XIII International Conference on Calorimetry in High Energy Physics 2008

(CALOR 2008)

Journal of Physics: Conference Series Volume 160

Pavia, Italy
26-30 May 2008

ISBN: 978-1-61738-374-8

Printed from e-media with permission by:

Curran Associates, Inc.
57 Morehouse Lane
Red Hook, NY 12571
www.proceedings.com

Some format issues inherent in the e-media version may also appear in this print version.

Printed by Curran Associates, Inc. (2010)

Institute of Physics
Dirac House, Temple Back
Bristol BS1 6BE UK

Phone: 44 1 17 929 7481
Fax: 44 1 17 920 0979

techtracking@iop.org

TABLE OF CONTENTS

OPERATING CALORIMETERS

CALORIMETRIC TECHNIQUES

ASTROPHYSICS AND NEUTRINOS

LHC

NEW TECHNIQUES

SIMULATION

PREFACE

The XIII International Conference on Calorimetry in High Energy Physics was held in Pavia, Italy, 26–30 May 2008, picking up the baton from the 2006 Conference in Chicago.

The Conference took place in the unique environment of the Theresian Room of the University Library. The attendees were surrounded by over 40 000 books of general interest and culture, and had the opportunity to see precious volumes written by such people as Galileo, Volta and Faraday.

The Workshop brought together more than 120 participants, including senior scientists as well as young physicists, confirming the central and ever-growing role of calorimeters in modern particle physics. The development of these detectors, as stressed by Professor Klaus Pretzl in his *lectio magistralis*, has made it possible to explore new frontiers in physics, and the present scenario is no exception to this rule.

With the LHC experiments almost completely installed and ready to take data, the Conference was an ideal chance to review the status of the different projects, whose development has been followed and discussed throughout the entire Calor series, and to show that they are capable of meeting the design specifications. Other highlights were the performance and physics results of calorimeters installed in currently operating experiments. In the session on astrophysics and neutrinos, the contributions confirmed the key role of calorimeters in this sector and demonstrated their growing application even beyond the field of accelerator physics. Considerable time was devoted to the state-of-the-art techniques in the design and operation of the detectors, while the session on simulation addressed the importance of a thorough understanding of the shower development to meet the demanding requirements of present experiments. Finally, on the R&D side, the particle flow and dual read-out concepts confronted the challenges issued by the next generation of experiments.

This complex material was reviewed in 83 presentations, now reported in these proceedings, that were debated in stimulating and fruitful discussions.

Outside of the Workshop, the participants were able to visit the historical Halls and Museum of the University, whose foundation dates back to the year 1361, and to enjoy a visit to the Certosa, a Carthusian monastery renowned for its exuberant architecture. Pavia welcomed the Conference participants by opening the doors of the Town Hall and offering a reception during which the Mayor's address underlined the importance of research and its applications in modern society.

The successful organization and the smooth running of the Conference is due to many people and Institutions. We gratefully acknowledge the financial support of the Istituto Nazionale di Fisica Nucleare (INFN), the Department of Nuclear and

Theoretical Physics and the University of Pavia, that made the Workshop possible, together with the contribution of our sponsors. The University also opened some rooms of the Chancellor's suite for the lunch and coffee breaks, and hosted the Conference Secretariat. We would like to express our deepest gratitude to the INFN and Department technical staff, who helped to prepare the Hall for the Conference and to provide computer services, and to the staff of the Theresian Library, who gave us access to the Room and organized a display of the many historical books from their vast and precious collections which are of interest to physicists.

Above all, the success of the meeting is due to the participants who animated it, and in particular to the speakers for their dedicated work in preparing their excellent talks and in providing the write-ups, and to the conveners for their essential role in shaping an interesting and well balanced scientific program.

Finally, we wish to thank the International Advisory Committee for their unfailing support and for offering us the opportunity to organize this Conference in Pavia.

Michele Livan
Chairman, Organizing Committee

International Advisory Commitee
M Danilov, ITEP Moscow
M Diemoz, INFN Roma I
A Ereditato, Bern
F Fabbri, INFN Frascati
T Kobayashi, ICEPP Tokyo
P Lubrano, INFN Perugia
S Magill, ANL Argonne
A Maio, LIP Lisbon
H Oberlack, MPI Munich
A Para, Fermilab
K Pretzl, Bern
Y Wang, IHEP Beijing
R Wigmans, TTU Lubbock
R Yoshida, ANL Argonne
R Zhu, Caltech

Local Organizing Committee
R Ferrari, INFN Pavia
M Fraternali, Università di Pavia
G Gaudio, INFN Pavia
M Livan, Università di Pavia (Chair)
P Pedroni, INFN Pavia
D A Scannicchio, INFN Pavia
V Vercesi, INFN Pavia

Session Organizers

Operating Calorimeters
W Sakumoto (University of Rochester)
D Schamberger (State University of NY at Stony Brook)

Calorimetric Techniques
C De La Taille (Université de Paris-Sud)
Paul Lecoq (CERN)
Frank Maas (GSI-Mainz University)
Jan Stark (LPSC Grenoble)

Astrophysics and neutrinos
I Gil Botella (CIEMAT)
A Vacchi (INFN Trieste)

LHC
P Bloch (CERN)
L Serin (Laboratoire de l'Accelerateur Lineaire/IN2P3/CNRS)

New Techniques
N Akchurin (Texas Tech University)
F Salvatore (Royal Holloway University of London)

Simulation
T Carli (CERN)
A Rimoldi (INFN e Università di Pavia)

Organization

Dipartimento di Fisica Nucleare e Teorica - Università degli Studi di Pavia
Istituto Nazionale di Fisica Nucleare – Sezione di Pavia
Università degli Studi di Pavia
Pragma Congressi, Corso Mazzini, 9 – Pavia

Sponsored by

CAEN
HAMAMATSU Photonis Italia
Iseg Spezialelektrinik GmbH
Wiener

Journal of Physics: Conference Series **160** (2009) 012001 doi:10.1088/1742-6596/160/1/012001

Physics with calorimeters

Klaus Pretzl

Laboratory of High Energy Physics, University Bern, Sidlerstr.5, CH 3012 Bern, Switzerland

E-mail: pretzl@lhep,unibe,ch

Abstract. Calorimeters played an essential role in the discoveries of new physics, for example neutral currents (Gargamelle), quark and gluon jets (SPEAR, UA2, UA1 and PETRA), W and Z bosons (UA1, UA2), top quark (CDF, D0) and neutrino oscillations (SUPER-KAMIOKANDE, SNO). A large variety of different calorimeters have been developed covering an energy range between several and 10^{20} eV. This article tries to demonstrate on a few selected examples, such as the early jet searches in hadron-hadron collisions, direct dark matter searches, neutrino-less double beta decay and direct neutrino mass measurements, how the development of these devices has allowed to explore new frontiers in physics.

1. Historical remarks

The American Astronomer Samuel Pierpot Langley has 1878 invented the bolometer in order to measure the infrared radiation of celestial objects [1]. His bolometer consisted of two platinum strips one of which was thermally isolated while the other was exposed to radiation. The exposed strip could be heated due to the absorption of electromagnetic radiation and thus would change its resistance. The two strips were connected to two branches of a Wheatstone bridge which allowed to measure the resistance change of the exposed strip. Therefor a bolometer behaves like a true calorimeter which measures the absorbed energy in form of heat. The name bolometer stems from the ancient Greec word $\beta o\lambda o\varsigma$ meaning ray and stands for ray-meter. With his bolometer Langley was the first to measure the energy flow in form of electromagnetic radiation from the sun. On Mount Whitney he determined the solar constant to 2.54 cal/ min cm^2 =1.77 kW/m^2, which was close to the more accurate average value of 1.37 kW/m^2 obtained today. Modern versions of Langley's bolometer are still most frequently used in astronomy to measure the energy flow of electromagnetic radiation from celestial objects and more recently also in cosmic microwave background radiation (CMB) experiments.

After the discovery of radioactivity by A.H.Becquerel in 1896, P.Curie and A.Laborde made a first attempt to measure the energy release in radioactive decays with a calorimeter [2]. At that time, a sharp difference between nuclei decaying into α particles which were emitted with the same energy and those decaying into electrons with energies distributed over a wide range was noticed. The beta decay seemed to violate the energy momentum conservation and turned out to be a big puzzle to the physics community. Micro-calorimeters were developed by C.D.Ellis and A.Wooster in 1927 [3] and independently by W.Orthmann and L.Meitner in 1930 [4] to determine the average energy of the electron in the beta decay of ^{210}Bi. With the differential calorimeter of W.Orthmann which is shown in Fig.1. and which is based on the same principle as Langley's bolometer heat transfers of the order of μW could be measured. He and L.Meitner [5] were able to determine the mean energy of the continuous β-spectrum in the ^{210}Bi decay to E=0.33MeV with an accuracy of 6%. These findings strongly supported the notion of a continuous β-spectrum which was leading to W. Pauli's neutrino hypothesis in 1930.

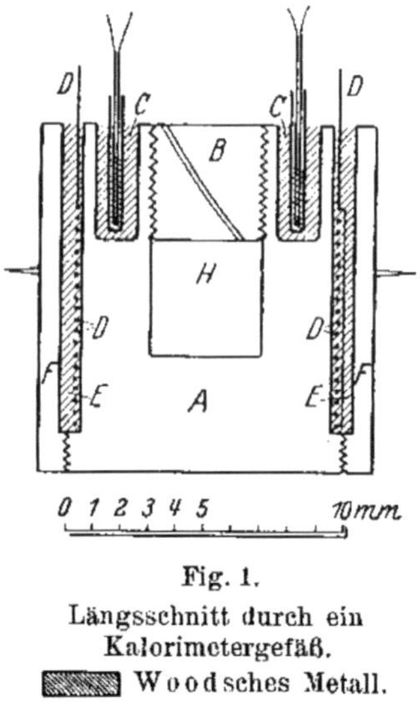

Figure 1. The calorimeter of W.Orthmann is shown.

In order to measure the spectrum of high energy cosmic ray particles N.L.Grigorov and collaborators introduced in 1954 a new calorimeter concept [6], the so-called sampling calorimeter. For this calorimeter they used ionization chambers (arrays of proportional counters) and scintillation counters sandwiched between thick iron absorber sheets. The visible energy of the interacting particles in the calorimeter was determined from the number of secondary particles in the shower and their energy loss in the ionization and scintillation counters. Because of the invisible energy loss in the absorber plates and in the detector sheets, sampling calorimeters need to be calibrated with particles of known energy, which turned out to be a problem in these early times due to the lack of high energy particle accelerators. Nevertheless, in 1957 N.L.Grigorov and his collaborators [7] constructed a sampling calorimeter in the Pamir mountains, at an altitude of 3860 m above sea level (Fig.2). Their calorimeter also employed $10m^2$ of emulsion sheets, which they placed between 3 lead sheets to study details of the primary interaction. In order to identify corresponding events in the emulsions and the calorimeter, two layers of emulsion sheets on top of each other were used: one fixed in space, the other moved by a certain amount in a certain time interval with respect to the other. The two emulsion film images were then matched and thus the time of the shower passage determined.

The development of calorimeters for high energy physics started in the late 60'ies. The discovery of the substructure of hadrons by the SLAC experiment in 1969 [8], the description of the results in a parton model of hadrons by R. Feynman and the introduction of the color octet of gluons by H.Fritsch, M.Gell-Mann, and H.Leutwyler were leading to a theory of strong interactions the so-called Quantum Chromo Dynamics (QCD). According to QCD partons would scatter via gluon exchange in central hadron-hadron collisions leading to two large p_T and two spectator jets (Fig.3).

These developments and the discovery of jet-like structure in e^+- e^- collisions at SLAC in 1975 [9] stimulated the community to introduce calorimeters with the advantage of having a better

Figure 2. The sampling calorimeter of Babayan, Grigorov, Shestoperov and Sobikyakov is shown.

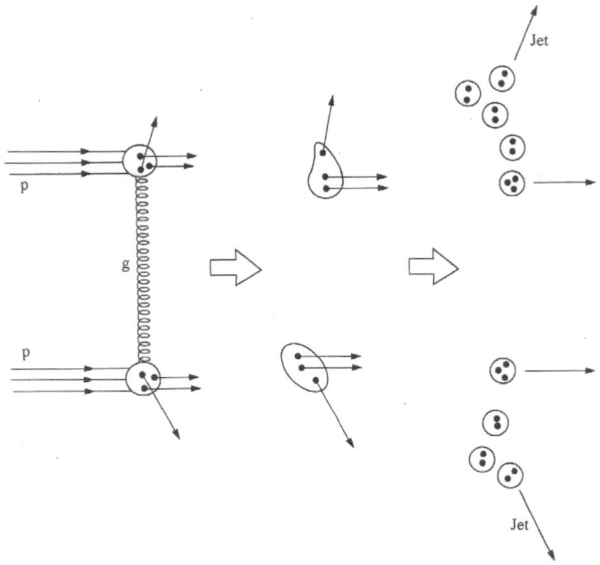

Figure 3. Central proton proton scattering process as described by QCD.

energy resolution than magnetic spectrometers and the capability of measuring the energy of charged as well as neutral particles. The discovery of neutral currents with GARGAMELLE at CERN in 1973 [10] also influenced the use of multi-ton calorimeters in neutrino experiments. A topical review of the development of calorimeters in astro and particle physics can be found in

[11].

In the following chapters some examples of calorimeter developments for new physics in the past and the presence are described: Early jet physics, dark matter searches, neutrinoless double beta decay and direct neutrino mass measurements.

2. The early days of jet physics

In the 1970'ies two high energy fixed target proton synchrotrons (the 400GeV synchrotron at Fermilab in 1972 and the 450 GeV SPS at CERN in 1976) as well as one proton-proton collider (the ISR at CERN in 1971) and two e^+- e^- collider (SPEAR at SLAC in 1972 and DORIS at DESY in 1974) started operation. These new facilities offered a unique playground for versatile research programs among which was the search for jets in hadron-hadron collisions. One of the questions raised was [12]: Do the cross sections for hard hadron-hadron scattering follow a scaling law

$$\frac{d\sigma}{dx_c} = \frac{1}{s} F(\theta_1^*, \theta_2^*, x_c) \tag{1}$$

with $x_c = \frac{E_{cal}}{E_{inc}}$ and θ_1^*, θ_2^* the minimum and maximum c.m.s. polar angles of the calorimeter? E_{cal} is the energy measured in the calorimeter and E_{inc} is the energy of the incident beam particle. This can be tested when keeping the c.m.s. angle of the calorimeter fixed while varying the incident beam energy. Such an experiment would need a calorimeter with the following features: segmentation in the polar angle θ and the azimuthal angle ϕ, large coverage in the central rapidity region, full coverage in ϕ in order to provide an unbiased jet trigger as well as a good jet energy resolution. The calorimeter should be movable on rails in order to change the distance to the target when changing the incident beam energy. The NA5 collaboration was among the first who constructed such a segmented calorimeter as shown in Fig.4.

Figure 4. The segmented NA5 calorimeter is shown.

Their calorimeter covers c.m.s. angles between $\theta_1^*=45^0$ and $\theta_2^*=135^0$. It consists of a electromagnetic part (scintillator/ lead) and a hadronic part (scintillator/iron) and has 24 segments in ϕ and 10 in θ. A very compact segmentation was obtained by introducing a novel wavelength shifter (WLS) readout system, which is shown in Fig.5.

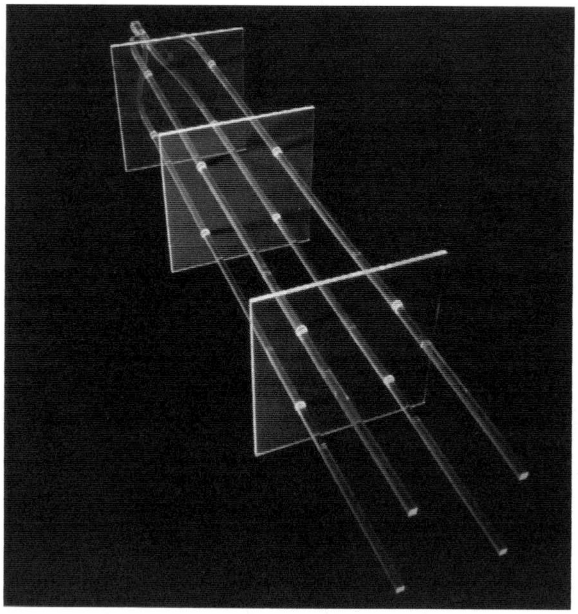

Figure 5. The novel wavelength shifter readout of the NA5 calorimeter is shown. See text for explanation.

By using a different WLS color (yellow) in the electromagnetic part of the NA5 calorimeter[13] than in the hadronic part (BBQ green), it was possible to guide the light of both colors in one rod to the two photomultipliers which, by using appropriate filters, were sensitive to either yellow or green light. This way the energy deposited in the electromagnetic and in the hadronic part of the calorimeter could be measured separately. The NA5 calorimeter was the first tower-structured scintillator sampling calorimeter in operation at CERN. After its first use in 1978 it has been employed in several followup experiments, like NA24, NA35, NA49, and after 30 years is still in operation in the North area of CERN. The central hole of the calorimeter was covered by a downstream calorimeter in order to be able to measure the total energy in the event. The construction parameters and the performance of both calorimeters are described in [14]. The ϕ segmentation enabled to set up and study calorimeter triggers with different ϕ acceptances like $\pi/2, \pi$ and 2π and compare the results with other experiments with limited calorimeter acceptance, like E260 at Fermilab [15]. The scaling behavior was studied by parameterising the cross section obtained with the azimuthal $\pi/2$ trigger by $\frac{d\sigma}{dp_T^2} \sim {p_T}^{-n} f(x_T)$ with $x_T = \frac{2p_T}{\sqrt{s}}$ and with the 2π trigger by $\frac{d\sigma}{d\Sigma|p_T|} \sim (\Sigma|p_T|)^{-n} f(x_T)$ with $x_T = \frac{\Sigma|p_T|}{\sqrt{s}}$. The scaling parameter n was derived from the cross sections σ_1 and σ_2 measured at 150 GeV/c and 300 GeV/c respectively for a fixed x_T and a fixed c.m.s. solid angle acceptance of the calorimeter by the relation $n = \ln \frac{\sigma_1}{\sigma_2} / \ln \frac{\sqrt{s_2}}{\sqrt{s_1}}$. The results are shown in Fig.6. Constituent scattering models in their simplest form predict a constant n=4 or 3 respectively. QCD models which include scale breaking effects, predict n to rise by 1 over the entire x_T region. The scaling parameters for the various triggers turn out to increase with x_T and are much larger than predicted by QCD and by the dimensional scaling of eq.(1). In looking for a jet-like event structure a planarity analysis of the NA5 data taken with the azimuthal 2π trigger was performed. The planarity P was calculated from a principal axis analysis of the transverse momentum distribution for each event as measured by the calorimeter. Particle momenta were derived from the energy clusters observed in the calorimeter. The planarity was defined as P=(a-b)/(a+b) with a (b) being the sum of the squares of the projected transverse momenta to the maximum (minimum) principal p_T axis. For an isotropic event structure one would expect P=0 and for pencil like jets P=1. The planarity distributions turned out to peak around P=0.5 and show practically no change when increasing the p_T trigger threshold. Also in p-p collisions at higher energies ($\sqrt{s} = 63GeV$) at the ISR, the R401 (Split Field Magnet) [16] and the R807 (Axial Field Spectrometer) [21] experiments show no dominant jet structure and no change of the mean circularity of 0.5 (circularity=1-planarity) with increasing trigger threshold. A detailed description of the NA5 results can be found in [17].

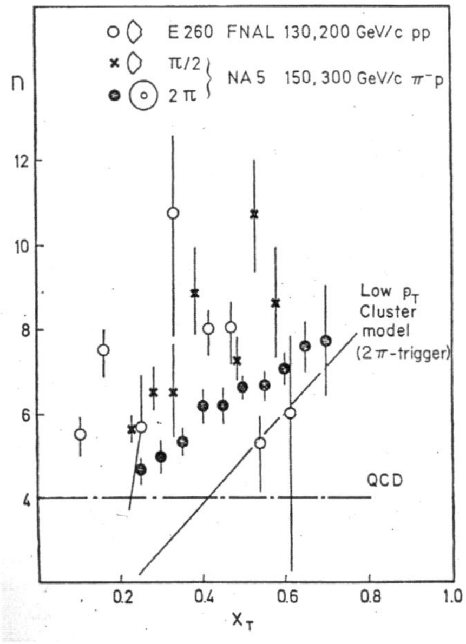

Figure 6. The scaling behavior of the cross sections measured with different calorimeter triggers is shown. For explanations see text.

Considering the expectations put into finding jets in hadron-hadron collisions these results turned out to be rather disappointing. However, as it became clear later the energies available at the SPS ($\sqrt{s} = 24GeV$) and the ISR ($\sqrt{s} = 63GeV$) were not sufficient to identify jets unambigously. Nevertheless, from the experimental techniques which were developed at low energies the next generation experiments at higher energies, like UA1 and UA2 at CERN and CDF at Fermilab, could profit a lot. In fact the central calorimeter of UA2 [18] is a direct transformation of the NA5 calorimeter from the laboratory to the c.m.s. system (Fig.7).

In 1982 the first unambiguous jets were seen by the UA2 [19] and the UA1 [20] experiments at the CERN $\bar{p}p$ collider ($\sqrt{s} = 540GeV$). In Fig.8 the differential cross section as a function of transverse energy ΣE_T as measured by the UA2 experiment is shown. It demonstrates a clear slope change at a transverse energy of around 60 GeV, where the dominant constituent scattering processes set in.

The measured jet angular distributions turned out to be consistent with the QCD predicted t-channel vector boson exchange and ruled out models with scalar gluons. In Fig.9 the results of

Figure 7. The central calorimeter of the UA2 experiment at the CERN $\bar{p}$p collider is shown.

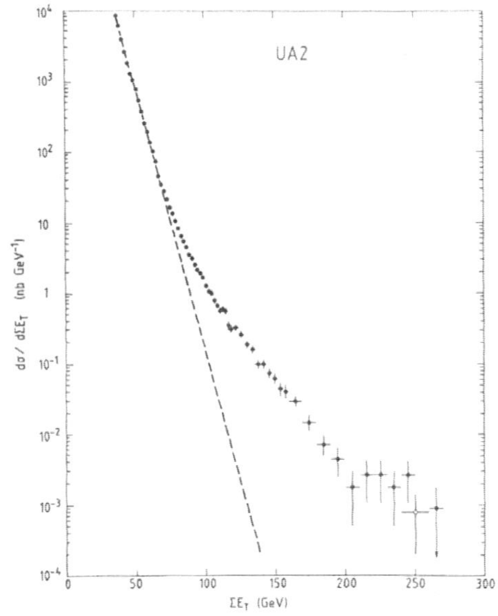

Figure 8. The differential cross section as a function of the transverse energy ΣE_T as measured by the UA2 experiment is shown.

UA2 and UA1 at the $\bar{p}p$ collider are compared to the results of R807 [21] at the ISR. Also shown is the LEGO-plot of a typical 2 jet event as seen by the calorimeter. It demonstrates again that constituent scattering processes were born out clearly only at very high energies. This turned out to be the case not only for jets but also for direct photons. Fortunately one could learn a lot about jets in the meantime such that one is ready to deal with them at LHC energies were they are produced abundantly.

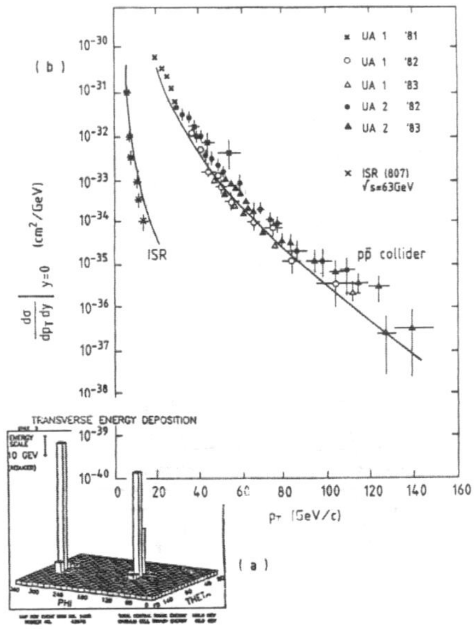

Figure 9. The differential cross sections as a function of transverse momentum p_T as measured by the UA2 and UA1 experiments at the $\bar{p}p$ collider and the R807 experiment at the ISR are shown. Also shown is the LEGO-plot of a typical 2 jet event as detected by the calorimeter.

3. Physics with cryogenic calorimeters

In the following chapters some examples are given where novel calorimeter techniques so-called cryogenic calorimeters are employed for the quest for the dark matter in the universe and the neutrino-less double beta decay as well as for the direct measurement of the neutrino mass. Although the basic idea of these techniques is going back to F.Simon [22] in 1935, its further development was recently very much stimulated by these searches. F.Simon claimed when cooling a small size calorimeter ($1cm^3$) down to liquid helium temperatures one could measure heat transfers of nW which is a gain of a factor 1000 in sensitivity compared to the W.Orthmann calorimeter. He argued that at low temperatures the heat capacity C of a micro-calorimeter is low and a small energy loss ΔE of a particle interacting in the calorimeter can lead to an appreciable temperature increase $\Delta T = \Delta E/C$. A typical cryogenic calorimeter is shown in Fig.10. It consists of an absorber with heat capacity C, a thermometer and a thermal link with heat conductance g to a heat reservoir with a constant bath temperature T_B. Particles interacting in the absorber cause a change in the resistance of the thermometer which is measured by a voltage drop V when passing a current I through the thermometer. Cryogenic calorimeters can be made from many different materials including superconductors, a feature which turns out to be very useful for different applications. They can be used as target and detectors at the same time. Most calorimeters used in high energy physics measure the energy loss of a particle in form of ionization or scintillation light. In contrast, cryogenic calorimeters are able to measure the total deposited energy in form of ionization and heat (phonons). This feature makes them very effective in detecting very small energy deposits with high resolution. There are several kinds of cryogenic detectors in use or under development. A topical review can be found in [23].

The most frequently employed calorimeters in the above mentioned searches are those which use transition edge sensors (TES) or semiconducting thermistors as thermometers. A TES sensor consists of a very thin super-conducting film or strip which is operating at a temperature in the transition between the super-conducting and the normal phase, where the resistance changes between zero and its normal value. The very strong dependence of the resistance change with temperature makes the TES calorimeter very sensitive to small input energies. A thermistor is a heavily doped semiconductor slightly below the metal insulator transition. Its conductivity at low temperatures can be described by a phonon assisted electron hopping mechanism between impurity sites. With these micro-calorimeters excellent energy resolutions of about 2 eV at 6 keV x-ray energies have been achieved. In Fig.11 a x-ray spectrum obtained with a state of the art Si(Li) solid state device (dashed line) and a cryogenic micro-calorimeter (solid line) using a Bi absorber and a TES sensor (Al-Ag) are compared [24].

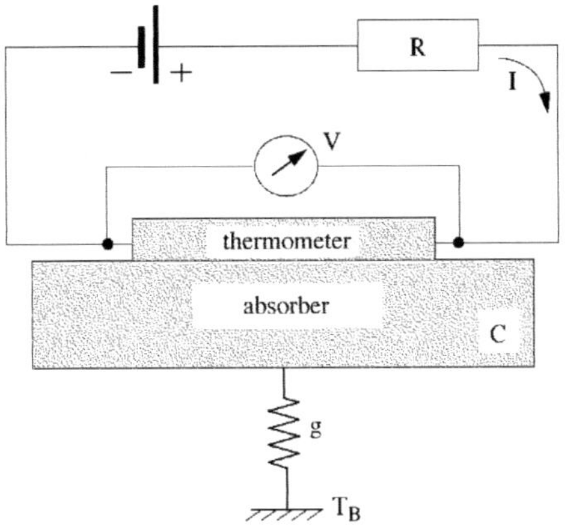

Figure 10. The principle of a cryogenic calorimeter is shown.

3.1. Direct dark matter searches

One of the outstanding mysteries in physics is the existence of dark matter and dark energy. From independent observations of the Cosmic Microwave Background radiation (CMB), the Large Scale Structure surveys (LSS) and the Supernova Type1 surveys (SN) [25] we have learned a lot about the matter budget of the universe. Combining these results one finds the universe contains 72±2% dark energy, 28±1% matter (including baryonic and dark matter), and 4.6±0.2% baryonic matter. However, the true nature of dark matter and dark energy which fills 95% of the universe is still unknown. While the dark energy is not directly detectable with todays technologies the dark matter, provided it consists of particles, can be searched for with earth bound detection systems owing to their large abundance in our galactic halo. There are several candidates for dark matter: massive neutrinos, weakly interacting massive particles, so-called WIMPs, and axions. Although neutrinos belong next to photons to the most abundant particles in the universe their mass turns out to be to small to significantly contribute to the missing mass. Neutrinos being relativistic at freeze out in the early universe are free streaming particles

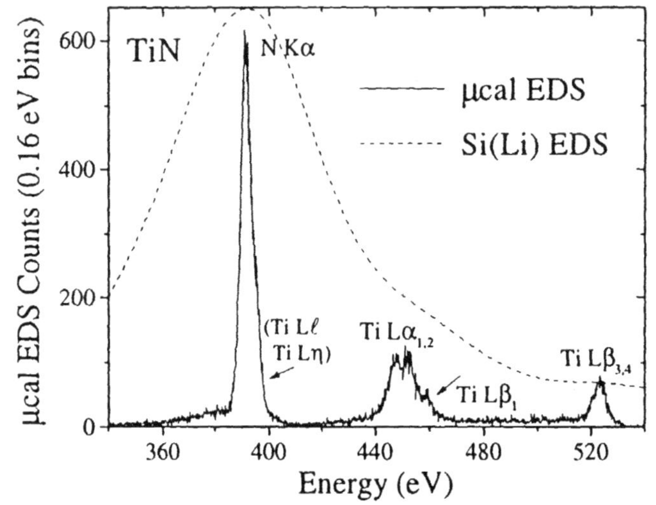

Figure 11. An x-ray spectrum obtained with a state of the art Si(Li) solid state device (dashed line) and a cryogenic micro-calorimeter (solid line) using a Bi absorber and a TES sensor (Al-Ag) are compared.

and would cluster only at very large scales. Therefore neutrinos would enhance large scale and suppress small scale structure formations. From the LSS power spectrum one can derive a ratio of the neutrino density to the matter density Ω_ν/Ω_m. From this and Ω_m determined from CDM and SN an upper limit for the sum of the neutrino masses $\Sigma m_\nu \leq 0.7eVc^{-2}$ was derived [25]. It is interesting to note that this upper limit of the neutrino masses is lower than the ones obtained from direct neutrino mass measurements so far. Axions which are named after a laundry detergent, are predicted by models to solve the CP violation problem in QCD and would be produced abundantly during the QCD phase transition in the early universe. The axion detection technique is based on Primakoff conversion in a magnetic field using microwave cavities. A topical review on dark matter searches including axions can be found in [26]. Today the most favored candidate for a WIMP turns out to be the lightest stable particle predicted by Super Symmetry (SUSY), the so-called neutralino. If the neutralino were to be discovered by the LHC at CERN it still would need to be confirmed as a dark matter candidate by direct detection experiments.

WIMPs can be detected via their elastic scattering with nuclei when penetrating a dark matter detector. In the case of neutralinos spin-independent as well as spin-dependent scatterings are possible. The expressions for the corresponding cross sections can be found in [27]. The WIMP mass can be derived from the characteristic nuclear recoil spectrum. Cryogenic detectors are well suited for this purpose since they are able to measure small recoil energies (order of eV) with high efficiency. They can also be made from many different materials. This turns out to be an advantage in order to verify a possible recoil signal with another detector material. However, cryogenic calorimeters are presently rather limited in their detector masses (kg). More conventional devices like LAr, LXe, Ge, and NaI allow eventually much larger detector masses (ton). This makes them also sensitive to annual modulations of the WIMP signal owing to the movement of the earth with respect to the dark halo rest frame. An effect which is claimed to be observed by the DAMA experiment in the Gran Sasso Laboratory. However, due to the quenching of the ionization and photon signals, conventional detectors have lower nuclear recoil detection efficiencies.

Dark matter detectors are located in deep underground laboratories to be protected from cosmic particles. Their sensitivity to WIMPs is very much limited by the background radiation from surrounding materials and rocks in underground laboratories as well as by the radioactivity of the detector material itself. Even the best local shielding and the use of radio-poor material for the detectors provides only limited effectiveness and is expensive. Nevertheless, orders of magnitude better sensitivities have been achieved when employing dual readout of the detectors. The CDMS and EDELWEISS collaborations use cryogenic calorimeters made of Si or Ge absorbers with thermistor or TES sensors. For each event they are recording the heat (phonon) signal as well as the ionization signal separately but simultaneously. The principle of this dual phonon-ionization readout is shown in Fig.12 . The CRESST collaboration is using $CaWO_4$ crystal absorbers and measures for each event the heat with a tungsten TES thermometer and the photon signal by a separate cryogenic detector, which consists of a silicon wafer with a tungsten TES thermometer (Fig.13). The dual phonon-ionization and phonon-photon readout of the detectors allows an active background recognition since for the same deposited energy in the absorber the ionization (or photon) signal from nuclear recoils is highly quenched compared to signals from background electrons (from Compton scattering). This feature provides a very effective separation of genuine nuclear recoils from electron background above 10keV recoil energies. This is demonstrated in Fig.14, taken from [28], which shows the energy equivalent of the pulse heights measured in the photon detector versus those in the phonon detector. The scintillating $CaWO_4$ crystal absorber was irradiated with photons and electrons using Cobalt and Strontium sources and with neutrons using an Americium Beryllium source. The upper band in Fig.14 shows the electron recoils (e) and the lower band the nuclear recoils (n).

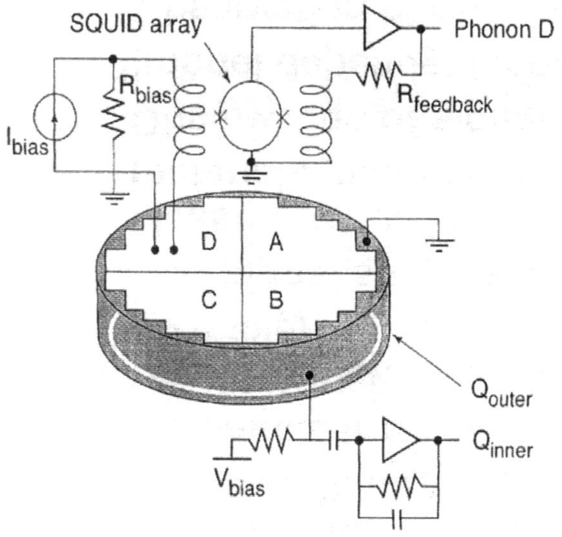

Figure 12. The principle of the dual phonon-ionization readout of the CDMS cryogenic calorimeter is shown.

Conventional WIMP detectors like the NaI crystals of DAMA and the LXe calorimeter of

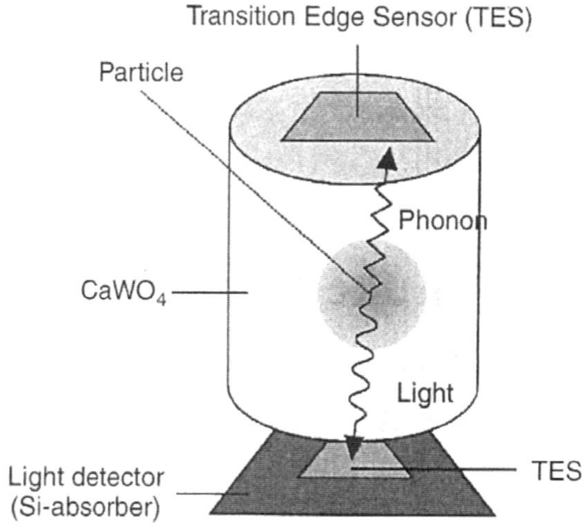

Fig. 1. Detector setup.

Figure 13. The principle of the dual phonon-photon readout of the CRESST cryogenic calorimeter is shown.

ZEPLIN I are using pulse shape analysis to differentiate between nuclear and electron recoils. The signals from nuclear recoils are faster due to the higher ionization density. A more effective background recognition is obtained by the XENON (LXe calorimeter) and the WARP (LAr calorimeter) detectors which use besides the timing information of the prompt photon signal also the ionization signal, which they obtain by drifting the ionization charges to the surface of the liquid calorimeter and by measuring the recombination photons emitted in the gas above the liquid. However, this very effective method works only at recoil energies above 50 keV. The use of LAr instead of LXe has many advantages, most importantly the lower price. However, LAr detectors have to deal with the radioactive ^{39}Ar isotope which is produced by cosmic radiation and which is limiting the sensitivity of these devices.

The experimental results are presented in Fig.15 as exclusion plots, which show the WIMP-nucleon cross sections (90% C.L.) for spin-independent interactions versus the WIMP mass. The sensitivities obtained with the cryogenic devices of CDMS II [29] (detector mass: 4.75 kg Ge and 1.1 kg Si) in the Soudan Underground Laboratory (2090 M.W.E.), USA, EDELWEISS [30] (0.96 kg Ge) in the Frejus tunnel (4800 m.w.e.), South of France, CRESST II [31] (0.6 kg $CaWO_4$) in the Gran Sasso Laboratory (3800 m.w.e.), Italy, and with the detectors XENON [32] (15 kg) and WARP [33] (2.6 kg) in the Gran Sasso Laboratory. Not shown in Fig.15 are the results of ZEPLIN I [34] (3.2 kg) in the Boulby mine , U.K., which are similar to WARP. Also shown are the early results of DAMA [35] (100 kg NaI) in the Gran Sasso Laboratory which claimed to see an annual modulation signal consistent with a WIMP mass $m_x = (52^{+10}_{-8}) GeV c^{-2}$. Although the other experiments are not in agreement with their results DAMA recently reconfirmed their findings with the DAMA/LIBRA detector (250 kg NaI) [36].

An extraction of the spin-dependent WIMP-nucleon cross section in a model independent way is not possible, since the nuclear and the SUSY degrees of freedom do not decouple from each other. Nevertheless, when using an "odd group" model which assumes that all the nuclear spin is carried by either the protons or the neutrons, whichever are unpaired, WIMP-nucleon cross

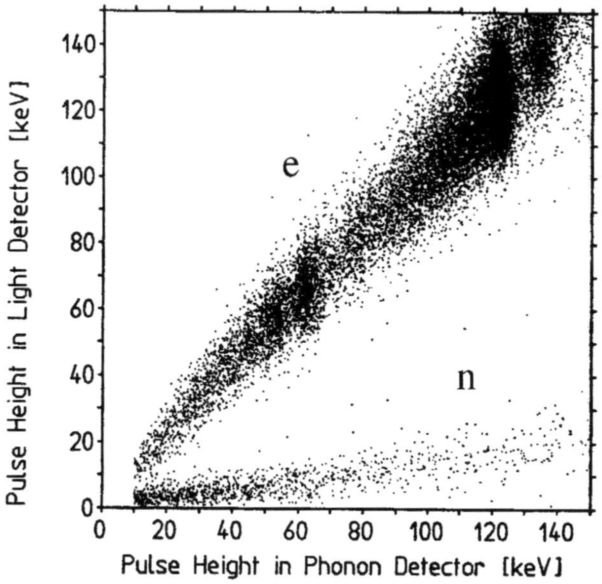

Figure 14. The energy equivalent of the pulse heights measured in the photon detector versus those in the phonon detector. The scintillating $CaWO_4$ crystal absorber was irradiated with photons and electrons using Cobalt and Strontium sources and with neutrons using an Americium Beryllium source. The upper band shows the electron recoils (e) and the lower band the nuclear recoils (n).

sections can be deduced. The CDMS collaboration gives a minimum upper limit of $2.7\text{x}10^{-38}$ cm^2 at 90% C-L. [37] for the spin-dependent WIMP-nucleon cross section, which is so far the best limit achieved with cryogenic calorimeters.

From Fig.15 it can be seen, that the experimental results start already excluding certain regions of SUSY model predictions [38]. New projects like EURECA (European Underground Rare Event search with Calorimeter Array) and Super CDMS with detector masses up to 1 ton are already on the drawing boards with the hope to be able to improve the WIMP detection sensitivity by several orders of magnitude.

3.2. Neutrino-less double beta decay

From neutrino oscillations we know that neutrinos have a mass and there is new physics beyond the Standard Model to be expected. However, oscillation experiments provide only mass differences of the neutrino flavors and direct neutrino mass measurements with novel detection techniques are in demand. There remains also the question whether neutrinos are Dirac or Majorana type particles. To answer this question one looks for the neutrino-less double beta decay. Double beta decay was first suggested by Maria Goeppert Mayer [39] in 1935. It occurs via spontaneous transition from a nucleus (A, Z) to its isobar (A, Z+2) and can proceed in two ways: the lepton number conserving process (A, Z)→ (A, Z+2) $+2e^-$ $+2\bar{\nu}_e$ where two electrons and two antineutrinos are emitted or the lepton number violating process (A, Z)→ (A, Z+2) $+2e^-$ with no neutrinos in the final state. In the neutrino-less double beta decay ($0\nu\beta\beta$) the energy spectrum of the decay would show a sharp peak with the energy sum of the two electrons. This process can only occur if the neutrino is a Majorana particle and if it has a non-vanishing mass. From the measured decay rate $(1/T^{0\nu}_{1/2})$ one can derive in principle its effective mass

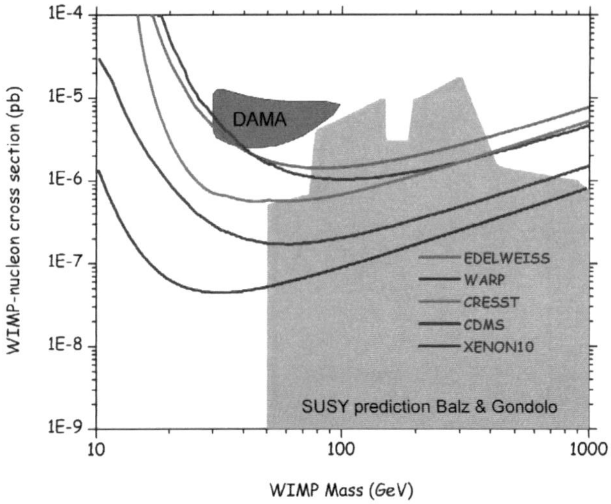

Figure 15. Exclusion plots for spin-independent WIMP interactions are shown.

$< m_\nu >$ or a lower limit of it:

$$(1/T^{0\nu}_{1/2}) = G_{0\nu}(E_0, Z) \mid M_{0\nu} \mid^2 < m_\nu >^2 \qquad (2)$$

where $T^{0\nu}_{1/2}$ is the lifetime of the decay, $G_{0\nu}(E_0, Z)$ is an accurately calculable phase space function and $M_{0\nu}$ is the nuclear matrix element, which is not very well known [40]. The calculated values of $M_{0\nu}$ can vary considerably, consequently the search for $0\nu\beta\beta$ should be made with several different nuclei in order to confirm an eventual discovery of this important process.

Because of their excellent energy resolution and their wide choice of $0\nu\beta\beta$ nuclei cryogenic calorimeters are very well suited for this research. This was first pointed out by E. Fiorini and T. Niinikoski in 1984 [41] who also pioneered the development of these detectors. A cryogenic $0\nu\beta\beta$ experiment with the name CUORICINO is currently taking data in the Gran Sasso Laboratory. It consists of an array of 62 TeO_2 crystals with a total weight of 40.7 kg. It is presently the largest cryogenic detector in operation. TeO_2 was chosen because of the high transition energy of 2528.8±1.3 keV and its large isotopic abundance of ^{130}Te with 33.8%. So far their results [42] show no evidence for $0\nu\beta\beta$ decay, but their lower limit of $T^{0\nu}_{1/2} \geq 1.8\text{x}10^{24}$ yr (90% C.L.) corresponding to a neutrino mass range of 0.1 to 0.9eV already has reached the sensitivity of the Heidelberg-Moskow experiment [43], which claimed evidence for $0\nu\beta\beta$ in ^{76}Ge. A review of present and future $0\nu\beta\beta$ decay experiments can be found in [44]. As a next step, the collaboration is constructing a 1 ton TeO_2 detector, named CUORE, which aims for a neutrino mass sensitivity between 8 and 45 meV. This is quite competitive to the goals of other projects under construction like EXO, GENIUS, GERDA, MOON, NEXT, SUPER-NEMO.

3.3. Direct neutrino mass measurements

The best upper limit for the electron neutrino mass of 2.2eV was obtained by the Mainz and the Troitsk experiments [45] from the electron spectroscopy of the tritium decay $^3H \rightarrow ^3He + e^- + \bar{\nu}_e$

with a transition energy of 18.6 keV. In the near future the KATRIN experiment using a much improved electron spectrometer is aiming to reach a sensitivity of 0.2 eV for the neutrino mass. A complementary approach was pioneered by the Genova group [46] by studying the beta decay of $^{187}Re \rightarrow^{187} Os + e^- + \bar{\nu}_e$ with a cryogenic micro-calorimeter. Rhenium is a superconductor with a critical temperature of 1.7K. Their calorimeter consisted of a Re crystal (2mg) coupled to a Ge NTD (neutron transmutation doping) thermistor. Re has the advantage of a low endpoint energy of 2.6 keV and a high natural abundance of 62.8% of ^{187}Re. In their first attempt they were able to reach an upper limit of the neutrino mass of 19 eV (90% C.L.) [47]. The Milan group [48] followed this method and has built an array of ten micro-calorimeters consisting of $AgReO_4$ crystals with masses between 250 and 350 μg. Their average energy resolution at the endpoint energy was 28.3 eV (FWHM). From a fit to the Curie plot of the ^{187}Re decay they obtained an upper limit for the neutrino mass of 15 eV.

The main advantage of this calorimetric approach is that the calorimeter measures the total energy of the decay including final state interactions, such as the de-excitation energy of excited atoms. However, in order to be competitive with the spectrometer experiments the cryogenic calorimeter has to have an excellent energy resolution (1 eV) and enough counting rate statistics at the beta endpoint energy. The latter may raise a problem for thermal calorimeters, since their signals are rather slow and limit the counting rate capability to several Hz. A new ambitious international project, named MARE (Micro-calorimeter Array for a Rhenium Experiment) [49], is prepared to deal with these problems and aims to reach a neutrino mass sensitivity of 0.2eV comparable to the KATRIN experiment. To reach their goal they need large arrays of detectors (10^4 detector elements) and energy and time resolutions of 1eV and 1μs respectively.

4. Outlook

In the coming years new physics (Higgs, SUSY etc.) is expected to be discovered with the LHC. The sophisticated calorimeters, as presented at this conference, will play thereby a central role. The decision on the construction of the future linear collider will also depend on the results obtained with the LHC. In the field of astrophysics and cosmology, large arrays of conventional and cryogenic calorimeters are under development or in construction for space-born x-ray and infrared observatories to study the formation and evolution of black holes and their role in galaxy formation, matter in extreme conditions (like in gravitational fields near black holes), supernova remnants, accretion powered systems and polarization in CMB. Large earth-bound calorimeters using the polar ice, the sea and the atmosphere as absorber are already in operation and larger ones are planned to search for high energy sources which emit particles (among them neutrinos and gammas) with energies beyond 10^{20} eV. Underground low background experiments are driving their sensitivity for the detection of the dark matter in the universe, the neutrino-less double beta decay and the neutrino mass with massive cryogenic or conventional calorimeters (on a ton scale) by several orders of magnitude beyond the presently achieved values. For the future, the development of new calorimeter techniques needs not only imagination but also continuous financial support for detector R&D.

4.1. Acknowledgments

The author wishes to thank Michele Livan and his collaborators for the excellent organization of this conference.

5. References

[1] Langley S P 1881 *Proceedings Of the American Academy of Arts and Sciences* Vol.**16** 342
[2] Curie P and Laborde A 1903 *Comptes rendus hebdomadaires des seances de l ' Academie des Sciences Paris* **136** 673
[3] Ellis C D and Wooster A 1927 *Proceedings of the Royal Society* **117** 109

[4] Orthmann W 1930 *Zeitschrift fuer Physik* **60** 10
[5] Meitner L and Orthmann W 1930 *Zeitschrift fuer Physik* **60** 143
[6] Grigorov N L *et al* 1954 *Zh.Eksp.Teor.Fiz.* **34** 5062*T*
[7] Murzin V S 1967*Progr. Elem.Part.Cosmic Ray Physi.* **34** 506
[8] Breidenbach M 1969 *Phys.Rev.Lett.* **23** 935
[9] Hanson G *et al* 1975 *Phys. Rev. Lett.* **15, 24** 1609
[10] Hasert F J *et al* 1973 *Phys. Lett.* B **46** 138
[11] Pretzl K 2005 *J.Phys.G: Nucl.Part.Phys.* **31** R133
[12] Ochs W and Stodolsky L 1977 *Phys.Lett.* B **69** 225
[13] Eckhardt V, Kalbach R, Manz A, Pretzl K P, Schmitz N, Vranic D 1978 *Nucl.Instrum. Methods.* **155** 389
[14] De Marzo C *et al* 1983 *Nucl. Instrum. Methods.* 217 405
[15] Bromberg C *et al* 1979 *Phys.Rev.Lett.* **43** 565
[16] Bsser F W *et al* 1973 *Phys.Lett.* B **46** 471
[17] De Marzo C *et al* 1983 *Nuclear Physics* B **211** 375
[18] Beer A *et al* 1984 *Nucl. Instrum. Methods in Physics Research* **224** 360
[19] Banner m *et al* 1982 *Physics Letters* B **118** 203
[20] Arnison G *et al* 1982 *Physics Letters* B **123** 115
[21] Akeson T *et al* 1983 CERN report Ep-83-130
[22] Simon F 1935 *Nature* **135** 763
[23] Pretzl K 2000 *Nucl. Instrum. Metods in Physics Research* A **454** 114
[24] Wollmann D A *et al* 2000 *Nucl.Instrum.Metods in Physics Research* A **444** 145
[25] Spergel D N *et al* 2007*Astrophys. J. Suppl. Series* **170** 377
[26] Pretzl K 2007 *Space Science Rev.* **130** 63
[27] Jungmann G, Kamionkowski M, Griest K 1996 *Phys. Rep.* **267** 195
[28] Meunier P *et al* 1999 *Appl. Phys. Lett.* B **75,9** 1335
[29] Ahmed Z *et al* 2008 *arXiv: 0802.3530, astro-phys.*
[30] Sanglard V *et al* 2005 *Phys.Rev.* D **71** 122002
[31] Angloher G *et al* 2005 *Astroparticle Physics* **23** 325
[32] Angle J *et al* 2008 *Phys. Rev. Lett.* **100** 021303
[33] Bennetti P *et al* 2008 *Astroparticle Physics* **28** 495
[34] Alner G J *et al* 2005 *Astroparticle Physics* **23** 444
[35] Bernabei R *et al* 2000 *Phys. Lett.* B **480** 23
[36] Bernabei R *et al* 2008 *arXiv: 0804.2741 astro-ph*
[37] Ahmed Z *et al* 2008 *arXiv: 0802.3530 astro-ph*
[38] Baltz E A and Gondolo P 2003 *Phys. Rev.* D **67** 063503
[39] Goeppert Mayer M 1935 *Phys.Rev.* **48** 512
[40] Vogel P 2001 *Double Beta Decay:Theory, Experiment and Implications, in Current aspects of Neutrino Physics, ed. Caldwell D O , Springer* 177
[41] Fiorini E and Niinikoski T 1984 *Nucl. Instrum. Methods in Particle Research* **224** 83
[42] Arnaboldi C *et al* 2005 *Phys.Rev. Lett.* **95** 142501
[43] Klapdor-Kleingrothaus H V *et al* 2004 *Nucl.Instr.Methods in Physics Research* A **522** 198
[44] Fiorini E 2007 *Europhysics News* **38, 4** 30
[45] Lobashov V M 2003 *Nucl.Phys.* A **719** 153
[46] Fontanelli F, Gatti F, Swift a, Vitale S 1996 *Nucl.Instrum. Methods in Particle Research* A **370** 247
[47] Gatti F *et al* 2001 *Nucl. Phys.* B **91** 293
[48] Sisti M *et al* 2004 *Nucl.Instrum. Methods in Particle Research* A **520** 125
[49] MARE http://mare.dfm.uninsubria.it

Journal of Physics: Conference Series **160** (2009) 012002 doi:10.1088/1742-6596/160/1/012002

Calorimeter Operations in RunII at DØ

Daniel Duggan

Florida State University, Tallahassee, FL.

Abstract. The operation and performance of the DØ calorimeter is presented. Calorimeter operation will be overviewed with specific focus on calibration techniques, hardware monitoring and stability, and overall data quality strategies.

1. Introduction

The DØ calorimeter is a hermetic liquid argon sampling calorimeter. It is housed in three separate crystostats, the central giving coverage of $|\eta| < 1.1$, and the entire calorimeter providing coverage of $|\eta| < 4.2$. The calorimeter was upgraded between RunI and RunII at Fermi National Laboratory, where it currently operates. Collisions are produced from the Tevatron using unpolarized proton and anti-proton beams with a center of mass energy of 1.96 TeV. The electronics were upgraded to optimize the system for the 396 ns bunch crossings and high instantaneous luminosities, which have exceeded $300 \times 10^{30}\ cm^{-2}s^{-1}$.

Although the electronics were upgraded, the crystostats have remained unchanged. In addition to original RunI calorimeter [1], both the Inner Cryostat Detector (ICD) and the preshower sytems are used in RunII for particle energy measurements. The ICD was installed to give additional coverage between cryostats, specifically in the region of $1.1 < |\eta| < 1.4$. The preshowers, both central and forward, were added to provide additional particle energy and position measurements as compensation for the added material in front of the calorimeter in RunII. More information about these systems can be found in Ref. [2].

2. Electronics

The DØ calorimeter is divided into four electromagnetic (EM) layers, four fine hadronic (FH) layers and up to three coarse hadronic (CH) layers to optimize the energy response for various types of particles. A schematic drawing of the entire calorimeter is shown in Fig. 1. The calorimeter uses a combination of metal absorbers and active liquid to accurately measure particles' energies that pass through it. Liquid argon (LAr) is the active sampling medium and the absorber layers, as well as the thicknesses, are listed in Table 1.

For the entire calorimeter, there are 47,364 active readout channels. All channels are finely segmented, 0.1×0.1 ($\Delta\eta \times \Delta\phi$), with increased segmentation for the maximal EM shower layer to 0.05×0.05. Up to 12 channels, or cells, form pseudo-projective towers in each $\eta - \phi$ position.

Each cell is comprised of a 2.3mm gap filled with LAr between an absorber plate and a G10 insulator board coated with an epoxy with high resistivity. Particles passing through the cell ionize the active medium and a potential of ~2 kV creates a current of electrons towards the resistive pad with an average drift time of ~450 ns. Copper pads etched into the board collect

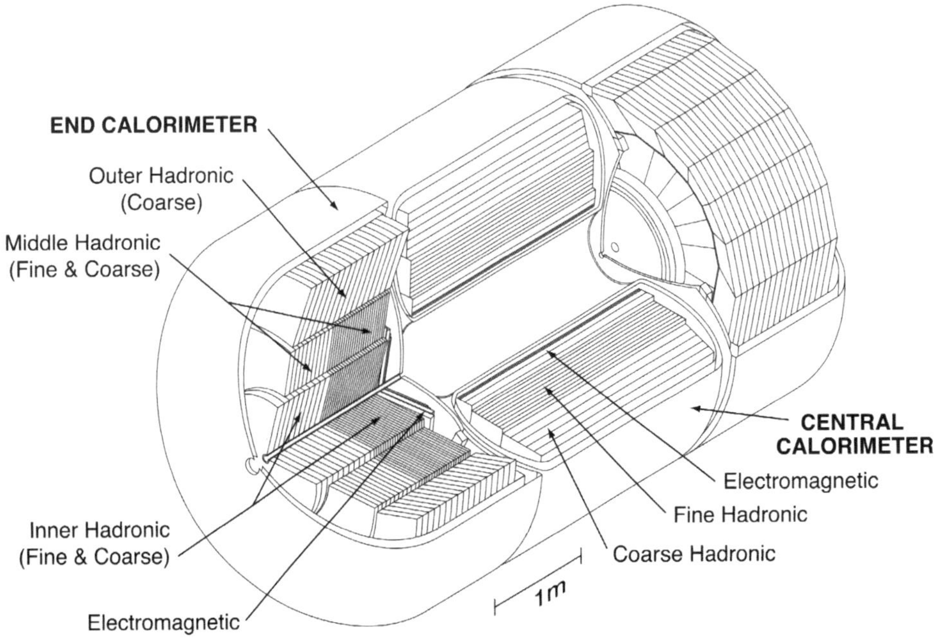

Figure 1. The three cryostat, hermetic liquid argon sampling calorimeter at DØ.

Table 1. Amount of material before each layer of the central (CC) and end cap (EC) calorimeters for the electromagnetic (EM), the fine hadronic (FH) and coarse hadronic (CH) layers. The thicknesses are listed in terms of radiation lengths (X_0) for the EM layers and in absorption lengths (λ_A) for the hadronic layers.

Layer	Region	Thickness	Material
EM	CC	2, 2, 7, 10	Uranium (3mm)
EM	EC	0.3, 3, 8, 9	Iron(1.4mm) + Uranium (4mm)
FH	CC	1.3, 1, 0.9	Uranium (6mm)
FH	EC	1.3, 1.2, 1.2, 1.2	Uranium (3mm)
CH	CC	3	Copper (46.5mm)
CH	EC	3, 3, 3	Iron (46.5mm)

the image charges produced, and pads are ganged together to form individual readout cells. Figure 2 shows a representation of the calorimeter cell design.

Once the charge is collected at the copper pads within the detector, they are sent via low impedance (30Ω) cables to the preamplification system. The preamplifiers are designed to match the cables' input impedance to prevent reflections in the signal. Fourteen types, or species, of preamplifiers are used to minimize cell-to-cell capacitance differences, which can be up to almost a factor of 10. This is done to create an integrated signal from the preamplification system that is independent of the input cells characteristics. This integrated signal is then sent to the Baseline Subtraction System.

The Baseline Subtraction (BLS) system receives the voltage signal from the preamplifier system and determines the energy deposited in each cell on an event-by-event basis. When the signal arrives to the BLS system, it is first intercepted from the Level 1 Calorimeter (L1Cal)

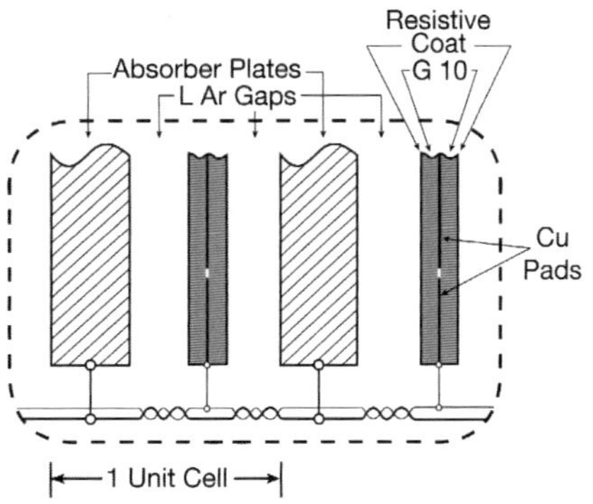

Figure 2. Schematic of a calorimeter cell.

trigger pick-off. The L1Cal sums the analog energy in towers of 0.2×0.2 ($\Delta\eta \times \Delta\phi$) for use in trigger decisions of event selection. The L1Cal was upgraded in 2006 to provide a more sophisticated trigger decision process [3]. The signal continues on to the signal shapers. The signal is shaped such that only the rising edge of the input signal is used in its energy estimation, which is approximately two thirds of the total signal. This is done because an integration of the entire signal is impossible given the timing constraints imposed by the Tevatron's 396 ns bunch crossings. The signal is then sampled at its newly shaped peak. In order to remove the remaining energy from previous events, the BLS system also samples the cell energy 396 ns before the peak of the signal. This excess energy is removed from the total measured signal energy and this modified signal energy is stored until a Level 2 accept is issued from the three-tiered trigger system. Due to the high volume of events processed and awaiting Level 2 decisions, Switched Capacitor Arrays (SCAs) are used for analog storage. The SCAs are also used for storage of signal information along the two calculated gain paths, $\times 1$ and $\times 8$. The use of two gain paths increases the dynamic range for the analog to digital conversion from 12 to 15 bit. The correct gain path is chosen by the BLS circuitry and once the event receives a Level 2 accept, it is sent to the Analog to Digital Converters.

The Analog to Digital Converters (ADCs) tranform the analog signal from the calorimeter cell into a digital energy value. It is here where signals are subject to a 2.5σ threshold in order to further suppress noise and minimum bias events. The ADC system also houses the timing and control operations of the entire calorimeter. It is synchronized to the accelerator clock and sends commands to the BLS system to regulate the signal and background sampling and subtraction. It also coordinates event processing among the ADCs and communications to the overall DØ detector framework. The digitized energy is sent to be used in event reconstruction.

3. Monitoring

Monitoring is an essential aspect to both efficient running and high quality calorimeter data. Monitoring is performed both online and offline for all parts of the system. This includes hardware readout of voltages, currents, temperatures as well as process statuses. Thousands of quantities are measured and monitored and if a problem does arise, alarms are connected to the data acquisition system to temporarily suspend data recording until the alarm status has been resolved. This is done to prevent bad quality data from being recorded. The hardware also contains failsafes so that no permanent damage would be done to the system due to malfunctions.

The monitoring also includes direct reconstruction of physics objects so that biases that are not observed as direct failures in the hardware can be possibly be seen in other ways. The reconstruction objects include missing transverse energy, as well as occupancy and energy, which

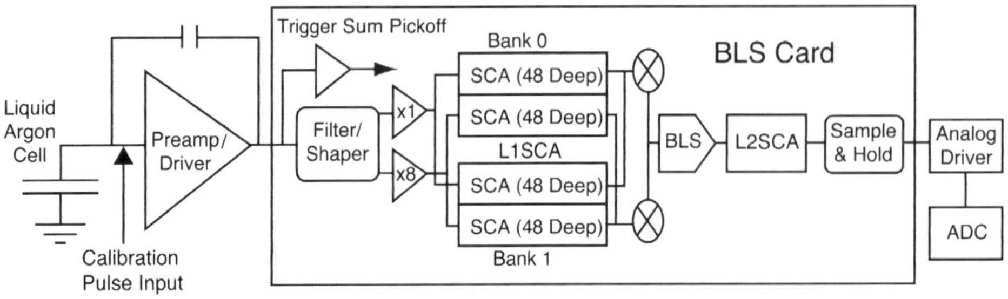

Figure 3. Schematic of the analog signal through the BLS circuitry using two independent gain paths.

are calculated at the cell and tower level. These quantities are reconstructed and monitored by a program that samples the data being recorded from the calorimeter. It has inline algorithms for both fast reconstruction of these objects and to search for signatures of failing hardware, unusually formed distributions and noise. Events containing any of these signatures are marked with data quality flags that used in the removal of bad events.

These tools are monitored by a trained shifter 24 hours a day, seven days a week in order to most efficiently address any potential problem that could arise with the calorimeter. Shifters have access to a large database of instructions, diagrams, and plots to aid in the proper identification and resolution of these problems. In addition to these resources, two pager carrying experts are available to answer and resolve any issues that are beyond the scope of the shifter.

Many times the information available for online quality criteria is not only used to help identify potential hardware failures, but they are also very sensitive to any noise that would possibly be introduced into the system. External noise has the potential to comprise good data, and extensive steps have been taken to remove it whenever it has appeared in the past. Safeguards were also developed to help prevent future instances, such as extensive grounding studies of the detector. Because of this proactive stance, the introduction of potential noise is unlikely. However, it is closely monitored and dedicated studies are performed to quickly identify and address whenever a candidate would arise. With constant monitoring and strict requirements on the these online hardware and physics readout values, we ensure the highest quality data and the most efficient running system possible.

4. Hardware Stability

Over 99.8% of the system is operational and performs optimally at all times. Almost all problems that do occur are well understood and thus the amount of downtime is minimized. However, access to a large part of the electronics systems is either restricted or, in some cases, not permitted. In these cases, efforts are made to migitate the downtime caused by these failures. As an example, it is advantageous to place the preamplification system within close proximity to the cryostat, but a power supply failure there results in a downtime of $8-10$ hours to access the area and repair the supply. In this case, redundant supplies are in place to be used in case of failures, thus avoiding extended periods without recording data for the affected section of the calorimeter.

Table 2 shows the most common hardware replacements for the calorimeter and maximum number of channels affected. As shown, most of the hardware replacements are in the BLS system. The BLS system is located directly under the detector in the collision hall, which means that access is restricted to times when there is no beam in the Tevatron. The replacement time itself is straightforward, and most often only single channels fail and have minimal impact on

Journal of Physics: Conference Series **160** (2009) 012002 doi:10.1088/1742-6596/160/1/012002

Table 2. Listed are the most common hardware replacements for the calorimeter. The interval of time in between replacements is given as well as the maximum number of affected channels. This number is expressed as a fraction of all channels in the system, as shown in the last column.

Hardware	Interval (weeks)	Repair time (hours)	Channels affected	Fraction of system affected
BLS Daughtercard	1	0.5	12	0.00025
BLS Motherboard	4	0.5	48	0.001
BLS Power Supply	16	2	768 (1536)	0.014 (0.028)
Preamp Power Supply	12	8-10	4608	0.085

the quality of recorded data. Finally, it is worth noting that problems due to overall aging of the detector have been minimal over its lifespan. Only one channel was permanently disabled because of effects consistent with aging, corresponding to less than 2×10^{-3} % of the system.

5. Calibrations

5.1. Online Calibrations

To obtain the most precise data possible, the calorimeter itself must be well-calibrated. This calibration is done using both online and offline techniques. The online calibrations are performed in two ways. The first is a pedestal measurement for all channels, which is performed daily. Pedestal calibrations measure the average ADC counts per channel *without beam* for both gain paths. These values are used during event energy calculations as the pedestals are subtracted from the integrated signal, as well as being used in the 2.5σ noise suppression. The stability of the recorded pedestal values from previous ones are used as a cross check as fluctuating pedestals are a direct sign of hardware failure.

The other main online calibration is a pulser calibration, which measures the electronics energy response channel by channel. This calibration is performed weekly, and also immediately after any calorimeter hardware replacement. The pulser is connected to the preamplification system where it inserts a known amount of charge. This generated signal is processed and the measured and inserted charges are compared to determine the response of the system. The input charges are varied in order to validate the entire dynamic range of the channels. The electronics' response is fit with a linear function that characterizes the energy response over this range. The fit, along with any non-linear corrections derived for it, are combined into calibration coefficients and are calculated individually and applied channel by channel. Together these calibrations are very effective tools to maintain optimal precision in the calorimeter, to aid in the identification of under-performing hardware, to minimize any and all cell to cell response variations, and to maintain a very robust and stable system.

5.2. Offline Calibrations

The online calibrations ensure a uniform and well understood response from all channels over their entire dynamic range. Calibrations complimentary to these are the ones performed offline. One main offline calibration is the ϕ-intercalibration of the calorimeter. In the absence of mechanical detector effects, the calorimeter cell response is independent of the cell's position in ϕ as the beams from the Tevatron come unpolarized. Thus all cells would have a uniform response. However, variations in the detector's mechanical structure can bias this response. This calibration is done to account for these effects in order to regain this uniformity. This is done by calibrating all cells in rings of ϕ throughout the detector. This normalization is performed

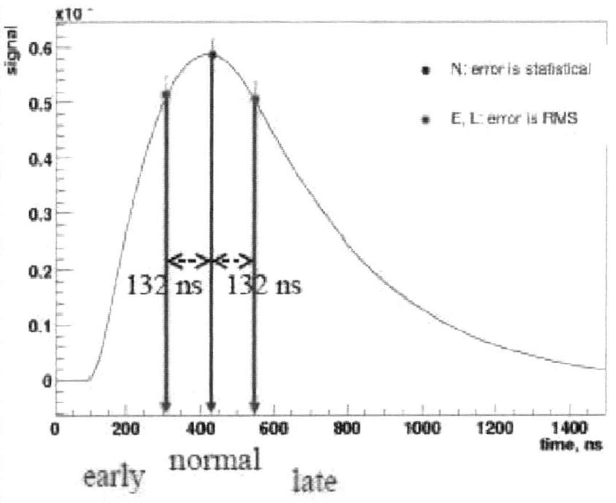

Figure 4. Triple timing studies use the information of the signal both before and after its peak. An optimal sampling point is one where the early sampling and the late sampling have the same values.

on data after all other calibrations have been taken into account. To obtain correction factors with very low uncertainty, these calibrations use ~5-15 million events that have been recorded parasitically from normal data taking. The original study was performed in 2006 and is explained in greater detail in Ref. [4]. As an example of the results from this study, the uncertainties in the central calorimeter are approximately 0.7%. Cross checks are performed on these calibration constants approximately every six months. The found variations in these time periods are very small, indicating that the overall system is very stable.

Additional studies are also performed on the processed calorimeter data to verify the quality and stability of calibrations in place. One such study is the triple timing study. It was originally performed as an independent measure of the calibration technique listed previously and was found to yield consistent results. This study uses information of the signal before and after its shaped peak in the BLS system. Because the signal is sampled every 132 ns, it is possible to measure the signal value during the rise and fall of the distribution as well as at the peak. Figure 4 shows the nominal sampling point of the peak. This signal shape is symmetric about this point up to where the early and late samplings occur, thus a difference of these measured values would indicate an undermeasured signal value. The size of this difference is used to estimate how much of a bias is present in the BLS sampling procedure. This study is performed using ~600,000 data events specially recorded to include this additional peak information. The study originally found that over 90% of all channels see less than a 0.5% percent diffence in these measured values. This procedure is repeated on the order of every six months as an independent verification of the sampling and timing stability of the BLS system and has found the system to be very stable with time with almost all channels showing variations on the order of 0.05%.

6. Conclusions

The DØ calorimeter has continued to perform efficiently and reliably throughout RunII at the Tevatron. It has maintained an efficiency greater that 90% over this entire period, and has seen improvement as RunII has progressed. Data quality is a top priority, and monitoring is performed online and offline for both hardware and reconstructed events to ensure the highest quality of recorded data, as well as preventative measures taken to counter potential noise problems. Online calibrations are performed as often as once a day for pedestal measurements and once a week for energy response measurements. These calibrations have helped to obtain a very stable system in which more than 99.8% of the system is at optimal performance. The problems that do occur typically affect less than 0.2% of all channels and are quickly fixed. The entire DØ collaboration looks forward much more high quality data and continued efficient running from the calorimeter throughout the entirety of RunII at the Tevatron.

Acknowledgments
We would like to extend our thanks to our DØ colleagues, as well as to Fermi National Laboratory for its continued support. I would also personally like to thank Walter Geist, Dean Schamberger, and George Ginther for all their help and guidance.

7. References

[1] S. Abachi *et al*, *Nucl. Instr. and Meth.* **A338**, 185 (1994).
[2] V.M. Abazov *et al*, *Nucl. Instr. and Meth.* **A565**, 463-537 (2006).
[3] M. Abolins *et al*, *Nucl. Instr. and Meth.* **A584/1**, 75-97 (2007).
[4] K. Peters, Proceedings of the 12th Calorimetry in High Energy Physics Conference, AIP Conference Proceedings Volume 867, 17-24 (2006).

Journal of Physics: Conference Series **160** (2009) 012003 doi:10.1088/1742-6596/160/1/012003

Belle Electromagnetic Calorimeter and its sBelle Upgrade

KEK, 1-1 Oho, Tsukuba, Ibaraki 305-0801, Japan

E-mail: isamu.nakamura@kek.jp

Isamu Nakamura (中村 勇)

Abstract. Electromagnetic calorimeter for the Belle experiment on KEKB asymmetric e^+e^- collider is being operated successfully for 9 years. In this presentation, systems and operation of the calorimeter are described. The belle detector will be upgraded to cope with higher luminosity running of KEKB collider aiming for accumulating 10 ab^{-1}. In this talk, a plan of electromagnetic calorimeter upgrade will also be shown.

1. Belle Calorimeter

The Belle experiment is for detailed measurement of B meson properties using the asymmetric e^+e^- collider, KEKB operated at $\Upsilon(4S)$ resonance. The Belle detector[1] is a general purpose detector consists of tracking devices, čerenkov detector and calorimeter placed inside the 1.5 T solenoidal magnetic field surrounded by muon detector instrumented in the iron yoke. The main purpose of the electromagnetic calorimeter is to detect γ and π^0 from B meson decay. Energy of such γ ranges from MeV to GeV. Energy resolution is very important because for some analyses one has to separate signal and background events only with mass distribution. To separate two photons from high momentum π^0, fine granularity is required. Hermeticity is also very important to study final states with neutrino. The Belle electromagnetic calorimeter consists of 8736 CsI(Tl) crystals with typical dimension of $6\times6\times30$ cm^3. The length corresponds to about 16 X_0. Figure 1 shows the schematic view of the calorimeter. The calorimeter has pointing geometry and covers polar angle from 12° to 155°. The scintillation light of the CsI(Tl) crystal has λ = 560 nm with rather slow decay constant of 1.3 μsec but abundant light output of about 50000 p.e./MeV. The light is read out by two independent sets of silicon PIN photodiodes [2] with sensitive area of 10×20 mm^2 connected to charge sensitive preamplifier. The signals are sent to the front-end electronics with cable of typically 10m long. Two signals from preamplifiers are first added and then processed with two shaper with 0.3(fast) and 1μsec(slow) shaping time. The fast signal are used to generate trigger signal while the slow signal is then sent to LeCroy MQT300A charge-to-time converter. The MQT converts 3 range/12 bit charge information into 15 bit timing information which corresponds to 18 bit dynamic range. The time encoded energy information is then recorded with LeCroy 1877S Fastbus TDC. The readout system consists of about 100 TDC boards in five crates read out by three VME systems. It takes about 30 μsec to read out one physics event. The trigger rate is about 300 Hz with instantaneous luminosity of 10^{34} $cm^{-2}sec^{-1}$, which corresponds to about 1% readout deadtime. Although the signal is recorded with TDC, only energy is recorded. No event timing information is available.

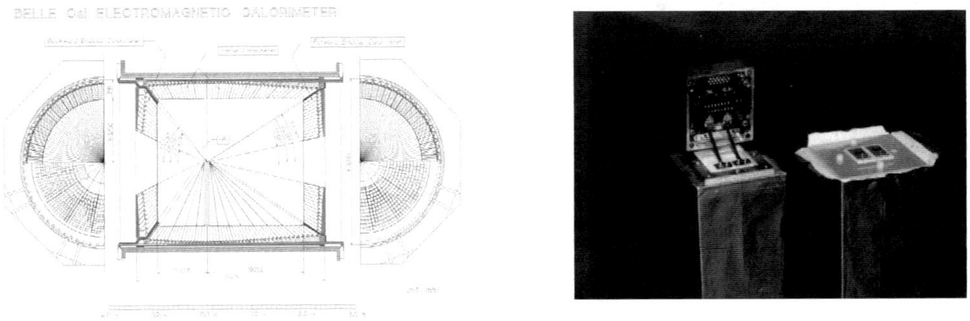

Figure 1. (Left) The Belle Electromagnetic Calorimeter. (Right) A counter.

2. Operation

Calorimeter energy is reconstructed from the TDC count (T) as

$$E = C \cdot \mathcal{G}(T - T_0), \tag{1}$$

where C, $\mathcal{G}$ and T_0 are time-to-energy conversion factor, electronics gain constant and pedestal, respectively. The gains and pedestals are monitored by dedicated calibration runs performed every morning. Calibration circuit is implemented in the frontend board so that test pulse can be injected to the preamplifiers. The conversion factor C is calibrated using bhabha events collected in typically one month period by minimizing chi squared defined as,

$$\chi^2 = \sum^{N_{\rm bb}} \frac{\left(E_{\rm exp} - \sum_{i \in {\rm cls}} C_i \cdot E_i\right)^2}{\sigma^2},$$

where $E_{\rm exp}$ is energy expected from the direction obtained from the track associated to the cluster, C_i is the conversion factor in equation 1, E_i is the energy measured in one crystal and σ is resolution. The sum runs all bhabha events. Taking derivatives of the χ^2 by each eight thousand C_is, minimization of χ^2 turns to the inversion of large 8000×8000 sparse matrix. Since tracking is not available in the very forward and backward regions, channels in those regions are calibrated with cosmic muon events which deposit mip signal to the counter.

To keep good performance it is important to keep environment stable. For this purpose, temperature and humidity are controlled by circulating water and dry air in the detector. Environment of the calorimeter is monitored by 312 temperature probes and 104 humidity probes. The bias current, temperature and current of power supplies are monitored and logged by every a few minutes, and are watched by safety shifters during operation.

3. Problems

3.1. Radiation Damage

Radiation damage is serious issue for experiments with high intensity accelerators. Radiation damages of crystals and photodiodes due to γ rays had been tested before the start of the experiment up to dose of several krad. However unexpected increase of dark current has been observed in the forward and backward regions of calorimeter. Figure 2 shows the dark current of the photodiode in barrel and forward endcap as a function of days after beginning of experiment. Dark current increase in the forward region is more than 100 nA while only several nA increase is observed in the barrel.Since the beams cross every several nsec, signal from beam induced γ become continuous current. Thus, γ ray does of the calorimeter can be measured by integrating the current in the photodiode in the following relation.

$$\int (I_{\text{on}} - I_{\text{off}})dt \sim \int E_{\gamma} dt,$$

where I_{on} and I_{off} are bias current with beam on and off, respectively. According to this evaluation γ ray dose in the forward endcap is estimated to be $\mathcal{O}(100)$ rad. If we translate the 100 nA increase of dark current using the radiation damage test using ^{60}Co performed before experiment, it corresponds to $\mathcal{O}(100)$ krad, that is 1000 times too high.

On the other hand it turned out that neutron background is rather high. Either spent electrons or γs from radiative bhabha hit accelerator element near the detector and back scattered neutron reach detector. Since radiation damage by neutrons have not been considered seriously, series of neutron radiation test ware performed for both crystals and photodiodes using the reactor 弥生 (YOYOI)[3]. Figure 3 shows the dark current increase as a function of neutron fluence. According to the result the neutron fluence at forward endcap is estimated to be $10^{11}/\text{cm}^2$, that is consistent with independent estimation by simulation. It is also confirmed that the crystals and photodiodes can be used up to $10^{13}/\text{cm}^2$ that is enough for future upgrade of KEKB accelerator[4].

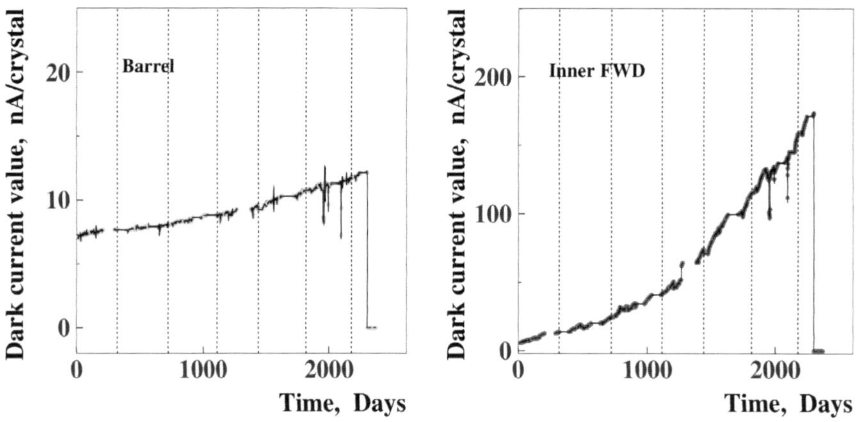

Figure 2. Darkcurrent in barral(Left) and foward endcap(Right) as functions of time after beginning of experiment.

3.2. Accererlator background

As the beam current increase, γs from accelerator background become visible. These background photons are studied in the random triggered events. Figure 4 show the sum of energies measured in the calorimeter as functions of total beam current. In the top figure energies are summed for all cells(crystals) with energy larger than 0.5 MeV while the bottom is for the sum of cluster energy. In the recent high current running, each event has roughly 2–3000 hit cells with E >0.5 MeV, total energy of 3 GeV from background photons. This is quite large as the beam energy of KEKB detector is 11.5(8+3.5) GeV. After applying clustering algorithm the effect is reduced to several background clusters and 500 MeV additional energy. Beam background produce not only additional clusters but also overlap to the clusters by the physics

Journal of Physics: Conference Series **160** (2009) 012003 doi:10.1088/1742-6596/160/1/012003

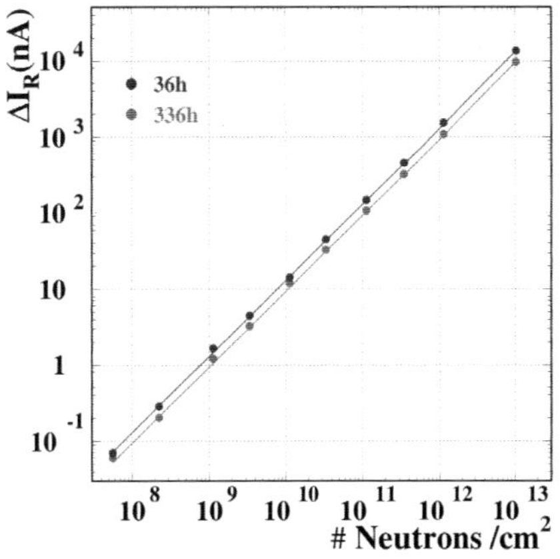

Figure 3. Darkcurrent increase as a function of neutron fluence.

events. The effect is illustrated in the Figure 4(Right). The Figure shows the additional energy caused by the background photons as a function of theta index of calorimeter with typical beam condition with luminosity of 10^{34}cm^{-1}sec^{-1}. One can see that each cluster has about 0.5 MeV additional energy in barrel and it increase in the endcaps to 1 MeV.

In the physics analysis, these background makes additional fake photons, shifts energy scales and deteriorate energy resolution. Photon reconstruction efficiency is also degraded because the background hits change cluster shapes that is used to identify low energy photons.

The above effect is roughly proportional to product of beam current and vacuum pressure. Since the vacuum is also proportional to current, background increase as current squared. We assume that the background level at the upgraded KEKB, which can provide luminosity of several $\times 10^{35}$cm^{-1}sec^{-1} with stored current of 13 A, becomes 20 times higher than current situation.

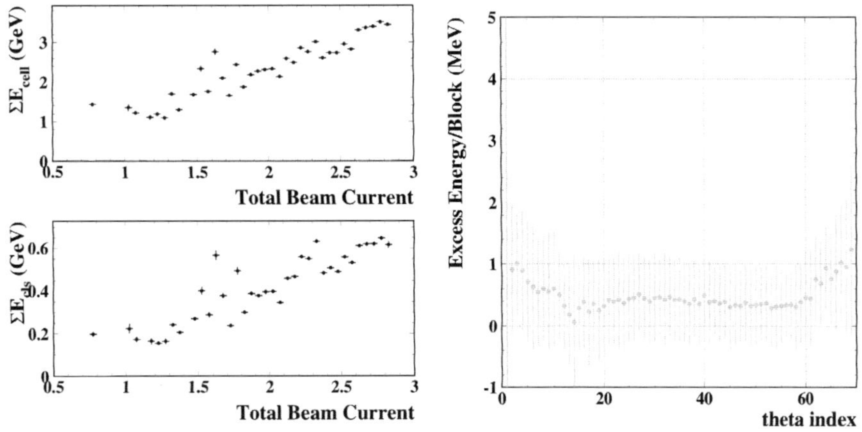

Figure 4. (Left) Total energy observed in random triggered events as functions of total stored current. (Right) Average additional energy in physics cluster due to background as a function of theta index of calorimeter.

Journal of Physics: Conference Series **160** (2009) 012003 doi:10.1088/1742-6596/160/1/012003

4. Upgrade Plan

With very successful operation of the KEKB accelerator and the Belle experiment, integrated luminosity recorded is approaching to 1 ab^{-1}. Now it is planned to improve further KEKB accelerator aiming at instantaneous luminosity of $8\times10^{35}cm^{-1}sec^{-1}$[4]. With that luminosity 10 ab^{-1} can be accumulated in several years. To achieve the luminosity beam current will be increased to more than 13 A. With such high current, background from accelerator will be 20 times higher than now and the trigger rate will be as high as 10 kHz. Current detector is not sufficient to be operated in the condition. For example, the dead time at 10 kHz is 30%. To cope with upgraded accelorator, calorimeter will also be upgraded.

To discriminate background events from physics events, timing information is useful. In order to record time stamps and eliminate readout deadtime, waveform is sampled with pipelined readout system. In the new readout electronics, the signals are shaped with shorter shaping time of 0.5 μsec and sampled by 18 bit ADC with 2 MHz frequency. The waveform is stored in pipe of 128 deep waiting for trigger. After trigger is initiated, 16 samples of waveform is fitted to extract timing and amplitude. With new electronics introduced, it is expected to obtain factor 7 reduction of background. Figure 5 shows expected timing distribution with new electronics obtained by simulation.

In the forward and backward regions where background is severe, replacement of current CsI(Tl) crystal to faster undoped CsI crystals are planned. Undoped CsI crystal has time constant of 10–30 ns which is much faster than that of CsI(Tl), though the light output is one tenth. To cope with small light output and 310 nm wavelength, photomultiplier(PMT) will be used for photodetection. The PMT has only three multiplication stages to make length short in order to fit in the current crystal containers and has finemesh structure to maintain gain of $\sim$ 50 in the 1.5 Tesla magnetic field. The signal from the PMT will be read out with 42 MHz sampling. With all these improvement more than 100 times background rejectioncan can be achieved.

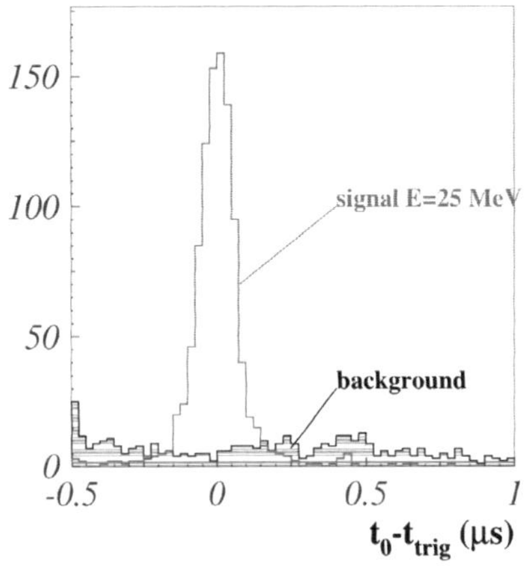

Figure 5. Expected timing distribution with new electronics.

References

[1] A. Abashian *et. al.*, Nuclear Instruments and Methods A479 117(2002)
[2] S2744-08, Hamamatsu Photonics, KK
[3] `http://www.tokai.t.u-tokyo.ac.jp/~rokan/`
[4] SuperKEKB Letter of Intent, KEK Report 04-4

Journal of Physics: Conference Series **160** (2009) 012004 doi:10.1088/1742-6596/160/1/012004

Performance and Operation of the *BABAR* Calorimeter

A. M. Ruland

Department of Physics, The University of Texas at Austin, 1 University Station C1600, Austin, Tx 78712-0264

E-mail: `ruland@physics.utexas.edu`

(For the BABAR calorimeter group)

Abstract. The performance and operation of the CsI(Tl) crystal calorimeter of the *BABAR* detector during the last years of operation is discussed.

1. Introduction

The *BABAR* detector is located at the PEP-II B Factory at the Stanford Linear Accelerator Center (SLAC). PEP-II is an asymmetric e^+e^--collider operating mainly at a center-of-mass energy of 10.58 GeV. This corresponds to the mass of the $\Upsilon(4S)$ resonance, which decays exclusively into $B^0\overline{B}^0$ and B^+B^- pairs. One main physics goal of the *BABAR* experiment was the measurement of *CP*-violating asymmetries in the decay of neutral B-mesons. Other goals of the experiment include precision measurements of the decays of bottom mesons to charm and τ leptons, as well as searches for rare decays utilizing the high luminosity delivered by the PEP-II accelerator. The *BABAR* detector (Fig 1) consists of 6 subdetectors. Starting from the interaction point and moving radially outwards there is a Silicon Vertex Detector, Drift Chamber, DRC (Cherenkov detector), an Electromagnetic Calorimeter, and an Instrumented Flux Return.

2. Calorimeter Goals

The calorimeter is designed for excellent efficiency, energy and angular resolutions over the energy range 20 MeV to 9 GeV. These requirements the choice for *BABAR* of a CsI(Tl) crystal calorimeter. The energy resolution of a homogeneous crystal calorimeter is given empirically as two terms summed in quadrature:

$$\frac{\sigma_E}{E} = \frac{a}{\sqrt[4]{E(\text{GeV})}} \oplus b, \tag{1}$$

where E and σ_E refer to the energy of a photon and its' RMS error in GeV. The energy dependent term, a, arises from fluctuations in photon statistics, electronics noise, and beam background generated noise. The constant term, b, arises from non-uniformity in light collection, leakage or absorption in the material between and in front of the crystals, and uncertainties in the calibration. The angular resolution is determined from the transverse crystal size and distance

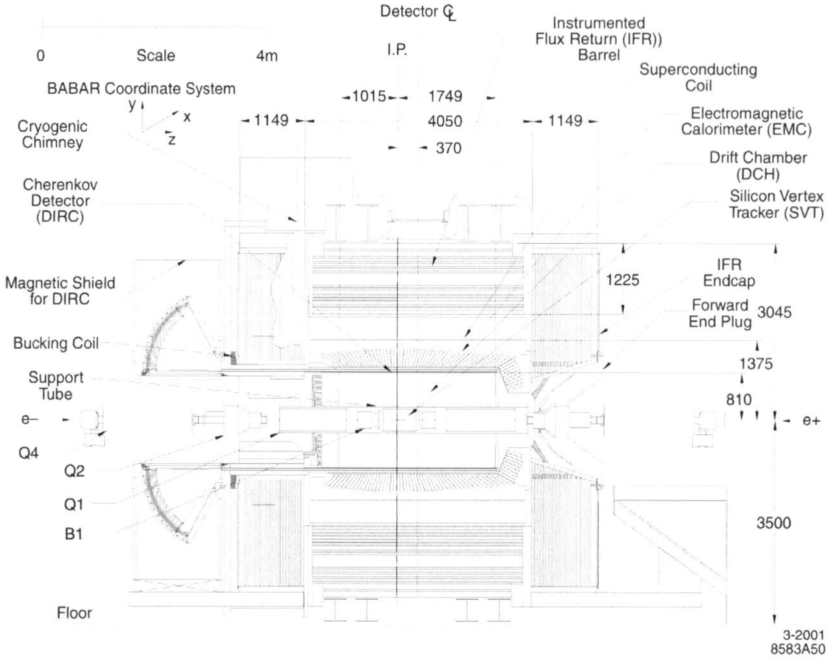

Figure 1. Longitudnal cross-section of the *BABAR* detector

from the interaction point. It can be empirically described by the equation

$$\sigma_\theta = \sigma_\phi = \frac{c}{\sqrt{E(\,\mathrm{GeV})}} + d, \tag{2}$$

3. Calorimeter Layout and Assembly

The *BABAR* calorimeter consists of 6580 CsI(Tl) crystals. The angular coverage is 126° in polar angle and 360° in azimuthal angle corresponding to a solid angle coverage of 90° in the center-of-mass (Fig 2). It is divided into two mechanically distinct sections: a barrel (5760 crystals) and an endcap (820 crystals). The barrel is subdivided into 48 polar angle rings with 120 crystals per ring. The endcap consists of 8 polar angle rings with the inner two rings containing 80 crystals each, the next 3 rings contain 100 crystals each, while the outermost 3 rings contain 120 crystals each. The geometry of the calorimeter is projective in ϕ, while in θ there is a non-projectivity of 14 mrad, except in the transition region between the barrel and endcap where it reaches 45 mrad. This non-projectivity minimizes energy losses through the spaces between the crystals. The 2 mm gap between the barrel and endcap is fully covered by the non-projectivity.

3.1. Crystal Assembly

The crystals are grown from CsI doped with 0.1% Thallium. They are trapezoidal in shape with a typical front(rear) dimensions of $4.7 \times 4.7 (6.1 \times 6.0)$ cm^2. The length of the crystals varies from 16 radiation lengths, X_0, in the backward part of the barrel to 17.5 X_0 in the forward part. For the details on the complete fabrication and assembly of the crystals see [1].

3.2. Mechanical Assembly

The individual crystals are assembled in carbon fiber modules. In the barrel, the modules contain 7×3 crystals in $\theta \times \phi$, except for the backward most module which contains 6×3

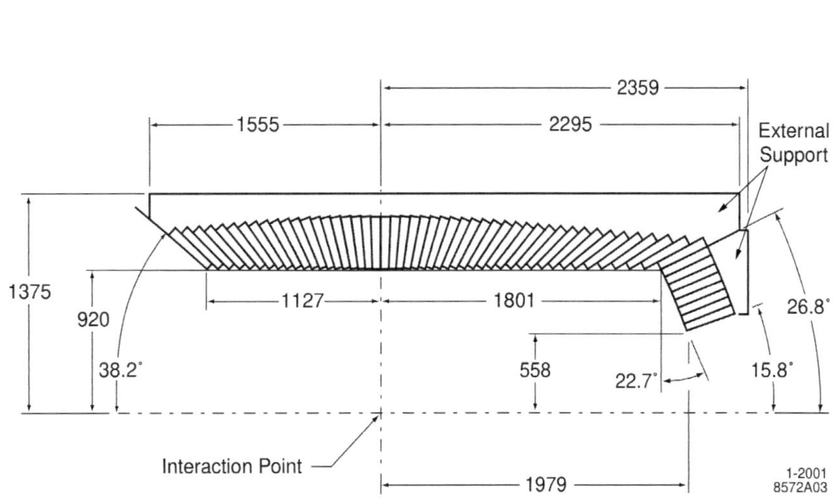

Figure 2. Longitudinal cross-section of the Calorimeter. Dimensions in mm.

crystals. The barrel consists of 280 modules which are inserted into an aluminum cylinder. The endcap contains 20 identical modules each with 41 crystals which are bolted to one of two semi-circular support structures.

4. Performance

4.1. π^0 and η mass and width

Fig 3a shows that two-photon invariant mass for hadronic events around the π^0 mass in data from 2001. Photons are required to exceed 30 MeV, while the π^0 energy is required to exceed 300 MeV. The reconstructed mass is measure to be 134.9 MeV/c^2, with a width of 6.5 MeV/c^2. The two-photon invariant mass for symmetric η's for $E_\eta > 1$ GeV is shown in Fig 3b. The reconstructed mass is 547 MeV/c^2, with a width if 15.5 MeV/c^2.

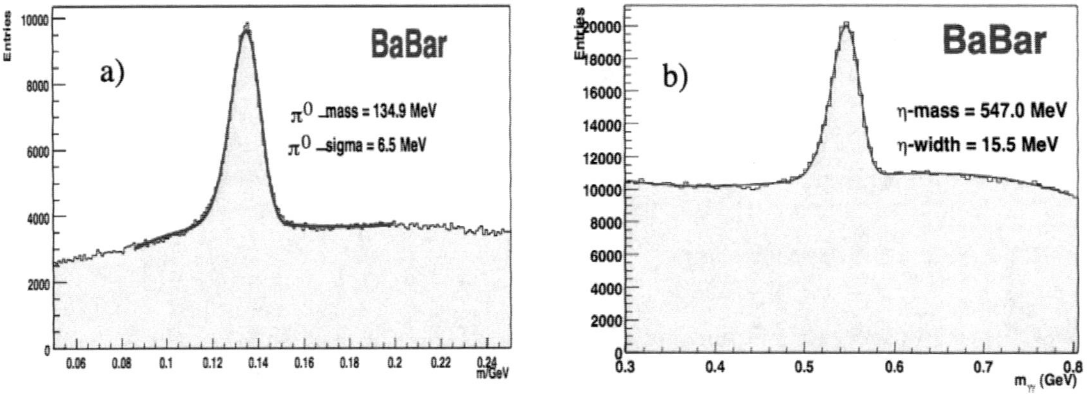

Figure 3. Invariant mass of the photons in hadronic events for π^0 mass (a) and η mass (b). The solid lines are fits to the data.

4.2. Resolution

Fig 4 shows the energy and angular resolution of the calorimeter derived from a variety of processes: radioactive source, symmetric π^0 and η decays, $\chi_{c1} \rightarrow J/\psi\gamma$, and Bhabha events. As the energy resolution of the π^0 and η is dependent on the calorimeter angular resolution, a simultaneous fit to energy and angular resolution was done for those cases, assuming an asymmetry of the photon energy distribution derived from Monte Carlo. A smearing of the MC energies is applied in order to have good agreement between Data and MC. The data from the radioactive source calibration shows a deviation form the fitted curve, as the photons from this process develop in single crystals and do not traverse material in front of the calorimeter. The angular resolution (Fig 4b) is derived from symmetric π^0 and η decays. Fits to the data yield

$$\frac{\sigma_E}{E} = \frac{(2.30 \pm 0.03 \pm 0.3)\%}{\sqrt[4]{E(\text{GeV})}} \oplus (1.35 \pm 0.08 \pm 0.2)\%, \tag{3}$$

$$\sigma_\theta = \sigma_\phi = \frac{(4.16 \pm 0.04)\text{mrad}}{\sqrt{E(\text{GeV})}} + (0.0 \pm 0.0)\text{mrad}. \tag{4}$$

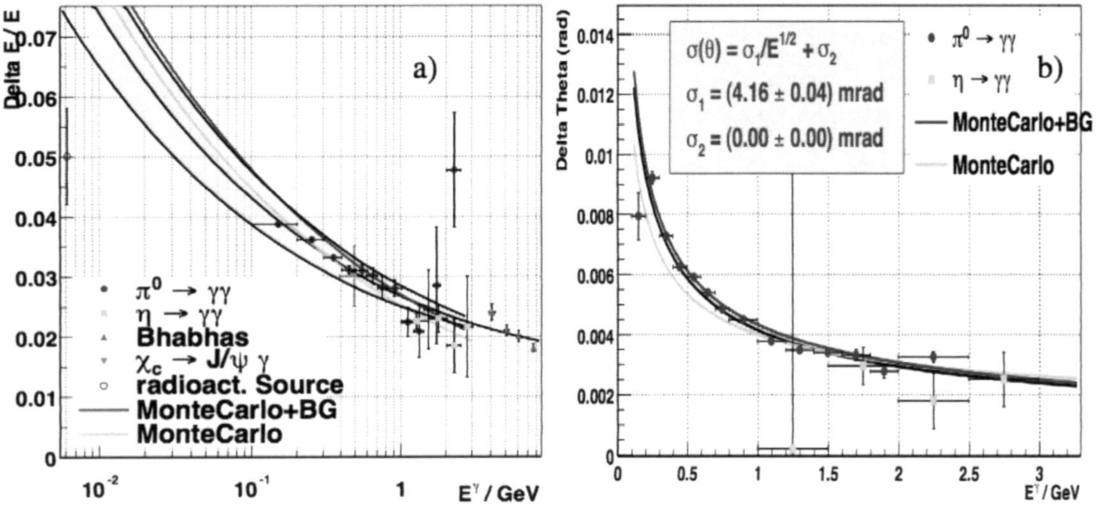

Figure 4. Energy resolution (a) and angular resolution (b). The central black line is a fit to the data. The outer black lines in (a) show the systematic uncertainty.

5. Calibrations

In spite of careful selection and tuning of the crystals, their light yield varies significantly and is non-uniform along the crystal axis. The light yield also changes over time due to backgrounds from beam-generated radiation. These differences in light yield must be calibrated away at different energies, corresponding to different average shower penetration. The calibration of the deposited energies is performed at two energies at the opposite ends of the dynamic range. The measurements are then combined by a logarithmic interpolation. At the low end of the range a 6.13 MeV photon from a radioactive source provides an absolute calibration, while at higher energies the relation between polar angle and energies of Bhabha events is used.

Corrections due to energy loss from shower leakage and absorption are performed as a function of polar angle and energy. For these corrections photons from π^0 decays are used with energies from 70 MeV to 2 GeV and photons from $e^+e^- \rightarrow \mu^+\mu^-\gamma$ are used with energies 400 MeV to 6 GeV. The details of this shower correction are given in these proceedings [2].

5.1. Radioactive Source Calibration

The radioactive source calibration uses 6.13 MeV photons produced in the reaction

$$^{19}F + n \rightarrow ^{16}N + \alpha, ^{16}N \rightarrow ^{16}O^* + \beta, ^{16}O^* \rightarrow ^{16}O + \gamma. \quad (5)$$

A fluid of polychlorotrifluoro-ethylene, activated by neutrons from a generator, circulates through a system of tubes in front of the crystals. ^{16}N has a lifetime of 7 seconds. All crystals on the calorimeter are calibrated with this method. The average resolution is 0.33%. Fig 5a shows the relative light yield change of the crystals in different calorimeter regions as function of the delivered integrated luminosity. This calibration is performed every 4 ± 1 weeks.

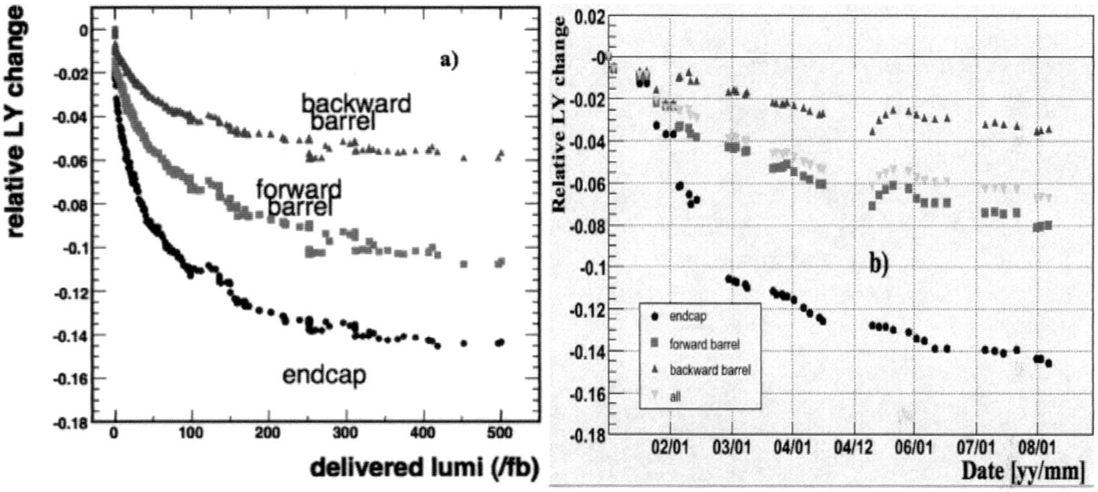

Figure 5. The relative light yield change for the radioactive source constants as a function of delivered luminosity (a) and the relative light yield change for Bhabha constants as a function of time (b).

5.2. Bhabha Calibration

At high energies, single crystal calibration is performed with a pure sample of Bhabha events. As a function of the polar angle of the $e^{\pm}$, the deposited cluster energy is constrained to equal that of a GEANT based Monte Carlo simulation. For a large number of energy clusters, a set of simultaneous linear equations relates the measured to the expected energy and thus permits the determination of a constant for each crystal. 200 $e^{\pm}$ hits per crystal result in constants with a statistical error of 0.35%. This calibration is done ~1 per month. Fig 5b shows the relative change in light yield of the crystals as a function of time for 3 different regions of the calorimeter as well as for the full set of constants.

6. Diagnostics and Monitoring

Real-time monitoring of the calorimeter hardware, currents, voltages, and temperatures is done using the EPICS software package [3]. EPICS is a set of open source software tools, libraries, and applications for real-time monitoring of large scientific experiments.

Journal of Physics: Conference Series **160** (2009) 012004 doi:10.1088/1742-6596/160/1/012004

Monitoring of the data is done in two separate passes. Initially, individual crystal parameters such as occupancies, hit timing, and multiplicity are extracted from the data as it is recorded. Automatic notification to the shifters is given when a parameter goes outside of preset boundaries. Secondly, fully reconstructed events are available within hours and provide higher level benchmarks such as π^0 mass and width, energy-momentum ratios, and track-cluster matching efficiencies.

6.1. Lighpulser Monitoring

Daily monitoring of the readout path from the crystals diodes to the DAQ system is performed using a lightpulser system. Spectrally filtered light is transmitted from a Xenon lamp to the rear of each crystal. The light pulse is similar in characteristics to scintillation light from CsI(Tl) crystals over the range of all but the lowest deliverable energies. This system allows for early detection of electronics issues that would affect data taking, as well as for monitoring relative lightyield changes in crystal response on a short time scale between calibrations.

6.2. Radiation Monitoring

Beam generated backgrounds are the major cause of reduction in the light yield of the crystals over time. In order to monitor this source of background, 116 RadFETs are placed in front of the calorimeter barrel and endcap crystals. These RadFETs are real-time integrating dosimeters based on solid-state MOS technology and are integrated into the EPICS monitoring system. As can be seen by Fig 6, the integrated dose is largest in the endcap which is closer the the beam line as well as more forward in polar angle making it more susceptible to beam generated background photons from small angle radiative Bhabha events in which an $e^{\pm}$ strikes a machine element.

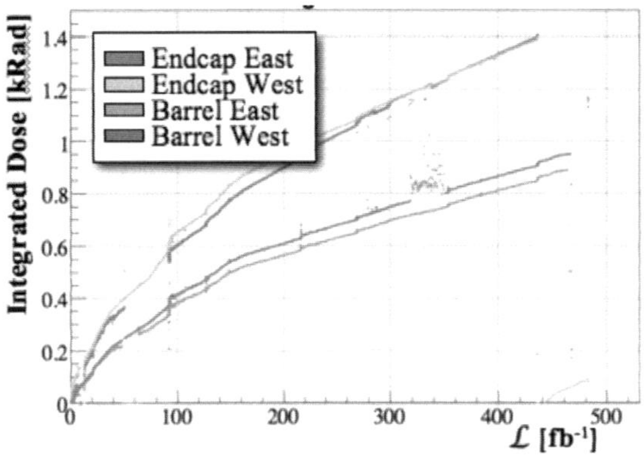

Figure 6. Average integrated dose for the barrel and endcap subdivided into east and west sides as a function of delivered luminosity.

7. Summary

The *BABAR* experiment concluded data taking in April 2008. For the past several years the Electromagnetic Calorimeter was recording data with high efficiency and excellent quality.

References

[1] *BABAR* Collaboration 2002 *Nucl. Inst. Meth.* **A479** 1.
[2] J.Marks, in these Proceedings.
[3] http://www.aps.anl.gov/epics/.

Journal of Physics: Conference Series **160** (2009) 012005 doi:10.1088/1742-6596/160/1/012005

Calibration of the *BABAR* CsI (Tl) Calorimeter

Jörg Marks for the Calorimeter Group of the *BABAR* Collaboration
Physikalisches Institut, Universität Heidelberg, Philosophenweg 12, D-69120 Heidelberg, Germany

E-mail: marks@physi.uni-heidelberg.de

Abstract. After nine years of operation, the *BABAR* experiment at the e^+e^- B factory PEP-II (Standford Linear Accelerator Center) stopped data taking in April 2008. An important part of the experiment is the electromagnetic calorimeter which consists of 6580 CsI crystals doped with thallium and read out by Si-PIN photodiodes. The light yield of the CsI crystals is changing in time due to radiation exposure. In addition to the changing light yield, passive material in front of and between the crystals as well as signal thresholds during the reconstruction influence the reconstructed energies. This requires a time-dependent calibration of the calorimeter. The calibration issues are reviewed and the calibration results obtained from various data samples are presented.

1. Introduction

The photon energy calibration of the *BABAR* electromagnetic calorimeter (EMC) [1] is performed in three main steps. The first two, the calibration of the electronic readout chain and the charge-to-energy conversion at the single crystal level were discussed in the previous presentation [2].

This presentation describes a third calibration step applied at the cluster level. The limited shower containment due to the total CsI crystal length and the dead material in front of the calorimeter (0.3 to 0.7 radiation length (X_0)) and between the CsI crystals lead to an energy loss of the total cluster energy measurement. Moreover, the noise suppression and the energy deposited in crystals not associated with the clusters during the reconstruction process cause an additional reduction. Therefore, the measured cluster energy has to be calibrated to the true energy of the primary electron or photon which is performed by the cluster energy calibration.

2. Cluster energy calibration

2.1. Cluster energy calibration scheme

The cluster calibration requires processes which allow a calorimeter-independent energy determination of the calibration signature (mostly photons). Their predicted energy has to be compared to the measured energy in the calorimeter over a range from well below 100 MeV up to the highest energies. Since no single process covers the whole energy range, photons from π^0 decays ($\pi^0 \rightarrow \gamma\gamma$) produced in multi-hadron events are used in the low-energy regime (70 MeV $< E_\gamma < 2$ GeV). The high-energy regime (400 MeV $< E_\gamma < 6$ GeV) is covered by photons from radiative muon events ($e^+e^- \rightarrow \mu^+\mu^-\gamma$).

The calibration function

$$C = f(E_\gamma, \theta_\gamma) \tag{1}$$

depends on the energy E_γ and the polar angle θ_γ of the photon. The emphasis of the calibration is to correct the measured cluster energies to the absolute energy scale and to describe the measured energy scale and their dependencies by the *BABAR* detector simulation based on GEANT4 [3].

Due to the poor statistical coverage of the full parameter space $(E_\gamma, \theta_\gamma)$, only the projections of the calibration function can be measured

$$C = g(E_\gamma) \cdot h(\theta_\gamma) \ . \tag{2}$$

In a first step, the polar-angle-dependent part $h(\theta_\gamma)$ is determined using radiative muon events. After applying the polar-angle-dependent part $h(\theta_\gamma)$, the energy-dependent part $g(E_\gamma)$ is extracted from both the π^0 and the radiative muon sample over the full energy range.

2.2. Calibration with radiative muon data

Since the initial state parameters of the process $e^+e^- \rightarrow \mu^+\mu^-\gamma$ are known and the muon energies are measured in a calorimeter-independent way, the reaction is kinematically over-determined. Thus, the energy and the polar angle of the photon can be determined by a kinematic fit independent of any calorimetric measurement. The predicted photon energy E_{fit} is compared to the photon energy measured in the EMC E^γ_{rec} by considering the ratio distribution $E^\gamma_{\text{rec}}/E^\gamma_{\text{fit}}$. The average values of the ratio distributions $\langle E^\gamma_{\text{rec}}/E^\gamma_{\text{fit}}\rangle$ are quantitatively evaluated by the peak position of a fit of a modified Log-normal distribution [4]. In the MC simulation, the cluster calibration function is extracted using the true photon energy E^γ_{true} instead of the predicted one.

The asymmetric line shape of the photon response distribution E^γ_{rec} causes systematic shifts of the ratio $E^\gamma_{\text{rec}}/E^\gamma_{\text{fit}}$ and therefore the data-over-MC-simulation ratio has to be used to extract the calibration function.

The angular part $h(\theta_\gamma)$ of the calibration function is determined in the different areas of the θ-modules (7 crystals each [2]) θ^i_γ. The following ratio yields the θ-dependent calibration function for data:

$$h_{\text{data}}(\theta^i_\gamma) = \frac{\langle E^\gamma_{\text{rec}}/E^\gamma_{\text{fit}}\rangle_{\text{data}}(\theta^i_\gamma)}{h_{\text{MC}}(\theta_\gamma) \cdot \langle E^\gamma_{\text{rec}}/E^\gamma_{\text{fit}}\rangle_{\text{MC}}(\theta^i_\gamma)} \ . \tag{3}$$

The function is measured for three different ranges of the photon energy (I. 0.5 − 1.0 GeV, II. 1.0 − 2.0 GeV, III. 2.0 − 4.0 GeV) depending on the polar angle of the crystals (θ crystal index) and is shown for the data taking period 2006 in Figure 1. The full lines represent the parametrizations.

In order to determine the energy-dependent calibration function $g(E_\gamma)$ for simulated data, the θ-dependent function $h(\theta_\gamma)$ is applied and $g_{\text{MC}}(E_\gamma)$ is obtained using the true photon energy.

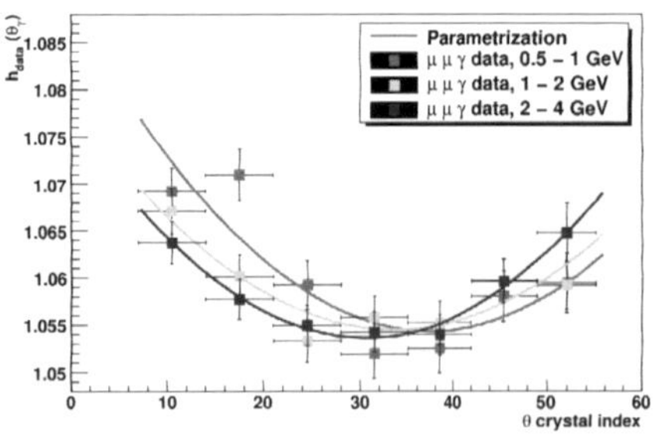

Figure 1. θ-dependent calibration in three energy ranges for $\mu\mu\gamma$ data of the data taking period 2006.

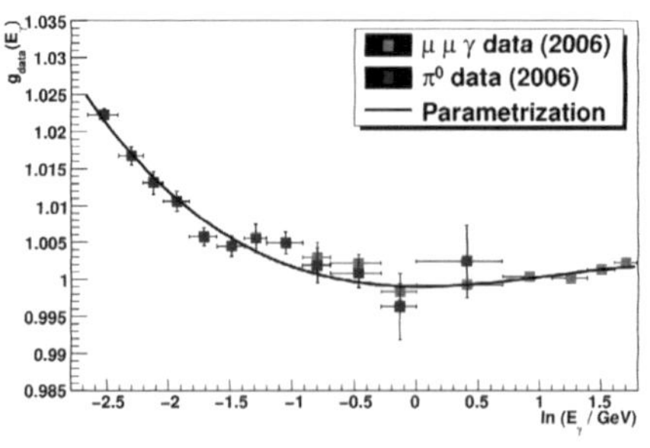

Figure 2. Energy-dependent calibration for π^0 and $\mu\mu\gamma$ data after applying the θ-dependent part (data taking period 2006).

The calibration function for data is extracted from

$$g_{\rm data}(E^i_\gamma) = \frac{h_{\rm data}(\theta_\gamma)\cdot\langle E^\gamma_{\rm rec}/E^\gamma_{\rm fit}\rangle_{\rm data}(E^i_\gamma)}{g_{\rm MC}(E_\gamma)\cdot h_{\rm MC}(\theta_\gamma)\cdot\langle E^\gamma_{\rm rec}/E^\gamma_{\rm fit}\rangle_{\rm MC}(E^i_\gamma)} \quad . \tag{4}$$

The red symbols in Figure 2 show the results obtained for the radiative muon data on a logarithmic E_γ scale.

2.3. Calibration with photons of π^0 decays

In order to measure $g(E^i_\gamma)$ in the low-energy regime, π^0 decays with the two decay photons in the same energy range (symmetric π^0's) are selected. The distribution of the invariant mass $m_{\gamma\gamma}$ of the π^0 candidates is shown as the blue distribution in Figure 3 for photons in the range 70 MeV $< E_{\gamma1}, E_{\gamma2} <$ 90 MeV. At low photon energies, peaking combinatorial background appears in the π^0 signal region. The combinatorial background contribution shown in Figure 3 (red histogram) is determined by combining photons from different events. Subtracting the background from the invariant mass distribution yields the π^0 signal distribution (Figure 4). The asymmetric line

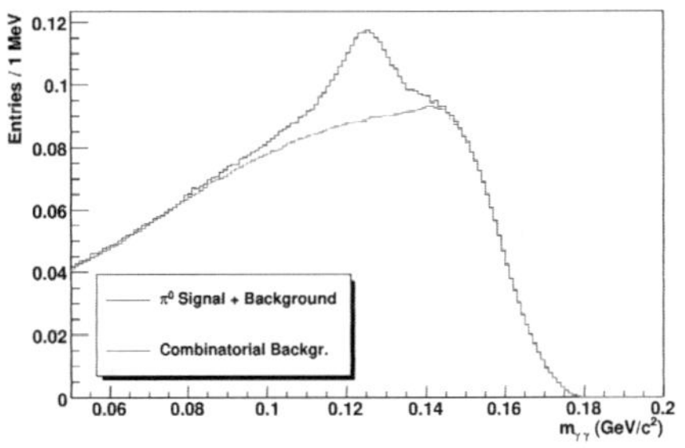

Figure 3. Invariant mass of all π^0 candidates (blue) and of the combinatorial background (red) for the data taking period 2006 (70 MeV $< E_{\gamma1}, E_{\gamma2} <$ 90 MeV).

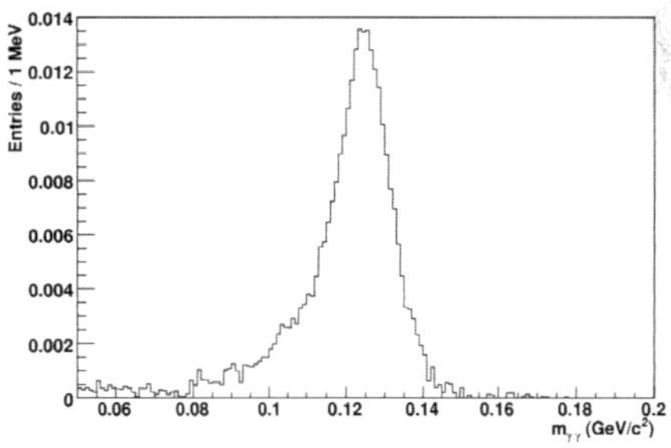

Figure 4. Invariant mass of the π^0 signal after subtraction of the combinatorial background contribution for the data taking period 2006 (70 MeV $< E_{\gamma1}, E_{\gamma2} <$ 90 MeV).

shape of the photon response causes the invariant mass $m_{\gamma\gamma}$ of the π^0 candidates to be shifted by $\Delta m = m(\pi^0) - m_{\gamma\gamma}$, [1] even if the photon response is perfectly calibrated. Δm is a function of the photon energy resolution, the tail of the photon response function and the position resolution of the photons. To account for these effects, the calibration function $g_{\rm data}(E_\gamma)$ in the low-energy regime is determined as the ratio of the average invariant mass values obtained with calibrated photon energies in the MC simulation and uncalibrated energies in the data,

$$g_{\rm data}(E^i_\gamma) = \frac{C_{\rm MC}(E_\gamma)\cdot\langle m_{\gamma\gamma}\rangle_{\rm MC}(E^i_\gamma)}{h_{\rm data}(\theta_\gamma)\cdot\langle m_{\gamma\gamma}\rangle_{\rm data}(E^i_\gamma)} \quad . \tag{5}$$

The average values of the invariant mass distributions $\langle m_{\gamma\gamma}\rangle$ are quantitatively evaluated by the determination of the peak position by a fit of a modified Log-normal distribution [4]. The blue symbols in Figure 2 show $g_{\rm data}(E^i_\gamma)$ on a logarithmic E_γ scale for the π^0 sample of the data taking period 2006. The full line represents the parametrization of the energy-dependent part of the photon calibration.

[1] The π^0 mass distribution is a convolution of the two asymmetric photon response functions.

This scheme implies a perfect description of the data by the MC simulation. Any deviation causes systematic shifts in the invariant mass and therefore shifts of the calibration function. Using typical discrepancies between data and MC simulation, the systematic uncertainties are estimated to be about 0.3 % in case of the π^0 sample. In case of the radiative muon sample, the systematic uncertainties decrease from 0.5 % at the lowest energy value to 0.1 % at the highest.

2.4. Cluster calibration results

The invariant mass ratio of data and simulation for the π^0 sample of the 2007 data taking period before (blue symbols) and after (red symbols) applying the calibration is shown in Figure 5. In case of the green symbols, the daughter photons of the π^0 decay originate from different energy ranges. Without cluster calibration, the deviations between data and MC simulations are as large as 1.5 % and $(E_\gamma, \theta_\gamma)$-dependent. After applying the calibration, within the uncertainties the data are well described by the MC simulation and are independent of E_γ and θ_γ. The data-

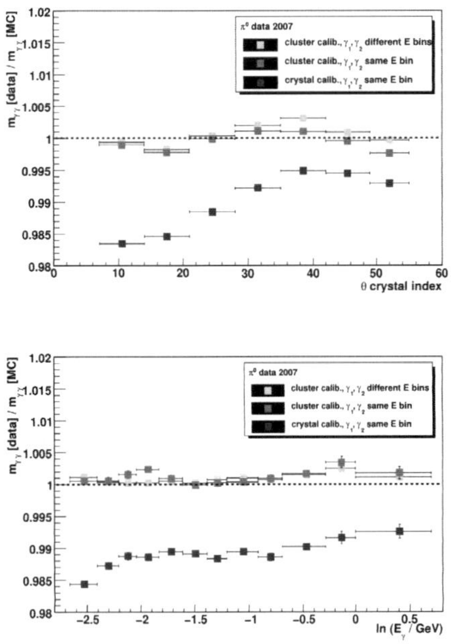

Figure 5. Invariant mass ratio of data and MC simulation for the π^0 sample of the 2007 data taking period.

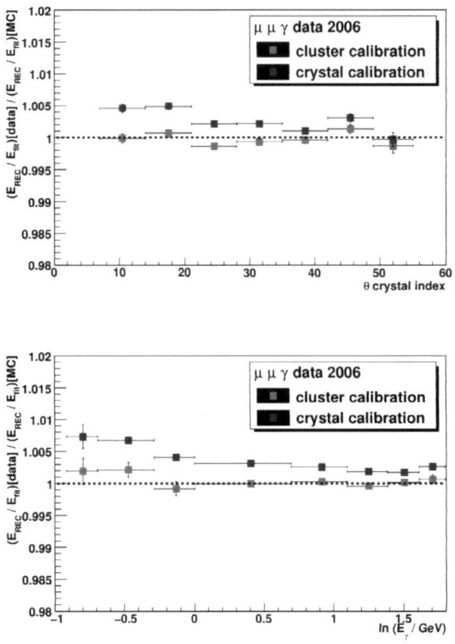

Figure 6. $E^{\gamma}_{\rm rec}/E^{\gamma}_{\rm fit}$ data-MC-simulation ratio for the radiative muon sample of the 2006 data taking period.

MC-simulation ratio of the quantity $E^{\gamma}_{\rm rec}/E^{\gamma}_{\rm fit}$ for the radiative muon sample of the 2006 data taking period is shown in Figure 6. As in case of the π^0 sample, after applying the calibration, the data are well described within the estimated uncertainties from the data-MC-simulation agreement. No remaining dependencies on E_γ and θ_γ are observed.

2.5. Validation of the cluster energy calibration

In the process $\Sigma \to \Lambda\gamma \to p\pi\gamma$ the photon energy is constrained and can be used to perform an independent test of the calibration at low photon energies (50 MeV $< E_\gamma <$ 400 GeV). The photon energy is calculated using tracking and particle identification information

$$E_\gamma^{\rm calc} = \frac{M_\Sigma^2 - M_\Lambda^2}{2(E_\Lambda - p_\Lambda \cos\theta_{\Lambda\gamma})} \,, \tag{6}$$

where M_X denotes the invariant mass, E the energy, p the momentum of the particles and $\theta_{\Lambda\gamma}$ the opening angle between Λ and γ. The ratio of the predicted to the measured photon energy in the projections of E_γ and θ_γ is measured and is found to be consistent with 1 within the total uncertainties of about 1 %.

In the high-energy regime, the photon energy scale can be probed by the process $B^0 \to K^{*0}\gamma \to K^+\pi^-\gamma$. The photon energy in this process ranges from 1.5 GeV to 3.5 GeV. The difference between the reconstructed energy $E^{*B^0}_{\mathrm{rec}}$ of the B^0 and the beam energy E^*_{beam} (both in the center of mass frame)

$$\Delta E = E^{*B^0}_{\mathrm{rec}} - E^*_{\mathrm{beam}} \tag{7}$$

is determined with quantities measured by the tracking system, particle identification and the photon energy. The measurements yield $\Delta E_{\mathrm{data}} = (-8.7 \pm 2.3)$ MeV for the data and $\Delta E_{\mathrm{MC}} = (-8.5 \pm 0.2)$ MeV for the MC simulation. This translates to a deviation of the photon energy scale of (-0.35 ± 0.09) % for data.

3. Smearing of the photon response in MC data

The line shape of the photon response is not fully described by the MC simulation. As a consequence, it is difficult to model selection criteria and efficiencies in physics analyses. The main goal is the improvement of the MC simulation description by tuning the MC model to data. Because of the large amount of MC data already simulated, a simplified approach is to introduce an additional term to the simulated photon response. It reflects the difference between the measurement and the MC simulation. Data of the process $e^+e^- \to \mu^+\mu^-\gamma$ allows to determine a smearing matrix (Students t distribution [5]) S_{ij} which connects the bin i of the distribution of the normalized photon response $E^\gamma_{\mathrm{rec}}/E^\gamma_{\mathrm{fit}}$ in MC simulation with the bin j of the data distribution. The smearing matrix depends on the photon energy and the calorimeter area.

After applying the smearing procedure to the MC data, the measured photon line shape in data is very well modeled. As an independent test, the ΔE distribution of $B^0 \to K^{*0}\gamma \to K^+\pi^-\gamma$ processes is compared in data and MC simulation. The ΔE line shape in data is very well described by the smeared MC simulation.

4. Monitoring of the EMC with muons

The signatures of minimum ionizing muons (from $e^+e^- \to \mu^+\mu^-$) measured in the EMC provide a stable signal in time with an energy loss in a CsI crystal of about 200 MeV depending on the muon energy. Selecting muons which fully pass one crystal and normalizing the muon signal by the crystal length allows to monitor the time stability of the crystal response. The average energy deposition of each crystal is determined as the peak position of a fit of a modified Log-normal distribution [4] to the muon signal.

The time stability over the full time of the BABAR operation can be seen by building the average over ϕ of the mean energy deposition of each crystal. The muon energy increases with decreasing θ (in forward direction). Therefore, the relativistic rise of dE/dx [6] leads to an increase of the average muon response in forward direction. The time dependence is illustrated in Figure 7 for the barrel part of the calorimeter. The variation over the data taking period from 2001 to 2008 is about ± 0.5 %.

The normalized muon response can be used to determine calibration factors for each crystal. The width of the measured cluster energy of the photons from radiative muons over the predicted photon energy $E^\gamma_{\mathrm{rec}}/E^\gamma_{\mathrm{fit}}$ judges the improvement in the resolution. The crystal calibration system considers the non-homogeneous light yield over the crystal length by using low energy and high energy signals (see [2]). However, in case of a calibration with muons this effect is not included and therefore the energy resolution degrades.

Journal of Physics: Conference Series **160** (2009) 012005 doi:10.1088/1742-6596/160/1/012005

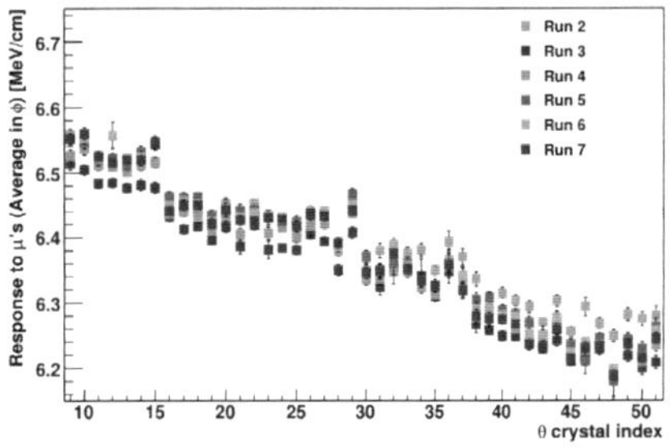

Figure 7. Normalized average muon response in ϕ as function of θ for different data taking periods.

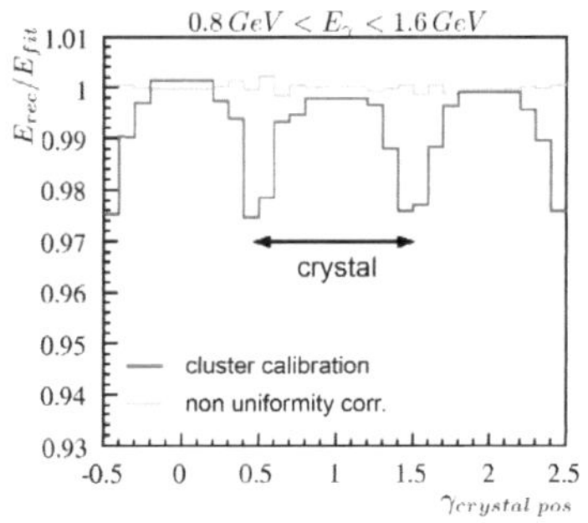

Figure 8. Normalized response $E^{\gamma}_{\text{rec}}/E^{\gamma}_{\text{fit}}$ across three CsI crystals before and after the lateral non-uniformity correction.

5. Lateral non-uniformity correction

Between the CsI crystals, dead material can not be avoided. As a result, a drop of the response $E^{\gamma}_{\text{rec}}/E^{\gamma}_{\text{fit}}$ at the crystal edges by about 2 - 3 % is observed (see Figure 8 red histogram). As in case of the cluster calibration, photons of the process $e^+e^- \rightarrow \mu^+\mu^-\gamma$ are used to obtain a correction function (lateral non-uniformity correction) depending on the impact position on the crystal surface. The correction function is determined in two dimensions and in five energy bins ($400\,\text{MeV} < E_\gamma < 8\,\text{GeV}$). Applying the lateral non-uniformity correction, the photon response is found to be independent of the impact position as shown in Figure 8 as green histogram. The energy resolution is measured to improve by about 10 %.

6. Association of EMC cluster and DIRC information

The dead material in front of the central part of the EMC varies between 0.3 and 0.6 X_0. The largest contribution is from the quartz bars of the *BABAR* particle identification system, an internally reflecting Cherenkov counter (DIRC) [1]. The fraction of photons with an electromagnetic shower starting in front of the EMC is about 8 %. These showers have a significant energy loss outside of the calorimeter. Therefore, the mean energy is smaller and a major contribution to the low-energy tail in the line shape of the photon response is observed. The additional fluctuations degrade the energy resolution. To study the calorimeter performance and the influence of resolution effects on data selection and efficiencies, it is of advantage to tag showers which started in front of the EMC. These early showers result in Cherenkov rings in the DIRC caused by the charged particles of the shower. Associating the Cherenkov rings and the EMC cluster allows for a selection of the early showering photons. In a sample of generic $B^0\bar{B}^0$ MC events, the fraction of true early showers in the number of reconstructed showers is 37 %. This number increases to 60 % for the $e^+e^- \rightarrow \mu^+\mu^-\gamma$ sample due to the smaller number of charged tracks. In data, 12 % of the π^0's have at least one daughter photon with a selected early shower. The average of the invariant mass distribution of these reconstructed π^0's is by about 1.5 MeV smaller than the invariant mass of π^0's without early showering photon.

7. Reference list

[1] B. Aubert et al. 2002 *Nuc. Instr. Meth.* A **479** 1-116
[2] A. Ruland et al. 2008 these Proceedings
[3] S. Agostinelli et al. 2003 *Nuc. Instr. Meth.* A **506** 250-303
[4] E. L. Crow and K. Shimizu 1987 *Stat.: Textb. and Monogr.* **88**
[5] M. Abramowitz and I. A. Stegun 1964 *Handb. of Math. Func.* Dover 0-486-61272-4
[6] H. Bethe and E. Bloch 1935 *An. d. Phys.* **337** Issue 3 325-400

The Crystal-Barrel/TAPS Experiment at ELSA
Current Status of the CsI(Tl) Calorimeters

Christoph Wendel for the CBELSA/TAPS Collaboration
Helmholtz-Institut für Strahlen- und Kernphysik, Universität Bonn, D-53115 Bonn, Germany
E-mail: cwendel@hiskp.uni-bonn.de

Abstract. The Crystal Barrel experiment at ELSA, as a medium-energy hadron physics experiment, is focused on baryon-spectroscopy and meson-photoproduction. The experimental setup includes three different, φ-symmetric calorimeters covering almost the complete solid angle. The detector system, together with a polarised target and a linearly or circularly polarised photon beam, allows in addition to the measurement of cross sections also the measurement of single- and double-polarisation observables. Baryon resonances up to masses of 2.5 GeV can be investigated.
In the course of a major upgrade of the CsI(Tl) calorimeters, two different techniques to detect the light generated in the crystals are compared with the known performance of the existing photodiode readout. The aim of the new techniques is to enable the calorimeter to be a first level trigger source while keeping a very good energy resolution comparable to the present photodiode readout.
The first solution is a complete redesign of the crystal readout, featuring two 10 mm × 10 mm APDs mounted directly on the crystals. With an integrated, fast preamplifier, this readout will allow both an energy measurement and the generation of a fast timing-/trigger-signal. The second solution is a modification of the existing setup. The photodiode will continue to provide the energy information while two additional SiPMs will create the timing-/trigger-signal.

1. Overview
To gain a detailed knowledge of the spectrum and the properties of baryons, photoproduction experiments are performed at the 3.5 GeV electron accelerator ELSA in Bonn [1]. The Crystal Barrel detector [2][3] with its almost 4π coverage is ideally suited to measure multi photon final states. The experiment is a fixed target experiment, it utilises a tagged beam of unpolarised, longitudinally or circularly polarised photons, created via bremsstrahlung from the (polarised) electron beam of the accelerator facility. As a reaction target either the Bonn Frozen Spin Target [4] or a pure liquid hydrogen target or a deuterium target can be used. Due to the forward boost of the reaction products, the forward angles are of special concern and three modular and interchangeable detectors were built to provide coverage of the region from 30° down to 2° depending on the physical demands.
In 2001 the Crystal Barrel calorimeter was modified to work with a wall of 528 TAPS BaF_2 detector modules [5], covering the solid angle from θ=30° to 6°. For the current measurement period, a new forward detector of 90 CsI(Tl) crystals was built to fit directly into the open forward cone, covering θ from 27.5° to 11.2°, while a smaller TAPS setup of 216 BaF_2 detector modules covers the area from 10.4° down to 1.6° (Fig.1).

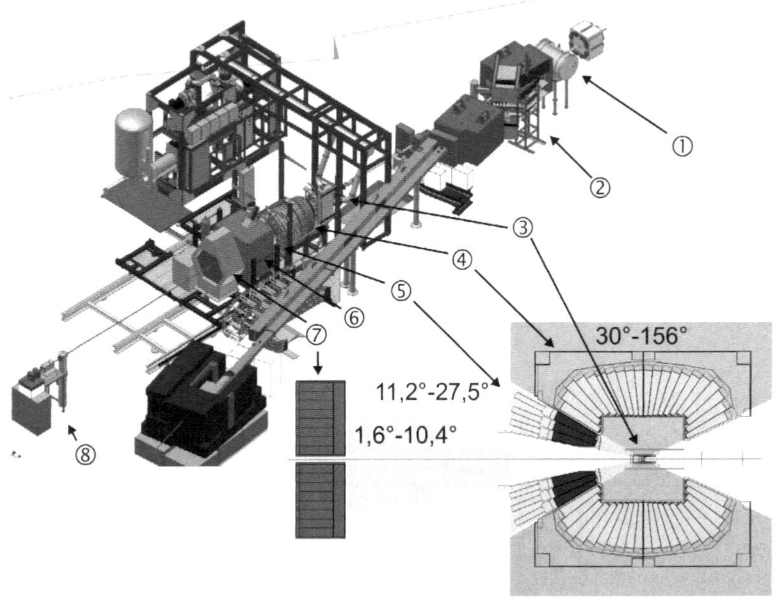

Figure 1.
(schematic) view of the Crystal Barrel experiment.

[1]Radiator Target
[2]Electron Tagging System
[3]Polarized Target Cryostat
[4]Crystal Barrel Detector
[5]Forward Detector
[6]CO_2-Čherenkow
[7]MiniTAPS
[8]Photon Beam Monitor

2. Current Status and Performance of the CsI(Tl) Calorimeters

In the present configuration, the Crystal-Barrel/TAPS experiment consists of three calorimeters, the BaF_2 calorimeter wall *MiniTAPS* and the two CsI(Tl) calorimeters *Crystal Barrel* and *CB Forward Calorimeter* (Fig. 2).

2.1. The Central Detector with the Crystal Barrel Calorimeter

The experiment's central detector consists of the Crystal Barrel as a main calorimeter and the Inner Detector for charge identification and first level trigger capability.
The 1230 CsI(Tl) crystals of the Crystal Barrel are arranged in 20 rings of 60 crystals each, covering the solid angle from 30° - 150° in θ (6° × 6° resolution) and one ring of 30 crystals covering 150° - 156° in θ (12° × 6°). The readout of the crystals is performed by photodiodes attached to wavelength shifting plates that cover the complete rear face of the crystals (Fig. 3-top). In each crystal's end cap enclosure is a preamplifier/driver electronic, providing amplification for 50 meter signal transmission to the ADC system. A special oil cooling system stabilises the temperature of the main calorimeter. With this readout, the Crystal Barrel reaches a measured energy resolution of $\sigma_{(E)}/E = 2.5\%/\sqrt[4]{E_{[GeV]}}$. The reconstructed angular resolution for the center of an electromagnetic shower is better than 1.5° [2]. In 1999, with the relocation to the ELSA facility, the original Fera-ADC readout was replaced by a FastBus 12 bit dual range ADC system [6]. Due to the long risetime of the shaped photodiode signals, the calorimeter cannot be used in a fast first level trigger scheme. For a second level trigger decision, a gated cellular cluster logic is used to calculate the event by event particle multiplicity within typically $4\,\mu s$ ($(1+N_{Part})\times 0.8\,\mu s$ [7]).
The Inner Detector as second part of the central detector system provides the missing first level trigger capability. It is a cylindrical fibre detector of 513 scintillating fibres arranged in three layers. Two of the three layers are wound around the detector in oposite directions, that way providing a redundant detection of charged particles and their direction. The reconstructed angular resolution is 0.1° in θ and 0.4° in φ, the detection efficiency for charged particles is measured to be 98% [3]. Matching of charged particles to hits within the Crystal Barrel is possible with a calibrated time resolution of 0.9 ns.

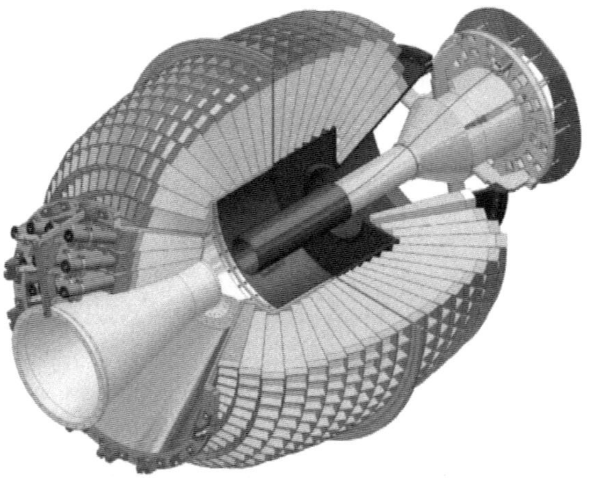

Figure 2. The *Crystal Barrel calorimeter* (center) with its *Inner Detector* (from the right) and the Crystal Barrel *Forward Detector* (from the left).

Figure 3. Photodiode readout of one of the main calorimeter crystals (top) and PMT readout of one of the forward calorimeter crystals (bottom).

2.2. The CB Forward Detector with the Forward Calorimeter

The CB Forward Detector was build in 2005 to cover the forward region between 30° and 10° in θ. 90 of the original Crystal Barrel crystals form the calorimeter part while a subdetector of 180 plastic scintillator tiles provides charge identification.

As in the original setup, the CsI(Tl) crystals have been arranged in three rings of 30 crystals each, with a covarege of 6° in θ and 12° in φ. Due to the needed holding structure for the detector, they now cover an area of 27.5° down to 10.2° with respect to the production target. To provide the desired first level trigger capability on photons, the crystals have been outfitted with lightguides and photomultipliers instead of the original photodiodes (Fig. 3-bottom). With fast, risetime compensating discriminators [8] a calibrated time resolution of 1.3 ns has been reached (Fig. 4). The first level trigger capability is achieved by a clusterfinder solution based on SRAM lookup units for the 90 crystals. The decision time for the whole system is 70 ns with an overall detection efficiency of 99,9% and a multiplicity accuracy of 95% [9].

For charged particle identification an additional subdetector based on plastic scintillator tiles was developed and mounted in front of the crystals (Fig. 9). The scintillation light is detected via thermoformed WLS fibres, which lead to multi anode PMTs outside of the detector. A time resolution for charged particles of $\sigma(t) = 1.7$ ns was reached, with a typical detection efficiency of 95%. The 180 scintillators match exactly the solid angle of the calorimeter crystals and allow, by a two layered, shifted setup, a spatial resolution of $6° \times 6°$ for the impact point of a charged particle [10].

2.3. Detector Trigger and DAQ system

The current data acquisiton consits of three parts.

The first part is the main trigger system. The trigger signals from all subdetectors are fed into a main trigger unit in which the desired trigger conditions can be selected to enhance specific reaction channels. Several trigger conditions can be used in parallel to optimise the use of beam time. After a fast first level trigger decision ($\sim$200 ns), the main calorimeter's cluster finder is gated and the final trigger decision, either experiment wide readout or fast reset, is performed. In parallel to the production trigger conditions, calibration and monitoring triggers are used for online monitoring of the detectors.

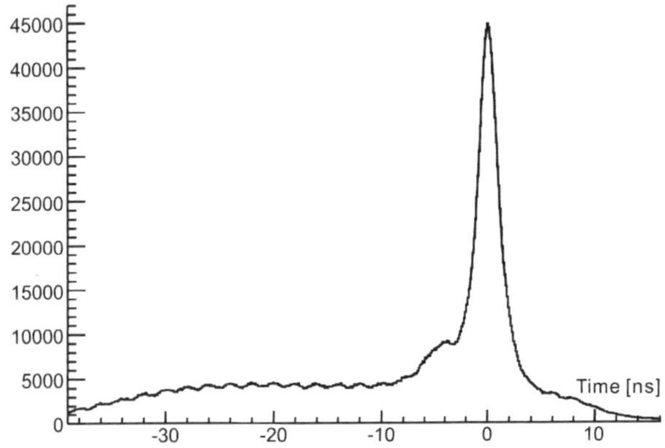

Figure 4. Calibrated time resolution for photons in the Forward Detector CsI(Tl) calorimeter. With a $\sigma(t)$ of 1.3 ns, the detector is able to resolve the bunch structure of the accelerator

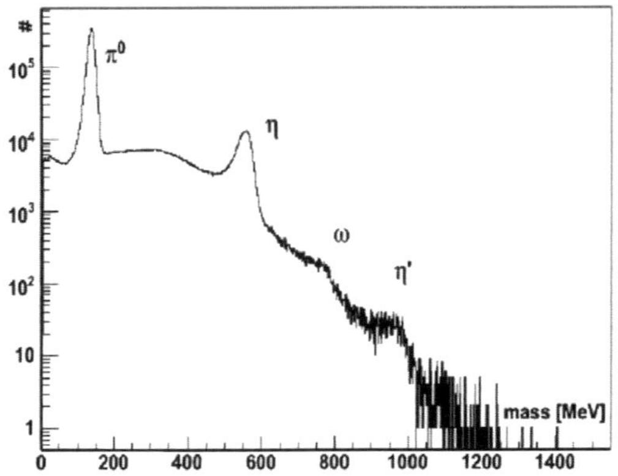

Figure 5. Online $\gamma\gamma$-invariant mass spectrum for events with three clusters (1 charged + 2 neutral) in the calorimeters.

The second part is the main DAQ system, it is based on a parallel readout scheme to minimize dead time due to data digitalisation and collection. Each subdetector is equipped with its own readout processor (local event builder), which collects the digitised data in a buffer as fast as possible. A global event builder collects and combines the data from all subdetectors and stores it on disk. The synchronisation of all local event builders is controlled over a special hardware synchronisation bus to guarantee data consitency event by event. The average event rate in the experiment is 1 kHz with a data rate written to disk of 8 MB/s.
The third part is a slowcontrol and online monitor system that constantly monitors all experimental parameters. The slowcontrol on the one hand monitors temperatures, voltages etc. and stores them in a database. It provides a web interface for accessing the database and changing detector parameters. An overview page warns the online crew in case of problems in the detectors. The online monitor on the other hand constantly copies events out of the incoming datastream. These events are instantantly displayed as raw spectra for all detectors. In parallel a preliminary calibration and analysis of the data is performed at a rate of 20 Hz. The display of these data provides the main tool to constantly monitor the experiment's performance while data taking. Problems within the experiment are instantly visible due to online spectra diverging from the norm and are marked accordingly. In fig. 5 shown is the one month sum of a typical online spectrum, in this case the $\gamma\gamma$-invariant mass of three particle events with one charged hit. In the logarithimc plot, the π^0 and η signal is clearly visible.

3. Future Readout Options

To improve the performance of the whole Crystal Barrel detector system, a full first level trigger capability on neutral and charged particles is desired. In addition the detection of charged particles will be improved by measuring the sign of charge and the momentum of the charged tracks in a new TPC, replacing the Inner Detector. For that, the whole calorimeter will again be placed in a magnetic field. The present readout with classical photomultipliers has to be replaced by a detection scheme working in 1.5 Tesla environment. Two different readout options are currently under investigation, either a complete redesign of the crystal readout based on avalanche photodiodes (APDs) or a modification of the current readout by adding silicon photomultipliers (SiPMs) to the present wavelength shifter.

Figure 6. Readout option with two 1 cm^2 APDs attached directly to the CsI(Tl) crystals.

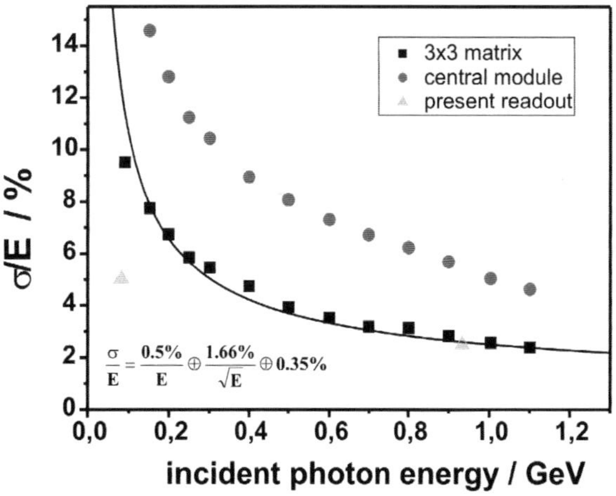

Figure 7. Energy resolution with the APD readout, compared to the current PD readout.

3.1. APD readout

For the APD readout, the current photodiodes would be replaced by two 1 cm^2 APDs of the PANDA type, attached directly onto the crystal rear face (Fig. 6). As with the present setup, a special preamplifier would be placed directly behind the APDs. First tests have been performed with an array of 3×3 crystals. With the PANDA amplifiers (thus not optimized for CsI(Tl) and our needs), the APDs readout already reached an energy resolution of $\sigma_{(E)}/E = 0.5\%/E \oplus 1.6\%/\sqrt{E_{[GeV]}} \oplus 0.35\%$, wich is similiar to the value known from the present photodiodes (Fig. 7). The time resolution of the APDs is with 1.8 ns close to the one measured with photomultipliers (1.3 ns). Currently under development is an optimized amplifier with a timing and energy output. Since the gain of APDs strongly depends on the temperature, methods for temperature stabilization are under investigation, the current design goal is a stabilisation of ±0.1 K at room temperature. Additionally, the prospects of a new readout system based on FADCs are currently investigated.

3.2. SiPM readout

The second readout option adds two SiPMs to the wavelength shifting plate of the present setup (Fig. 8). The energy readout and thus the known, good energy resolution of the photodiode is not touched. The SiPMs add the trigger capability to the calorimeter. The signals of both SiPMs are added and amplified on an additional board within the crystal endcap. First measurements proved the feasability of this method, with the presently tested SiPMs (MEPhI/PULSAR, 9 mm^2, 5kPixel) a trigger threshold of 27 MeV could be reached. Recent measurements on a 3×3 crystal array with an improved readout board suggest, that the threshold could be lowered to 20 MeV. Currently, the optimization of the SiPMs and amplifier boards for the desired trigger threshold of less than 20 MeV is under investigation. To reach this goal, an appropriate type of SiPM has to be chosen corresponding to the peak wavelength of the WLS (625 nm). Since dark counts are the main noise contribution, the size of the SiPM and number of pixels has to be optimized for the expected light yield at trigger level to suppress the SiPM inherent noise.

Figure 8. Readout option with two $9\,\text{mm}^2$ SiPMs additionaly attached to the WLS of the current setup.

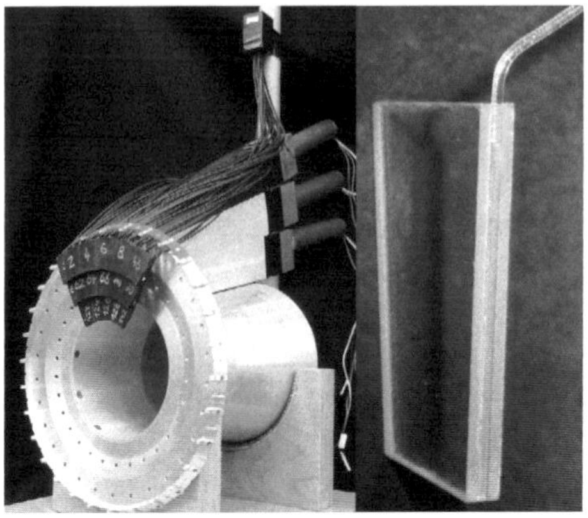

Figure 9. New subdetector for charge identification in the Crystal Barrel Forward Detector (left), based on 3 mm plastic scintillators with thermoformed fibre readout (right).

4. Summary

The Crystal Barrel calorimeter with its photodiode readout is now running for almost 20 years. In 1999, with its relocation to the ELSA facility, the detector system has been modified to fit the new experimental requirements. A new inner detector has been built and the calorimeter's readout system redesigned. In 2005, the forward region was transformed into a separate forward detector with a fast PMT readout. Since 1999, the calorimeters have been used in three different measurement periods and were a reliable source for competative and new data on baryon spectroscopy and meson photoproduction.
With a major redesign of the crystal readout, a first level trigger capability will be reached, extending the performance of the detector system substantially. New reaction channels will be made accessible over the full solid angle.

References

[1] W. Hillert 2006 EPJA 28 139-48
[2] E. Aker et al. 1992 NIM A321 69.
[3] G. Suft et al. 2005 NIM A538 416
[4] H. Dutz 1999 NIM A436 430–42
[5] R. Novotny 1998 Nucl. Phys. B 61B 137–42
[6] A. Ehmanns 2000 PhD-Thesis Univ. Bonn
[7] H. Flemming 2001 PhD-Thesis Univ. Bochum
[8] P. Marciniewski et al. 2002 IEEE Nucl. Sc. V49 312–17
[9] Ch. Funke 2008 PhD-Thesis Univ. Bonn
[10] Ch. Wendel 2008 PhD-Thesis Univ. Bonn

XIII International Conference on Calorimetry in High Energy Physics (CALOR 2008) IOP Publishing
Journal of Physics: Conference Series **160** (2009) 012007 doi:10.1088/1742-6596/160/1/012007

Operation and Performance of the CDF Calorimeters

Giovanni Pauletta

for the

CDF collaboration

Dipt di Fisica, Universita' di Udine, Via delle Scienze, Udine, UD, Italy

pauletta@ud.infn.it

Abstract. All electromagnetic and hadronic components of the The CDF run II calorimeter system are scintillator - based sampling calorimeters employing both older (Run I) and newer fiber - based techniques for light extraction. The system has now been operating successfully for several years and is presently taking data at high luminosity thanks to its design characteristics and the upgrade of its frontend daq and trigger electronics. The composition and operation of the calorimeter system is described with due consideration for calibration and maintenance techniques required for the preservation of data quality and stability and its performance will be summarized with reference to design expectations.

1. The CDF calorimeters: design characteristics

All CDF calorimeters (fig. 1a) are scintillator-based sampling calorimeters. The central region which extends to a pseudorapidity (η) of about 1 is spanned by an 18 - radiation length (X_0) lead (1/8 inch) - scintillator (5 mm. polystyrene) em calorimeter (CEM) [1] followed by a 4.7 - interaction length (λ) iron (2.5 cm) - scintillator (1cm PMMA) hadron calorimeter (CHA)[2]. The basic structural units of both calorimeters are 15° wedges (fig. 1b) which are subdivided in 10 projective towers, each subtending 0.1 units of pseudorapidity and 15° of azimuth. Light from all active scintillator elements of each tower is collected on each side of the wedge by wavelength-shifter (wls) and transported to the photomultiplers (PMTs) by means of plexiglass guides which are routed through the "cracks" between wedges which end up spanning ~5% of the azimuth. Strip chambers, located at 5.9 X_0 (see fig. 1b), measure transverse shower development at shower maximimum with orthogonal strips and wires.

CEM light-output is ~100 photoelectrons (p.e.) per GeV per PMT and design energy resolution $\sigma(E)/E$ for electrons is $13.5\%/\sqrt{E\sin\theta}$ (E in GeV). The large sampling fraction of the CHA was chosen for good resolution ($33\%/\sqrt{E\sin\theta}$ with a 4% constant term) for 50 GeV pions but this design performance was compromised at higher energies because of the limited thickness due to financial and structural limitations.

The CHA is complemented by the Wall Hadron Calorimeter (WHA) [2] (see fig. 2a) which extends its η-range to ~1.1. This is also an iron (5 cm) – scintillator (1 cm PMMA) but the sampling fraction is smaller than for the CHA because higher forward energies require more absorber and its resolution is correspondingly worse. Preliminary studies [2] indicated that light output from both Hadron calorimeters might be expected to decrease because of the adverse effects of radiation on the scintillator attenuation length. On the other hand the magnetic field has been observed to increase scintillator output by ~4% [2].

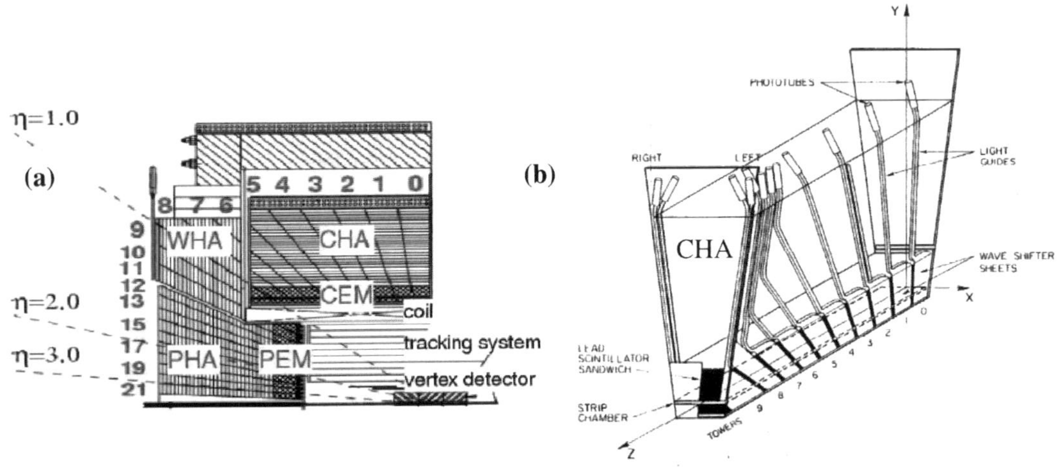

Figure 1 (a) A quadrant of the CDF detector showing the positions of the calorimeters. (b) A central calorimeter wedge showing the light-collection system for the inner CEM. An analogous wls – acrylic light pipe system (not shown) collects a routs light for the CHA.

In anticipation of the Run II luminosity upgrade, the configuration of the Run I forward calorimeters was modified to accommodate the shorter bunch spacing. The resulting Plug configuration is shown in fig.2a. It includes a lead-scintillator em calorimeter (PEM) [3] followed by an iron-scintillator hadron calorimeter (PHA) [4]. These calorimeters span the range $1.1 < |\eta| < 3.64$. A "miniplug" calorimeter [5] for forward diffraction studies will not be described here. The design of the plug calorimeters is shown in more detail in fig. 2. The plug calorimeters were the first large-scale application of the "scintillator tile" + wls fiber configuration. It is illustrated in fig. 2b. The PEM alternates 4 mm polystyrene (Kuraray SCSN 38) scintillator with 4.5 mm lead plate lined with 0.5 mm stainless steel in 22 layers for a total of 21 X_0 (corresponding to 1 λ). The PHA uses thicker (6mm) tiles of the same scintillator which it alternates with the 5.08 cm slabs of iron inherited from the Run I Plug calorimeter, for a total of 7 λ. This iron was incremented (darker shading in fig. 2b) so as to extend the polar angle covered by the calorimeter to smaller angles as a substitute for forward gas-based calorimeters used in Run I. A stack of tiles alternated by absorber constitutes a projective tower and the light collected from all tiles in a tower is routed to the photocathode of a single PMT (Hamamatsu R4125) for each calorimeter. The PEM is preceded by al pre-radiator (PPR) comprising a single layer of 10 mm PVC scintillator (BC408) [6] subdivided into the same tower-based tile structure but viewed separately by dedicated PMTs.

Transverse shower development is measured by a shower max counter (PES) located behind the 4th layer of the PEM, at ~ 6 X_0 (see fig 2a). The PES [7] comprises 2 layers of 5mm by 6mm PVC (BC404) scintillator strips of varying length, read out by 0.83 mm wls fiber (Kuraray Y11) embedded longitudinally in the strip. The strips are are arranged in 8 U/V planes, each spanning 45^0 of azimuth.

Design performance was $16\% / \sqrt{E} \oplus 1\%$ for the PEM with 5 p.e./ minimally-ionizing particle (mip)/ tile corresponding to a total light output of 400 p.e./GeV. For the PHA, it was $70\% / \sqrt{E} \oplus 4\%$ with 5 p.e./mip/tile, corresponding to a total light output of 40 p.e./GeV.

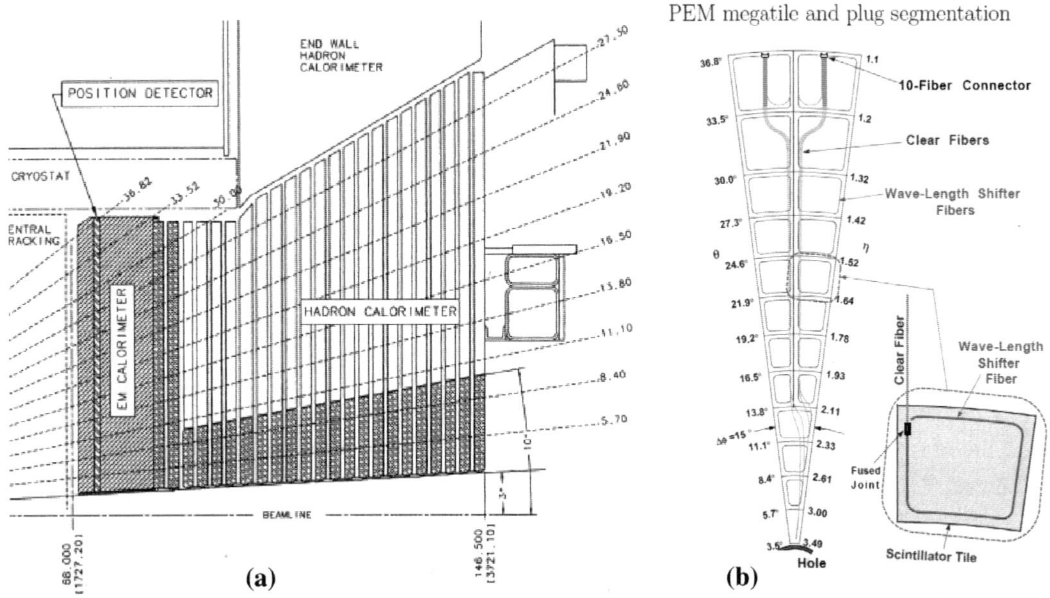

Figure 2. (a) Plan of a CDF plug calorimeters showing their coverage in polar angle and the location of the shower max (PES) detector. (b) A 15-degree azimuthal section (a "megatile") of PEM showing how it is subdivided into "tiles" together with an enlargement of one of these tiles showing how light is collected by wls fiber embedded in the tile.

All calorimeters were read out by the same frontend (f.e.) electronics [8] based on the QIE (Q Integrator and Encoder) custom-built 8-range (I/2-I/256) current-splitter/integrator/encoder developed to accommodate the run II requirement for increased dynamic range. The f.e. CAFÉ cards, containing the QIE chip, a 10-bit FADC, Q-injection circuitry and FRAM are mounted on the f.e. VME boards known as the ADMEMs (ADC + MEMory) which provide the Level 1 trigger with transverse energy sums for each tower, obtained using Xilinx FPGA's.

Calibration and monitoring instrumentation [1-4,9,10] is an essential part of the CDF calorimeter system. It is needed both to transfer test-beam calibrations and to monitor stability and functionality at run time. There are two essential elements: (1) lasers or flashers which inject pulsed light at the PMT photocathode and (2) radiation sources which are used to stimulate the scintillator. The first element is used to monitor the signal collection and amplification chain, beginning with the PMT, whereas the radiation sources test the entire chain, including the light generation and transmission process in the scintillator. Together, they enable one to disentangle problems and instabilities originating in the light generation and transmission system from those originating in the chain beginning with the PMTs. (e.g. magnetic field effects on scintillator response [2] and deterioration in plug PMT response discussed later). The light-injection systems also incorporate LEDs for injection of low-level continuous light at the photocathode in order to reduce rate-dependent gain fluctuations. The charge-injection circuitry incorporated in the f.e. CAFÉ cards also allow for monitoring of the signal digitization stability by the QIEs.

All CDF calorimeter are equipped with a radioactive source calibration system [11] employing either Cs^{137} or Co^{60} sources, the former being preferred because less penetrating radiation produces more localized effects. The CEM calorimeter is equipped with a Xenon flasher system and LEDs [9] whereas the CHA and WHA rely on a Nitrogen UV laser for light injection [2]. A Nitrogen laser system is also used to distribute light pulses to both plug calorimeters.

2. Measuring the calibrations

In this section we outline the procedure whereby the energy scale of the calorimeters was determined and transferred

2.1. General procedure

The following procedure was used to set the energy scale for all CDF calorimeters

- Relative channel gains are derived from radioactive source and/or laser measurements
- Absolute calibrations are obtained from test beam (TB) measurements of E/p. All HVs were set for the same Nominal Gain of all towers (e.g. 2 pc/GeV).
- Dimensionless correction factors, called LERs (Linear Energy Response), where *LER=Nominal Gain/Real Gain*, were determined for the correction of residual differences.
- After application of the LERs a unique Scale Factor (SCL) was determined for all channels so that $E\ (GeV) = count_i^{raw} \times LER_i \times SCL(GeV / count)$.
- Calibrations were transferred to the detector by repeating the radioactive source runs in situ and applying the test beam constants adjusted for the change in readout electronics, eventual differences in radioactive sources and the "brightening" effect of the CDF magnetic field on the scintillator output.

2.2. An example

Some procedural details are given here for the plug calorimeters by way of illustration (a more detailed account of the calibration procedure is given in refs [3] [4]). A replica of the real plug calorimeters, spanning 45^0 for the PEM and 60^0 for the PHA, was constructed and set up for data taking at the MT6 TB using the Run I (RABBIT) electronics. All towers of the replica module were exposed to e^+ and π^+ and μ^+ beams whose momenta were varied between 5 and 230 GeV/ and tagged with a resolution of between 1.1% and 1.6% at lower and upper ends of the momentum range, respectively. LERs and SCLs were determined as a function of energy from measurements with electrons and pions while the energy deposited by muons was used to check the calibration model.

The energy resolution for the PEM is shown in Fig. 3 where the design resolution is also shown for comparison. Non–linearity for the PEM was better tha 1% and non-uniformty less that 2% .

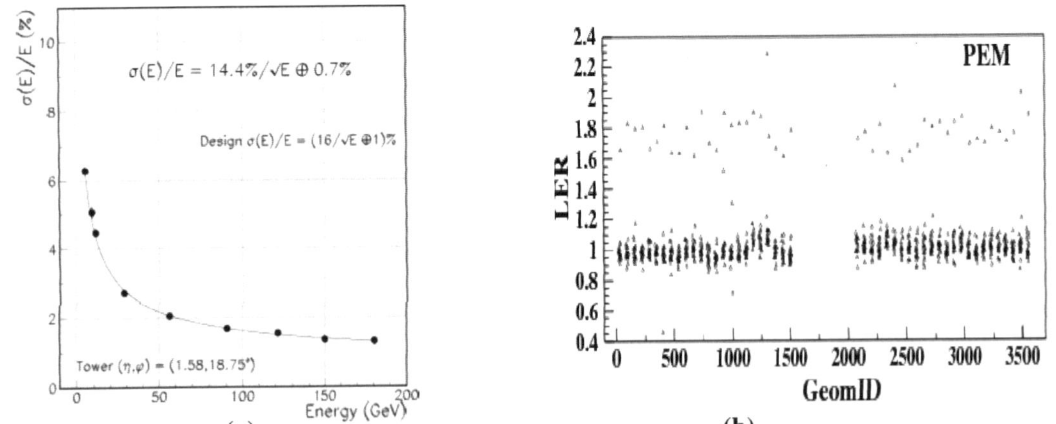

Fig 3 (a) The PEM resolution for a tower of the PEM replica as measured at test beam. (b) The LERs for the PEM calorimeters at CDF, versus geometrical ID. Points on the left are for the west plug.

Radioactive source runs were also taken at TB and, on the basis of the measurements of beam momentum, energy deposited in the calorimeter, current generated by the radioactive sources, calibration transfer constants (in pC/GeV/μA) were determined with the beam centered on all towers. These were then corrected for differences in the radioactive sources (Co^{60} at CDF vs. Cs^{137} at TB), a shorter (132ns vs 2.2 μs) charge integration gate at CDF and enhancements ("brightening") in the scintillator light output [2] due to the CDF magnetic field. Radioactive source runs were repeated at CDF and an example of the resulting LERs, for all towers of both PEMs is shown in fig. 4b.

3. Keeping the calibrations: run-time calibration and monitoring

The response of the calorimeters can vary with time for a variety of reasons ranging from component failure to gradual deterioration and it is therefore of critical importance to monitor all channels and update calibrations as often as possible. In this section we outline the procedures employed for the CDF calorimeters

3.1. Organization

CDF data is expected t be calibrated and run through (offline) production within 2 months of being taken. This sets the overall time scale and it is the responsibility of the "Physics coordinator" to ensure that calibration updates are in the data base every ~200 pb^{-1} of integrated luminosity. A "calibration coordinator" oversees a "calibration group" of "system" experts, each responsible for a detector subsystem. For the calorimeter systems experts are nominated for each of the CEM. CHA. WHA, PEM, PHA, PES, HAD/EM timing. System experts are expected to update the on-line calibration data-base on a weekly basis.

3.2. Operations

One can distinguish between two types of calibration:

- Detector-level hardware calibrations which are initiated by the shift crew at fixed intervals (e.g. daily). These are automated to the degree that it is possible.
- Higher-level automated calibrations: a calibration executable, that runs continuously, strips out, stores (e.g. in ntuples) and processes events (e.g.di-muon/electron events, minimum bias events, jet events) for calibration purpouses.

Monitoring the results of calibration runs is greatly facilitated by the calibration data base and monitoring software (DBANA). Monitoring takes place at different levels: at calibration time by the shift crew; on a weekly basis by the above-mentioned system experts who are responsible for identifying and solving problems and for updating calibrations on a weekly basis; by the experts over longer periods in preparation for offline production.

Examples of typical detector-level calibrations are the QIE and plug laser calibrations. In the QIE calibration sequence, known charge is injected into the QIE (using the charge-injection circuitry on the CAFÉ card, mentioned previously) for calibration of the electronic gain of all calorimeter channels. The procedure also identifies bad or noisy electronic channels. Calibration data are processed by online "consumer" programs which store pedestals, means, r.m.s. deviations and gains in the data base where they can be monitored using DBANA. Shift crew monitor the data right after the calibration and signal eventual deviations from reference data for inspection by the experts.

The plug laser system is used to distribute light from each laser pulse to all plug calorimeter channels and to stable PIN diodes which monitor the intensity fluctuations of each laser pulse to better that 1%. Ratios of calorimeter-to-PIN signals are therefore expected to be stable to that level and eventual deviations are indicative of malfunction or of gradual variations in the channel calibration beginning a the channel PMT. Thanks to this system, a progressive degradation of the plug PMT responses was brought to light at an early stage. Laser data are also processed on-line by a

Journal of Physics: Conference Series **160** (2009) 012007 doi:10.1088/1742-6596/160/1/012007

"consumer" program which stores the results in the calibration data base where they may be monitored using DBANA and used to correct for the degradation in PMT gain.

An example of a higher-level calibration is the procedure used to track and update the CEM calibrations: 8 GeV e.m. "objects", selected by a "calibration trigger", are stored and subsequently used for E/p in situ calibration. This procedure evidenced a ~3% annual drop in calorimeter response due to scintillator degradation. The monitoring of variations in the Z peak also showed luminosity-dependent variations at the <1% level. Response variations are also monitored by tracking the minimum-bias rate at fixed η.

CHA LERs are also monitored and updated using E/p measurements since problems developed in the CHA radioactive source drive mechanism: J/PSI muon momenta are compared to the muon energy measured by the calorimeter. When muon statistics are insufficient, monitoring of the minimum–bias rates at fixed η is used as a fall-back procedure These procedures confirm previous observations of response degradation due to deterioration in the scintillator attenuation. The degradation rate is shown in fig. 4 (left).

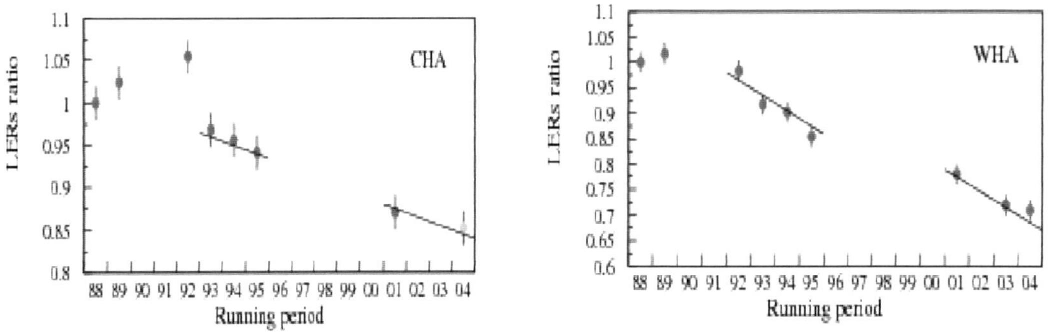

Figure 4. LER variation for the CHA and WHA calorimeters vs year.

The rate of degradation is more pronounced for the WHA as seen in fig. 4. The WHA radioactive source system continues to operate satisfactorily and periodic source runs are usd to track the response variations.

Response variations in the plug calorimeters (see fig. 5) are due to PMT gain decreases, as already mentioned above..

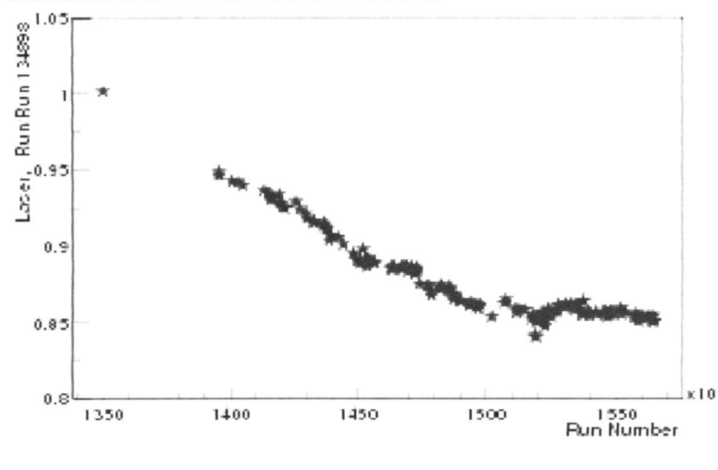

Figure 5. The response of a channel of the west PHA vs run number

They increase with η and range between 1% and 8%. The source of the PMT gain degradation is not understood

4. Jet energy scales

.

Calorimeter responses are major ingredients of the Jet energy scales which are needed to evaluate the energies of partons resulting from underlying physics processes. As such they are essential for most analyses and are the major source uncertainty in the top quark mass. Given their importance, they are the responsibility of a dedicated "Jet Energy Resolution group".

The parton energy is distributed amongst all elements of the parton shower and its evaluation depends on an understanding of the shower evolution and on the "jet energy corrections" such as the those deriving from the dependence of calorimeter response on the type of particle, non linearities in the calorimeter response, energy losses due to acceptance and energy losses from the jet clustering algorithm. Amongst the major responsibilities the jet energy resolution group is the evaluation of energy corrections and the improvement of jet clustering algorithms. Tuning the calorimeter simulation MC , based on GFLASH [12], is an essential element of this work. Test beam data (for p>20GeV) and in situ data (for p<20GeV) were used to tune the MC for single particle calorimeter response. As a result, the jet scale uncertainty has recovered from the upgrade changes and is now slightly better than the final Run I estimate [13,14]

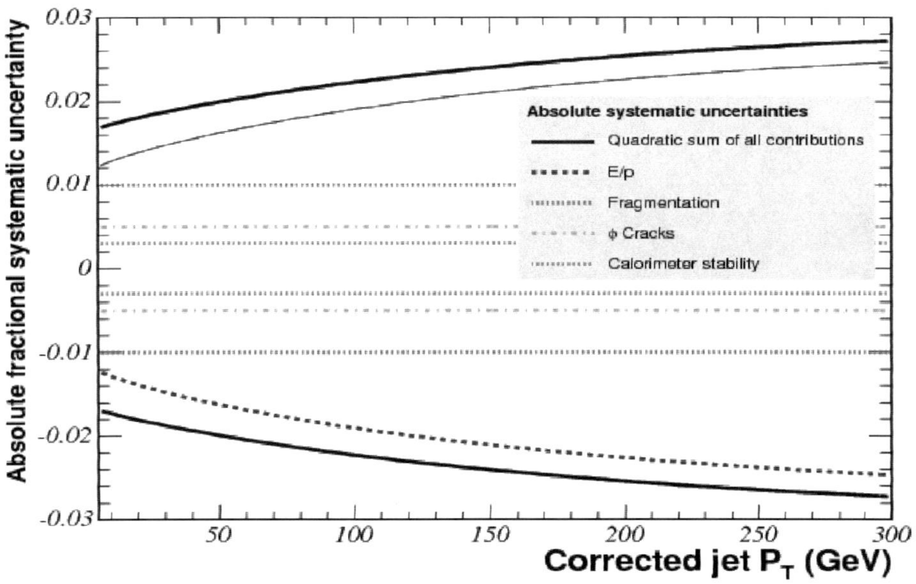

Figure 6. Uncertainties in the absolute jet energy scle

5. Coclusions

The CDF calorimeters are operating at better than design expectations and are expected to survive to the end of Run II without significant further loss of performance. Operational procedures have now settled down for the home stretch. Further improvements in the jet energy scale are expected as analysts hone their tools and exploit all the bells and whistles.

Journal of Physics: Conference Series **160** (2009) 012007 doi:10.1088/1742-6596/160/1/012007

References

[1] L. Balka et al.1988 *Nucl. Instrum. Methods* **A267** 272
[2] Bertolucci S et al. 1988 *Nucl. Instrum. Methods* **A267**
[3] Albrow M et al. 2002 *Nucl. Instrum. Methods* **A480** 524
[4] De Barbaro P, in *Proceedings of SCIFI97* (AIP Conf. Proceedings, 1997) FERMILAB-CONF-98/057-E; Jinbo Liu, in *Proceedings of CALOR97*, World Scientific 1998
[5] Goulianos K et al. 2004 *Nucl. Instrum. Methods* **A518** 42
[6] Albrow et al., *Nucl. Instrum. Methods* **A431** 104
[7] Apollinari G, Goulianos K, Melese P and Lindgren M 1998 *Nucl. Instrum. Methods* **A412** 515
[8] Shar T, Nelson C and Wesson T 2000 *IEEE Trans. On Nucl. Sci.* **47** 1834 Nelson C, Shaw T, 2002 in *Proceedings of CALOR2002* 644
[9] Hahn SR et al.1988 *Nucl. Instrum. Methods* **A267**
[10] Cauz D et al. 1996 in Proceedings *of the Sixth International Conference on Calorimetry in High Energy Physics*, Frascati Italy, Frascati Physics Series 1996 p369
[11] Barnes V et al., 1998 in *Proceedingd of SCFI 97*, Notre Dame, IN, USA 1998 p. 420
[12] Grindhammer G, Rudowicz M and Peters S 1990 *Nucl. Instrum. Methods* **A290** 469
[13] Bhatti A, Canelli F, Heineman B. Et al., 2006 *Nucl. Instrum. Methods* **A566**
[14] Hatakeyama K in *Proceedings of TOP2006*, PoS TOP2006:014,2006

STAR Calorimetry

W W Jacobs, for the STAR Collaboration

Indiana University Cyclotron Facility and Department of Physics, 2401 Milo B. Sampson Lane, Bloomington IN 47408 USA

jacobsw@indiana.edu

Abstract. The main STAR calorimeters comprise a full Barrel EMC and single Endcap EMC plus a Forward Meson Spectrometer. Together they give a nearly complete coverage over the range -1 < pseudorapidity < 4 and provide EM readout and triggering that help drive STAR physics capabilities. Their description, status, performance and operations (and a few physics anecdotes) are briefly presented and discussed.

1. Introduction and overview

Many thanks to the organizers for the invitation to this excellent conference held on this occasion in a lovely/historical setting (and with lots of great food and discussions!).

STAR is one of two large experiments sitting on the RHIC rings at Brookhaven Laboratory located on Long Island in New York. RHIC (the Relativistic Heavy Ion Collider) as its name suggests, provides collisions of heavy ion species with center-of-mass energies up to $\sqrt{s_{NN}} = 200$ GeV. It also provides polarized proton beams, with a capability for either transverse or longitudinal polarization at the experiments, with collision energies $\sqrt{s} = 200$ GeV, and in future years 500 GeV.

The main STAR operating calorimeters (pictured in figure 1), are the Barrel EMC (BEMC), the Endcap EMC (EEMC) and the Forward Meson Spectrometer (FMS). Taken together they provide nearly compete EM coverage for pseudorapidity $-1 \leq \eta \leq +4$. They total ~ 6800 "towers", ~ 43,000 "SMD" and ~7000 "Pre- and Post-Shower" readout channels. There are other smaller calorimeters at STAR, notably the Zero Degree Calorimeters (ZDC), used for monitoring collider luminosity, and versions of a Forward Pion Spectrometers (FPD) preceding FMS, that won't be discussed further here.

Unlike the configuration planning of many of the other detector projects we have heard presented at this conference, the BEMC, EEMC and FMS calorimeters at STAR were not among its "baseline" detectors. This accounts for a somewhat different "modus operandi" we have followed with incremental installation, commissioning, calibration, upgrade and maintenance activities. As well, the various calorimeters were driven by different physics interest subgroups: the BEMC designed largely for heavy ion studies with full tracking coverage from the Time Projection Chamber (TPC); the EEMC driven primarily according to spin physics (and having only partial TPC coverage but additional tracking envisioned); and the FMS which addresses some "heavy ion" issues along with emphasis on future spin physics. The first BEMC patch install was in 2001, with full install/all

Figure 1. The operating STAR calorimeters as discussed herein: (left) the Barrel EMC shown during insertion of one of the 60 modules/side around the TPC, (center) the Endcap EMC as seen removed from STAR via the west poletip/carriage and (right) the Forward Meson Spectrometer which views the interaction region through the opening in the EEMC/poletip.

tower readout/trigger in '06 (full electronics in '07). Similarly the EEMC test install occurred in '02, with full install/all tower readout/trigger in '03 (full readout in '05). The FMS on the other hand, went quickly from final prototype to full detector installation and readiness for run 8, although the first FPD prototype (a section of an EEMC test module) was deployed already in '02.

The point of this brief chronology is to emphasize the incremental nature of the efforts and note also the small group involved in the various activities from installation to commissioning. As well, one should add that the "politics" of such an implementation demands that there be an increased capability and physics output for each yearly running period. In fact the latter was accomplished on a pretty regular basis starting with measurement of the forward π^0 inclusive single spin asymmetry and cross section from '02 pp running (FPD prototype); '03 π^0 inclusive cross section from d+Au, '03-4 polarized pp inclusive jet results for ΔG, '04 Au+Au heavy flavor x-sec (all with 1/2 BEMC), '06 polarized pp di-jet "Sivers" with EMC trig data (full EEMC/BEMC), and recently much more spin and heavy-ion physics. Presently the STAR EMC's dominate triggering for spin physics as well as in recent years the heavy-ion data bandwidth; EMC data are now also becoming a major part of STAR physics analysis and output (recall STAR's baseline "heart" is the TPC). Overall, things have gone well and the EMC performance/operations, driven by available money and manpower vs. physics output has worked (maybe not perfectly but) well. There certainly has been great productivity!

Before presenting details of the individual calorimeters in the next sections, it is useful here to mention briefly some common organizational aspects of EMC equipment, analysis and calorimeter support procedures we have found helpful at STAR. In particular for the BEMC and EEMC, while both are sampling calorimeters with similar structures and features, there are also differing implementations (some significant) and approach. These issues are brought together in a STAR EMC "operations and detector analysis" based group (EMC^2) for which I have had oversight and whose weekly phone meetings (and some collaboration meeting sessions) I have coordinated over the past ~ 3 years. Discussed issues include detector equipment support and detector performance; calibrations (all kinds); run status tables/data base of detector components (for physics replay); detector based analysis algorithm development (from physics work groups); software, including the EMC subsystems coordinators reporting to the STAR software group. This group-like structure of contributors with somewhat diverse yet common interest seems to have worked reasonably well.

For illustration, we take the tower FEE digitizer as a common hardware issue. Nominally the FEE comprises a 32-input 9U digitizer with a 12bit ADC per channel, 10 MHz beam crossing rate and pipeline architecture (for EEMC the "magnetic" noise suppression components are removed and higher PMT gain used on poletip); trigger section outputs high tower (HT) and trigger patch (TP) information. In early running and then subsequently, "ghost pedestals", n*256 noise, and header corruption were traced to clock issues (problem then jumped from EEMC to BEMC and back, etc.). Besides the "fix", this spawned many common diagnostic tools to monitor/catch these detrimental hardware behaviors. Separately, in another hardware related symbiosis, the EEMC's required remoted power supplies (due to the high poletip magnetic fields), were found to be the final real solution to serious and continued BEMC supply failures. There are many other examples of various types.

2. STAR barrel EMC

The Barrel Electromagnetic Calorimeter (BEMC) completely surrounds the STAR Time Projection Chamber (TPC), covering an area of nearly 60 m^2. Figure 2 highlights several BEMC components.

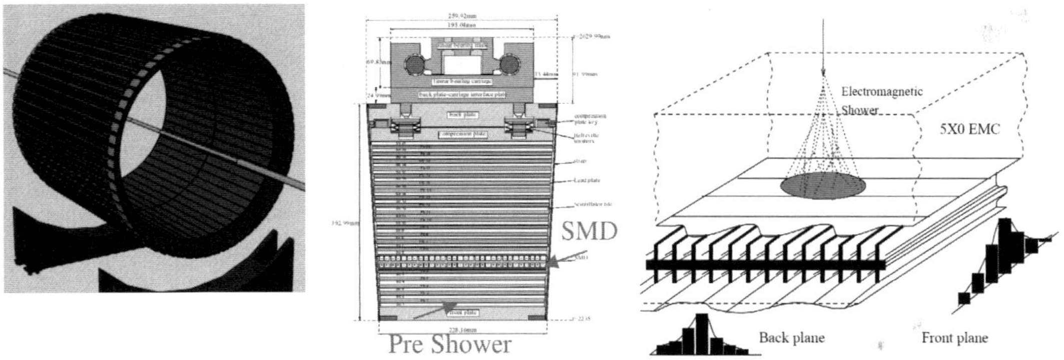

Figure 2. BEMC components: (left) schematic of 120 modules surrounding the beam pipe, (center) "end view" drawing of module cross section showing mechanical structure and location of the SMD and preshower, and (right) schematic of SMD detector response.

The nominal BEMC parameters are (see also reference [1]):

- $-1.0 < \eta < 1.0$; full azimuthal coverage; $dE/E \sim 14\%/\sqrt{E}$
- 4800 towers; 36k SMD strips; 4800 preshowers
- 120 modules (40 towers/module); 60 west and 60 east
- $(\Delta\eta, \Delta\varphi)$module ~ (1.0, 0.1); depth = 21 X_o
- $(\Delta\eta, \Delta\varphi)$tower ~ (0.05, 0.05)
- Shower Max Detector (SMD) at a depth 5 X_o
- $(\Delta\eta, \Delta\varphi)$ ~ (0.007, 0.007)
- Preshower: first 2 scintillator layers (depth 2 X_o) combined/shared with tower readout

Overall the BEMC has worked performed well in producing lots of physics as mentioned above. However, of interest to this session of the conference are some details of operation. I have space to mention briefly only a few items from the talk regarding calibration and component performance next.

For pre-run checkout and online tower calibrations several techniques have been and still are employed. A LED pulser system can be used to adjust individual/aggregate HV for the tower tubes (for latter, e.g., moving all gains from 28 to 60 GeV full scale to match the EEMC for beginning of run 5). Slope equalization from minimum bias events is used at beginning of run to fine-tune the HV and is especially useful for getting new tubes/bases tuned in. This procedure is fast (turnaround in less than a day) but is only relative in the various η "rings". During the run start, gains are checked using MIP's. Data production is requested in fastOffline mode and gives early confirmation that things are reasonable for the proper scale and triggering. For analysis and offline, the full calibration uses the STAR μDST framework. MIPs set relative gains for each tower; while electron E/p tracked with TPC are used to normalize η-rings. An artifact of particle isolation issues at the largest η are presently corrected to a flat response as suggested by other calibration techniques. Beginning in 2006, additional statistics in the data sets allowed for a tower-by-tower electron E/p measurement. Currently higher p_T electrons as well as calibration with π^0's is being revisited.

The BEMC has had some significant component performance issues that have required effort to remedy. For example, for the towers, the original tower FEE crate design was a compact assembly including integrated power supply, controls and LV for the tower CW bases and etc. The solution to continued supply failures (as high as one every few weeks when running) was to remote the supplies to racks away from the magnet backlegs. As well the coupling of CW base LV to FEE power reduced flexibility and hard resets the CW when FEE's cycled following FPGA bit upset, so the solution here was to go to a completely separate supply as well as separate HV control functions. This separation has proven more flexible, reliable and easier on the components. For the towers themselves, as of 9/07 ~ 2% (96 towers) were not working with a plan to bring back about 40 towers (rest likely dead "forever" with broken optical fibers?). This to be compared to situation of '05 when 10-15% of the towers were dead! Similarly the SMD for the BEMC has taken some work to improve following initial damage due to a bad HV supply. The number of chambers lost during a running period has gone from 8-10/run for period '03-'05 to ~1 per run period the last few years. In all, a total of 53 out of 3600 (1.5%) anode wires have had to be cut permanently. The failure of SMD FEE cards remains fairly high (20-15 per run period) although recently reduced on the west side which has a different batch of storage capacitor array (SCA) chips; amplifier and shaper (15 chns) has a failure systematic on the top FEE card whereas SCA (150 chns) problems are systematic on east side. For the readout crates one had lost 50% by end of '05 run; now reduced to 1 (out of 8 total) at end in '07. For the preshower which uses similar electronics/crates, ~ 10-15% of channels not working at run 8 end ... usefulness of signal in commissioning stage and being exploring in physics analyses at end of run 7 and into run 8.

3. STAR Endcap EMC

The Endcap Electromagnetic Calorimeter (EEMC) is mounted on the inside of the west STAR magnet poletip. The nominal EEMC parameters are (see also reference [2]):

- 2 half annuli – $1 < \eta < 2$; SS structure; 23 layers Pb/SS laminate; depth = 21 X_0
- 720 projective towers (comprising 24 layers)
    - 6° and 12° megatile; ~ 100% coverage
    - ($\Delta\eta$, $\Delta\varphi$) tower ~ (0.057 to 0.099, 0.01)
    - dE/E ~ $16\%/\sqrt{E}$
- Depth segmentation
    - 2 separate preshower (1,2) layers (1440 chns)
    - high position resolution SMD @ ~ 5 X_0
    - -- 6912 triangular-shaped scint strips; base 10 mm, height 7 mm
    - -- "u", "v" stereo planes in 30° sectors w/ overlap; ~ 280 cm to Interaction Region
    - postshower 24th layer (720 chns)

For the readout, of particular interest is the digitization of all individual channels of the SMD and pre-and post-showers. This is accomplished with an exceedingly compact design of Multi-Anode PMT

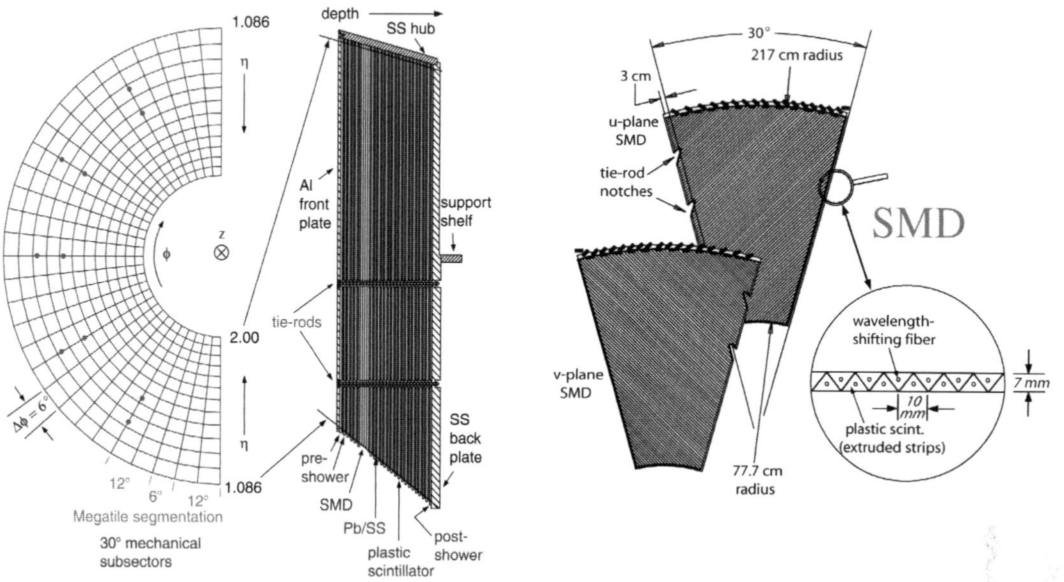

Figure 3. Schematic representation of EEMC components: (left) plan and elevation view of tower structure and layers, (right) illustration of stereo SMD planes of triangularly shaped scintillator.

(MAPMT) and associated CW base and FEE assembly (4 FEE cards with 4 chns x 12 bit ADC/each for every MAPMT). These assemblies are housed in magnetic shielding boxes, 12 assemblies = 192 chns/box. This has proved to be an elegant/robust system for individual readout of the ~ 9.2k channels.

In general the EEMC performance has been "remarkably good", but we can touch herein on only a few aspects. Regarding EEMC configuration and repair: as a special constraint, all PMT and FEE are located on back of the STAR west poletip in order to minimize light fiber pathlengh and localize electronics (only 54 fibers go to data collectors off the poletip). The ~ 1kG fields requires all the PMT's to be in shielding "boxes"; similarly, the MAPMT/FEE compact assembly as just described. In terms of maintenance, access to the poletip requires a manlift, and if running, the STAR magnet ramped down and locked out (e.g., latter means a minimum of a few hour turn around so rely mainly on scheduled/bi-weekly ~ 1 shift access periods for repairs/maintenance during runs). However, also during RHIC shutdowns, EEMC access limited when poletip is retracted for work on other systems (TPC, TOF, BSMD, etc.), hence, require ~ 2-3 weeks with poletip attached to do maintenance every year (the single box repair turnaround is ~ 1 day, which includes "piggyback" of box to neighbor box leaving light fibers attached). The good news is that EEMC electronics/readout failure rates are relatively low and declining! For the tower tubes and bases we have seen at peak ~ 1% /yr; for MAPMT it was 2% the first year but now <0.5%. The failure rate for MAPMT FEE boards is similarly declining to from < 1% to ~ 0.1%. Ancillary problems are with the HV controller units (ship for repair) and issues with the cooling for the MAPMT boxes (bubbles in the chiller and compressor fail/shutdown problems). A nuisance factor is that ~ 5% of the readout boxes require de-fibering for maintenance work due to fiber runs being miscalculated. Offline QA with pedestal monitoring catches many problems including complete failure, "fat pedestal" diagnostic, stuck bits, data base swap, etc. But, we have a recent example of low tower bits being stuck 50% of the time which eluded all diagnostics. The moral is to be always looking for the next problem which is presently ignored!

Calibrations for the Endcap at run startup are similar as described for the BEMC (slopes, etc.). However, a main difference is the fact there is only partial coverage by the STAR TPC tracking. In its place was "invented" a MIP identification using a SMD u x v plane hit isolation pattern

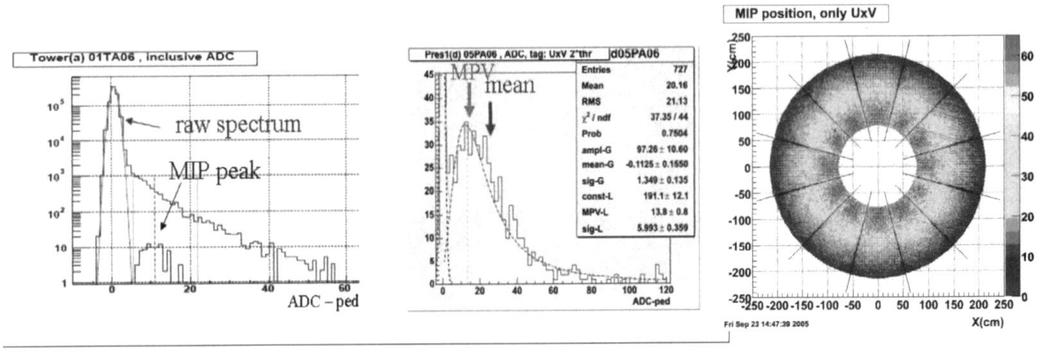

Figure 4. EEMC MIP calibration using SMD "tracking" isolation: (left) tower response, (middle) 1st preshower layer response, (right) SMD u x v MIP response. See text for details.

0000000 -- X[X] – 0000000 along with the MIP-like response of all the other the detector components (2 preshowers, tower and postshower). This 26-fold condition is highly selective and allows MIP calibration of the various EEMC components by excusing the detector components from the coincidence in turn. The ideal tower gains are set to 60 GeV E_T for full scale (4096 chns); the RMS spread from this procedure is ~ 9%. We find from run 5 to 6 (for example) that the gains are stable to ~ 1%, so there is no indication so far of fiber, scintillator or PMT degradation for the EEMC.

A performance issue for the EEMC SMD is that of π^0/γ separation for which the triangularly shaped scintillator strip design with individual readout were envisioned. The physics interest is in investigation of the gluon contribution to the proton spin, probed in γ-jet double spin measurements in longitudinally polarized pp collisions via partonic qg Compton scattering. The expectation was that a factor ~ 3 reduction of π^0's was needed at p_T ~ 10 GeV/c (more required to go down in p_T) for reasonable extraction of the direct γ-jet signal. Extensive analysis is ongoing with the limited run 6 data sample and includes use of SMD shower shape discrimination analysis (e.g., a maximum sided fit residual in both u and v planes) along with detailed response of the rest (e.g., and including both preshower layer energies) of the EEMC detector components. For the SMD we have found that the initial simulations are not realistic in producing details of the transverse shower shape. At present we have embarked on using a library of SMD response shapes of real photon showers indentified from η particle decay (shapes replace Monte Carlo according to SMD sector, φ, η and photon energy). First look at sided residual discrimination with these "data driven" shapes in pythia for background and signal slices look encouraging although there remains considerable work to do for a robust extraction.

4. Level 0 High Tower, Jet Patch and higher level triggering in STAR

The STAR calorimeters can trigger at Level 0 (L0) on High Tower (HT), Trigger Patch (TP) or Jet Patch (JP). Generally set HT: ≥ 1 (of 4800 BEMC or 720 EEMC) if tower ET > threshold; JP: ≥ 1 (of 12 BEMC or 6 EEMC) hard-wired jet patches (sum of several TP's) if total Σ ET > thresh [note: there is also a minbias condition]. Triggering on something as large (Δη =1; Δφ =1) as a jet patch requires careful trigger signal processing to avoid noise issues. The tower digitizer trigger 12 bit ADC is reduced to 10 bits at the FPGA. A special "ped_4" (to put pedestal in range 16-31 channels is then subtracted). For the trig patch (already produces a 6 bit HT) the 2 least significant bits are dropped and Σ 8bits => Trig Patch sums without noise (single channel pedestals that sum in bigger jet patches for triggering). The above information is also passed on to a higher level 2 (L2) trigger that can accept or abort events. As example 2006 JP rate was ~ 150 Hz with L2 trigger used to fit limited bandwidth (~ 2.5 Hz for this trigger). Run 9 "DAQ1000" for the TPC will relax these rate limits – but data volume still a concern! Issues in running these EMC triggers are "hot towers", particularly for low threshold HT (HI physics); for spin physics where the JP is a workhorse due to reduced trigger bias on jet type, the issues are keeping good pedestals, tracking stuck trigger bits on the digitizer boards, etc. For the upcoming run 9 we plan L0 wiring/FPGA upgrades => smooth out present trigger response in φ and η.

It is worth mentioning that a byproduct (and an extremely valuable one) of the L2 triggering is QA feedback from L2 trigger algorithms. In particular, L2 samples all events (STAR "online" just ~1% and random) and can easily be trigger sorted and/or other diagnostics performed and summarized. We typically monitor pedestal residual per tower, HT and TP frequency/trigger, di-jet correlations, etc. In fact in run 6 a physics measurement ("Sivers" A_N measurement for di-jets [3]) was made directly out of L2 by identifying and triggering on 2 localized clusters, $\eta\Delta \times \phi\Delta = 0.6 \times 0.6$, with E_TEMC > 3.5 GeV, $|\phi\Delta| > 60°$ from which the cosine of the di-jets was reconstructed => sign of net k_{Tx} for each event. Of course trigger data can be replayed (fast) offline to optimize cuts, etc. ... a novel use of L2!

5. STAR EMC calorimetry operations & Procedures

In addition to (and complementing some of) the detailed comments in the above sections, we summarize here the main overall operational protocol and associated personnel:

- While the collider runs yearly there is generally a "summer" maintenance period (no running due to weather, electricity costs, etc.) during which the EMC priorities are sequenced for repair and upgrades directly impacting performance in next run period (mostly EMC expert).
- At run startup: collisions timing checks, pedestals (triggering), pulsers, etc. (EMC expert)
- With beam and starting with minbias trigger => QA detector response and initial calibrations; then activate EMC HT, JP, level 2 (L2) triggering and tune cuts, rates, etc. (EMC expert)
- Monitoring in running mode: online QA-plots (a fraction of events/all triggers) saved per run (shift crew and EMC expert); L2 trigger diagnostics and monitoring plots, ratios, etc. available few minutes after each run and saved to web (~ EMC expert)
- Problems: shift crew reset supplies, reload FPGA (from radiation upsets, etc.); EMC experts mask hot channels, deal with stuck trigger bits, and all other problems
- Special runs: minbias/fill for status tables; background (HT, JP w/o minbias); other special
- Access: ~ bi-weekly (1 shift); or on demand "under fill prep"/machine probs/emerg (.25-1hr)

6. Forward Meson Spectrometer (FMS)

The FMS is a hermetic electromagnetic calorimeter, built from finely segmented lead-glass detectors, positioned west of the interaction point at STAR. Some FMS attributes are:

- 2-m square, 1264 cell Pb-glass array at 7.5 m from center of STAR
- Small cells / 476 total: 3.8 cm square x 60.2 cm; π^0 vs. γ to 60 GeV
- Large cells / 788 total: 5.8 cm square x 45 cm (Scott F2) π^0 vs. γ to 40 GeV

The detector is actually mixture of recycled/built parts (a very "green" detector!); e.g., Pb-glass from E831/FNAL (large) and IHEP + JLAB (small); bases and tubes also some borrowed/built. However, all the Pb-glass was re-wrapped and extensively QA/tested. There is a north-south split at the beam pipe and halves movable by carriage since at this forward position there is strong x_F and p_T correlation which can then be mapped. An advanced custom "QT" board readout for FMS cells was built. Salient features of these 32 analog input (~ 80 ns capture time), 12-bit ADC /channel boards, include 5-bit TDC / channel (5 ns time stamp background reduction) and five FPGA for data and trigger. For each trigger 32 bits of information are produced for each RHIC crossing and shipped on fiber to PCI receivers; DAQ is linux based. The trigger treats groups of 16 crystals: for HT trigger (implemented) – e.g., highest of 16 ADC values; for "overlapping patches", select a contiguous group

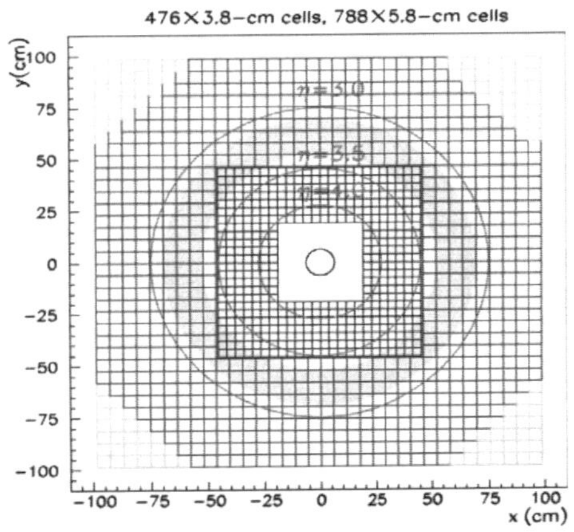

Figure 4. A schematic of the STAR forward meson spectrometer as seen from the interaction point. The Blue beam penetrates at the center of the matrices. The array is 'stacked for physics" in forward direction; the yellow shaded area represents a conservative fiducial volume used for detection of direct γ candidates, with the remainder of the FMS used as a veto of photons arising primarily from decay.

of 6 bits sum /6 bits HT => a dynamic clustering and ~ movable trigger patch (to be commissioned). Regarding initial Run 8 FMS performance and operations: using adapted shower shape codes from FPD, and preliminary calibration/efficiency, see clear evidence for π^0 events in both large and small tiles. Detector capabilities were commissioned as needed for dAu, then polarized pp running and included a LED gain tracking system. At runs's end there were only ~ 10 "problem channels" overall.

7. Future Directions and Developments for STAR Calorimetry

The need for better EMC operations and performance as well as the push to address new physics goals has defined a number of near and longer term tasks. A summary of the notable items by category is:

- Readout and Trigging: upgrades to trigger control/driver units => more EMC thresholds; continue with further development of EMC L2 and FMS trigger algo's for efficient triggering and monitoring; parts of EMC readout now slow! re: new TPC "DAQ1000" => rework barrel SMD/preshower RDO crate FPGA; all receivers to DDL transfer (ALICE) and PCI/linux
- Tracking: forward GEM tracker (FGT) => STAR upgrade plan for ~2010 (e.g., charge sign separation of W decay e's for flavor dependent spin PDF determination); optimize (meantime) forward tracking algo's and new DAQ hardware for maximum TPC reach
- Shielding, material and etc. Fight "extra" material associated with central tracking and other upgrades; tunnel shielding for remaining EMC "hot spots" => bandwidth/data integrity
- Calibrations, algorithm development and etc. improve tower calibrations (e.g., π^0's and e's where possible) and other components for next tier of physics output; improve π^0/γ and e/h discrimination algos (γ -jets currently of interest for spin, HI and QCD)

8. Summary

STAR Calorimetry was installed and commissioned over several years and serves a variety of both heavy ion and spin physics needs At present, all EMC components are running well and increasingly contributing to STAR physics. Examples in triggering and EMC response analysis were presented and selected operations, procedures, component performance issues reviewed along with anticipated upgrades. We look forward to many future years of successful operations and physics output!

References

[1] M. Beddo *et al.*, 2003 Nucl. Instr. Meth. **499**, 725.
[2] C.E. Allgower *et al.*, 2003 Nucl. Instr. Meth. **499**, 740.
[3] B.I. Abelev *et al.* 2007 PRL **99**, 142003.

Journal of Physics: Conference Series **160** (2009) 012009 doi:10.1088/1742-6596/160/1/012009

Status of Electro-Magnetic Calorimeter in BESIII

Li ZHOU Junguang-LU Tao-HU Xiao-CAI Mingyi-DONG

Jian-FANG Boxiang-YU Zhigang-WANG Zhenghua-AN

Institute of High Energy Physics, Chinese Academy of Sciences

19 Yuquanlu Road, Shijingshang District Beijing 100049, CHINA

zhoul@mail.ihep.ac.cn

1. Introduction

The Electro-Magnetic Calorimeter (EMC) plays an important role in the BESIII detector, whose primary function is to measure precisely energies and positions of electrons and photons. High detection efficiency and good resolutions for lower energy photons are very important for the electromagnetic calorimeter of BES III.

The general physics requirements of the BESIII detector leading to the design of the EMC based on CsI(Tl) crystals, with the following expected performance:

- The measurable energy range of electron or photon is from 20 MeV to 2 GeV. The energy resolution is about $2.5\%/\sqrt{E(GeV)}$, good energy resolution is important especially in the energy region below 500 MeV.
- The position resolution for EM shower is $\sigma_{x,y} \leq 6mm/\sqrt{E(GeV)}$.
- A neutral(γ) energy trigger is provided.
- A good e/π separation in the energy region above 200 MeV is expected.
- Crystals have a fine granularity and signal readout so that overlapping showers can be reconstructed, especially for high energy π^0's. The electronics noise for each crystal should be less than 220 KeV.

2. Construction of the EM Calorimeter

The CsI(Tl) calorimeter are composed of one barrel and two endcap sections . The barrel has an inner radius of 94 cm and a length of 275cm, covering the polar angle of 146.5° - 33.5°($\cos\theta \sim 0.83$). The endcaps with inner radius of 50cm are placed at $z = \pm 138$ cm from the collision point, covering the polar angle of 32.5° - 21.3°. The total acceptance is 93 % of 4π . A small gap of about 5cm between endcaps and the barrel is reserved for mechanical supporting structure, cables and cooling pipes. in the barrel ,There are 44 rings of crystals along the z direction., 120 crystals in Φ.All crystals with a small tilt angle of 1.5° in the Φ direction and 1°～3° in the θ direction. Each endcap consists of 6 rings and split into two half circles. The entire calorimeter have 6240 CsI(Tl) crystals with a total weight of about 24 tons.

The shape and dimension of barrel crystals are illustrated in Fig.2-1. the section of crystal. Crystals typically have a front face of about 5cm × 5cm and a rear face of about 6.5cm × 6.5cm. The length of all crystals is 28 cm(15X_0). For easy manufacture, two out of four side faces of crystals are perpendicular to the front and the rear faces. The shape of endcap crystals is more complicated due to varying radius of the rings, and the dimension of the crystal close to the boundary of half circle may be different from the inner ones. There are 31 types of crystals and a total of 960 crystals for the endcap.

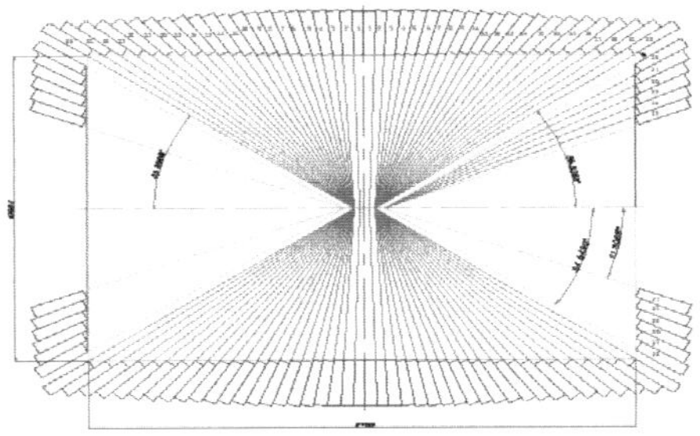

Fig.2.1Configuration of the electromagnetic calorimeter (left) One of Cs(Tl) crystal detector(right)

3. Quality Control

All crystal was manufactured by France Sanit –Gobain, Shanghai Institute of Ceramics, Beijing Hamamatsu and 3000, 1920, 1320 piece respective. CsI(Tl) crystals from mass production should satisfy a number of requirements: relatively light output>42%, light uniformity<7%, The tolerance of the crystal dimension is defined as +0, -200 μm for all transverse dimensions, ± 1mm for length. and radiation hardness <20% decrease of light output after 1000rads (Small sample) or < 9% decrease of light output after 100rads(real size) .

3.1 light output and Uniformity

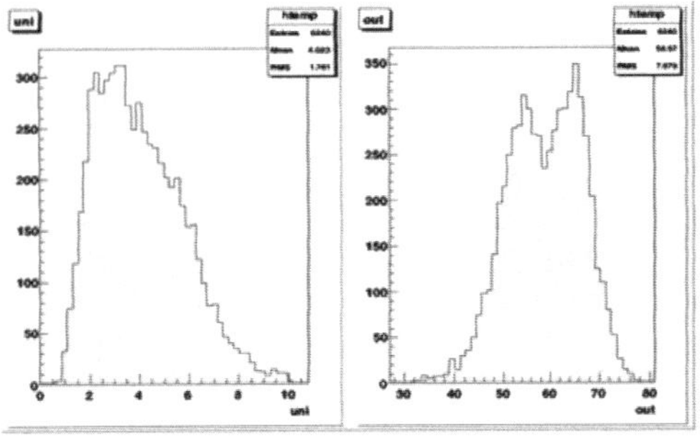

Fig.3.1 Relations the light output

and the uniformity

The light output of a crystal is defined as the number of photoelectrons per MeV. When a crystal is wrapped by 2 layers of 130 μm thick Tyvek, coupled to two photodiodes (S2744-08), readout by an amplifier with 1 μ s shaping time, the light output should be more than 5000 photoelectrons per MeV. A 2-inch Bialkali photomultiplier, PMT is used for testing of crystals. Each crystal is tested by measuring the position of 662 KeV γ ray photoelectric peak using a ^{137}Cs source, at 9 points along the long direction of crystal. There is a 0.3mm air gap between the crystal and the PMT. The light output of a crystal is defined as the average of the 9 measurements and will be normalized to the light output of a small reference CsI(Tl) crystal with 1 inch in diameter and 1 inch in length. Result is shown in Fig.3.1, it include barrel and endcape crystal total are 6280 piece.

3.2 Radiation Hardness

With the expected dose of the BESIII calorimeter of about 300 rad/year, the decrease of the light output of CsI(Tl) is required to be less than 10%/Krad in normal operation, and less than 20%/Krad after CsI(Tl) crystals are irradiated for short time by the ^{60}Co γ source with a high radiation dose. Total 482 crystal are rejected hereinto 4 is dimension problem.

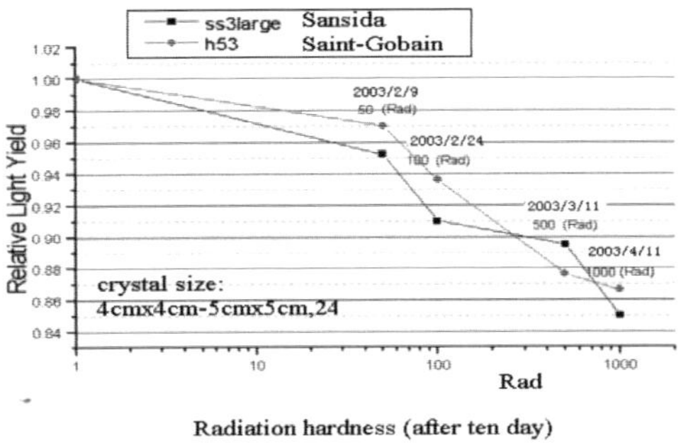

a

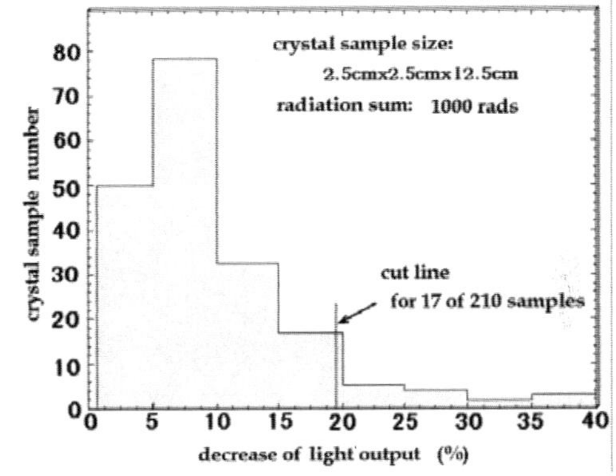

Fig.3.1Radiation hardness of the large CsI(Tl) crystals from different manufacturers(left) . 17 pieces of 210 samples have not passed (right)

3.3 Photodiodes and Charge Sensitive Preamplifiers

Hamamatsu S2744-08 is selected as the photodiode to be used. Its exterior dimension is 14.2mm× 27mm. The photosensitive area is 10cm x 20cm and the thickness is 300μm. 13088 pieces of photodiodes were burned in 65 degree for 150 hours at a reverse voltage of 70V in IHEP Beijing. Photodiodes, crystals and preamplifiers are selected and grouped together, so as to make the gains of all calorimeter modules similar. Measured the noise of each photodiode and the 60 KeV X-ray peak position of the 241Am X-ray source, the noise level of the photodiode and the relative gain of each electronic channel can be calibrated.

4. The Construction of CsI(Tl) Modules

A CsI(Tl) crystal module consists of one CsI(Tl) crystal and two photodiodes. The crystal (except the rear surface) are firstly wrapped in a 260μ thick Tyvek sheet, and then wrapped in a 50μ laminated sheet with 25μ aluminum and 25 μ Mylar. The aluminum layer works for shielding against the electronic noise, and the Mylar is used to isolate each crystal electrically from others and also isolate the detector from the support structure. The rear end of the crystal is the window for collecting scintillation light. Two pieces of type S2744-08 photodiodes are glued at the central part of the crystal rear surface via a 3mm thick Lucite with a size of $2.1cm \times 2.5cm$. Epoxy glue is used both for CsI-lucite and Lucite-photodiode joints. Before gluing, except the area for Lucite-photodiode, the rest part of the rear surface is covered by a 500μ thick Tyvek reflection sheet. Four taped holes are drilled on the centerlines of four sides of the rear face by a special technique. An 8 mm thick aluminum base plate with a cut area for photodiodes is fixed to the crystal rear end by four self-tapping screws. An aluminum shield box covering two preamplifiers is fixed to the aluminum base plate with other four screws. The electrical ground of the preamplifier is connected to the shield box and the aluminum shield layer to avoid external noise interference. There is a rectangle hole on the shield box for settling the monitoring light fiber. Through this hole and the prefabricated holes of aluminum base plate and the Tyvek reflection sheet, one end of the fiber with metal sheath contacts the rear surface of the crystal to inject monitoring light. All processes of the assembly are kept in a dry room with humidity less then 10% and at a constant temperature

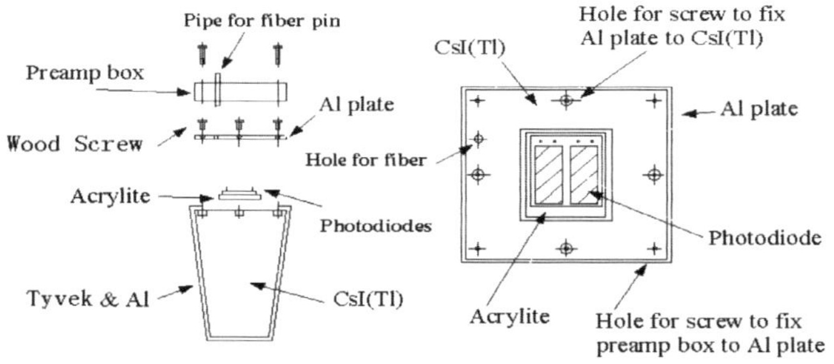

Fig.4.1Assembly of a CsI (Tl) crystal module

5. Cosmic-ray test

After the detector modules are assembled, the differences in their gain should be evaluated by using cosmic-ray, before installing the crystals into the calorimeter structure. When a cosmic-ray muon passes through the crystal being laid horizontally, about 25 MeV is deposited. If the accuracy of the track length of the incident particle in the crystal reaches about ±3mm and a crystal are measured at 9 points longitudinally, with a total of about 4000 events, nine distributions of pulse heights per cm path can be obtained, with an energy resolution of about 10% and the accuracy of the pulse height peak of about 2%. Fig.5.1 are Relative light out test result, all crystal are content to the request.

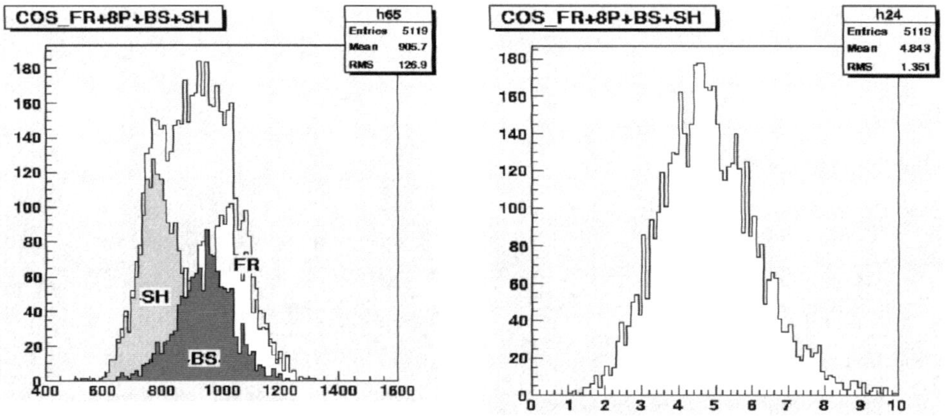

Fig.5.1 The light output and the uniformity, FR is France Sanit –Gobain,

SH is Shanghai Institute of Ceramics, BS is Beijing Hamamatsu

The cosmic-ray test setup could load 60 crystal modules simultaneously for test, and each test takes 72 hours. In addition, this setup also includes random calibration pulse and LED fiber monitoring system. The former is used to calibrate the electronic noise and pedestal and the latter is used to calibrate the relationship between the light intensities and the energies of cosmic ray muon. The data of the cosmic-ray tests are as the basic data of each crystal module.

6. Support Structure of Crystal

The mechanical structure of the calorimeter consists of a barrel and two endcaps. They should support all the inner detectors such as TOF and drift chambers, and attached to the main structure (iron yoke) of the detector. Each endcap consists of two half rings, which has to be movable so that inner detector can be accessed. The container of the barrel calorimeter is a slat structure and is divided into 60 (in Φ direction) compartments, each contain two rows of crystals (88 crystals). The 60 stainless steel bars are connected to the two stainless steel end-rings with screws. The inner edge of the stainless steel bars are connected to the stainless steel opening-cut girder and its two sides through two stainless steel reinforce bars are joined with the two adjacent stainless steel bars. Thus the whole support frame is joined together. In order to be disassembled easily for fixing inner detectors, each endcap consists of two half rings. When BES III inner detectors need to be repaired, the iron yoke of endcap can be moved away about 1430mm along the beam direction (east-west direction) and the two half rings are moved separately in the south-north direction. The available space is enough to examine and repair the main drift chamber (MDC).

7. Conclusion

BESIII Barrel Electro-Magnetic Calorimeter has been sitting in super conduct loop in this autumn, all detector module passed test by cosmic ray and LED fiber system check, we can expect that we will enjoy more physics in future.

Fig.7.1 installation EMC, left is barrel right is endcap

References

[1] BESIII design report 2003

[2] Bell collaboration, Nucl. Instrum. Meth. A479, 117-232 (2002).

[3] KEK Progress Report 1995 April-1996 Mach

[4] BABAR collaboration, Nucl. Instrum. Meth. A479, 1 (2002).

[5] BarBar Collaboration, A systematic Study of Radiation Damage to Large Crystals of CsI(Tl) in the BarBar Detector, Calor2002 report

Calorimeters for absolute luminosity at upgraded DAΦNE

V.Romano, M.Schioppa
Universita' della Calabria, Cosenza, Italy

F.Bossi, B.Buonomo, G.Mazzitelli, F.Murtas, P.Raimondi
Laboratori Nazionali di Frascati dell'INFN, Frascati, Italy

N.Arnaud, D.Breton, P. Roudeau, A.Stocchi, A. Variola, B.Viaud
LAL, Orsay, France

P.Valente
INFN Roma "La Sapienza", Roma, Italy

P. Branchini
Sez. INFN Roma Tre, Roma, Italy
E-mail: `Paolo.Branchini@roma3.infn.it`

Abstract. This paper describes the *LUMI* project which aims at providing fast, reliable and absolute luminosity measurements at the modified DAΦNE interaction point in Frascati for testing the new "crabbed waist" scheme. We present a description of the experimental setup (two luminometers, *LUMI*1 and *LUMI*2), the simulation framework developped for this project and summarize the results and performances.

1. Introduction

The LNF Accelerator Division (LNF-DA) is currently modifying the DAΦNE interaction region. The new layout is based on the *crabbed waist* scheme [1] which is expected to increase the collider luminosity by up to an order of magnitude. The study of the machine performances in this configuration is of utmost importance, both for the DAΦNE experiment and for a future SuperB-factory whose luminosity should exceed 10^{36} cm^{-2}s^{-1} [2]. The upgraded machine has started operations in january 2008, it is going to deliver beam to the *SIDDHARTA* experiment for about six months.

SIDDHARTA [1] will estimate the machine luminosity from the kaon pairs rate seen by the detector. However, given the importance of the experiment, additional measurements of the

[1] The SIDDHARTA (SIlicon Drift Detector for Hadronic Atom Research by Timing Application) experi ment will be devoted to precision measurements on kaonic hydrogen and kaonic deuterium

luminosity based on more standard calorimetric techniques are needed. Therefore, we have built and installed two luminometers around the interaction point (IP). The first one, *LUMI1*, is a double-arm lead-scintillator calorimeter operated in coincidence with a thin GEM tracker to measure Bhabha scattering events. The second one, *LUMI2*, is a crystal forward calorimeter, aiming at measuring the rate of single bremmstrahlung events.

2. Description of the experimental setup

In this section, we describe the experimental setup designed and built for the *LUMI* project: the two luminometers *LUMI1* and *LUMI2*, its electronics and the associated readouts. A careful characterisation of the detector performances has been obtained with dedicated calibrations with test-beam data at the BTF.

2.1. Conventions

The (horizontal) beam axis is labelled z, with the positrons (electrons) traveling towards $+z$ ($-z$) respectivel y. y is the vertical axis (orientated upwards) and x is chosen so that (x, y, z) is a direct orthonormal frame. In the DAΦNE IP, the beams cross each other with a 25 mrad horizontal angle whose effect is to boost the tracks to the $-x$ direction.

2.2. LUMI*1*

This luminometer is symmetrical with respect to the plane containing the IP and perpendicular to the beam axis (z) ; its main components are a lead-scintillator calorimeter and a GEM tracker.

The lead-scintillator calorimeters surround the QD0 final quadrupole magnets and are located at 32.5 cm from the IP.

Their inner and outer radius (respectively 104mm and 341mm) have been chosen to cover the polar angle range between $18 - 27°$ which corresponds to the GEM angular acceptance. In the plane transverse to the beam axis they are segmented in twelve 30 degrees-wide sectors. Two of them intersecting the horizontal ($x - z$) plane (on each side of the beampipe) are not instrumented, both for engineering/mechanical reasons and because the background is expected to be very high in these areas (horizontal plane).

Each instrumented trapezoidal sector is built by sandwiching eleven lead slabs between between trapezoidal plastic scintillator tiles. The first eight (last three) slabs are 5 mm (10 mm) thick while the scintillator tiles are 1 cm thick. The total thickness of the calorimeter, 17.5 cm, corresponds to $\sim$ 12.5 X_0. The expected energy resolution at 510 MeV (nominal beam energy for the Φ-factory) is about 25%.

Lead and scintillator elements are inserted inside a properly machined aluminum frame providing mechanical support. The tiles are connected to three wave-length shifting fibers (WLS), Bicron BCF-92, collecting the light and conveying it a Photonis-Philips XP 2262B photomultiplier (PMT). Each PMT integrates the light coming from all the tiles of a given sector: therefore, the luminometer is not sensitive to the longitudinal development of the shower. In front of each calorimeter, at a distance of 18.5 cm from the IP, a ring of triple-GEM detectors is installed around the beampipe. Each unit of the GEM tracker has an half-moon shape; the top (bottom) half covers azimuthal angles between 14 and 166° (194 and 346°) respectively. A detailed description of the GEM detector can be found in [3]. In principle, Bhabha events are tagged by the coincidence of two back to back high energy ($\sim$500 MeV) deposits in the calorimeters, properly combined with the signals from the GEM detectors. We keep however open the possibility of determining the rate of Bhabha events with the two systems *independently*. At a luminosity of 5×10^{32} cm^{-2}s^{-1}, this rate is $\sim$150 Hz in the angular region of interest.

Table 1. Expected rate of events at $10^{33}\text{cm}^{-2}\text{sec}^{-1}$

$e^+ e^- \to e^+ e^-$	$e^+ e^- \to e^+ e^- \gamma$	$e^+ e^- \to e^+ e^- \gamma\ \gamma$
312.5 Hz		
$18 < \theta < 27°$	95% below 1.7mrad	15% below 1.7mrad

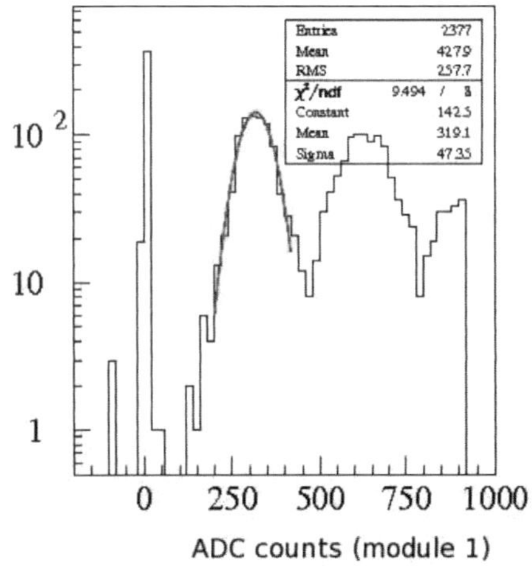

Figure 1. Energy distribution in the first module

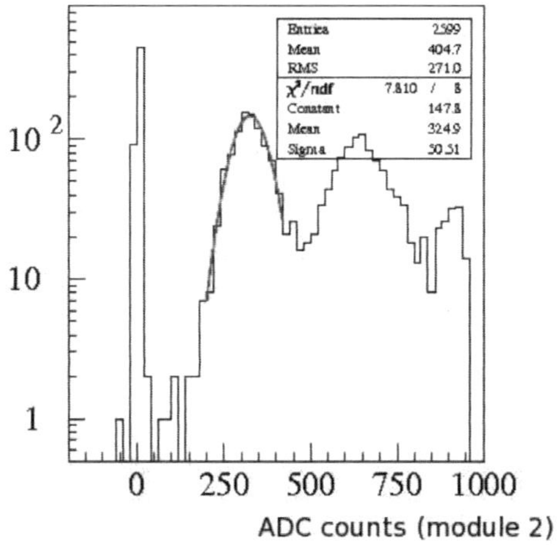

Figure 2. Energy distribution in the second module

2.3. LUMI*2*

A second luminometer, *LUMI2*, is located 170 cm away from the IP on the 'electron' ($-z$) side of the detector just after the beam lines are split. It is made of four $PbW0_4$ crystals (squared section of $30 \times 30\text{mm}^2$ and 110mm heght) are put in series along z corresponding to about $\sim 13\ X_0$). The detector are equipped with Hamamatsu R7600 PMT; in this configuration the electronic stays at about 110mm from the horizontal plane. The aim of these detectors is to measure the rate of the photons emitted by $e^+e^- \to e^+e^-\gamma$ processes. They occur with a cross section of 85 mb for photon energies above 100 MeV, and 95% of the radiated photons can be found inside a cone of 1.7 mrad aperture.

Because of the boost which pushes towards the $-x$ direction the trajectory of the photon, *LUMI2* is placed along the beampipe at x=-5 cm and rotated by 4 degrees in the horizontal plane with respect to the beam axis. The two modules of the calorimeter were tested at the *beam test facility* (btf) in Frascati. The energy distribution for a 500 MeV imping electron are shown in figure 1 and 2 for the two different modules. The different peaks of the energy distribution correspond to the average number of electrons present in the beam. The resolution we get is about 15%.

3. *LUMI*1 performances

Before installation each module of the *LUMI1* calorimeter has been tested and estensively calibrated using the btf in Frascati. Equalization was performed setting the beam energy at 510 MeV and varying the photomultiplier (PM) high voltage (HV). The ADCs peak values for a typical PMT are shown in figure 3 as a function of the HV. The performances of the calorimeter were studied in terms of linearity and resolution using the same setup we have later

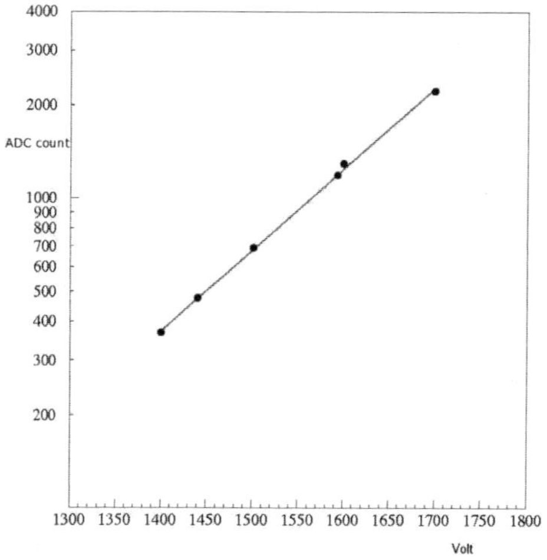

Figure 3. ADC's average distribution value as a function of the PMTs HV for a fixed beam energy.

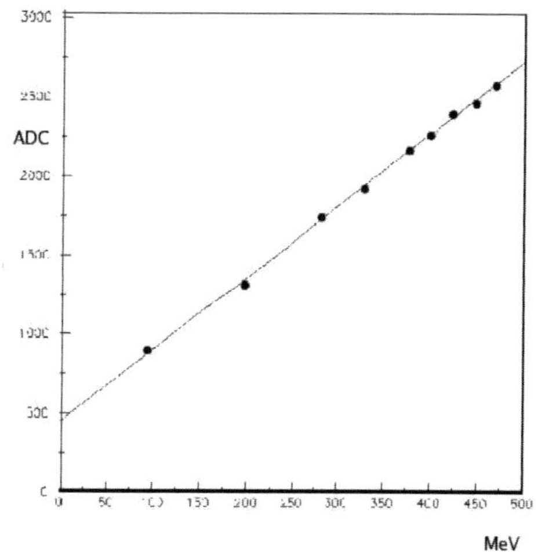

Figure 4. ADC's average distribution value as a function of beam energy for a selected HV value.

on installed on DAΦNE. Figure 4 shows the reconstructed energy in arbitrary units without pedestal subtracion as a function of the beam impinging energy on the calorimeter. A straight line is superimposed to guide the eye.

GEANT3 has been used to simulate the shower initiated by the electrons and positrons in the calorimeter, and to estimate the energy lost by all the corresponding particles while they cross the scintillating tiles. The parameters used as inputs to the simulation were: the photon yield produced by the scintillator per unit energy, the photon attenuation in this material and in the WLS optical fibers and finally the gain of the PMTs to which these fibers drive the photons. Fig. 5 shows the number of reconstructed photons found in the calorimeter central sector and in its two closest neighbours. The average number of photons correspond to the nominal energy of 510 MeV. We expect therefore an energy resolution close to 25% for a 510 MeV electron. Fig. 6 shows the number of collected photons as a function of the azimutal track angle. The ϕ-dependence is clearly due to the calorimeter segmentation and to the correlation between this angle and the distance between the hit location in the tile and the fibers collecting the light. The same phenomenon is responsible for the θ dependence shown in 7.

To correct for this effect, a correction matrix

$$C_{i,j} = N_{i,j}/N_{ideal} \tag{1}$$

is constructed where $N_{i,j}$ is the number of photons collected in a given (θ_i, ϕ_j) cell and N_{ideal} the number of photons in the best (θ, ϕ) cell (in principle located at the bottom and in the middle of the tile). The θ information is based on the GEM radial segmentation. The azimutal angle ϕ is estimated via an weigthed average of the ϕ values of the calorimeter sectors. The weights used to compute this barycenter are the number of photons in each of the calorimeter sector. Figure 8 shows the energy distribution obtained for selected Bhabha on data after correction (red cross) together with Monte Carlo prediction (black line). The correction used improves the calorimeter resolution from 25% to 20%.

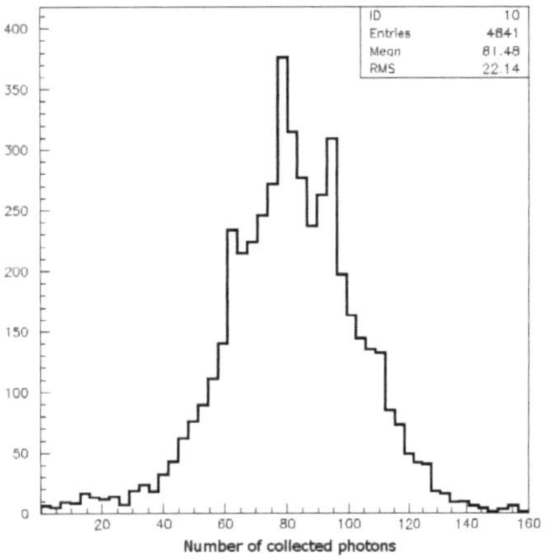

Figure 5. Number of reconstructed photons in the calorimeter central sector.

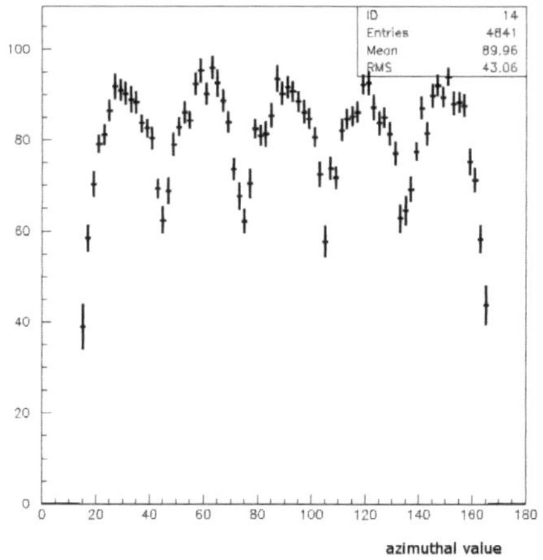

Figure 6. Number of reconstructed photons as a function of the true azimutal track angle.

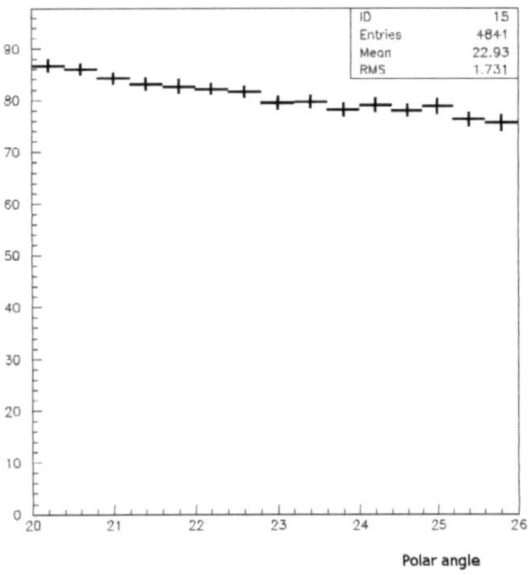

Figure 7. Number of reconstructed photons as a function of the true polar track angle.

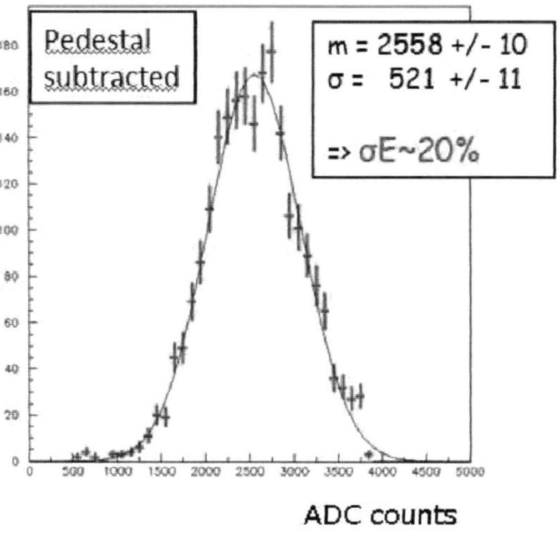

Figure 8. Reconstructed energy distribution.

3.1. Trigger

The trigger system is based on local energy deposits in the calorimeter. Analog signals delivered by the phi sector of a single module are added. NIM equipment has been used to set a threshold for the single module and perform the coincidence with the module on the opposite side. A Bhabha was defined as two energy deposits of more than 200 MeV each occurring in two opposite modules within a time window of 50 ns. A two level scheme has been adopted in order to produce a first trigger signal (T1) to start the calorimeter FEE digitisation while the second level trigger signal (T2), delayed by 4 μs with respect to the T1 starts the front-end electronics (FEE) read-out. The time between bunch crossing in DAΦNE is 2.7 ns, therefore the trigger operates in continuos mode. To avoid the intrinsic jiitter of the trigger signal formation the first level trigger is synchronized within 50 ps with the machine radiofrequency before being delivered to the FEE.

3.2. Electronics and Readout
Most of the FEE and data acquisition (DAQ) components, as well as the HV supplies are borrowed from KLOE [4]. The signals coming from the *LUMI*1 photomultipliers are amplified, inverted and delivered to a 3-stage splitter. One stage consists of a constant fraction discriminator which delivers a current signal to the FEE KLOE TDCs [5]. Another stage consists of a 3-pole Bessel filter followed by a KLOE ADC. The third stage sums the signal over up to five pmts (the top or bottom half of one calorimeter side) and this signal is used to trigger the experiment.

The KLOE FEE modules are build around a custom bus, the AUX-bus [4], which only uses the VME standard for initialization purposes. The AUX interface allows crate event building and data synchronization checks. A trigger-driven read out controller starts the data taking process from the FE boards and delivers data frames to the control manager through C-bus. The control manager board is acquired using the KLOE procedures via VME by a MVME6100 processor. This CPU is in charge of the data framing and delivers data frames to a PC computer via a gigabit ethernet link. The data are finally written to disk by the PC. A simple graphical user interface (GUI) develloped in JAVA performs the run control functions. The entire acquisition chain has been installed in the DAΦNE IP pit.

4. Conclusions
The detector has been installed in February 2008. It has been running continuosly since then giving feed-back to the DAΦNE team.

5. References
[1] D. Alesini et al., LNF-06/33 (IR), (2006).
[2] SuperB Collaboration, SuperB CDR arXiv:0709.0451.
[3] F. Sauli, Nucl. Instrum. Meth. A386, (1997) 531-534.
[4] A. Aloisio et al., Nucl. Instrum. Meth. A516 (2004) 288-314.
[5] M. Passaseo, E. Petrolo, S. Veneziano, Nucl. Instrum. Meth. A367 (1995) 418-421.

Journal of Physics: Conference Series **160** (2009) 012011 doi:10.1088/1742-6596/160/1/012011

The MEG liquid xenon calorimeter

Giovanni Gallucci [1]

Istituto Nazionale di Fisica Nucleare Sezione di Pisa, Largo Bruno Pontecorvo 3, 56127 Pisa, Italy

E-mail: giovanni.gallucci@pi.infn.it

Abstract. The MEG experiment at PSI, starting data taking this year, searches for the muon decay into one electron and one photon with a sensitivity to branching ratios around 10^{-13}, two orders of magnitude better then the present best experimental limit. To reach this goal a new kind of large acceptance, large mass (roughly 2.2 Tons) calorimeter based on liquid xenon scintillation light was developed. Several calibration techniques developed to monitor the calorimeter behavior during all the experiment data taking will be shown together with the experimental resolutions obtained.

1. The MEG experiment at PSI

1.1. Motivation

The MEG experiment [1] searches for the $\mu^+ \rightarrow e^+ + \gamma$ decay. In the Standard Model (SM) framework lepton flavor is preserved, therefore this decay is forbidden. If the present neutrino masses and mixing are considered, the probability of this transition is different from zero but negligible (BR$\sim 10^{-55}$). However all SM extensions enhance the rate through mixing in high energy sector of the theory and in particular, various supersymmetric grand-unified theories (SUSY-GUT) predict a branching ratio in the range $\sim 10^{-12} \div 10^{-14}$ [2]. The signature of a $\mu \rightarrow e + \gamma$ decay at rest is a simultaneous emission of the two daughter particles, having opposite directions and having the same energy equal to half of the muon mass $E_\gamma = E_e = m_\mu/2$ = 52.8 MeV. The main background is given by the accidental coincidence of a positron from the normal muon decay with a high energy photon coming either from radiative muon decay, positron brehmsstraalung or annihilation-in-flight. Given the experimental resolutions on the measurement of the positron energy ΔE_e, of the photon energy ΔE_γ, of their relative timing and angle ($\Delta t_{e\gamma}$ and $\Delta\theta_{e\gamma}$ respectively) and the muon stopping rate R_μ the probability of misidentifying an accidental coincidence as a signal evidence is proportional to:

$$BR_{Acc} \propto R_\mu \, \Delta E_e \, \Delta E_\gamma^2 \, \Delta t_{e\gamma} \, \Delta\theta_{e\gamma}^2 \tag{1}$$

It is apparent from equation (1) that superior energy, position and timing resolutions on both the electron and photon sides are essential in setting stringent limits for this decay. The present experimental limit is set to BR($\mu \rightarrow e + \gamma$) $< 1.2 \times 10^{-11}$ by the MEGA experiment [3]. The MEG experiment aims at reaching a sensitivity of 10^{-13} by using a novel liquid xenon scintillation calorimeter to measure the photon four-momentum.

[1] For the MEG Collaboration

1.2. Muon beam and sub-detectors
In the MEG setup, muons from the presently highest intensity muon beam, the $\pi E5$ beam line at Paul Scherrer Institut, are stopped in a thin target. Positron momentum and trajectory are measured by a magnetic spectrometer (COBRA) composed by a super-conducting non-uniform magnet and 16 drift chambers. The inhomogeneous field quickly sweeps positrons towards plastic scintillating bars that measure their timing. Photons reach the liquid xenon calorimeter where their energy, position and timing are measured (see Figure 1).

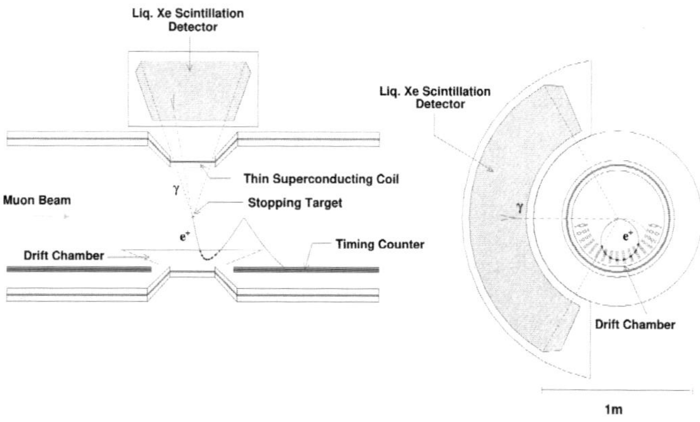

Figure 1. The set-up of the MEG experiment. The C-shaped liquid xenon calorimeter is visible on both views

2. The liquid xenon calorimeter
2.1. Structure
The MEG electromagnetic calorimeter was operated for two months at the end of 2007. It is a 0.8 m^3 C-shaped volume filled with about 800 liters of liquid xenon cooled (T = 165 K) by a pulse-tube refrigerator and by auxiliary liquid nitrogen lines mounted on the internal part of the cryostat. The cryostat structure is made of steel. Its front part is made of a thin steel window (0.5 mm) supported by an aluminum honeycomb and carbon fibers, to reduce the photon interaction probability between the production target and the liquid xenon (see Figure 2). The internal and external radii of the active volume are 65 cm and 112 cm respectively, for a 17 X_0 thickness, and angular extension $\pm 60°$. Inside, mounted on an aluminum and plastic (peek, only inner face) structure, there are about 846 2" UV-sensitive photomultiplier tubes (HAMAMATSU Photonics Inc.) immersed in the liquid xenon. They have a compact structure with twelve amplification stages in order to operate also in a moderate magnetic field ($\approx$ 50 gaus).

2.2. Proprieties of Xenon as a scintillation medium
There are two different ways to produce scintillation light in xenon: ionization and excitation of the atoms. Both ways lead to form an excimer, a Xe_2 molecule existing only in an electronic excited state, which eventually de-excites giving rise to a VUV photon (λ_{peak} = 178 nm, $\Delta\lambda$ = 14 nm FWHM). The two mechanisms are different and lead to different characteristic times. Further, the ratio between ionization and excitation is different for different particles. Liquid xenon light yield is very high ($\approx$ 40000 phe/MeV, $\approx$ 80% of NaI) and it has a very fast response time (4 ns, 20 ns, 45 ns decay components). The high density (2.95 g/cm^3) and atomic number (Z = 54) of liquid xenon allow to make a compact detector. Thanks to the

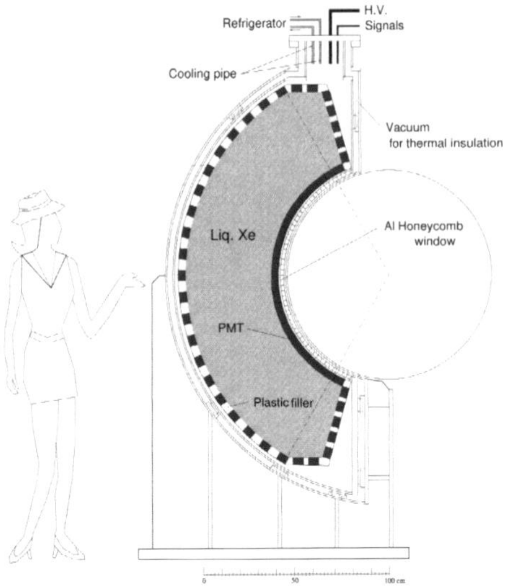

Figure 2. View of liquid xenon calorimeter

above mentioned scintillation mechanism (excimer production) self-absorption should be small and light attenuation is dominated by Rayleigh scattering and absorption processes due to impurities in the liquefied gas; in particular water vapor and molecular oxygen have absorption lines in the range of interest (VUV). In order to purify xenon, a gas-phase purification system with water, oxygen and particle filters runs in parallel to a similar liquid-phase purification system with a fluid pump to increase purification efficiency and speed.

2.3. Data Acquisition system

The signals from all photomultipliers are read from two different digitizer types. The first is a 100 MHz digitizer for triggering purposes. The second one is a 2 GHz digitizer called Domino Ring Sampler [4] which consists of a circular pipeline of 1024 capacitors sequentially cleared and opened to sample the incoming signals. This is necessary to obtain a timing resolution of 100 ps by bin interpolation. In figure 3 we can see alpha and gamma waveforms for the two different types of digitizer.

2.4. Calibrations

In order to ensure that the required performances are reached and maintained during the MEG run, we developed several complementary and redundant methods to calibrate and monitor the behavior of the calorimeter.

2.4.1. LED and alpha sources

36 LEDs are mounted on the calorimeter lateral faces in twelve different positions, able to lighten all photomultipliers. Assuming a linear response of PMTs and a Poisson distribution for photoelectrons, the charge distribution variance is proportional to the photomultiplier gain according to following expression:

$$\sigma^2 = g\,(q - q_0) + \sigma_0^2 \tag{2}$$

where σ is the variance, g is the gain, q is the charge and q_0 and σ_0 are the charge and the variance of the pedestal respectively. By impulsating the LEDs at different voltages and using

Journal of Physics: Conference Series **160** (2009) 012011 doi:10.1088/1742-6596/160/1/012011

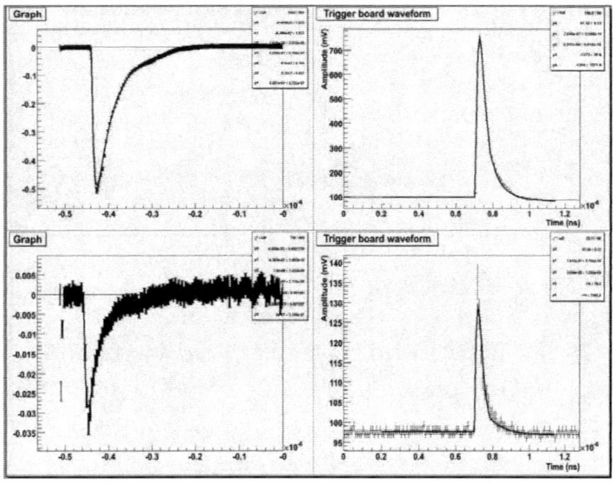

Figure 3. Alpha and γ waveform for DRS (negative) and trigger (positive) digitizer

equation (2), it is possible to obtain a good evaluation of photomultiplier gains (see Figure 4). Five tungsten wires, with a diameter of 100 μm, are mounted inside the calorimeter along the

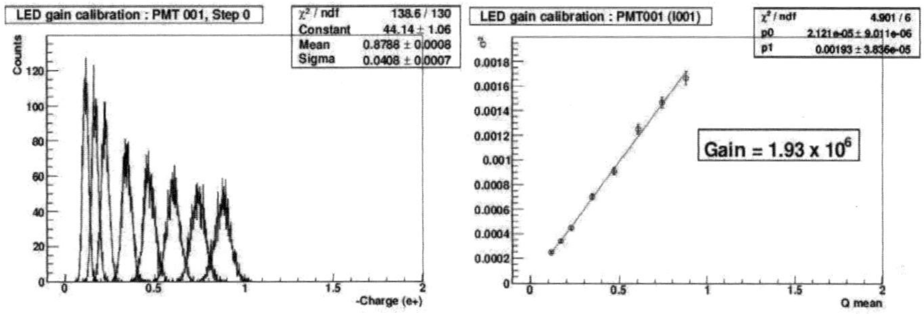

Figure 4. Example of a gain fit for a photomultiplier

beam direction and fixed on lateral faces. On each wire there are five ^{241}Am dots which emit alpha particles releasing their energy in a region about 40 μm around the source. The energy and spatial distribution of these events are known and can be easily simulated. The comparison between the simulated and the real number of photoelectrons of an alpha event provides a good evaluation of the photomultiplier quantum efficiency. The exponential fit of the relation between the ratio of simulated and real photoelectrons number versus the distance of each photomultiplier provides an evaluation of the liquid xenon absorption length (see Fig. 5 (b)).

2.4.2. Cockcroft-Walton accelerator

Normally available radiative γ-ray sources produce photons with energies much below what execpeted for $\mu \rightarrow e + \gamma$ decay (52.8 MeV). The MEG experiment bought therefore a 1MeV Cockcroft-Walton proton accelerator which is used to produce 17.6 and 14.6 MeV lines (see Fig. 5 (a)), by means of the resonant reaction

$$p + {}^{7}_{3}Li \rightarrow {}^{8}_{4}Be + \gamma \qquad (3)$$

allowing a simple, precise and fast energy calibration of the calorimeter. The reaction

$$p + {}^{11}_{5}Bo \rightarrow {}^{12}_{2}Be + \Sigma\gamma \qquad (4)$$

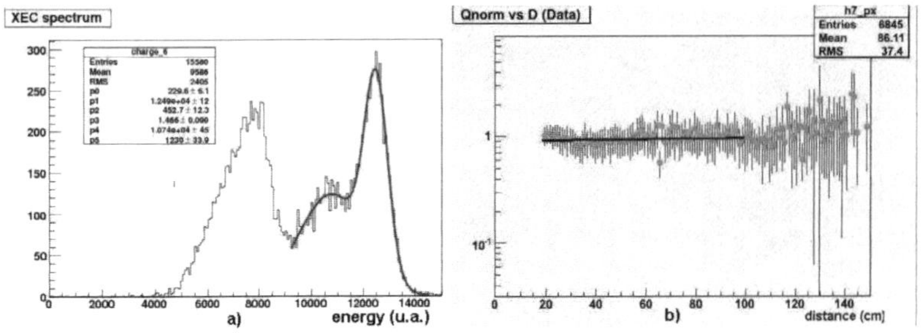

Figure 5. a) Litium peak. b) A distribution of the ratio of simulated and real photoelectrons number versus the distance of each photomultiplier to evaluate Xe purity

with simultaneous emission of two photons (4.4 and 11.7 MeV) useful to study the coincidence between calorimeter and timing counter is also used.

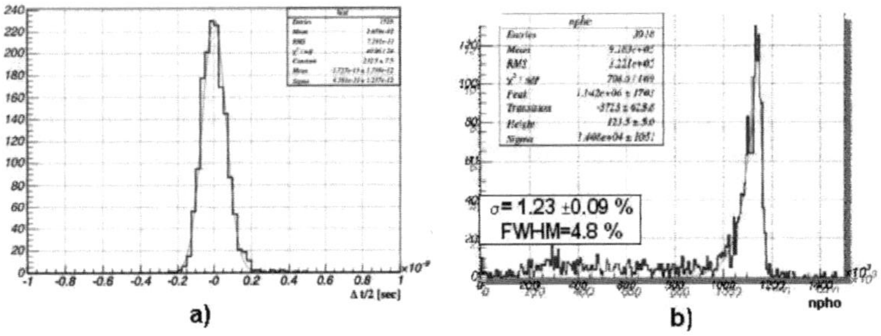

Figure 6. Time (a) and energy (b) resolution on 52.9 MeV photons

2.4.3. Charge exchange reaction

A negative pion beam is stopped in a liquid hydrogen target, when the charge exchange reactions:

$$\begin{aligned} \pi^- + p \rightarrow \quad & \pi^0 + n \\ & \hookrightarrow \gamma + \gamma \end{aligned} \tag{5}$$

The energy spectrum of the two γs from the π^0 decay is flat between $54.9 < E_\gamma < 82.9$ MeV and there is a correlation between their energy and opening angle: in particular when two photons are emitted at 180°, they have 54.9 and 82.9 MeV energies respectively. It is therefore possible to study the calorimeter response at an energy close to interesting one. In particular to select only 54.9 and 82.9 MeV photons, we use an auxiliary segmented NaI detector positioned at the opposite side of the calorimeter, so that only back to back γ pairs are selected. In order to scan the whole ϕ angle, the NaI is mounted on a mechanical structure that allows to move it in different positions.

2.5. Performances of calorimeter

With the aforementioned calibration methods the calorimeter performances have been checked. In particular the energy, position and timing resolution and the linearity of calorimeter response have been measured. In figure 7 (a) the detector linearity is showed: the first four points are from Cockcroft-Walton reactions, the others two from pion charge exchange reactions. In figure

6 the timing and energy resolution obtained with 54.9 MeV photons are showed. For spatial resolution some holes and edge of a lead collimator have been used (see Fig. 7 (b)). In table 1 these results are summarized.

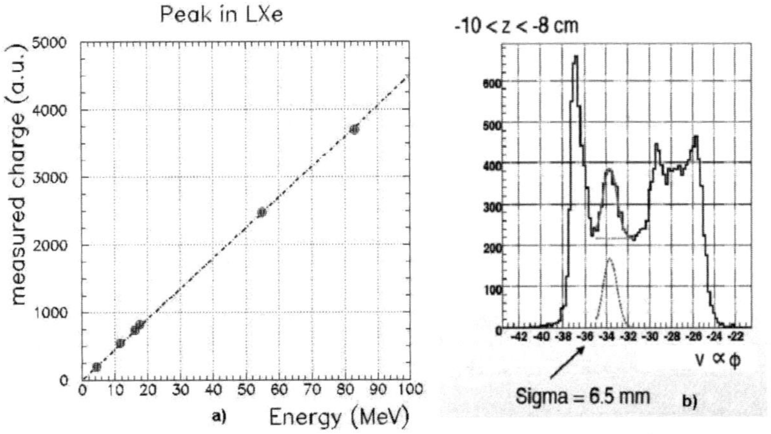

Figure 7. a) Checked charge at different photon energies in the calorimeter. b) Spatial resolution obtained by 52.9 MeV photons using a lead collimator

Table 1. Summary of liquid xenon detector resolution

Measurement	Resolution (FWHM)
γ Energy (on 55 MeV)	4.8
γ Position (mm)	15.0
γ Time (nanosecond)	0.15

3. Conclusions

The MEG calorimeter detector is the biggest liquid xenon scintillation in the world. In order to calibrate and monitor that the required performances are reached and maintained, several complementary and redundant methods are developed. The calorimeter was successfully installed and operated for 2 months in 2007 when, using the calibration methods, it showed resolutions in agreement with what proposed to reach MEG goals. The data taking period of the MEG experiment will start in July 2008.

References

[1] A. Baldini *et al.*, Research Proposal to INFN 2002.
[2] R. Barbieri *et al.*, 1995 , *Nucl. Phys.* B **445** 215.
J. Hisano *et all.*, 1997 *Phys. Rev.* B **391** 341.
R. Ciafaloni *et al.*, 1996 , *Nucl. Phys.* B **458**.
[3] M. Ahmed *et al.*, [MEGA collaboration] 2002 *Phys. Rev.* D **65** 112002.
[4] S.Ritt, 2004 *Nuclear Science Symposium Conference Record*, IEEE.

Journal of Physics: Conference Series **160** (2009) 012012 doi:10.1088/1742-6596/160/1/012012

The ALICE Electromagnetic Calorimeter Project

F. Ronchetti, on behalf of the ALICE collaboration.

Istituto Nazionale Fisica Nucleare,
Laboratori Nazionali di Frascati
Via E. Fermi 40
00044 Frascati (Roma)

Federico.Ronchetti@LNF.INFN.IT

Abstract. The ALICE Experiment (A Large Ion Collider Experiment) aims to study the properties of quark-gluon matter using Pb-Pb collisions at a center of mass energy (per nucleon pair) of $\sqrt{s_{NN}}$ = 5.5 TeV with the Large Hadron Collider (LHC) at CERN. The EMCal consists in a large area electromagnetic calorimeter able to extend the measured momentum range of photons and electrons by over an order of magnitude. In addition, the EMCal will enhance the capability of the overall ALICE setup to perform better jet reconstruction by measurement of the neutral energy component of jets, photons and neutral pions. The EMCal will also produce a fast high-p_T trigger: the anticipated minimum bias average Pb-Pb interaction rate is very high (around 8 kHz), thus a fast high-p_T trigger will provide an enhancement in high p_T events in central collisions. The EMCal covers a geometrical region from $-0.7\leq\eta\leq0.7$ (in pseudo-rapidity η) and 120° in the azimuthal angle ϕ. In particular, the ϕ-coverage has been chosen to allow the detection of γ-jet events in coincidence with the other ALICE complementary calorimeter, the PHOS. The EMCal is a modular sampling calorimeter: it can measure showers up to 20 radiation lengths. Each module is composed by 4 towers of a Pb-scintllator sandwich (shashlik). The shape of the basic module is tapered to allow a projective geometry of the final assembly with respect to the interaction point. An assembly of 12×24 modules is called a super-module. The complete EMCal is a high granularity detector containing 11 super modules for a total of 12.672 towers. An independent optical readout of each tower is provided using wavelength shifting fibers coupled to an APD (Avalanche Photo Diod). The APD readout was chosen to allow the operation in the high B-field environment created by the ALICE solenoidal magnet. The gain of the APD is monitored using a LED activated scintillator installed on into each module.

1. Introduction

1.1. Physical motivation

The ALICE experiment at LHC is made out of a wide array of detector systems for measuring hadrons, leptons, and photons. ALICE is designed to carry out comprehensive measurements of high energy nucleus-nucleus collisions, in order to study QCD matter under extreme conditions and to study the phase transition between condensed matter and the Quark-Gluon Plasma (QGP) [1,2]. The interaction and energy loss of high energy partons in matter provides a sensitive tomographic probe of the medium generated in high energy nuclear collisions (jet quenching) [3-6]. Jet quenching measurements have played a key role at the Relativistic Heavy Ion Collider (RHIC) and will be central to the study of nuclear collisions at LHC. The EMCal enhances ALICE's capabilities for jet quenching measurements. The addition of the EMCal into ALICE enables triggering on high energy jets, reduces significantly the measurement bias for jet quenching studies, improves jet energy resolution, and augments existing ALICE capabilities to measure high momentum photons and electrons. Combined with ALICE's excellent capabilities to track and identify particles from very low p_T to high p_T, the EMCal enables an extensive study of jet quenching at the LHC.

1.2. The ALICE Electromagnetic Calorimeter

ALICE is designed for measurements in the high multiplicity environment of heavy ion collisions and is well suited for jet quenching studies. It has excellent momentum resolution for charged particles from 100 MeV/*c* to 100 GeV/*c*, covering nearly the full range of fragment momentum for the highest energy jets accessible in heavy ion collisions. The EMCal will complete ALICE's capabilities to measure jet quenching. The most important features of the EMCal are an efficient and unbiased fast trigger (Level 0/1) for high energy jets, and measurement of the neutral portion of jet energy. The EMCal will also improve jet energy resolution, and enhance ALICE capabilities to measure high p_T photons, neutral hadrons, and electrons.

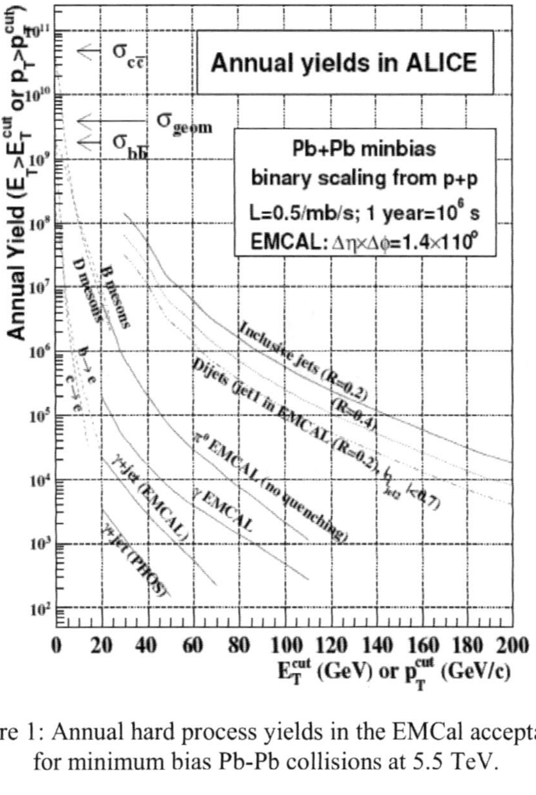

Figure 1: Annual hard process yields in the EMCal acceptance for minimum bias Pb-Pb collisions at 5.5 TeV.

Fig. 1 shows the annual yield for various hard processes in the EMCal acceptance, for minimum bias Pb-Pb collisions at nominal luminosity. The EMCal kinematic reach for inclusive jets extends beyond 200 GeV, while for di-jets with a trigger jet in the EMCal and the recoiling jet in the TPC acceptance it is about 170 GeV. The γ-jet rate is statistically robust for $p_T \leq 40$ GeV/*c*, while the yield for inclusive electrons from semi-leptonic decays of *b* and *c* extends to $p_T \sim 25$ GeV/*c*.

ALEPH [7] and STAR [8] have shown that jet measurements based on electromagnetic calorimetry and charged particle tracking have similar energy resolution to e.m. and hadronic calorimetry. In fact, charged particle tracking is superior to hadronic calorimetry for suppressing backgrounds to jet measurements in the high multiplicity environment of heavy ion collisions. Full exploitation of jets as a probe of QCD matter at the LHC requires both broad kinematic reach of jet energy and detailed measurement of jet structure, from the hardest hadronic fragments to very soft fragments. Much of the interesting physics may indeed be carried by low p_T hadrons, which have the greatest sensitivity to the jet interaction with the medium. From the experience of RHIC measurements [9,10], particle identification is expected to be critical in elucidating the physics of jet quenching. The EMCal acceptance, triggering and measurement capabilities, combined with the excellent tracking and particle identification capabilities of ALICE, enable the most extensive measurements of jet quenching at the LHC.

2. Detector design

2.1. Design overview.

The overall design of the EMCal is heavily influenced by its integration within the ALICE magnet. The EMCal is to be located inside the large room temperature solenoidal magnet of ALICE within a cylindrical integration volume approximately 112 cm deep in the radial direction, sandwiched between the ALICE spaceframe and the ALICE magnet coils. Due to the presence of the PHOS carriage below the ALICE TPC and the HMPID above the ALICE TPC, the EMCal covers a region of about 110 degrees in azimuth above the TPC adjacent to the HMPID and its acceptance is well matched to ALICE physics goals.

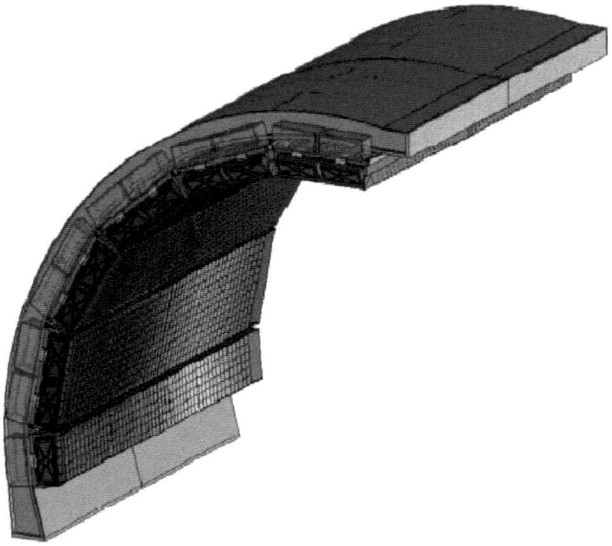

Figure 2: The array of super modules shown in the installed position on their support structure.

The conceptual design of the electromagnetic calorimeter for the ALICE experiment is based on the Shashlik technology as implemented in the PHENIX experiment [11] at RHIC, HERA-B at DESY or LHCb [12] at CERN. The scope and basic design parameters of the proposed calorimeter have been chosen to match the physics performance requirements of the high p_T physics program. Fig. 2 shows the EMCal super modules mounted in the installed position on their support structure. A continuous arch of super modules, each spanning ~20 degrees in azimuth, is indicated. The EMCal is positioned to provide partial back-to-back coverage with the ALICE Photon Spectrometer (PHOS) calorimeter. Small azimuthal gaps (~ 3.0 cm) are provided between super modules to facilitate installation and alignment. These gaps are positioned in line with the TPC sector boundaries. The chosen technology is a layered Pb-scintillator sampling calorimeter with a longitudinal pitch of 1.44 mm Pb and 1.76 mm scintillator with longitudinal Wavelength Shifting Fiber (WLS) light collection. The full detector spans η = -0.7 to η = 0.7 with an azimuthal acceptance of $\Delta\phi$ = 110. The detector is segmented into 12,672 towers, each of which is approximately projective in η and ϕ to the interaction vertex. The towers are grouped into super modules of two types: full size which span $\Delta\eta$ =0.7 and $\Delta\phi$ = 20 and half size which span $\Delta\eta$ = 0.7 and $\Delta\phi$ = 10. There are 10 full size and 2 half size super modules in the full detector acceptance (Fig. 2). The super module is the basic structural units of the calorimeter. These are the units handled as the detector is moved below ground and rigged during installation.

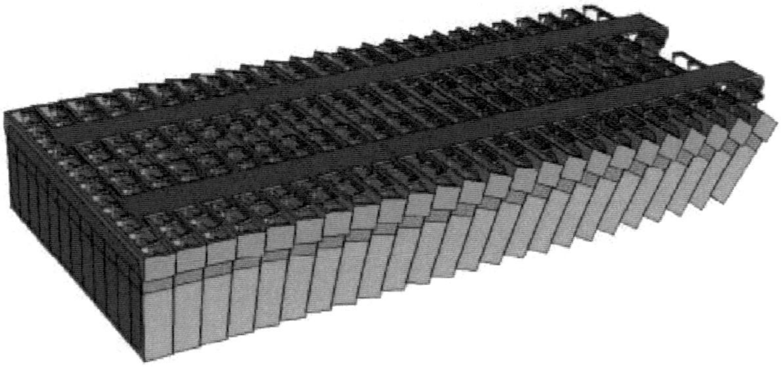

Figure 3: ALICE EMCal super module concept.

Fig. 3 shows a super module with its external mechanical structure stripped away to illustrate the stacking of modules within the super module. This figure shows a full size super module with 12× 24 modules configured as 24 strip modules of 12 modules each. The supporting mechanical structure of the super module has been removed so that the strip module stacking into a nearly projective geometry can be seen. The electronics integration pathways are illustrated. Each full size super module is assembled from 12 × 24 = 288 modules arranged in 24 strip modules of 12 modules each. Each module has a rectangular cross section in the ϕ direction and a trapezoidal cross section in the η direction with a full taper of 1.5°. The resultant assembly of stacked strip modules is approximately projective with an average angle of incidence of less than 2° in η and less than 5° in ϕ. A single module and an assembled strip module is indicated schematically in Fig. 4.

2.2. Module Design

The smallest building block of the calorimeter is the individual module illustrated in Fig. 4. Each individual module contains 2 × 2 = 4 towers built up from 77 alternating layers of 1.44 mm Pb and 1.76 mm polystyrene, injection molded scintillator. White, acid free, bond paper serves as a diffuse reflector on the scintillator surfaces while the scintillator edges are treated with TiO_2 loaded reflector to provide tower to tower optical isolation and improve the transverse optical uniformity within a single tower. The Pb-scintillator stack in a module is secured in place by the static friction between individual layers under the overall load of ~350 kg. The module is closed by a skin of 150 μm thick stainless steel screwed by flanges on all four transverse surfaces to corresponding front and rear aluminum plates. This thin stainless skin is the only inert material between the active tower volumes. The internal pressure in the module is stabilized against thermal effects, mechanical relaxation and long term flow of the Pb and/or polystyrene by a customized array of 5 non-linear spring sets

Journal of Physics: Conference Series **160** (2009) 012012 doi:10.1088/1742-6596/160/1/012012

(Bellville washers) per module. In this way, each module is a self supporting unit with a stable mechanical lifetime of more than 20 years when held from its back surface in any orientation as when mounted in a strip module.

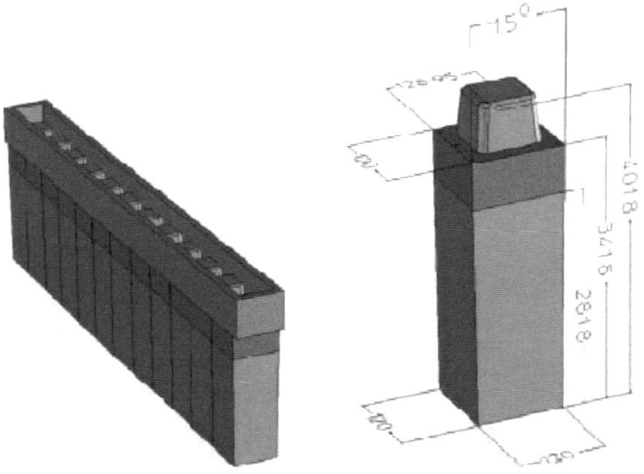

Figure 4: A single 1.5° taper module (right hand side) with the dimensions of the prototype shown in mm. The left hand figure shows a single strip module comprised of 12 EMCal modules integrated onto a single strong back.

Fig. 5 below shows a cut away view of the back end of a single module illustrating the internal components used to sustain the module compression and a segment of the strip module strong back.

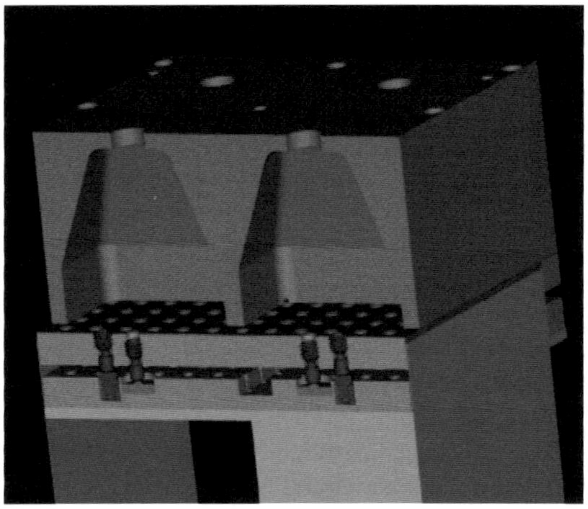

Figure 5: Cut away view of the back end of a single module showing the components that maintain the module's compression. The Bellville washers act between the two containment Al plates to keep the module internal load. A segment of the strip module strong back is also shown.

All modules in the calorimeter are mechanically and dimensionally identical. The front face dimensions of the towers are 6×6 cm^2 resulting in individual tower acceptance of $\Delta\eta\times\Delta\phi$=0.014×0.014 at η=0.

2.2.1. *Sampling fraction.* The present conceptual design incorporates a moderate detector average active volume density of ~5.68 g/cm^3 which results from a 1:1. 22 Pb to scintillator ratio by volume. This results in a compact detector consistent with the EMCal integration volume at the chosen detector thickness of 20.1 radiation lengths.

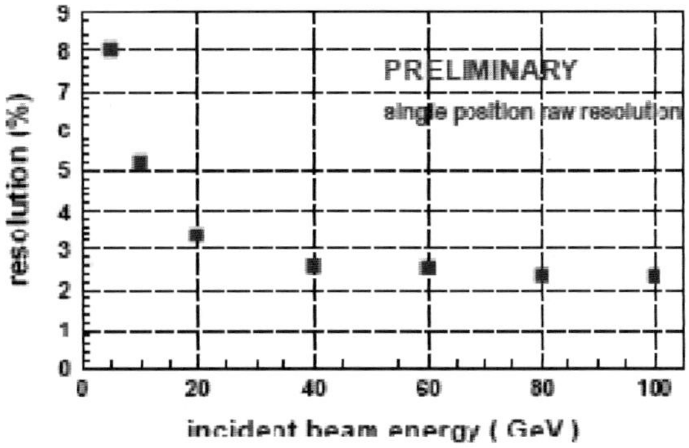

Figure 6: EMCal II generation prototype test beam results (Oct `07) for the energy resolution (no final calibration and no beam resolution subtraction).

Practical considerations, including the total assembly labor cost, suggest reducing the total number of Pb/scintillator layers thus decreasing the sampling frequency. Fig. 6 shows the measured energy resolution curve from the 16-module II generation prototype from the CERN SPS test beam (October 2007). The result was obtained using temporary calibration constants and does not include the subtraction of the beam momentum spread.

2.2.2. *Optical and Photo Sensors.* Scintillation photons produced in each tower are captured by an array of 36 Kuraray, Y-11, double clad, WLS fibers that run longitudinally through the Pb/scintillator stack (Fig. 7). Each fiber terminates in an aluminized mirror at the front face end of the module and is integrated into a polished, circular group of 36 at the photo sensor end at the back of the module. The fiber bundles are pre-fabricated and inserted into the towers after the module mechanical assembly is completed. The 36 individual fibers are packed into a circular array 6.8 mm in diameter and held in place inside a custom injection molded grommet by Bicron BC-600 optical cement. An optical quality finish is applied to the assembled bundle using a diamond polishing machine. At the other end of the bundle, individual fibers are similarly polished and mirrored with a sputtered coat of aluminum thick enough to ensure the protection of the inner mirror. The response of the Al-coated fiber is considerably flatter with an overall increase in efficiency in the range of about 25% in the vicinity of shower maximum (i.e. the location of the highest energy deposition for an electromagnetic shower). This number accounts for material immediately in front of the detector; which ranges between 0.4 and 0.8 radiation lengths, and assumes 5.5 - 6.0 radiation lengths for shower maximum for 10 GeV photons. At this depth in the detector, the mirrored fiber response is very uniform and contributes nothing significant to the non-linearity of the detector as a whole. Other factors which can significantly impact the electromagnetic performance of the calorimeter, include scintillator edge treatment and the density of the wavelength shifting fiber readout pattern and the material chosen for the interlayer diffuse reflector. For scintillator edge treatment and fiber density, advantage was taken from the extensive studies made by the LHCb collaboration for their ECAL [12]. In particular, a diffuse reflector edge treatment was adopted, such as that obtained with Bicron Titanium Dioxide loaded white paint (BC622A) with a total fiber density of about one fiber per cm^2. In the case of the interlayer diffuse reflector, a white, acid free, bond paper was used in place of the Teflon based commercial TYVEK. While TYVEK produces slightly better surface reflectivity, its coefficient of friction is too low to permit its use in this design where the module's mechanical stability depends somewhat on the interlayer friction.

Figure 7: Fibre bundles with attached APD and preamplifier of four towers of an EMCal module.

The 6.8 mm diameter fiber bundle from a given tower connects to the APD through a short light guide/diffuser with a square cross section of 7 mm × 7 mm that tapers slowly down to 4.5 mm × 4.5 mm as it mates (glued) to the 5 mm × 5 mm active area of the photo sensor. Fig. 7 shows also 4 pre-fabricated fiber bundles inserted into the towers of a single prototype module. In this picture all of the module rear enclosing and structural elements are omitted so the convergence of the wavelength shifting fibers may be seen as they converge to the light guide (inside the black plastic tube) and finally to mate with the APD and charge sensitive preamplifier (CSP). The selected photo sensor is the Hamamatsu S8664-55 Avalanche Photo Diode. This photodiode has a peak spectral response at a wavelength of 585 nm compared to an emission peak of 476 nm for the Y-11 fibers. However, both the spectral response and the quantum efficiency of the APD are quite broad with the latter dropping from the maximum by only ~5% at the WLS fiber emission peak. At this wavelength, the manufacturer's specification gives a quantum efficiency of 80%.

3. **Module Integration to Strip Modules and Super Modules.**

As described above, the super module is the basic building block of the calorimeter. Starting with 288 individual modules which are rather compact and heavy, the main engineering task is to create a super module structure which is rigid, with small deflections in any orientation yet does not require extensive, heavy external stiffening components that would reduce the volume available for the active detector. The solution adopted for the ALICE EMCal is to develop a super module "crate" which functions not as a box for the individual modules but rather an integrated structure in which the individual elements contribute to the overall stiffness. The super module crate is effectively a large I-beam in which the flanges are the long sides of the crate and the 24 rows of strip modules together. This configuration gives to the super module good stiffness for both the 9 o'clock and 10 o'clock locations. For the 12 o'clock location, the I-beam structure of the super module is augmented by a 1 mm thick stainless steel forward sheet (traction loaded), which controls the bending moment tending to "open" the crate main sides, and helps to limit deflection of strip modules. Ridges are provided on the interior surfaces of the crate to allow precision alignment of the strip modules at the correct angle. The stiffness given by this I-beam concept allows the use of non-magnetic light alloys for main parts of the super module crate. Parts of the super module crate will be made mainly from laminated 2024 aluminum alloy plates. The two main sides (flanges of the I-beam) of the crate will be assembled from 2 plates, 25 mm and 25 mm thick, bolted together and arranged so as to approximately follow the taper of the 20 degree sector boundary. Each of the 24 rows of a super module contain 12 modules as described above. Each of the modules is attached to a transverse beam by 3.4 mm diameter stainless steel screws. The 12 modules and the transverse beam form a strip module. The strip module is 1440 mm long, 120 mm wide, 410 mm thick. The total weight of the strip module is approximately 300 kg and like module, it is a self supporting unit. The transverse beam, which is the structural part of the strip module, is made from cast aluminum alloy with individual cavities along its length where the fibers emerging from towers are allowed to converge. The casting process is well suited to forming these cavities and the overall structure, saving considerable raw material and machining time. Fig. 4 (left panel) shows the overall layout and dimensions of a strip module. In addition to functioning as a convenient structural unit which offers no interference with the active volume of the detector and forming the web of the I-beam structure of the super module, the transverse beam of the strip module provides protection for the fibers, a structural mount for the light guide, APD and charge sensitive preamplifier and a light tight enclosure for these elements.

4. Conclusions

According to the very important but not final RHIC results, jet quenching studies are the ultimate tool to probe QCD dense matter. In order to go further there is a need to measure the jet structure modification over a huge kinematic range, while having an efficient jet triggering and particle tracking and, last but not least, a detailed particle identification down to low p_T. The ALICE detector augmented with the EMCal can fully meet these demands.

References

[1] F. Carminati et al. (ALICE Collaboration), J. Phys. G30 1517 (2004).
[2] ALICE Collaboration, Physics Performace Report, Vol 2, J. Phys. G (2006).
[3] R. Baier, D. Schiff and B. Zakharov, Ann. Rev. Nucl. Part. Sci. 50, 37 (2000).
[4] M. Gyulassy, I. Vitev, X. N.Wang and B.W. Zhang, Quark Gluon Plasma 3, 123 (ed. R. Hwa and X. N. Wang, World Scientific, Singapore); nucl-th/0302077.
[5] A. Kovner and U. Wiedemann, Quark Gluon Plasma 3, 192 (ed. R. Hwa and X. N. Wang, World Scientific, Singapore); hep-ph/0304151.
[6] P. Jacobs and X. N. Wang, Prog. Part. Nucl. Phys. 54, 443 (2005).
[7] D. Buskulic et al. (ALEPH), Nucl. Instr. Meth. A360, 481 (1995).
[8] M. Miller et al. (STAR), proceedings of PANIC05, hep-ex/0604001.
[9] S. S. Adler et al. (PHENIX), Phys. Rev. C69, 034909 (2004).
[10] J. Adams et al. (STAR), nucl-ex/0601042.
[11] L. Aphecetche et. al., (PHENIX), Nucl. Instrum. Methods A499 (2003) 521.
[12] LHCB Collaboration, LHCb TDR2, CERN/LHCC 2000-36, 6 September 2000.

Studies of the effect of charged hadrons on lead tungstate crystals

Francesca Nessi-Tedaldi
Swiss Federal Institute of Technology (ETH), CH-8093 Zürich, Switzerland
E-mail: Francesca.Nessi-Tedaldi@cern.ch

Abstract. Scintillating crystals are used for calorimetry in several high-energy physics experiments. For some of them, performance has to be ensured in difficult operating conditions, like a high radiation environment, very large particle fluxes and high collision rates. Results are presented here from a thorough series of measurements concerning mainly the effect of charged hadrons on lead tungstate. It is also shown how these results can be used to predict the effect on crystals due to a given flux of particles.

1. Introduction

The effect of large hadron fluxes on $PbWO_4$ crystals has become an important subject of investigation with the construction of the Large Hadron Collider (LHC) at CERN and of experiments using such crystals. The question had to be studied, whether large hadron fluences cause a specific, possibly cumulative damage, and if so, what its quantitative importance is and whether it only affects light transmission or also the scintillation properties of lead tungstate. It should be noticed that all damage observed in earlier tests [1, 2] could always be ascribed to the ionising dose associated with the hadron flux, apart from some indication of a hadron-specific damage in BGO, which can be extracted [3] from existing data, as discussed later herein. However, none of the previous tests on lead tungstate had been extended to the full integrated fluences expected at the LHC. Therefore, a thorough series of irradiation tests on lead tungstate crystals [4, 5, 6] has been performed over several years, with exposures reaching up to the highest integrated fluence expected at the LHC after 10 years of running. In this presentation, an overview is given on the main results obtained through these tests, and the understanding reached is summarised.

2. Proton and γ irradiation studies

In a first study to explore the dependence on the integrated fluence of a possible hadron-specific damage [4], eight CMS [7] production crystals of consistent quality [7] were irradiated in a 20 to 24 GeV/c proton flux of either $\phi_p = 10^{12}$ p cm^{-2}h^{-1} or $\phi_p = 10^{13}$ p cm^{-2}h^{-1} at the CERN PS accelerator [8]. The maximum fluence reached was $\Phi_p = 5.4 \times 10^{13}$ p cm^{-2}. The crystals were nearly parallelepipedic, with dimensions $2.4 \times 2.4 \times 23$ cm^3. To disentangle the contribution to damage from the associated ionising dose, complementary ^{60}Co γ-irradiations were performed at a dose rate of 1 kGy/h on further seven crystals, since a flux $\phi_p = 10^{12}$ p cm^{-2}h^{-1} has an associated ionising dose rate in $PbWO_4$ of 1 kGy/h. The maximum dose reached, of 50.3 kGy, is just $\sim$ 50% below the one reached in proton irradiations. The measurements of transmission

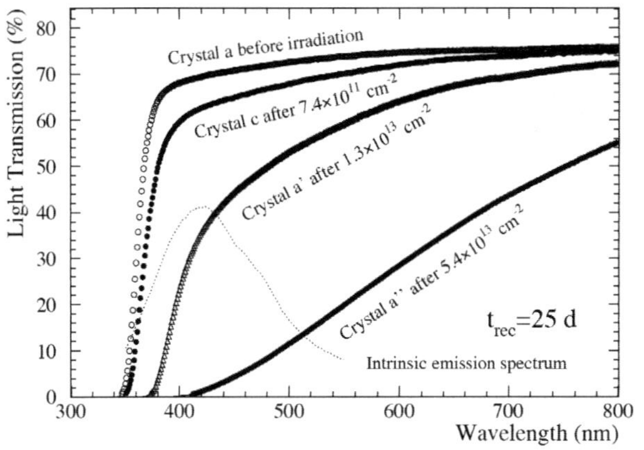

Figure 1. Light transmission curves for crystals with various degrees of proton-induced radiation damage [4]. The emission spectrum units [9] are arbitrary.

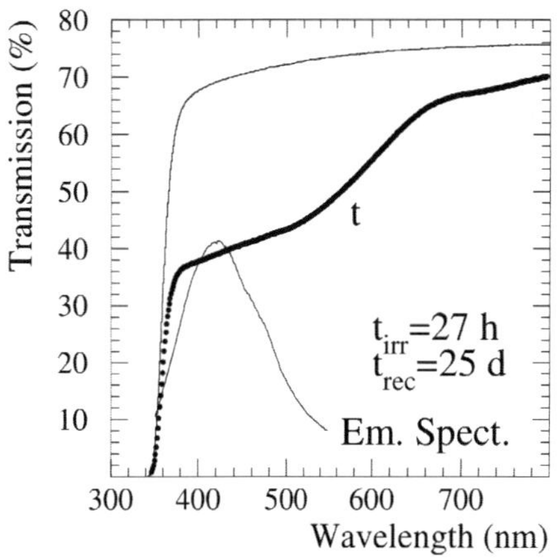

Figure 2. Transmission curves for a γ-irradiated crystal [4] prior to (thin) and after (thick line) irradiation. Emission spectrum as in figure 1.

damage revealed [4] that proton irradiation decreases the light transmission for all wavelengths and moves the ultra-violet transmission band-edge to longer wavelengths, as can be seen in figure 1. In γ-irradiated crystals, the transmission band-edge does not shift at all, even after the highest integrated dose: one only observes the well-known absorption band [10] around 420 nm that can be appreciated in figure 2. These results demonstrate the qualitatively different nature of proton-induced and γ-induced damage.

The damage is usually quantified through the induced absorption coefficient for a given wavelength λ, which for the Longitudinal Transmission, measured through the length ℓ of a crystal, is defined as

$$\mu_{IND}^{LT}(\lambda) = \frac{1}{\ell} \times \ln \frac{LT_0}{LT} \tag{1}$$

where LT_0 (LT) is the Longitudinal Transmission value measured before (after) irradiation through the length of the crystal, and analogously for Transverse Transmission (TT). The data for μ_{IND}^{LT} versus light wavelength for proton-irradiated crystals in figure 3 show a λ^{-4} dependence, which is absent in γ - irradiated crystals. This is an indication of Rayleigh scattering from small centres of severe damage, as they might be caused by the high energy deposition of heavily ionising fragments along their path, locally changing the crystal structure. Since the crystals contain heavy elements, Pb and W, it was argued in [4] that a hadron-specific damage can in fact arise from the production, above a $\sim$20 MeV threshold, of heavy fragments with up to 10 μm range and energies up to $\sim$100 MeV, causing a displacement of lattice atoms and energy losses, along their path, up to 50000 times the one of minimum-ionising particles. The observed behaviour confirms this qualitative understanding. A further evidence evidence of Rayleigh scattering is the complete polarisation of scattered light, as observed using Polaroid films.

As it can be seen in figure 4, where the evolution over time of damage values measured at the peak of scintillation emission wavelength, $\mu_{IND}^{LT}(420 \text{ nm})$, is plotted, only a small fraction of the hadron damage recovers on a time scale comparable to the expected duration of LHC running.

The correlation between the induced absorption coefficient at 420 nm and fluence [4] shown in figure 5, is consistent with a linear behaviour over two orders of magnitude, showing that proton-induced damage in $PbWO_4$ is predominantly cumulative, unlike γ-induced damage,

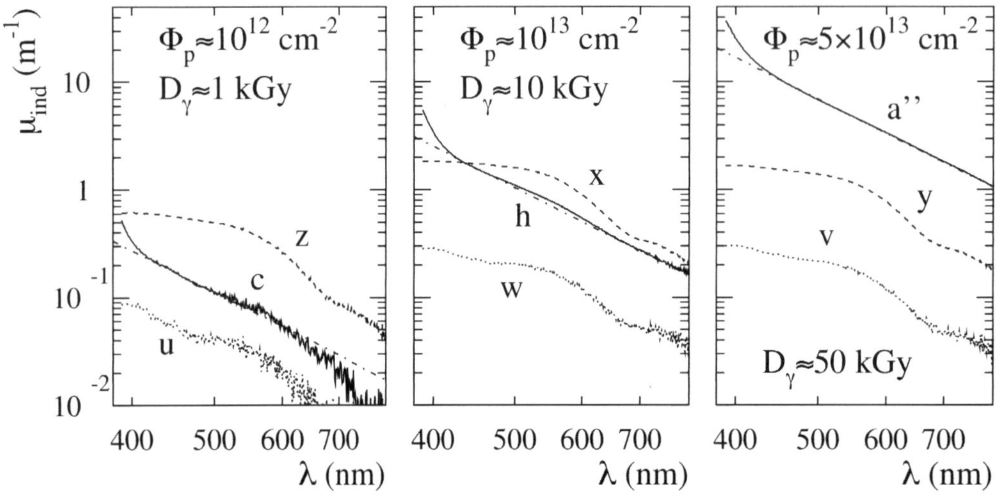

Figure 3. Plot of $\mu_{IND}(\lambda)$ versus λ for proton-irradiated crystals (a'', c, h) and for γ-irradiated ones (u, v, w, x, y, z) [4]. The dot-dashed line shows λ^{-4}, fitted to the proton-damage data.

which reaches equilibrium [4, 10]. No flux dependence was observed.

Damage values from protons and γ's were also compared in a study performed on BGO[11].

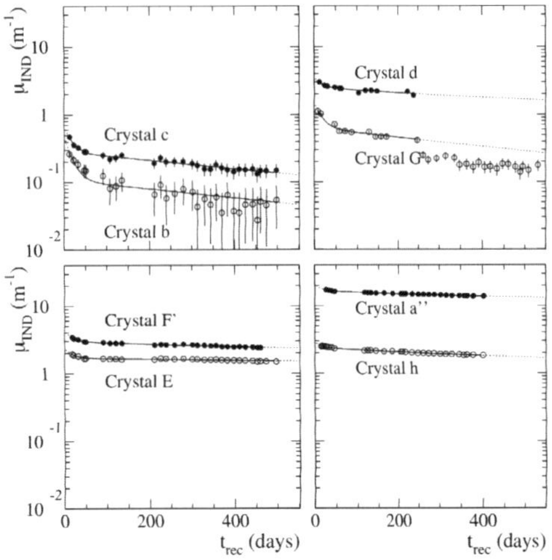

Figure 4. Evolution of longitudinal induced absorption $\mu_{IND}^{LT}(420\ \text{nm})$ over time for proton irradiated crystals [4].

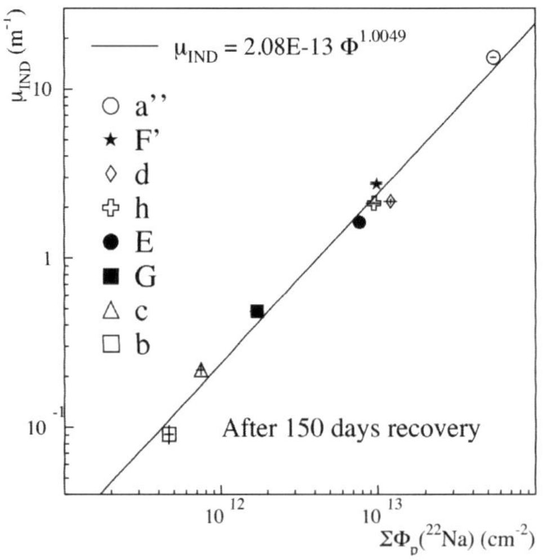

Figure 5. Induced absorption $\mu_{IND}(420\ \text{nm})$ as a function of cumulative proton fluence [4].

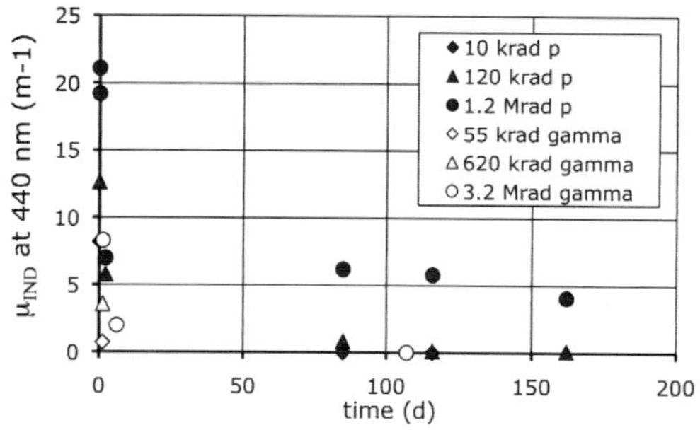

Figure 6. Evolution over time of the Longitudinal induced absorption, $\mu_{IND}(440\text{ nm})$, for proton- (full symbols) and for γ-irradiated (white symbols) BGO crystals, extracted from data in [11].

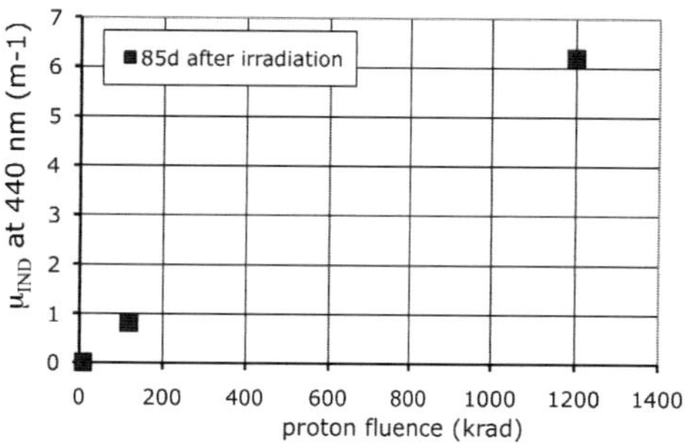

Figure 7. Correlation between the absorption $\mu_{IND}(440\text{ nm})$ induced by proton irradiation and proton fluence in BGO, extracted from data in [11].

The changes in band-edge are similar to what is seen in $PbWO_4$, and long enough after irradiation, when the ionising-radiation damage contribution has recovered (figure 6), a remaining proton-induced damage can be extracted, that behaves linearly with fluence, as visible in figure 7.

The relevant quantity for detector operation being the scintillation light output, any effect of hadrons on this quantity is of great importance. Detector calibration through a light injection system, as foreseen e.g. by CMS [7], is based on the assumption that changes in light output are all due to changes in light transmission. The study published in [5] compares the correlations between Light Output loss and induced absorption in proton- and γ-irradiations for the same set of crystals above. Those correlations (figures 8 and 9) show that, in the explored range of proton fluences, no additional, hadron-specific damage to the scintillation mechanism could be observed in lead tungstate within the measurement's precision.

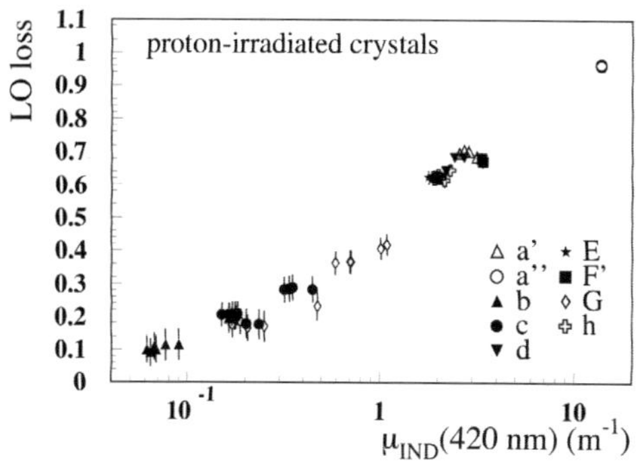

Figure 8. Correlation between induced absorption, $\mu_{IND}(420\text{ nm})$, and Light Output loss for proton induced damage in $PbWO_4$ [5].

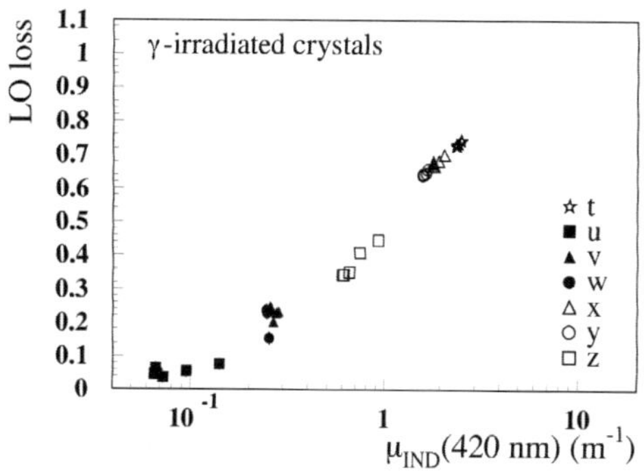

Figure 9. Correlation between induced absorption, $\mu_{IND}(420\text{ nm})$, and Light Output loss for γ-induced damage in $PbWO_4$ [5].

3. Comparative proton and positive pion irradiation studies

Crystals used in high-energy physics detectors will typically be exposed to hadrons – mostly charged pions – with different energies. In the CMS experiment at the LHC for example [7], the large hadron fluxes are due to particles - mostly pions - whose energies rarely exceed 1 GeV. To understand how to scale proton-damage values to the ones caused by such pion fluences, a previously γ irradiated crystal (w in [4] and in figure 3) was cut into three, 7.5 cm long, sections ($w1$, $w2$ and $w3$) after thermal annealing of its γ-induced damage.

The middle section, $w2$, was irradiated with 290 MeV/c positive pions at the Paul Scherrer Institute in Villigen, Switzerland. Details about the irradiation are found in [6]. The crystal was irradiated up to a total fluence $\Phi_\pi = (5.67 \pm 0.46) \times 10^{13}$ cm^{-2}, while the average flux on the crystal was $\phi_\pi = 4.13 \times 10^{11}$ cm^{-2} h^{-1}.

The first, *w1*, and last, *w3*, sections were irradiated with 24 GeV/c protons. They were placed at the same time, $w1$ in front of $w3$, into the proton-irradiation facility [8] of the CERN PS accelerator, with the beam entering through the small *w1* face, so that the hadronic cascade could develop through both crystals, as illustrated in figure 10. The procedure was totally analogous to the one described in [4]. The proton fluence on the $w1$ crystal front face was $\Phi_p = (1.17 \pm 0.11) \times 10^{13}$ cm^{-2}.

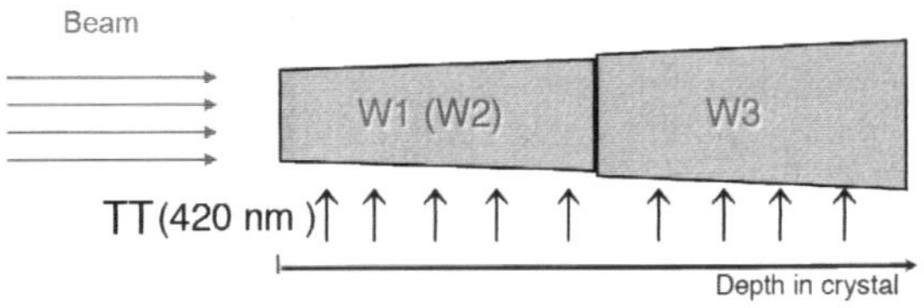

Figure 10. Illustration of $w1$ and $w3$ ($w2$) crystal placement during proton (pion) irradiation, and of the TT measurement direction after irradiation.

After it was demonstrated, in [5], that hadron damage only affects light transmission in $PbWO_4$, the study of pion damage was focussed on this observable.

At first sight (figure 11) the light transmission measured through the length of the crystal as a function of wavelength shows, after irradiation, the same behaviour for positive pions and protons. The band-edge shift is observed, which is characteristic for hadron-irradiated lead tungstate and BGO. It should be pointed out that similar damage amplitudes for the three crystals were obtained on purpose through a suitable choice of fluences.

The evolution of damage with time for all three crystals is shown in figure 12, where it is evident that pion damage follows the same pattern as observed for proton-irradiated crystals [4].

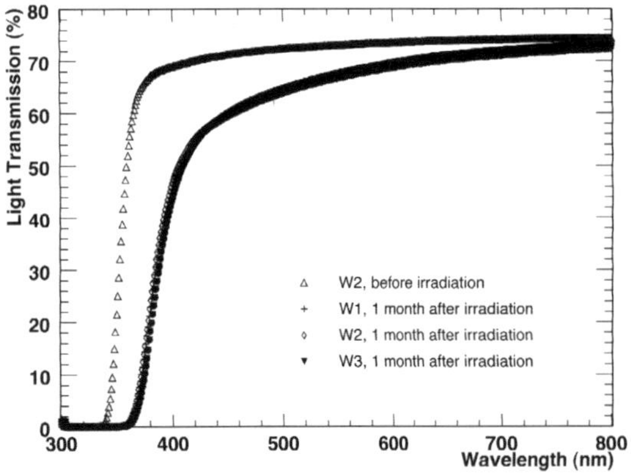

Figure 11. Longitudinal light transmission curves for the three sections of crystal w, one month after hadron irradiation, in comparison to the values before irradiation [6].

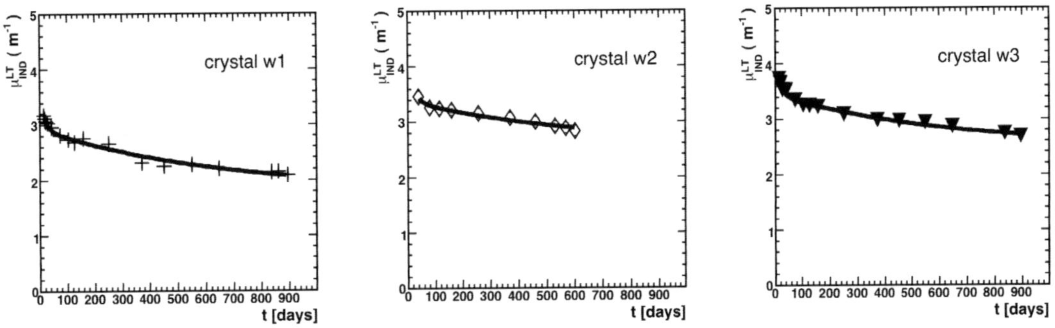

Figure 12. Evolution of the longitudinal induced absorption $\mu_{IND}^{LT}(420\text{ nm})$ over time, showing recovery, for the three sections of crystal w [6].

The data, as indicated by the lines, are well fitted by the same function used in [4], namely a sum of a constant plus two exponentials with time constants $\tau_1 = 17.2$ days and $\tau_2 = 650$ days. A similarly good fit is obtained through the form

$$\mu_{IND}^{LT,\,j}(420\text{ nm}, t_{\text{rec}}) = B_0^j\left(e^{-t_{\text{rec}}/\tau_2} + B_2\right) + B_1^j e^{-t_{\text{rec}}/\tau_1} \qquad (2)$$

where t_{rec} is the time elapsed since the irradiation and j, $(j = 1, 2, 3)$ is the crystal index. This parametrisation yields, for $t >> \tau_1$, a damage amplitudes ratio between crystals which is independent from the elapsed time. The fits being almost indistinguishable, the comparison between crystals remains independent of the selected fit function.

A deeper insight into the damage amplitude ratio between positive pions and protons was gained in [6] by looking at the induced absorption profile along the depth of the crystals. The transverse induced absorption coefficients $\mu_{IND}^{TT}(420\text{ nm})$ were obtained from measurements of the Transverse Light Transmission (TT) performed sideways at various positions along the crystals length, as indicated in figure 10. The profiles for this quantity are plotted in figure 13.

It was advocated in [4] that damage could be proportional to the density of stars, i.e. of inelastic hadronic interactions caused by a projectile above a given threshold energy. To verify

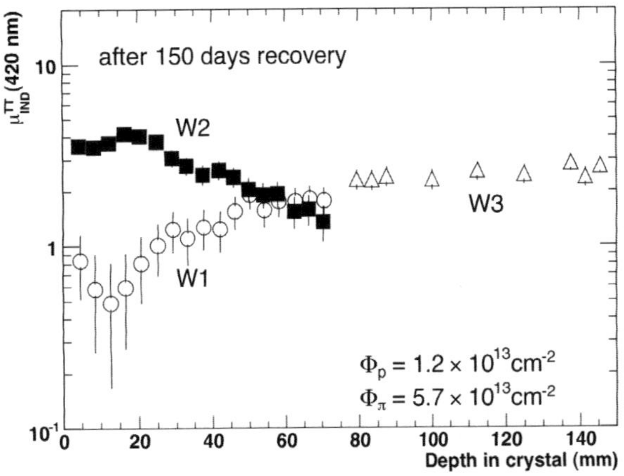

Figure 13. Transverse induced absorption, $\mu_{IND}^{TT}(420\text{ nm})$ as a function of position along the crystal [6]. Data for $w1$ and $w3$ are shown according to the crystals' position during irradiation.

Journal of Physics: Conference Series **160** (2009) 012013 doi:10.1088/1742-6596/160/1/012013

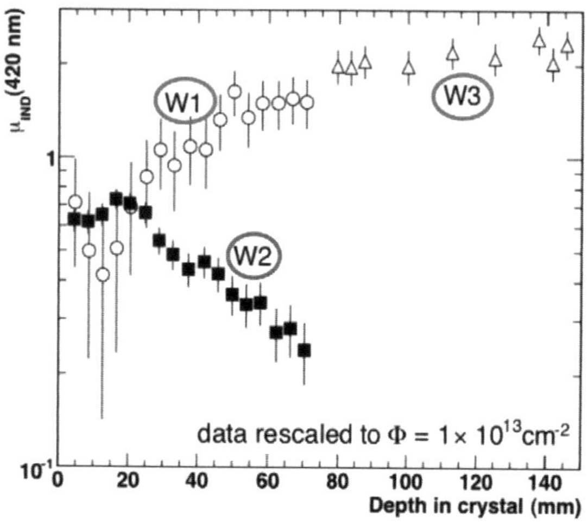

Figure 14. Same data as in figure 13, rescaled to a fluence of $10^{13}\mathrm{cm}^{-2}$.

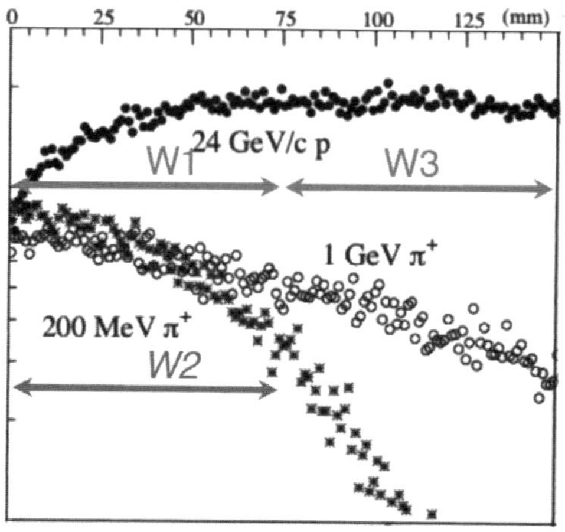

Figure 15. Star densities from FLUKA Monte Carlo simulations, as a function of depth in the crystal [4],[12].

this hypothesis, the damage profiles from figure 13 have been rescaled first of all to an identical fluence value, of $10^{13}\mathrm{cm}^{-2}$, as it is justified due to the cumulative nature of the damage. The so-obtained profiles are shown in figure 14. In figure 15, the star densities from FLUKA Monte Carlo simulations are shown as a function of depth in the crystal, where arrows indicate the lengths covered by the crystals during irradiation. One notices the fast drop, due to absorption, for 200 MeV π^+, while the density rises and stays high for 24 GeV/c protons throughout the depth corresponding to $w1$ and $w3$. The comparison of the damage profile from protons in *w1* and *w3* and pions in *w2* from figure 14 with the simulated star densities profiles in figure 15 shows an agreement that confirms in a striking way the hypothesised mechanism. A direct comparison is then performed in figure 16. The ratios of rescaled transverse absorption coefficients measured for positive pions in *w2* and for protons in crystal *w1* are plotted as a function of depth. The corresponding ratio of star densities obtained from the FLUKA simulations described in [4] and [12] is also plotted. The measured ratios and the star densities ratios are in agreement throughout, within the experimental uncertainties.

This comparison establishes that the damage amplitudes caused by 24 GeV/c protons and

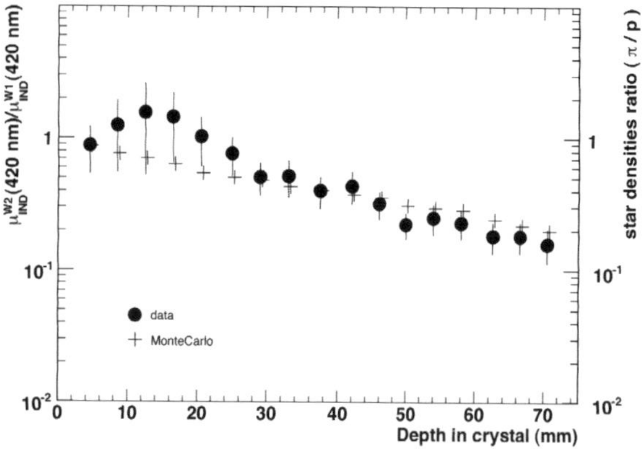

Figure 16. Ratio of μ^{TT}_{IND}(420 nm) results for $w2$ and $w1$ crystals from figure 14, compared to the corresponding simulated star density ratios from figure 15 [6].

290 MeV/c positive pions in lead tungstate scale according to the star densities caused by those particles in the crystal as calculated by Monte Carlo simulations.

4. Conclusions and outlook

Irradiation studies with 20 to 24 GeV/c protons and ^{60}Co photons in a fluence regime as expected at the LHC demonstrate that hadrons cause a specific, cumulative damage in lead tungstate, which solely affects the crystal light transmission while its scintillation properties remain unaffected. Therefore, it is possible to monitor the damage through light injection. All characteristics of the damage are consistent with it being due to an intense local energy deposition from heavy fragments. Since this damage mechanism should be absent in crystals with elements below Z=71 [13], a hadron damage test in such crystals should confirm the present understanding. A positive pion irradiation shows that the damage amplitudes caused by 24 GeV/c protons and 290 MeV/c pions in lead tungstate scale like the star densities those particles cause in the crystals according to simulations. The results can thus be used to estimate the expected damage for different experimental conditions.

References

[1] Batarin V A et al 2004 *Nucl. Instr. and Meth.* A **530** 286
[2] Auffray E et al 1998 CMS-NOTE-1998-069 (Geneva: CERN)
[3] Nessi-Tedaldi F 2006 *Proc. 9th ICATPP Conf. on Astroparticle, Particle, Space Physics, Detectors and Medical Physics Applications* ed M. Barone et al (Singapore: World Scientific) p 857-863
[4] Huhtinen M, Lecomte P, Luckey D, Nessi-Tedaldi F, and Pauss F 2005 *Nucl. Instr. Meth.* A **545** 63-87
[5] Lecomte P, Luckey D, Nessi-Tedaldi F and Pauss F 2006 *Nucl. Instr. Meth.* A **564** 164-168
[6] Lecomte P, Luckey D, Nessi-Tedaldi F, Pauss F and Renker D 2008 *Nucl. Instr. Meth.* A **587** 266-271.
[7] The CMS Collaboration 1997 *The ECAL Technical Design Report* CERN/LHCC 97-33, CMS TDR 4, (Geneva: CERN).
[8] Glaser M et al, 1999 *Nucl. Instr. Meth.* A **426** 72
[9] Zhu R Y 2004 *Trans. Nucl. Sci.* **51** 1560-1567
[10] Zhu R Y 1998 *Nucl. Instr. and Meth.* A **413** 297
[11] Kobayashi M et al 1983 *Nucl. Instr. Meth.* **206** 107-117
[12] Huhtinen M, *private communication* and http://indico.cern.ch/conferenceDisplay.py?confId=a03642 (Geneva: CERN)
[13] Iljinov A S et al, 1989 *Phys. Rev.* C **39** 1420-1424

The CMS-HF quartz fiber calorimeters.

A. Penzo [a] and Y. Onel [b]
a) INFN, Sezione di Trieste, Trieste, Italy
b) University of Iowa, Iowa City, USA
(for the CMS Collaboration)

Abstract: Two hadronic forward (HF) calorimeters extend the acceptance of CMS at large rapidities and are built with radiation-hard components (steel absorbers and quartz fibers) to resist the severe radiation levels in the forward regions. Very high energy jets can be measured in HF, by detecting Cherenkov light emitted by shower particles in the quartz fibers. The HF calorimeters are now installed in the underground CMS cavern; after commissioning, the detectors are being prepared for beam. Progress in calibration work and current plans for the HF calorimeters during the initial LHC runs are summarized.

Keywords: Quartz, Cherenkov, calorimeter, radiation.
PACS: 07.20.Fw, 29.40.Vj, 42.70.Ce

1. Introduction

The Large Hadron Collider (LHC) will reach unprecedented high energies ($\sqrt{s}$ =14 TeV) and luminosities (L = 10^{34} cm^{-2} s^{-1}), opening vast new opportunities for particle physics research and discoveries, but producing extreme radiation conditions for experiments in the underground caverns. Each LHC apparatus needs to resist high radiation levels, without compromising its excellent performance. For the CMS detector, among many other crucial issues, an almost hermetic calorimetry coverage is required to measure missing E_T and forward high-energy jets, potential key signals of new physics [1].

Two hadron forward ($HF^{\pm}$) calorimeters on both sides of the CMS interaction point (IP), at about ±11 m, cover the very forward angles of CMS, in the pseudorapidity range (3.0 < $|\eta|$ < 5.0), leaving only clearance for the beam pipes (Fig.1). To resist the high LHC radiation levels in this region, they are built with radiation hard components: steel absorbers with longitudinal quartz fibers, which should survive very high doses (of the order of Grad). About 1000 km of quartz fiber have been used for the two HF modules, which weight about 250 tons each. These devices can measure hadronic jets up to few TeV energies, detecting the Cherenkov light, emitted by the shower particles in the quartz fibers, and transported through the same fibers to photomultiplier tubes (PMT), located on the downstream side of the calorimeter modules (Fig. 1). These detectors have become a benchmark reference for large-scale applications of quartz fiber calorimeters [2].

The two modules have been assembled at CERN in Building 186, where both were also calibrated with radioactive sources. Some HF wedges had been earlier tested and calibrated with electrons and hadrons in the SPS H2 beam line. The two modules were then transported from Building 186 to the CMS assembly hall SX5 at LHC Point 5 [3] and subsequently lowered in the underground UXC5 cavern [4].

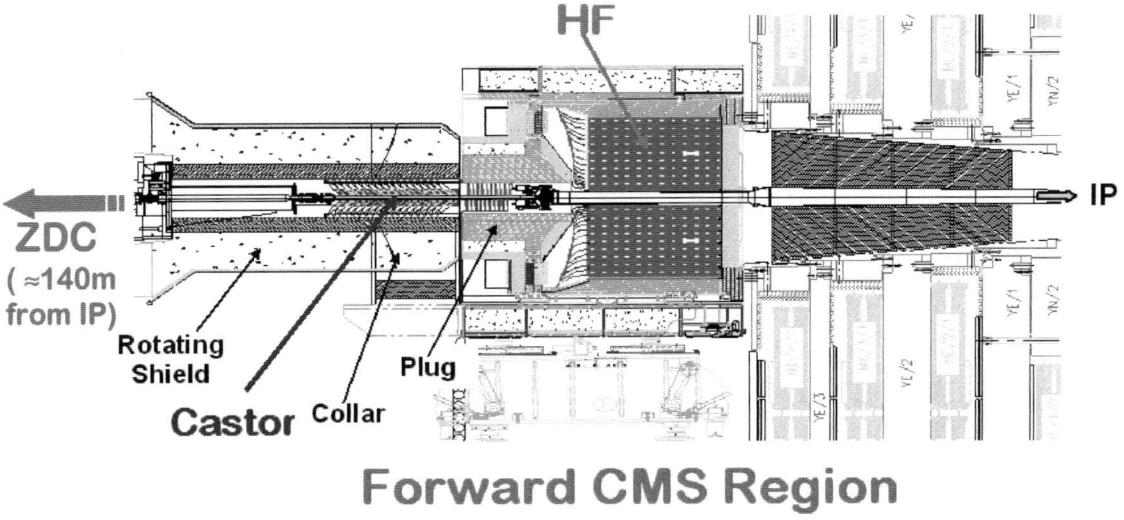

Fig. 1 – The forward region on one side of the IP in CMS: the quartz calorimeters HF (fibers) and Castor (plates) cover respectively pseudorapidity ranges ($3.0 < |\eta| < 5.0$) and ($5.0 < |\eta| < 7.0$); at 140 m further from IP, the ZDC (fibers) calorimeter measures high-energy forward neutrals.

2. Quartz Fiber Cherenkov Calorimetry

Cherenkov radiation has been frequently used to measure the energy of electromagnetic showers, for instance in lead-glass calorimeters [5]: now it becomes increasingly useful for hadronic calorimetry as well, thanks to quartz fibers, a new type of active material for calorimeters, providing very high radiation hardness, excellent optical quality and fine granularity. Valuable properties of Cherenkov calorimeters [6] are well recognized:

- Cherenkov light is strongly correlated with the showering particles' trajectories, providing directional information;
- Cherenkov radiation emission is essentially instantaneous, thus giving a fast response, useful for high rate applications;
- Cherenkov radiation is produced by charged particles with velocity above a certain threshold: these detectors are essentially blind to low energy particles and neutrons, and preferentially sensitive to the large amount of $e^{\pm}$, produced in EM processes, and emitting Cherenkov radiation in quartz down to $E \geq 190$ keV;
- A sizable fraction of hadronic cascades in an absorber is transformed into γ rays (mainly through π^0 decays), giving $e^{\pm}$ as final products; in a Cherenkov calorimeter, hadronic showers are essentially visible through their EM components, with lateral width much narrower than for real hadron cascades (as appearing in ionizing or scintillating calorimeters) and concentrated near the core of the shower, thus making jet isolation easier;
- Quartz fibers maintain good transparency to visible light up to Grad doses [7].

Quartz fibers for calorimetry were introduced by P. Gorodetzky and co-workers [8], for detectors exposed to very high radiation levels, typically Grads (10^9 rad =10 MGy): for instance "Zero Degrees Calorimeters" (ZDC) at very forward angles in fixed-target heavy-ion scattering experiments [9]. A number of detectors based on quartz fibers have been constructed in order to profit from the above listed merits [10]. There are however also some shortcomings:

- The light output of this type of calorimeters is quite low due to several reasons:
    - The energy of showering particles is mainly dissipated through ionization losses; only a minute fraction is released in the form of Cherenkov radiation;
    - The fraction of quartz fibers in the calorimeter is rather low (about 1%) to keep the overall cost of the calorimeter at an affordable level;
    - Only a small faction of produced photons is trapped in the fibers and transported to photodetectors;
- As, contrary to visible light, fibers' transparency to UV deteriorates under high radiation at LHC, there is no advantage in using UV-sensitive photodetectors;
- This type of calorimeters is highly non-compensating, i.e. the detector is essentially sensitive only to the electromagnetic shower core and the so-called e/h ratio is quite large (e/h $\approx$ 5), implying that the signal for high-energy hadrons (for instance pions), is considerably lower (by a factor of about 1.5) than the signal for electrons (or photons) of the same energy.

3. Structure of the HF Calorimeters

The forward calorimeter is essentially a cylindrical steel structure with outer radius 130 cm, inner radius 12.5 cm (to accommodate the beam pipe) and length 165 cm ($\approx$10 λ_I). The absorber of each HF module weights 108 t and is azimuthally subdivided in 20-degree modular Fe wedges assembled in 2 half-cylinders; thirty-six such wedges (18 on either side of the interaction point) make up the HF calorimeters. Each absorber wedge has a matrix of longitudinal grooves spaced by 5 mm horizontally and vertically; in each groove, quartz fibers are inserted. The fibers have a 600 μm diameter fused-silica core, 630 μm with the polymer hard-clad, and 800 μm with protective acrylate buffer. Half of the fibers run over the full depth of the absorber while the other half starts at a depth of 22 cm from the front of the detector. These two sets of fibers are read out separately. This arrangement makes it possible to distinguish showers generated by electrons and photons, which deposit a large fraction of their energy in the first 22 cm, from those generated by hadrons, which produce signals in both calorimeter segments. We refer to the long fiber section as L (it measures the total signal), and the short fiber section as S (it measures the energy deposition after 22 cm of steel). Long and short fibers alternate in nearby grooves spaced by 5 mm inside the absorber. The packing fraction by volume (fiber/total) in the first 22 cm is 0.57% and it is twice as large beyond that depth.

The fibers extend behind the back face of the absorber (Fig. 2), and are grouped in bundles separately for L and S fibers, corresponding to towers in ($\eta - \phi$), with a typical segmentation of 0.175 x 0.175 ($\Delta\eta x \Delta\phi$); the fiber bundles are routed from the back of the calorimeter to air-core light guides which penetrate through a steel-lead-polyethylene shielding matrix, protecting the PMTs housed in the readout boxes. Each wedge has 2 readout boxes (RBX), each housing 24 PMTs (8-stage, with bialkali photocathode and borosilicate glass window; the field of the CMS magnet at the HF position is sufficiently low to have negligible effect on the HF PMTs, which are individually shielded.

Fig. 2 : HF wedges during assembly and fiber stuffing in the steel absorber, in Build. 186

Each HF module is divided into 36 segments in azimuth and 12 segments in pseudo-rapidity, resulting in 432 towers per side, each of them corresponding to a physical channel, and sampled by fibers of 2 different lengths, (L) and (S) readout by separate PMTs. The L fibers are sensitive both to photons/electrons and hadrons, while the S fibers miss most of the EM showers (developing and dissipating most of their energy in the front part of the absorber) and are therefore preferentially sensitive to hadrons.

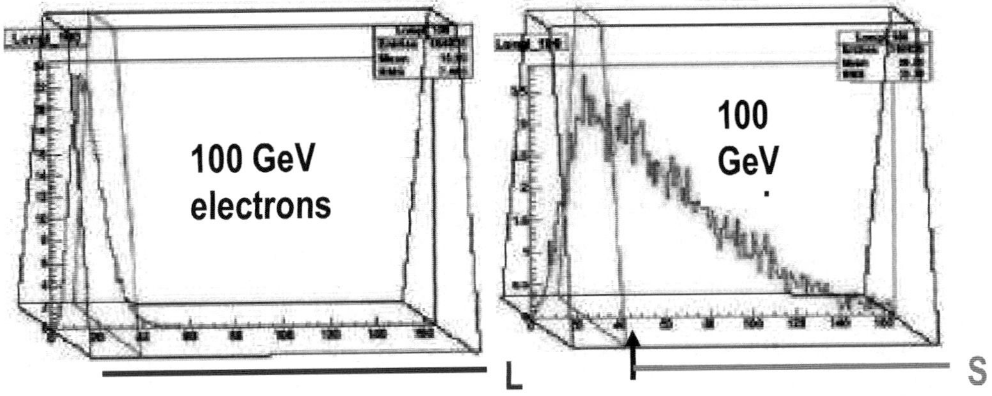

Fig. 3 : The development of EM and HAD showers in one HF wedge

Using this longitudinal segmentation of the calorimeter, where the sampling fraction in the front (EM) section is lower with respect to the rear (HAD) section it is possible to achieve a partial equalization of the HF response to electrons/photons and hadrons, despite the strong non-compensating nature of this type of calorimeter.
Signals from the PMT are digitized on a bunch by bunch basis using a dedicated chip, and then routed to a set of 36 Trigger and Readout (HTR) boards each servicing 24 HF physical towers.

4. Performance of the HF Calorimeters

The HF calorimeter is based on quartz fibers where Cherenkov radiation forms the basis of signal generation. Thus, the detector is essentially sensitive only to the electromagnetic shower core and is highly non-compensating (e/h $\approx$ 5). This feature is also manifest in narrow (and relatively short) showers compared to similar calorimeters based on ionization. The choice of fused-silica optical fibers as active material is dictated by its exceptional radiation hardness. The electromagnetic energy resolution is dominated by photoelectron statistics and can be expressed as $\Delta E/E = A/\sqrt{E} + B$. For electrons the stochastic term A $\approx$ 198% and the constant term B $\approx$ 9%. The hadronic energy resolution is largely determined by the fluctuations in neutral pion production in the shower development: when expressed as in the electromagnetic case, A $\approx$ 280% and B $\approx$ 11%.
The light yield of the HF calorimeters is approximately 0.25 phe/GeV for electrons in L fibers, thus 1 phe corresponds to about 4 GeV; high energy muons traversing the absorber module, give approximately a similar value (~ 3.5 GeV in average); if a muon impacts a PMT, it generates on average a signal equivalent to 30 phe (~ 120 GeV).
The transverse width of the shower seen via Cherenkov light is rather narrow: in the case of electrons, ~ 98% of the visible energy is contained in 5 cm around the shower axis (~ 85% within 1 cm). In the case of hadron showers, the transverse shower profile is wider: 70% within 1 cm (90% within 5 cm).

5. Commissioning and Calibration of the HF calorimeters

The two HF modules have been assembled at CERN in Building 186, where both were also calibrated with radioactive sources and with LED in 2006. Some HF wedges had been earlier tested and calibrated with electrons and hadrons in the SPS H2 beam line. The two modules were then transported to the CMS assembly hall SX5 at LHC Point 5 [2] (Fig.4a). The HF calorimeters are now installed in the underground CMS cavern; most of installation and commissioning was done in 2007 and beginning of 2008; soon the calorimeters will be prepared for beam (Fig. 4b).
The HF modules were the first CMS pieces to be lowered into the experimental hall UXC5, at the end of 2006; in the beginning 2007 HF^+ was the first detector to be connected through the cable chains to the read-out electronics situated in USC5 cavern and have the front-end electronics powered-up: therefore HF^+ commissioning started underground, obtaining, after intensive work, to have all channels functional and properly mapped into the readout. HF^+ was the first active detector included in the Global Runs (GR), a series of monthly week-long integration exercises [12]. Since the end of May 2007 (GREM), HF^+ was in regular use for electronic checks, and also for software development and testing, giving an excellent opportunity for debugging electronics and software in realistic conditions [13] (Fig. 5).

Fig. 4: a) One of the HF modules after transport to SX5, before lowering underground to UXC5; b) HF^- being brought to the beam level in UXC5: the collar shield behind HF^- is placed on the platform attached to the HF^- support, sitting on top of the yellow "raiser" structure[11].

It was soon followed by the other CMS subsystems (Fig. 6) and joined early in 2008 by HF^-. With the approaching of beams in LHC, the activities around HF are focusing on precision tuning and calibration of the detectors, as well as preparation for initial operations with beams.

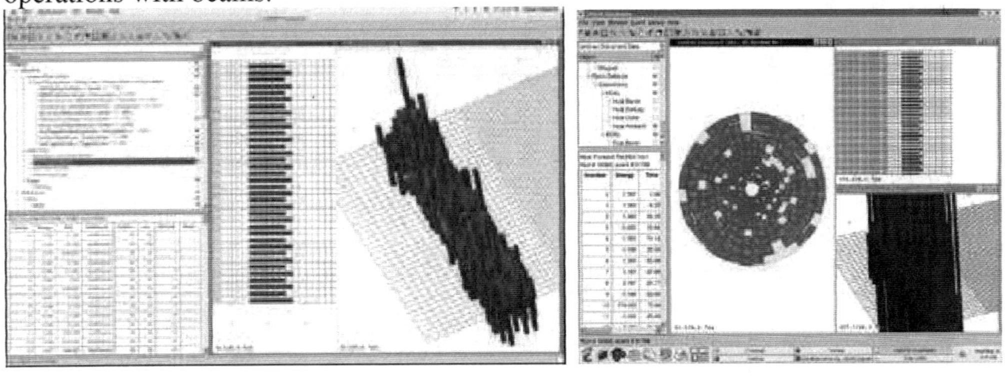

Fig. 5 : a) Quasi-online event display shows the HF^+ reconstructed hits from the global run 9965; b) Events display of the HF^+ calibration data (by Ianna Osborne) [13].

The calibration of the HF is based on three independent hardware systems: Laser Gain Monitor, Light Emitting Diodes, and Moving Radioactive Source. These systems should deliver the required calibration resulting in an overall energy scale uncertainty of approximately 3%. The chain of calibration steps starts from the detailed studies of the PMT properties performed at reception of the 2000 R7525HA PMTs from Hamamatsu: the gains (G) of all PMTs have been determined and subdivided in groups of sufficiently close values [14].

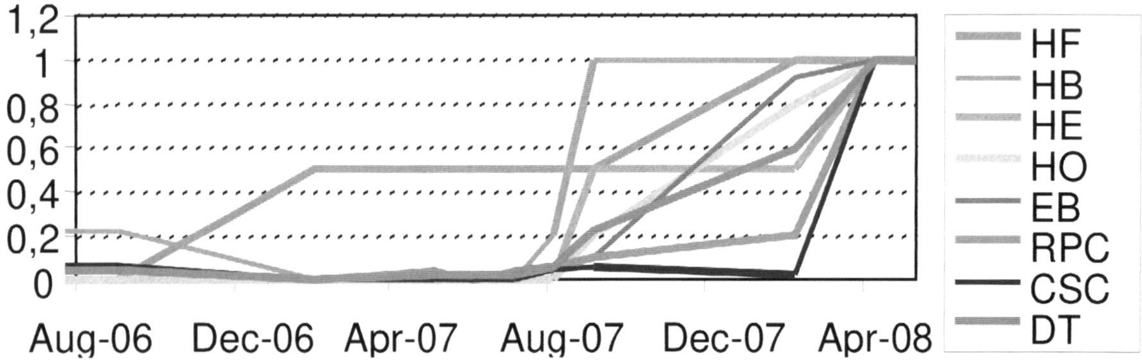

Fig. 6 . Fraction of CMS detectors active in the Global Runs.

In 2004, during test beam studies, several slices of the HF were tested with electron, pion, and muon beams at energies varying from 30 to 300 GeV. The response to these different particles and energies were compared with results of simultaneous runs with a Co^{60} wire-source [15]. A systematic source scan of all HF wedges was performed in Bldg 186 on fully HF assembled modules, before transport to SX5. During these tests, data were separately taken with the wire-source remote and inserted into the detector. The charge collected as a function of the source position inside the absorber provides information on the energy calibration of the detector (Fig. 7). There is a clear difference in the number of counts between the wire-source positions inside the detector and at remote park position. This sourcing campaign was planned in order to transfer the beam energy calibration (100 GeV electrons) from the few wedges tested on the beam to the totality of the wedges in both HFs.

Additionally LED runs have been taken at different levels in order to estimate the gains G of PMTs (from SPE spectra [16]) and the dependence of G on the HV of the PMTs. Using these measurements it was found that the relative gains measured at the University of Iowa test station and the gains obtained from the SPE data are highly correlated [17]. These relative gain values were used to group the PMTs going to different tower positions (the lower gains closer to the beam). It also confirms the reliability of using the SPE peaks in monitoring the gains with the LED system.

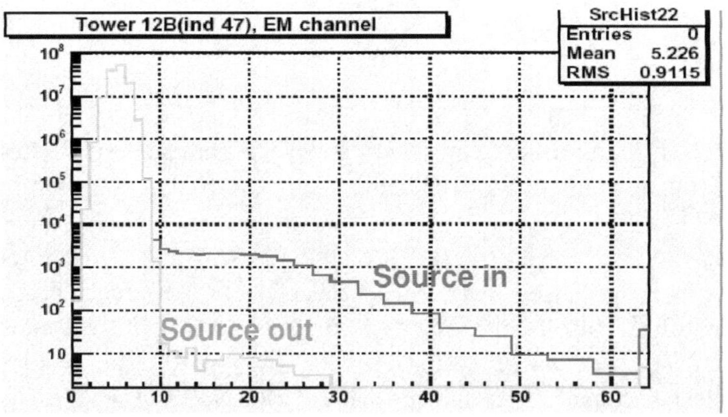

Fig. 7: The spectra from an L fiber, for source outside or inside HF absorber.

The energy calibration in terms of the amount of charge collected per unit energy of a particle is done with the help of the wire sourcing and the measured value is about 0.21 and 0.34 GeV/fC for L and S fibers respectively [18]. These studies help to understand the operation of the QIE elements, pedestals, and the noise in the system. The three calibration factors: fC/phe, GeV/fC and phe/GeV should be consistent with 1, when multiplied together: a value of 0.25 phe/GeV means a consistency within 10%.

6. Initial operation with beams and first results

The basic role of the HF calorimeters at high luminosity [1] will consist in improving missing-energy (E_T) determination and tag vector-fusion events with forward jets (Fig. 8) for physics and luminosity monitoring as a service to LHC and CMS.

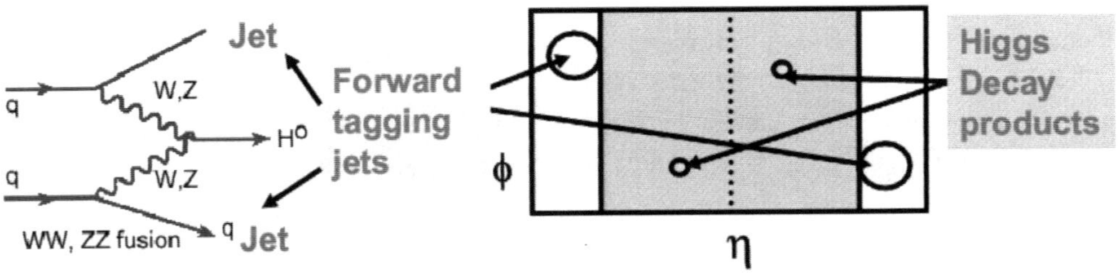

Fig. 8 – The VBF diagram for Higgs production, and the distribution of reaction and Higgs decay products, showing the tagging role of high energy forward jets.

During initial operation of LHC (at low luminosities) the HF calorimeters will be crucial for start-up luminosity studies, and may collect some data on forward physics. For instance HF will be able to measure forward jets for studies of QCD and diffractive processes, as well as dijets with large $\Delta\eta$ gaps ($\Delta\eta > 6$), coming from processes, where forward energetic "spectator" jets in high cross section hadronic interactions mimic somehow the role of "tagging" jets in the EW VBF process. For Mueller-Navelet dijets [19], the statistics in a low luminosity pp run ($L \sim 1\ pb^{-1}$) may reach 10^4, enough to start studies of low-x QCD [20].

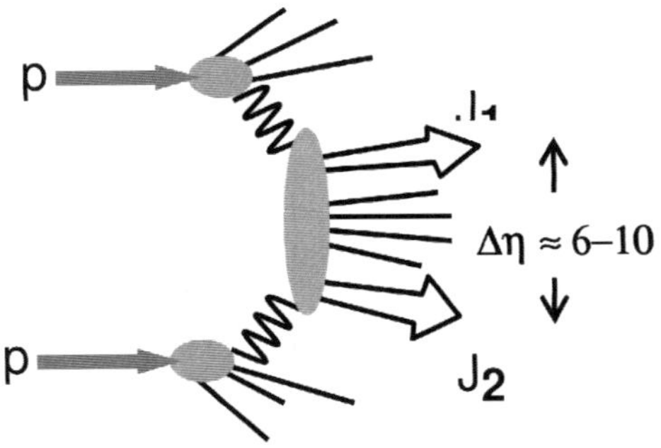

Fig. 9 . The Mueller-Navelet diagram for dijets production process [19, 20].

The online luminosity measurement [21] in CMS is based on HF, which is sensitive to a large fraction of the inelastic cross section. The design goal for the real time measurement is to determine the average luminosity <L> in less than one second. Possible techniques for extracting the real time luminosity signal have been studied:

- Zero-counting, based on the probability p(0) of beam-crossings (BX) without interactions: $\mu = -\ln p(0)$ (μ is the average number of interactions per BX);
- Hit counting, based on the number of towers above threshold for each BX; this quantity is linearly related to <L>, if multiple hits are negligible;
- Total transverse energy E_T measurements, adapted to the highest luminosity: both N_{ch} and E_T should be proportional to μ (hence to <L>).

In order to extract the luminosity signal from the HTR an additional mezzanine board (HLX) is mounted on each HTR board. Two identical servers collect data from all 36 HLX and transfer the data to a server interfacing to various clients.

7. Conclusions: the fate of HF calorimeters

In ten years of LHC operation at nominal luminosity, the parts of HF closest to the beampipe ($4.5<\eta<5$), will experience radiation doses close to 1 Grad (10^7 Gy) and large neutron fluxes leading to the activation of the absorber, at levels of ~10 mSv/h in the inner parts (1-10 μSv/h on the periphery). Soon after extensive LHC operation, the HF modules will become off-limits and access, maintenance and repair may become almost impossible. It is expected however that active elements of HF (quartz fibers + PMTs) survive these levels of radiation with limited deterioration: fibers' optical transmission is expected to be degraded by ≤ 50% after 10 years of LHC operation and quartz fibers are practically insensitive to neutrons and to low energy particles from decays of activated radionuclides (no Cherenkov light).The PMTs are properly shielded against radiation emerging from the absorber region and maintenance of read-out boxes will be performed with the help of semi-robotic extractor tools. HF is equipped with radiation monitors located at the periphery of the detector, and with a system (Raddam) to measure the transmission properties of few reference quartz fibers embedded in the absorber, as a function of the absorbed doses [22]. Such a monitoring system will be extremely useful to follow-up the degradation of the calorimeter response and provide the appropriate corrections, to maintain the quality of the HF calibration. In conclusion, with increasing integrated LHC luminosities, the body of HF will become inaccessible, thus making maintenance problematic, but the detector should maintain its design functionality, and provide high quality data for many years of LHC operation.

References

[1] N. Kauer *et al.*, H → W W as the discovery mode for a light Higgs boson. Phys. Lett. B503, 113 (2001).
N. Akchurin *et al.*, Study of low mass Higgs using pp → qqH at CMS. FERMILAB FN-0714/FN-0723 (2002); CMS NOTE 2002/016 (2002).
E. Yazgan *et al.*, Search for a standard model Higgs boson in CMS via vector boson fusion in the H → WW → l ν l ν channel, Eur.Phys.J.C53:329-347,2008.

[2] A.Ferrando *et al.*, A compensating quartz fiber calorimeter for small angle calorimetry at the LHC, Nucl.Instr.Meth. A390 (1997) 63.
J.Merlo, CMS hadronic forward calorimeter, Nucl.Phys 61B (1998) 41.
V.Gavrilov, CMS quartz fiber calorimeter, Nucl.Instr.Meth. A453 (2000) 24
G. Bayatian *et al.*, Design, Performance and Calibration of the CMS Forward Calorimeter Wedges, Eur. Phys. Jour. C53, 139 (2008)

[3] See http://cmsinfo.cern.ch/outreach/CMStimes/2006/07_31/index.html for details and photographs (see also CERN Courier, Vol. 46, N. 7, September 2006, p. 7).

[4] See http://cmsinfo.cern.ch/outreach/CMStimes/2006/11_06/index.html for details and photographs (see also CERN Bulletin, N. 47/48 November 2006, p. 4).
[5] C. A. Heusch, Use of Cherenkov Techniques for Total Absorption Measurements – CERN - EP/84 - 96 (1984). Seminar on Cherenkov Detectors and their Applications in Science and Technology in honor of P.A. Cherenkov for his 80th birthday.
[6] R. Wigmans - Review of Cherenkov Calorimetry, Lisbon 1999: Calorimetry in High Energy Physics", Ed. G.Barreira *et al.*, World Scientific, Singapore (2000) 335
[7] U. Akgun and Y. Onel, Radiation-Hard Quartz Cerenkov Calorimeters, AIP Conference Proceedings of XII International Conference on Calorimetry in High Energy Physics, p 282 (2006).
J. P. Merlo and K. Cankocak, Radiation Hardness Studies of high OH-Content Quartz Fibers Irradiated with 24 GeV Protons," Proceedings of 9th ICATTP Conference, World Scientific (2006) 838, CMS CR- 2006/005, (2006).
V.Gavrilov *et al.*, Study of quartz fiber radiation hardness, CMS TN-94-324 (1994).
V.Hagopian, Radiation damage of quartz fibers, CMS CR 1999/002
[8] P. Gorodetzky *et al.*, Quartz fiber calorimetry, Nucl. Instr. Meth. A 361 (1995) 1.
A.Contin *et al.*, Development of Quartz Fiber Calorimetry, CERN-DRDC/94-4, 1994
[9] See, for instance: R.Arnaldi *et al.*, Quartz fiber ZDCs at CERN SPS and LHC, Nucl .Instr. Meth. A409 (1998) 608
S.N.White, Forward measurements in RHIC and LHC heavy ion collisions, Nucl. Instr. Meth. A409 (1998) 618
[10] N. Akchurin and R. Wigmans - Quartz fibers as active elements in detectors for particle physics, Rev. Sci. Instr. 74 (2003) 2955-72.
G. Mavromanolakis, Quartz Fiber Calorimetry and Calorimeters, HEP-Cavendish/04/34 - UA-NPPS/06/2003, Cavendish Lab, Cambridge, UK.
[11] See http://cmsinfo.cern.ch/outreach/CMStimes/2007/10_01/index.html
[12] See http://cmsinfo.cern.ch/outreach/CMStimes/2008/01_28/index.html
[13] See http://cmsinfo.cern.ch/outreach/CMStimes/2007/06_04/index.html
[14] U. Akgun *et al.*, Complete tests of 2000 Hamamatsu R7525HA phototubes for the CMS-HF Forward Calorimeter, Nucl. Instr. And Meth. A550, 145(2005).
[15] N. Akchurin *et al.*, Calibrating the forward calorimeter of the CMS detector using a Co^{60} radioactive source, Instr.Exp.Tech. 50 (2007) 744.
[16] E. Gülmez *et al.*, Using single photoelectron spectra in the calibration of the CMS-HF Calorimeter, Proc. 9th ICATTP Conference, World Scientific (2006).339..
M. Baarmand *et al.*, Single Photoelectron Spectra Analysis of the HF Calorimeter, CMS IN 2007 – 018 (2007)
G. Safronov *et al.*, Single photoelectron peak measurement and its application for HF calibration and PMT gain monitoring, CMS IN 2008/000 (2008)
[17] K. Dindar *et al.*, Comparison of PMT relative gains measured at the University of Iowa and at CERN CMS IN-2005/045 (2005).
[18] E. Gülmez, The CMS-HF Forward Calorimeter Commissioning, CMS-CR 2007-071, Presented at the 10th ICATPP Conference, Oct. 8-12, 2007, Como, Italy
[19] A.H.Mueller, H.Navelet, An Inclusive Minijet Cross-Section and the Bare Pomeron in QCD ,Nucl. Phys. B282 (1987) 727
[20] S. Cerci and D. d'Enterria, Low-x QCD studies with forward jets in pp at 14 TeV, CMS-CR 2008-023, 28 May 2008; presented at Quark Matter 2008, Jaipur, India, 4-10 Feb 2008.
[21] N. Adam *et al.*, The CMS Luminosity System, CMS-IN 2007-030 (2007)
[22] K. Cankocak *et al.*, Radiation-hardness measurements of high OH-content quartz fibres irradiated with 24 GeV protons up to 1.25 Grad., CMS Note 2007/003 (2007).

Quartz Plate Calorimeter as SLHC Upgrade to CMS Hadronic EndCap Calorimeters

U. Akgun[a], P. Bruecken[a], K. Cankocak[a], L. Cremaldi[b], F.Duru[a], I. Dumanoglu[d], J. Merlo[a], A. Penzo[e], Y. Onel[a], D. Winn[c]

[a]University of Iowa, Iowa City, IA 52242, USA
[b]University of Mississippi, University, MS 38677, USA
[c]Fairfield University, Fairfield, CT 06430, USA
[d]Cukurova University, Adana, Turkey
[e]University of Trieste and INFN, Trieste, Italy

(for CMS Collaboration)

Abstract. Due to an expected increase in radiation damage under super-LHC conditions, we propose to substitute the scintillator tiles in the original design of the hadronic endcap (HE) calorimeter with quartz plates. Quartz is proved to be radiation hard by the radiation damage tests with electron, proton, neutron and gamma beams. Using wavelength shifting fibers, it is possible to collect efficiently the Cherenkov light generated in quartz plates. This paper summarizes the results from various test beams, bench tests, and Geant4 simulations done on methods of collecting light from quartz plates, as well as radiation hardness tests on quartz material.

Keywords: Quartz, Cherenkov, calorimeter, radiation.
PACS: 07.20.Fw, 29.40.Vj, 42.70.Ce,

INTRODUCTION

The Large Hadron Collider (LHC) is designed to provide 14 TeV center of mass energy with proton proton collisions every 25 ns. After a few years of running, the LHC will be upgraded to super-LHC (SLHC), that will operate at 10 times higher luminosity ($L = 10^{35}$ cm^{-2} s^{-1}), thereby allowing new physics discoveries. In the current design, the hadronic endcap (HE) calorimeter of the compact muon solenoid (CMS) detector uses Kuraray SCSN81 scintillator tiles, and Kuraray Y-11 double clad wavelength shifting (WLS) fibers. These materials have been shown to be moderately radiation hard up to 2.5 MRad . The scintillation photons are collected by wavelength shifting fibers lined close to the edge of the scintillators, which we defined as the HE-shape [1]. Under SLHC conditions the lifetime radiation dose in the HE calorimeter region will increase from 2.5 MRad to 25 MRad. The scintillator tiles used in the current design of the HE calorimeter will lose their efficiency due to radiation damage.

As a solution to the radiation damage problem in the SLHC era, we propose to substitute scintillators by quartz plates and carry out the light via UV absorbing, blue emitting WLS fibers. We performed radiation hardness tests on various types of quartz material in the form of fiber, under electron, proton, neutron, and gamma radiation. Results show that quartz will not be affected by the radiation dose at the SLHC. In the first part of this report we summarize these different radiation hardness studies performed by our group. Radiation damage to WLS materials will be studied at a later date.

However, when quartz plates are used, the detected photons come from Cherenkov radiation, which yields 100 times less light than the scintillation process. The focus of this study is to find an efficient way to collect light from quartz plates. At the University of Iowa and Fermi National Accelerator Laboratory, we tested and simulated different sizes of quartz plates with different fiber geometries embedded in them to obtain maximum light. To make up for the reduction in light production, use the far-UV region to collect Cherenkov photons since the number of generated Cherenkov photons increase as $1/\lambda^2$ [2]. The second part of this report presents the results from the studies performed on finding an efficient way to collect light from quartz plates.

RADIATION DAMAGE STUDIES ON QUARTZ FIBERS

Quartz is known to be radiation hard in general. However, not all the quartz types have the same amount of radiation hardness. Among the different types of quartz it is important to find the best option to replace the CMS HE calorimeter tiles. To investigate different fiber radiation damage properties under various radiation types the following tests have been performed;

Quartz Radiation Damage with Electrons

The quartz radiation damage test with an electron beam has been tested on nine different high OH^- quartz fibers, with hard plastic cladding (qp) and quartz cladding (qq) using 500 MeV electrons from the Linac Injector of LEP (LIL) at CERN. The transmission of Xe light was measured *in-situ* in the 350-800 nm range. The induced attenuation at 450 nm found to be 1.52±0.15 dB/m for a 100 Mrad absorbed dose. We observed some darkening on the fibers, but the radiation did not change the tensile strength of the quartz fibers [3].

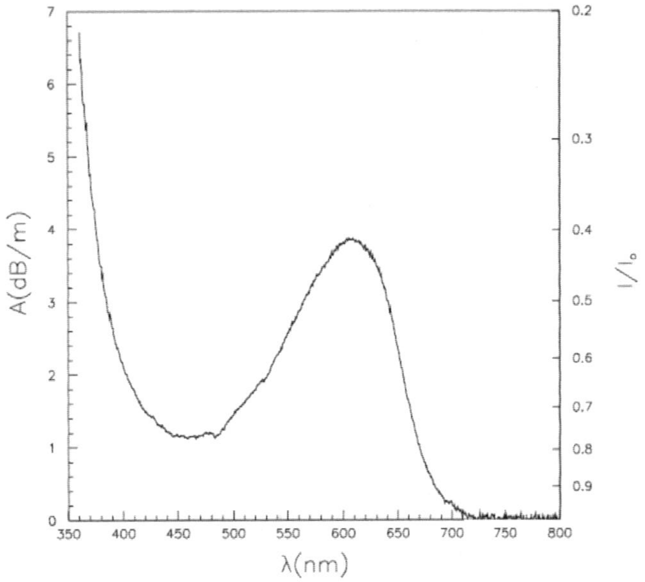

Figure 1: Typical behavior of induced attenuation versus wavelength for Polymicro qp fiber with 300 micron core, irradiated to 54 Mrad.

We achieved a set of mechanical and optical tests for the properties after irradiation, of nine types of fused-silica core fibers. We measured in situ the darkening and the recovery of two fibers at the same time. This arrangement allows us to make a direct comparison of the optical properties of two different fibers, independently of dose, injected light or electron beam fluctuations. The main results are: (1) there is no significant difference in optical properties between the polymer-clad (qp) fiber drawn by Polymicro Technology and the fluorine doped silica clad fibers drawn by Polymicro or Hesfibel from Heraeus performs under irradiation up to 100 Mrad. (2) Tensile strength measurements show that Polymicro fibers exhibit highest tensile strength with minimum RMS. There is no significant radiation effect on the tensile modulus, which remains at 4.6 ± 0.4 GPa. (3) The quality of the core material strongly affects the fiber performance. We measured a very different attenuation in the fibers drawn, by a single company (Polymicro), where the preforms came from two different suppliers located in Germany and in Russia. (4) The attenuation and recovery as a function of time are well represented by two-parameter fits. (5) The measurement of quartz fiber darkening by attenuation of light transmission 14.4 s after the beam stop is insensitive to short-lifetime luminescence induced by the Cherenkov UV emission or dose accumulation.

Three bands of luminescence have been observed in fused silica at 280, 470 and 650 nm with lifetimes around 4 ns; 2–10 ms; and 20 ms; respectively. The characteristics of the luminescence, mainly in the 470 nm band, in the fibers have to be further measured in order to fully evaluate the effects of this band on the performance of the quartz plate calorimeter.

Quartz Radiation Damage with Protons

To complete our knowledge of radiation damage in quartz fibers and to investigate their possible use at the Super LHC (SLHC) we initiated high level irradiation with 24 GeV protons at CERN. About 10^{16} protons/cm^2 were sent onto 1.20 m of quartz fibers, corresponding to a dose of 1.25 Grad [4]. The fiber radiation damage induced by protons exhibited the same well known behavior as with electrons: high light attenuation below 380 nm and in the band 550–680 nm, moderate attenuation in the band 400–520 nm and practically no attenuation above 700 nm. The damage varies exponentially with dose; fast in the first hours and slow after. Above 0.6 Grad we observe a new phenomenon: the radiation damage is not recoverable in the range 580–650 nm and below 380 nm. The two fibers tested, 0.6 mm quartz core diameter, one with hard plastic cladding (qp) and the other with quartz cladding (qq), were supplied by the US firm Polymicro Tech. Inc. (PT).

The fibers were installed, with their light support, on a remote controlled table at the CERN facility IRRAD. Three loops of qq and qp fibers (40 cm long) were wrapped together and placed with a slope of 2.5% relative to the beam. The total irradiated length (L) was about 1.2 m. No material was inserted in front of the fibers as in the case of electrons. The 24 GeV proton beam at IRRAD was delivered as 3–4 sub-cycles in a (19.6 ± 3.0) s long supercycle. This beam delivers 10^{11} protons/cm^2/burst in a spot of 2x2 cm^2. The sub-cycle sharing changed during the irradiation. The number of protons delivered was recorded and the dose calibration was done using thin Al samples. We regularly checked the proton beam stability in position, width and time.

The attenuation measurements were performed *in-situ*. We tested quartz fibers with quartz cladding (qq) and with plastic cladding (qp). The light attenuation in the fiber is defined as

$$A(\lambda, D) = \alpha(\lambda)\left[\frac{D}{D_s}\right]^{\beta(\lambda)} \qquad (1)$$

Where α and β are parameters for qq and qp fibers that are determined by fitting the ratios as a function of wavelength and dose.

$$\frac{I(\lambda,D)}{I(\lambda,0)} = e^{\left[-\left(L/4.343)\alpha(\lambda)\left(\frac{D}{D_s}\right)^{\beta(\lambda)}\right)\right]} \quad (2)$$

Choosing a scale factor D_s = 100 Mrad, α is the attenuation at 100 Mrad, in dB/m, L being expressed in m.

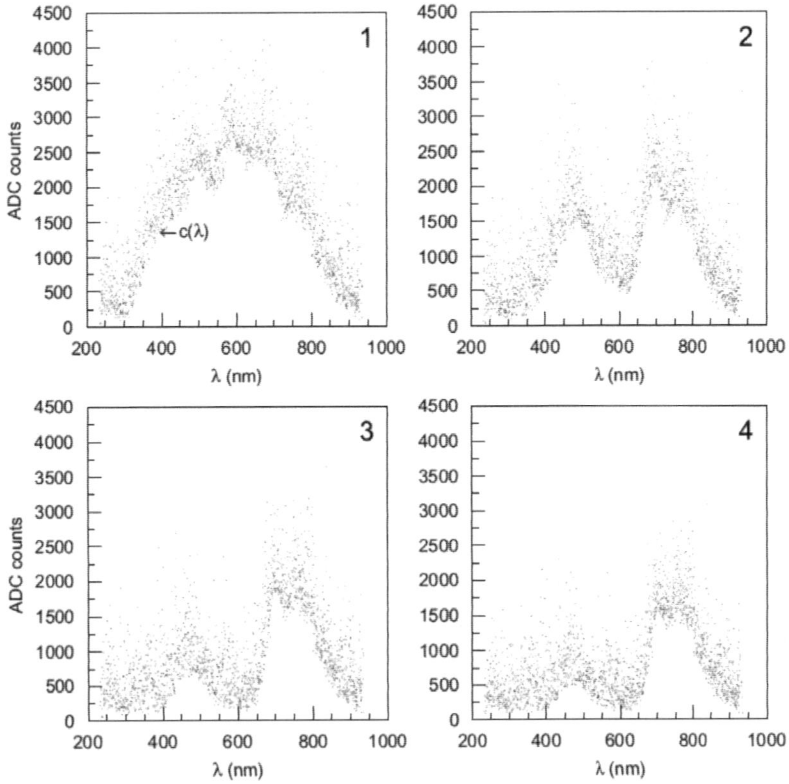

Figure 2: Evolution of the transmitted light spectra from a 1.25m of qq fiber as a function of dose 0 rad (1), 240 Mrad (2), 0.8 Grad (3) and 1.25 Grad (4).

Radiation damage due to protons and damage recovery are in agreement within errors with our previous results of electron irradiations with the same type of quartz used by Polymicro Technology Inc. The two types of irradiation were quite different: 24 GeV protons compared to 0.5 GeV electrons, in different geometries. 1.25 Grad with protons compared to 50Mrad with electrons.

As in electron irradiations we did not observe significant differences in proton irradiation between qq and qp fiber up to 1 Grad. Near 450 nm the observed difference appears in the fits at high dose and above 1 Grad there is a steep decrease of the transmitted signal in qp fiber.

For doses above 0.5 Grad there is no recovery of damage below 380 nm and in the range of 580–650 nm, the quartz becomes opaque. The PMTs detecting Cherenkov light are sensitive to the blue light (about 450 nm) emitted from the fibers where the effects are maximum.

For a 2 m qp fiber we observed a decrease of 30% in light transmission resulting from a dose accumulation of 20 Mrad, as well as an increase in light transmission by 22% 10 days after an irradiation of 100 Mrad in 200 days.

Quartz Radiation Damage with Neutron and Gamma

We also performed radiation damage tests to measure the degradation of quartz optical fibers under neutron and gamma radiation, which is expected to be present in the CMS hadronic calorimeters at the LHC and even more during SLHC.

For this purpose we selected seven different types of quartz material in the form of fiber from Polymicro Technologies. The selected quartz types were FVP 300-315-345, FSHA 300-330-350, FDP 300-315-345, FBP 600-660-710, FVP 600-660-710, FVP 600-660-710 UVM, and FSHA 600-630-800 [5]. The fibers were tested for light transmission degradation after a large radiation dose. They were bombarded with pulses of high-energy neutrons produced by the Intense Pulsed Neutron Source (IPNS) at Argonne National Laboratory for 313 hours. Seven sets, of five fibers each, were placed in an irradiation tube about 25 cm away from the IPNS target. These fibers were irradiated for a two-week period during which the integrated current delivered to the IPNS target was 4456 μA-hrs. The fibers were exposed to a total of 17.6 MRad of neutron and 73.5 MRad of gamma radiation.

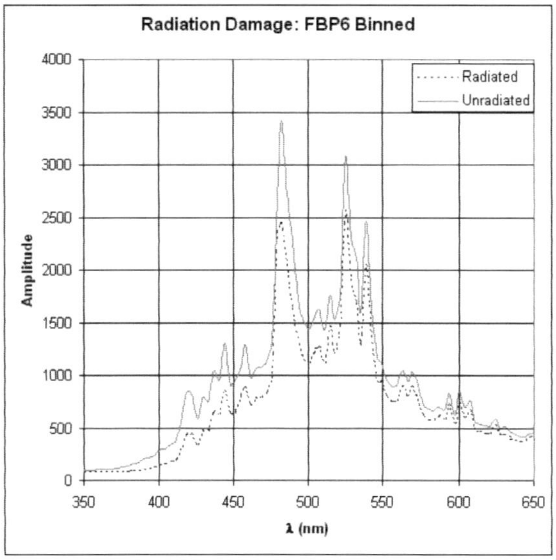

Figure 3: FBP 600-660-710 spectra before (red) and after (black) radiation.

The optical transmission of the fibers was then measured and compared to the baseline measurements. An Ocean Optics PX-1 Xenon flash lamp and an Ocean Optics SD2000 spectrometer were used for these measurements. The light was sent into a bifurcated optical fiber and was split into two channels. The tests show that a special radiation-hard solarization resistant quartz fiber (FBP 600-660-710) gives the best results. Solarization is defined as a phenomenon where a temporary change occurs to a material due to exposure to high energy electromagnetic radiation. The response of this fiber to the radiation can be seen in Figure 2, where the solid line is the response of the quartz fiber before being subject to the radiation and the dotted line is the response after irradiation.

Journal of Physics: Conference Series **160** (2009) 012015 doi:10.1088/1742-6596/160/1/012015

CHERENKOV LIGHT AND A QUARTZ PLATE CALORIMETER

Cherenkov Light Collection from Quartz Plates

As stated before, our proposal is to collect the Cherenkov photons from the far-UV range to help make up for the deficiency against scintillators [6]. For this purpose we selected Saint Gobain BCF-12 WLS fibers, which can absorb photons down to 280 nm, and emit at 435 nm. In the current design of the HE plates, fibers collect the scintillation photons from the edges of the plates. This simple fiber geometry works well for the scintillators since the scintillation photons are generated in random directions. However, the Cherenkov photons are generated at a fixed angle with respect to the momentum of the charged particle. Since the photons are scarce we cannot afford to make them propagate to the edges of the plates. The fibers should be placed close to the photons for efficient light collection. We investigated the most uniform and efficient fiber embedding geometry to collect the Cherenkov photons. For this purpose, various fiber embedding geometries were considered. Here we report the results for the following geometries: Bar-shape, HE-shape, Y-shape, and S-shape (see Figure 4).

All of the plates were wrapped with Tyvek, which is a very strong, synthetic material. The University of Iowa bench tests showed that Tyvek is as good a reflective material as aluminum and Mylar in both the UV and visible wavelength region [7]. In all the tests reported below Hamamatsu R7525-HA photomultiplier tubes (PMTs) [8] were used to measure the light at the end of the fibers.

The calculations and simulated model for our system show that the amount of Cherenkov radiation in a quartz plate is around 1% of the scintillation photons from the same size scintillator tile. To collect the light in a more efficient way, we worked on different plates with different sizes and with different fiber geometries embedded in them. After many beam tests and bench tests we came to the conclusion that by using a quartz plate with the bar-shape fiber geometry, we can collect almost 70 percent of the light that the original HE tile would yield.

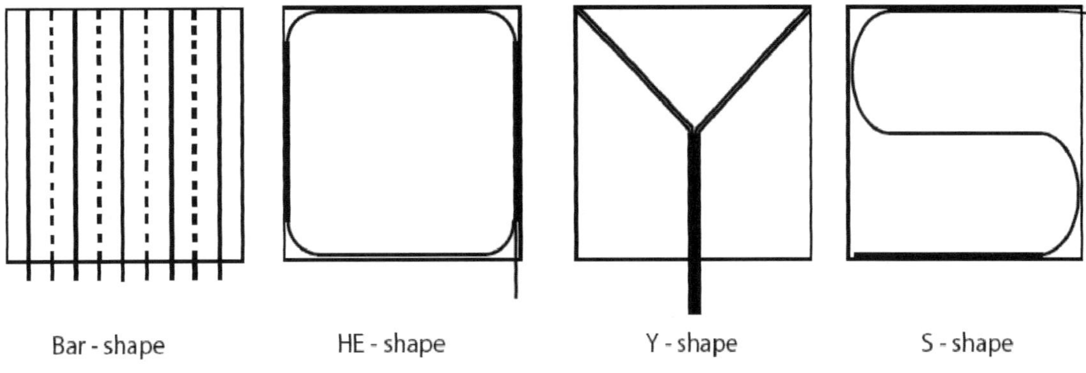

Figure 4: Fiber geometries embedded into quartz plates. The dotted lines of the bar-shape represent the fibers on the other side of the plate.

The plate with the barshape fiber geometry is shown to be very uniform since the fibers are distributed uniformly throughout the surface. To improve the light collection, a thick quartz plate with a small area, and many WLS fibers embedded in it should be used. It should be remembered that the space between the absorber layers of HE calorimeter (9 mm) limits the thickness of the quartz plates. Simulations show that the surface non-uniformity of light output is around 26% for the bar-shape and the ratio of collected light with respect to HE scintillator is about 70%. The analyses of the mean arrival time showed that the light collection is extremely fast (< 5 ns) which makes quartz a good candidate for the SLHC era. Even if the 10 times higher luminosity is obtained by decreasing the bunch crossing to 12.5 ns, the photon arrival time for quartz plates will be well within the gate.

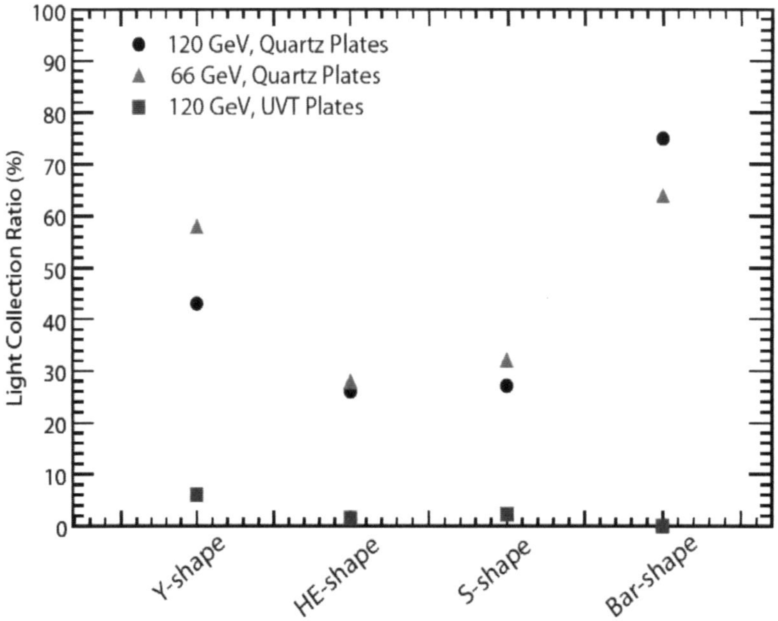

Figure 5: The light collection ratios of quartz plates with different fiber geometries to the original HE scintillator. The variations in the response for different beam energies are due to lack of statistics at 66 GeV beam energy. The test results of the UV Transmitting Acrylic (UVT) are also marked.

Quartz Plate Calorimeter Prototype – I (QPCAL-I)

Based on the R&D studies described above, we have built a quartz plate calorimeter prototype (QPCAL-I). The calorimeter consists of 20 layers of quartz plates with 20cm x 20cm x 5 mm dimensions. The Cherenkov light was collected by Saint Gobain BCF-12 WLS fibers, with "bar shape" geometry. Hamamatsu R7525-HA PMTs were used to collect the light from WLS fibers.

The prototype was tested at the CERN H2 test area with 2 different configurations; hadronic configuration with 7 cm iron absorbers between each layers, and electromagnetic (EM) configuration with 2 cm iron absorbers. The hadronic and EM configurations were tested with various energies of pion and electron beams, respectively.

QPCAL-I showed superior response linearity and uniformity (see Figure 6). QPCAL-I has a preliminary hadronic resolution of 18%. The preliminary hadronic and EM resolutions are shown in Figure 7.

Journal of Physics: Conference Series **160** (2009) 012015 doi:10.1088/1742-6596/160/1/012015

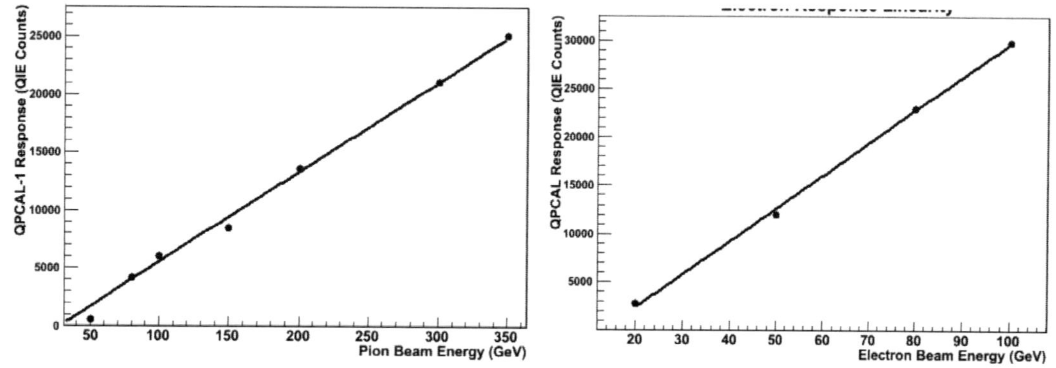

Figure 6: Hadronic (Left) and EM (Right) response linearity of QPCAL-1, versus beam energy.

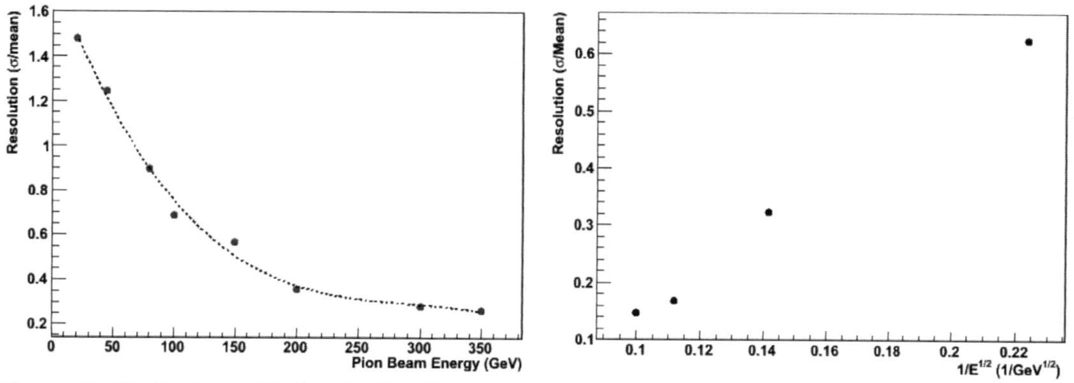

Figure 7: Preliminary Hadronic (Left) and EM (Right) energy resolution of QPCAL-1, versus beam energy.

ACKNOWLEDGMENTS

We would like to thank D. Green, J. Freeman, and A. Skuja of HCAL management, Iowa–Turkish Universities Raddam Group. This research is funded by DOE (DE-FG02-91ER40644), Iowa-Fermilab MOU for CMS construction and M&O projects, and The University of Iowa office of VP of Research and Graduate College.

REFERENCES

1. The HCAL Technical Design Report, CERN/LHCC 97-31, CMS TDR 2, 20 June 1997.
2. G. F. Knoll, Radiation Detection and Measurement, 1989.
2. I. Dumanoglu *et al.,* Nucl. Instrum. Meth. A 490, 444-455, 2002.
3. K. Cankocak *et al.,* Nucl. Instrum. Meth. A 585, 20-27, 2008.
5. P. Bruecken *et al.,* CMS IN-2006/014, 2006.
6. F. Duru *et al.,* IEEE Trans. Nucl. Sci., 55, 734-740, 2008.
7. U. Akgun *et al.*, IEEE Trans.Nucl.Sci.53:1547-1550,2006.
8. U. Akgun *et al.*, Nucl. Instrum. Meth. A550:145-156,2005.

Journal of Physics: Conference Series **160** (2009) 012016 doi:10.1088/1742-6596/160/1/012016

New crystal technologies for novel calorimeter concepts

Paul Lecoq

CERN, Geneva, Switzerland

paul.lecoq@cern.ch

Abstract. Present calorimetric systems give a global information on the total energy deposit at a given time in large detector cells but provide no details on the cascade mechanism of this energy deposition in space and time, as well as on the physics of the signal generation.

In the domain of High Energy Physics (HEP) high-precision measurement of hadrons and jets is one of the detector challenges at future high energy colliders. It has been shown that higher segmentation of the calorimter and/or the simultaneous recording of the scintillation light produced in an active medium, which is proportional to the total energy deposited by the shower particles, and the Cherenkov light, which is only produced by the charged, relativistic shower particles, can significantly improve the performance of present hadron calorimeters.

At low energy, for instance for medical imaging devices, the detailed recording of the whole Compton-photoelectric interaction chain would have a strong impact on the spatial resolution, energy resolution and sensitivity of the imaging cameras.

Recent progress in heavy scintillating crystal production methods as well as in nanotechnologies applied to diffractive optics introduce interesting perspectives for he development of innovative strategies for an homogeneous but finely structured calorimeter. It is shown here how a new class of metamaterials based on these technologies can open the way to new calorimeter concepts allowing to simultaneously record with high precision the maximum of information on the shower such as its direction, the spatial distribution of the energy deposition and its composition in terms of electromagnetic, charged and neutral hadron contents.

1. Introduction

The present paper proposes to take advantage of new scintillator production processes and of recent advances in nanotechnologies, applied in particular to the domain of photonic crystals and diffractive optics, to develop novel detector concepts capable of delivering much richer information than usual in a X- or Gamma - ray detector and more generally in a calorimeter.

Homogeneous crystal calorimeter are well known to give excellent electromagnetic energy resolution at low energy, as was first demonstrated by the spectacular charmonium spectroscopy study at SPEAR with the NaI(Tl) Crystal Ball in the seventies (Figure 1a). Large crystal based calorimeters build in the last 25 years (L3 with BGO, Babar and Belle with CsI, CMS with PWO) have demonstrated that systematics can be controlled with high precision on several thousands or even tens of thousands of crystals to give excellent energy resolution also at high energy (Figure 1b).

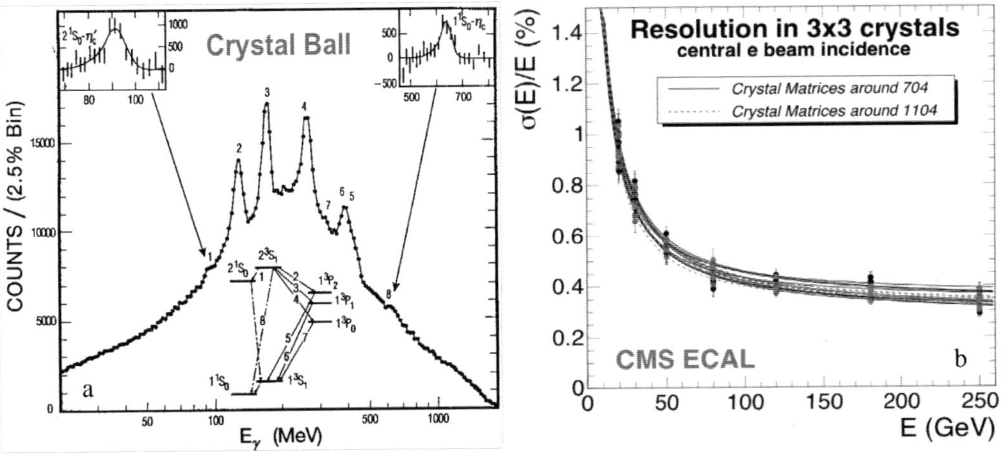

Figure 1: Charmonium spectroscopy measured by the Crystal Ball (a), and test beam resolution curve for several crystals of the CMS ECAL (b)

However scintillating crystals are considered to have poor performance, when associated to hadron calorimeters for hadron and jet calorimetry. This is due to the fact that homogeneous calorimeters are by nature non compensating. Moreover quenching mechanisms limit the scintillation efficiency in high ionizing density regions, giving rise to different responses for minimum ionizing and heavily ionizing particles. As a result the e/h ratio is generally much larger than 1 limiting therefore the hadronic energy resolution, as long as corrections on an event to event basis cannot be made to take into account the fluctuations of the electromagnetic fraction in hadronic showers.

Indeed in the domain of High Energy Physics (HEP) high-precision measurement of hadrons and jets is one of the detector challenges at future high energy colliders. A global (integral) approach for jet calorimetry cannot achieve a better jet resolution than 60-70%/√E. Whatever the technical approach, the requested jet resolution of 30%/√E requires high granularity analysis of jet showers and/or a precise determination of the different shower components (electromagnetic, charged hadronic, neutral).

The present paradigm with associated intense R&D in Europe and US is the one of particle flow, which amounts to develop tracking calorimeters with extremely high granularity such as silicon tungsten sandwiches aiming at the identification of all particles in jets and the tracking of individual showers This implies a highly segmented calorimeter with a huge number of channels and complex engineering problems to extract the signals.

An alternative approach, much less explored at the level of R&D, is to integrate over the shower development and aim to measure independently the scintillation light produced in an active medium, which is proportional to the total energy deposited by the shower particles, and the Cherenkov light, which is only produced by the charged, relativistic shower particles. Since the latter are almost exclusively found in the electromagnetic (em) shower component (dominated by π^0s produced in hadronic showers), a comparison of the two signals makes it possible to measure the energy fraction carried by this component, f_{em}, event by event. As a result, the effects of fluctuations in this component, which limit the performance of non-compensating calorimeters (non-linearity, poor energy resolution, non-Gaussian response function), could be eliminated or at least reduced to improve substantially the energy resolution of hadronic showers in general and of jets in particular. This would lead to an important improvement in the hadronic calorimeter performance.

This concept, developed by R. Wigmans and others with the DREAM calorimeter [1] suffers however some limitations depending on the way it is implemented. In its original design the basic element of this calorimeter is an extruded copper rod, which acts as the absorber. This 4x4mm^2 rod is hollow, the central cylinder has a diameter of 2.5 mm. In this hole is inserted the active material made of several plastic scintillating fibers acting as ionizing detectors, mixed with undoped fibers, intended for detecting Cherenkov light. This is therefore a sampling calorimeter with a resolution limited by sampling fluctuations inherent to this kind of calorimeter.

Another approach, promoted by the same group is based on the separation of the scintillation and Cerenkov signal produced in homogeneous detector blocks made of scintillating crystals such as Bismuth Germanate (BGO), Lead Tungstate (PWO) or others. This solution is of course free of sampling fluctuations. There is however a non negligible coupling between the Cerenkov and scintillating signals in a bulk scintillating crystal. Indeed there is a strong overlap between the continuous Cerenkov emission spectrum (varying as $1/\lambda^2$) and the scintillation excitation spectrum, usually in the high frequency part of the optical window of the crystal, where the Cerenkov emission is the highest. Cerenkov emission is one of the channels of scintillation excitation mechanism because of the high UV photo-absorption cross-section in the scintillation excitation region. In other words it means that part of the Cerenkov light is absorbed by the luminescent centers of the crystal and reemitted at longer wavelength as scintillation light, as shown in Figure 1 for Luthetium Orthosilicate crystals (LSO).

We propose an alternative to these approaches, combining their relative merits and limiting their drawbacks. It relies on recent impressive development in the domain of crystal growth and nano-photonics opening the way to homogeneous calorimeter designs with high granularity 3D imaging as well as particle idendification capability.

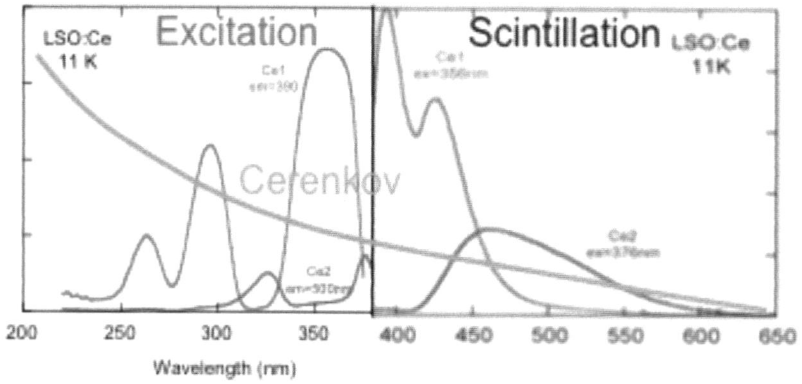

Figure 2: Excitation (left) and emission spectra (right) of he two LSO scintillation centres. The Cerenkov spectrum is superimposed and shows the strong overlap with the excitation spectrum2

2. The micro-pulling-down crystal fiber production technology

The analysis of the potential of different crystal growth techniques from the melt shows that pulling-down technology from a shape-controlled capillary die gives the possibility to produce elongated crystals with dimensions that are not accessible using traditional cutting and polishing of bulk crystals grown by the more standard Czochralski or Bridgeman methods. The size of the melting zone in the pulling-down technique is up to one order of magnitude smaller than that observed in the Czochralski method. Therefore, it is believed that the pulling-down process can be considered as a good way to achieve stationary pulling conditions and can facilitate the growth process, allowing for instance much faster growth and higher concentration of doping ions, even for those with high segregation coefficient.

A group of institutes and companies closely linked to the Crystal Clear colaboration [2] (CERN, VUB Brussels, University Claude Bernard LPCML, company Fibercryst Lyon, company Cyberstar, Grenoble) is presently involved in the development of this new crystal growth technology, which allows the fast and cheap production of heavy crystal scintillating fibers. High quality fibers of LuAG (Lutetium Aluminum Garnet) have been grown in collaboration with the company Fibercryst (Lyon) and the Laboratoire de Physico-Chimie des Matériaux Luminescents (LPCML) from University Claude Bernard (Lyon). Other well known heavy scintillating crystals such as BGO, LSO, LYSO [3], YAP and LuAP have also been grown in fibers of different sizes and PWO is under study. These high quality scintillating fibers are characterized at CERN and LPCML and can open attractive possibilities for the design of future detectors for HEP or other applications. The most relevant parameters of these crystals are summarized in Table 1.

Properties **Crystal**	***Light yield (ph/Mev)***	***Decay time (ns)***	***Density (g/cm3)***	***Lumi-nescence wave-length (nm)***	***Expe-rience of fiber pulling***	***Easy to grow crystal***	***Easy handling and polishing***	***Other remarks***
YAG:Ce	20 000	70	4.57	550	Strong	YES	YES	up to 2m
LYSO:Ce	25 000	40	7.40	420	Strong	NO	YES	30cm up to now
BGO	8 000	300	7.10	480	Strong	NO	YES	up to 2m
LuAG:Ce	20 000	40-70	6.73	535	Medium	YES expected	YES	1 to 2m
LuAG:Pr	20 000	20	6.73	290-350	Medium	YES expected	YES	Should behave as a garnet (like YAG)
GSO:Ce	14 000	60	6.71	460	Weak	Medium	NO	Brittle and tends to cleave at polishing
YSO:Ce	14 000	37 and 82	4.45	420	Weak	YES expected	YES	A few cm

Table 1: Some parameters of crystalline fibers grown by Fibercryst

In the present state of the art the pulling down technique allows to grow fiber (rod) shaped crystals with a controlled diameter between 0.3 and 3.0mm and up to 2 m in length. By modifying the shape of the capillary die it is also possible to produce elongated crystalline materials with more complex non-cylindrical cross-section (square, rectangular, hexagonal) for easier integration of the crystal in complex detectors. The procedure based on pulling-down technology was recently improved at the LPCM Laboratory and Fibercryst Company in Lyon to grow both single crystal fibers and shaped bulk crystals. The crystals are produced from the melt obtained at the capillary die positioned at the centre bottom of a cylindrical Iridium crucible as illustrated in Figure 3. Once the melt drop is formed, the growth process is initiated after connection of the seed with the drop at the bottom of the crucible (capillary die). Then the seed is pulled down continuously with a pulling rate ranging from 0.1 to 0.5 mm/min (about 10 times faster than Czochralski and 50 times faster than Bridgeman). Ongoing developments are going along a multiple capillary die crucible allowing to grow several fibers in parallel. The industrial optimization and cost effectiveness study of this process still needs to be made but this should lead to comparable costs per unit volume of grown crystal than with standard crystal growth approaches.

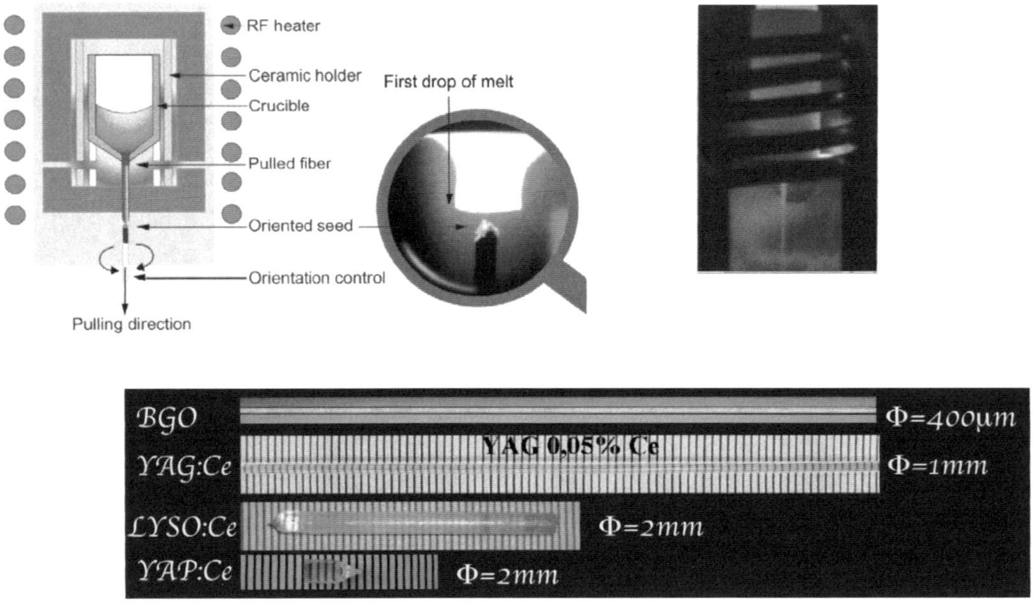

Figure 3: The micro-pulling down crystal growth technology (Courtesy Fibercryst)

3. The metamaterial concept

The basic concept of this proposal is to structure the standard detector block or pixel used so far in such a way as to extract more information than the total energy deposit in the block. In electromagnetic calorimeters the dimensions are typically 1 Moliere radius in the lateral and 25 radiation lengths in the longitudinal directions.

With the proposed approach it is possible to conceive detector blocks made of trunks of "cables" constructed from crystal fibers. Various scintillators can be selected to build these cables, having different emission wavelength, different scintillation yield and decay time, so that a direct encoding of the light is made as a function of the point of emission. A single photodetector can then decode this complex signal and derive a much more complete information about the shower process. For high energy calorimeters materials with different UV transmission cut-off can be selected so that the part of the Cerenkov emission in the detected light can be determined. Another approach is to choose scintillating materials activated by a doping ion instead of self activated scintillators (BGO, PWO).

Crystals such as Cerium doped LSO or LYSO for instance (Lutetium-Yttrium orthosilicate), both very fast (40nsec decay time) and dense (7.4g/cc) or LuAG (Lutetium Aluminum garnet, 70ns decay time and 6.73 g/cc density), or LuAP (Lutetium Aluminum Perovskite, 17nsec decay time and 8.34 g/cc density) are excellent candidates for mixing Cerium doped fibers, which would then behave as scintillators, with undoped fibers of the same material, which would only produce Cerenkov light. One could then obtain a very homogeneous, dense and compact system capable of disentangling the electromagnetic from the hadronic part in a shower in a calorimeter with a uniform radiation length, Moliere radius and interaction length in the whole volume of the detector. Moreover the assembly of the fibers can be organized in a flexible way, allowing a multitude of detector geometries.

If the detection of neutrons proves to substantially improve the overall jet energy resolution (requires simulation work) neutron sensitive scintillation fibers can also be inserted in the cables. No attempts have been made yet to grow such fibers but scintillators like Lithium Tetraborate LBO ($Li_2B_4O_6$), Lithium fluoride based materials like LiCAF ($LiCaAlF_6$) or more generally the Cerium doped elpasolite family ($Cs_{2-x}Rb_xLiMX_6$, where X=Sc, Y, La, Lu and X=Cl, Br, I) are very attractive candidates because of the presence of high neutron capture cross section Lithium and Boron and of their low density (2.42 for LBO), which makes them rather insensitive to gamma conversions.

It must be noticed that, although monocrystalline, the fibres are flexible enough to be assembled and twisted as in a cable in order to avoid straight gaps aligned with the particle trajectory. The exact number and distribution of the different type of fibers (Cerenkov, scintillating, neutron sensitive) as well as their diameter will be optimized by Monte Carlo.

At the extremities of each metablock a diffractive optics plate, patterned in such way as to match the cable structure will concentrate the light emerging from all fibers of the same type to an avalanche photodiode or SiPMT. Recent developments in nanotechnologies have allowed considerable progress in the tailored design and manufacture of such systems, also called MOEMS (Micro-Opto-Elecro-Mechanical-Systems), which are extensively used for adaptive optics in astronomy [4]. A sketch of this concept is shown on Figure 4.

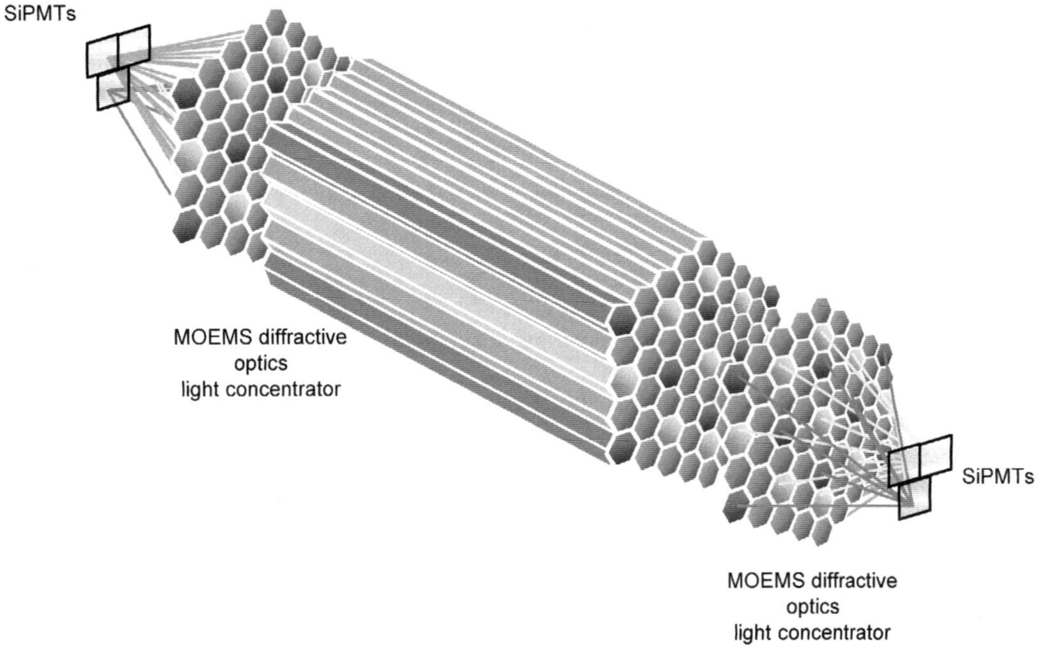

Figure 4: Principle of the metacable with 3 types of fibers and diffractive optics light concentrators

4. Material selection and first test results

We have selected LuAG as the most adequate material to design a proof of concept detector. This is a garnet of Lutetium and Aluminum with crystallographic properties rather close to YAG, a material already developed with the micro-pulling-down technology for the growing market of fiber lasers. Indeed excellent quality LuAG fibers, up to 2m in length have been grown in a consistent way. LuAG is an excellent and versatile host for a scintillator with a density of 6.73g/cm3, a radiation length of 1.41cm and an interaction length of 23.3cm. Moreover, it can easily be doped with Cerium giving rise to a bright (20'000 photons/MeV) and fast (70ns) scintillation peaked at 535nm, matching well the quantum efficiency maximum of solid state photodetectors. Would a faster scintillation signal be needed LuAG could also be doped with Praseodynum with an emission peak around 300nm and a decay time of 20ns only.

Another advantage of LuAG is its fundamental absorption edge at 250nm in the UV combined with a very large transparent optical window. Combined with a high index of refraction (1.84, to be compared to 1.55 for quartz) This makes LuAG an excellent Cerenkov radiator with a Cerenkov energy threshold of 97KeV only, to be compared to 190KeV for quartz.

The most important physico-chemical and optical properties measured so far are summarized in table 2.

Physico-chemical properties

Structure / Space group	Cubic / Ia3d
Density (g/cm^3)	6.73
Zeff	62.9
Radiation length X_0 (cm)	1.41
Interaction length (cm)	23.3 *LuAP: 19.8* *Fe: 17*
Hardness (Mohs)	7.5 *PWO: 3* *BGO, glass: 5*
Fracture toughness (Mpa.m$^{1/2}$)	1.1
Cleavage plane / H_2O solubility	No / No
Melting point (°C)	2260
Thermal expansion @ RT (°K^{-1})	8.8 10^{-6}
Thermal conductivity @ RT (W/m°K)	31

Optical properties

Light yield: Ce or Pr doped (ph/MeV)	20'000 *1/2 NaI(Tl)*
d(LY)/dT	?
Emission wavelength (nm): Ce doped Pr doped	535 290, 350
Decay time (ns): Ce doped Pr doped	70 20
Refractive index @ 633nm (isotropic) n^2= 3.3275151 - 0.0149248 λ^2	1.842 *Quartz: 1.55*
Fundamental absorption undoped (nm)	250
Max. Cerenkov 1/2 angle	57°
Total reflexion 1/2 angle	33°
Cerenkov threshold e energy (KeV)	97

Table 2: Physico-chemical and optical properties of Lutetium Aluminum garnet (LuAG) crystals

The small attached movie shows the flexibility of a 1m long, 350 micron thick LuAG:Ce fiber, as well as several 1mm diameter fibers arranged in a matrix of an alpha detector.

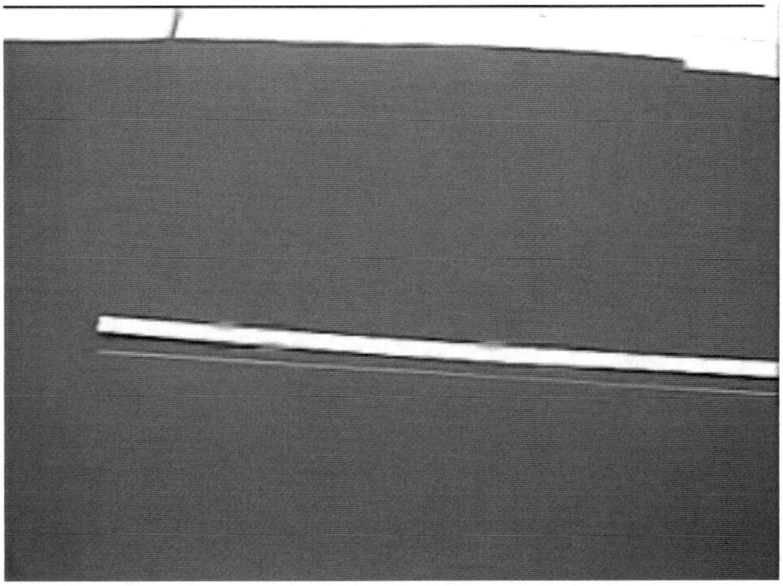

5. Conclusion

We are exploring the potential of recent progress in the field of crystallogenesis and photonics crystals to develop a new concept of calorimeter based on metamaterials to simultaneously record with high precision the maximum of information of the cascade conversion process such as its direction, the spatial distribution of the energy deposition and its composition in terms of electromagnetic, charged and neutral hadron contents. The main objective is to combine the merits of a finely segmented imaging calorimeter and of the dual readout approach.

References

[1] Akchurin N, Atramentov O, Carrell K, Gümüs K Z, Hauptman J, Kim H, Paar H P, Penzo A, Wigmans R, Separation of scintillation and Cherenkov light in an optical calorimeter, *NIMA 550 (2005) 185–200*

R. Wigmans, *these proceedings*

[2] R&D proposal for the study of new fast and radiation hard scintillators for calorimetry at LHC, Crystal Clear Collaboration, *CERN/DRDC/P27/91-15, project RD 18*

[3] Anfré P, Dujardin C, Fourmigué J M, Hautefeuille B, Lebbou K, Pedrini C, Perrodin D, Tillement O, Evaluation of Fiber-Shaped LYSO for Double Readout Gamma Photon Detection, *IEEE TRANSACTIONS ON NUCLEAR SCIENCE, VOL. 54, NO. 2, APRIL 2007*

[4] Ambs P, Sheng Y, Diffractive optics, *Opt. Eng., Vol. 43, 2503 (2004)*; DOI:10.1117/1.1813443

Crystal Calorimeters in the Next Decade

Ren-yuan Zhu

California Institute of Technology, Pasadena, CA 91125, USA

E-mail: zhu@hep.caltech.edu

Abstract. Crystal calorimeter has traditionally played an important role in precision measurement of electrons and photons in high energy physics experiment. Recent interest in calorimeter technology extends its application to measurement of hadrons and jets with dual readout. Potential application of new generation scintillating crystals of high density and high light yield, such as LYSO, in high energy physics experiment is described. Candidate crystals for the proposed homogeneous hadronic calorimeter concept is also discussed.

1. Introduction

Total absorption shower counters made of inorganic crystal scintillators have been known for decades for their superb energy resolution and detection efficiency [1]. In high energy and nuclear physics, crystal calorimeters have been constructed, and their use has been a key factor in the successful physics programs of many experiments. The use of crystals for electromagnetic calorimeter in high energy and nuclear physics experiment is well understood. With proper calibration and monitoring, crystal calorimeters achieve their designed resolution for photon and electron measurement [2]. Recent interest in homogeneous crystal hadronic calorimeter extends its application to measurement of hadrons and jets with high resolution [3]. This concept adapts dual readout for both Cherenkov and scintillation light, which is extensively studied recently by the Dream collaboration [4]. Section 2 of this paper describes optical and scintillation properties of heavy crystal scintillators commonly used in particle physics experiment. Fast and bright crystals discovered in the last two decades, such as cerium doped lutetium oxyorthosilicate ($Lu_2(SiO_4)O$ or LSO) [5], cerium doped lutetium yttrium oxyorthosilicate ($Lu_{2(1-x)}Y_{2x}SiO_5$, LYSO) [6] and cerium doped lanthanum tri-halides,e.g. $LaCl_3$ and $LaBr_3$[7] are also covered. The expected performance of a LYSO electromagnetic calorimeter is elaborated in Section 3. Section 4 discusses candidate crystals for the homogeneous crystal hadronic calorimeter concept.

2. Properties of Crystal Scintillators

Table 1 lists basic properties of heavy crystal scintillators: NaI(Tl), CsI(Tl), BaF_2, CeF_3, bismuth gemanade ($Bi_4Ge_3O_{12}$ or BGO), lead tungstate ($PbWO_4$ or PWO) and LSO [8]. All, except CeF_3, have either been used in, or actively pursued for, high energy and nuclear physics experiments. LSO and LYSO crystals are widely used in the medical industry. Mass production capabilities exist for all these crystals.

Figure 1 is a photo showing twelve crystal samples. In addition to samples listed in Table 1 pure CsI, CsI(Na), LYSO as well as $LaCl_3$ and $LaBr_3$ are also shown in this photo although the last two are not yet in mass production stage. Samples are arranged in an order of their

Table 1. Properties of Heavy Crystal Scintillators with Mass Production Capability

Crystal	NaI(Tl)	CsI(Tl)	BaF_2	CeF_3	BGO	$PbWO_4$	LSO(Ce)
Density (g/cm^3)	3.67	4.51	4.89	6.16	7.13	8.3	7.40
Melting Point (°C)	651	621	1280	1460	1050	1123	2050
Radiation Length (cm)	2.59	1.86	2.03	1.70	1.12	0.89	1.14
Molière Radius (cm)	4.13	3.57	3.10	2.41	2.23	2.00	2.07
Interaction Length (cm)	42.9	39.3	30.7	23.2	22.7	20.7	20.9
Refractive Index^a	1.85	1.79	1.50	1.62	2.15	2.20	1.82
Hygroscopicity	Yes	Slight	No	No	No	No	No
Luminescence^b (nm)	410	560	300	340	480	425	420
(at Peak)			220	300		420	
Decay Time^b (ns)	245	1220	650	30	300	30	40
			0.9			10	
Light Yield^{b,c}	100	165	36	7.3	21	0.30	85
			4.1			0.077	
d(LY)/dT^{b,d} (%/°C)	-0.2	0.4	-1.9	~0	-0.9	-2.5	-0.2
			0.1				
Experiment	Crystal Ball	CLEO *BaBar* BELLE BES III	TAPS	-	L3 BELLE	CMS ALICE PrimEx Panda	SuperB

a At the wavelength of the emission maximum.
b Top line: slow component, bottom line: fast component.
c Relative light yield of samples of 1.5 X_0 and with the PMT quantum efficiency taken out.
d At room temperature.

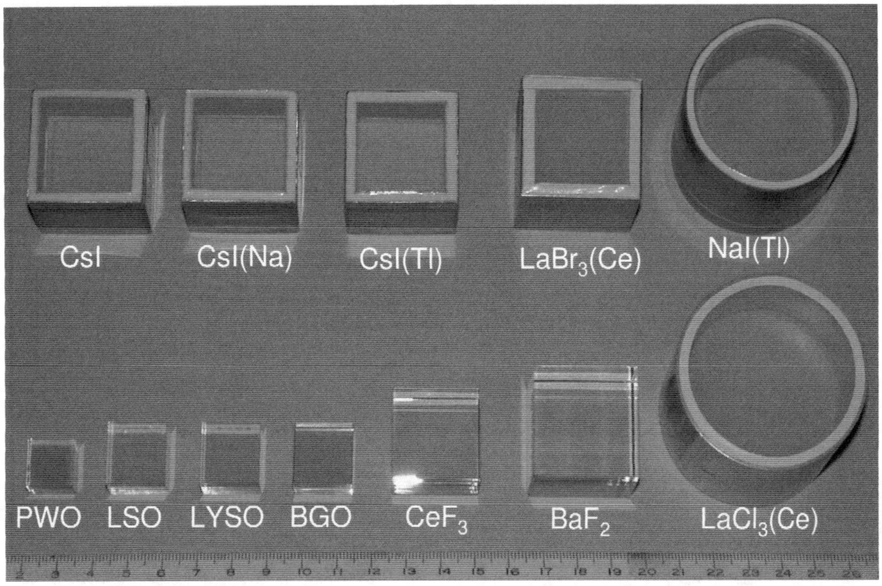

Figure 1. A photo shows twelve crystal scintillators with dimension of 1.5 X_0.

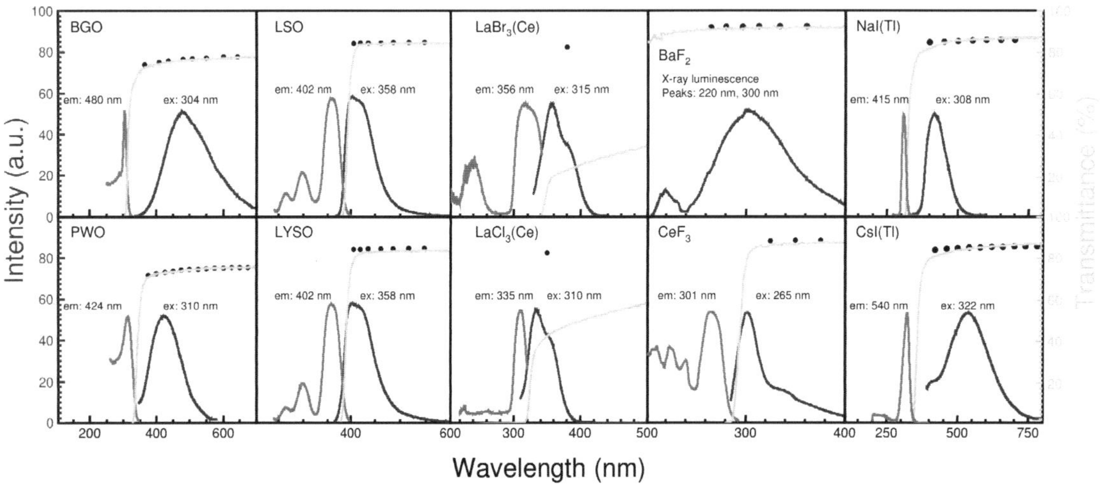

Figure 2. The excitation (red) and emission (blue) spectra (left scale) and the transmittance (green) spectra (right scale) are shown as a function of wavelength for ten crystal scintillators. The solid black dots are the theoretical limit of the transmittance.

density, or radiation length. All non-hygroscopic samples are wrapped with white Tyvek paper as reflector. Hygroscopic NaI, CsI, $LaBr_3$ and $LaCl_3$ are sealed in package with two ends made of quartz windows of 3 or 5 mm thick to avoid surface degradation. To minimize uncertainties in light output measurement caused by the sample size dependence all samples have a cubic shape of $1.5 \times 1.5 \times 1.5\ X_0^3$, except NaI(Tl) and $LaCl_3$ which are a cylinder with a length of 1.5 X_0 and areas at two ends equaling to $1.5 \times 1.5\ X_0^2$ to match the 2 inch diameter of the PMT cathode.

Figure 2 shows a comparison of the transmittance, emission and excitation spectra as a function of wavelength for ten samples. The solid black dots in these plots are the theoretical limit of the transmittance, which was calculated by using corresponding refractive index as a function of wavelength taking into account multiple bounces between the two parallel end surfaces and assuming no internal absorption [9]. Most samples, except $LaBr_3$ and $LaCl_3$, have

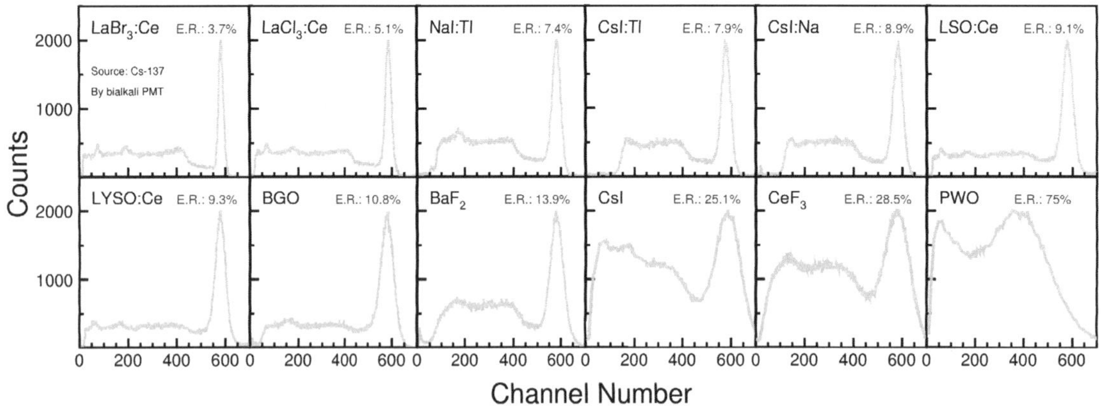

Figure 3. ^{137}Cs γ-ray pulse hight spectra measured by a Hamamatsu R1306 PMT are shown for twelve crystal samples. The numerical values of the FWHM resolution (E.R.) are also shown in the figure.

Journal of Physics: Conference Series **160** (2009) 012017 doi:10.1088/1742-6596/160/1/012017

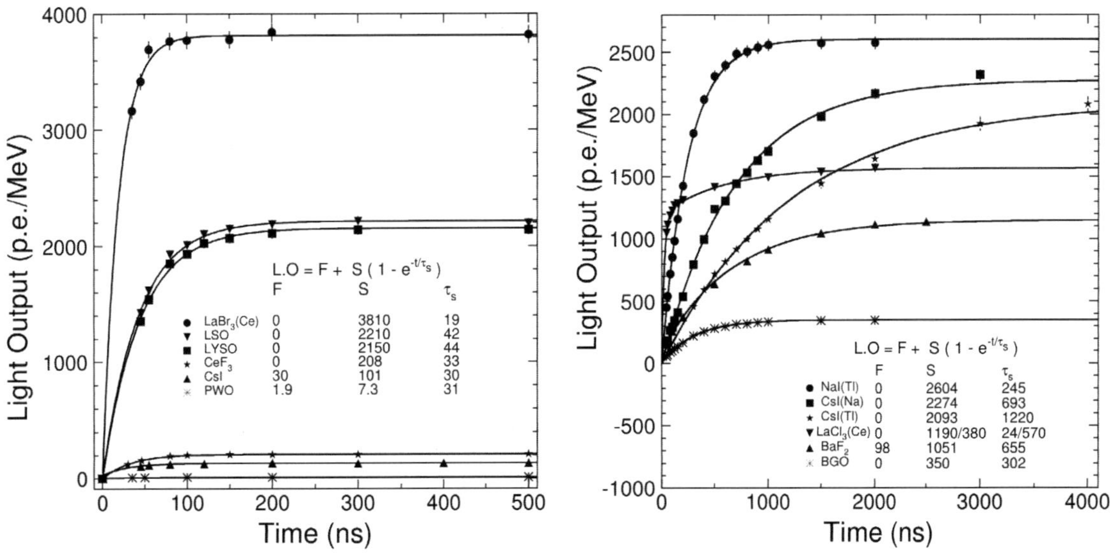

Figure 4. Light output measured by using a XP2254b PMT is shown as a function of integration time for six fast (Left) and six slow (Right) crystal scintillators.

their transmittance approaching the theoretical limits, indicating negligible internal absorption. The poor transmittance measured for $LaBr_3$ and $LaCl_3$ samples is probably due to scattering centers inside these samples.

It is interesting to note that BaF_2, BGO, NaI(Tl), CsI(Tl) and $PbWO_4$ have their emission spectra well within the transparent region showing no obvious self-absorption effect. The UV absorption edge in the transmittance spectra of LSO, LYSO, CeF_3, $LaBr_3$ and $LaCl_3$, however, cuts into the emission spectra and thus affects crystal's light output. This self-absorption effect is more seriously in long crystal samples used in high energy and nuclear physics experiment as extensively discussed for LSO and LYSO crystals [10].

Figure 3 shows the ^{137}Cs γ-ray pulse hight spectra measured by a Hamamatsu R1306 PMT with bi-alkali cathode for twelve crystal samples. Also shown in these figures are the corresponding FWHM energy resolution (E.R.). γ-rays spectroscopy with a few percents resolution is required to identify isotopes for the homeland security application. It is clear that only $LaBr_3$ approaches this requirement. All other crystals do not provide sufficient energy resolution at low energies.

Figure 4 shows light output in photo-electrons per MeV energy deposition as a function of the integration time, measured by using a Photonis XP2254b PMT with multi-alkali photo cathode,

Table 2. Emission Weighted Quantum Efficiencies (%)

Emission	LSO/LYSO	BGO	CsI(Tl)
Hamamatsu R1306 PMT	12.9±0.6	8.0±0.4	5.0±0.3
Hamamatsu R2059 PMT	13.6±0.7	8.0±0.4	5.0±0.3
Photonis XP2254b	7.2±0.4	4.7±0.2	3.5±0.2
Hamamatsu S2744 PD	59±4	75±4	80±4
Hamamatsu S8664 APD	75±4	82±4	84±4

Journal of Physics: Conference Series **160** (2009) 012017 doi:10.1088/1742-6596/160/1/012017

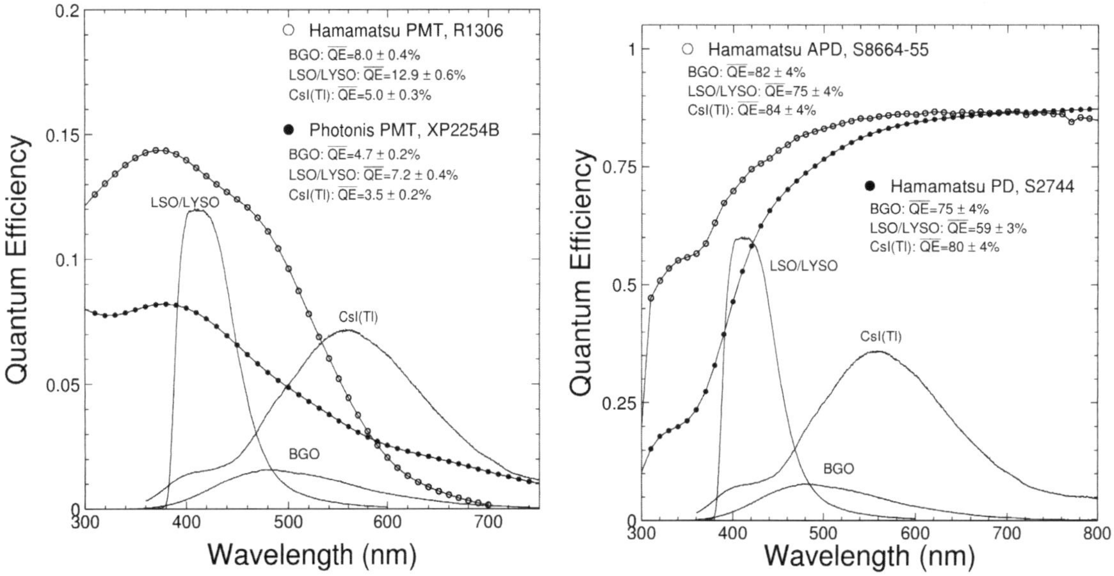

Figure 5. Left: Quantum efficiencies of a Hamamatsu 1306 PMT with bi-alkali cathode (open circles) and a Photonis 2254B PMT with multi-alkali cathode (solid dots) are shown as a function of wavelength together with the emission spectra of the LSO/LYSO, BGO and CsI(Tl) samples, where the area under the emission curves is proportional to their corresponding absolute light output. Right: The same for a Hamamatsu S8664 Si APD (open circles) and a Hamamatsu S2744 Si PIN diode (solid dots).

for six fast crystal scintillators (Left): $LaBr_3$, LSO, LYSO, CeF_3, undoped CsI and $PbWO_4$ and six slow crystal scintillators (Right): NaI(Tl), CsI(Na), CsI(Tl), $LaCl_3$, BaF_2 and BGO. The corresponding fits to the exponentials and their numerical results are also shown in these figures. The undoped CsI, $PbWO_4$, $LaCl_3$ and BaF_2 crystals are observed to have two decay components. Despite its poor transmittance the cerium doped $LaBr_3$ is noticed by its bright fast scintillation, leading to the excellent energy resolution for the γ-ray spectroscopic applications.

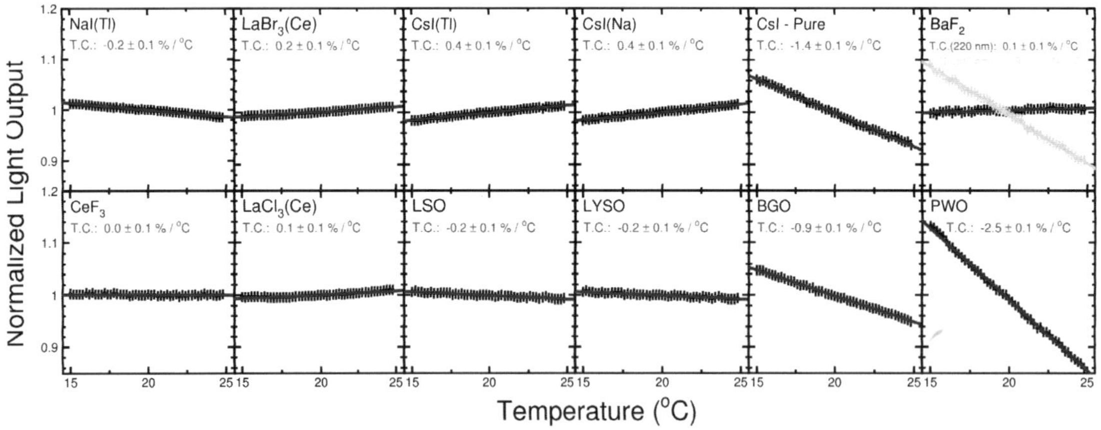

Figure 6. Light output temperature coefficient obtained from linear fits between 15°C and 25°C for twelve crystal scintillators.

The LSO and LYSO samples have consistent fast decay time (~40 ns) and photo-electron yield, which is 6 and 230 times of BGO and $PbWO_4$ respectively.

Since the quantum efficiency of the PMT used for the light output measurement is a function of wavelength, it should be taken out to directly compare crystal's light output. Figure 5 shows typical quantum efficiency as a function of wavelength for a PMT with bi-alkali cathode (Hamamatsu R1306) and a PMT with multi-alkali cathode (Photonis 2254B), a Si APD (Hamamatsu S8664) and a Si PIN PD (Hamamatsu S2744). The emission spectra of LSO/LYSO, BGO and CsI(Tl) crystals are also shown in these figures. Table 2 summarized numerical result of the emission weighted average quantum efficiency for several readout devices. The light output values in Table 1 are listed with the PMT quantum efficiency taken out. The light output of LSO and LYSO crystals is a factor of 4 and 200 of that of BGO and $PbWO_4$ respectively.

Scintillation light yield from crystal scintillators may also depends on the temperature. Fig 6 shows light output variations for twelve crystal samples between 15°C and 25°C. The corresponding temperature coefficients obtained from linear fits are also listed in the figure. The numerical result of these fits is also listed in Table 1.

3. LYSO Crystal Electromagnetic Calorimeter

Because of their board application in medical industry large size LSO and LYSO crystals with consistent optical and scintillation properties have been routinely grown [10]. Figure 7 shows four long crystal samples of $2.5 \times 2.5 \times 20$ cm^3. Figure 8 shows the spectra of 0.51 MeV γ-rays from a ^{22}Na source observed with coincidence triggers [10]. The readout devices used are a Hamamatsu R1306 PMT (Left) and 2 Hamamatsu S8664-55 APDs (Right). The FWHM resolution for the 0.51 MeV γ-ray with the PMT readout is about 12% to 13% for these long samples, which can be compared to 15% for the BGO sample. With APD readout, the γ-ray peaks are clearly visible. The energy equivalent readout noise was less than 40 keV for these long LSO and LYSO samples.

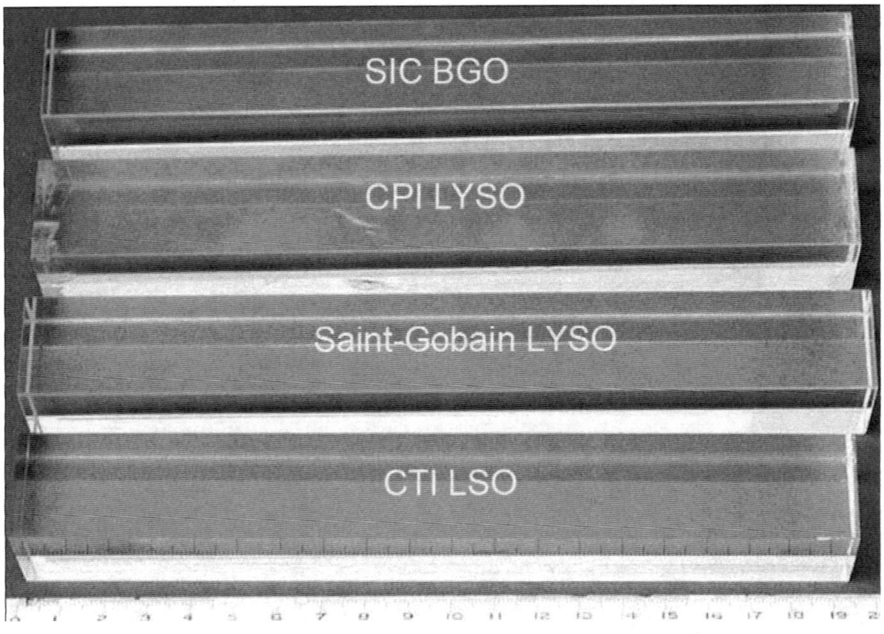

Figure 7. A photo shows four long crystal samples with dimension of $2.5 \times 2.5 \times 20$ cm^3.

Journal of Physics: Conference Series **160** (2009) 012017 doi:10.1088/1742-6596/160/1/012017

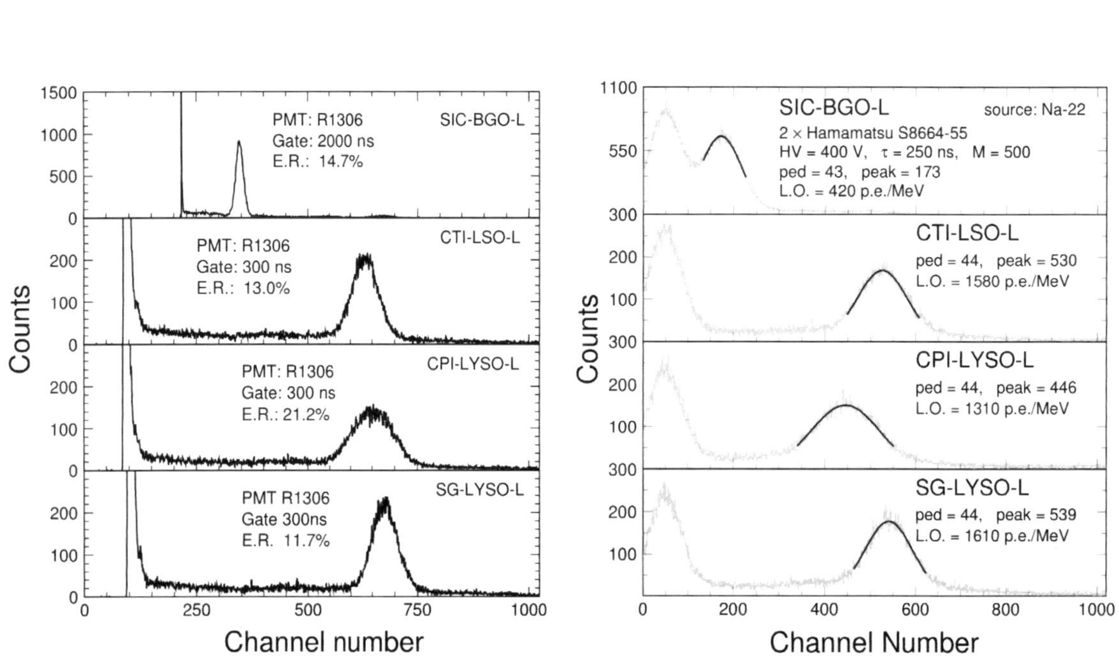

Figure 8. The 0.511 MeV γ-rays spectra from a ^{22}Na source measured with a coincidence trigger using a Hamamatsu R1306 PMT (Left) and two Hamamatsu S8664-55 APDs (Right) for long BGO, LSO and LYSO samples of $2.5 \times 2.5 \times 20$ cm^3 size.

LSO/LYSO crystals is also found to be much more radiation hard than other crystals

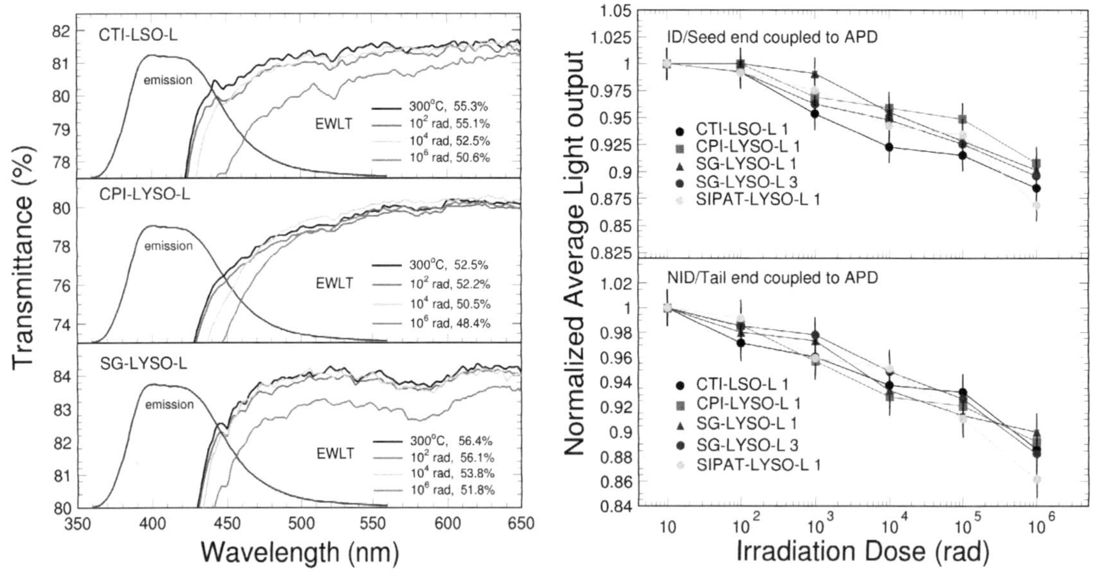

Figure 9. Left: Transmittance spectra are shown as a function of wavelength in an expanded scale together with the photo-luminescence spectra for three long LSO and LYSO samples before and after the irradiation with integrated doses of 10^2, 10^4 and 10^6 rad. Right: Normalized light output with ID (top) and NID (bottom) end coupled to the readout device of two S8664-55 APDs is shown as a function of the integration dose for five long LSO and LYSO samples.

commonly used in high energy and nuclear physics experiment, such as BGO, CsI(Tl) and $PbWO_4$ [11]. Their scintillation mechanism is not damaged by γ-ray irradiation. Radiation damage in LSO and LYSO crystals recovers very slow under room temperature but can be completely cured by thermal annealing at 300°C for ten hours. The γ-ray induced readout noise was estimated to be about 0.2 MeV and 1 MeV equivalent respectively in a radiation environment of 15 rad/h and 500 rad/h for LSO and LYSO samples of $2.5\times2.5\times20$ cm^3. Figure 9 (Left) shows an expanded view of the longitudinal transmittance spectra for three samples before and after several steps of the γ-ray irradiation with integrated dose of 10^2, 10^4 and 10^6 rad. Also shown in the figure is the corresponding numerical values of the photo-luminescence weighted longitudinal transmittance ($EWLT$). Figure 9 (Right) shows the normalized average light output as a function of integrated dose for five long samples from various vendors. It is interesting to note that all samples show consistent radiation resistance with degradations of both the light output and transmittance at 10 to 15% level after γ-ray irradiation with an integrated dose of 1 Mrad.

Assuming the same APD based readout scheme as the CMS $PbWO_4$ calorimeter, the expected energy resolution of an LSO/LYSO crystal based electromagnetic calorimeter would be

$$\sigma_E/E = \mathbf{2\%}/\sqrt{E} \oplus \mathbf{0.5\%} \oplus 0.001/E, \tag{1}$$

which represents a fast calorimeter over large dynamic range with low noise. Such calorimeter would provide great physics discovery potential for high energy physics experiments in the proposed super B factory [12] as well as the proposed International Linear Collider (ILC) [13]. Because of its fast scintillation and good radiation hardness LYSO crystals are also proposed for the CMS $PbWO_4$ crystal endcap calorimeter upgrade at SLHC [14].

4. Crystal Hadronic Calorimeter

Crystals have recently been proposed to construct a homogeneous calorimeter, including both electromagnetic and hadronic part [3]. This homogeneous hadronic calorimeter concept removes the traditional boundary between ECAL and HCAL, so eliminates the effect of dead materials in the middle of the hadronic shower development. It takes advantage of recently implemented dual readout approach to measure both Cherenkov and scintillation light to achieve good energy resolution for hadronic jets measurement [4]. Because of the unprecedent volume (70 to 100 m^3) foreseen for such calorimeter [3], the crystal material must be dense (to reduce the volume),

Figure 10. Left: A photo shows three crystal samples of $5 \times 5 \times 5$ cm^3 investigated for the homogeneous hadronic calorimeter concept. Right: The transmittance spectra of PbF_2 (green), BGO (blue), PWO (red) and UG11 (black) are shown as a function of wavelength. Also shown in this figure are the Cherenkov emission spectrum (dashed blue) and the normalized figure of merit for the Cherenkov light measurement with the UG11 filter.

UV transparent (to effective collecting the Cherenkov light) and allows a clear discrimination between the Cherenkov and scintillation light.

Figure 10 (Left) shows samples of three $5 \times 5 \times 5$ cm^3 crystal samples: PbF_2, BGO and PWO. Crystals of this size can be seen as typical building block for a crystal hadronic calorimeter. All material are dense (PbF_2 has of a density of 7.7 g/cm^3) with a nuclear interaction length about 22 cm. Mass production capability exists for all three candidate materials with cost among the lowest for materials of such density. Figure 10 (Right) shows the transmittance spectra of PbF_2 (green), BGO (blue), PWO (red) and a UG11 filter (black) as a function of wavelength together with the Cherenkov emission spectrum (dashed blue). The UG 11 filter can be used to select the Cherenkov light with small or no scintillation contamination. Also shown in this figure is the normalized figure of merit for the Cherenkov measurement (TWEM) by using the UG11 filter, which is defined as the transmittance weighted Cherenkov emission spectrum. Their numerical values are 1.0:0.53:0.21, which would be 1.0:0.82:0.75 without using the UG11 filter. Among these materials PbF_2 is the most effect in collecting the Cherenkov light because of its good UV transmission.

Effective discrimination between Cherenkov and scintillation light can be realized by using optical filter. Figure 11 (Left) shows a set-up used to investigate Cherenkov light collection and its separation from the scintillation light. Cosmic-rays were triggered by two finger counters with coincidence. The Cherenkov and scintillation light pulses generated by cosmic-rays were measured simultaneously by two Hamamatsu R2059 PMT coupled to the sample through optical filters UG11 and GG400. GG400 is a low-pass filter with cut-off at 400 nm. The UG11 filter is used to select the Cherenkov light as shown in Figure 10 (Right). The GG400 filter is used to select the scintillation light with small contamination of the Cherenkov light. The output of these two PMTs were digitized by an Agilent 6052A digital scope. Figure 11 (Middle and Right) shows the front edge of the scintillation light pulse from BGO and PWO, observed through the GG400 filter. Their delay from the trigger (t_d) and rise time (t_r) are identical with numerical values of 6.2 ns and 1.9 ns respectively. Figure 12 shows the Cherenkov light pulse shape observed for PbF_2 (Left), BGO (Middle) and PWO (Right) through the UG11 filter. All pulses have consistent time structure in the delay (6.1 ns), the rise time (1.8 ns), the fall time (4.2 ns) and the FWHM width (3.0 ns). It is interesting to note that there is no difference observed

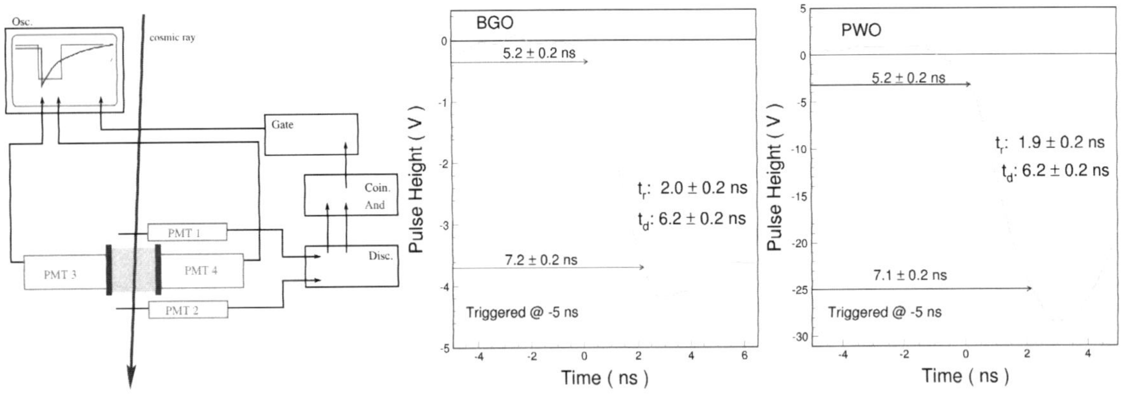

Figure 11. Left: A schematic showing a simple set-up used to measure cosmic-ray generated Cherenkov and scintillation light simultaneously by using two Hamamatsu R2059 PMT. The light pulses are recorded by an Agilent 6052A digital scope. Digital scope traces of the scintillation light front edge measured by a Hamamatsu R2059 PMT with GG400 filter for the BGO (Middle) and PWO (Right) samples.

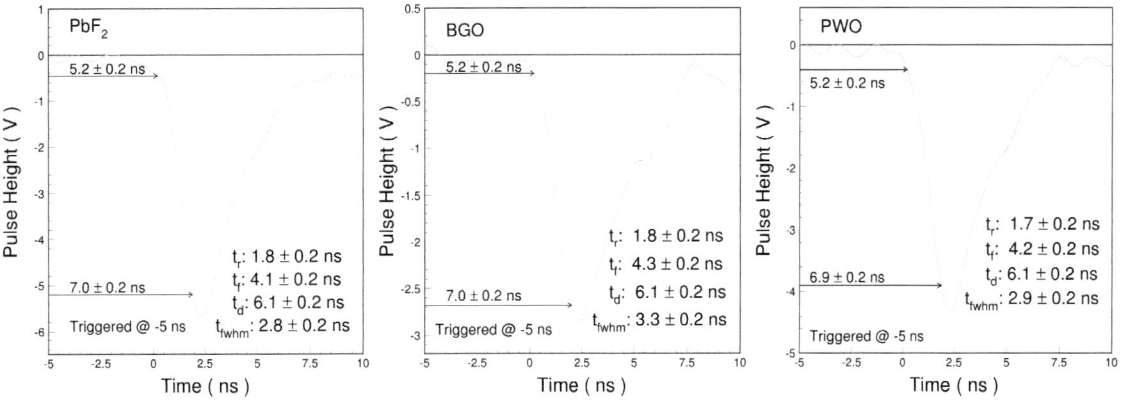

Figure 12. Digital scope traces of Cherenkov light pulse measured by a Hamamatsu R2059 PMT with UG11 filter for the PbF_2 (Left), BGO (Middle) and PWO (Right) samples.

in the delay and rise time between the Cherenkov and scintillation light, indicating that only the light pulse width and fall time are useful for the discrimination between the Cherenkov and scintillation light. A slow scintillator may help this discrimination.

The ratio of Cherenkov versus scintillation light was measured to be 1.55% and 22% for BGO and PWO respectively. These values are consistent with the scintillation light yield shown in Table 1, the emission weighed quantum efficiency of bi-alkali cathode of the Hamamatsu R2059 PMT shown in Figure 5 (Left) and the TWEM values shown in Figure 10 (Right).

Development of cost-effective material is crucial for the homogeneous hadronic calorimeter concept. While BGO is the best material to be used for such calorimeter, R&D is actively pursued by the high energy physics community for additional materials. One approach is to develop PWO crystals with slow scintillation emission . Green (560 nm) and slow emission with a few μsec decay time was observed by selective doping in PWO crystals [15]. Such crystals were reported to have a factor of ten more light than yttrium doped PWO crystals used in high energy physics experiment. This slow and green scintillation would be desirable for this application.

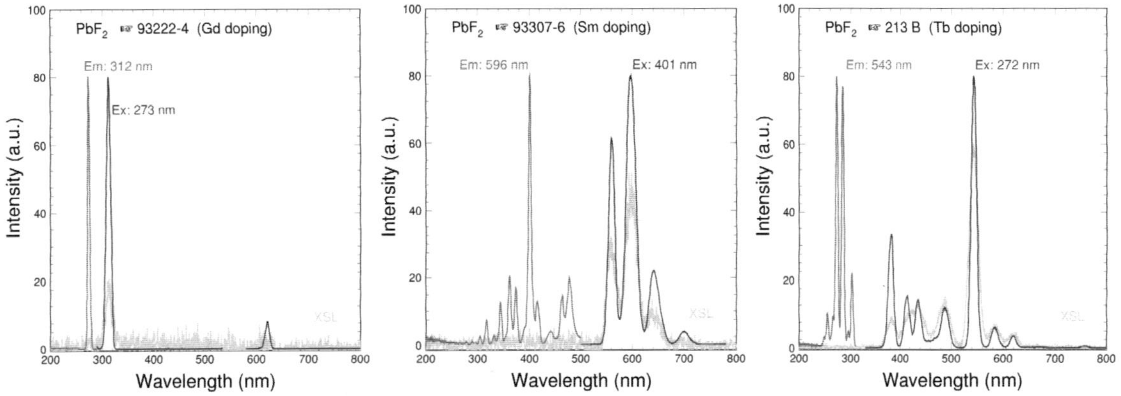

Figure 13. UV excitation (red), photo-luminescence (blue)and x-luminescence (green) spectra are shown as a function of wavelength for Gd (Left), Sm (Middle) and Tb (Right) dopes PbF_2 samples.

Another approach is to develop scintillating PbF_2 crystals by selective doping. Observations of fast scintillation in Gd or Eu doped PbF_2 crystals were reported [16, 17]. Our investigation shows that rear earth doping introduces scintillation in PbF_2, but not in the level can be measured by using γ-ray source. Figure 13 show the excitation, photo-luminescence and x-luminescence spectra for Gd, Sm and Tb doped PbF_2 crystal samples. It is noted that the scintillation of Sm and Tb doped PbF_2 samples is between 500 to 600 nm, which is desirable for Cherenkov/scintillation discrimination. Investigation is continuing aiming at developing cost-effective materials for this concept.

5. Summary

Precision crystal electromegnetic calorimeters have been an important part of high energy physics detector. In addition to the Panda PWO electromagnetic calorimeter at Fair [18] the availability of mass production capability of large size LSO and LYSO crystals provides an opportunity to build a LSO/LYSO crystal electromegnetic calorimeter with unprecedent energy resolution over a large dynamic range down to MeV level. Such calorimeter, if built, would greatly enhance the physics discovery potential for high energy and nuclear physics experiments in the next decade.

Recent interest in high energy physics community to pursue homogeneous hadronic calorimeter with dual readout opens a new area of crystal calorimetry to achieve good energy resolution for hadronic jets in the next decade. The main challenge for this concept is to develop cost effective heavy crystal scintillators with good UV transmission and excellent Cherenkov/scintillation discrimination.

Acknowledgments

This work is partially supported by the U.S. Department of Energy Grant No. DE-FG03-92-ER40701 and the U.S. National Science Foundation Award PHY-0612805 and PHY-0516857.

References

[1] G. Gratta, H. Newman and R.-Y. Zhu, *Annu. Rev. Nucl. Part. Sci.* **44** 453 (1994).
[2] R.-Y. Zhu, in *Proceedings of XII International Conference on Calorimetry in Particle Physics*, Ed. S. Magill and R. Yoshida, AIP Conference Proceedings Volume 867 (2006) 64.
[3] A. Para, http://ilcagenda.linearcollider.org/contributionDisplay.py?contribId=15&sessionId=3&confId=2474.
[4] R. Wigmans, in *Proceedings of VII International Conference on Calorimetry in Particle Physics*, Ed. E. Chen *et al.*, World Scientific (1998) 182, and N. Akchurin *et al.*, *Nucl. Instr. and Meth.* **A537** (2005) 537. Also, see papers in these Proceedings.
[5] C. Melcher, V.S. Patent 4958080 (1990) and 5025151 (1991).
[6] D.W. Cooke, K.J. McClellan, B.L. Bennett, J.M. Roper, M.T. Whittaker and R.E. Muenchausen, *J. Appl. Phys.* **88** (2000) 7360 and T. Kimble, M Chou and B.H.T. Chai, *2002 IEEE NSS Conference Record.*
[7] E. Loef, P. Dorenbos, E. Eijk, K. Kraemer and H. Guedel, *Nucl. Instr. and Meth.* **A486** (2002) 254.
[8] R.H. Mao, L.Y. Zhang and R.-Y. Zhu, *IEEE Trans. Nucl. Sci.* **NS-55** (2008).
[9] D.A. Ma and R.-Y. Zhu, *Nucl. Instr. and Meth.* **A333** (1993) 422.
[10] J.M. Chen, R.H. Mao, L.Y. Zhang and R.-Y. Zhu, *IEEE Trans. Nucl. Sci.* **54** (2007) 718.
[11] J.M. Chen, R.H. Mao, L.Y. Zhang and R.-Y. Zhu, *IEEE Trans. Nucl. Sci.* **54** (2007) 1319.
[12] SuperB Conceptual Design Report, **INFN/AE-07/2, SLAC-R-856, LAL 07-15**, March (2007).
[13] R.-Y. Zhu, *A LSO/LYSO Crystal Calorimeter for the ILC*, talk presented in 2005 ILC Workshop, Snowmass. http://nicadd.niu.edu/cdsagenda//fullAgenda.php?ida=a0561.
[14] R.-Y. Zhu, *Development of LYSO Crystals for CMS at SLHC*, talk presented in CMS ECAL SLHC Workshop, CERN. http://www.hep.caltech.edu/ zhu/talks/ryz_080415_SLHC.pdf.
[15] R.H. Mao *et al.*, in *Proceedings of IX International Conference on Calorimetry in Particle Physics*, Ed. B. Aubert *et al.*, Franscati Physics Series Vol. XXI (2000) 709-720.
[16] D. Shen *et al.*, *Jour. Inor. Mater.* **Vol 101** (1995) 11.
[17] C. Woody *et al.*, in *Proceedings of SCINT95*, Delft University Press, Delft, The Netherlands, (1996).
[18] Philippe Rosier, in these Proceedings.

Recent results from the DREAM project

Richard Wigmans

Department of Physics
Texas Tech University
Lubbock, TX 79409-1051, U.S.A.

E-mail: wigmans@ttu.edu

Abstract. The aim of the DREAM project is to develop calorimeters that are able to measure the four-vectors of all fundamental constituents of matter, including fragmenting quarks, with a precision of 1% or better. To achieve this, the factors that limit the performance of the present generation of calorimeters are eliminated one by one, in the order at which these factors dominate. In this talk, I give an overview of the results achieved so far, and outline our plans for the future.

1. Introduction

High-precision jet spectroscopy will be increasingly important in future high-energy accelerator experiments, particularly at a Linear e^+e^- Collider. DREAM (Dual REAdout Method) calorimeters seem to be well suited for this task. The key aspect of DREAM detectors is the simultaneous measurement of scintillation light and Čerenkov light generated in the shower development process. By comparing these two signals, the electromagnetic shower fraction can be measured event by event, both for single hadrons and for jets, and the detrimental effects of fluctuations in this fraction can be eliminated.

The merits of this technique were first illustrated with a calorimeter in which the two signals are provided by two different types of optical fibers. More recently, we have been concentrating on crystals ($PbWO_4$ and BGO), which have the potential of eliminating (or at least reducing) the contributions of the next two important sources of fluctuations: photoelectron statistics and sampling fluctuations. I will describe the techniques used to unravel the signals from these crystals into Čerenkov and scintillation components. The detailed time structure measurements we performed for these studies also make it possible to measure the contributions of neutrons to the signals. This would help to reduce the effects of fluctuations in nuclear binding energy losses, which is the last frontier in the quest for ultimate hadronic calorimeter performance. There will be detailed talks about all these aspects at this conference. Gabriella Gaudio will present the evidence for the contribution of Čerenkov light to the signals from $PbWO_4$ crystals, and report on the temperature dependence of this contribution. Cecilia Voena will discuss the precision of event-by-event measurements of the Čerenkov fraction in crystal calorimeters. Davide Pinci will show results of dual-readout calorimetry performed with simple crystal electromagnetic calorimeters used in conjunction with our DREAM fiber module, and John Hauptman will talk about the neutron contributions to the signals. Finally, Roberto Carosi will describe a new system intended to measure the time structure of all our signals with nanosecond resolution. In this talk, I will provide context and mention some highlights.

Journal of Physics: Conference Series **160** (2009) 012018 doi:10.1088/1742-6596/160/1/012018

2. Dual-Readout Calorimetry

The DREAM Collaboration tries to improve the performance of hadron calorimeters by eliminating or reducing the factors that limit that performance. Since the resolution of calorimeters is determined by fluctuations, eliminating or reducing the (effects of the) dominant fluctuations is the key to improving it. In non-compensating calorimeters, the hadronic energy resolution is dominated by fluctuations in the em shower fraction, $f_{\rm em}$. DREAM has shown that (the effects of) these fluctuations can be eliminated by simultaneously measuring both the visible deposited energy (dE/dx) and the Čerenkov light generated in the shower absorption process.

Calorimeters that use Čerenkov light as signal source are, for all practical purposes, only responding to the em fraction of hadronic showers [1]. This is because the electrons/positrons through which the energy is deposited in the em shower component are relativistic down to energies of only 200 keV. On the other hand, most of the non-em energy in hadron showers is deposited by non-relativistic protons generated in nuclear reactions [2]. Such protons do generate signals in other types of active media (scintillator, LAr). By comparing the relative strengths of the two signals, the em shower fraction can be determined and the total shower energy can be reconstructed using the known e/h value(s) of the calorimeter. This can be seen as follows. The measured calorimeter signal can be written as the sum of an em term and a non-em term. The latter is affected by the e/h ratio:

$$\begin{aligned} S &= E\left[f_{\rm em} + (h/e)_S(1-f_{\rm em})\right] \\ Q &= E\left[f_{\rm em} + (h/e)_Q(1-f_{\rm em})\right] \end{aligned} \tag{1}$$

The scintillator (S) and Čerenkov (Q) signals are different because of differences in the e/h ratios. When these e/h ratios are known, these equations can either be solved for the em shower fraction ($f_{\rm em}$) or for the shower energy (E).

2.1. *Fiber-based detectors*

This principle was first demonstrated with a calorimeter that used two active media, hence the name DREAM (Dual REAdout Method): scintillating fibers measured the visible shower energy (S), while clear fibers measured the generated Čerenkov light (Q). Using the ratio of the two signals, the value of $f_{\rm em}$ could be determined *event-by-event*, since

$$\frac{Q}{S} = \frac{f_{\rm em} + 0.21\,(1-f_{\rm em})}{f_{\rm em} + 0.77\,(1-f_{\rm em})} \tag{2}$$

where 0.21 and 0.77 represent the h/e ratios of the Čerenkov and scintillator calorimeter structures, respectively.

The merits of this method are clearly illustrated by Figure 1, which shows the overall Čerenkov signal distribution for 100 GeV π^- (a), as well as distributions for subsamples selected on the basis of their $f_{\rm em}$ value (b) [3], determined by Equation 2. Each $f_{\rm em}$ bin probes a certain region of the overall signal distribution, and the average value of the subsample distribution increases with $f_{\rm em}$.

Once the value of $f_{\rm em}$ was determined, the signals could be corrected in a straightforward way for the effects of non-compensation. In this process, the energy resolution improved, the signal distribution became much more Gaussian and, most importantly, the hadronic energy was correctly reproduced. This was true both for single pions as well as for jets The results for 200 GeV "jets" are shown in Figure 1c-e. These "jets" were in fact not fragmenting quarks or gluons, but multiparticle events created by pions interacting in a target placed upstream of the calorimeter. Using only the *ratio* of the two signals produced by this calorimeter, the resolution for these "jets" was improved from 14% to 5%, in the Čerenkov channel. It was shown that this 5% resolution was in fact dominated by fluctuations in side leakage in this (small, only 1030 kg instrumented volume) detector. Eliminating such fluctuations led to a further considerable improvement (Figure 1e).

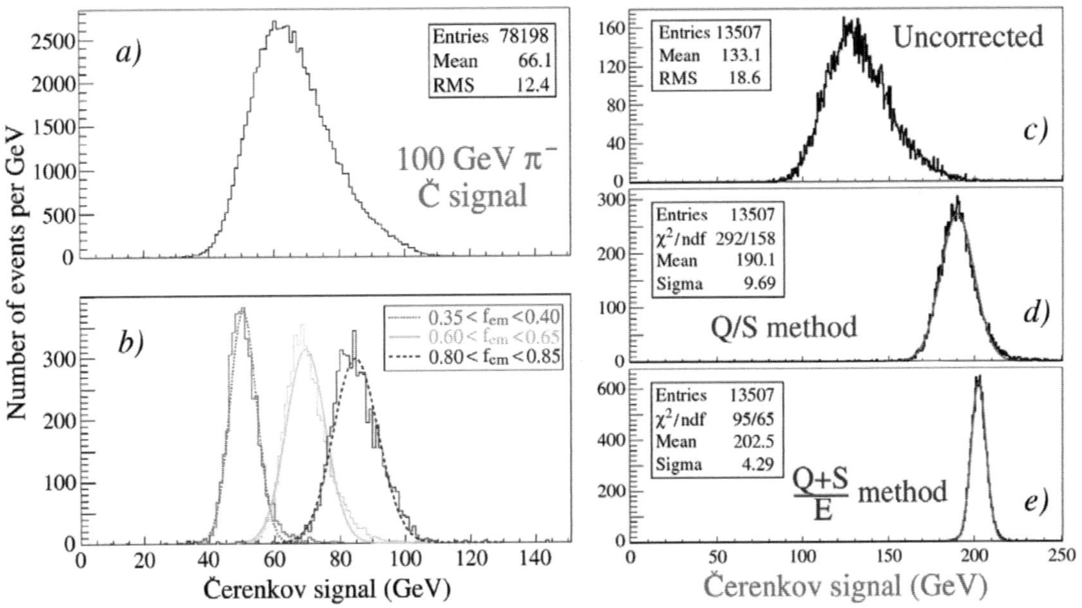

Figure 1. Čerenkov signal distribution for 100 GeV π^- (a) and distributions for subsamples of events selected on the basis of the measured $f_{\rm em}$ value, using the Q/S method (b). Signal distributions for high-multiplicity 200 GeV "jets" in the DREAM before (c) and after (d) corrections on the basis of the observed Čerenkov/scintillator signal ratio were applied. In diagram e, energy constraints were used, which eliminated the effects of lateral shower fluctuations that dominate the resolution in d [3].

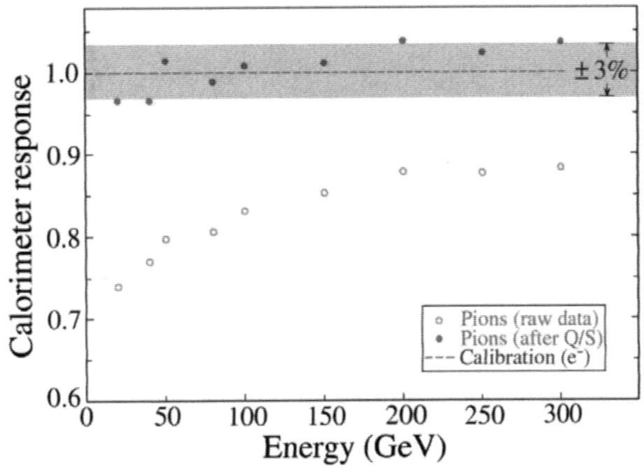

Figure 2. The calorimeter response (average signal per unit energy) for pions before and after the corrections based on a comparison of the scintillator and Čerenkov signals were applied [3].

The shower energy could also be determined directly (*i.e.,* without first measuring $f_{\rm em}$) by solving Equations 1 for E:

$$E = \frac{S - \chi C}{1 - \chi}, \quad \text{with } \chi = \frac{1 - (h/e)_S}{1 - (h/e)_C} \sim 0.3 \text{ for this calorimeter} \tag{3}$$

Figure 2 illustrates that the hadronic signal non-linearity (a notorious problem for non-compensating calorimeters) was indeed more or less eliminated with such a procedure, and that the calorimeter response to pions became within a few percent equal to that for electrons, with which the instrument was calibrated. Also the "jet" energy was well reconstructed in this way. Whereas the raw data gave a mean value of 133.1 GeV for 200 GeV "jets" (Figure 1c), the described procedure increased it to 190.1 GeV (Figure 1d). Any remaining discrepancies are most likely due to the effects of side leakage, and could thus be eliminated in a sufficiently large detector.

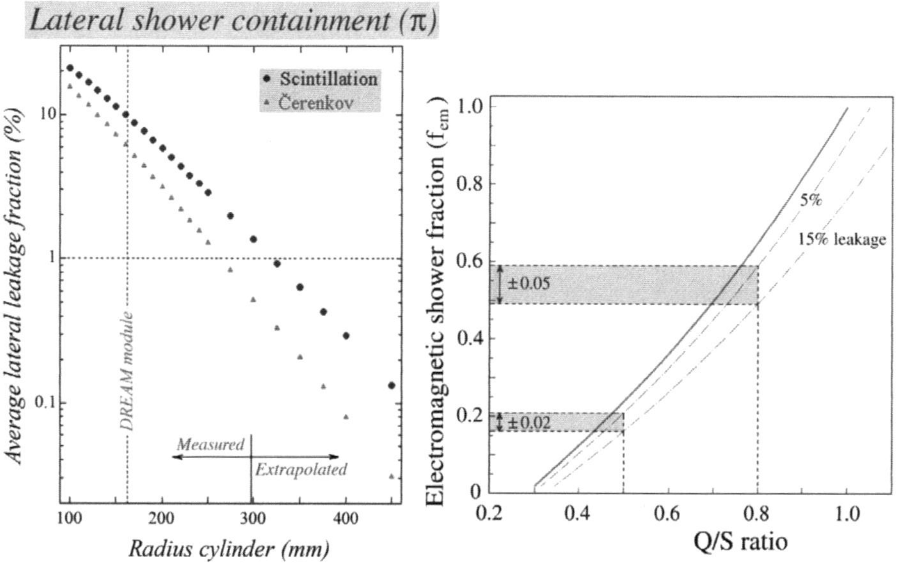

Figure 3. Leakage outside of a DREAM calorimeter cylinder, as a function of the radius of that cylinder. Results are given for 100 GeV π^+ as a fraction of the total signal in an infinitely large calorimeter, separately for scintillation and Čerenkov light (a). The relationship between the em shower fraction needed for the correction procedures and the measured Q/S signal ratio. This relationship is given for a perfect calorimeter and for calorimeters that are too small to fully contain the hadron showers (b) [4].

The importance of shower leakage in this relatively small calorimeter is illustrated in Figure 3. Figure 3a shows that for showers induced by 100 GeV pions, on average $\sim 90\%$ of the scintillator signal and $\sim 95\%$ of the Čerenkov signal that would be recorded in an infinitely large instrument of this type is observed in our detector. The effects of this leakage on the calorimeter performance are illustrated in Figure 3b, which shows the relationship between the em shower fraction ($f_{\rm em}$) which one needs to know for the correction procedure, and the Q/S signal ratio which is measured. Event-to-event fluctuations in shower leakage lead to an uncertainty in the value of $f_{\rm em}$ derived from these measurements, and this uncertainty affects the energy resolution and the value of the reconstructed energy. Figure 3a also shows that in order to reduce the average energy leakage fraction from 10% to 1%, the radius of the detector would have to be roughly doubled.

2.2. *Homogeneous DREAM calorimeters*

Once the effects of the dominant source of fluctuations are eliminated, the resolution is determined and limited by other types of fluctuations. In the case of the DREAM detector, these fluctuations include, apart from fluctuations in side leakage which can be eliminated by making the detector sufficiently large (see Figure 1e), *sampling fluctuations* and fluctuations in the *Čerenkov light yield*. The latter effect alone (8 Čerenkov photoelectrons per GeV) contributed $35\%/\sqrt{E}$ to the measured resolution. Both effects could be greatly reduced by using crystals for dual-readout purposes. Certain dense high-Z crystals ($PbWO_4$, BGO) produce significant amounts of Čerenkov light, which may be separated from the scintillation light by exploiting differences in time structure, spectral properties and directionality.

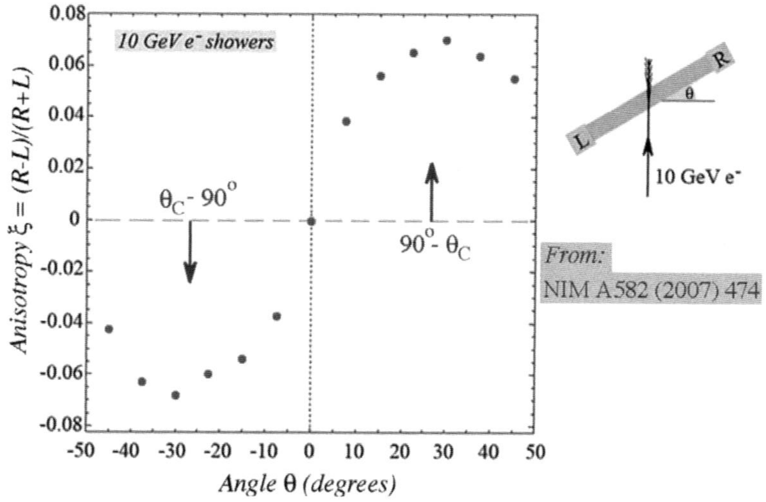

Figure 4. Left-right response asymmetry measured for 10 GeV electrons showering in a $2.5X_0$ thick $PbWO_4$ crystal, as a function of the orientation of the crystal (the angle θ) [5].

Figure 4 provides proof that a significant fraction of the response from a scintillating $PbWO_4$ crystal is indeed due to Čerenkov light. The figure shows an angle-dependent asymmetry between the signals from the two PMTs (L and R) located at opposite ends of a crystal exposed to a beam of 10 GeV electrons. This asymmetry indicates the presence of a directional component in the light produced in this crystal. This component is largest when the crystal is oriented at such an angle that the Čerenkov light emitted by the particles strikes the PMT that records it perpendicularly, *i.e.*, at $\theta \sim 27°$ for PMT R and at $\theta \sim -27°$ for PMT L (the Čerenkov angle in $PbWO_4$ is 63°) [5].

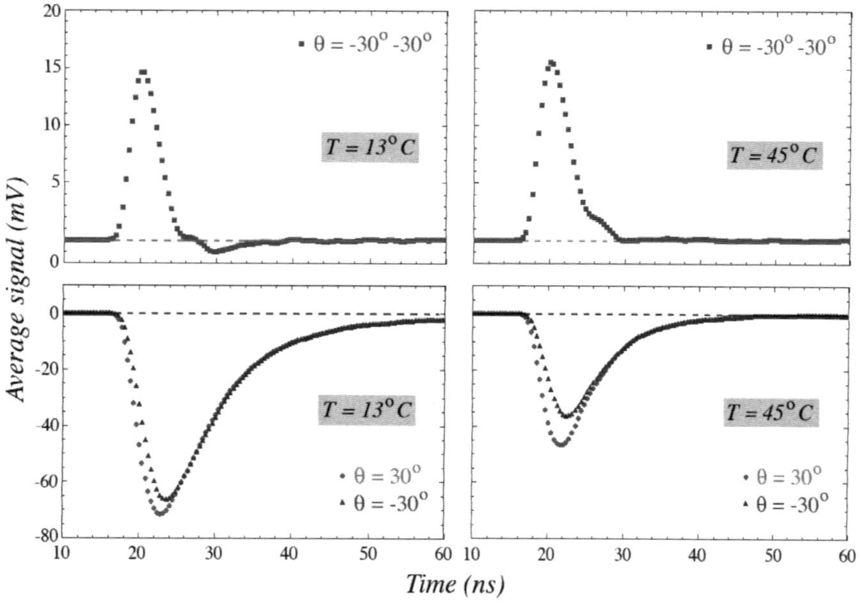

Figure 5. Average time structure of the signals from PMT R (see Figure 4) measured for 50 GeV electrons traversing the $PbWO_4$ crystal at $\theta = 30°$ and $\theta = -30°$ (*bottom* plots), as well as the difference between these two signals (*top* plots), measured for two different temperatures: 13°C (*left*) and 45°C (*right*) [6].

Figure 5 shows the time structure of the signals from a lead tungstate crystal traversed by a beam of 50 GeV electrons, for two different angles and two different temperatures. Each of the bottom graphs depicts the average time structures of the signals measured at $\theta = 30°$ and $\theta = -30°$ with PMT R. The difference between the signals recorded at these two angles is shown in the top graphs, separately for each of the two temperatures. At $\theta = 30°$, Čerenkov light produced in the showers initiated in the crystals by the high-energy electrons is preferentially detected in PMT R, since this light is emitted at an angle of $\arccos(1/n) = 63°$ by the charged relativistic shower particles traversing the $PbWO_4$ crystal (which has a refractive index $n = 2.2$). This Čerenkov light manifests itself as an additional prompt component, superimposed on the scintillation light that constitutes practically the entire signal measured at $\theta = -30°$. As a result, the signals measured at $\theta = 30°$ are larger, and rise steeper than those measured with the same PMT at $\theta = -30°$. For PMT L, the opposite effect was observed. As expected, here the signals at $\theta = -30°$ were measured to be larger and steeper than those at $\theta = 30°$.

Interestingly, the Čerenkov component was not affected by the change in temperature. However, the intensity of the scintillation component decreased by about a factor of two when the temperature was increased from 13 to 45°C. As a result, the contribution of Čerenkov light is much more evident from the time structure measured at the higher temperature [6].

Details of the time structure of the $PbWO_4$ signals have been successfully used to measure the Čerenkov/scintillation ratio for individual events [7], and the em fraction of hadron-induced showers in a calorimeter system with a $PbWO_4$ em section [8, 9].

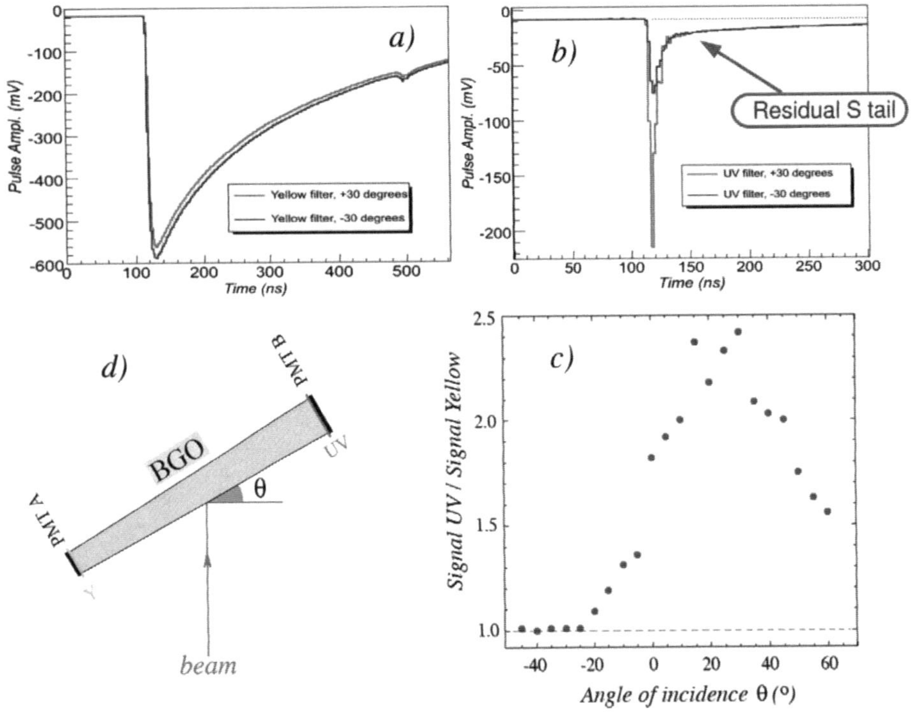

Figure 6. Average signals produced by 50 GeV electrons traversing a BGO crystal at angles of $\pm 30°$ with the longitudinal axis (d), as measured with a UV filter (b), or with a yellow filter (a). The ratio of these two signals, as a function of the angle of incidence of the beam electrons (c) [7].

Even more interesting features were observed in bismuth germanate ($Bi_4Ge_3O_{12}$, or BGO) crystals. Even though Čerenkov radiation represents a much smaller fraction of the light produced by these crystals, it is easier to separate and extract it from the signals. The much longer scintillation time constant and the spectral difference are responsible for that. Figure 6 shows recent results obtained by the DREAM Collaboration. The different spectra for the scintillation and Čerenkov light are evident from Figures 6a and b, which show the time structures of signals recorded with two different filters. The "prompt" component observed in the ultraviolet signal is due to Čerenkov light. This can be concluded from Figure 6c, which shows the ratio of both signals as a function of the angle of incidence of the beam electrons. This ratio peaks when the photons emitted at the Čerenkov angle ($\theta_C = \arccos 1/n \approx 63°$) travel along the longitudinal crystal axis: $\theta = 27°$ (Figure 6d).

Figure 6b shows that also a small fraction of the scintillation light passes through the UV filter. This offers the possibility to obtain all needed information from that one signal. An external trigger opens two gates: one narrow (10 ns) gate covers the prompt component, the second gate (delayed by 30 ns and 50 ns wide) only contains scintillation light. The latter signal can also be used to determine the contribution of scintillation to the light collected in the narrow gate. In this way, the Čerenkov/scintillation ratio can be measured event-by-event on the basis of one signal only [7]. We have used this technique to assess the possibilities for dual-readout calorimetry with a crystal em calorimeter section [9].

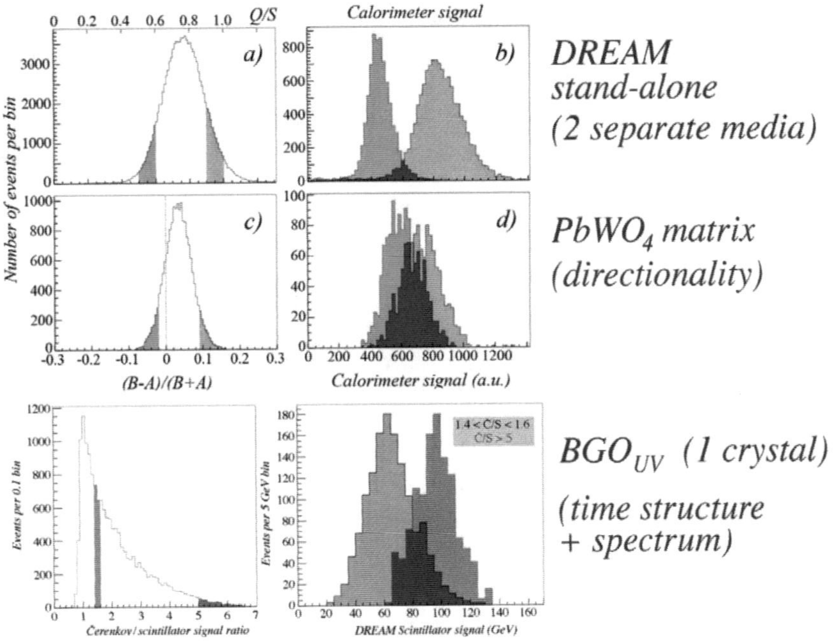

Figure 7. Total hadronic signal distribution in the DREAM fiber calorimeter, for two event samples with a different electromagnetic fraction (b, d, f). The em fraction was determined on the basis of the Q/S signal ratio in the fiber calorimeter itself (a) [5], from the directionality of the signals measured in a $PbWO_4$ crystal matrix installed upstream (c) [8]), or from the time structure of the signals from an upstream BGO crystal (e) [9].

In Figure 1, we showed how the measurement of the em shower fraction, $f_{\rm em}$, (from the Čerenkov/scintillation ratio) for individual events was crucial for improving the hadronic calorimeter performance: event samples with different $f_{\rm em}$ values represented different regions of the total signal distribution. As illustrated in Figure 7, we have observed similar effects for calorimeters in which one or more crystals served as an em section. Figure 7c,d shows results of tests in which the fiber calorimeter was used in conjunction with a matrix of 19 $PbWO_4$ crystals which served as an em calorimeter section, read out by two PMTs. The asymmetry in the signals from these two PMTs was used as a measure of the em content of the developing shower. Figure 7d shows that the total signal in the combined calorimeter system depended indeed in the expected way on this measured asymmetry [8]. Figure 7e,f concerns the results of similar tests with a BGO crystal serving as the em calorimeter section. In this case, the em shower fraction was derived from the time structure of the signals from this crystal (see also Figure 6b).

Figure 7f shows that the total calorimeter signal is indeed very sensitive to the em fraction measured in this way: a large em fraction leads to a large total signal, a small em fraction to a small total signal, just as in the measurements of the fiber calorimeter in stand-alone mode, which are shown for reference purposes in Figure 7a,b [9]. In these measurements, the Čerenkov signals from the crystals were found to comprise a substantially larger number of photoelectrons than those from the quartz fibers: > 30 photoelectrons/GeV.

2.3. Further improvements

The features described above offer the potential to improve the already impressive DREAM results significantly. Should one succeed to eliminate, or at least greatly reduce the contributions of sampling fluctuations and photoelectron statistics to the hadronic energy resolution in this way, then the last hurdle toward ultimate performance is formed by the fluctuations in visible energy, *i.e.*, fluctuations in the energy fraction used to break up atomic nuclei. It has been demonstrated that the kinetic energy carried by the neutrons produced in the shower development process is correlated to this invisible energy loss [2]. Efficient neutron detection, a key ingredient for compensating calorimeters, not only brings e/h to 1.0, but also greatly reduces the contribution of fluctuations in invisible energy to the hadronic energy resolution. It has been demonstrated that this reduces the ultimate limit on this resolution to $\sim 13\%/\sqrt{E}$ [10], in compensating lead/plastic-scintillator calorimeters. This would translate into mass resolutions of $\sim 1.5\%$ for hadronically decaying W and Z bosons, much better than the requirements of a Linear e^+e^- Collider experiment.

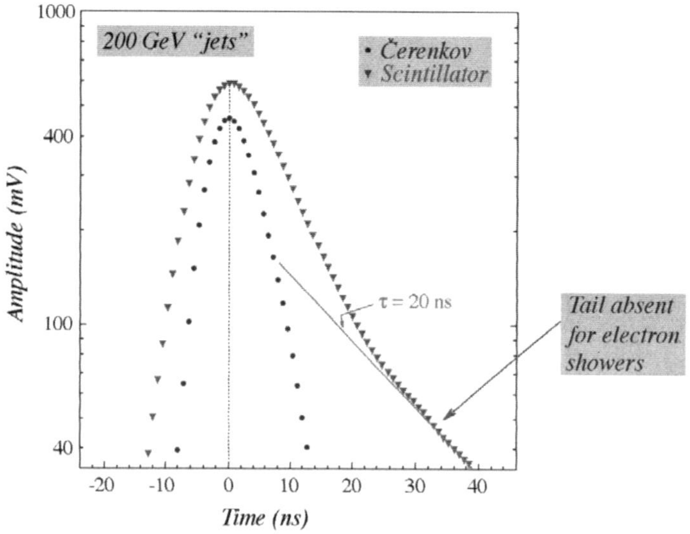

Figure 8. Average time structure of the Čerenkov and scintillation signals recorded for 200 GeV "jets" developing in the DREAM calorimeter. The scintillation signals exhibit a tail with a time constant of about 20 ns, which is absent in the Čerenkov signals. [11].

Detailed measurements of the time structure of the calorimeter signals, examples of which are given in Figures 5 and 6, make it also possible to measure the contribution of neutrons to the shower signals. Figure 8 illustrates this with data recently taken with the DREAM fiber calorimeter [11]. The figure shows the average time structure of Čerenkov and scintillator signals measured for 100 GeV π^+ showers developing in this calorimeter. The scintillator signals exhibit an exponential tail with a time constant of ~ 20 ns. As we established earlier [12], this tail has all the characteristics expected of a (nonrelativistic) neutron signal and is thus absent in the time structure of the Čerenkov signals.

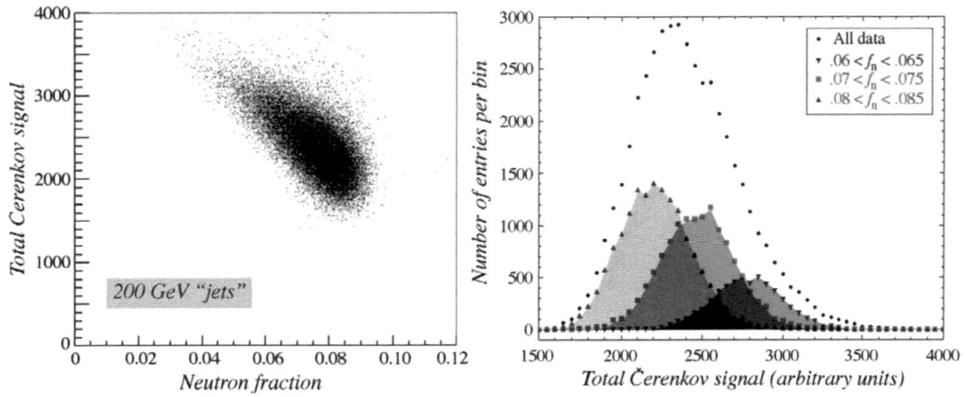

Figure 9. Scatter plot for 200 GeV "jets. For each event, the combination of the total Čerenkov signal and the fractional contribution of neutrons to the total scintillator signal is represented by a dot (left diagram). Distribution of the total Čerenkov signal for 200 GeV "jets" and the distributions for three subsets of events selected on the basis of the fractional contribution of neutrons to the scintillator signal (right) [11].

The new data (obtained for the entire detector, instead of individual towers) made it possible to measure the relative contribution of neutrons to the scintillator signals, f_n, event-by-event. Figure 9 shows that this neutron fraction is clearly anti-correlated to the total Čerenkov signal (left diagram), and that a subsample of events with (approximately) the same f_n value corresponds to a certain region of the total signal distribution (right diagram), in much the same way as the f_{em} fraction probes the total signal distribution (Figure 1b). And just as for the f_{em} measurements, this information could be used to

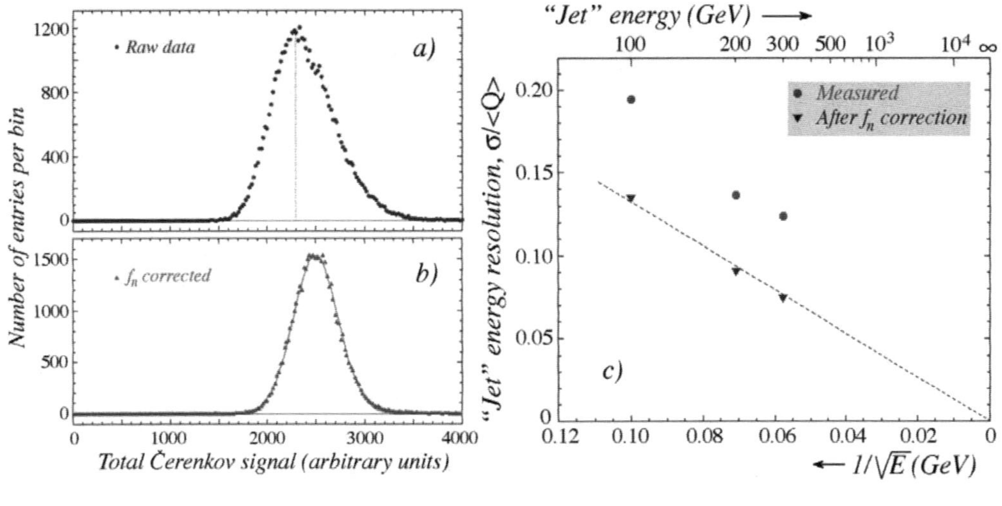

Figure 10. Distribution of the total Čerenkov signal for 200 GeV "jets" before (a) and after (b) applying the correction based on the measured value of f_n, described in the text. Relative width of the Čerenkov signal distribution for "jets" as a function of energy, before and after a correction that was applied on the basis of the relative contribution of neutrons to the scintillator signals (c) [11].

improve the hadronic performance of the fiber calorimeter. This is illustrated in Figure 10, which shows that the asymmetric signal distribution characteristic for a non-compensating calorimeter (Figure 10a) could be transformed into a narrower, Gaussian distribution (Figure 10b), and that in this process any non-Poissonian contributions to the energy resolution were eliminated (Figure 10c).

Journal of Physics: Conference Series **160** (2009) 012018 doi:10.1088/1742-6596/160/1/012018

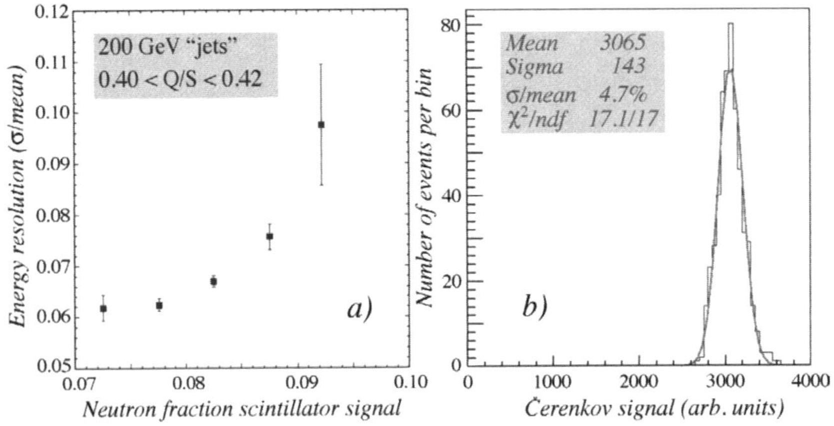

Figure 11. The energy resolution for 200 GeV "jets" with the same em shower fraction, as a function of the fractional neutron contribution to the scintillator signals (a). Čerenkov signal distribution for 200 GeV "jets" with $0.70 < Q/S < 0.75$ and $0.45 < f_n < 0.65$, together with the results of a Gaussian fit (b) [11].

We would like to point out that these improvements were achieved by using the information from only *one* type of fiber, *i.e.,* the time structure of the signals from the scintillating fibers. This in contrast to the improvements resulting from the measurements of $f_{\rm em}$, which require two different types of fiber. However, the (anti-)correlation between f_n and $f_{\rm em}$ is not perfect. The elimination of fluctuations in $f_{\rm em}$ takes care of the effects of the *average* contribution of invisible energy. However, for a given value of $f_{\rm em}$, the invisible energy fluctuates about this average, and a measurement of f_n may track these fluctuations. Therefore, measurements of f_n provide information *complementary* to that obtained from the Q/S signal ratio.

This is illustrated in Figure 11a, which shows that the energy resolution of a subsample of 200 GeV "jet" events with (approximately) the same em shower fraction is clearly affected by the relative contribution of neutrons to the signals. As f_n increases, so does the fractional width of the Čerenkov signal distribution. A larger $f_n$values means that the average invisible energy fraction is larger. This in turn implies that the event-to-event fluctuations in the invisible energy are larger, which translates into a worse energy resolution, even in signals to which the neutrons themselves do not contribute. Figure 11b illustrates the quality of the response function that was achieved with the combined information on the em shower fraction and the contribution of neutrons to the signals. This Čerenkov signal distribution concerns 200 GeV "jet" events with a Q/S value between 0.70 and 0.75 and a fractional neutron contribution to the scintillator signals between 0.045 and 0.065. The distribution is very well described by a Gaussian fit, with an energy resolution of 4.7%. The resolution was further reduced, to 4.4%, when the neutron fraction was narrowed down to 0.05 - 0.055. As a reminder, we mention that these results were achieved in a calorimeter with an instrumented mass of only about 1 ton. The mentioned resolutions are most likely strongly dominated by the contributions of (Čerenkov) photoelectron statistics (estimated at 2.5%) and side leakage. Eliminating these remaining factors, by means of more efficient detection of the Čerenkov signal in a sufficiently large detector may bring the resolution close to the theoretical limits.

3. Future plans

High-quality energy measurements will be an important tool for accelerator-based experiments at the TeV scale. There are no fundamental reasons why the four-vectors of all elementary particles could not be measured with a precision of 1% or better at these energies. However, reaching this goal is far from trivial, especially for the hadronic constituents of matter. Unfortunately, little or no guidance is provided by hadronic Monte Carlo shower simulations in this respect. In the past 30 years, all progress in this domain therefore has been achieved through dedicated R&D projects such as DREAM, and this is still the way to go today.

The dual-readout R&D project concentrates strongly on experimental tests of the validity of the principles on which improvement of hadronic calorimeter performance is based, and will continue to do so. We will continue to determine the factors that limit the performance of our prototypes and try to eliminate these factors, or least reduce their effects to insignificant proportions. Until this process has been completed, we will tend to ignore issues concerning the incorporation of this type of detector into a 4π experiment, and simulations in general.

So far, the dual-readout approach has been remarkably successful. It combines the advantages of compensating calorimetry with a reasonable amount of design flexibility. Since there is no limitation on the sampling fraction, the factors that limited the energy resolution of compensating calorimeters (SPACAL, ZEUS) to $\sim 30\%/\sqrt{E}$ can be eliminated, and the theoretical resolution limit of $\sim 15\%/\sqrt{E}$ seems to be within reach. Dual-readout detectors thus hold the promise of high-quality calorimetry for *all* types of particles, with an instrument that can be calibrated with electrons.

References

[1] Akchurin N *et al.* 1997, Nucl. Instr. and Meth. **A399**, 202.
[2] Wigmans R 2000, *Calorimetry, Energy Measurement in Particle Physics*, International Series of Monographs on Physics, Vol. 107, Oxford University Press (2000).
[3] Akchurin N *et al.* 2005, Nucl. Instr. and Meth. **A537**, 537.
[4] Akchurin N *et al.* 2008, Nucl. Instr. and Meth. **A584**, 304.
[5] Akchurin N *et al.* 2007, Nucl. Instr. and Meth. **A582**, 474.
[6] Gaudio G 2008, Contribution to these Proceedings.
[7] Voena C 2008, Contribution to these Proceedings.
[8] Akchurin N *et al.* 2008, Nucl. Instr. and Meth. **A584**, 273.
[9] Pinci D 2008, Contribution to these Proceedings.
[10] Drews G *et al.* 1990, Nucl. Instr. and Meth. **A290**, 335.
[11] Hauptman J 2008, Contribution to these Proceedings.
[12] Akchurin N *et al.* 2007, Nucl. Instr. and Meth. **A581**, 643.

XIII International Conference on Calorimetry in High Energy Physics (CALOR 2008) IOP Publishing
Journal of Physics: Conference Series **160** (2009) 012019 doi:10.1088/1742-6596/160/1/012019

Detection of K$^+$ Mesons in Segmented Electromagnetic Calorimeters

D.I. Glazier, T.C. Jude for the CrystalBall@MAMI and A2 collaborations.

School of Physics, University of Edinburgh, Kings' Buildings, Edinburgh, EH9 3JZ, UK

E-mail: dglazier@ph.ed.ac.uk

Abstract. The combination of the CrystalBall and TAPS electromagnetic calorimeters were installed in the MAMI A2 hall in 2003. Here they are able to detect the reaction products from photo-induced reactions in combination with the Glasgow photon tagger. In the last two years the MAMI facility was upgraded from 885 MeV to 1.5 GeV, the A2 photon tagger underwent a similar upgrade crossing the threshold for strangeness photoproduction. For the CrystalBall this created a new challenge, to identify K^+ mesons above the large background from other charged hadrons, in a situation where the detector setup does not benefit from a magnetic field to help separate particle species. These proceedings outline a novel technique which uses the decay products of the K^+ as a strangeness tag.

1. The CrystalBall at MAMI

Since 2003 the CrystalBall and TAPS electromagnetic calorimeters have been the central detectors facilitating experiments in the tagged photon hall (A2) at the MAMI electron accelerator facility. The close to complete acceptance for photons make it an ideal site for investigations of the photoproduction of neutral mesons. In addition, complimentary detector systems have successfully extended the range of particles that can be analysed in the calorimeters allowing a more diverse range of photoproduction experiments to be performed.

1.1. MAMI

The Mainzer Microtron (MAMI) [1, 2], is a well established electron accelerator providing an extremely high quality electron beam. An energy upgrade was completed in 2007 increasing the beam energy from 885 MeV (MAMI-B) to 1.5 GeV (MAMI-C). Crucially this crossed the threshold for associated strangeness production where kaons are produced in tandem with a hyperon (Λ or Σ).

1.2. The Tagged Photon Hall (A2)

In the A2 tagged photon hall the MAMI e^- beam is converted into a real photon beam via bremsstrahlung. The residual e^- from this process are momentumanalysed in a magnetic spectrometer, developed by the University of Glasgow. This tagger can be used to analyse up to $10^8 \gamma s^{-1}$ and photon energies between 5-92 % of the incident electron beam energy. The 352 scintillators in the focal plane provide a photon energy resolution of around 2 MeV with a 1.5 GeV beam. An additional "microscope" can be installed to improve the resolution to sub MeV for precision experiments. The photon beam can be circularly polarised up to 80 % through transfer from longitudinally polarised electrons, or, linearly polarised through coherent bremsstrahlung in an aligned diamond radiator also up to 80 %. In parallel with the MAMI-C upgrade the tagger underwent a successful upgrade of its own [3]. The magnetic field was increased to allow the bending of 1.5 GeV electrons while the focal plane detectors were replaced and electronics revamped.

The A2 collaboration have available a number of target systems including cryogenic Hydrogen, Deuterium and ^{3}He, a wide range of solid nuclear targets such as Carbon, Calcium and Lead and will soon have a number of polarised targets for experiments on Hydrogen, Deuterium and ^{3}He.

1.3. CrystalBall

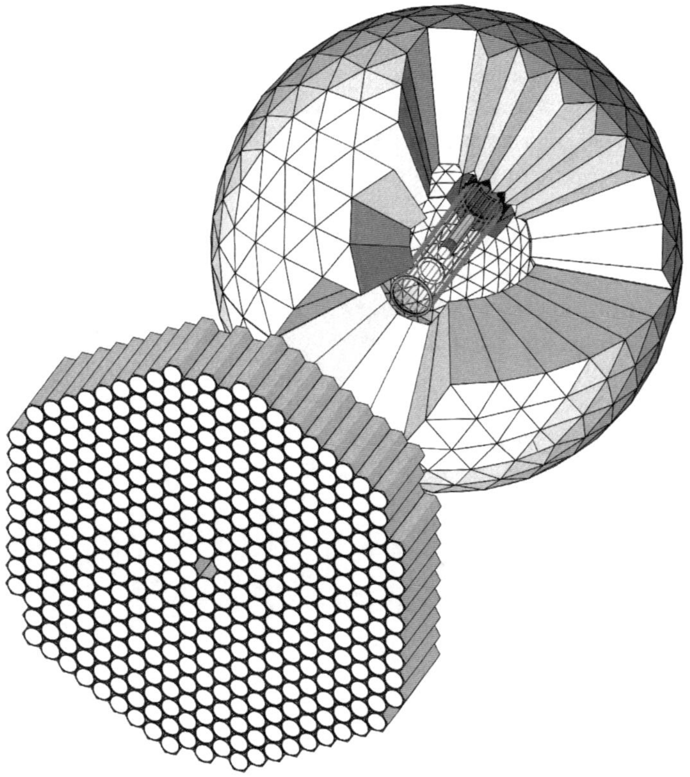

Figure 1. The CrystalBall and TAPS calorimeters as arranged in the A2 hall. Shown inside the ball are the PID detector and Hydrogen target.

The CrystalBall is a well traveled calorimeter [4]. Conceived and built at SLAC in the 1970s it has been the central detector system for experiments at SPEARS, DORIS and BNL prior to arriving at MAMI. It consists of a segmented array of 672 NaI crystals spherically arranged around the target. Each crystal is a 41 cm long truncated pyramid which is equivalent to 15.7 radiation lengths. This arrangement covers 93 % of phase space leaving two holes at the front and back for the incident beam to pass through. Photons traversing into the crystals will produce an electromagnetic shower contained typically in less than 13 neighbouring crystals, analysis of which provides the energy and direction of the initial photon. The granularity provides angular resolution of approximately 2° and the NaI crystals light output characteristics enable an energy

Journal of Physics: Conference Series **160** (2009) 012019 doi:10.1088/1742-6596/160/1/012019

resolution of $\sigma_E = 2.0\% E^{0.36}$. By considering multiphoton final states neutral mesons such as the π^0 and the η can be identified and their 4-momentum reconstructed.

Each crystal is fitted with a photomultiplier the output of which is split 3 ways, the first to the event trigger, the second to a multi-hit TDC and the third to a 40 MHz sampling ADC. As well as sampling the main signal the ADCs are recorded before and after and event trigger to allow subtraction of remnant light [5, 6]. The timing resolution consists of a 3 ns component from the trigger timing and a 2 ns component intrinsic to each crystal. For the technique outlined in this paper it is the relative time resolution of 2 ns which is of importance.

1.4. TAPS

For fixed target experiments such as those performed in the A2 hall the forward angular region is particularly important. The CrystalBall has a hole in this region covering approximately 0-20°, so to increase the acceptance in this crucial region of phase space the TAPS electromagnetic calorimeter has been installed. The TAPS detector has featured prominently in previous A2 experiments as well as spending time at ELSA, GSI, Ganil and CERN. Since the upgrade to MAMI-C, TAPS has been positioned 1.45 m upstream from the target and consists of 384 BaF_2 crystals. Each crystal is hexagonally shaped and 25 cm, or 15 radiation lengths long. The energy resolution $\sigma_E = 2.5\% E$ while the angular resolution is around 1°. The fast component of the BaF_2 crystals coupled to constant fraction discriminators facilitate a timing reolution of 1 ns.

1.5. Additional Detector Systems

As mentioned previously the CB and TAPS calorimeters are excellent γ detectors. However to increase their potential two additional detector systems have been added. Firstly 2 cylindrical MWPCs have been installed inside the CB. These provide accurate track information ($\sigma_\theta \simeq 1°$) as well as a vertex position.

Secondly a particle ID detector (PID) [8] is installed inside the wire chambers. This consists of 24x50cm long plastic scintillators orientated parallel to the beam direction. The scintillator thickness is 4 mm and it thus provides a ΔE signal which allows a separation of charged pions, protons and deuterons.

2. Outline of Physics Programme

Such a versatile detector system covering almost all accessible phase space coupled to an intense high resolution photon beam allows for an ambitious and diverse programme of experiments. The main themes of the MAMI programme include, precision determination of nucleon resonance properties, tests of fundamental symmetries and test of low energy theorems. A few experiments from the programme are described below.

The provision of polarised beam and targets provides an excellent basis for measurements of meson photoproduction. In addition the development of large acceptance recoil nucleon polarimetry will allow for the first time sufficient observables for 'model independent' measurement of the photoproduction amplitudes. This will aid our understanding of the nucleon excitation spectra as well as the underlying reaction mechanism.

The upgrade to 1.5 GeV beam energy also makes MAMI-C a competitive η factory. The large acceptance for photon detection allows precision measurements of the subsequent decay, in particular rare decays, providing stringent tests of fundamental symmetries. In addition decay to $3\pi^0$s provides sensitivity to the $\pi\pi$ scattering length.

The multiparticle final state detection, coupled with the high luminosity photon tagger, enables measurements where a very accurate and comprehensive determination is required to extract the important physics, for example radiative resonance decay. In particular, measurement of the $N\pi\gamma$ final state with polarised beam is sensitive to the magnetic moment of the $\Delta^+(1232)$, providing a route to measure this poorly known quantity.

There is also many physics opportunities that will be investigated with nuclear targets such as medium modifications of the ω meson and multinucleon knockout experiments. The first publication from the collaboration used a nuclear target and the ability of the CB to measure nuclear decay γs to investigate incoherent π^0 production on Carbon [8], which allows access to transition matter form factors and the $N\Delta$ interaction. A similar measurement of coherent π^0 production on Lead will allow an extraction of the neutron skin thickness which is a crucial quantity for nuclear physics and the physics of neutron stars.

3. K$^+$ Photoproduction

K$^+$ photoproduction is an important tool in baryon spectroscopy as it is naturally sensitive to resonances with strong strangeness decay branches. This may included a number of missing resonances predicted by the quark and other phenomenological models but as yet unseen in partial wave analysis of pion scattering data. A number of experiments have already been completed at ELSA, CLAS and GRAAL although some cross section results are inconsistent [9]. Another aspect of K$^+$ production that is particularly suited to study at the upgraded MAMI-C is threshold production. In this regime accurate data will provide a test of Low Energy theorems such as Chiral Perturbation Theory.

4. K$^+$ Detection

One of the crucial requirements of any detector system is being able to distinguish between different particle species. Commonly this is done through measurement of two variables from momentum, energy or velocity therefore determining the mass of the particle. For the CB the short distance between the target and crystals as well as the considerable length of the crystal preclude a sufficiently accurate extraction of velocity from time of flight. In practise the CB can only directly determine the kinetic energy of the particle through calorimetry and a combination of the CB and PID detector energy signals allows separation of protons and pions using the standard ΔE-E analysis.

This method however is not sufficient for separating K^+ due to a number of factors. Firstly detector resolution and hadronic interactions in the CB crystals make it impossible to separate the kaons from the several orders of magnitude greater incidence of protons in the CB. Secondly the K^+ will likely decay within the ADC sampling time making the energy deposited in the ball invalid for ΔE-E analysis. However this second factor can instead be used to our advantage to actually tag the presence of the K^+ as outlined in the following two sections.

4.1. K^+ Decays

The K^+ undergoes a Weak decay with a lifetime of 12 ns, significantly larger than the CB resolution of 2 ns, but well within the standard TDC sampling time used of 200 ns. For the majority of cases the K^+ will stop in the CB crystals after around 1ns and then decay at rest some time later. The two main decay channels are,

- $K^+ \rightarrow \mu^+\nu_\mu$ 63.51% (muonic)
- $K^+ \rightarrow \pi^0\pi^+$ 21.16% (pionic)

In the case of the muonic decay the neutrino will leave undetected while the muon will deposit the 152 MeV kinetic energy from the decay kinematics. This energy allows the muon to travel 25 cm through the CB and pass through around 5 crystals. The muon itself will also decay but its lifetime of 2 ms means its decay will generally not register in the CB TDCs. In the pionic decay all of the kinetic energy and mass of the π^0 can be contained inside the CB, whereas the π^+ will deposit just its kinetic energy of 109 MeV plus 3 MeV from its own muonic decay. This results in up to 360 MeV ($M_K - M_\pi$) being deposited in a large number of crystals around the original K^+ hit crystal.

4.2. K^+ Tagging Algorithm

The procedure developed for identifying a K^+ through its decay is as follows,

(i) Construct as large a cluster of crystal hits as possible through combining neighbouring hits.

(ii) Split this cluster into separate side clusters according to the timing of the hits. i.e. each side cluster contains hits with timing signals within 8 ns of each other.

(iii) The main cluster that occurs first in time is identified as the incident K^+, while the second side cluster is due to its decay.

(iv) Cut on the second side cluster energy ($> 100MeV$),average time of hits relative to the K^+ ($> 10ns$), length from the K^+ cluster to furthest crystal in decay cluster ($> 11cm$) and number of crystals (> 1) to select the K^+ muonic and pionic decay candidates.

(v) If the second side cluster passes the cuts check the first cluster passes a ΔE-E analysis.

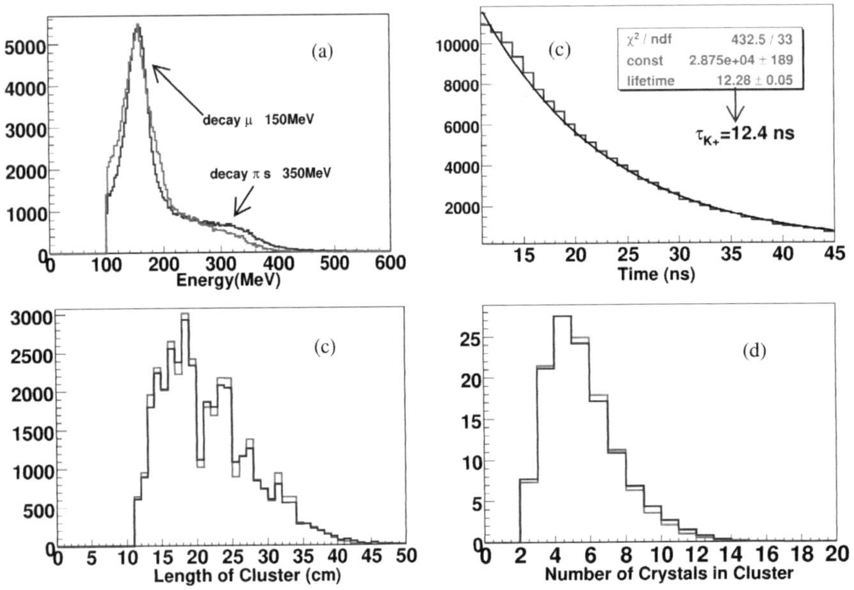

Figure 2. The plots show the energy, time, length and number of crystals in the second clusters. The second cluster cuts and ΔE-E analysis have been applied. The blue line shows the plots from real data samples, whereas the red line shows the result from a GEANT4 simulation of K^+ photoproduction events. The 150 MeV μ^+ peak is clearly seen in the energy spectra while the time spectra (real data only) shows the K^+ lifetime.

5. Data Analysis

This technique was developed using both real experimental data and a detailed GEANT4 [10] simulation of the A2 CB detector systems [11] using K^+ photoproduction events. The simulation shows the technique works well for both the CB and TAPS calorimeters, however as data has only been collected under trigger conditions that allow K^+ to be detected with the CB the following analysis will concentrate on only that detector. The plots in figure 2 show the comparison of the experimental data and simulation for the characteristics of the second side cluster after the algorithm of the previous section has been applied. The simulation gives a very good overall description of the data. The differences in the energy spectra, fig. 2a are attributable to acceptance effects arising because of the difference in angular distribution of K^+ in the phase

Journal of Physics: Conference Series **160** (2009) 012019 doi:10.1088/1742-6596/160/1/012019

space simulation compared to the data. The μ^+ energy of 150 MeV is particularly prominent in the energy spectra, with the pionic decay events forming the higher energy shoulder. The timing spectra, fig. 2b, show an exponential decay consistent with the K^+ lifetime, this is the same for the simulated data. The distribution of decay cluster length, fig. 2c, and multiplicity, fig. 2d, also show very good agreement with the simulation.

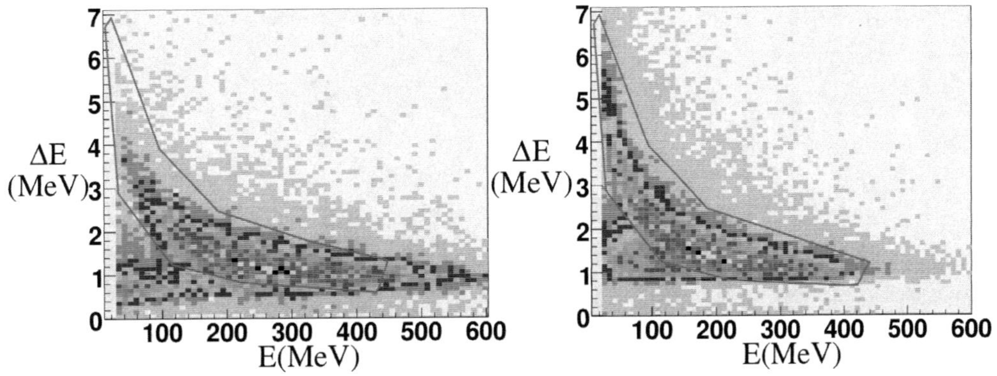

Figure 3. The real (left) and simulated (right) ΔE-E analysis. The red region shows the events selected as K^+

The additional constraint on the ΔE-E analysis is shown in figure 3 where the K^+ cluster energy is plotted versus the PID signal. A ridge consistent with that expected for K^+ is clearly visible. The red cut region is effective for suppressing any residual π^+ background. With all cuts applied the K^+ detection efficiency is around 10 % for K^+ up to 330 MeV.

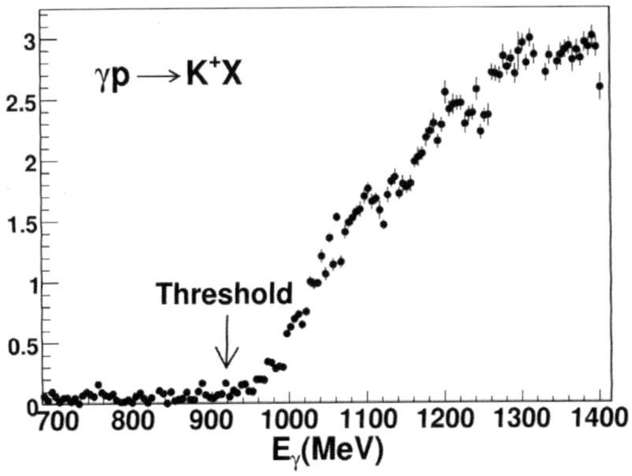

Figure 4. The raw K^+ excitation spectra taken from real experiment showing the reaction threshold of 920 MeV.

6. Some Preliminary Photoproduction Results

Already some data has been taken with the 1.5 GeV circularly polarised beam and the hydrogen target. Presented here are the very preliminary first results from analysis of this data with the new technique. Figure 4 shows the K^+ yield as a function of photon beam energy, uncorrected for acceptances. The main point to observe here is the spectrum shoots up at the $K^+\Lambda$ threshold of around 920 MeV photon energy, with little background below this supporting the validity of the technique for selecting the kaons. In addition, the missing mass of the beam, target and K^+ system has been reconstructed in fig.5. The two most probable reactions in our energy range are $K^+\Lambda$ and $K^+\Sigma^0$ production and the masses of these two particles are clearly seen in the missing mass spectra. The left panel shows the real data with peak width and position consistent with the simulation on the right, note that an equal number of Λ and Σ^0 were assumed in the simulation. The separation is already at a reasonable level, but the missing mass resolution will be improved further when the wire chambers are fully implemented giving more accurate track information for the K^+. An energy loss correction, not yet implemented, will further increase the separation. The two channels may also be separated through the radiative decay of the Σ^0 to a Λ by detecting the decay photon with energy of the order of 100 MeV, in coincidence in either the CB or TAPS.

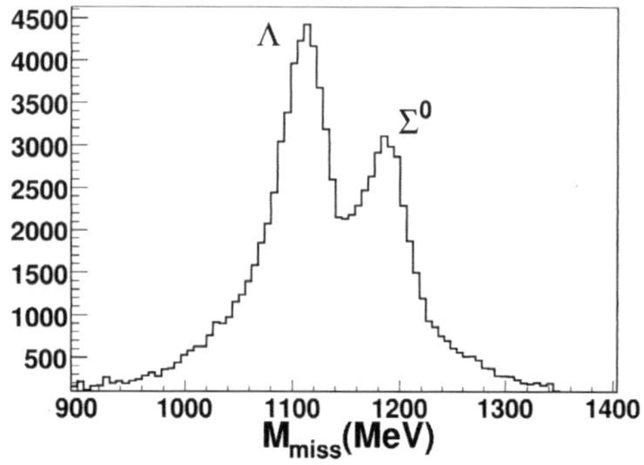

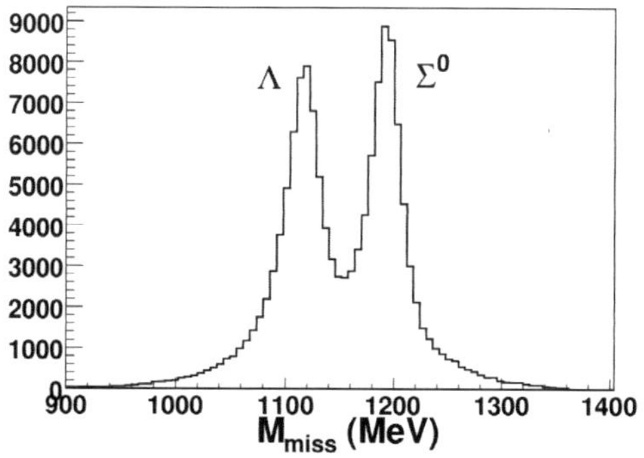

Figure 5. The missing mass of the beam+target-K^+ system, showing the masses of the Λ and Σ^0 associated with the K^+ production. The left hand panel is real data, the right simulation assuming an equal number of Λ and Σ^0 events.

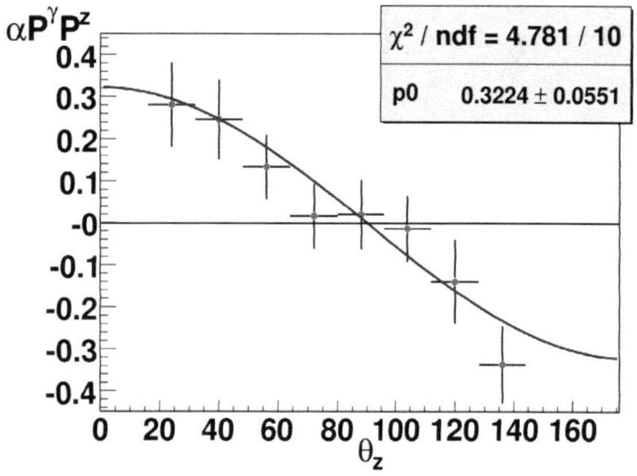

Figure 6. The asymmetry of the decay neutron with respect to the Λ momentum direction and beam helicity.

The angular distribution of the Λ weak decay is sensitive to its polarisation. With our detector we are able to detect either the decay to $\pi^0 n$ (36%) or $\pi^- p$ (64%). In these decays the polarisation P^i, with respect to a given axis i, is related to the angular distribution of the nucleon with respect to that axis via,

$$N(\theta_i) = N_0 \left(1 + \alpha P^i \cos\theta_i\right)$$

where $\alpha = 0.64$ is the Λ asymmetry parameter. Flipping the direction of the polarisation via the beam helicity allows us to form an asymmetry out of this distribution, we can thus deduce the polarisation transferred from the incident circularly polarised photon beam, P^γ, to the recoiling Λ,

$$A(\theta_i) = \frac{N^+ - N^-}{N^+ + N^-} = \alpha P^\gamma P^z \cos\theta_i$$

Figure 6 shows this asymmetry for an axis defined along the Λ momentum direction and the decay to $\pi^0 n$. This preliminary analysis already shows a clear asymmetry related to the polarisation transfer.

References

[1] H. Herminghaus, K.H. Kaiser, H. Euteneuer 1976 *Nucl. Instrum. Methods.* A **138** 1
[2] T. Walcher 1990 *Prog. Part. Nucl. Phys.* **24** 189
[3] J.C. McGeorge *et al.* 2008 *Upgrade of the Glasgow photon tagging spectrometer for Mainz MAMI-C Eur. Phys. J.* A i2007-10606-0
[4] A. Starostin *et al.* 2001 *Phys. Rev.* C **64** 055205
[5] D. Krambrich, Ph.D. thesis, Universitat Mainz, 2006
[6] D.P. Watts 2004 *Calorimetry in Particle Physics: Proceddings of the 11th Int. Conf., Perugia, Italy* (Singapore, World Scientific) edited by C. Cecchi *et al.* p560
[7] R. Novotny 1991 *IEEE Trans. Nucl. Science* **38** 379
[8] C.M. Tarbert *et al.* 2008 *Phys. Rev. Letts.* **100** 132301
[9] J.W.C. McNabb *et al.* 2004 *Phys. Rev.* C **69** 042201
[10] S. Agostinelli *et al.* 2003 *Nucl. Inst. Meths.* A **506** 250-303
[11] http::/www.ph.ed.ac.uk/nuclear/G4

Journal of Physics: Conference Series **160** (2009) 012020 doi:10.1088/1742-6596/160/1/012020

The NA62 rare Kaon decay experiment Photon Veto System

V. Palladino for the NA62 Photon Veto Working Group. *

July 15, 2008

Abstract

The NA62 experiment at CERN SPS has been proposed with the purpose to measure the Branching Ratio for the ultrarare decay $K^+ \to \pi^+ \nu\bar{\nu}$. The photon veto system will have to provide a rejection better than 10^{-8} for π^0 decays. The system is composed by several detectors. The larger ones constitute the Large Angle Veto (LAV) system, and will cover an angular region up to 50 mrad with respect to the incident beam. In order to optimize the cost/performance ratio we have tested three different technologies for the LAV system using a low energy electron beam at the LNF Beam Test Facility. A lead-scintillating fibers prototype has been built on purpose; a lead-scintillating tiles prototype was on loan by former CKM collaboration; a set of lead glass blocks from the OPAL barrel calorimeter have also been tested. We present preliminary results on detector performances and compare the three solutions.

1 The NA62 experiment

The Branching Ratio (BR) $K^+ \to \pi^+ \nu\bar{\nu}$ can be related to the CKM element V_{td} (the less well known one), the precise theoretical estimation in the Standard Model (SM) and SUper SYmmetry (SUSY) will allow us to have a probe of the flavour sector or the evidence of physics beyond the Standard Model, if deviation from the predicted SM value will be observed.

The SM prediction is $(8.22 \pm 0.84) \times 10^{-11}$. Until now only three events have been collected and the experimental value is $1.47^{+1.30}_{-0.89} \times 10^{-10}$[1]. The NA62 experiment (a.k.a. P326) has been proposed to detect 100 events of $K^+ \to \pi^+ \nu\bar{\nu}$ with a S/B ratio of $10:1$ [2].

*A. Antonelli, E. Capitolo, P.S. Cooper, L. Iannotti, M. Moulson, M. Raggi and T. Spadaro, from Laboratori Nazionali di Frascati dell'INFN, Frascati, Italy; P.S. Cooper is a visitor fron Fermi National Laboratory, Batavia IL, USA; F. Ambrosino, M. Napolitano, V. Palladino, A. Romano and G. Saracino, from Dipartimento di Scienze Fisiche and Sezione INFN, Napoli, Italy; R. Fantechi, G. Lamanna, I. Mannelli and S. Venditti, from Dipartimento di Fisica and Sezione INFN, Pisa, Italy; E. Leonardi, F. Perfetto, M. Serra and P. Valente, from Dipartimento di Fisica dell'Universitá "la Sapienza" and Sezione INFN, Roma, Italy; Speaker e–mail: vito.palladino@na.infn.it.

The experimental layout is proposed in 1.

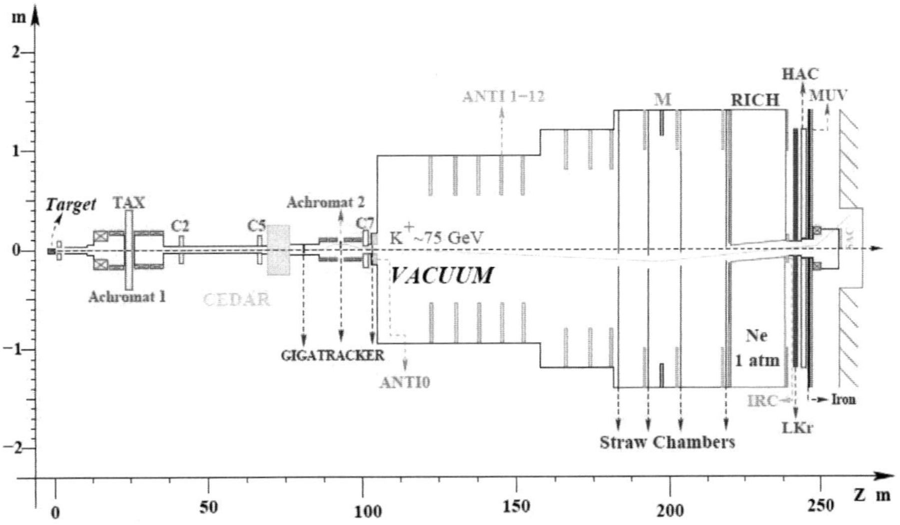

Figure 1: NA62 experimental setup.

The experiment will make use of a 75 GeV unseparate positive secondary beam. The total beam rate is 800 MHz providing $\tilde{5}0$ MHz of K^+'s. The fiducial decay volume (in vacuum) begins 102 m from the production target position. The K^+ decay rate in the 120 m long fiducial decay will be about 10 MHz.

The success of the experiment depends crucially in obtaining the required level of background rejection. To provide this, many different detectors have been proposed for both kinematic based and particle identification based rejection.

The decay fiducial volume is surrounded by 12 LAV (Large Angle Veto) stations ring shaped, that cover the angular range from 8.5 to 50 mrad. The last 35 m of the decay region host a dipole spectrometer with four straw-traker stations operated in vacuum. The spectrometer, together with the RICH system and the dedicated muon detector downstreamm, will provide the required level of π/μ separation.

The NA48 Liquid Krypton calorimeter (LKr) and some other small detectors are used to veto small angle (high energy) photons.

The apparatus must be able to reject background from, e.g. $K^+ \rightarrow \pi^+\pi^0$ decays at level 10^{-12}. Kinematical cuts on the K^+ and π^+ tracks will provide a factor 10^{-4} and ensure 40 GeV of electromagnetic energy in the photon veto; this energy must be detected with an inefficiency better than 10^-8. The LAV system will be able to detect photons as low as 200 MeV with inefficiency better than 10^-4. In addiction, the LAV system must have a good energy and time resolution in order to define a precise energy threshold and to use the system in the trigger logic. In order to minimize the amount of material crossed by photons

Journal of Physics: Conference Series **160** (2009) 012020 doi:10.1088/1742-6596/160/1/012020

produced in π^0 decays and reach such a high efficiency, the LAV system will be operated in vacuum.

2 LAV system

The LAV system will be made by 12 ring shaped subdetectors. The detectors at the first five stations have inner radii of 60 cm and outer radii of 96 cm. The other stations have inner and outer radii which match the shape of the vacuum chamber. The largest covers the range from 90 cm → 140 cm.

The LAV requirements are:

- *Photon Inefficiency* at 10^-4 level or better in the energy range 200 MeV → 1 GeV ;
- *Mechanical design* optimized in order not to spoil the vacuum inside the decay volume ;
- *Time Resolution* better than 1 ns in order to use the system in the trigger logic ;
- *Energy Resolution* at level 10% or better for 1 GeV photons , in order to define a precise energy threshold.

The overall front surface is about 30 m^2. We now describe the three protoypes we have tested at the Frascati Beam Test Facility (BTF).

2.1 Scintilating Fiber Protoype

This prototype is based on scintillating fibers [4]: the structure consists of 1 mm diameter scintillating fibers sandwiched between 0.5 mm thick lead foils. The fibers are arranged orthogonal to the incident particle and are read out at both ends. Two Ushaped modules form a station. We have built one U shaped module from scratch. The prototype, built in Frascati, has the inner radius of 60 cm and the outer of 72.5 cm. These dimensions are 35% of the dimensions for an upstream module, this thikness was chosen in order to reduce the prototype costs; it should be sufficient for the trasverse shower containment of low-energy electrons incident half-way between inner and outer edge along the radial direction. In order to reduce the costs the fiber density is not uniform in the beam direction, this of course slightly spoils the energy resolution which however is not a crucial feature for a veto system. The read out is provided by a PMTs matrix coupled with the fibers by Winston-cones light concentrators.

2.2 Scintillating Tiles Protoype

The second possible technology choice is based on a sandwich of scintillating tiles (5 mm thick) and lead foils (1 mm thick) with wavelength shifter (WLS) fibers read out [?]. An assembly of wedge-shaped modules forms the veto station. A partial prototype had been built by CKM collaboration at FermiLab. The prototype was tested with 1.2 GeV electrons at Jefferson Lab, the tests confirmed that the detector inefficiency was at most 3×10^{-6}. We have obtained CKM prototype on loan for further test at lower energies, and comparison. The structure is fully described in the reference [?].

2.3 Lead glass blocks

The last possible choice is an original reuse of the leadglass blocks from the OPAL barrel calorimeter. Blocks will be arranged in five staggered sub-rings, this disposition is able to ensure at least three layers (corresponding to about 20 radiation lengths)involved in the particle detection.

3 BTF setup and test

The Frascati Beam Test Facility (BTF) [5] allows tests with low multiplicity beams of electrons with energies ranging from 100 MeV to 750 MeV. The beam pulse width is 10 ns with a repetition rate of 50 Hz. The BTF beam is generated using electrons and positrons coming from DAΦNE LINAC, which after impacting on a target are converted in electrons in a wide energy spectrum. The energy selection is then provided using a 45° dipole. The target with double twin slit reduces the beam multiplicity from 10^9 to < 1. The beam direction may be tuned using another 45° dipole and two quadrupoles.

4 The BTF tagging System

The selection used to tag single electrons is shown in Figure 2. The trigger counters are all made with 10 mm thick plastic scintillator:

F1 is a paddle of area $60 \times 85 mm^2$, it is positioned few centimeters from the end of the vacuum beam pipe,

H1 is a paddle of area $200 \times 130 mm^2$ with a 14 mm diameter hole in the center, it is postioned 10 mm downstream of F1,

H2 is a paddle of area $330 \times 100 mm^2$ with a 14mm diameter hole in the center, it is positioned 90 cm downstream of H1,

F2 is a paddle of area $60 \times 85 mm^2$, it is positioned as little as 10 mm upsteam of the prototype to be tested.

The trigger logic for one electron events selection can be summarized as $F1\ \overline{H1}\ \overline{H2}\ F2$, where F1 anf F2 refer to the presence of charged signal on the paddles consistent with the passage of a single electron, and $\overline{H1}$ and $\overline{H2}$ refer to no signal in the hole paddles.

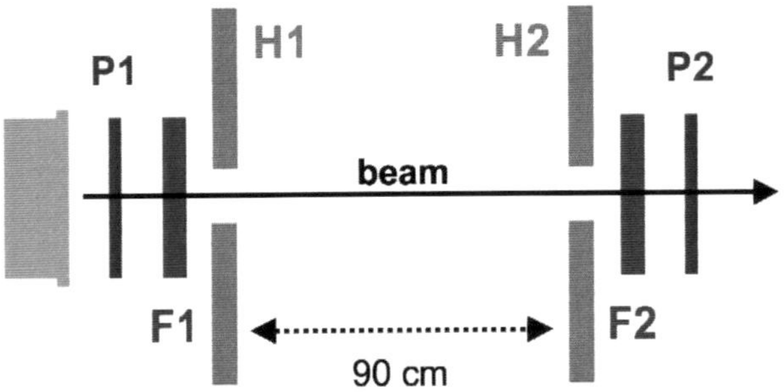

Figure 2: A scheme of BTF tagging system.

The thikness of the paddles has been chosen to allow efficient identification of events with exactly one electron in the paddle within a 10 ns linac pulse. The hole dimension has been chosen in order to help rejecting events with straight beam particles present.

The mistag probalility has been monitored and correspond to a false rate of $< 2 \times 10^-6$ at 90% CL.

5 Technology choice

During 25-day run in June-July 2007 we testes all three technologies [4]described above. For the lead glass block test we used four blocks positioned orthogonally wrt the beam axis, hit approximately in their center by the beam.

The inefficiency results are shown in the Table 5.

Beam Energy[MeV]	Tagged Events	Event Below 50MeV	Inefficiency
Scintillating Fiber			
203	68 829	5	$7.3^{+4.1}_{-3.3} \times 10^{-5}$
350	207 385	3	$1.4^{+1.1}_{-0.9} \times 10^{-5}$
483	371 633	1	$2.7^{+4.7}_{-1.7} \times 10^{-6}$
Scintillating Tiles - Preliminary			
203	651 65	2	$3.1^{+3.5}_{-1.9} \times 10^{-5}$
350	221 162	3	$1.4^{+1.0}_{-0.9} \times 10^{-5}$
483	192 412	1	$5.2^{+9.1}_{-3.3} \times 10^{-6}$
Leadglass - Preliminary			
203	25 069	3	$1.2^{+0.9}_{-0.8} \times 10^{-4}$
483	91 511	1	$1.1^{+1.9}_{-0.7} \times 10^{-5}$

We can see that the effiency is above the requirements for all solutions. The availability of a large amount of OPAL leadglass blocks has driven our choice in favour of this technology.

6 Cosmic test and LG characterization

After the definitive choice of the LGs as the baseline technology, we started a systematic characterization of them. With this aim in Naples we arranged a test station composed by a darkbox and a tower tracker. The procedure consists in measuring, for each LG, the gain and the Photoelectron Yield (PeY).

6.1 PMT Gain

PMT gain is obtained for each High Voltage (HV) applied to the PMT, by a linear fit of the observed variance of the response versus the response itself, when the block is exposed to different intensity LED pulses inside a dark box. The values are corrected for the contribution of the gain fluctuations. The gain curve is then obtained by fitting the measured gain versus the HV value, and the PMTs are all equalized to a HV such to get a gain $G = 10^6$.

6.2 Photoelectron yield

Photoelectron Yield (PeY) is the number of photoelectrons produced per 1 MeV deposited energy by muons travelling trasversally in a leadglass block. The measurement of PeY has been made in Naples with a tracker tower that enable us to track cosmic muons and select those traversing the blocks with the desired inclination. The PeY is then defined as :

$$PEY = \frac{Q}{G\,e\,77MeV}$$

where Q is the collected charge, G is the PMT gain, e is the elementary charge and 77 MeV is the mean energy released by a muon traversing a 10 cm block. The observed PeY for 40 blocks chosen at random among the available OPAL blocks, is about 0.4 pe/MeV, and the distributions shows a spread in PeY values at level 10

7 Lead glass prototype studies

After the single block characterization we have assembled in Naples, using 25 blocks, a small prototype mounted in a linear shape 3. The geometrical staggered disposition is the same of the final one, in order to study edge effects and measure a more realistic inefficiency value.

7.1 Data set

The data referred to in this section were collected during a two week test at the BTF in February 2008. Data were taken at various energies and hitting the detector in different geometrical configurations. In this paper we report a preliminary analysis of part of the data collected at energy 471 MeV using some different beam impact positions. We will focus our attention on comparing beam

Figure 3: The linear prototype used for February 2008 test.

impacting on the center of one block (Central Beam), and beam impacting on the lateral (longer) edge of one block (Edge Beam).

7.2 Intercalibration and Energy Calibration

Response of each block is characterized by its PeY as measured with cosmic rays.One of them was chosen as reference (PeY_{ref}) and the intercalibration costants were found dividing each PeY to PeY_{ref}. In order to set an absolute energy scale we defined a clustering algorithm. Given the energy of incident electrons and the intercalibration constants we can measure the energy scale as the ratio between the number of ADC counts and the electrons energy. Clustering has been made using a variable threshold channel by channel, optimized on the background level observed in each cell of the prototype. The background was dominated by soft photons in the BTF hall; the integral of this background on all 25 cells was monitored during data taking and was typically at level of about 7 MeV. The energy scale on runs with the same energy was stable at level of 3%, this effect is still to be investigated; it could be related to hysteresis effects in the BTF magnets used to drive the beam to the experimental hall.

7.3 Cluster shape and Energy Resolution

For central beam events the average number of cells for a cluster is 5 while it gets to 6 for edge beam events. In the longitudinal (along the beam) direction, on average a central beam cluster hits 3 planes of the prototype, and less than 10% of events hits the last plane, while a edge beam cluster hits the fifth plane in about 30% of the cases. In both cases 80% of the energy is released in the first plane. The energy resolution observed is about 9.5 % at 471 MeV for central beam events and about 11 % for edge beam events.

7.4 Time Resolution

The time response, measured with a 12 bit TDC, has been corrected for each block for the time walk by means of the appropriate slewing corrections, evaluated as described in [4]. For a block in the most upstream plane, which is hit always in the same position by the beam, the typical time resolution is at level 200 ps. Resolution is instead of about 1 ns for a block in the second plane, dominated thus by the distribution of the impact points on the block. We also defined an energy weighted time average for a cluster; its resolution is about 560 ps.

7.5 Inefficiency

A preliminary analysis of part of our data set gives the following results for the LG prototype:

Beam type (E=471 MeV)	Tagged Events	Event Below 50MeV	Inefficiency
Central	47 252	2	$4.2^{+4.8}_{-2.6} \times 10^{-5}$
Edge	9 711	1	$1.0^{+1.8}_{-0.7} \times 10^{-4}$

8 Conclusions and future plans

An intensive R&D program has been carried out to assess the technology and evaluate the performances of a Large Angle Veto system for the NA62 experiment. The lead glass blocks formerly used for the OPAL barrel calorimeter are our baseline choice. A prototype with a geometry very similar to the one foreseen for the final ring shaped detectors has been tested and its performances meet the requirements for the LAV system. Further tests will be made, in vacuum, on a secondary beam extracted from the SPS in Octber 2008. The desing of a complete ring is almost completed; construction of the first complete ring will likely start in year 2009.

References

[1] E949 Collaboration, V. Anisimovsky *et* al., CERN, Geneva, Tech. Resp. CERN/SPSC 2005-013, 2005.

[2] G. Anelli *et al.*, CERN, Geneva, Tech. Resp. CERN/SPSC 2005-013, 2005.

[3] E. Ramberg, P. Cooper and R. Tschirhart, *IEEE Trans. Nucl. Sci.*, vol. 51, p2201, 2004.

[4] M. Moulson *et al.*, Presented at 2007 IEEE Nuclear Science Symposium and Medical Imaging Conference (NSS MIC 2007), Honolulu, Hawaii, 28 Oct - 3 Nov 2007.

[5] G. Mazzitelli *et al.*, *Nucl. Instrum. Meth. A*, vol. 515, p. 524, 2003.

Journal of Physics: Conference Series **160** (2009) 012021 doi:10.1088/1742-6596/160/1/012021

Test beam study of the PANDA shashlyk calorimeter prototype

D A Morozov, S K Chernichenko, A A Derevschikov, V Y Kharlov, Y A Matulenko, V V Mochalov, A V Ryazantsev, P A Semenov, A P Soldatov, O P Yuschenko and A N Vasiliev

Institute for High Energy Physics, Pobeda Street 1, Protvino, 142281, Russia

E-mail: Dmitry.Morozov@ihep.ru

Abstract. The Shashlyk calorimeter prototype for the PANDA experiment has been constructed at IHEP and experimentally tested using the 1-19 GeV electron beam with high precision momentum tagging at the IHEP accelerator. Results of the first measurements for the fine-segmented Shashlyk calorimeter prototype in the wide energy range up to 19 GeV are presented. Fair energy and position resolutions having been obtained are in a good agreement with the Monte-Carlo simulations. Detection inefficiency due to holes for straight light fibers has turned out to be negligible for PANDA. Plans for the near term future including approaching test beam study of π^0 reconstruction capability of shashlyk prototype are also discussed.

1. Introduction

The physics program of the PANDA project at the international FAIR facility at GSI (Germany) is based on a state-of-the-art universal detector for strong interaction studies at high intensity cooled antiproton beam with an energy up to 15 GeV. This program relies heavily on the capability to measure photons with excellent energy and position resolution. For this purpose PANDA has proposed to employ electromagnetic calorimeters using two different technologies: a compact calorimeter around the target based on lead tungstate crystals and a fine-segmented Shashlyk-type calorimeter in the forward region (Figure 1). PANDA physics program requires 4π angular coverage of photon detection, which is achieved mainly by the target spectrometer EMC. However, to measure everything in the forward region with angles less than 5° in vertical plane and 10° in horizontal plane, forward spectrometer EMC is used. The target spectrometer EMC is going to be built from the lead tungstate scintillating crystals (PWO). It has been already realized in the CMS, ALICE and PRIMEX electromagnetic calorimeters [1, 2, 3]. The fine-segmented shashlyk type calorimeter proposed for KOPIO experiment showed a good performance in the energy range up to 1 GeV [4] - energy resolution was about $3\%/\sqrt{E}$. Since PANDA requires much wider energy region calorimeter, test beam measurements of the shashlyk calorimeter prototype were carried out at IHEP with the use of electron beam in the energy region from 1 to 19 GeV. The results of energy and position resolution measurements as well as the Monte-Carlo simulation predictions are presented.

In this report we describe a fine-sampling electromagnetic calorimeter prototype with lead absorber, where the thickness of the absorber is significantly smaller than radiation length of lead. Such a small thickness of the absorber layers results in a small interaction probability of

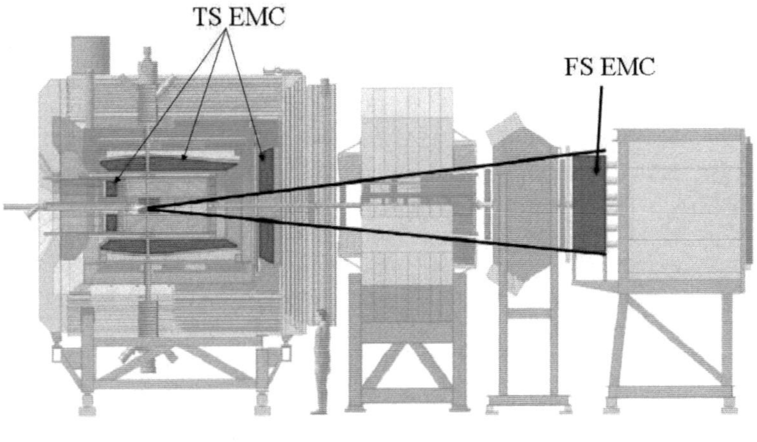

Figure 1. Panda detector schematic view. Target spectrometer calorimeter (TS EMC) and forward spectrometer calorimeter (FS EMC) are shown in red.

the secondary shower particles. A design of this prototype is close to the KOPIO prototype including the same lateral size of cells. The results of miscellaneous studies of parameters of the prototype in the energy range from 1 to 19 GeV are presented in this report.

2. Shashlyk modules structure

The PANDA forward spectrometer EMC has some specific requirements. Since it is situated outside the target region, seven meters from the interaction point beside the dipole magnet, geometrical size constrains are not so strict as well as position (angular) resolution requirements (see Figure 1). The EMC should cover $\sim 4m^2$ and measure mainly relatively high energy particles in the forward direction. However, the EMC has to keep a good energy resolution in the range of up to 10 GeV, a rather low detection threshold of 10-20 MeV and a fast signal to prevent pile-up effect. One of the most interesting solutions which satisfies all the above requirements and has another advantage of low cost is so called shashlyk calorimeter with fine segmentation.

The modules were assembled from 380 alternating layers of lead and scintillator plates. Lead plates were doped by 3% of antimony to improve their rigidity. Scintillator plates were made of polystyrene doped by 1.5% of paraterphenyle. Scintillator was manufactured at IHEP scintillator workshop by the molding technology (see Figure 2). The physical properties of the modules are presented in Table 1.

Scintillation light was collected by the wave-length shifting (WLS) optical fibers BCF-91A with a diameter of 1.2 mm. Each fiber penetrated through the module along its longitudinal axis twice, forming a loop in front of the module. The radius of the loop was 28 mm. The 72 such looped fibers penetrated through the modules in the longitudinal direction with the spacing of 9.3 mm forming the grid of 12×12 fibers per module. All 144 fiber ends were assembled into a bundle of a diameter about 10 mm, glued, cut and polished in order to deliver the scintillating light to the photodetector attached at the downstream end of the module. A Hamamatsu photomultiplier R5800 has been chosen as a photodetector for the prototype. The diameter of the photocathode was 25.4 mm, the number of dynodes was 10, the applied high voltage was about 1100 V. Each photomultiplier was monitored by the LED light carried to the photocathodes by a clear polystyrene fiber. The purpose of the LED monitoring system was to check whether the photodetector was working, and to study long-term variations of the LED amplitude.

Journal of Physics: Conference Series **160** (2009) 012021 doi:10.1088/1742-6596/160/1/012021

Table 1. Shashlyk module properties.

lead plate thickness	0.275 mm
scintillator plate thickness	1.5 mm
number of layers	380
effective radiation length, X_0	34 mm
total radiation length	$20X_0$
effective Moliere radius, R_M	59 mm
module size	$110\times110\times675$ mm^3
module weight	18 kg

Figure 2. Shashlyk modules production at IHEP scintillator workshop.

Figure 3. 3×3 matrix of shashlyk modules on movable platform.

3. Prototype test beam setup

The prototype consisted of 9 fine-sampling modules. The modules were assembled into 3×3 matrix installed on the x, y-moving table (see Figure 3). The beam channel 2B of the U-70 accelerator was used to study the performance of the calorimeter prototype. The secondary beam of negatively charged particles of momenta from 1 to 19 GeV/c contained more than 70% of electrons with some fraction of muons, and hadrons (mainly π^- and K^-). Electron beams available at IHEP have momentum spread of several percent. In order to test high precision calorimeters a dedicated beam tagging setup was constructed at the beam output building [5]. The tagging system is illustrated in Figure 4 and was based on the dipole magnet M and four sets of 2-coordinate drift chambers DC1 to DC4, the bending angle of the magnet was 55 mrad. A trigger of the experimental setup used the coincidence of 4 scintillator counters - S1, S2 and S3 installed upstream before the first drift chamber DC1, and S4 installed after the last drift chamber DC4 in front of the calorimeter prototype (ECAL). This setup was used to measure beam particle momentum with a precision of 0.13% at high energies where multiple scattering was not dominating.

An amplitude information was measured by 15-bit charge sensitive ADC modules LRS2285 over 150 ns gate with a sensitivity of 30 pC/count. To read out a time information from the drift chamber stations, the TDC LRS337 CAMAC modules have been used. Remotely controlled the x, y-moving table positioned the prototype across the beam with a precision of 0.4 mm. Data acquisition system included two crates with ADC and TDC modules as well as control modules to synchronize a read out process. VME crate with CAMAC parallel branch driver and PCI-VME bridge linked all the electronics into the complete system. Detailed description of the data

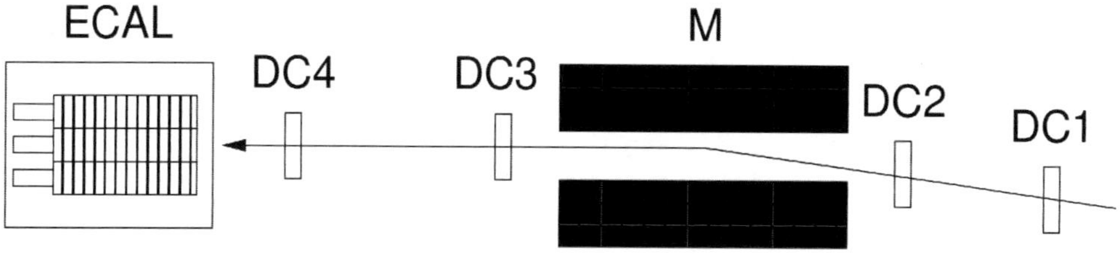

Figure 4. Beam tagging system for the shashlyk prototype studies.

acquisition system and front-end electronics can be found in [6].

4. Monte Carlo simulations

The relevant simulation tools were developed. These tools, at the first stage, are intended mainly for cross-check of experimental results as well as for tuning of the reconstruction algorithms. Proving consistency of Monte Carlo and real data, we plan to use these tools for further optimization of module design and reconstruction algorithms to provide better performance of the photons and π^0s reconstruction. Simulation studies were performed with GEANT3 [7] as a Monte Carlo engine with detailed description of materials and module geometry. The developing shower produces light which originates from two different sources:

- scintillation in plastic plates due to continuous energy losses when charged particles pass the active calorimeter material,
- Cherenkov radiation when charged particles pass the WLS fibers.

The simplified technique results in counting energy deposition in the active material (with some corrections to take into account the light attenuation in the fibers) and ignoring Cherenkov radiation inside the fibers. This method is very fast while can not reproduce all the details of the calorimeter response such as non-uniformity due to fibers and cell borders. For these studies, the detailed light propagation was applied taking into account optical properties of materials, reflections at plate borders, light capture by fibers with account of the cladding and the Cherenkov light production and propagation inside the fibers. It was assumed that attenuation length in the scintillator is 70 cm and in the fiber is 400 cm, scintillator refraction index is 1.59, total inner reflection efficiency at large scintillator faces is 0.97 and reflection of diffusion type was assumed at side scintillator faces with the same probability. The mean deposited energy for one optical photon production in the scintillator was assumed to be 100 eV and the Cherenkov photons were generated by GEANT.

5. Calibration

The modules were calibrated by a 19-GeV/c beam. Each module was exposed to the beam using the x, y-moving table. The energy spectrum from one module (Figure 5) shows a sharp peak at 19 GeV corresponding to the energy deposited by electrons. Another peak at low energies is due to minimum ionization loss of muons and charged hadrons. A wide distribution of the deposited energy between the two peaks is due to hadron showers. The best relative calibration coefficients were found by equalizing minimum ionizing particle (MIP) signals, while the absolute calibration was obtained by setting the total measured energy in the 3×3 matrix to 19 GeV. Events where only one module has an energy above the threshold of 100 MeV were selected for the MIP calibration. The energy distribution around the MIP peak (Figure 6) has two contributions, one is caused by the Landau distribution of the ionization energy loss, and another one is due to the finite energy resolution of the calorimeter at low energy. The MIP peak was fitted by the

Journal of Physics: Conference Series **160** (2009) 012021 doi:10.1088/1742-6596/160/1/012021

Landau distribution, and the most probable value of the fitting function served for the relative calibration.

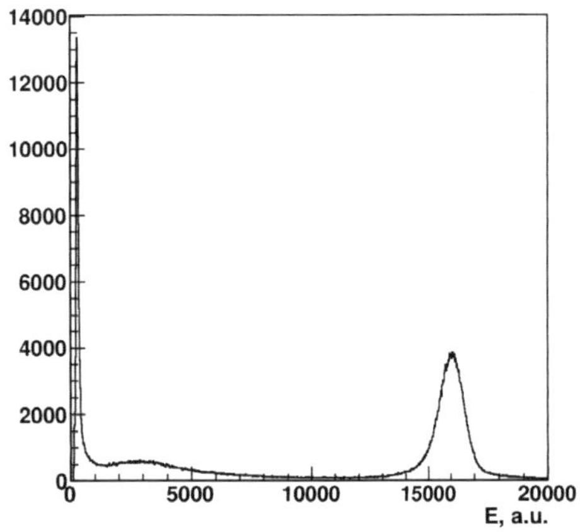

Figure 5. Energy deposited by the 19-GeV/c beam in one module.

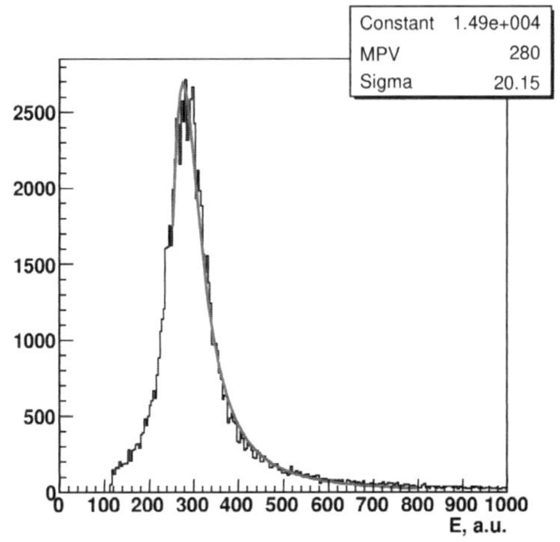

Figure 6. Energy spectrum near MIP peak position. Fitting function is the Landau distribution.

6. Energy and position resolution results

After dedicated calibration runs, where each module was exposed to the 19-GeV/c beam, the ECAL prototype was fixed so that the beam hit the central module, and it was exposed to beams at momenta of 1, 2, 3.5, 5, 7, 10, 14 and 19 GeV/c. For each beam momentum, the magnetic field in the spectrometric magnet M was adjusted to provide the same bending angle of the beam. The momentum of the beam particle p was measured by the magnetic spectrometer, and the energy E measured in the calorimeter prototype is linearly correlated with the momentum p, as illustrated by Figure 7. Therefore, in order to obtain a true energy resolution, the measured energy should be corrected by the beam momentum, or the energy resolution can be represented by the width of the distribution of the E/p ratio (Figure 8). The energy resolution is obtained from the Gaussian fit of the right peak around $E/p = 1$. The energy resolution σ_E/E measured by electrons at the energies from 1 to 19 GeV are shown in Figure 9. Black stars represent the experimentally measured points, and the solid lines (green for experimental and red for MC values) are results of the fit by the function

$$\frac{\sigma_E}{E} = \frac{a}{E} \oplus \frac{b}{\sqrt{E}} \oplus c, \tag{1}$$

where E is in GeV. The obtained parameters are summarized in Table 2.

There is a good agreement of our experimental results with the MC data and with the previous studies at lower energies [4], where a stochastic term was 2.9%.

A position resolution has been measured by comparison of exact impact coordinate of the beam particle, measured by the last drift chamber DC4, and the center-of-gravity of the electromagnetic shower developed in the calorimeter prototype. Figure 10 shows the dependence of the measured coordinate on the true one. The spatial resolution in the middle of the module is shown in Figure 12. Again, the black stars stand for experimental data points, the curves - for fit by the function:

$$\sigma_x = \frac{a}{\sqrt{E}} \oplus b[mm], \tag{2}$$

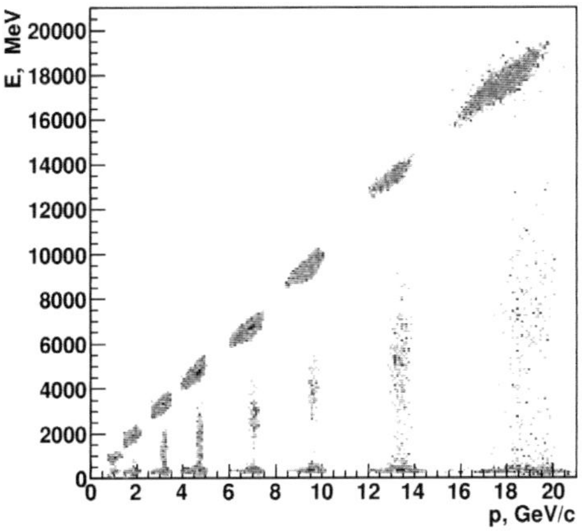

Figure 7. Correlation between the measured energy and the beam momentum.

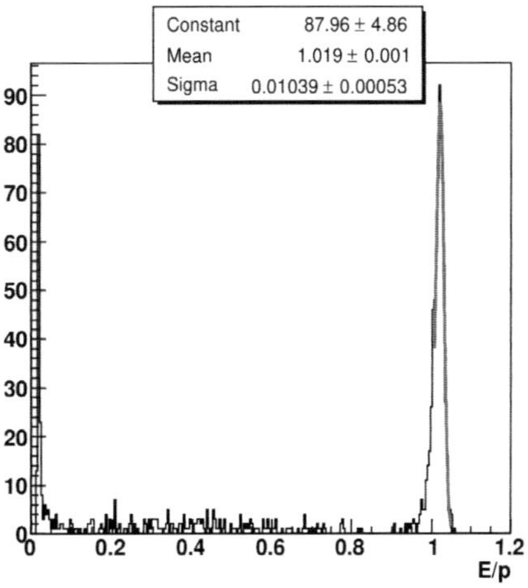

Figure 8. Ratio of the energy E to the momentum p at 19 GeV/c.

Table 2. Parameters describing energy resolution of the prototype [in %].

Experiment data fit	MC data fit
$a = 3.5 \pm 0.3$	$a = 3.3 \pm 0.1$
$b = 2.8 \pm 0.2$	$b = 3.1 \pm 0.1$
$c = 1.30 \pm 0.04$	$c = 1.23 \pm 0.02$

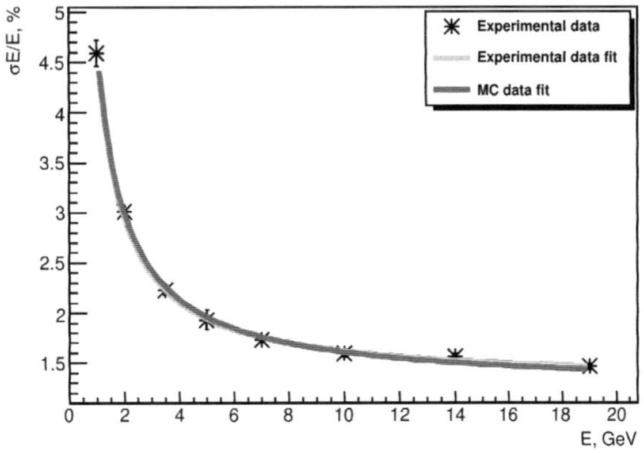

Figure 9. Energy resolution dependence on energy.

where E is in GeV. Parameters of the fit are shown in Table 3.

Note that this data is for the worst case position resolution at the module center. Usually resolution near the module edge is 3 times better. The MC result is in good agreement with the experimental data.

Table 3. Parameters describing position resolution of the prototype in the center of the cell [in mm].

Experiment data fit	MC data fit
$a = 13.1 \pm 1.0$	$a = 14.1 \pm 1.0$
$b = 4.0 \pm 0.8$	$b = 3.7 \pm 0.9$

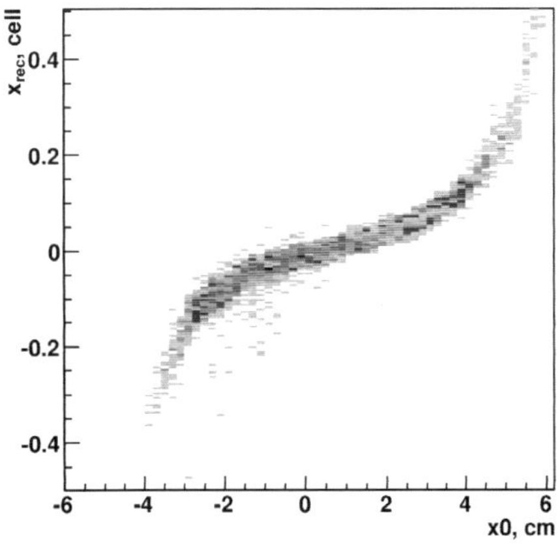

Figure 10. Energy resolution dependence on energy.

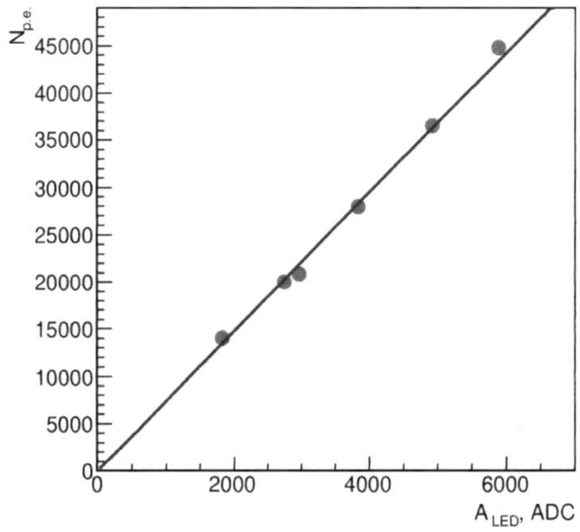

Figure 11. Light yield vs amplitude for single module.

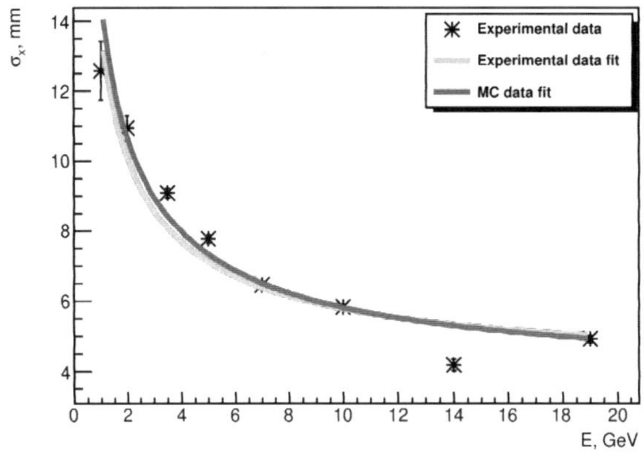

Figure 12. Position resolution dependence on energy.

7. Light output measurement

Light yield was measured with the highly stable LED light. The fluctuations of the measured amplitude are due to statistics of detected photoelectrons only. For the measured LED amplitude A and the dispersion of its distribution σ_A, one has a number of detected photoelectrons expressed by the equation:

$$n_{p.e.} = \left(\frac{1.2}{\sigma_A/A}\right)^2. \quad (3)$$

A set of runs with six different LED amplitudes has been taken. The dependence of the photoelectron statistics on the LED amplitude for one cell is shown in Figure 11. This plot was fitted by the linear function, and the slope represents the number of photoelectrons per one ADC count. Being divided by the calibration coefficient, one obtains the number of photoelectrons detected by the photomultiplier, when the amount of deposited energy in the module is 1 MeV. The mean value of that number for all 9 modules is $N_{p.e} = 5.3 \pm 0.2$ per 1 MeV.

8. Detection inefficiency

Due to various mechanical inhomogeneities of the modules one can expect to observe the dependence of the energy E deposited in the calorimeter on the hit coordinates (x, y). The "hot" zones, if any, should be seen at the WLS fiber positions, at the steel strings, and at the boundaries between the modules. The possible lateral non-uniformity of the energy response was studied with the data collected in the 19-GeV/c beam. The last drift chamber DC4 was used to measure the coordinate of the beam particle incidence onto the calorimeter surface. As the beam contained several particle species which interact differently with the calorimeter medium (see Figure 5), the mean deposited energy was measured as a function of (x, y) for two energy intervals, $E < 0.5$ GeV and $16 < E < 22$ GeV corresponding to the MIP peak and that of the electromagnetic shower, respectively. Within the available statistics, lateral non-uniformity of the energy response is observed not more than 1-2 %.

9. Conclusions

The measurements of energy and position resolutions of the PANDA forward electromagnetic calorimeter prototype of fine-sampling type have been carried out at the IHEP test beam facility at the Protvino 70 GeV accelerator. The prototype was designed and assembled at IHEP scintillator workshop. The results for 110×110 mm^2 cell size are the following. Energy resolution: $\sigma_E/E = 2.8/\sqrt{E} \oplus 1.3[\%]$, position resolution: $\sigma_x = 13.1/\sqrt{E} \oplus 4.0[mm]$. Monte-Carlo model predictions describe data very well. The results are also consistent with the previous measurements at lower energies [4].

The number of photoelectrons detected by photomultiplier in our measurements is $N_{p.e} = 5.3 \pm 0.2$ per 1 MeV deposited within one module.

The non-uniformity of the energy response of the prototype due to holes for straight fibers studied with the use of electrons and MIPs has turned out to be negligible. Monte-Carlo simulations are in a good agreement with the obtained experimental results.

The features determined for our calorimeter prototype meet the design goals of the PANDA experiment. However, the final conclusion on lateral sizes of the cells as well as on Shashlyk longitudinal sampling structure could be done only after study of reconstruction efficiency of π^0-mesons of different energies. We plan to have a shashlyk test beam run in 2008 with a prototype of 8x8 smaller cells (55×55 mm^2).

This work has been partially supported by the Contract with Rosatom number H.4d.47.03.08.118, by the INTAS grants with Ref. No. 06-1000012-8845 and 06-1000012-8914, and by the RFBR grant 05-02-08009.

References

[1] Dafinei I 2005 Performance of lead tungstate crystals for the CMS electromagnetic calorimeter *Nucl. Instr. and Meth.* A **537** 368

[2] Aleksandrov D V *et al.* 2005 A high resolution electromagnetic calorimeter based on lead-tungstate crystals *Nucl. Instr. and Meth.* A **550** 169

[3] Kubantsev M *et al.* 2006 Performace of the PrimEx electromagnetic calorimeter *Preprint* JLAB-PHY-06-528
[4] Atoian G S *et al.* 2008 *Nucl. Instrum. Meth.* A **584** 291
[5] Batarin V A *et al.* 2003 Development of a momentum determined electron beam in the 1-45 GeV range *Nucl. Instr. and Meth.* A **510** 211
[6] Batarin V A *et al.* 2003 Precision measurement of energy and position resolutions of the BTeV electromagnetic calorimeter prototype, Nucl. Instrum. and Meth. *Nucl. Instr. and Meth.* A **510** 248261
[7] Brun R *et al.* 1987 GEANT3 reference manual *Preprint* DD/EE/84-1 CERN

Construction and tests of a fine granularity lead-scintillating fibers calorimeter

P Branchini[1], F Ceradini[2], G Corradi[3], B Di Micco[1] and A Passeri[1,4]
[1] INFN Sezione Roma Tre, via della Vasca Navale 84, 00146 Roma, Italy
[2] Università Roma Tre, via della Vasca Navale 84, 00146 Roma, Italy
[3] INFN Laboratori Nazionali di Frascati, via E Fermi 40, 00044 Frascati, Italy.

E-mail: antonio.passeri@roma3.infn.it

Abstract. We report the construction and the tests of a small prototype of the lead-scintillating fiber calorimeter of the KLOE experiment, instrumented with multianode photomultipliers to obtain a 16 times finer readout granularity. The prototype is 15 cm wide, 15 radiation lengths deep and is made of 200 layers of fibers 50 cm long. On one side it is read out with an array of 3×5 multianode photomultipliers Hamamatsu type R8900-M16, each segmented with 4×4 anodes, the read out granularity being 240 pixels of 11 × 11 mm^2 corresponding to about 64 scintillating fibers each. These are interfaced to the 6 × 6 mm^2 pixeled photocathode with truncated pyramid light guides made of Bicron BC-800 plastic to partially transmit the UV light. Each photomultiplier provides also an OR of the 16 last dynodes that is used for trigger. The response of the individual anodes, their relative gain and cross-talk has been measured with the light (440 nm) of a laser illuminating only few fibers on the side opposite to the read-out. We finally present the first results of the calorimeter response to cosmic rays in auto-trigger mode.

1. Introduction

The lead-scintillating fibers calorimeter of the KLOE experiment at the DAFNE φ-factory has taken data for many years and has proven to provide excellent performances as tracking calorimeter and time of flight detector at low energy [1,2]. It is made by stacks of about 200 grooved 0.5 mm thick lead foils alternated with layers of cladded 1 mm diameter scintillating fibers. The fibers are arranged in a periodic equilateral triangle array whose side is 1.35 mm, as shown in fig. 1. The final result is an almost homogeneous compound of lead:fibers:glue with relative volume fractions 42:48:10. The average density is ~5 g/cm^3 and the radiation length ~1.5 cm. The scintillating fibers have a fast scintillating decay time of ~2.2 ns and the light attenuation length along them is around 4 m. Due to the very high scintillator content such detector is extremely well suited both for energy and time measurements. The measured energy resolution for photons in KLOE is $5.7\%/\sqrt{E(GeV)}$ with a negligible constant term, while the intrinsic time resolution of the calorimeter is $54\ ps/\sqrt{E(GeV)} + 50ps$, the constant term being due to residual miscalibrations between cells.

In the KLOE calorimeter the spatial reconstruction of the energy depositions is obtained, for the coordinate along the fiber length, by measuring the arrival time difference of the scintillating light at

[4] To whom any correspondence should be addressed.

both ends of the fibers, which provides a resolution of ~1 cm when the energy is released at a distance of 2 m from the photomultiplier. The reconstruction precision transverse to the fibers is determined by the size of the light readout cells, which are squares of 4.4 x 4.4 cm^2 at both sides of the calorimeter.

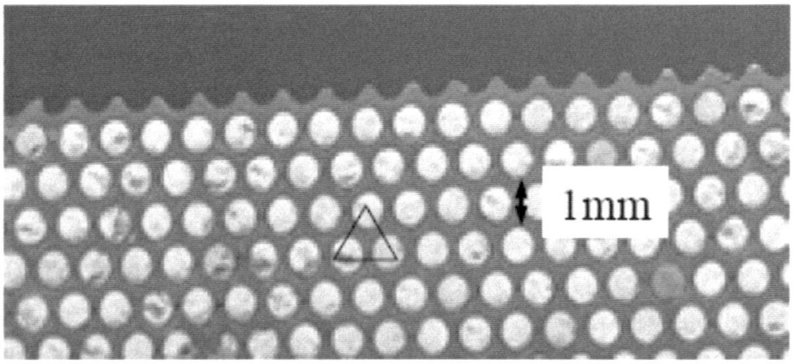

Figure 1. Lead- scintillating fibers structure in the KLOE calorimeter.

This segmentation yields a resolution of ~1.3 cm (4.4/√12) for isolated showers, which matches the average shower size. However, such coarse granularity does not allow to disentangle two nearby energy deposits, which in KLOE may occur quite frequently and have a high probability to be merged if the distance between their centroids is less than 20 cm. This limitation is particularly nasty in all physics analyses in which the photon counting is crucial to separate signal and background. Moreover, the coarse granularity does not allow to track the details of the energy deposition development, which are essential to identify the different kind of particles entering the calorimeter.

A detailed FLUKA simulation of one KLOE calorimeter module [3] showed that, increasing the readout granularity by a factor of 16 (i.e. making 1.1 x 1.1 cm^2 elementary readout cells), couples of particles impinging into the calorimeter with a separation down to few centimeters can be easily disentangled, and that the energy deposition shape for electrons and muons can be reconstructed to a level that allows efficient particle identification on an event by event basis.

Here we present the practical implementation of this concept on a small prototype of the KLOE calorimeter, having the standard thickness of about 15 radiation lengths, length of 50 cm and width of 13.5 cm. On one side a matrix of 3 x 5 standard KLOE light guides and PMs had been already installed. We designed a completely new readout structure for the other side, where the 3 x 5 matrix of KLOE cells is matched by 15 macro-cells, each segmented into 16 cells 11 x 11 mm wide and read out by a multi-anode photomultiplier with 16 anodes.

2. Multi-anode photomultipliers characterization

We have chosen the Hamamatsu R8900 M16 multi-anode photomultiplier [4], the latest development of conventional multi-anode technology based on weak electrostatic focusing and implementing bi-alkali photo-cathodes. This device has a sensitive surface which is 83% of its physical area (25.7 x 25.7 mm^2) and provides a compact readout solution for our detector. The photocathode is segmented into a 4 x 4 matrix where each cell has a nominal active surface of 6.0 x 6.0 mm^2 and a dead zone 0.3 mm wide is present between any two adjacent cells. Beside the 16 cells output signals, the device provides also in output the sum of its 16 last dynodes, which is very useful for triggering purposes. Finally, this device exists also in an improved quantum efficiency version: we used three of such PMs for the readout of the first plane of the prototype.

It is well known that multi-anode PMs suffer by wide response non-uniformity between different cells, which have to be accounted for in data reconstruction. We studied such effect in laboratory by means of the Hamamatsu Laser pulser PLP-10 at 440 nm, which provides light pulses of 100 ps duration. We illuminated each single cell and we studied its charge response as a function of the supply high voltage. The response at a generic high voltage V is normalized to the response at V=500 V. In each cell the behaviour is well described by an exponential function, with similar slopes

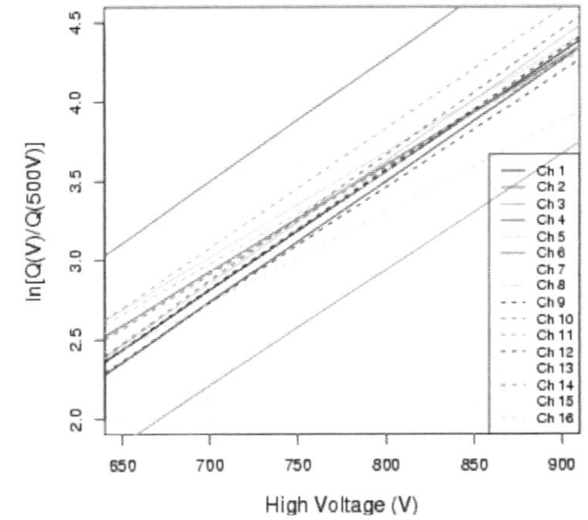

Figure 2. Charge response as a function of high voltage for the 16 anodes of a single R8900 M16. Only the fits are shown to save space.

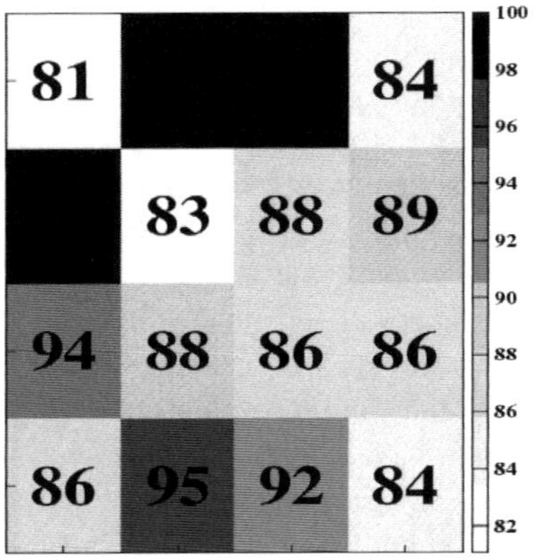

Figure 3. Charge response variation at HV=800 V in a typical R8900.

but with wide offset variations, as it is shown in fig.2 for an individual PM. Figure 3 shows the charge response level in the cells of a single R8900 at a working point of 800 V. Typical response variations are in the range of 20% , which in some individual cell may occasionally reach 30-35%.

A second important issue for multi-anode PMs is the cross talk between different cells due to the internal structure of the device. We studied also this effect by illuminating a single cell and measuring the charge response of all the others. In this way we build a cross talk matrix for each R8900, of which an example is shown in fig.4. The cross talk level is generally around few per cent, apart from single cases in which only the cell adjacent to the illuminated one may reach 20-30%.

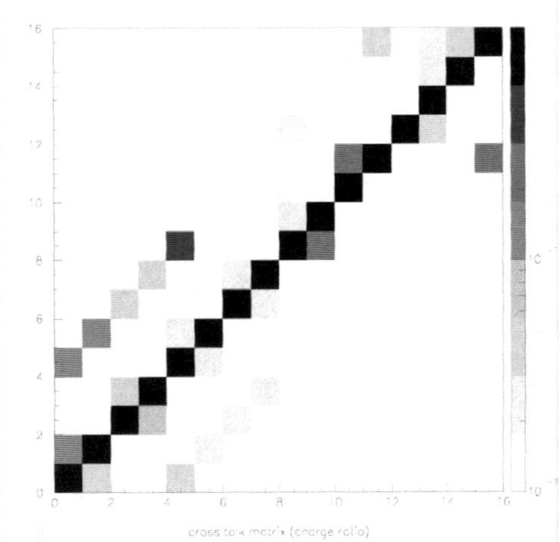

Figure 4. Cross talk matrix for a single R8900 at 800 V. Illuminated cell on the x

axis, responding cell on the y axis.

3. Light guides and mechanics

A 11 x 11 mm^2 readout cell corresponds to about 64 scintillating fibers. To collect the light emitted by such fibers and to concentrate it into one 6.0 x 6.0 mm^2 photocathode cell of the R8900-M16 a special set of light guides has been designed and realized. An overall truncated pyramid structure maps the 16 elementary cells corresponding to one KLOE cell into one multi-anode PM. Such structure is actually segmented into 16 individual light guides which map each single readout cell into a 5.7 x 5.7 mm^2 square in the centre of one multi-anode cell (we prefer not to use the edges of the photocathode cells which are usually less efficient). Individual guides have three slightly different shapes depending whether they are at the corners, at the centre or at the sides of the macro-cell. A length of 6 cm is enough to keep all the guide faces well below the minimum reflection angle. The UV transparent BC800 plastics from Bicron has been used for all the guides. An aluminium grid (anodized in black) keeps the guides in place on the cathode side. Individual guides are only separated by a thiny slice of air (the plastic-air contact surface being the most advantageous for light reflection) and they touch each other only on the calorimeter surface. A picture showing details of the light guides structure is presented in figure 5.

Figure 5. View of two blocks of 16 light guides, one seen from the side to be glued to the calorimeter, the other seen from the side to be put in optical contact with the R8900.

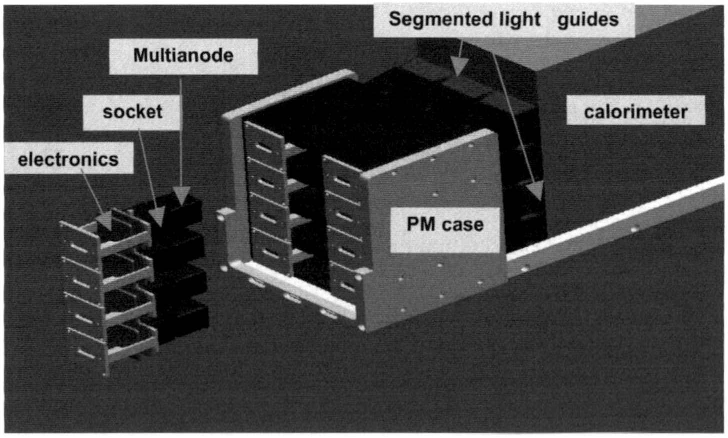

Figure 6. Mechanical sketch of the readout support structure (here only 4 PM rows are showed, instead of the final 5)

A mechanical structure, sketched in fig.6, has also been designed to support the PMs and their electronics and to allow easy insertion and extraction of individual PMs.

4. Electronics and data acquisition

Both for the front-end electronics and for the data acquisition we exploited as much as possible the existing spare boards of the KLOE experiment. The only part that required a completely new design was the pre-amplification stage, for various reasons: a higher amplification factor is needed in individual channels due to the reduced light input, a compact solution is required to process the 16+1 signals produced by each multi-anode and finally the signals have to be inverted to be fed into the KLOE calorimeter electronic chain, which works only with positive signals. Fig.7 shows the electrical scheme of a single inverting pre-amplifier stage providing a factor of 10 amplification. A small board hosting 17 such channels has been produced, and plugged to each multi-anode PM.

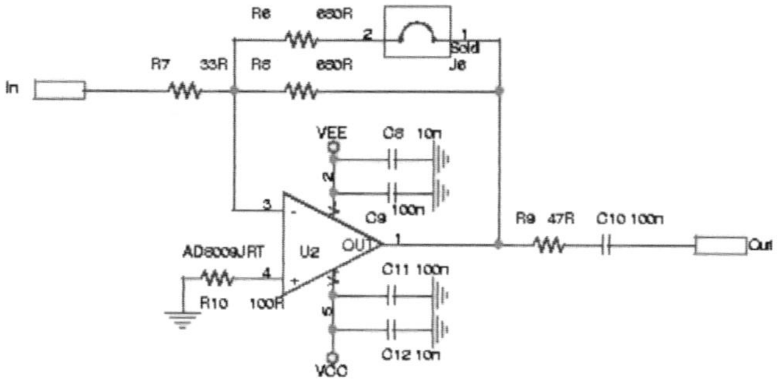

Figure 7. Inverting preamplifier stage for a single multi-anode channel.

Signals are then processed into the so called SDS boards [1], where they are split in 3 copies: one is fed into the KLOE calorimeter ADC [5], the second is first discriminated with constant fraction technique and then fed into the KLOE calorimeter TDC [5], and the third one can be fast-summed to up to 4 neighbouring cell signals for triggering purposes.

The data acquisition system is a replica of the KLOE one [6], with the only important difference that the second level real time CPU is now a Motorola MVME 6100.

Overall, the fine granularity light readout includes 255 channels (16+1 for each of the 15 multi-anodes PM), each producing both an ADC and a TDC output, for a grand total of 510 electronic channels to be acquired. All ADC channels have been acquired with random trigger and all pedestals have been evaluated. An offline threshold is put to each ADC channel at 3 times the rms of its pedestal distribution.

5. Optical cross talk determination

The possibility of optical cross talk between the small light guides that we installed on the calorimeter has to be carefully investigated. To measure the amount of this effect in a clean way we dismantled the standard light readout system on the opposite calorimeter side and from there we injected a laser pulse in individual scintillating fibers. We studied the response of the fine granularity channels while scanning the opposite calorimeter face with the laser pulse. Illuminating the 16 cells of the same multi-anode PM we found a level of cross talk very similar to that observed during the R8900

Journal of Physics: Conference Series **160** (2009) 012022 doi:10.1088/1742-6596/160/1/012022

characterization and described in section 2. An example is shown in fig.8, where a cross talk at the level of few % is observed in the cells adjacent to the illuminated one, while all the others are at the per mil level. This means that all the cross talk sources arising from the light guides system and from the full electronic connection are negligible with respect to the intrinsic R8900 cross talk.

Optical cross talk can finally be isolated by checking the response of cells adjacent to the illuminated one but belonging to a different multi-anode PM. We observed a cross talk level below the per mil, as it is shown in fig. 9, and a slight under-fluctuation (positive signals going negative) in the other cells of the multi-anode, which is probably an electronic effect. We can then safely conclude that no sizeable optical cross talk is introduced by the fine granularity readout.

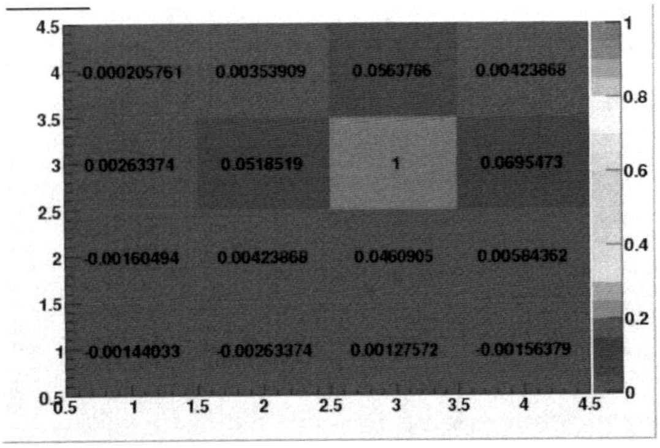

Figure 8. Typical response pattern when a single cell is illuminated in an individual multi-anode PM. On both axis the scale is simple cell numbering.

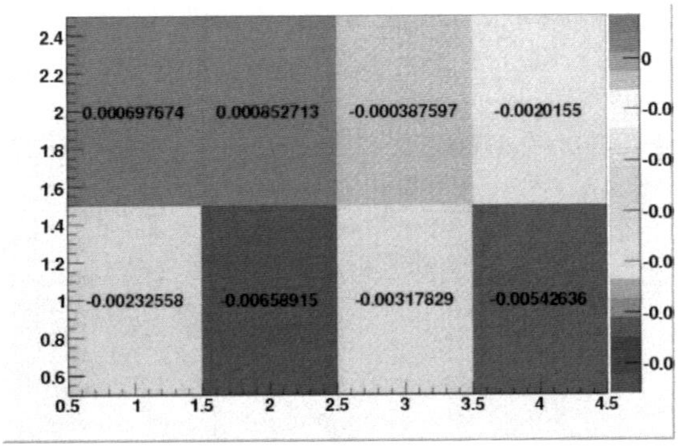

Figure 9. Response pattern on half multi-anode when one adjacent cell of the neighbour PM is illuminated.

6. Cosmic rays test

Cosmic rays have been acquired in auto-trigger mode, using the signals of the last dynodes sum provided by the R8900. The logic OR of the 3 sum signals of the top row of multi-anodes has been put in coincidence with the OR of the 3 sum signals of the bottom multi-anode row. Two typical cosmic rays event displays are shown in fig.10, from where the tracking potentialities of our detector become evident.

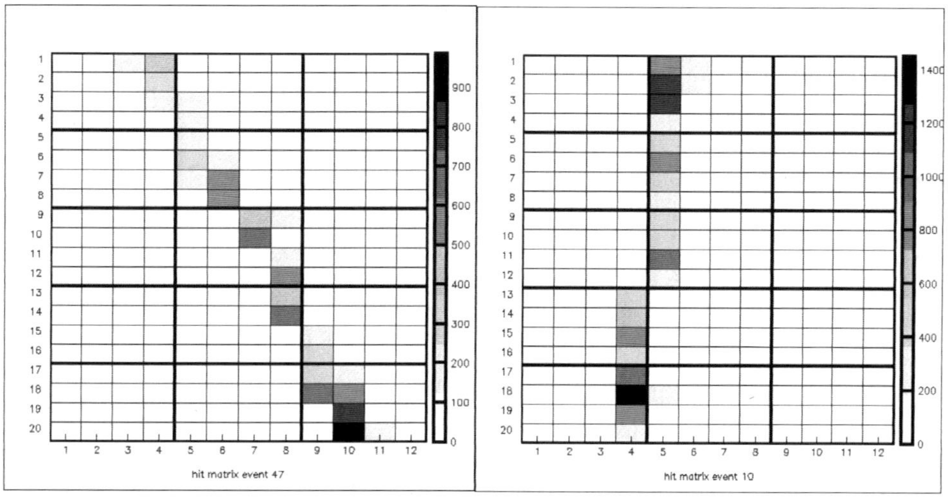

Figure 10. Two typical cosmic rays event displays

A rough equalization of the multi-anode response has been performed by setting the high voltage of each R8900 at a value that fixes at ~3000 counts the peak of the distribution of the sum of the cells above threshold in the PM. A nice Landau shape is then obtained for the distribution of the sum of all the cells above threshold in the detector, as shown in fig.11. Individual cell spectra show a Landau peak too, even if superimposed to a low charge background due to residual external light contamination (which will disappear by improving the shielding), as can be seen from the example shown in fig.12.

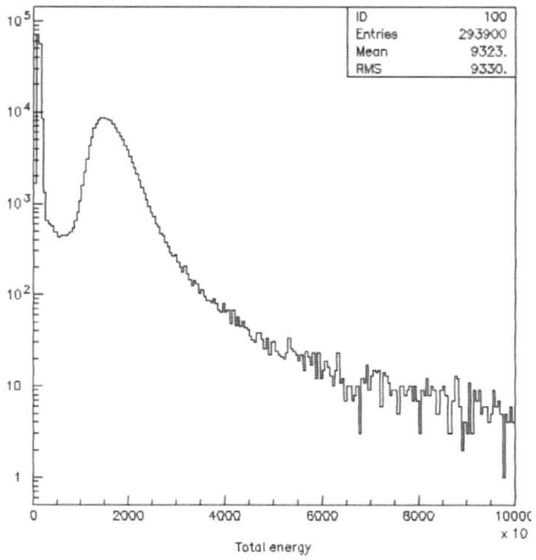

Figure 11. Offline sum of all the cells above threshold in the calorimeter over 15000 cosmic rays events.

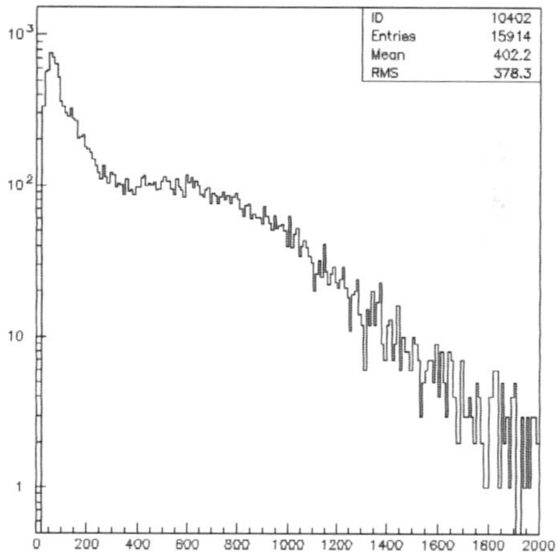

Figure 12. Single cell spectrum for cosmics rays events.

7. Conclusions

We have built a fine granularity light readout system for a KLOE calorimeter prototype using multi-anode photomultipliers, with negligible optical cross talk. Cosmic rays tests of the full detector shows good overall performances and very promising tracking potentialities. A test beam with electrons has already been performed and data are being analyzed. Further tests with neutrons and pions are being planned.

Acknowledgments
This work is partly supported as a PRIN project by the Italian Ministero dell'Istruzione, Università e Ricerca Scientifica.
All this work would have been impossible without the very qualified and patient work of our mechanical technicians. In particular we warmly thank A.Iaciofano and his collaborators M.Capponi and V.Penna. For the mechanical design we thank M.Lobello. For the delicate and long cable preparation we thank F.Budano, G.Pellizzoni and A.Rossi. For the general electronic support we thank R.Lomoro. A very special thanks goes to the student Paul Mekhedjian (now at University of California) for the careful work of multi-anode PMs characterization. We thank also the students S.Troisi and V.Bortolotto who took part to the optical cross-talk measurement and to the cosmic rays test. Finally we thank our polish collegues P.Moskal and J.Zdebik of the Jagellonian University in Cracow who helped us in the realization of the event display program.

8. References

[1] Adinolfi M et al. 2002 The KLOE electromagnetic calorimeter *Nucl. Inst. Meth.* A **482** 364
[2] Ambrosino A et al. 2004 Data handling, reconstruction and simulation for the KLOE experiment *Nucl. Inst. Meth.* A **534** 403
[3] Di Micco B, Branchini P, Ferrari A, Loffredo S, Passeri A and Patera V 2006 A new FLUKA simulation of the KLOE calorimeter *Nucl. Phys. B (Proc. Suppl.)* **172** 243
[4] Kawasaki Y et al. 2006 Performance of a multi-anode photomultiplier employing a weak electrostatic focusing system (Hamamatsu R8900 series) *Nucl. Inst. Meth.* A **564** 378
[5] Antonelli M et al. 1998 The ADCs and TDCs for the KLOE electromagnetic calorimeter *Nucl. Inst. Meth.* A **409** 675
[6] Aloisio A et al. 2004 Data acquisition and monitoring for the KLOE detector *Nucl. Inst. Meth.* A **516** 288

Journal of Physics: Conference Series **160** (2009) 012023 doi:10.1088/1742-6596/160/1/012023

Measurement and simulation of the neutron detection efficiency with a Pb-scintillating fiber calorimeter

M.Anelli[a], G.Battistoni[b], S.Bertolucci[a], C.Bini[c], P.Branchini[d], C.Curceanu[a], G.De Zorzi[c], A.Di Domenico[c], B.Di Micco[d], A.Ferrari[e], S.Fiore,P.Gauzzi[c], S.Giovannella[a], F.Happacher[a], M.Iliescu[a,f], M.Martini[a,h], S.Miscetti[a], F.Nguyen[d], A.Passeri[d], A.Prokofiev[g], P.Sala[b], B.Sciascia[a], F.Sirghi[a].

Institutions:
(a) Laboratori Nazionali di Frascati, INFN, Italy,
(b) Sezione INFN di Milano, Italy,
(c) Università degli Studi "La Sapienza" e Sezione INFN di Roma, Italy,
(d) Università degli di Studi "Roma Tre" e Sezione INFN di Roma Tre, Italy,
(e) Fondazione CNAO, Milano, Italy,
(f) IFIN-HH, Bucharest, Romania,
(g) The Svedberg Laboratory, Uppsala University, Sweden.
(h) Dipartimento di Energetica dell'Universitá "La Sapienza", Roma, Italy.

E-mail: `matteo.martini@lnf.infn.it`

Abstract. We have measured the overall detection efficiency of a small prototype of the KLOE Pb-scintillating fiber calorimeter to neutrons with kinetic energy range [5,175] MeV. The measurement has been done in a dedicated test beam in the neutron beam facility of *The Svedberg Laboratory*, TSL, Uppsala. The measurement of the neutron detection efficiency of a NE110 scintillator provided a reference calibration. At the lowest trigger threshold, the overall calorimeter efficiency ranges from 28 % to 33 %. This value largely exceeds the estimated $\sim$ 8 % expected if the response were proportional only to the scintillator equivalent thickness. A detailed simulation of the calorimeter and of the TSL beam line has been performed with the FLUKA Monte Carlo code. The simulated response of the detector to neutrons is presented together with first data to Monte Carlo comparison. The results show an overall neutron efficiency of about 35 %. The reasons for such an efficiency enhancement, in comparison with the typical scintillator-based neutron counters, are explained, opening the road to a novel neutron detector.

1. Introduction

Detection of neutrons with energies from a few to hundreds MeV is usually performed with organic scintillators. The high concentration of hydrogen atoms provides a proton target and the elastic scattering of neutrons produces recoil protons which produce a visible response. Typical efficiency is of the order of 1 % per cm of scintillator thickness [1]. On the other hand the extended range rem counters used in radiation protection [2, 3] are based on the idea of adding one layer of a medium-high Z material to the organic scintillator. This enhances the neutron response through inelastic processes that result in an abundant production of secondary neutrons.

The KLOE calorimeter [4] is a high sampling lead scintillating fiber calorimeter. Energy and time resolutions are

$$\sigma_E/E = 5.7\%/\sqrt{E\ (\mathrm{GeV})}$$

and

$$\sigma_t = 57\ \mathrm{ps}/\sqrt{E\ (\mathrm{GeV})} \oplus 100\ \mathrm{ps}$$

.

The calorimeter was primarily designed to detect photons, the detection efficiency is $\sim 90\%$ at $E = 20$ MeV and reaches 100% above 70 MeV.

During a KLOE study, hints of a high efficiency for low energy neutrons (< 20 MeV) were observed. In this exercise we use charged kaons events and we measure an average efficiency that is between three and four times higher than the one expected for the equivalent scintillator thickness. This result was confirmed also by the official Monte Carlo simulation of the KLOE experiment.

To understand the underlying physical mechanisms which produce this difference, and to make a more systematic study of the calorimeter response, we planned a test beam to expose the calorimeter to a dedicated neutron beam and we developed a complete simulation of the detector geometry and response.

Apart from the possibility to develop compact, inexpensive, fast and highly efficient neutron counters, this work has also been motivated by the prospects of the search for deeply bound kaonic nuclei [5] and of the neutron electromagnetic form factor measurements in the time-like region [6]. These are two fundamental items for the DAΦNE upgrade under study at the Laboratori Nazionali di Frascati of INFN (LNF). In both cases the neutron detection efficiency is a key point for the measurements.

2. Measurement of the neutron detection efficiency at TSL

The KLOE calorimeter prototype used in this measurement is composed of ~ 200 layers of 1 mm diameter blue scintillating fibers glued inside grooved lead layers of 0.5 mm thickness. The final structure has a fiber:lead:glue volume ratio of 48:42:10 resulting in a density of ~ 5 g/cm^3. The total external dimensions are $13 \times 24 \times 65$ cm^3 where the second value is the calorimeter depth and the third one is the fiber length. The calorimeter is readout at both fiber ends in order to reconstruct this coordinate by time difference. The readout is organized in four planes along the calorimeter depth for the first 16.8 cm. Each plane is then divided in 3 columns along the horizontal coordinate originating cells of 4.2×4.2 cm^2. Larger readout elements are used in the rear part of the calorimeter. The small elements are coupled to 1-1/8 inch diameter standard bialkali photomultipliers, PM, through light guides.

When running with neutron beam, the calorimeter was positioned with fibers running vertically, the lower (upper) end is called side A (B). Each PM signal has been split in three replicas. The first two signals allow measurements of charge and time while the third one is used to summing up the PM charges of the first four planes of each side. The discriminated signals obtained, SA and SB, are combined in overlap coincidence, $SA \cdot SB$, to trigger on the calorimeter.

A reference counter for the efficiency measurement was built with a 5 cm thick bulk of NE110 organic scintillator, of transversal dimensions 10×20 cm^2, by coupling it at the two ends to two EMI9814 PM's.

A "beam position monitor" is also used. This detector was made by an array of seven scintillating counters 1 cm thick to check the beam shape during run.

Journal of Physics: Conference Series **160** (2009) 012023 doi:10.1088/1742-6596/160/1/012023

Figure 1. Experimental setup used during test beam. (1) Calorimeter prototype, (2) beam position monitor and (3) NE110 reference counter.

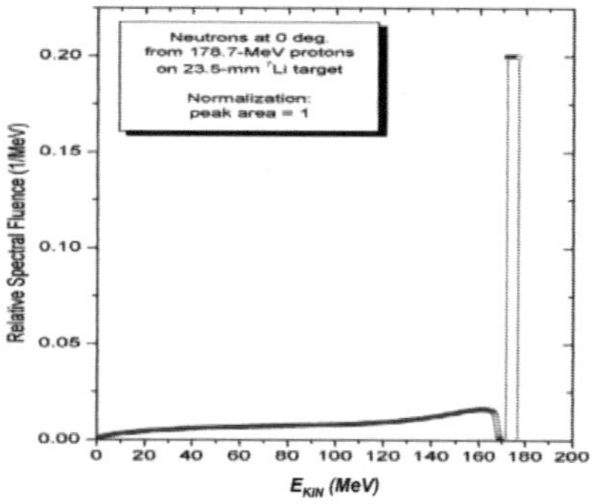

Figure 2. Neutrons energy spectrum at the TSL facility.

The experimental setup is shown in Fig.1, where we have the calorimeter prototype, the reference counter and the beam position monitor used to detect beam dimensions.

When running with the beam, the scintillator was positioned with its longest dimension along the horizontal coordinate. To trigger on the scintillator, the PM signals, $S1$ and $S2$, were discriminated and an overlap coincidence, $S1 \cdot S2$, was formed.

We ran our experiment at the "The Svedberg Laboratory" (TSL) high-energy neutron facility[7]. At this facility, protons from the Gustaf Werner cyclotron were directed on a ^{7}Li target generating a neutron beam which is then geometrically shaped by an iron collimator block with a 2 cm diameter cylindrical hole. The energy and angular distributions of neutrons are defined by the differential cross section of the ^{7}Li(p,n) reaction at forward angles [8]. The neutron energy spectrum is dominated by a peak at a few MeV below the primary proton energy, and a long tail down to thermal neutrons (see Fig.2). In our settings, we performed two different test beams, using high energy neutrons (178.7 MeV) and low energy neutrons (21 and 46 MeV). The neutron beam time structure was in phase with the cyclotron RF which had a period, T_{RF}, of $\sim$ 45 ns. The beam was emerging from the collimator at 3 m distance from the target. In order to ensure full beam acceptance we ran at a distance of 5.1 m from the target and centered the beam in the middle of our detectors. Very low intensity neutron beams, from 1.5 kHz/cm^2 to 6 kHz/cm^2, have been required in order to minimize the probability of counting more than a neutron per event. The neutron rate has been measured with a ionization-chamber monitor (ICM), with an absolute accuracy estimated at the level of 10 %.

All discrimination and trigger signals were formed with NIM logic while the DAQ was based on VME standard. Data have been acquired with a simplified version of the KLOE data acquisition system, composed by 40+40 ADC and TDC channels and able to write on disk at a rate of 1.7 kHz. The effective ADC gate was 300 ns wide, while the TDC worked in common start mode with the start provided by the trigger and the stop given by the discriminated signals delayed of 150 ns by a monostable inside the board. The trigger signal was either $SA \cdot SB$ or $S1 \cdot S2$ after phase locking with the cyclotron RF replica. Typical runs consist of $\sim 0.5 - 1.5 \times 10^6$ events.

The detector efficiency to the overall neutron spectrum, ϵ_D, has been determined taking the ratio of the counted detector rate, R_D, with the rate of the neutron beam estimated by the ICM, R_n. For a given trigger threshold, assuming full beam acceptance and no background, R_D is the ratio between the acquired rate, R_{DAQ}, and the fraction of DAQ live time, F_{live}. The incoming neutron rate, R_n, is instead given by:

$$R_n(\text{kHz}) = 9.09\,\pi \cdot R_{ICM} \cdot F_{peak}$$

where R_{ICM} is the ICM rate in Hz counted in each run and F_{peak} is the fraction of neutrons at peak energy.

3. Calibration and measured efficiencies

For the scintillator, the horizontal scale has been calibrated fitting the energy spectra obtained with a reference β source. For the calorimeter, each cell has been calibrated using minimum ionizing particles and then applying the energy scale determined with electron and photon beams [9] in the same prototype.

The effective trigger thresholds cut-off have been evaluated from the data looking at the charge integrated by all the cells of the calorimeter at different thresholds. The ratio of these distributions normalized to the tail integrals has been fit with a Fermi-Dirac functions to extract the effective threshold value.

A clustering procedure is used to reconstruct the timing in the whole calorimeter and derive the kinetic energy of the primary neutron from time of flight, TOF. A first data-Monte Carlo comparison of TOF has been made at cell level. In the cells of the central column more than 90 % of the beam is contained. Data reconstructed clusters, with a single fired cell, show a ratio lateral/central fired cells higher then what expected from the simulation. Moreover, lateral cells show a flatter time distribution compared with Monte Carlo. We interpreted this effect as a presence of background due to low energy neutrons that form an halo around the central beam core. An overall good agreement is observed when taking into account also the shape of the halo contribution; this is obtained from data looking at the TOF distribution of the outer cells of the calorimeter where the background contamination is evident. We have estimated, for the run at 174 MeV, that halo events contribute to $\sim 30\%$ of the total number of events. In Fig.3, we show the TOF distribution for the central and the outer cells of the first calorimeter plane.

Concerning low energy run, we are still completing the TOF study to evaluate the correct halo fraction. As a preliminary evaluation, we rely on halo measurement carried out by TSL beam experts. They performed a scan of the area near the collimator using a fission monitor counter. This measurement has been confirmed by our background counters. By integrating over the whole calorimeter plane, we estimate a fraction of halo contamination: $F_h = (20 \pm 10)\%$.

The measured efficiencies for the scintillator and for the calorimeter are shown in Fig.4 and Fig.5, as a function of the trigger threshold in MeV, after correcting for the halo fraction. In Fig.5, the calorimeter efficiency is also compared with the expected result obtained considering the scintillator equivalent thickness.

These results show that at the lowest trigger threshold the neutron detection efficiency of the calorimeter ranges from 28 % to 33 %, depending on the beam intensity. For comparison, the efficiency of the 5 cm thick NE110 scintillator ranges from 4 % to 10 %, for values of the trigger threshold below 5 MeV of electron equivalent energy, in good agreement with the available

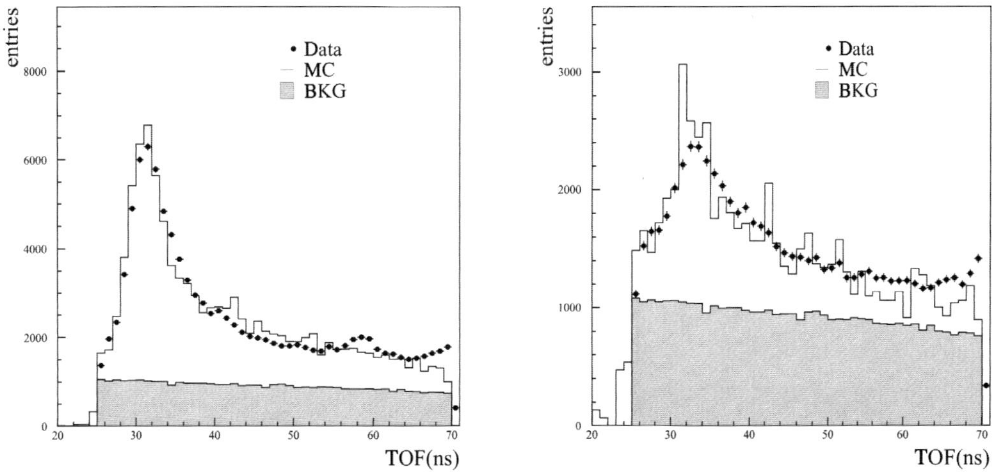

Figure 3. Data-Monte Carlo comparison of time of flight in the central calorimeter cell for the first calorimeter plane with 174 MeV neutrons. Dots are data, histograms are the simulation (solid) and background (green) contributions. (left) central cell, (right) outer cell.

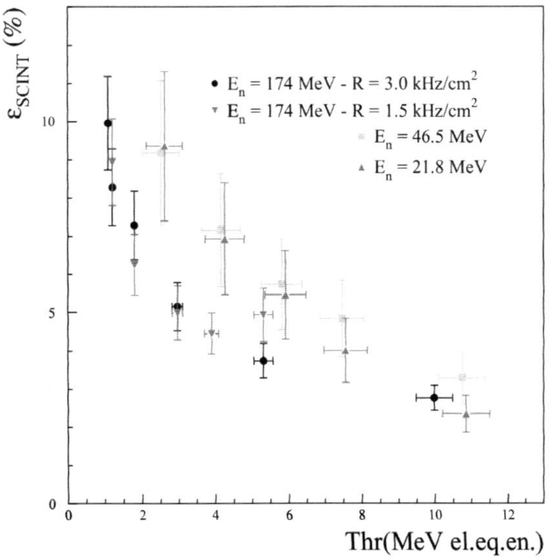

Figure 4. Dependence of ϵ_{scint} on the applied trigger threshold. The horizontal scale is in MeV of electron equivalent energy. The systematics is 10 % on both horizontal and vertical scale. A comparison with available measurements in literature [1] is also shown.

measurements in literature. This indicates that the measured calorimeter efficiency is sizeably enhanced with respect to the expected 8−10% based on the amount of scintillator only.

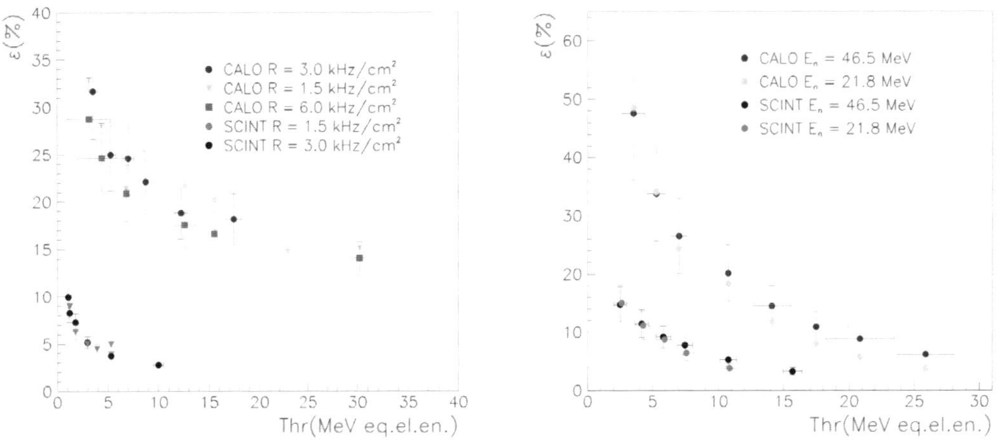

Figure 5. Dependence of ϵ_{calo} on the trigger threshold. The horizontal scale is in MeV set for electron response. (left) Run at 174 MeV, (right) run at 21 and 46 MeV.

4. Monte Carlo simulation and comparison with data

The Monte Carlo code FLUKA [10, 11] has been used for a detailed simulation of the calorimeter structure. The TSL experimental beam-line, from the neutron source to the collimated beam, has been also simulated, in order to have a reliable characterization of the neutron beam impinging on the detector (see Fig.6). Using the tool "LATTICE" the whole calorimeter module has been designed. All the compounds have been carefully simulated: for the fibers, an average density between cladding and core has been used, for the glue we have taken into account the right fraction of epoxy resin and hardener. FLUKA computes the energy deposits in the scintillating fibers, taking into account the signal saturation due to the Birks law. For each energy deposit, the average number of photoelectrons is estimated and then attenuated to the calorimeter ends with the proper attenuation length. The photoelectron statistics and the generation of the discriminated signal are also simulated, while the trigger effect has not yet been included.

The primary reason for the observed efficiency enhancement appears to be the huge inelastic production of neutrons on the lead planes. For neutrons in the high energy peak (175 MeV), the probability to have an inelastic interaction is 31.4 % on the lead, compared to 7.0 % on the fiber and 2.2 % on the glue. The secondary particles generated in such inelastic interactions are on average 5.4 per event, counting only the secondary neutrons above 19.6 MeV. Among the produced secondaries, 62 % are neutrons, 27 % photons, 7 % protons while the remaining 4 % are nuclear fragments. Typical inelastic reactions on lead are:

n Pb $\rightarrow$ xn + $y\gamma$ + Pb
n Pb $\rightarrow$ xn + $y\gamma$ + p + residual nucleus
n Pb $\rightarrow$ xn + $y\gamma$ + 2p + residual nucleus

and so on. Low-energy neutrons (below 19.6 MeV) are transported in FLUKA with a multi-group algorithm, that uses a neutron cross-section library derived from the most recently evaluated data. These secondary neutrons give also a sizeable contribution to the response: due to the larger inelastic cross section the neutron shower-like effect increases and originates on average about 100 secondaries per event, out of which $\sim$ 5 protons and $\sim$ 1 photon directly contribute in generating a visible response.

The high sampling frequency of the calorimeter appears to be a crucial point in the efficiency enhancement. First of all, the protons and the electromagnetic energy produced on lead in the

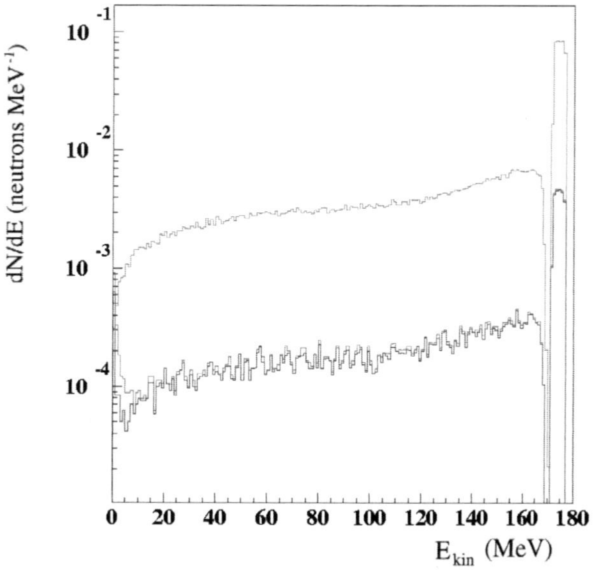

Figure 6. Neutron energy spectra as computed with the FLUKA simulation. From the top: at the source, at the collimator exit and on the calorimeter entrance.

inelastic processes can be detected by the nearby fibers down to very low energies. Moreover, secondary neutrons are produced on following lead planes with decreasing energies, thus having larger probabilities to produce ionizing particles on the nearby fibers. The isotropic distributions, which characterize the inelastic processes, also play a role: the backscattered neutrons contribute to increase the collision density in the first calorimeter planes, so containing the neutron shower depth.

A good agreement between data and simulation is observed when taking into account also the halo contribution as shown in Fig3 for the TOF and in Fig.7 for the collected charge.

In Fig.8 we show the dependence of ϵ_{calo} as a function of the trigger threshold compared to the FLUKA simulation after correcting R_{DAQ} for the halo contribution estimated with the TOF. The comparison is done for the run at 174 MeV. A pretty good data/simulation agreement is observed. Comparison between data and Monte Carlo for the low energy run is in progress.

5. Conclusion and prospects

Using an high sampling Pb-scintillating fibers calorimeter, the first measurement of the detection efficiency for neutrons between 20 and 180 MeV has been performed at the "The Svedberg Laboratory" of Uppsala. The calorimeter efficiency, integrated over the whole neutron energy spectrum, ranges between 32-50% at the lowest trigger threshold. This result is between three and four times larger than what expected considering the equivalent scintillator thickness of the prototype. Cross-check measurement of the neutron efficiency has been done using a NE110 scintillator. This result agrees with previous published results.

Monte Carlo simulation shows that the origin of such enhancement is related both to the shower-like effect due to the inelastic processes in the calorimeter structure and to the high sampling fraction of the detector. Preliminary comparison between data and Monte Carlo is

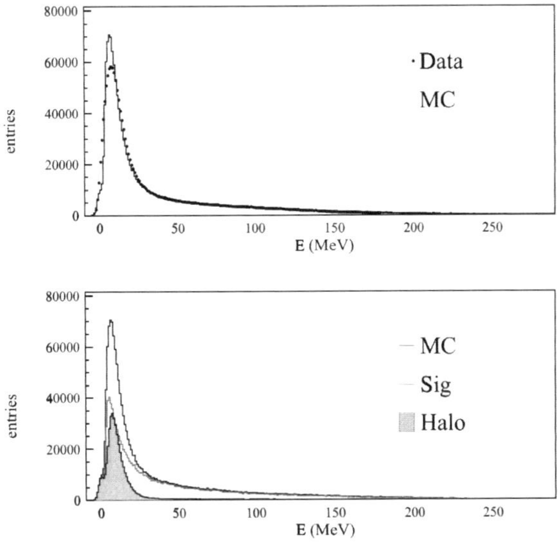

Figure 7. Comparison between data and Monte Carlo for the collected charge. Halo contribution has been included in the simulation (bottom plot).

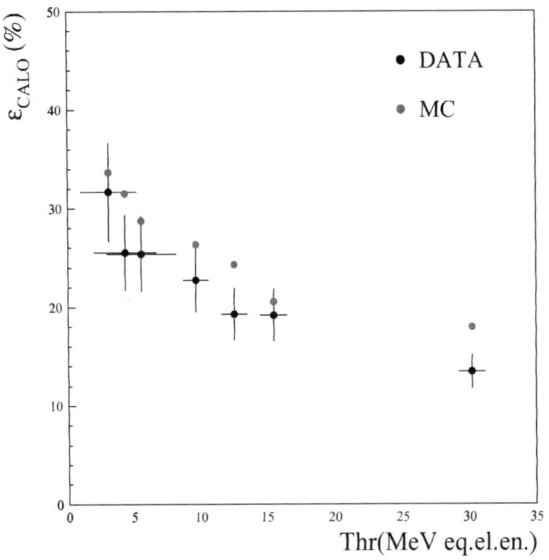

Figure 8. ϵ_{calo} comparison between data (bullet) and simulation (open circle) as function of the applied trigger threshold for the run at 174 MeV.

satisfactory. Full simulation of the calorimeter, of the beam line and of the experimental hall (together with local shielding) is in progress.

Further test foreseen for fall 2008 at TSL. In this new data taking campaign, we plan to use some new detectors to explain the physical effects observed. To better describe the beam halo, we are building a new beam position monitor with $(1 \times 1)\ cm^2$ granularity read out using multianodes photomultipliers. We are preparing a high granularity calorimeter prototype to study the processes inside the modules with best resolution. To this last test beam, we are also building a small calorimeter with the same structure of the used prototype but with different sampling fraction. In particular, this last detector has more lead in its composition, and this is important to better study the efficiency enhancement that is predicted by Monte Carlo has due to the secondary interactions of neutron with the high Z material.

Acknowledgment
We thank M. Arpaia, G. Bisogni, A. Cassarà, A. Di Virgilio, U. Martini, A. Olivieri and all the mechanical LNF division for the help in the construction of the NE110 reference counter and the overall mechanical supports. We also acknowledge A. Balla for the help in the setting of the electronic chain. We cannot forget to thank P. Caponera and M. Rossi, which have boldly transported all materials in the long trip from LNF to TSL. Moreover we want to warmly acknowledge all the TSL staff for the help during the data taking. This work was partially supported by “Transnational access to Research Infrastructure” TARI - The Svedberg Laboratory, HadronPhysics I3, Contract No. RII3-CT-2004-506078.

References

[1] A. Del Guerra, NIM **135** (1976) 319-330.
[2] C.Birattari et al., NIM **A 297** (1990), 250-257.
[3] C.Birattari et al., NIM **A 338** (1994), 534-543.
[4] The KLOE collaboration, NIM **A 482** (2002) 364-386.
[5] LOI_MARCH_AMADEUS.pdf
in http://www.lnf.infn.it/lnfadmin/direzione/roadmap/.
[6] loi_06.pdf in http://www.lnf.infn.it/conference/nucleon05/FF.
[7] A.V. Prokofiev et al., PoS (FNDA2006) 016 (2006).
[8] A.V. Prokofiev et al., Journal of Nuclear Science and Technology, Supplement 2 (2002), 112-115.
[9] A. Antonelli et al., NIM **A 354** (1995), 352-363.
[10] A. Fassò et al., CERN 2005-10 (2005), INFN/TC_05/11, SLAC-R-773.
[11] A.Fassò et al., eConf C0303241 (2003), arXiv:hep-ph/0306267.

Journal of Physics: Conference Series **160** (2009) 012024 doi:10.1088/1742-6596/160/1/012024

Jet energy scale calibration in the D0 experiment

Jeroen Hegeman

Nikhef, Kruislaan 409, 1098 SJ Amsterdam, The Netherlands

E-mail: jhegeman@nikhef.nl

Abstract. Using the Run IIa data set the D0 experiment has reached a jet energy calibration precision on the level of 1–2% over a wide kinematic range. This paper presents the methods used and the results obtained. Special attention is paid to the remaining systematic uncertainties.

Most physics analyses at hadron colliders involve multiple jets, making the calibration of the jet energy scale an important input. Since many of the results based on the Tevatron Run IIa data set are limited by systematic uncertainties, any improvements in jet energy scale translate directly into improved precision of the published results.

1. The D0 detector

The D0 detector [1, 2] is a general purpose detector designed for the study of high-p_T effects in proton-antiproton collisions. Situated in the Fermilab Tevatron collider ring, during Run IIa it collected approximately 1 fb^{-1} of data at $\sqrt{s} = 1.96$ TeV.

The D0 detector uses a uranium-liquid argon sampling calorimeter system (figure 1) consisting of a central calorimeter covering the pseudorapidity region $|\eta| \lesssim 1.1$ and two endcap calorimeters extending the range to $|\eta| \lesssim 4.2$. Both central- and endcap calorimeters are housed in individual cryostats. The region between central- and endcap cryostats ($1.1 \lesssim |\eta| \lesssim 1.4$) is instrumented with a scintillator based inter-cryostat detector. The cell based readout provides a transverse segmentation of $\Delta\eta \times \Delta\phi = 0.1 \times 2\pi/64$ up to $|\eta| = 3.2$. More forward the cell size increases to avoid too small cells. The electromagnetic calorimeter consists of four layers with a total thickness of $\approx 20\ X_0$. The third EM layer, around the approximate electromagnetic shower maximum, is segmented twice as finely in both η and ϕ directions. The hadronic calorimeter contains three (central) or four (endcaps) fine hadronic layers followed by a coarse hadronic layer (using copper and stainless steel absorbers instead of uranium). The total depths of the central- and endcap hadronic calorimeters are $\approx 7.2\ \lambda$ and $8.0\ \lambda$ respectively. A scintillating fiber preshower detector covers the range $|\eta| < 1.3$. Preshower information is used both in EM cluster reconstruction and in photon identification. Intercalibration of the D0 calorimeter in rings of ϕ has been performed using $Z \to e^+e^-$ and QCD di-jet events [3]. The D0 detector simulation is based on the GEANT package.

2. Jet reconstruction

Jets are reconstructed using the 'Run II midpoint cone algorithm' [4]. Cells with the same η/ϕ coordinates are combined into pseudo-projective towers (figure 1b). Starting from seed towers, calorimeter towers within a cone of radius $\Delta\mathcal{R} = \sqrt{(\Delta y)^2 + (\Delta\phi)^2} < \mathcal{R}_{\text{jet}}$ around the seed are clustered. The jet axis is determined by four-vector addition of the clustered towers. This process

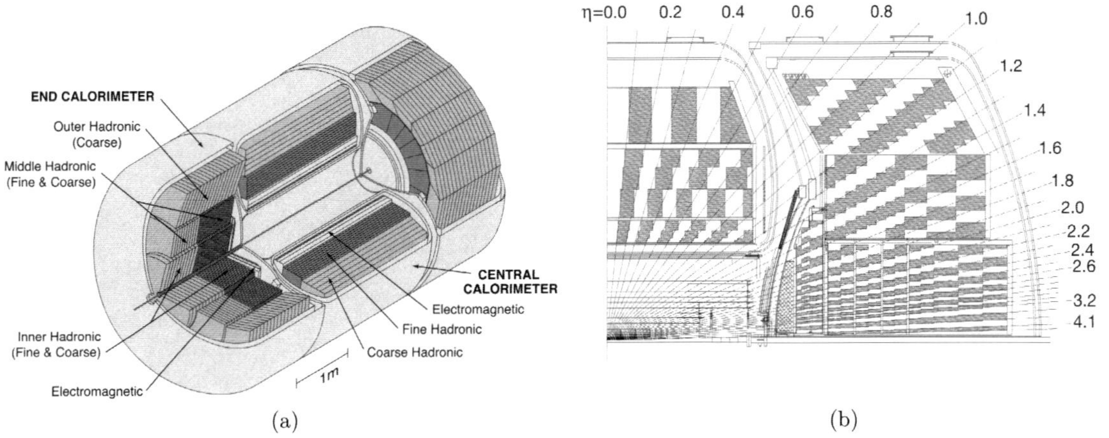

Figure 1: (a) Isometric view and (b) beam-parallel view of a portion of the D0 calorimeter system. The shading patterns indicate the calorimeter segmentation.

is repeated until the jet axis is stable. The midpoints between jet pairs are used as additional seeds to improve infrared safety. Identical solutions are removed and, depending on the amount of overlap, overlapping jets are either split or merged into a single jet. Two different cone sizes are used: 0.5 and 0.7 and the jet reconstruction threshold is $p_T \geq 6$ GeV/c. By using stable Monte Carlo particles instead of calorimeter towers the exact same jet reconstruction algorithm can be applied in simulation.

3. Jet energy scale corrections

Typically, physics analyses depend on data-to-Monte Carlo comparisons of reconstructed physics objects. To allow for such comparisons the jet energy scale corrects the jet energy as measured in the calorimeter back to the jet energy as reconstructed at the stable-particle level.

To avoid depending on the possibly imperfect detector simulation, especially considering the modified Run II detector geometry compared to the pre-Run I testbeam data, a data-driven calibration procedure is employed. Any Monte Carlo used employs a tuned version of the PYTHIA event generator.

The subcorrections making up the jet energy scale correction (eq. (1)) are (i) the offset subtraction O removing spurious energy, (ii) the absolute (R) and η-dependent (F_η) response and (iii) the out-of-cone showering corrections S correcting for lateral energy leakage out of the jet cone. Each subcorrection is calibrated individually and possible method biases are taken into account explicitly where necessary.

$$E_{\rm jet}^{\rm ptcl} = \frac{E_{\rm jet}^{\rm meas} - O}{F_\eta \cdot R \cdot S} \qquad (1)$$

It should be noted that the underlying event, consisting of the beam remnants and their connections with the hard scatter constituents, is an integral part of the physics process and neither can nor should be corrected for in the energy calibration procedure.

3.1. Offset subtraction

The offset correction subtracts all energy from the jet cone that is not related to the hard scatter event. Contributions from (both electronics and uranium) noise and pile-up energy from previous

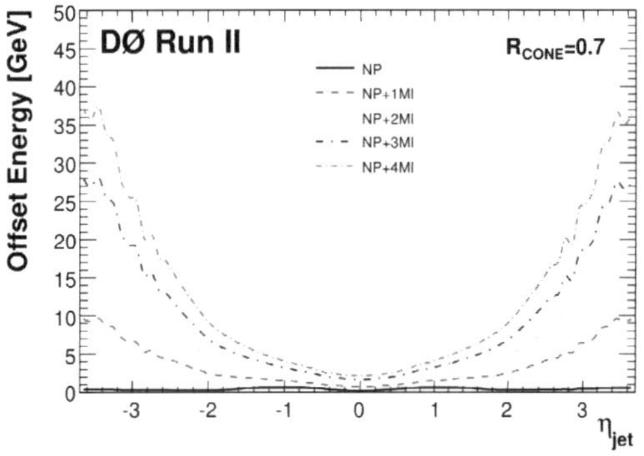

Figure 2: The noise/pile-up (NP) and multiple interactions (MI) contributions to the offset energy as a function of jet pseudorapidity for $\mathcal{R}_{\text{jet}} = 0.7$ cone jets.

bunch crossings is separated from the effects of possible additional proton-antiproton interactions. The noise/pile-up contribution is estimated from zero-bias data, data triggered on the presence of a bunch crossing, vetoed for inelastic interactions. Energy depositions due to multiple interactions are estimated from minimum-bias data: data triggered on the presence of a (possibly inelastic) interaction. The overall energy difference between minimum-bias events with N_{PV} reconstructed vertices and single vertex events is considered as the energy originating from $N-1$ additional interactions:

$$\text{MI}(N_{\text{PV}}) = \text{MinBias}(N_{\text{PV}}) - \text{MinBias}(N_{\text{PV}} = 1)$$

The 396 ns bunch crossing separation together with the high instantaneous luminosities delivered by the Tevatron result in a Poisson-distributed average number of up to five simultaneous proton-antiproton collisions per beam crossing. This implies that most events contain at least one interaction in addition to the hard scatter under study.

Both noise/pile-up and multiple interactions corrections are estimated as average energies over rings in ϕ for each η-ring of towers. This is translated to a per-jet offset correction (figure 2) based on the position of the jet axis and the cone size.

Calorimeter information is zero-suppressed: cells with negative energies or with a signal within 2.5 σ from their noise pedestals are dropped from the readout. The presence of jet energy thus locally increases the noise survival level leading to an underestimation of the offset energy inside jets. This effect is largely counteracted by the fact that photons, due to their small shower sizes, are less than jets affected by the zero-suppression, leading to a slight underestimate of the response. These two effects were studied in detail in Monte Carlo. The remaining effect is of the order of 1.5% below $p_{\text{T}} \approx 50$ GeV/c and is incorporated in the combined jet energy corrections.

3.2. Response corrections

The presence of the inner trackers, solenoid and preshower detector corresponds to a significant amount of material in front of the calorimeter: $\approx 4.0\ X_0$ at normal incidence, rising to $\approx 4.4\ X_0$ in the endcaps at $\eta = 2$. To reduce pile-up effects signal integration is restricted to the first ≈ 260 ns of each crossing. Compared to the ≈ 450 ns typical drift time in the liquid argon gaps this means only two thirds of the charge is collected. The result is a strongly non-compensating calorimeter: $e/h > 1$, making the response corrections by far the largest contribution to the overall jet energy scale.

3.2.1. The MPF method The MPF (missing E_{T} projection fraction) method uses the fact that in back-to-back tag-and-probe events tag and probe are perfectly balanced at the particle

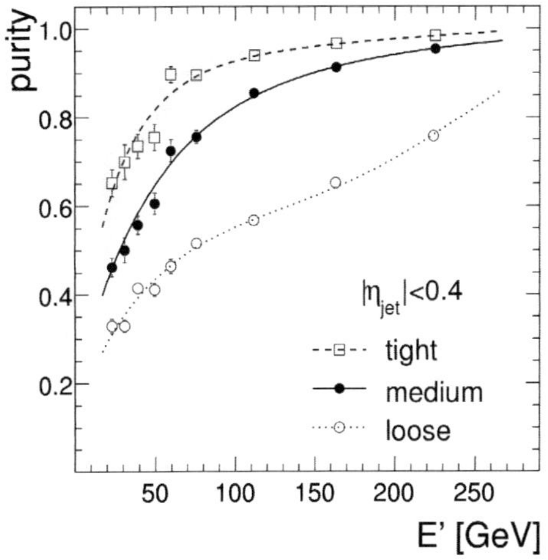

Figure 3: The purity of the photon+jet data sample used for response corrections estimated for three different photon selection criteria (loose, medium and tight). The band surrounding the curve corresponding to the tight photon selection corresponds to the total uncertainty on the photon purity.

level. Starting from a calibrated tag, any miscalibration of the probe will be reflected in missing transverse energy. The detector response to the probe can be expressed as:

$$R_{\text{probe}} = 1 + \frac{\vec{E}_{\text{T}} \cdot \vec{p}_{\text{T, tag}}}{\vec{p}^{\,2}_{\text{T, tag}}}.$$

To suppress the effects of the jet energy resolution all response corrections are parameterized versus the expected jet energy $E' \equiv p_{\text{T, tag}} \cosh \eta_{\text{jet}}$.

3.2.2. Absolute calorimeter response A photon+jet event sample using central photons ($|\eta| < 1.0$) is used to estimate the response of central jets ($|\eta| < 0.4$). The tag photon and probe jet are required to be tightly back-to-back: $\Delta\phi > 3.0$. The photon selection criteria include stringent cuts on the fraction of energy deposited in the EM calorimeter, shower shape information in the third EM layer and the preshower detector, as well as both track- and calorimeter based isolation cuts. Nevertheless the photon+jet sample contains a significant fraction of QCD di-jet contamination, typically due to jets with leading neutral pions. The photon purity was estimated for three different photon selection criteria (figure 3) in both Monte Carlo and data (using a template based method). Measured purities in data and MC were found to be in excellent agreement [5]. Corrections for di-jet contamination were studied in all three samples, bringing the response back to the same value in all cases. The final response correction (figure 4a) is based on the tight photon selection, giving the highest purity and the smallest purity correction. The uncertainties on the response (figure 4b) are dominated by the photon purity (low p_{T}) and the photon energy calibration (above $p_{\text{T}} \gtrsim 30$ GeV/c). It should be noted that the response corrections are known to a precision of $\approx 1.3\%$ up to the highest jet energies.

The MPF method balances the full hadronic recoil against the tag photon under the assumption that the jet fully contains the hadronic recoil. It is thus sensitive to additional energy in the event, e.g. jets below the reconstruction threshold. The effects of the reconstruction threshold, the choice of the $\Delta\phi$ back-to-backness cut and different underlying event models were studied in Monte Carlo where individual energy depositions can be compared to the true energies. The residual correction is of the order of $\approx 2\%$ and is incorporated into the combined jet energy scale corrections.

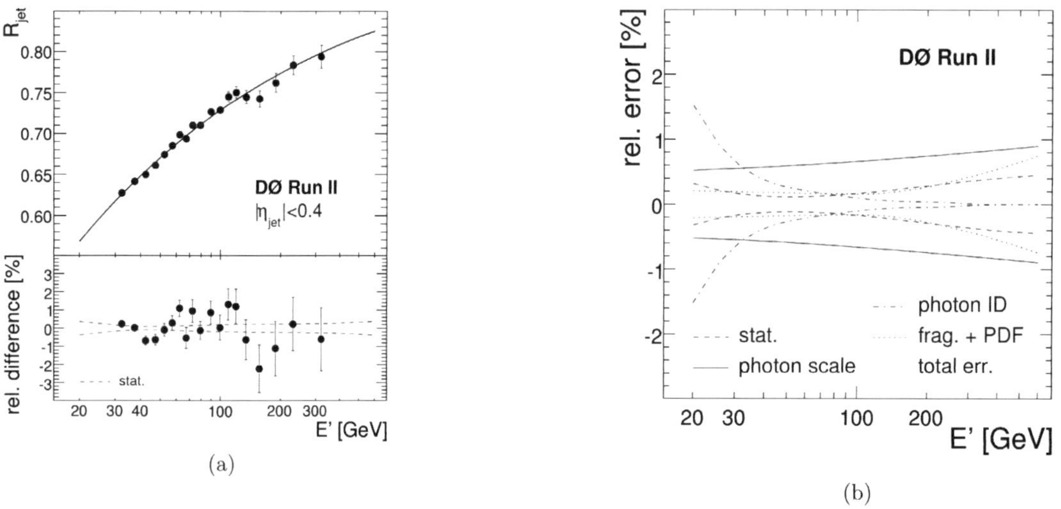

Figure 4: (a) The fitted absolute response for central $\mathcal{R}_{\text{jet}} = 0.7$ cone jets. The bottom inset shows the fit residuals and the statistical uncertainty on the fit. (b) The systematic uncertainties affecting the absolute response ($|\eta| < 0.4$). Except at low p_T the propagation of the photon energy calibration is the dominant contribution.

3.2.3. η-Dependent response corrections While the D0 calorimeter is fairly uniform within the central- and endcap regions the gaps between the cryostats are instrumented differently. Also, the amount of dead material depends on jet pseudorapidity.

The relative response correction calibrates forward jets with respect to central jets. Together with the absolute detector response this provides the response correction for forward jets.

The relative response is measured using the MPF method on both photon+jet and di-jet samples. The tag photon (jet) is required to be within $|\eta| < 1.0$ ($|\eta| < 0.4$), the probe jet is unrestricted in pseudorapidity. Tag and probe are required to be back-to-back: $\Delta\phi > 3.0$. The di-jet sample reaches higher in jet transverse momentum and is used to determine the shape of the response corrections at high p_T.

The different quark/gluon compositions of jets in photon+jet and di-jet events lead to different response corrections. Whereas photon+jet events at low p_T are dominated by quark jets, the gluon jet fraction increasing with increasing p_T, di-jet events show the opposite behavior. Gluon jets tend to be wider, containing more particles with a correspondingly lower average particle energy. Both the tracker magnetic field as well as the presence of dead material lower the response of low energy particles, resulting in an overall lower response for gluon- than for quark jets (figure 5). Since forward jets traverse more material before reaching the calorimeter, this effect becomes stronger for jets in more forward regions.

Figure 6 shows a summary of the relative response corrections for $\mathcal{R}_{\text{jet}} = 0.7$ cone jets in both photon+jet and di-jet events for a wide range of jet energies. The largest corrections are needed in the inter-cryostat regions: $0.5 \lesssim |\eta| \lesssim 1.8$. The influence of dead material increases at larger angles, making the relative response drop strongly in the forward direction.

The dominant systematic uncertainties result from imperfections in the parameterizations of the energy dependence and the η-interpolation.

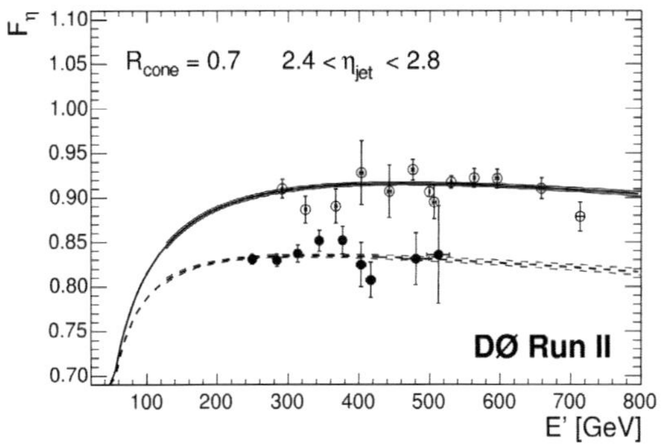

Figure 5: Relative response in data for $\mathcal{R}_{\text{jet}} = 0.7$ cone jets within $2.4 < |\eta| < 2.8$ for both photon+jet (bottom curve) and di-jet (top curve) events. Compared to di-jet events photon+jet events contain more gluon jets, resulting in a lower response.

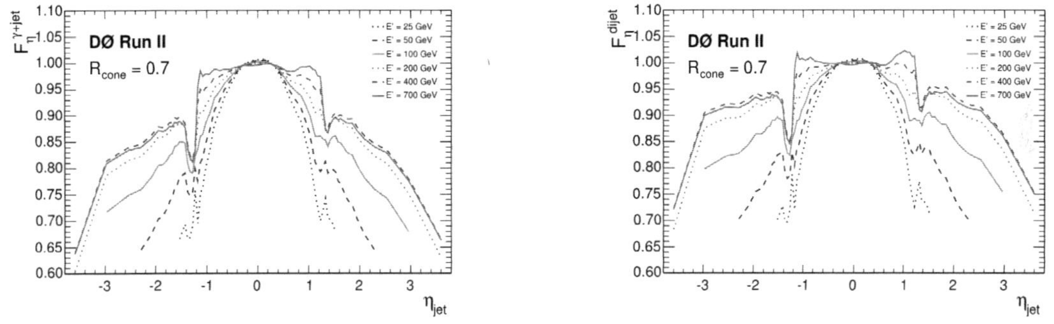

(a) (b)

Figure 6: Relative response corrections F_η in data (a) for photon+jet events and (b) for di-jet events.

3.3. Out-of-cone showering correction

The cone jet algorithm reconstructs the jet energy as the sum of all energy present within the jet cone. Due to detector effects like particles bending in the magnetic field as well as parts of the shower leaving the jet cone in the calorimeter, the energy inside the cone may differ from the true jet energy. The out-of-cone showering correction corrects for the net energy difference due to these detector showering effects.

Using back-to-back photon+jet events in Monte Carlo without any noise and/or multiple interactions the energy deposited by each particle can be tracked in the detector simulation. Distinguishing between particles assigned to the particle level jet and other particles allows for the creation of jet- and non-jet energy deposition profiles as a function of the radial distance $\Delta\mathcal{R}$ away from the jet axis. After offset subtraction a template fit to the data is used to estimate the true energy distributions from particles inside/outside the jet (figure 7). The net showering correction can be obtained from the jet- and non-jet energy profiles after the fit:

$$S = \frac{\int_0^{\mathcal{R}_{\text{jet}}} (E(\text{jet}) + E(\text{non-jet}))}{\int_0^{\infty} E(\text{jet})}$$

Since the out-of-cone showering correction is dominated by edge effects the correction becomes larger for smaller jet cone sizes. Similarly the correction becomes more prominent for jets at

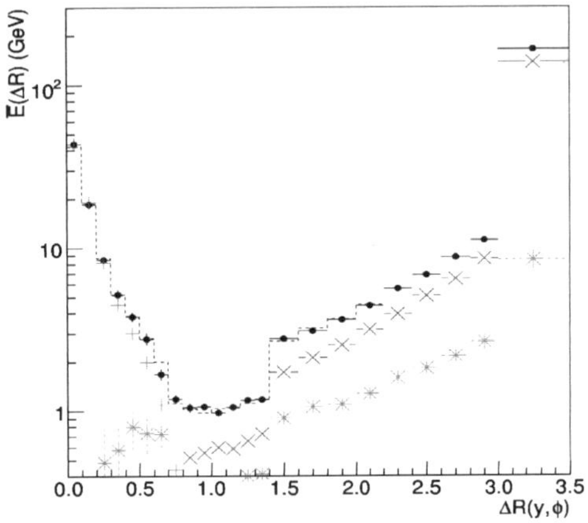

Figure 7: The jet- (+) and non-jet (×) energy distributions fitted to data for central $\mathcal{R}_{\rm jet} = 0.7$ cone jets ($|\eta| < 0.4$) of $100 < p_{\rm T} < 130$ GeV/c. The full circles show the data, the dashed line the template the sum of the fitted energy templates. The shaded band around the fit results reflects the statistical fit uncertainty. The offset energy profile (∗) is kept fixed in the fitting.

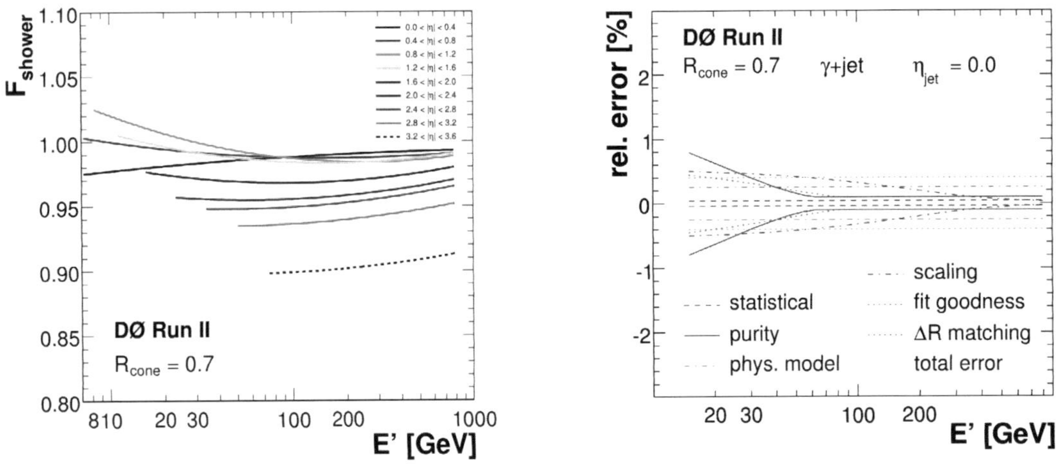

(a) (b)

Figure 8: (a) Out-of-cone showering corrections for $\mathcal{R}_{\rm jet} = 0.7$ cone jets in different pseudorapidity ranges and (b) the systematic uncertainties for central jets. The dominant uncertainties are all related to the shape of the jet energy distribution profile and it's agreement with the data. The uncertainty marked 'scaling' reflects the effect of down-scaling the single pion response in Monte Carlo to match the low-$p_{\rm T}$ jet response between data and Monte Carlo.

larger pseudorapidities where the physical size of jets decreases (figure 8a). The dominant systematic uncertainties to the out-of-cone showering corrections (figure 8b) are all related to the dependability of the jet energy profile. Impurities in the photon+jet sample modify the relative quark/gluon content which is translated directly into the shape of the jet profile. The same holds true for the underlying physics models for fragmentation and hadronisation as well as for the underlying event modeling. Uncertainties due to these effects were studied in Monte Carlo. A uniform uncertainty was added to account for the differences in fit agreement over the different pseudorapidity ranges.

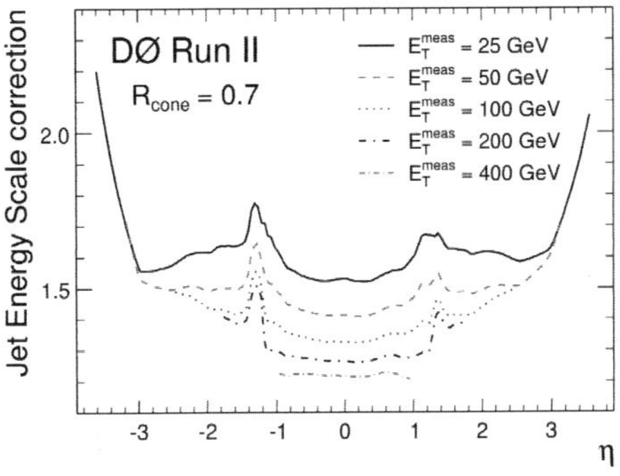

(a)

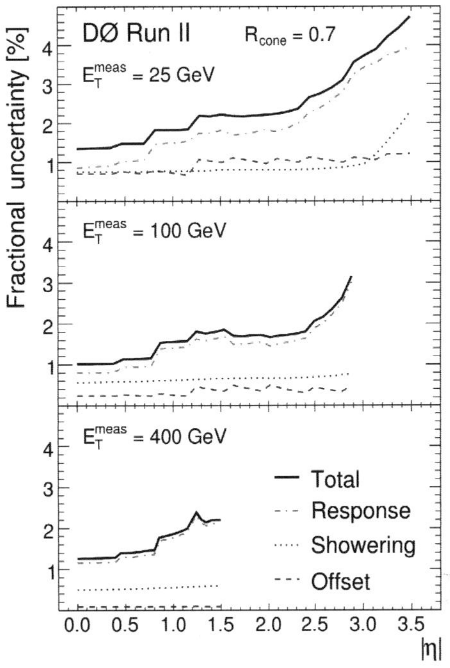

(b)

Figure 9: Total energy scale corrections for $\mathcal{R}_{\text{cone}} = 0.7$ cone jets figure 9a and the corresponding relative uncertainties figure 10b (total: solid line) split into contributions from offset (dashed line), response (dash-dotted line) and showering (dotted line) corrections

4. Combined jet energy scale corrections

Figure 9 shows the combined jet energy scale corrections and the corresponding uncertainties. Over a wide kinematic range the uncertainties are of the 1–2% level. All subcorrections are known to the same level of relative precision. Due to the size of the response correction this becomes the dominant uncertainty on the combines jet energy scale. At low p_T the uncertainty is dominated by the purity of the photon+jet sample. At the high p_T end the uncertainties are affected by decreasing statistics in the di-jet sample. Overall the limiting factors are the photon purity and the electromagnetic energy calibration.

The high precision jet energy calibration is directly visible in physics results. An example is the recent measurement of the inclusive jet cross section to a precision of $\approx 10\%$ [6].

Acknowledgments

The results presented here are the outcome of many long term efforts of the D0 jet energy scale group and would not have been possible without the support of the whole D0 collaboration. The author would also like to thank the Stichting voor Fundamenteel Onderzoek der Materie (FOM) in the Netherlands, as well as the organizers of the CALOR 2008 conference.

References

[1] Abachi S *et al.* (D0) 1994 *Nucl. Instrum. Meth.* **A338** 185–253
[2] Abazov V M *et al.* (D0) 2006 *Nucl. Instrum. Meth.* **A565** 463–537 (*Preprint* physics/0507191)
[3] Peters K 2006 *AIP Conf. Proc.* **867** 17–24 (*Preprint* physics/0608002) URL http://www.slac.stanford.edu/spires/find/hep/www?eprint=physics/0608002
[4] Blazey G C *et al.* 2000 (*Preprint* hep-ex/0005012) URL http://www.slac.stanford.edu/spires/find/hep/www?eprint=hep-ex/0005012
[5] The D0 collaboration *Paper in preparation for Nucl. Instrum. Meth.*
[6] Abazov V M *et al.* (D0) 2008 *Submitted to PRL* (*Preprint* 0802.2400)

Particle Flow Algorithm and calorimeter design
Journal of Physics: Conference Series

Jean-Claude Brient
Laboratoire Leprince-Ringuet
CNRS-IN2P3/Ecole polytechnique

E-mail: brient@llr.in2p3.fr

Abstract. The boson tagging at ILC, based on di-jets masses reconstruction, has lead to consider the reconstruction of multijet events as a major goal for the detector on this machine. One technique to optimise the multijet reconstruction is called Particle Flow Algorithm (PFA) and consists in software compensation and individual particle reconstruction. The main direction of the R&D studying calorimetry optimised for PFA, is concentrating on fine granularity calorimeters with a high degree of longitudinal segmentation. These studies include comparison of simulation models with data to measure their degree of agreement, the technical issues of building a detector optimised for PFA calorimetry, and development of algorithms for software compensation and particle flow reconstruction. Several technologies for electromagnetic and hadronic calorimeters are pursued with prototypes in test beam and new generation prototypes , very close to ILC like detector are under development with a very high degree of integration.

1. Introduction
The boson tagging is essential to fulfil the physics program at the International Linear Collider, which is mainly related to W, Z and Higgs boson production studies. The majority of these boson decays to quark-antiquark, leading to dijet decays of the boson. The optimisation of the dijet mass resolution will improve the tagging of the different bosons. A technique called PFA for Particle Flow Algorithm has been proposed [1] to optimise the dijet mass resolution. The detector could therefore be optimised to achieve the best possible performances in the PFA, that is to reconstruct individually each particle of the final state. It is possible to classify the particles in a jet in 3 "species": The charged tracks, the photons and the neutral hadrons (neutrons, neutral kaons). The short lived neutral hadrons (K°short and Λ are reconstructed in the tracker device and are no longer considered as neutral hadrons). The charged tracks are reconstructed from the tracker device, the photon are reconstructed in the electromagnetic calorimeter (ECAL) and the neutral hadrons in the ECAL and hadronic calorimeter (HCAL). This is supposing that it is possible to disentangle for each particle the deposited energy in the calorimeter. This is the goal of the calorimeter design to optimise the PFA performances. In this case, the energy resolution can be written like $\sigma^2 \text{jet} = \sigma^2\text{ch.} + \sigma^2\gamma + \sigma^2 h^0$. With a typical charged track resolution on momentum ($\Delta p/p \sim$ few 10^{-4}), an ECAL energy resolution ($\Delta E/E \sim 12\%$) and a typical energy resolution in HCAL of $\Delta E/E \sim 45\%$, the jet energy resolution could be as good as $\sigma_{jet}=0.14\sqrt{E_{jet}}$, which is by far much better than what was obtained up to now with calorimetric approach

2. Starting from Physics

An important part of the physics programme at ILC is based on multi boson final state. It is summarised in the following table 1.

Table 1. main physics processes to be studied at ILC

Bosons only	Fermions-bosons
ZH	e+e-H, e+e-Z
WW	νν H, νν Z
ZZ	tt H , νν HH
ZHH	e ν W
ZWW	ννWW, ννZZ

The Z and W bosons decay predominantly to quarks pair, with a ratio of branching fraction BR(Boson →q qbar)/BR(Boson →Leptons) of 7 for the Z boson, 2.1 for the W boson and about 5.7 for a standard model Higgs with a mass of 120 GeV/c². The situation of the Higgs could be more complicated if the Higgs is non-standard. Therefore the multijets final state could be used to tag the boson using their reconstructed masses. It is possible only because of the signal to background ratio (i.e. the background of qqbar production for multijets final state) at the cross section level which is favourable (less few order of magnitude). It is illustrated in figure 1

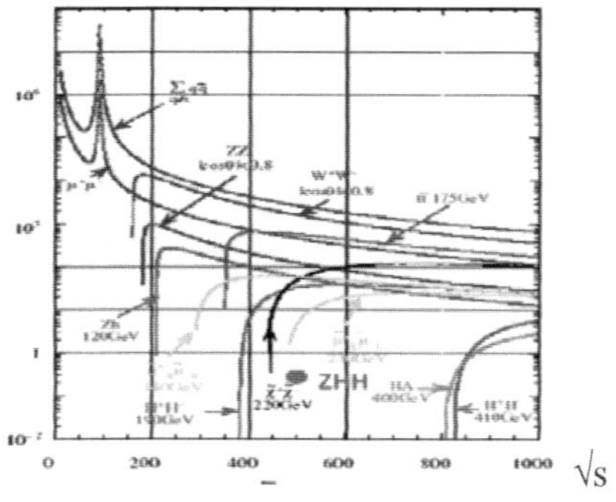

Figure 1. Some interesting cross section in e+e- interaction from Z peak to 1 TeV

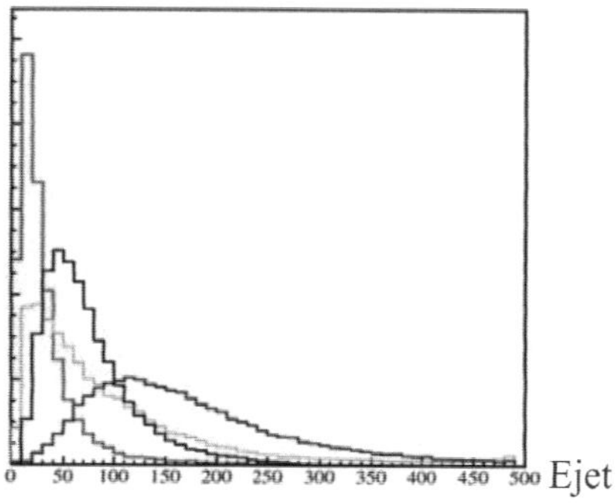

Figure 2. Jets energy distribution for ννH (2jets) in red, ttbar (green) and WW (blue) at 1 TeV and ZH(with 120GeV mass) at 500 GeV

The energy distribution of these jets shows a peak at low energy, with most of the distributions below 250 to 300 GeV. Figure 2 shows some examples of these distributions.

3. The jet energy resolution for the ILC physics programme

The energy resolution on the jets can be written as follows: $\Delta E = a \sqrt{E} + b\, E + c$, where E is the jet energy. Studies have been performed [2] showing that a=0.3 and b=0 fit the requirement for the physics programme at ILC. It is illustrated on figure 3, with the ZZ/WW separation based on the two dijet masses. The two peaks at different masses (W,Z) is clearly seen for a=0.3 while it is largely diluted for a=0.6. Similar studies have been done for Higgs mass measurement, for Higgs couplings, but also for SUSY, for Higgs CP violation decays, analysed with tau decays, etc...

4. Particle Flow Algorithm

One way to obtain the value of a=0.3 and b=0., could be to use the method called Particle Flow Algorithm (PFA), which consists in reconstructing each particle individually [1,3]. This method has been partially already used at LEP, for example by ALEPH. The idea is to reconstruct the charged tracks in the tracker device, then to use the ECAL for the photon(s) and the ECAL and HCAL for the other neutral particle(s) (K°,neutron). An important point to note is the difference with the energy flow method, where balance of energy is made between momentum measurement given by the tracker device and the energy deposited in the calorimeter.

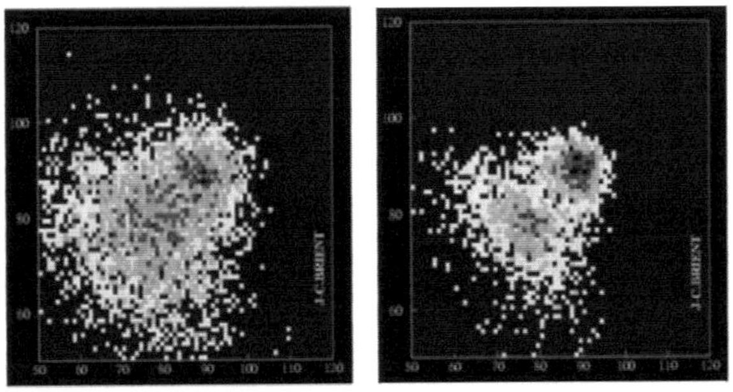

Figure 3. WW and ZZ separation in the 4jets events, with the two dijets masses. Right for a=0.6 and left for a=0.3 (a is defined in the text)

For PFA, the tagging of particle is given by topological separation between showers from charged tracks and from neutral particles. Such an approach can be compared to the calorimetric approach for what concerns the resolution. It is summarised in the table 2.

Table 2. a,b, and c for a jet energy resolution given by $\Delta E = a \sqrt{E} + b E + c$

Detectors	**a**	**b**	**c** (GeV/c²)
ALEPH *(real data)*	0.6	0.	0.6
ATLAS central region	0.6	0.03	0.
H1 (*real data)*	0.5	0.05	0.
Goal for ILC detector	0.3	0.	0.5

The topological reconstruction leads to a kind of recommendation's list for the detector and more precisely for the calorimeter.

A good lateral granularity gives a better separation. It is mandatory to have a good longitudinal segmentation, in order to avoid showers mixing (helping for good pattern recognition of the showers). The calorimeter has to be at a large distance from the interaction point giving a better opening of the jets, making easier to separate the showers. A large magnetic field will also help because of_the bending of the charged particles. Eventually, in order to have a good efficiency to reconstruct all the low energy particles, the coil has to be put outside the calorimeter, avoiding having too large matter in front of the calorimeter. From the recommendation list, we can extract some logical consequences for a calorimeter dedicated to PFA performances:

- A calorimeter with a very large number of channels ("high number of pixels for the shower picture), leading to a unprecedented level of integration
- A very compact calorimeter , which reduces the radius of the coil

- Prototype in test beam to debug such a device (essential for such brand new view of this type of device)

Among the consequences, we can list (non-exhaustive): the VFE electronics has to be embedded in the detector, leading to the problem of power dissipation, the local zero suppress has to be done locally, a very strict care of the common mode has to be performed in order to maintain a very good signal over noise ratio at the mip level, etc... and eventually , the active device must be very stable in time but also in temperature, etc...

5. The expected performances with PFA

A large amount of work has been devoted to develop a good PFA reconstruction program in the last 2 years. Interesting results based on full GEANT4 simulation and full reconstruction has been obtained. The simulated detectors were similar, with gas tracker device but with different types of calorimeters. However, both reconstruction programs are based on PFA approach. Figure 4 shows the jet energy resolution curve as a function of the jet energy, for calorimetric approach from ATLAS and H1, and ALEPH for PFA. Moreover, on the figure, there are results from M.Thomson [4] for silicon diode ECAL and results from A.Miyamoto et al [5] for scintillator ECAL and HCAL. The results shows a value of the parameter a better than 0.3 and b very small as long as the jet energy is in the good region for the ILC physics program (below 250-300 GeV).

The calorimetric approach is less interesting even with a good stochastic term because of the constant term which is inherent to this approach and due to dead zone, inter-calibration, non-linearity, electronics response, etc... The constant term per pixel could be relatively large in PFA approach without playing a role because of the large number of pixels per GeV and per particle. A simulation with 5% constant term per pixel has been shown to be reduced to negligible effect in PFA approach.

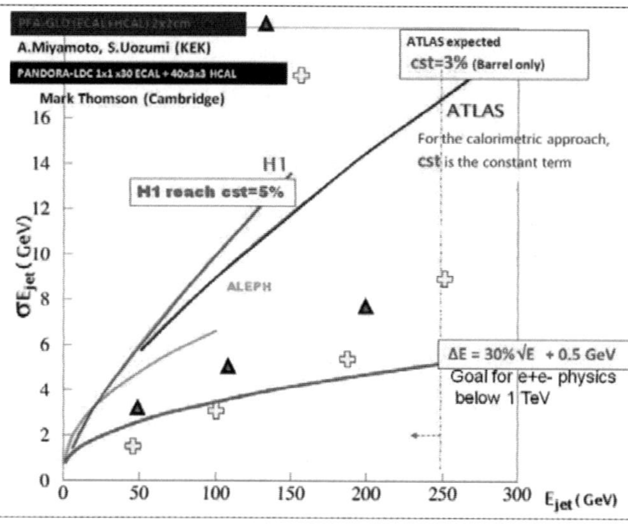

Figure 4. Jet energy resolution for calorimetric approach (H1, ATLAS) and for ALEPH (partial PFA) and for fully simulated GEANT4 events, reconstructed by two different PFA programs

5.1. Questioning the expected performances

The quality of the description of the hadronic shower in GEANT4 is relatively poor, especially when speaking about width and depth at such a level of precision (cm size). A first attempt has been made to change the physics list in order to observe the impact on the expected performances. The results of the study are shown in Table 3. Between these 2 models, the hadronic shower width differs by about 20%, therefore, as a first conclusion; it can be said that the jet energy resolution based on PFA depends only marginally on the hadronic shower model.

The second possible problem could be the real condition for a detector with a very large number of readout channels. The noise, the dead zone of the geometry, the intercalibration, etc.. which could cause large degradation in the performances could be tested in real condition during the test beam period

Table 3. Variable a as function of energy and hadronic shower model used in the detector simulation. (studied by M.Thomson [4])

Hadronic physics list in GEANT4	Jet energy (GeV)	Value of a
QGSP_BERT	45	0.226
LHEP	45	0.232
QGSP_BERT	100	0.293
LHEP	100	0.302

6. The CALICE collaboration

A large collaboration, called CALICE for **CA**lorimeter for **LI**near **C**ollider **E**xperiment [6], is working on such device since 2002. The collaboration is made of about 280 physicists/engineers from 45 institutes and 12 countries. Electromagnetic calorimeters (ECAL) as well as hadronic calorimeters (HCAL) are studied in the collaboration. Three different types of ECAL are proposed to reach the goal. Two are more or less classical, first a silicon diode–tungsten sampling calorimeter, then a scintillator strip-tungsten sampling calorimeter but read by MPPC, the new photo counter in fashion. The third one is based on MAPS, (CMOS Active Pixel Sensors) with digital readout of pixel with size of 50x50 μm^2. The last one is still at the level of first demonstration of feasibility, while the two "classical" are already prototyped with test beam for 2 years at CERN and now at Fermilab. On the side of hadronic calorimeter (HCAL), the classical solution is proposed as a sampling device made of stainless steel and very small scintillator tiles (3x3 cm^2) read by silicon photomultiplier. In addition, a brand new type of calorimeter is proposed with gas devices (RPC, GEM or Micromegas) with pad size of about 1x1cm^2 and read only on 1 to 3 bits (DHCAL). In this last case, the hadronic energy would be given by the shower development (size, depth, area and volume) and believing GEANT4, with an energy resolution better than the one with tile and analogue readout. This last point has clearly to be demonstrated and the concept has to be validated

7. Prototypes in test beam

Test beam has been performed in 2006 at DESY with electrons from 1 to 6 GeV and at CERN H6b with electrons, pions and protons, with energy from 8 GeV to 120 GeV. The figure 5 hereafter shows the ECAL prototype silicon-tungsten in front of a prototype of 1m3 of AHCAL (tile-iron) behind and finally a tail catcher (TCMT), made of strip of scintillator , used to monitor the longitudinal leakage of high energy shower. The level of complexity of assembling and running is illustrated by figure 6, which shows the PCB and the readout cables for the ECAL prototype with its 9720 channels in an active volume of 18x18x18 cm^3. The test beam has allowed checking the level of noise, the stability of the response, the readout and DAQ system capability and more generally the quality of the detector during the test beam period.

Figure 5. Test beam area at CERN H6b, with the prototypes of ECAL, of Tile HCAL and the one of TCMT behind.

Figure 6. The silicon tungsten ECAL prototype. The number of readout lines and cables can be noted.

8. Test beam for PFA

In addition to the standard goal for test beam, we want to use the test beam data to study the issue of the hadronic shower. Among the related question, we can list:

- The separability of the clustering on the hadronic shower of the PFA by superposing real interaction taken in our prototypes.
- The test of the hadronic shower model in the GEANT4 simulation (depending of particle type and energy)
- The energy resolution of each device AFTER clustering, which the relevant parameter for PFA approach

The test at low energy in progress at Fermilab is an important part of this programme. It is illustrated by figures 7 and 8, showing the on-line event display of typical events of this test beam data.

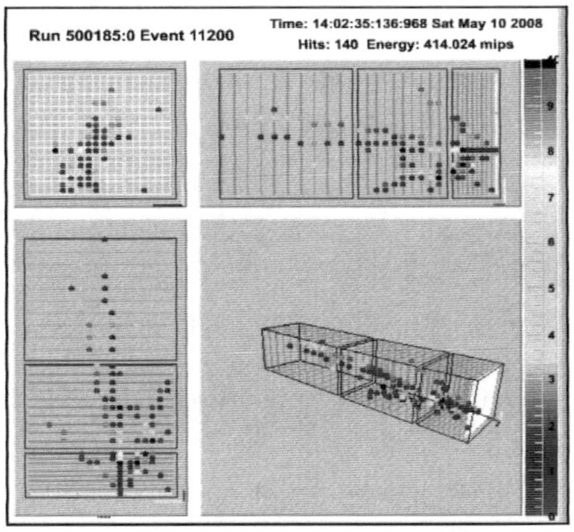

Figure 7. 8 GeV pion interacting in ECAL at the MTBF-FNAL test beam spring 2008

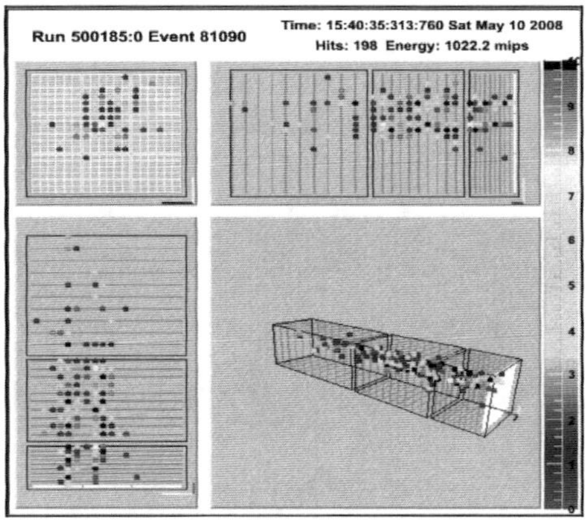

Figure 8. Simultaneous arrival at the FNAL test beam. Visual inspection confirms easy separation between the 2 showers

The data taken from 2006 to 2007 at DESY and CERN shows at least a satisfactory global behaviour of the detector. Important points were discovered during this test beam, such as the capacitive coupling between floating guarding and external pixels in the silicon wafers for the ECAL or the need for very precise monitoring of the data condition for the use of silicon photomultiplier. The analyses are in good progress and preliminary results will be shown at this workshop.

9. Conclusion

To optimise the use of the luminosity for an ILC machine, the decays of the bosons to jets have to be used for physics analysis. It is possible, because the signal to background ratio of the cross section e^+e^- to multijets final state is favourable, that is below few order of magnitude before selection. A new technique, called the Particle Flow Method, has been introduced to optimise the jets reconstruction in the jet energy region below 300 GeV. Results of full GEANT4 simulation and reconstruction with this technique have shown performances such as the ones needed for the ILC programme up to a centre of mass energy of 1 TeV. These results are based on simulation of ultragranular calorimeters, which give the best performances for PFA.

The CALICE collaboration designs prototypes of such ultragranular device, gaining expertise in this brand new domain for calorimeter, where the key word is reconstruction and not the energy resolution of single particle. Test beam data has been taken with the first generation prototype, at DESY and CERN. New test beam period, devoted to very low energy hadron, is in progress at Fermilab. Several tens of millions of interactions of electrons, positrons, and different types of hadrons with the first generation prototype have been registered and analyses are in good progress, with preliminary results shown at this workshop and papers to come.

REFERENCES

[1] J.C. Brient *Summary of the calorimeter session.* . Oct 2000. 5th International Linear Collider Workshop (LCWS 2000), Fermilab, Batavia, Illinois, 24-28 Oct 2000. Published in *Batavia 2000, Physics and experiments with future linear e+ e- colliders* 893-903

[2] T.Barklow *Physics Requirements for Jet Energy and Di-Jet Mass Resolution Physics Requirements for Jet Energy and Di-Jet Mass Resolution,* Presented at ECFA-LC and ILC08, Warsaw, June 2008.
J.C. Brient *Measurement of the Higgs decays into W W* at future e+ e- linear colliders.* LC-PHSM-2004-002, 2004. 12pp.

[3] J.C. Brient *Improving the jet reconstruction with the particle flow method: An introduction.* Mar 2004. 7pp. 11th International Conference on Calorimetry in High-Energy, Perugia 28 Mar - 2 Apr 2004. Published in *Perugia 2004, Calorimetry in particle physics* 445-451

[4] M.Thomson, *Pandora performances*, presented at the ECFA-LC workshop, Warsaw, June 2008

[5] A.Miyamoto et al, *PFA performances*, presented at the TILC08 workshop Sendai, March 2008

[6] CALICE collaboration, http://llr.in2p3.fr/activites/physique/flc/calice.html

Results of the CALICE Scintillator ECAL beamtest at DESY

Daniel Jeans, for the CALICE collaboration
Department of Physics, University of Kobe, Kobe, Japan
E-mail: jeans@people.kobe-u.ac.jp

Abstract. A prototype ILC scintillator-Tungsten electromagnetic calorimeter was constructed, with active layers consisting of scintillator strips, individually read out by MPPC devices. This prototype was exposed to positron beams at DESY in March 2007. Preliminary results of the analysis of the collected data are presented.

1. Introduction

A future detector at the ILC requires excellent jet energy reconstruction to maximise the physics output of the accelerator. The ability to distinguish the hadronic decays of Z and W bosons is one of the most stringent requirements, leading to a required relative jet energy resolution of around 3%. One way to achieve this resolution is by means of a particle flow approach, which uses the tracking system to estimate the charged jet energy, and the calorimeter to measure only the neutral energy component. The success of this approach depends on the ability to distinguish the energy deposited in the calorimeter by individual particles within a jet, allowing neutral energy deposits to be reliably identified. This leads to the requirement of a calorimeter with small Molière radius to reduce the overlapping of showers, and a calorimeter with finely segmented readout.

In this paper we describe a prototype of an ECAL designed to satisfy these requirements. A small sampling calorimeter was constructed, with active layers consisting of small scintillator strips, individually read out by photon detectors, and absorber layers of tungsten, respectively giving the required high granularity and small Molière radius.

This prototype was exposed to positron beams with momenta between 1 and 6 GeV/c at the DESY-II electron synchrotron. The preliminary results of the analysis of the collected data are presented in this paper.

2. Scintillator ECAL prototype detector

The ECAL prototype consisted of 26 layers of scintillator and tungsten, as shown in figure 1. Each active layer consisted of 18 1×4.5 cm^2 strips of 3mm thick scintillator, arranged in two rows, giving a total cross–section of 9×9 cm^2. A photon detector was attached to the end of each strip to read out the produced light. The photon detector used was the Multi Pixel Photon Counter (MPPC) produced by Hamamatsu Photonics [1]. This has a high photon detection efficiency, very compact size, and a reasonable cost, making it a suitable candidate for such a calorimeter at the ILC. The strips were orientated to be perpendicular in adjacent layers, giving an effective granularity close to 1×1 cm^2.

The tungsten absorber layers had a thickness of 3.5 mm, and a gap of around 1 mm was used to route the cables to extract the MPPC signals.

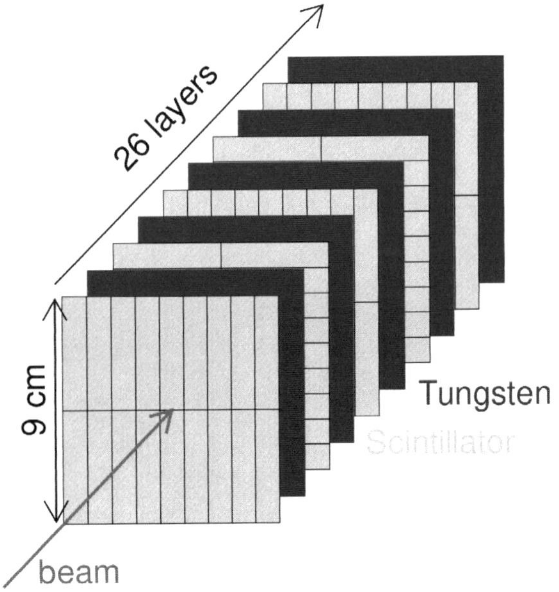

Figure 1. Cartoon showing the geometry of the first few detector layers. In total, 26 9×9 cm^2 layers, each consisting of 18 $4.5 \times 1.0 \times 0.3$ cm^3 scintillator strips, were interleaved with 0.35 cm thick tungsten plates. The strip orientation was alternated in adjacent layers.

Three types of scintillator strip were tested. Two types were based on a "megastrip": a $9 \times 4.5 \times 0.3$ cm^3 tile of Kuraray scintillator, as shown in figure 2. The tile was machined with eight pairs of thin grooves, into which white PET film was placed, giving a total of nine strips with reasonable optical isolation. One type of megastrip also had a 1mm hole drilled along the length of the strip, into which a wavelength–shifting fibre (WLSF) was inserted. The megastrip tiles were covered with 3M reflector film.

Scintillator strips manufactured by an extrusion method were also tested. These were produced by the Misung Chemical Company (Korea) in association with Kyungpook National University. These strips had dimensions of $4.5 \times 1 \times 0.3$ cm^3, and were produced with a co–extruded coating of TiO_2 (for light shielding) and a central hole with 1mm diameter (into which a WLSF was placed). A photograph is shown in figure 3. The cost of producing such extruded strips is significantly less than that of the megastrip tiles.

Three half–detectors of 13 layers were constructed, each using a different type of scintillator. Different arrangements of these half–detectors were tested: WLSF readout megastrip upstream of direct readout megastrip; direct readout megastrip upstream of WLSF readout megastrip; and extruded scintillator upstream of WLSF readout megastrip. Since the maximum of EM showers occurs at around layer 6 or 7 in this detector, the upstream half–detector measures most of the deposited energy.

The MPPC signals were read out using the CALICE readout electronics and DAQ developed by our CALICE colleagues at DESY, LAL-Orsay, and in the UK. They are essentially the same as those used by the CALICE Analogue–HCAL group.

3. Beamtest

This detector prototype was exposed to beams of positrons at the DESY laboratory in March/April 2007. Positron beams with a tunable momentum between 1 and 6 GeV/c were

Figure 2. Photograph of megastrip tile made of Kuraray scintillator, showing the machined grooves for optical isolation, and central holes for the WLSF.

Figure 3. Photograph of a Misung C.C. extruded scintillator strip, with inserted WLS fibre.

provided by the DESY-II electron synchrotron. The beamline was instrumented with four drift chambers to precisely measure the beam position on an event-by-event basis, and various scintillator trigger and veto counters were used to trigger on and select interesting events. The detector prototype was placed on a movable stage, allowing the beam to be scanned over the front face of the detector.

4. Detector calibration

To calibrate the detector, the Tungsten absorber plates were removed from the calorimeter, and data were taken with a 3 GeV/c positron beam. The signals on the trigger and veto counters and the drift chambers were required to be consistent with the passage, without showering, of a single positron along the beamline. Events in which the positron had passed through a particular strip were selected by considering the signals on the scintillating strips in the two up– and down–stream layers with the same strip orientation, and requiring them to be consistent with a single positron passing through the strip, parallel to the beamline. Such a calibration sample was defined for each of the strips in the detector. An example of the MPPC signal, after pedestal subtraction, in one of these samples is shown in figure 4. The distribution is well fitted by a Gaussian–convoluted Landau function. These signal distributions were used to extract the calibration constant for each of the scintillator strips.

Since the MPPC properties vary significantly with temperature, the dependence of these calibration constants on temperature was measured. Several calibration runs were taken during data–taking, allowing their dependence on the detector temperature to be extracted. All readout channels showed a similar temperature dependence of their calibration constant of around -2% per Kelvin.

The uniformity of the MIP response along the length of the scintillator strip was also measured. The beam was scanned over the entire front face of the detector, illuminating all regions of the strips. The drift chambers were used to determine the precise position at which the positron intersected the strips. The response to single positrons was measured as a function of the distance along the strip axis (i.e. the distance from the MPPC position); the results for a number of strips are shown in figure 5. The most uniform (and also the largest) light yield was shown by the WLSF read out megastrip, with a variation of around 80-90% along the strip length. The direct read out megastrips showed a slightly larger non-uniformity, of around 70-80%, due to light attenuation inside the scintillator. The extruded strips showed a significantly larger non-uniformity, of around 40-50%.

The reasons for the non-uniformity of the extruded strips was investigated in a separate

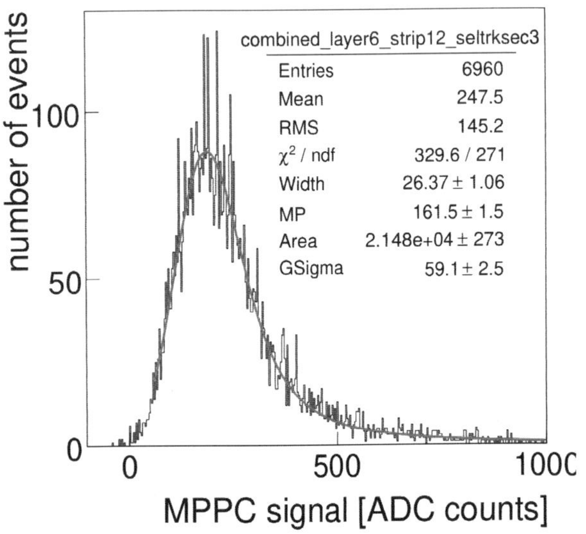

Figure 4. An example of the measured MIP signal in a single scintillator strip, used to calibrate the detector response. The measured distribution (black histogram) is fitted by a Landau distribution convoluted with a Gaussian (red curve).

beamtest. Non-perfect matching between the WLSF and the MPPC, and scintillator with a relatively short attenuation length were identified as the major causes. Improved samples of scintillator are under development and will be used in the next prototype.

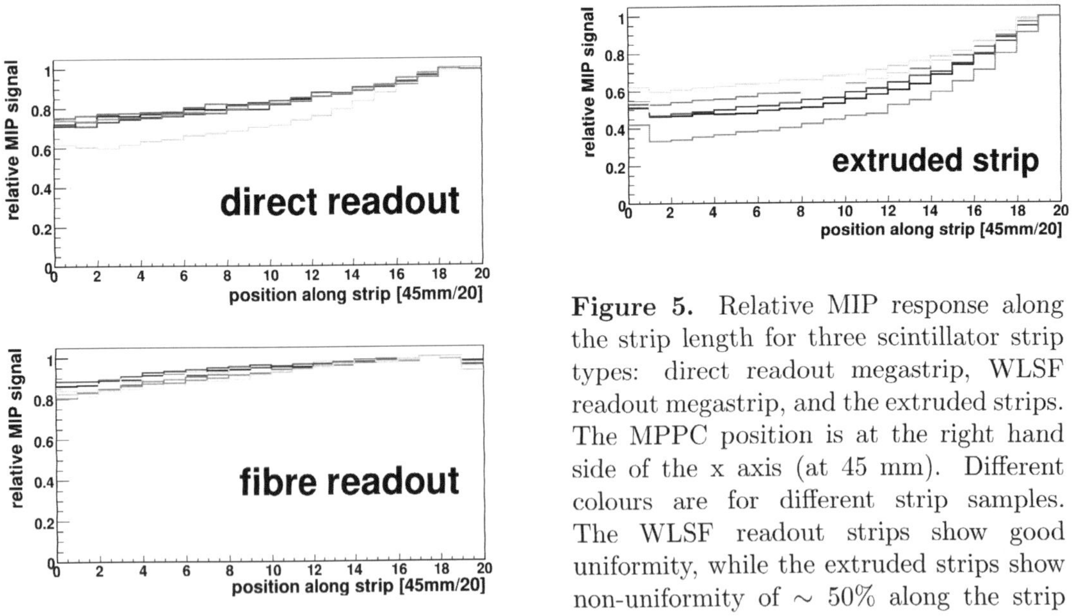

Figure 5. Relative MIP response along the strip length for three scintillator strip types: direct readout megastrip, WLSF readout megastrip, and the extruded strips. The MPPC position is at the right hand side of the x axis (at 45 mm). Different colours are for different strip samples. The WLSF readout strips show good uniformity, while the extruded strips show non-uniformity of $\sim 50\%$ along the strip length.

The same events used for the calibration were also used to measure the optical cross-talk across the boundaries between adjacent strips. The average signal recorded when a positron passed through the neighboring strip was compared with the signal when the strip itself was hit. Examples of these distributions in the megastrips and extruded strips are shown in figures 7. The measured cross-talk in the "megastrip" layers was around 10% in the transverse direction (perpendicular to the strip axis), and rather smaller ($\sim$ 1%) in the longitudinal direction (i.e. between the two megastrip plates in the same layer). The cross-talk in the extruded strips was much smaller in the transverse direction ($\sim$ 1%). The system of grooves in the megastrip tile does not achieve perfect optical isolation, while the white co-extruded covering of the extruded strips does achieve good isolation. A method to correct for the cross-talk was developed, and was applied in the analysis of the EM shower data.

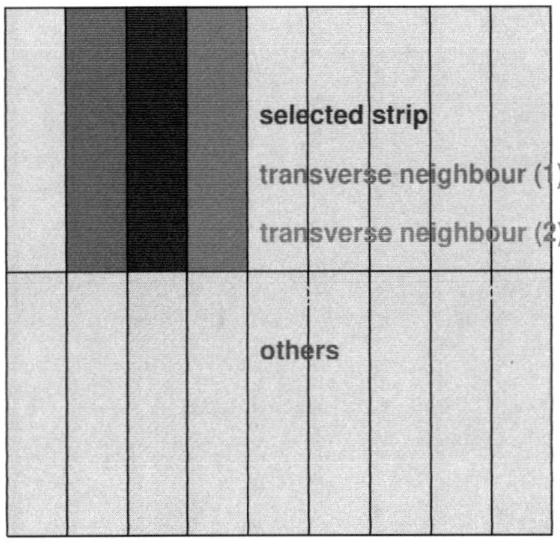

Figure 6. Definition of transverse and longitudinal neighbours for the cross-talk measurement.

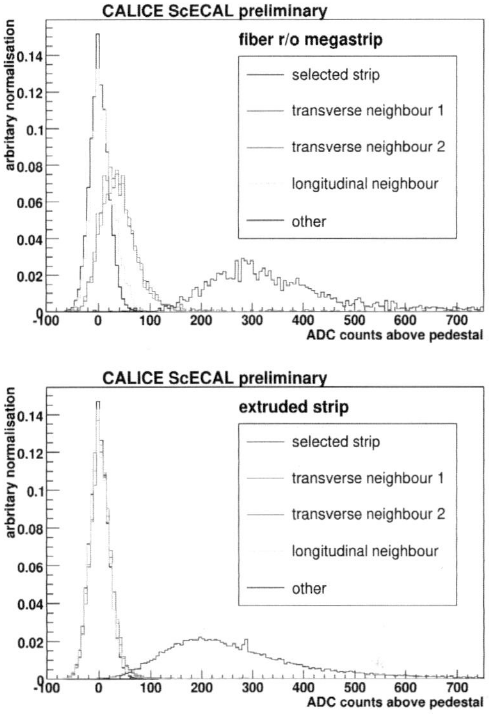

Figure 7. Cross-talk in megastrip with WLSF readout (top) and extruded strips (bottom). The measured signal in a strip when hit by a positron (blue histogram), when a neighbouring strip in transverse (red, pink) and longitudinal (cyan) directions is hit, and when a non-neighbouring strip is hit (black).

5. Energy linearity and resolution

The response of the calorimeter was measured using runs in which the Tungsten absorber plates were inserted. The response of the calorimeter to positrons in the momentum range $1 \rightarrow 6$ GeV/c was measured. The measured signals from the MPPCs were calibrated using the measured calibration constants, and corrected for the effects of temperature variation and optical cross-talk.

The detector performance was measured in two regions of the detector. "Central" events was defined as those in which the positron was measured (by the drift chambers) to have hit the front face of the calorimeter in the central 1×1 cm^2 region of the detector. In this region the lateral leakage of showers is minimised. However, in the central region the shower energy is mostly deposited at the ends of the scintillating strips, either close to or far from the MPPC position. The effects of any response non-uniformity along the strip are therefore maximised in this central region.

To investigate these effects, events in the "quarter" regions were also considered. These quarter regions are defined as the four 1×1 cm^2 regions whose centres are at $x, y = \pm 22.5$ mm with respect to the detector centre, as shown in figure 8. In these events, most of the energy is deposited close to the centre of the scintillating strips, minimising the effects of any non-uniform response along the strip length. However, the lateral shower leakage out of the calorimeter module is somewhat larger than for events in the central region.

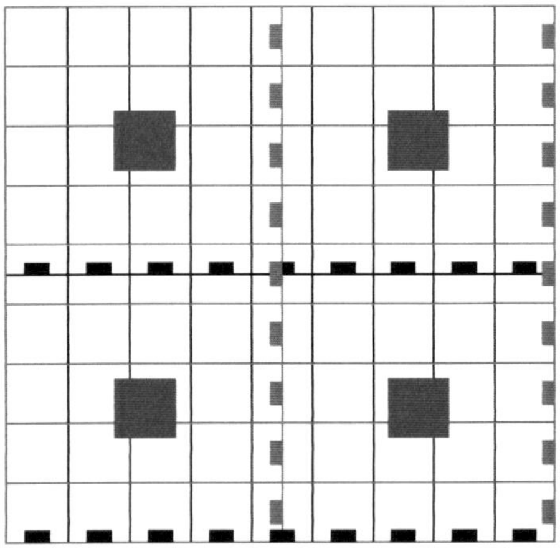

Figure 8. Cartoon of the front face of the calorimeter prototype. The strip and MPPC positions in the X and Y layer orientations are shown in red and black. The four 1×1 cm^2 quarter regions are shown as magenta squares.

Figure 9 shows the reconstructed energy, in terms of MIP equivalents, at the six beam momenta of 1, 2, 3, 4, 5 and 6 GeV/c, for the first detector configuration (WLSF readout megastrip module upstream of the direct readout megastrip module). These energy distributions are well fitted by Gaussian distributions. The mean values of the fitted Gaussian functions are used to measure the detector's energy response, and the fitted widths to estimate the energy resolution.

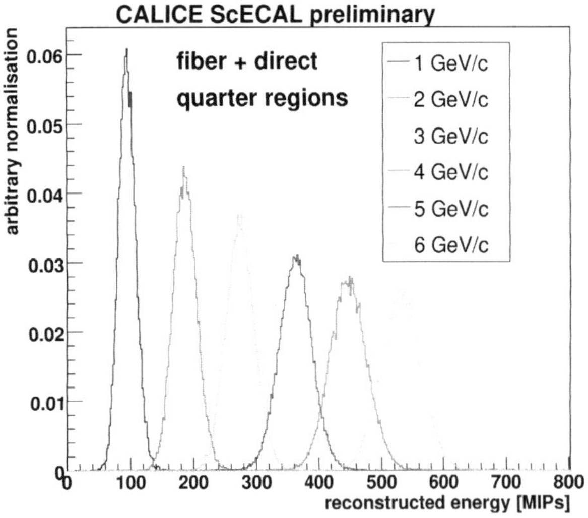

Figure 9. The reconstructed energy in the first (fibre+direct) detector configuration. Only events in which the beam hit the "quarter regions" are included. The six histograms show the reconstructed energy for beam momenta between 1 and 6 GeV/c, and are fitted by Gaussian functions.

The mean energy response for centrally injected events of the three detector configurations is shown in figure 10. The configurations have very similar response to EM showers, and the response is linear to within $\sim 4\%$ in this energy range.

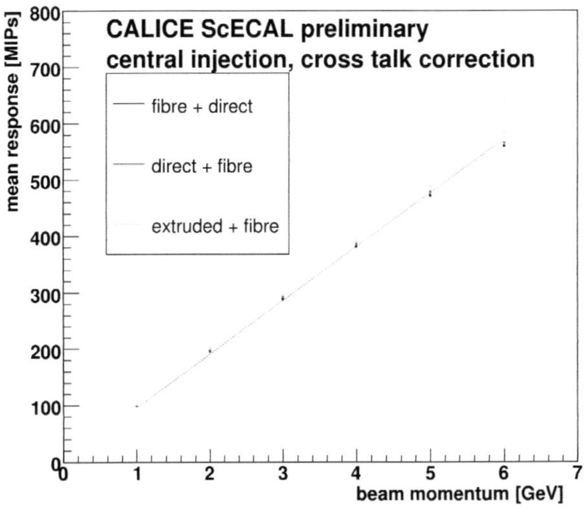

Figure 10. The mean reconstructed energy as a function of beam momentum, for centrally injected events. The three lines show the results for the different detector configurations.

The measured energy resolution in the two detector regions are shown in figures 11 (for the quarter regions) and 12 (in the central region). In the quarter region, the energy resolution of the three configurations is rather similar, the "extruded+fibre" configuration having only slightly worse resolution than the others. In the central region, however, the energy resolution of the "extruded+fibre" configuration is significantly worse than that of the other two. This difference is due to the effects of the strip response non–uniformity, which is largest in the extruded strips, and whose effects are maximised in the central detector regions.

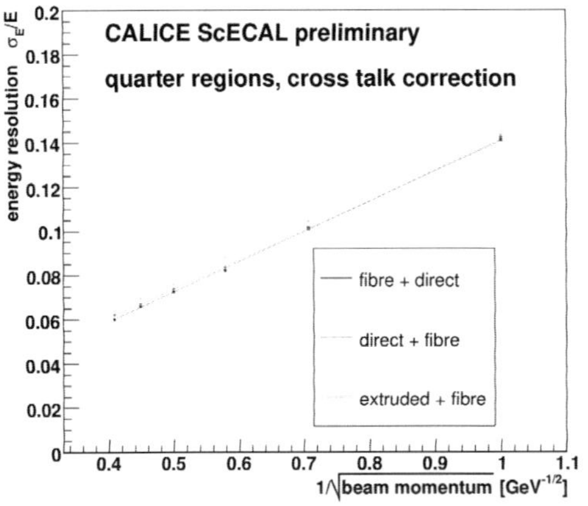

Figure 11. The measured energy resolution of the three detector configurations, in the quarter detector regions.

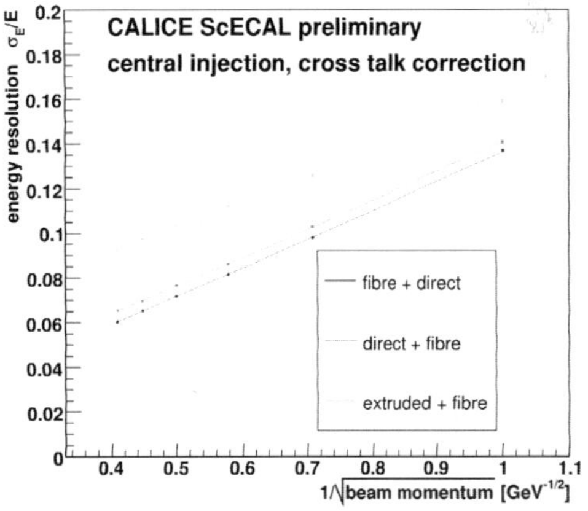

Figure 12. The measured energy resolution of the three detector configurations, in the central detector region.

A fit of the measured energy resolutions to the quadratic sum of a stochastic and constant term give the values shown in table 1. The stochastic term is similar in all configurations and regions, between 13.5 and 14.5%, however the constant term varies significantly between the

Journal of Physics: Conference Series **160** (2009) 012026 doi:10.1088/1742-6596/160/1/012026

different configurations, particularly in the central region, again due to the non–uniformity of the strip response.

Table 1. Results of the fits of the measured energy resolutions to the quadratic sum of a stochastic and constant term. Results are shown for the central and quarter regions of the three detector configurations. Errors are statistical only.

	Quarter		Central	
	stochastic	constant	stochastic	constant
fibre + direct	13.98 ± 0.07	1.96 ± 0.12	13.39 ± 0.05	2.57 ± 0.07
direct + fibre	13.83 ± 0.07	2.58 ± 0.09	13.70 ± 0.06	3.39 ± 0.05
extruded + fibre	14.61 ± 0.08	2.35 ± 0.12	14.52 ± 0.09	7.26 ± 0.05

6. Future plans

Now that several issues have been identified and resolved, a larger detector prototype is being constructed. It will have larger transverse size ($18 \times 18\text{cm}^2$), and more longitudinal layers (30). Improved extruded scintillator strips will be used. This detector will be taken to Fermilab later in 2008, and exposed to beams of electrons with a significantly higher range of momenta and also to beams of other particle types. Data will be collected together with the CALICE Analogue-HCAL detector, which will be positioned downstream of the ECAL.

7. Summary

A sampling calorimeter prototype for an ILC detector was designed and constructed. Active layers consisted of scintillator strips individually read out by MPPC devices, while absorber layers were made of Tungsten, giving a calorimeter with small Molière radius and highly segmented readout, suitable for the particle flow approach to jet energy measurement. Three different types of scintillator strips were tested: cast scintillator "megastrips" with and without wavelength-shifting fibre readout, and extruded scintillator strips.

This detector was exposed to positron beams in the momentum range $1 \rightarrow 6$ GeV/c at the DESY-II synchrotron. Three detector configurations, using different types of scintillator strips, were tested. The detector was calibrated, and the non–uniformity of strips response to MIPs was measured, as was optical cross-talk between adjacent strips. The "megastrip" scintillators showed good uniformity, but also optical cross-talk between neighbouring strips of around 10%. The extruded strips showed significant non-uniformity of around 50% along the strip length, but good optical isolation.

The response of the detector to electromagnetic showers was measured in the two detector regions where the effects of strip non–uniformity were maximised and minimised. In the region where effects of non-uniformity are minimised, the three configurations showed similar performance: good linearity, and energy resolution with a stochastic term of around 14% and a constant term of around 2%. In the region with significant effects of non–uniformity, the constant term rose to around 7% for the detector configuration with largest non-uniformity.

A larger prototype is presently being designed and constructed. Improved samples of extruded scintillator will be used. This prototype will be tested at Fermilab in Fall '08, in conjunction with other CALICE calorimeters.

References

[1] http://jp.hamamatsu.com/products/sensor-ssd/4010/index_en.html

Calibration and Monitoring of a Scintillator HCAL with SiPMs CALICE Scintillator HCAL

Angela Lucaci-Timoce (on behalf of the CALICE collaboration)
DESY, Notkestr. 85, 22607 Hamburg, Germany
E-mail: angela-isabela.lucaci-timoce@desy.de

Abstract. The operational experience with a highly-granular analogue hadronic calorimeter (AHCAL) consisting of 7608 individual scintillator tiles readout via Silicon-Photo-multipliers (SiPM) is presented. The calibration of each cell is based on minimum ionizing particle signals for which in general a muon beam is used. In addition, a correction for the non-linearity introduced by the finite number of pixels (1156) in the SiPM is applied. The aspects of temperature and voltage dependence of SiPM are addressed, and monitoring and calibration procedures are discussed. Such procedures are essential for the extrapolation of calibration factors over several days of data taking with the calorimeter. For this purpose a versatile UV-LED light distribution system was developed, capable of delivering light to all tiles with intensity from a few photo-electrons to the saturation of the SiPM. The procedures are tested using data collected with the AHCAL at the CERN SPS test beam.

1. Introduction

The CALICE collaboration built a high granularity calorimeter prototype for a future high energy electron positron linear collider. In view of testing shower simulation and validating the particle flow approach, data were taken in the years 2006 and 2007 at CERN SPS test beam. For a discussion of the physics results, see [1]. The technical solutions for a realistic calorimeter, like mechanical design and DAQ, are presented in [2].

The response of SiPMs depends on temperature and voltage variations. These dependencies are monitored with an LED system, with twelve LEDs per HCAL module. Each LED illuminates 18 SiPMs and one PIN photodiode to monitor the LED signal. The LED system is used to obtain the individual SiPM gain in runs with low LED light intensity (see section 2.3). Another feature of the system is to supply reference pulses monitored by PIN diodes. In addition, it provides the full dynamic range for checking the SiPM response function.

The temperature variations are monitored by five temperature sensors aligned vertically in the center of each AHCAL module.

In case of long data taking periods of several days, the calibration factors extracted in the beginning may be extrapolated at the end of the data taking, with the help of the monitoring system.

2. AHCAL Calibration Chain

The analog signal of the SiPMs is given by the total number of fired pixels. But due to the limited number of pixels (1156) and the finite pixel recovery time (20-500 ns), the SiPMs get saturated when the numbers of produced photoelectrons is of the order of the total number of pixels. As a result, a non-linear SiPM response curve is obtained (see Figure 1).

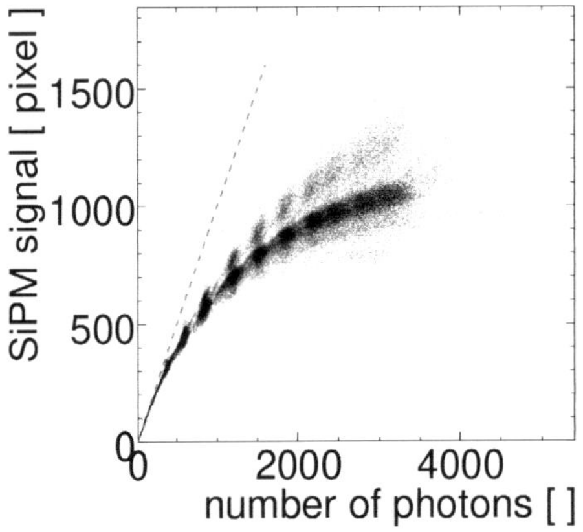

Figure 1. Dependence of the SiPMs signal (in pixels) on the number of produced photoelectrons, measured at ITEP (Russia). The dashed line indicates the linear behaviour. The measurements were done with 'bare' SiPMs (i.e. before being installed on the scintillator tiles) on the test bench.

Each calorimeter cell is calibrated based on signals from minimum ionising particles (MIPs), for which muons are used in a first approximation. The calibration of one cell i is done according to:

$$E_i[MIP] = \frac{A_i[ADC]}{A_i^{MIP}[ADC]} \cdot f_{resp}(A_i[pixels]) \quad (1)$$

where $A_i[ADC]$ is the amplitude registered in cell i, in units of ADC channels, and $A_i^{MIP}[ADC]$ is the MIP amplitude in cell i (see section 2.1). The SiPM response function $f_{resp}(A_i[pixels])$ corrects for the non-linearity of the SiPM response. It acts on the amplitude expressed in number of pixels, and returns the saturation correction factor which needs to be applied to linearize the amplitude in MIPs.

The amplitude in pixels is obtained by dividing the amplitude of a cell by the corresponding SiPM gain $G_i[ADC]$:

$$A_i[pixels] = \frac{A_i[ADC]}{G_i[ADC]}. \quad (2)$$

The procedure to obtain the individual SiPM gain is discussed in section 2.3.

2.1. AHCAL MIP Calibration with Muons

As already mentioned, muons are used as minimum ionising particles in a first approximation. The amplitude spectra of the 216 tiles of each HCAL modules in muon events are fitted with a convolution of Gaussian and Landau distribution (see Figure 2). For reference, the noise distribution obtained with random trigger events is overlayed. The MIP calibration factor is defined as the difference between the most probable value of the minimum ionising peak and the mean of the noise peak.

To reject noise, a cut of 0.5 MIPs is applied. As a result, the MIP uncertainties affect the reconstructed energy and the noise level.

The MIP signal to noise ratio is defined as the ratio between the MIP value and the σ of the noise distribution, and it is a measure of the separation power between the noise and the minimum ionising peak. The distribution of the signal to noise ratio for the data taken in 2006

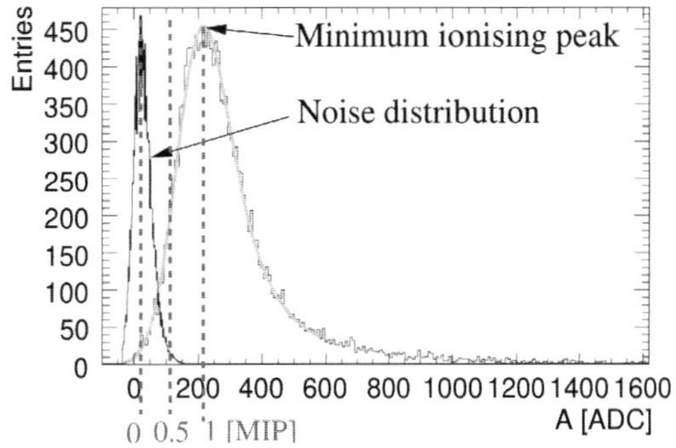

Figure 2. Example of amplitude spectra for an HCAL tile in muon events.

and 2007 is shown in Figure 3. On average, a signal to noise ratio of 9-10, and of 12, was obtained for the year 2007, and 2006, respectively.

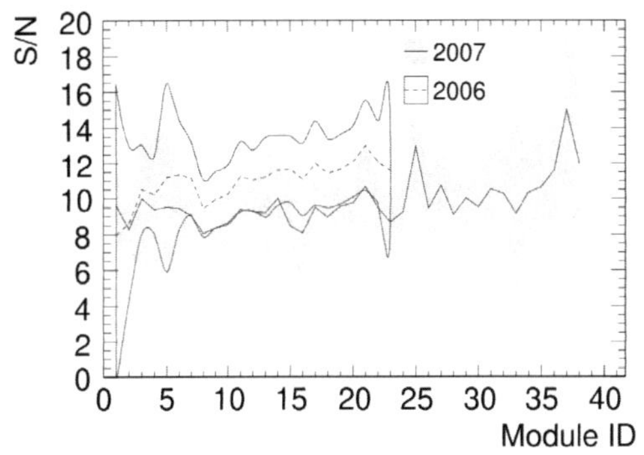

Figure 3. Distribution of the average MIP signal to noise ratio in each AHCAL module for the 2006 and 2007 data taking periods. The error band indicates the spread (RMS) between all cells in one module [3].

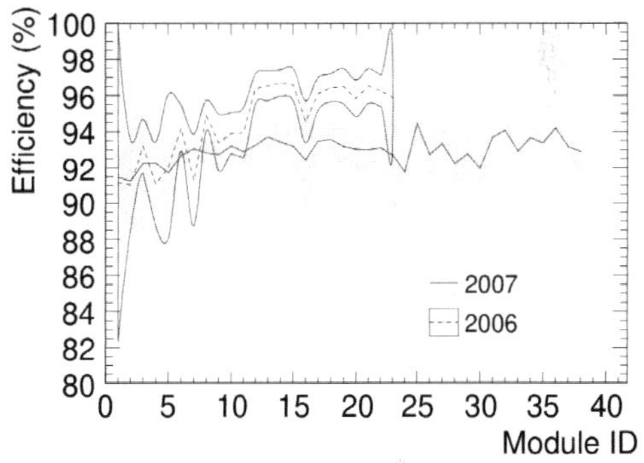

Figure 4. Distribution of the average MIP detection efficiency in each AHCAL module for the 2006 and 2007 data taking periods. The error band indicates the spread (RMS) between all cells in one module [3].

The MIP detection efficiency, defined as the ratio between the MIP value and the σ of the MIP distribution, is presented in Figure 4, for a threshold cut of 0.5 MIPs. An average of 93% and of 96% was obtained in 2007, and 2006, respectively. The difference is due to a decrease of the light yield in the year 2007, due to operation at higher temperatures.

2.2. MIP Calibration with Hadrons

In case of a deep-site detector, which is a possibility for the ILC, cosmic muons may not be enough to calibrate the calorimeter. Therefore, solutions have been looked for of using other types of particles as MIPs instead of muons. One idea, proposed in 2004 by A. Raspereza, was to use hadrons as MIPs. Since their interaction length is high ($\lambda_i \sim 17$ cm, which corresponds to about 8 HCAL layers), long tracks within hadron showers are abundant, and they are good MIP candidates.

The idea was followed by the CALICE group in Munich. Track-like clusters in hadron events were selected based on the deep analysis algorithm of V. Morgunov [7]. The amplitude spectra of single tiles were fitted with an exponential convoluted with a Landau distribution. For tiles close to the beam axis, a clear MIP peak was observed (see Figure 5), whereas for tiles outside the beam axis, the background is predominant (see Figure 6).

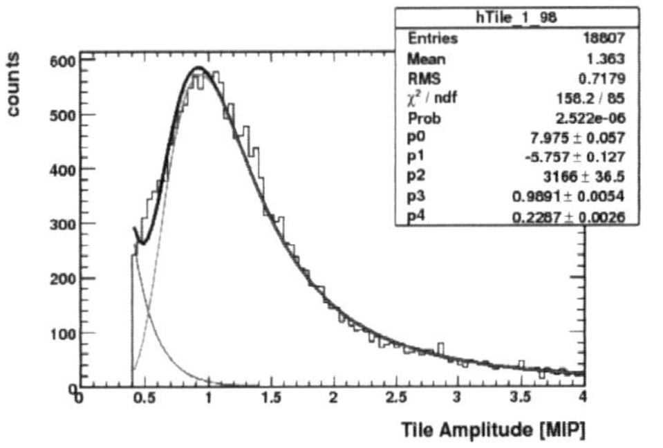

Figure 5. Amplitude spectra for an AHCAL tile close to the beam axis, in hadron events [6].

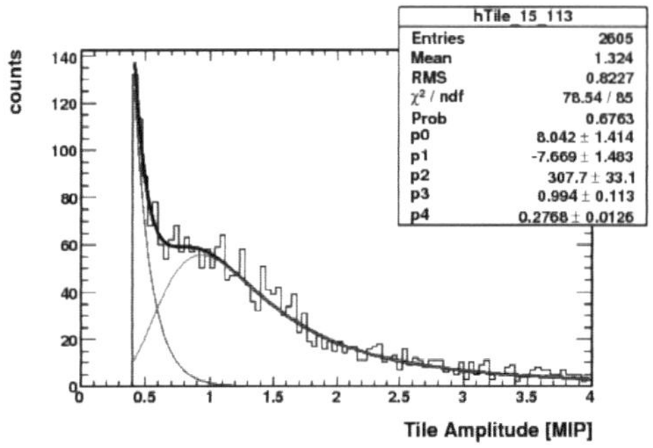

Figure 6. Amplitude spectra for an AHCAL tile outside the beam axis, in hadron events [6].

2.3. Gain Calibration

Since the saturation behaviour of the SiPMs appears at the pixel scale, one needs a way to obtain this scale. This is done via gain calibration, which in addition offers the possibility of looking directly at the SiPMs and monitoring their performance.

The gain calibration is done with the LED system. For all HCAL channels, single photon spectra are taken at low intensity LED light, and fitted with a multi-Gaussian function. An example of such a spectra is shown in Figure 7. The gain value is obtained as the difference between the means of two consecutive single photon peaks. The efficiency of such a calibration is high. For example, for the CERN test-beam data, 96.9% of the channels were calibrated. The rest of 1.7% had the LEDs off, and 1.4% had missing calibration, but these channels were recalibrated at the end of the data taking period, once the detector was back at DESY.

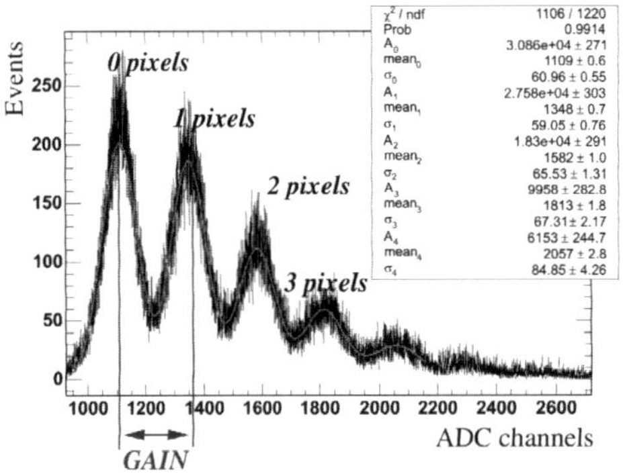

Figure 7. Example of a single photon spectra for an AHCAL channel.

2.4. Saturation Correction

The response function of a SiPM can be approximated by the function:

$$N_{pixel} = N_{total} \cdot \left[1 - \exp\left(-\frac{N_{photoelectrons}}{N_{total}}\right)\right], \tag{3}$$

where N_{total} is the number of pixels at which the SiPMs saturate.

By comparing measurements performed on the 'bare' SiPMs (ITEP) and on the SiPMs mounted on a tile (CERN), the CALICE Bergen group showed that the mounted SiPMs saturate at lower values that 'bare' ones (see Figure 8). This result was later confirmed by the DESY group and by ITEP measurements. The reason is due to geometrical misalignment. The mounted SiPMs do not get completely illuminated by the light cone from the wave length shifting fiber, such that less effective pixels contribute to light detection. To correct for this, a scaling factor $N_{tot}(CERN)/N_{tot}(ITEP)$ was introduced ([5]).

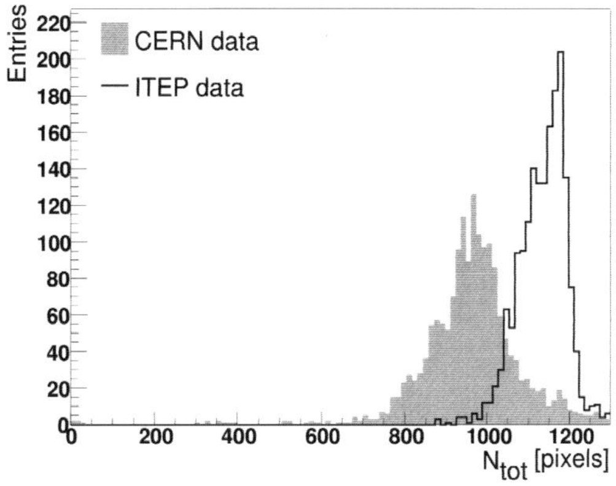

Figure 8. Distribution of the SiPM saturation values: comparison between 'bare' SiPMs (ITEP measurement) and mounted SiPMs (CERN measurement) [4].

3. Time Dependence of the SiPM Response

The SiPMs are operated in Geiger mode, in which the bias voltage is higher than the breakdown voltage:

$$U_{bias} = U_{breakdown}(T) + \Delta U. \tag{4}$$

Typical values for the U_{bias} are 40-70 volts.

The parameters of the SiPMs depend primarly on the overvoltage $\Delta U = U_{bias} - U_{breakdown}(T)$, and indirectly on temperature, via the breakdown voltage.

The SiPM temperature and voltage dependencies can be expressed as:

$$\begin{aligned} G &= G(T, U) \\ \epsilon &= \epsilon(T, U) \end{aligned} \tag{5}$$

where G is the SiPM intrinsic gain, and ϵ is its photo-detection efficiency.

3.1. SiPM Dependencies on Voltage

The gain and amplitude dependencies on voltage were measured at ITEP, during SiPM production. Depending on the operation point, two groups of SiPMs were observed: group A, operated at about 3 V above breakdown, and group B, operated at about 2 V above breakdown (see Figure 9). The dependencies were well reproduced at the CERN test beam, although a different test set-up was used (see Figures 10 and 11).

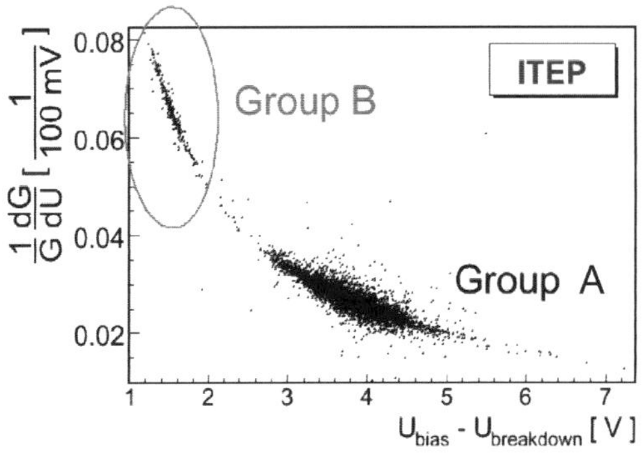

Figure 9. Dependence of the relative SiPM gain on the overvoltage $\Delta U = U_{bias} - U_{breakdown}$ (ITEP measurement).

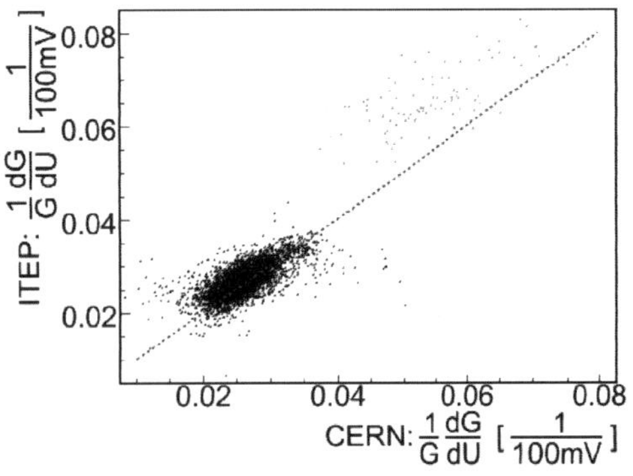

Figure 10. Relative dependencies of **gain** on voltage: comparison between ITEP and CERN measurements. The black points correspond to group A of SiPMs and red to group B (see text for details).

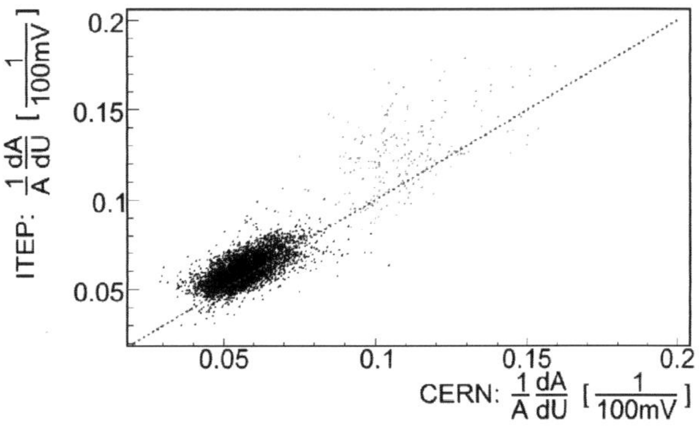

Figure 11. Relative dependencies of **amplitude** on voltage: comparison between ITEP and CERN measurements. The black points correspond to group A of SiPMs and red to group B (see text for details).

3.2. SiPM Response on Temperature Dependence

The temperature variations are measured in test-beam only (i.e. CERN). During one month of data taking, temperature variations of 3-4 degrees are measured. Since the gain depends inversely linear with temperature (see Figure 12), this translates into a variation of $-1.7\% \cdot K^{-1}$ for the gain of the SiPMs from group A. A larger variation, of $-4.0\% \cdot K^{-1}$ for group A, is measured in case of the SiPM amplitude. This is due to the fact that the amplitude is the product of SiPM gain and photo-detection efficiency, both sensitive to temperature variations.

3.3. Voltage Adjustment

The temperature effects on the SiPM gain can be compensated for via voltage adjustement. In a first step, the gain dependence of voltage is fitted with a straight line, and the slope dG/dU is determined. Next, the bias voltage is adjusted to correct for gain deviations:

$$\Delta U = \frac{G - G_0}{dG/dU}, \tag{6}$$

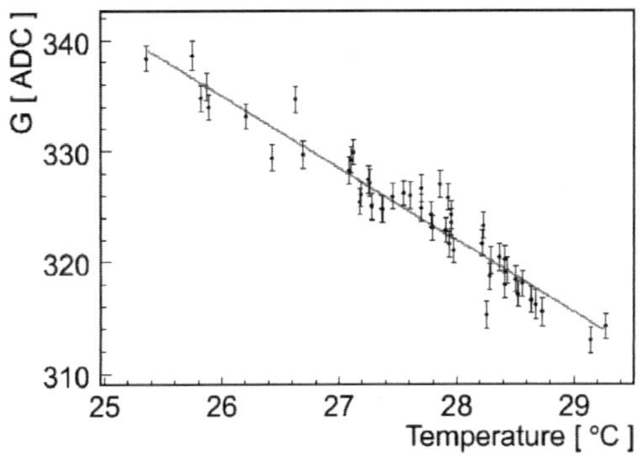

Figure 12. The SiPM gain dependence on temperature.

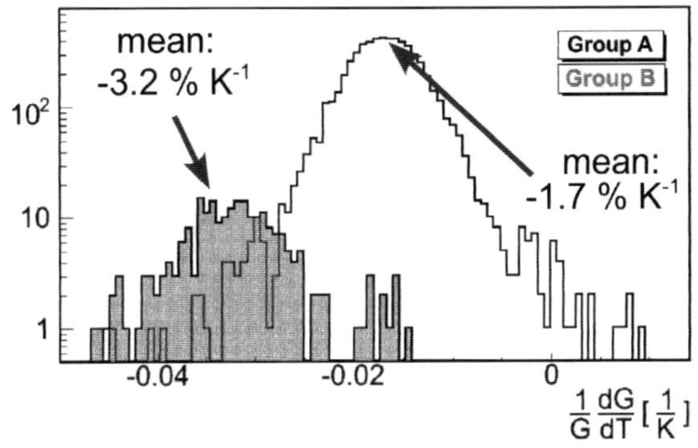

Figure 13. Distribution of the relative gain dependence on temperature.

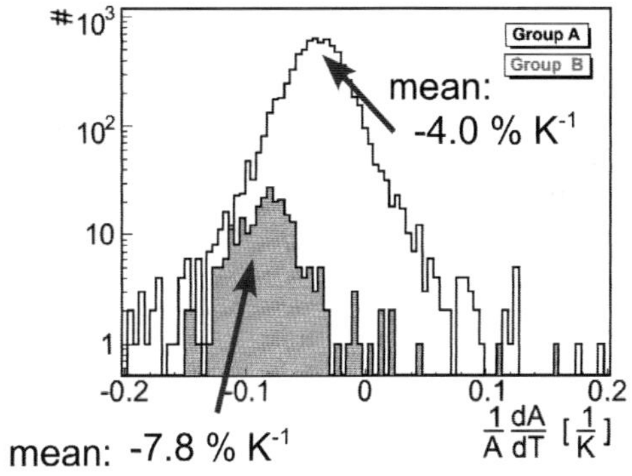

Figure 14. Relative dependence of amplitude on temperature.

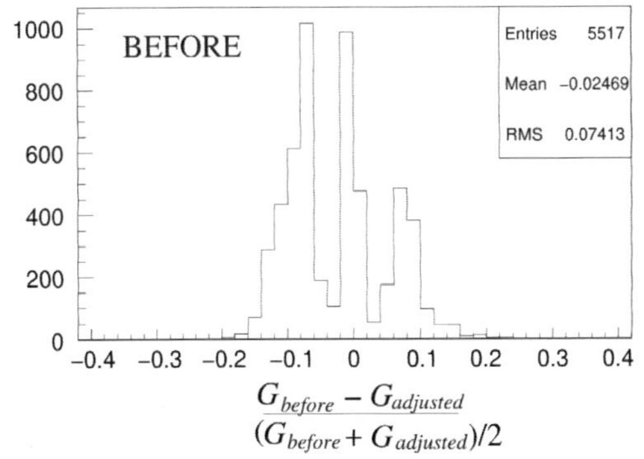

Figure 15. Deviation of the gain from the mean gain, before voltage adjustment.

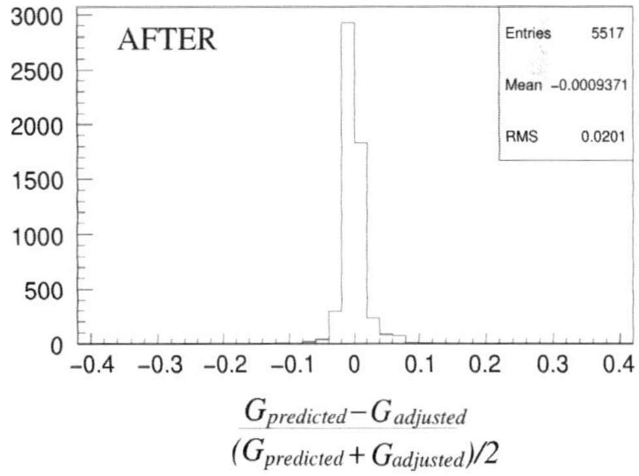

Figure 16. Deviation of the gain from the mean gain, after voltage adjustment.

and the gain $G_{adjusted}$ is measured. The method is checked by comparing the theoretically calculated gain:

$$G_{predicted} = G_{before} + \Delta U \cdot \frac{dG}{dU}, \tag{7}$$

where G_{before} is the gain measured before voltage adjustments, with $G_{adjusted}$.

Figures 15 and 16 show the deviations of the gains difference from the mean, before and after voltage adjustment. It can be seen that the corrected gain has the expected behaviour.

3.4. Conclusions
The CALICE analog scintillator HCAL is the first experiment to handle large samples of SiPMs (7608 channels in total). The signal of an HCAL cell is calibrated using minimum ionising particles and corrected for the non-linearity of the SiPM response curves. The SiPM performance can be monitored directly via gain calibration. The dependence of the SiPM signal on temperature variations can be corrected for, although the benefits for the data analysis are still to be demonstrated.

3.5. Acknowledgments
This work is supported by Deutsches Elektronen Synchrotron Hamburg.

[1] E. Garutti, these proceedings
[2] F. Sefkow, these proceedings
[3] CALICE collaboration, "The reconstruction of the energy lost by a 120 GeV muon in the highly granular hadronic calorimeter for the International Linear Collider", CALICE note, *CAN-009*
[4] CALICE Bergen group, private communication
[5] CALICE collaboration, "The scintillator ECAL beam test at DESY, 2007", CALICE note, *CAN-012*
[6] F. Simon, "Temperature correction for hadron runs", talk given at the CALICE collaboration meeting in Argonne, USA, 17-19 March, 2008
[7] V. Morgunov and A. Zhelezov, "Deep analysis of hadronic showers", talk given at ILC Software and Tools Workshop, Orsay, 2-4 May, 2007

Journal of Physics: Conference Series **160** (2009) 012028 doi:10.1088/1742-6596/160/1/012028

2nd generation ASICs for CALICE/EUDET calorimeters

F. Dulucq, J. Fleury, C. de La Taille, G. Martin-Chassard, L. Raux, N. Seguin-Moreau.

Orsay Micro Electronics Groups Associated (OMEGA) Laboratoire de l''Accélérateur Linéaire. Université Parsi XI. F 9198 ORSAY Cedex

taille@lal.in2p3.fr

Abstract. Imaging calorimetry depends heavily on the development of high performance, highly integrated readout ASICs embedded inside the detector which readout the millions of foreseen channels. Suitable ASICs prototypes have been fabricated in 2006-2007 and show good preliminary performance.

1. Introduction

After several years of data taking on 10 000 channels of electromagnetic and hadronic calorimeters, the CALICE collaboration has turned to addressing the technological challenge of a highly integrated, highly granular calorimeter that can be scaled to a full size detector [1].

For this purpose, 2nd generation readout integrated chips (ASICs) are needed (the first generation ASICs are those used for the current test-beam prototype). They incorporate the digitization and the sequenced readout handling inside the ASIC on top of the previous analog signal processing so that everything can be integrated inside the detector and having only a few digital lines coming out.

The chips have been developed, starting by the Digital Hadronic calorimeter RPCs which was more urgent and simpler as it is semi-binary with a chip called HaRDROC described in section 2. This chip has allowed to validate the digital readout sequence of a set of chips daisy-chained. The second chip was SPÏROC for the analog hadronic calorimeter silicon PMS

2. DHCAL technological prototype : HARDROC1 ASIC

HARDROC readout is a semi-digital readout with two thresholds (2 bits readout) which allows both good tracking and coarse energy measurement, and also integrates on chip data storage.

The 64 channels of the ASIC are made of:

- Fast low impedance preamplifier with 6bits variable gain (tuneable between 0 and 4)
- Variable shaper (50-150ns) and Track and Hold to provide a multiplexed analog charge output up to 10pC.
- Variable gain fast shaper (15ns) followed by two low offset discriminators to autotrigger down to10 fC. The thresholds are loaded by two internal 10 bit- DACs.
- A 128 deep digital memory to store the 2*64 discriminator outputs and bunch crossing identification coded over 24 bits counter.

HARDROC1 has been fabricated in AMS SiGe 0.35µm technology in September 2006. It has an area of $16mm^2$ and is packaged in CQFP240 package.

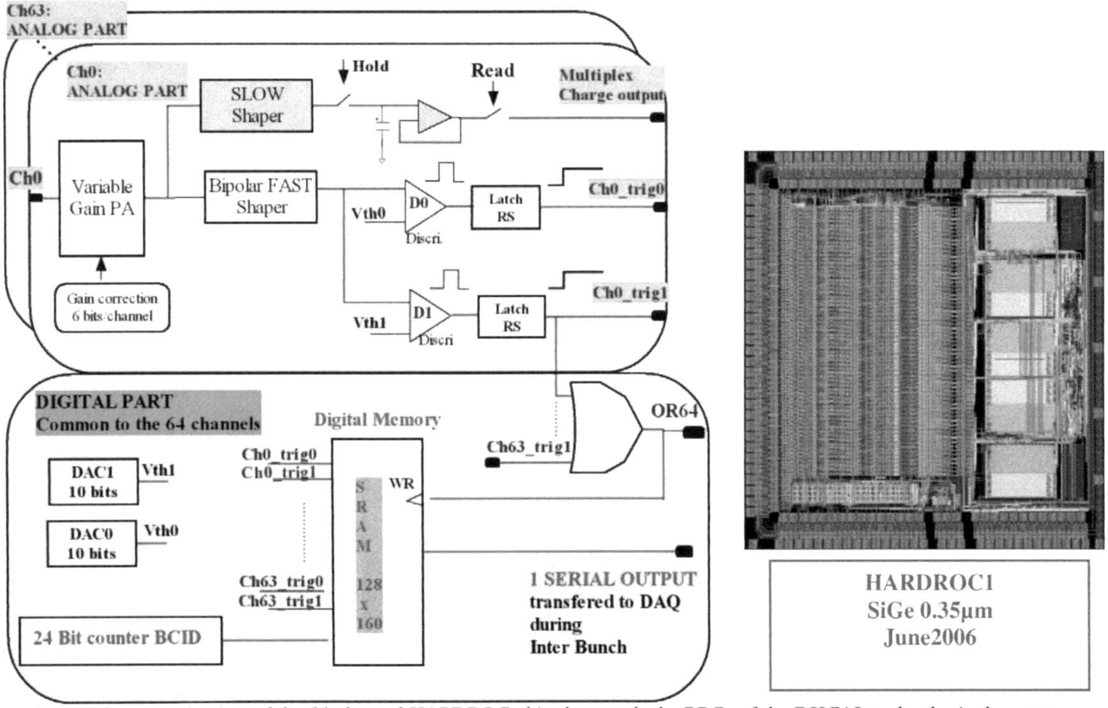

Figure 1 : synoptic view of the 64-channel HARDROC chip that reads the RPCs of the DHCAL technological prototype

2.1. Chip design

The block diagram of the ASIC is given in Figure 1. Each input signal is first amplified thanks to a variable gain current-sensitive preamplifier which exhibits low noise and low input impedance to minimise crosstalk. It allows accommodating the gain depending of the detector choice, up to a factor 4 to an accuracy of 6% with 6 bits.
The amplified current feeds then a slow shaper combined with a Sample and Hold buffer to store the charge in 2pF and provide a multiplexed charge output (5MHz) up to 10pC.
In parallel, trigger outputs are obtained via fast channels made of a fast (15 ns) shaper followed by two low offset discriminators. The discriminator thresholds are set by two internal 10 bit DACs.
Each trigger output is latched to hold the state of the response until the end of the clock cycle. The trigger1 outputs (corresponding to Vth1 <Vth0) are OR wired to generate an internal trigger used to start the memorization of the 128 trigger outputs as well as the Bunch Crossing Identification delivered by a 24 bit counter, needed to associate hits in the DAQ to bunch crossing identification (BCID). It is also possible to capture event data using an external trigger provided from outside the chip. All the bias currents are programmed through the Slow Control and the chip is power pulsed to decrease the power consumption. 10µW/channel as targeted with a 1% beam duty cycle.

2.2. Testbench measurements

HaRDROC has been extensively measured from December 2006 to December 2007. In particular, it was the first time that the power pulsing could be tested as well as the readout scheme of the calorimeter with built-in memory, token ring and daisy chain mechanism.

The analog part, largely reused from another ASIC (MAROC worked as foreseen. The bipolar fast shaper gain is about 3.5mV/fC and its peaking time is equal to 15ns. The Slow Shaper gain is about 50 mV/pC and its peaking time can be tuned from 100ns to 150ns. The crosstalk (Fig.3) has been measured by sending 100fC in one channel and looking to the direct neighbours. This 2% crosstalk is well differentiated and located on the input.
The linearity of the two 10 bits integrated DACs used to generate the thresholds of the discriminators, has been measured and shows residuals within ±5 mV for a 2.6V dynamic range which corresponds to an Integral Non Linearity of 0.2% (2LSB). The slope is 2.5mV per DAC unit.
The trigger sensitivity (s-curves) has been measured on the 64 channels of the chip. The quite large non uniformity between channels (±25 %) is explained by current mirror mismatch (small transistor size to optimize speed at low current) and can be corrected using the gain tuning of the input preamp as shown in Figure 2. The s-curves also show a total noise around 1fC.

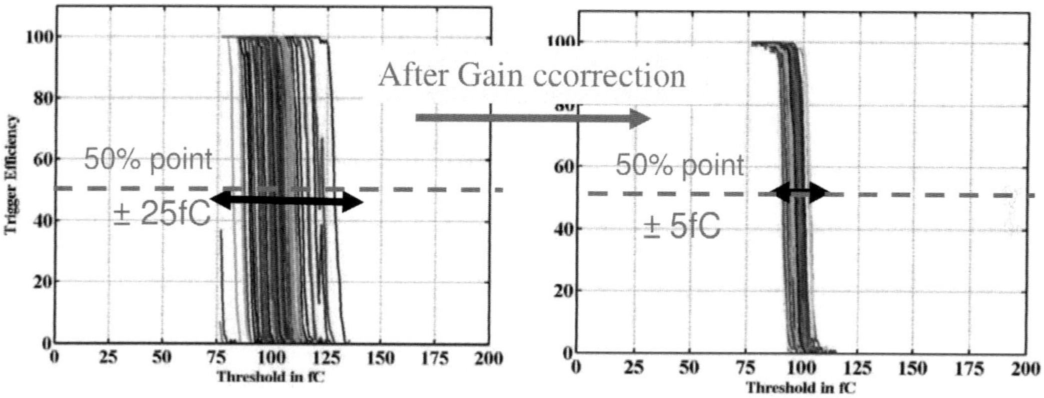

Figure 2: s-curves of the 64 channels of HARDROC1 before and after gain correction

As can be seen in Figure 3 , the dispersion between the channels is low enough to allow to bring the threshold down to 10fC. In particular, the pedestal is uniform within 1 DAC count (2.5mV or 1fC in high gain) which shows the excellent offset control obtained with this technology.

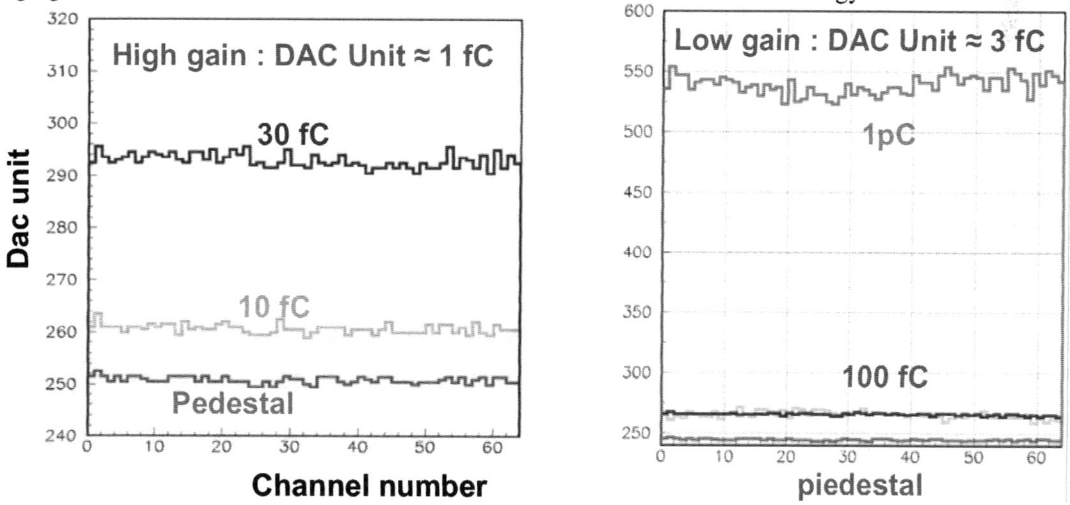

Figure 3: threshold for 50% trigger efficiency for the 64 channels

Last, but not least, the power pulsing capability has been tested on HARDROC1.There are 3 independent signals of power-on : analog, digital and DAC. Each stage can be forced on by slow control overruling the power-on pulse. As can be seen in Figure 4, it takes only 2 μs for the analog part to be "awake" and provide a discriminator output ("trigger") when an input charge of 100fC is sent 2µs after the power-on signal. As could be expected, the DAC is much slower to settle (this is the reason why it had a separate power-on signal) and in it needs 25µs to reach its nominal value within a few mV.

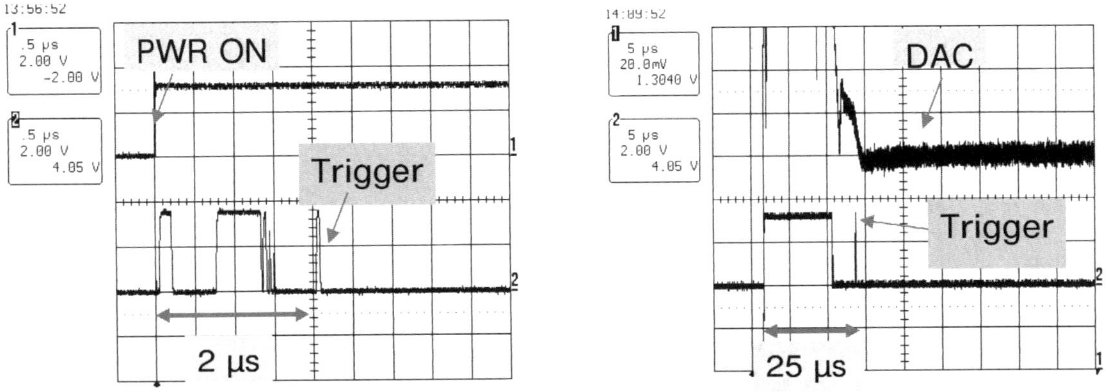

Figure 4: power-on of the analog part (left) and of the DAC (right)

2.3. HaRDROC digital part:

Because of the very high number of electronic channels foreseen in the final detector, chips will be embedded inside the detector and are designed to be daisy chained without any external circuitry, to limit to a bare minimum the number of output lines on the detector. A memory has been integrated in HARDROC to store during the bunch train the 2bits trigger outputs of each channel as well as the BCID, and this for every hit.
The data format is 128(depth)*[2 bits * 64 ch + 24 bits (BCID) +8 bits (header)] =20kbits. There is one serial output which is transferred to the DAQ during the inter-bunch.

The auto trigger mode which is crucial for the chip functionality has been checked successfully down to 10fC. The trigger crosstalk has been measured by injecting a charge in one channel and measuring the trigger efficiency on the direct neighbours. There is no crosstalk for injected charge lower than 1.6pC, corresponding to a crosstalk value of 1%.
A PCB hosting four HARDROC chips (4X64 channels) has also been designed to study the signal connection between the different chips before extracting it through a USB device. The PCB board will be associated to both RPC and µMEGAS detectors in order to validate the whole concept through exposure first to cosmics and then to beam test at CERN.

3. AHCAL technological prototype and SPIROC

As for the ECAL and DHCAL, the AHCAL must also prove that it is scalable to large modules. There, the main issues are the integration of the electronics inside the detector with its power dissipation and readout issues (similar to the ECAL and DHCAL) and mechanical assembly of the system. Also the thickness is a very important issue, although not as critical as in the ECAL.
Once again, the ASIC is a central, critical issue of the whole detector usual coined as "no chip, no detector !". For this purpose, a fully integrated ASIC named SPIROC (for Silicon Photomultiplier Integrated Read Out Chip) has been developed.

3.1. SPIROC ASIC architecture

Journal of Physics: Conference Series **160** (2009) 012028 doi:10.1088/1742-6596/160/1/012028

The SPIROC ASIC that reads 36 SiPMs is an evolution of the FLC_SiPM used in the physics prototype. As for the ECAL, it keeps most of the analog part, adds an analog memory to record up to 16 events of a train and adds the auto-triggering capability. The digitization is brought inside the chip as well as the data handling.

The first prototype has been fabricated in june 2007 in AMS SiGe 0.35μm. The chip layout is shown in Figure 5, its area is 35 mm^2 for 36 channels and it is packaged in a CQFP240 package

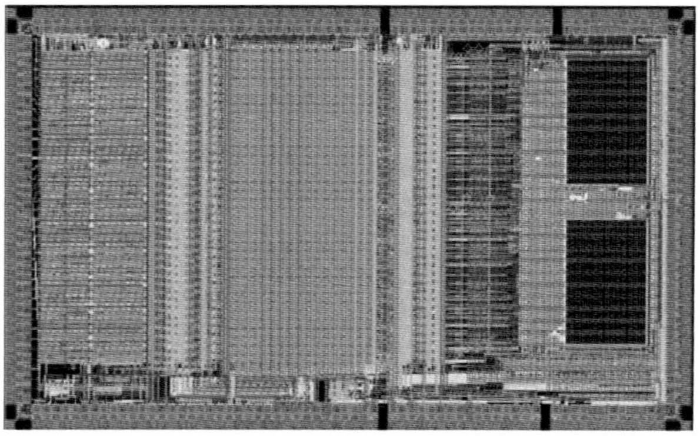

Figure 5: layout of SPIROC1

The schematic diagram of one channel is shown in Figure 6.

- The 8bit input DAC has been conserved, however its power dissipation has been brought down by 3 order of magnitude to 1μW/channel as it is not power pulsed. The DAC also has the particularity of being powered with 5V whereas the rest of the chip is powered with 3.5V.
- The voltage amplifier architecture with variable gain has also been kept, with a gain variable on 4 bits. However, the high gain/low gain separation splitting is now done at the preamp level by having two preamps on parallel on the input.
- The charge is measured on both gains by a "slow" shaper (50-150ns) followed by an analog memory with a depth of 16 capacitors.
- The auto-trigger is taken on the high gain path with a high-gain fast shaper followed by a low offset discriminator. All these blocks are new. The discriminator output is used to generate the hold on the 36 channels. The threshold is common to the 36 channels, given by a 10 bit DAC similar to the one from HARDROC with a subsequent 4bit fine tuning per channel.
- The discriminator output is also used to store the value of a 300ns ramp in a dedicated analog memory to provide time information with an accuracy of 1ns
- A 12bit Wilkinson ADC is used to digitize the data at the end of the acquisition period.

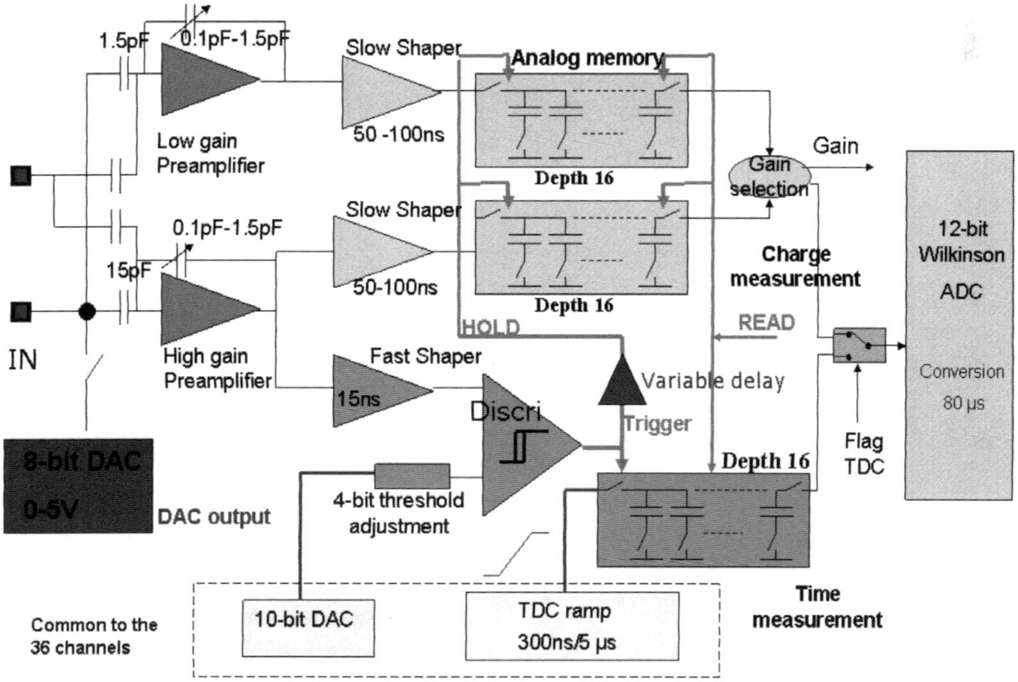

Figure 6: schematic diagram of one channle of SPIROC1

The digital part is complex is it must handle the SCA write and read pointers, the ADC conversion, the data storage in a RAM and the readout process.

3.2. measured performance

The chip has been received in October 2007 and underwent a long series of tests both in Orsay and Hamburg. The analog performance has shown good results but an inverted polarity in the ADC discriminators has unfortunately precluded any measurements with the internal ADC.

- Input DAC : the span goes from 4.5V down to 0.5V with a LSB of 20 mV. The default value is 4.5V in order to operate the SiPM at minimum over-voltage when the DAC is not loaded. The linearity is ±2% (5LSB), just enough for the SiPM operation but consistent with the allocated area. Also, the dispersion between channels, although not fundamental could also be improved. The power dissipation is well within the specs and the 100nA bias current to Vdd makes the chip difficult to measure without special precautions.
- Threshold and gain selection 10bit DACs behave similarly to the HARDROC.
- Charge path : the photoelectron to noise ratio of 8 allows to nicely resolve the single photoelectrons peaks
- Auto-trigger and discriminators. The ability to self-trigger on single photoelectrons or on the MIP is a new feature of SPIROC.

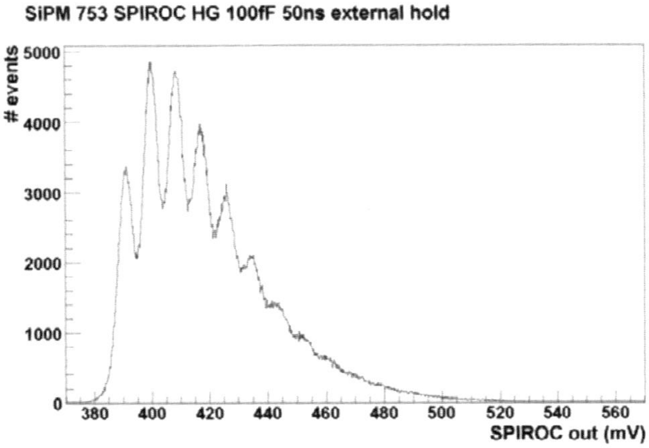

4. Conclusion

The first AISCs of 2nd generation for CALICE have been prorotyped and tested with good performance that validated the choices made. They will go in production in first half of 2009 to equip the technological EUDET modules.

References

[1] JC Brient : proceedings of this conference and references therein
[2] I Laktineh : proceedings of this conference and references therein
[3] F. Sefkow : proccedings of this conference and references therein

Journal of Physics: Conference Series 160 (2009) 012029 doi:10.1088/1742-6596/160/1/012029

CALICE: status of a data acquisition system for the ILC calorimeters

Valeria Bartsch for the UK CALICE collaboration

Department of Physics and Astronomy, University College London, Gower Street, London, WC1E6BT, UK

E-mail: bartsch@hep.ucl.ac.uk

Abstract. A data acquisition system is described which will be used for the next generation of prototype calorimeters using particle flow algorithms for the International Linear Collider. The system could also be the basis for the final detector. The design is sufficiently generic such that it should have applications elsewhere, be they either ILC detectors or within high energy physics in general. The data acquisition system will be implemented using FPGAs and built using off-the-shelf components and networking hardware. The EUDET ECAL technical prototype will be used to demonstrate the feasibility of this approach. The design philosophy, the current status of the project and its aims are presented giving an overview of the effort.

1. Constraints on the Data Acquisition System

The concept of particle flow algorithms and the beam structure of the planned ILC accelerator put constraints on the data acquisition (DAQ) system which are described in the following.

Particle flow algorithms are a widely accepted approach to improve the energy resolution at the International Linear Collider (ILC). Particle flow uses the high segmentation of about 1 cm×1 cm in the detection layer of a sampling calorimeter to track particles through the calorimeter. The segmentation results in about 24 million readout pads in the electromagnetic calorimeter (ECAL) of a planned ILC detector [1]. It is estimated that about 250 GB of raw data per bunch train need to be handled depending on the very front end electronics (VFE) and the detector type chosen for the final calorimeter. Because of the high number of readout channels it is necessary to minimise the cost of the electronics. This is achieved by using standard networking chipsets and protocols. Already at an early stage of the calorimeter prototypes the DAQ system has been designed for scalability. In the current detector plans the space for the electronics and cooling inside the calorimeters is very small. In the EUDET prototype the DAQ components are not yet minituarised to allow some flexibility for the test beams.

The ILC accelerator will generate bunch trains with a length of about 1 ms and a repetition rate of 200 ms resulting in an inter-train gap between bunch trains of about 199 ms which can be used for the readout of the calorimeters. It is vital that no data is lost through buffer overflows during the bunch train phase and that the data get time stamps for the event building which is done at a later stage.

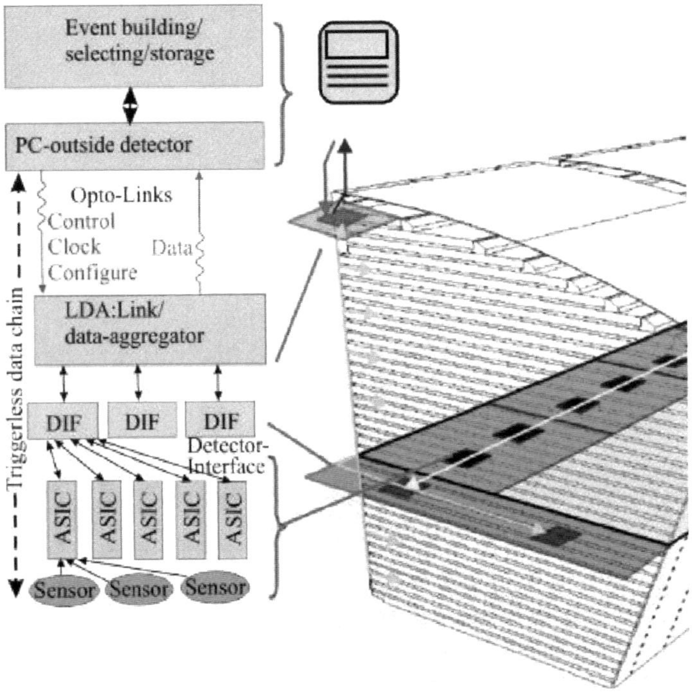

Figure 1. Integration of the DAQ components in the EUDET module

2. Overview over the proposed Data Acquisition System

2.1. Detector Interface Board

Fig. 1 shows a schematic of a calorimeter module and its DAQ system. The detector signals are buffered and digitised by the Very Front End (VFE) ASICs situated on the detector slabs. The VFE ASICs depend on the detector used in the prototypes. The interface between the VFE and the DAQ system is provided by the detector interface (DIF) board which is located at the end of the calorimeter slabs. The DIF board consists of a customised part which translates the signals of the ASICS into more generic signals and a generic part of the DIF which passes the data on and sends configuration on to the VFE electronics. Thus providing a common interface to the DAQ components further away from the detector. The connections between the DIFs have been designed in order to give an additional level of redundancy as can be seen in Fig. 2. The DIF is connected via HDMI cabling to the rest of the DAQ system. HDMI [2] is a home entertainment system standard with small cabling and connectors which are commercially available, at low cost. They are rated at more than 300 Mb/sec for data.

2.2. Link Data Aggregator

In order to accumulate the data sent from the DIF a link data aggregator (LDA) collects the data from several DIFs and sends it further to the off detector receiver (ODR). The number of DIFs aggregated in the LDA depends on the detector type chosen in the detection layer. The optimal number of DIFs/LDA must take into account the number of available pins on the FPGA, the bandwidth per DIF and the cost effective maximum bandwidth of the optical link. The current LDA prototype can host up to eight links to the DIFs, but plans for a new

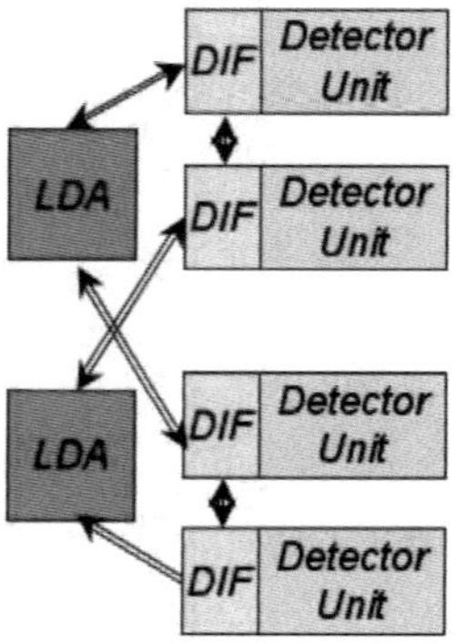

Figure 2. Connection redundancies between DIFs and LDAs

prototype with a higher number of DIF connections are already made. Physically the LDA has banks of HDMI connectors for connections to the DIFs and a small form-factor pluggable (SFP) connector for the optical link off-detector. The prototype version of the LDA is built on a commercial development board with a Xilinx Spartan3-200 FPGA and two purpose specific add-on boards: one providing the SFP and serialiser chipset for the optical link, and the second hosting eight working HDMI connectors with clock fan-out hardware.

2.3. Off Detector Receiver

The LDA is situated at the edge of the detector, the ODR however is located in the counting room connected to the on-detector DAQ system by an optical fibre. It sends control and configuration data to the LDA for distribution to the DIFs and finally sends the data off the detector to the event building. The ODR is realised as a PCI-Express card and can serve up to four LDAs per card, one PC can host up to two ODRs. For the proposed DAQ system currently a Xilinx Virtex 4 FPGA is used, thereby using a commercial FPGA board. At present, the data stream which will come from the detector in the EUDET test beam is simulated by an internal data generator in the firmware, or data received via an on-board ethernet interface.

The user interface to the ODR card contains two parts: a custom driver - mainly tasked with mapping card memory to the user space and providing direct memory access (DMA) support and a client program. The client retrieves data from the ODR card memory and stores it on the local disk. The client is a fully programmable multi-threaded application with separate threads for data transfer from the ODR to the host memory and IO threads for storing to the local disk. The performance of the prototype ODR has been investigated using the Ethernet interface to provide an input data stream which is coped to host memory and is currently higher than 500 MByte/sec without any disk writes as can be seen in Fig. 3. The number of DMAs has been optimised for this figure.

3. Clock and Control Handling

3.1. Clock and Control Board

For the event building a good clock is essential. It is understood that a machine clock will be used which will be fed into the ODRs and fanned out to the LDAs and DIFs. The requirements on the clock are a low jitter and a fixed latency between the machine clock and the clock in the DIFs. Commercial networking hardware is not suited for this task as it is most efficient when it can buffer data and provides no guarantees on delivery times. Similarly networking hardware built into modern FPGAs suffers from varying latency. Therefore a clock and control board has

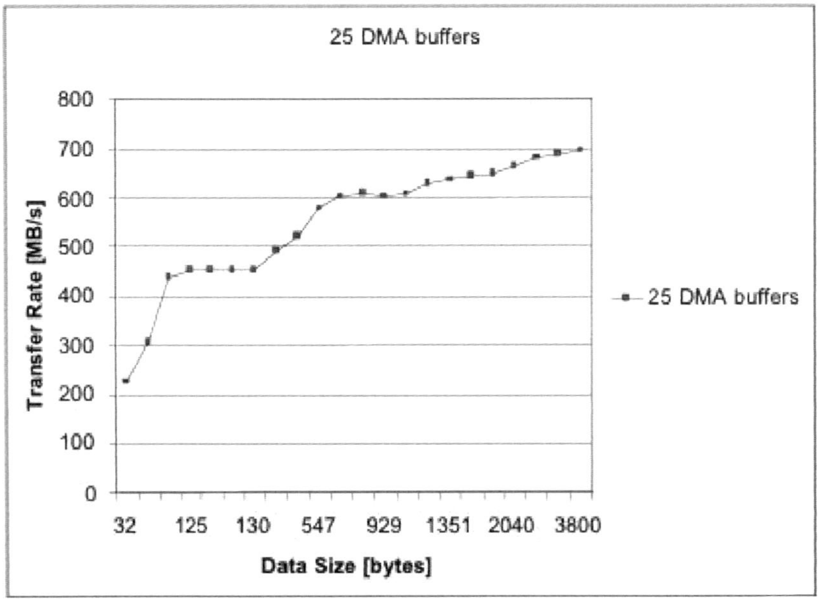

Figure 3. Data transfer rates of the off detector receiver depending on the data size.

been custom designed with links to eight ODRs.

The clock and control module must interface with the machine and provide stand-alone signal and clock generation. It will also receive a busy signal from the very front end electronics. The HDMI cables configuration used for the DIF to LDA link will be reused here. Control data will be sent over the same HDMI type cabling.

3.2. Resets due to Single Event Upsets

Single event upsets (SEUs) can occur in the electronics if a particle (typically a neutron, proton or pion above a certain threshold energy which is about 20 MeV) traverses it. The traversing particle transfers energy to a nucleus in the sensitive volume which creates a large ionization along its short path. If the deposited energy in the sensitive volume is higher than the threshold energy the electronics changes its state, in other words an upset has been triggered which needs to be reset by the control data. A study has been made to estimate the frequency of resets needed and therefore to estimate the impact of SEUs on the control stream.

For this study which is described in more detail in [3] simulations using PYTHIA and MOKKA have been performed for physics events and beamstrahlung remnant events which occur frequently according to the TESLA TDR. WW, QCD, $t\bar{t}$, Bhabha, beamstrahlung pair production and $\gamma\gamma \rightarrow$ hadrons events were simulated. A spectrum of particles traversing the ECAL front-end electronics is shown in Fig. 4. The $\gamma\gamma \rightarrow$ hadrons machine background and QCD events dominate the spectrum. By comparing the particle spectrum with SEU cross sections of several FPGAs as found in the literature the SEU rate has been estimate to lie between 14 minutes to 12 hours for all of the FPGAs in the TESLA ECAL. This requires the FPGAs in the ECAL to be reset at a higher rate to mitigate these effects.

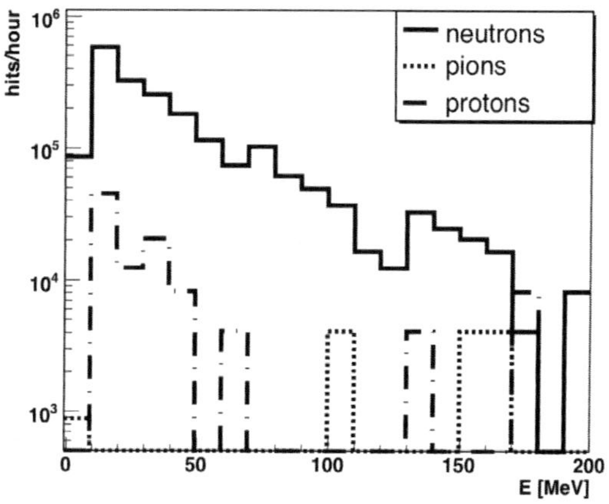

Figure 4. Energy spectrum at the FPGAs for various processes

4. DAQ Software

As framework of the DAQ and control software DOOCS [4] has been chosen. DOOCS provides a three layered modular framework. The lowest layer being hardware interfaces also called device servers, the middle layer provides an interface to a finite state machine, a naming service, a DAQ server and web service e.g. for alarm handling, the upper layer provides user interfaces to programs like ROOT, MATLAB or LabVIEW and DOOCS specific user interfaces. Several network protocols can be used like remote procedure calls, TINE and channel access.

The device server handles all properties of a particular device, e.g. an ODR, LDA, DIF or ASIC. These servers then create as many device instances also called locations as needed of its type. A client server is an independent program which receives that data and sends control messages to the servers. The system is distributed, because the device definition is done on the server side only and is transparent to the clients. Whenever a server is started, new device instances are created or new properties are added, these changes are immediately available on the network; there is no central database to hold the items. The names of the device server (IP-Addresses) for the clients are resolved by an Equipment Name Server (ENS). For the EUDET prototype an ENS naming convention has been chosen. The device servers however have not yet been implemented.

5. Conclusion and Outlook

The DAQ system for the EUDET detector will be a hierarchical DAQ system which minimises the area needed on the detector and uses commercial components to minimise the cost of the DAQ system. In order to make the development for the test system independent of the detector type tested a DIF board will make the readout generic. An LDA card will concentrate the data from the DIF and send it off to the ODR which is located off the detector in a control room. The task of the ODR is to store the data. At the time of writing each hardware component of the DAQ system exists and tests are ongoing. For the LDA the needed firmware still needs to be implemented. The component tests will be finished this autumn with the integration tests starting this winter. It is envisaged that the hardware interfaces of the DAQ software to each hardware components will be finished this autumn. This means that the hardware and software developments are synchronised. The goal is to have the whole DAQ system ready for the EUDET test beam in 2009.

Journal of Physics: Conference Series **160** (2009) 012029 doi:10.1088/1742-6596/160/1/012029

6. References

[1] TESLA Technical Design Report, Part IV, A Detector for TESLA, T. Behnke et al., 2001.
[2] HDMI- High Definition Multimedia Interface Consortium homepage, http://www.hdmi.org
[3] V. Bartsch, et al., Nucl. Instr. and Meth. A (2008)
[4] DOOCS homepage, http://tesla.desy.de/doocs/doocs.html

Semi-Digital Hadronic Calorimeter for Future High Energy Physics Experiments

Imad Laktineh

Université de Lyon, Université de Lyon 1, CNRS/IN2P3, IPNL 4 rue E.Fermi
69622 Villeurbanne cedex
E-mail : laktineh@in2p3.fr

Abstract. A new concept of high granularity hadronic calorimeter based on a semi-digital readout for future ILC experiments is presented. The aim of this concept is to provide the HCAL with a tracking capacity in addition to a good energy measurement resolution. The sensitive medium of this HCAL is made of very thin gas detectors. The readout is based on detector-embedded electronic boards equipped with low consumption daisy-chained 64-channel chips. The nice results obtained with a slice test made of small detectors show that the concept is successful and can be used to build a 1m3 semi-digital HCAL prototype.

1. Introduction

One of the most promising techniques that the next generations of high energy physics experiments are likely to exploit is the one based on particle flow algorithms. The concept uses the fact that better jet energy resolution can be achieved when jet particles are tracked down in the different sub-detectors. The idea behind this is that if the particle is tracked in all the sub-detectors, then its energy can be estimated in the most suitable one obtaining thus the best possible measurement.

To achieve this goal the different sub-detectors should be able to have a tracking capacity in addition to their own specific job. This means that future calorimeters should be modified to allow tracking particles in addition to the energy measurement. The statement is valid for both the electromagnetic and the hadronic calorimeters [1]. Following this idea we present here a new generation of hadronic calorimeters with high granularity in both the transverse and the longitudinal directions offering tracking capacities.

To cope with the tremendous number of electronic channels associated to this kind of calorimeter, a new generation of electronics has been developed. The new electronics is detector-embedded with daisy connection scheme. It provides a semi-digital readout while reducing the power consumption by a factor of 100 thanks to the power pulsing mechanism it offers. Different gas detectors are proposed to provide the sensitive media for this new generation of hadronic calorimeter. Glass Resistive Plate Chamber is one of those candidates which associate high performance to low cost production.

In this paper we first introduce the semi-digital hadronic calorimeter concept. We then present the new electronics and the readout system that was developed for this purpose. We then describe the GRPC detector that we used as a sensitive medium to build our test setup. Finally we describe the setup and the first results obtained with the new concept.

2. Semi-Digital Hadronic CALorimeter concept:

The standard hadronic calorimeter using scintillators as sensitive media can perform high precision energy measurement [2]. However, it may suffer from its poor transverse granularity due to the size of

Journal of Physics: Conference Series 160 (2009) 012030 doi:10.1088/1742-6596/160/1/012030

scintillators that can be used in this kind of detector. This may become an important limitation factor when it comes to the PFA application.
To overcome this limitation a digital hadronic calorimeter with high granularity was proposed within the CALICE collaboration [3] with the aim to attain high transverse granularity (1 cm^2) To achieve this, the analogue readout usually used in the standard HCAL is traded against a binary one and the scintillator is replaced by gas detectors adequate for high granularity sensitive media realisation.

Although this digital HCAL with its high granularity structure is well suited for the PFA application, one may think that the binary readout can affect the energy resolution the HCAL is able to provide. This is not true at low energy. Indeed, up to few GeVs, the simple counting of the hits left by the hadronic shower in the HCAL gives very good estimation of the primary particle energy. The energy resolution obtained in this case is even better than in the case of the analogue readout which suffers from the Landau fluctuation. When the energy becomes high either in the case of a single particle or a jet the particles density in the shower central region becomes higher and a simple counting of the hit numbers does not allow one to establish a good correlation between this number and the number of shower particles and hence prevents to achieve good energy measurement resolution
A way to improve on the high energy resolution while keeping the high granularity of such a calorimeter can be obtained by going from binary readout to a semi-digital one. Replacing one-bit readout by two-bit readout can be of big help. It allows one to distinguish between the simple case with one particle going throw one pad of the detector and the cases where few or many particles going through a detector pad.
A work realized by a KEK group [4] has shown that with such a readout one can have three thresholds which may be fixed for instance to (0,5, 10 and 100 mip energy). The energy measurement resolution obtained in the case using scintillator sensitive medium is as good as the one obtained with the analogue readout.
In order to prove the validity of the semi-digital HCAL concept in the context of the future international linear collider experiments, we conceived and realized a new ASIC. We also conceived and built electronic boards capable to host four ASICs. The boards were then associated to GRPC detectors. A readout system was also developed and the whole system was tested using a cosmics bench.

3. Sensitive medium for the DHCAL

Increasing the granularity of the HCAL necessitates a new gender of sensitive media with a transverse segmentation as small as 1 cm^2. In addition, the thickness of such a medium should be very small to have the most compact HCAL. This is motivated by the fact that the HCAL is expected to be inside the coil of the future ILC experiments. Different gas detectors can satisfy those requirements like the Glass Resistive Plate chamber [5], the Gas Electron Multiplication chamber [6] and the MICRO MEsh GASeous chamber [7]. The thickness of such detectors can be of only few millimetres. Many efforts are underway to adapt them to the DHCAL concept. Here after we describe the GRPC that was used to validate the semi-digital HCAL concept.

The GRPC proposed for the semi-digital HCA is made of two thin glass plates separated by spacers which can be of different shapes. Fishing lines are the standard spacers. An ionising gas is enclosed inside the two plates. High voltage is applied to the two plates through a very thin layer of resistive painting on the outer side of the plates. The resistive painting used is usually made of graphite (less than 1M Ω/□). A thin Mylar layer separates one of the glass plates from the electronics card which contains on the inner face the 1 cm^2 cooper pads.
When a charged particle crosses the chamber, it ionises the gas and a shower is produced. The electrons of the shower are attracted to the glass connected to the anode. They induce opposite charges on the pads of the electronics card producing thus the signal.
According to the nature of the ionising gas and the high voltage value one can obtain an avalanche mode or a streamer mode. The streamer mode produces an important quantity of charge whose

evacuation from the glass plate takes more time and thus reduces the detector rate capability (not more than few Hz/cm^2).
The avalanche mode is more appropriate for the future ILC hardonic calorimeter. The amount of produced charge is a factor of 100 less which implies a detection rate that can go beyond 100 Hz/cm^2. The gas which is usually used to obtain the avalanche mode is a mixture of the TFE (93%), the isobutene (5%) and the SF_6 (2%). The last two gases are used as photon and electron quencher respectively to keep the signal localised around the crossing article.

4. Electronics readout system

The high granularity HCAL necessitates also to readout a tremendous number of channels.
This leads to two major technical challenges that the electronics readout has to tackle. The first is the power consumption. With up to fifty millions channels needed to readout a hadronic calorimeter like the one expected for the future ILC experiments [8] it is of a primary importance to reduce the power consumption to the lowest level. The second challenge resides in the number of connection needed for this enormous number of channels. This involves an unthinkable quantity of cables to be used and whose effect on the detector geometry can result in many cracks and holes preventing the detector from being hermitic which renders the PFA application much less efficient. To solve these two problems a new electronic chip (ASIC) called HARROC [9] was developed. It can deal with 64 channels each of them having two independent comparators to provide two thresholds. The thresholds can be fixed by software to integer values in the interval between 0 and 1023 (DAQ units). An important feature of this chip is the power pulsing. It allows the chip to be active just before the bunch crossing and off just after. This permits to reduce the power consumption by a factor of 100 in the case of the ILC cycle machine which is supposed to last 200 ms of which only one millisecond is reserved for the bunch crossings. The chip is conceived to be very tiny so it can be embedded on the detector. It has the advantage to be connected to other chips through the daisy chain mechanism so that the chips of the same detector plan can communicate with each other. Slow control parameters as well as data collected in the detector can then circulate among the chips while the communication with the outside acquisition system is hugely reduced. In addition to these features, the chip has an internal memory capable to contain up to 128 events. The aim of this memory is to record the events which occur during the bunch crossing and then read them out during the idle time. This means that during the bunch crossing the chips are able to record all the signals and hence no event can be lost because of online selection. Of course this is possible due to the low occupancy rate expected in the HCAL of the future ILC experiments. Another feature of this ASIC is the possibility to change the gain of each channel in the interval [0-4] with a 6-bit precision. This is a useful tool to render the response of the different channels to the same signal as identical as possible. The cross-talk between two channels of one ASIC was measured by injecting an electric signal equivalent to the one produced by the average charge left by a mip (1 pc) in a GRPC through one channel using an appropriate test board. The signal observed in the other channels was found to be less than 2%. In figure1 a synoptic presentation of the HARDROC chip is shown.

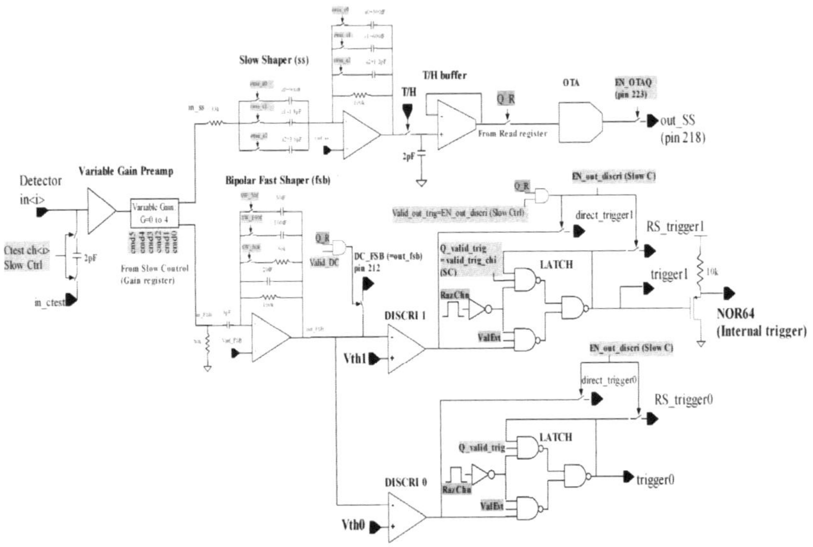

Figure1: Synoptic scheme of the HARDROC chip

To validate the concept of a semi-digital hadronic calorimeter with high granularity, a Printed Circuit Board was conceived to host 4 hardroc chips and to provide the connection between adjacent chips as well as the connection of the first chip to the readout system. For simplification reasons the readout system using FPGA device was also implemented on the same PCB as well as a USB device responsible of the communication between the FPGA and a computer. The PCB is an 8-layer circuit with two internal layers reserved for the digital signal transfer. The routing was optimised to reduce the cross-talk among adjacent pads. The thickness of the PCB is only 800 microns to keep the effective detector thickness as small as possible. On one of the two PCB faces, 256 cooper pads of 1X1 cm^2 were printed. The distance between two adjacent pads was chosen to be 500 microns. This was intent to allow the charged particles which cross the detector between two pads to be detected at least by one of the two pads since the spread of the produced shower at the anode level is expected to be around 1 mm^2. The pads are connected internally to the ASIC channels through the PCB structure. The cross-talk among adjacent pads was tested before other electronics components were fixed on the PCB by injecting a charge of 1 pc on one pad using an appropriate probe. The charges induced on the adjacent pads were then measured and found to be less than 0.3 % of the injected one which means that the cross-talk due to the PCB is negligible with respect to the ASIC one which is itself very small. Acquisition software was also developed. It permits to download the different slow control parameters to the different ASICs and to collect data from these ASICs through the FPGA device. Two readout modes were implemented. The first one is the ILC-Like one where events are recorded during the bunch crossings and the readout takes place after. The other mode was conceived for cosmics and beam test studies. In this mode the acquisition and data taking is stopped when an external trigger occurs. The memory of the different ASICs is then read out. In both modes each event is associated with a time stamp. In the external trigger mode the time difference between the external trigger and the last recorded event is also given. This determines the time occurrence of each event with respect to the external trigger one.

5. Experimental setup

Few GRPC detectors of 33.55 X 8.35 cm^2 each were built and associated to the PCB hosting the ASICs as well as the readout system. Three of the fully equipped detectors were placed in a mechanical structure and oriented horizontally so they can detect cosmics. A system of three scintillators associated with PhotoMultipliers was built to form a trigger system for the cosmics crossing the detectors (figure2). A software graphic interface based on LABVIEW was elaborated. The interface was used to calibrate the lower threshold of the 256 channels of each of the three electronics board. Indeed a charge of 100 fc was injected through internal capacitors to each channel one by one for different values of gains. Each step was repeated 100 times to accumulate significant statistics. For each gain value the response of the studied channel was recorded and an efficiency curve called (Scurve) was obtained by varying the lower threshold by a step of 1 DAQ unit. An automatic procedure using a fit with a Dirac-Fermi function leads then to determine the inflexion point (corresponding roughly to the 50% efficiency level) for each channel and for each gain. The correlation between the gain and the inflexion point DAQ values was then found to be linear and was fitted by a straight line. The operation was repeated for the 256 channels of each detector. The average value of the 256 inflexion points obtained with gain =1 was then selected as a reference value. Then for each channel the gain was chosen according to the aforementioned study so that the new inflexion value was as close as possible to the reference value. Once all the gains are corrected, the Scurves were produced again. Figure3 shows the dispersion of the inflexion points obtained before and after the gain correction. After the gain correction the dispersion of the 256 channels is less than 1.3 DAQ units. This represents a dispersion of less than 2.6 fc in charge values. Indeed, to determine the correlation between the DAQ unit and the injected charge, charges with different values were injected and the inflexion point value determined at gain =1. This allows one to find the correspondence between the charge and the DAQ units which appears to be: 1 fc = .48 DAQ units.

Figure2: The test setup

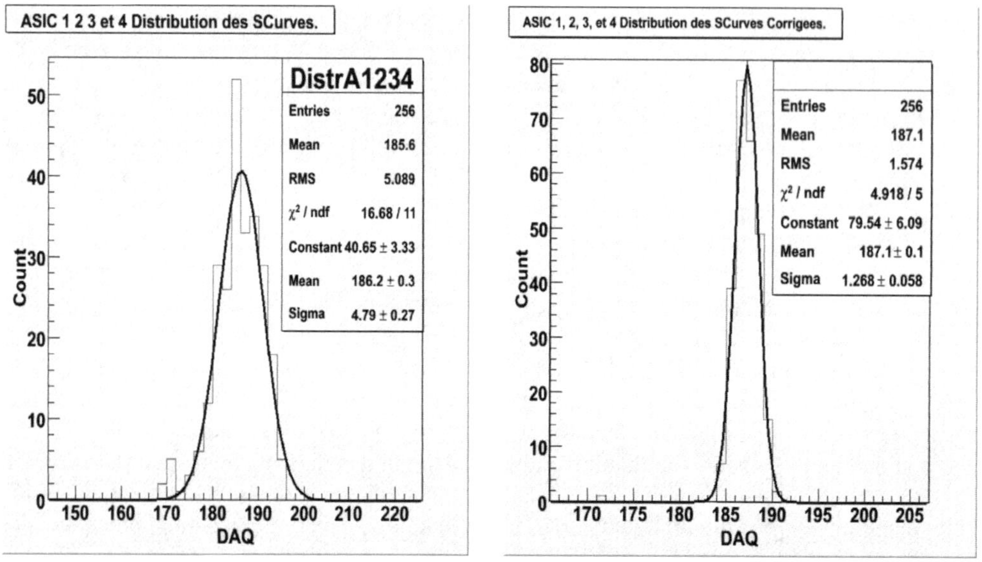

Figure3: The DAQ dispersion before (left) and after (right) the gain correction

After the gain corrections applied to the three PCBs the whole system was mounted. An external trigger provided by the scintillator-PM system permits to stop the acquisition of the three of them when a cosmic was signalled. The data recorded in the 12 ASICs is then read and the events (hits) which occurred in the time window corresponding to the signal time arrival with respect to the external scintillator are considered for efficiency and multiplicity studies. The time window was estimated using the measured time difference between the analogue signal obtained from the cathode side and the trigger signal. It was then confirmed by the results obtained on the anode side. The width of the time window is taken to be of 200 ns which is the resolution given by the internal ASIC clock of 5 MHz.

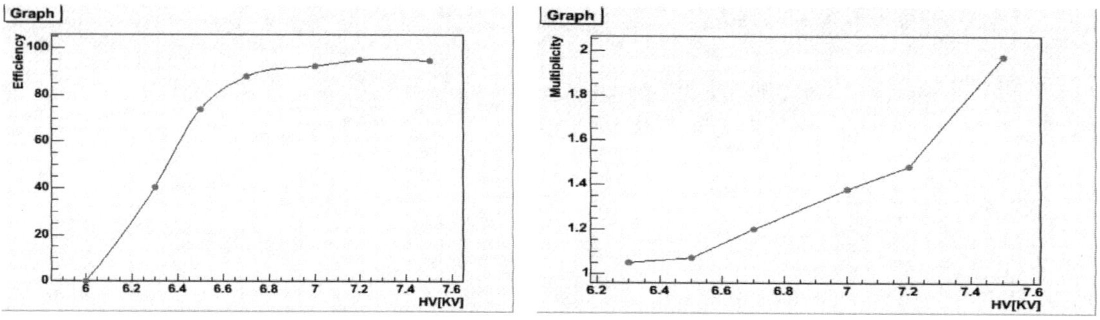

Figure4: Efficiency (left) and multiplicity (right) versus H.Voltage values of a GRPC with graphite painting (threshold =100 fc).

The efficiency and multiplicity was first obtained by observing the presence of hits in the time window mentioned above and their number for different values of the high voltage. The results are shown in figure4 for a standard GRPC detector (graphite painting and fishing line spacers) which are in agreement with the results found using standard electronics. Other detectors using more resistive painting (LICRON) and more sophisticated gas distribution inlet as well as ceramic microspheres as spacers were studied using the same dispositive. These modifications lead to an important reduction of the multiplicity as shown in figure5. It goes down from 1.6 to 1.3 at 7.4 kV. This is an important result since the multiplicity of a mip should be close to one in order to have the best tracking precision. In order to study the noise effect on the previous results the same study was realised using shifted time windows. The results obtained indicate that the noise contribution to the measured efficiency is negligible. The noise was also measured independently by counting the number of recorded hits by time interval in each channel for different high voltage values. In the worse scenario which takes place for high voltage values (>7.5 KV) and low threshold value (< 100 fc) the noise is less than 1 Hz/cm^2 which is ridiculously irrelevant when considered within the 300 ns time interval during which one bunch crossing is expected to take place in the future ILC and during which the acquisition system is active.

Although the two thresholds were present for the recorded events, only the first threshold was considered for the time being. The hits obtained using the second threshold, were however checked to be included in the set of those obtained with the lower threshold and their number decreases with the increase value of the threshold. The information provided by the second threshold will be soon used when the setup is exposed to a beam at CERN during this summer in order to study its effect on the tracking precision as well as the energy resolution.

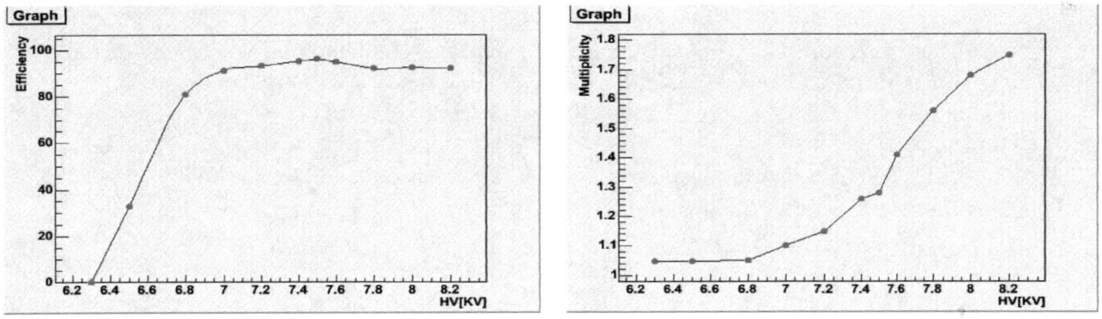

Figure5: Efficiency (left) and multiplicity (right) versus H.Voltage values of a GRPC with LICRON painting (threshold =100 fc).

The efficiency and the multiplicity versus high voltage studies for each of the GRPC detectors were redone using the information from the other two GRPCs. This allows one to obtain more precise results since the hits position in the studied detector can be predicted from the ones recorded in the other detectors. The results obtained confirmed the previous ones

6. Conclusion

The semi-digital HCAL concept proposed for the future ILC can be a powerful tool to apply the PFA techniques through the high granularity it offers. The use of 2-bit readout rather than a binary one (1-bit) can help to keep the energy resolution close to the one of the analogue one especially when the charge spectrum associated to a mip in the GRPC is less spread. This can be in fact achieved by going from simple to multigap GRPC [10].

The electronic readout system was designed and realised. A cosmics bench test was built and fully equipped GRPC detectors were tested. The first results show that the concept is successful. Yet, many efforts are underway to improve on both the detector and the electronics performance. A mini hadronic calorimeter made of five GRPC detectors separated with iron slabs of 2 cm each is already built. Beam tests programmed this summer at CERN will permit to provide more hints on the way to improve the performance of the semi-digital concept.

The success of the tests already realised and those to come will lead to the construction a hadronic calorimeter prototype of $1m^3$ in the coming two years.

References

[1] Innovations in ILC detector design using a particle flow algorithm approach Stephen R Magill 2007 *New J. Phys.* **9** 409.

[2] Hadron Calorimetry at the LHC, J. Proudfoot 34th SLAC Summer Institute On Particle Physics (SSI 2006), July 17-28, 2006

[3] Calice collaboration web page: http://llr.in2p3.fr/activites/physique/flc/calice.html

[4] Software studies of GLD calorimeter, MATSUNAGA, Hiroyuki, LCWS2006 Bangalore proceedings.

[5] Operation of narrow gap RPC with tetrafluoroethane based mixtures, V. Koreshev, V. Ammosov, A. Ivanilov, Yu. Sviridov, V. Zaets and A.Semak, *Nucl. Instr. Meth.*, vol. A456, pp. 46-49, 2000.

[6] The Gas Electron Multiplier (GEM),Bouclier, R.; Capeans, M.; Dominik, W.; Hoch, M.; Labbe, J.-C.; Million, G.; Ropelewski, L.; Sauli, F.; Sharma, A. Nuclear Science, IEEE Transactions on Volume 44, Issue 3, Jun 1997 Page(s):646 - 650

[7] MICROMEGAS : A high-granularity position-sensitive gaseous detector for high particle-flux environment, Y. Giomataris, Ph. Rebourgeard, J.P. Robert and G. Charpak, Nucl. Instr. Meth. A376(1996)29-35.

[8] International Linear Collider Reference Design Report 2007 online at http://www.linearcollider.org/cms/

[9] HARDROC, HAdronic Rpc Detector Read-Out Chip. N.Seguin-Moreau,TWEPP 2007, PRAGUE et IEEE HAWAI 2007, proceedings:

[10] A new type of resistive plate chamber. The multigap RPC, Cerron-Zaballos et al., NIM,1996 vol. 374 n°1.

XIII International Conference on Calorimetry in High Energy Physics (CALOR 2008) IOP Publishing
Journal of Physics: Conference Series **160** (2009) 012031 doi:10.1088/1742-6596/160/1/012031

Calorimeter R&D for the SuperNEMO Double Beta Decay Experiment

Matthew Kauer - on behalf of the SuperNEMO Collaboration
Dept. Physics and Astronomy
University College London
Gower Street
London, UK
WC1E 6BT

E-mail: kauer@hep.ucl.ac.uk

Abstract. SuperNEMO is a next-generation double beta decay experiment based on the successful tracking plus calorimetry design approach of the NEMO3 [1] experiment currently running in the Laboratoire Souterrain de Modane (LSM). SuperNEMO can study a range of isotopes, the baseline isotopes are ^{82}Se and possibly ^{150}Nd. The total isotope mass will be 100–200 kg. A sensitivity to a $0\nu\beta\beta$ half-life greater than 10^{26} years can be reached which gives access to Majorana neutrino masses of 50–100 meV. One of the main challenges of the SuperNEMO R&D is the development of the calorimeter with an unprecedented energy resolution of 4% FWHM at 3 MeV ($Q_{\beta\beta}$ value of ^{82}Se).

1. Introduction

The recent observation of neutrino oscillations and the resulting measurements of the neutrino mass differences has motivated experimental searches for the absolute neutrino mass. Neutrinoless double beta decay ($0\nu\beta\beta$) is the only practical way to understand the nature of neutrino mass and one of the most sensitive probes of its absolute value. Ettore Majorana proposed that neutrinos could be their own anti-particles [2], and this lead to Furry's conclusion [3] that neutrinoless double beta decay (Figure 1B) is possible via neutrino exchange if the neutrinos are Majorana particles and have non-zero mass.

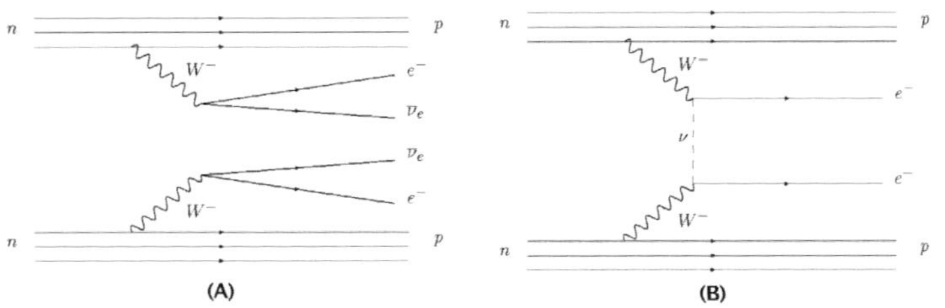

Figure 1. Feynman diagrams of a $2\nu\beta\beta$ (A) decay allowed in Standard Model, and $0\nu\beta\beta$ (B) decay allowed if neutrinos are massive and Majorana particles.

The effective Majorana neutrino mass $\langle m_{\beta\beta}\rangle$ is proportional to the square root of the $0\nu\beta\beta$ decay half-life $T_{1/2}^{0\nu}$ in equation (1), where $G^{0\nu}$ is the kinematic phase-space factor and $M_{0\nu}$ is the nuclear matrix element. The experimental signature of $0\nu\beta\beta$ is two electrons with the energy sum equaling the $Q_{\beta\beta}$ of the decay. There are other mechanisms to explain neutrinoless double beta decay [4], but the above mechanism is the most favored due to the minimal required modifications to the Standard Model.

$$[T_{1/2}^{0\nu}]^{-1} = G^{0\nu}|M_{0\nu}|^2\langle m_{\beta\beta}\rangle^2 \tag{1}$$

2. SuperNEMO Detector

SuperNEMO is ~100 kg source isotope (^{82}Se or ^{150}Nd), tracker + calorimeter detector with a projected neutrinoless double beta decay half-life sensitivity of 10^{26} years (~ 50 meV effective Majorana neutrino mass). The SuperNEMO baseline design calls for 20 modules ($\sim 4\times2\times1$ m), each holding 5 kg of source isotope. Both sides of the source foil have 9 layers of Geiger mode drift cells enclosed by the calorimeter walls. Each module will hold ~600 8" PMTs.

The project is currently in a 3 year design study and R&D phase and the collaboration comprises over 90 physicists from 12 countries. The R&D program focuses on four main areas of study: isotope enrichment, tracking detector, calorimeter, and ultra-low background materials production and measurements. The goals of the R&D are summed up in Table 1.

Table 1. SuperNEMO Parameters and Goals

Parameters	Goals
Isotope	^{82}Se (or ^{150}Nd)
Mass	100–200 kg
$0\nu\beta\beta$ Detection Efficiency	30%
Energy Resolution FWHM at 3 MeV	4%
^{214}Bi Source Purity	$< 10\ \mu$Bq/kg
^{208}Tl Source Purity	$< 2\ \mu$Bq/kg
Operation Time	5 years
$T_{1/2}^{0\nu\beta\beta}$ Sensitivity	10^{26} years
Effective Majorana Mass ($\langle m_{\beta\beta}\rangle$)	50–100 meV

The significance of energy resolution is best illustrated by the half-life sensitivity formula [5]. This formula (2) has limitations in accurately predicting the sensitivity of the specific SuperNEMO detector, but does demonstrate the significance of energy resolution. The energy resolution ΔE factors in with equal importance as isotope mass M, runtime t, and number of background events N_{bkg}. Factors N_A and A are Avogadro's number and atomic mass of the isotope and ε and κ_{CL} are the detector efficiency and the confidence level on the half-life sensitivity $T_{1/2}$. The dominating background to $0\nu\beta\beta$ is the irreducible $2\nu\beta\beta$ channel, therefore the energy resolution of the calorimeter becomes the dominating parameter determining the detector's overall sensitivity to neutrinoless double beta decay.

$$T_{1/2} \propto \ln 2 \cdot \frac{N_A}{A} \cdot \frac{\varepsilon}{\kappa_{CL}} \cdot \sqrt{\frac{M \cdot t}{N_{bkg} \cdot \Delta E}} \tag{2}$$

Simulations done for ^{82}Se with a projected calorimeter energy resolution of 12% and 8% FWHM at 1 MeV and normalized to 10^{26} year $0\nu\beta\beta$ half-life, clearly displays the importance of energy resolution for this experiment (Figure 2). At 12% energy resolution, the high energy tail from the $2\nu\beta\beta$ energy spectrum overlaps the $0\nu\beta\beta$ peak, but at 8% energy resolution there is separation.

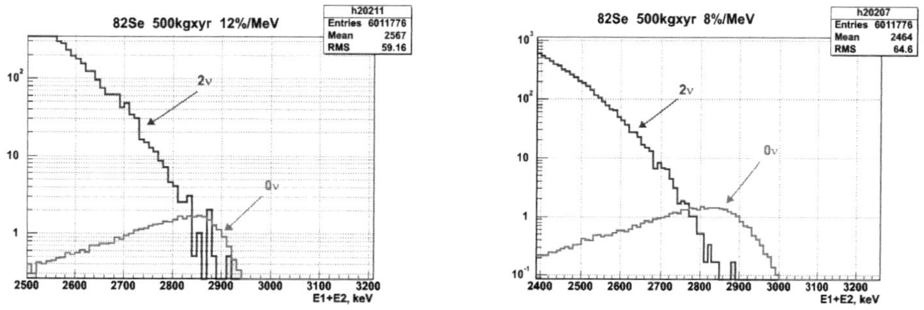

Figure 2. Simulations for 500 kg·yr ^{82}Se. The $0\nu\beta\beta$ half-life (RED) is normalized to 10^{26} years. Expectations for energy resolutions 12% (left) and 8% (right) $\frac{\Delta E}{E}$ FWHM at 1 MeV.

3. Calorimetry Goals of SuperNEMO

The calorimeter R&D is subdivided into three main groups: energy and time resolution studies, calibration, and PMT radio-purity. The energy resolution R&D is the main focus of this report. As with all PMT based calorimeters, PMT gain stability and linearity must be both intrinsically good and experimentally well understood to ensure the accurate reconstruction of data. Conventional LASER/LED configurations prove difficult with many channels. A promising alternative method is one photo-electron peak monitoring [6] because the PMT gain can be extracted independent of light amplitude. The R&D also investigates the use a low activity alpha source embedded into the plastic scintillator as a means to monitor the gain.

Specific to low background counting experiments, ultra-pure materials must be used throughout the detector. The PMTs are one of the main sources of contamination with emphasis on the purity of the cathode glass which is closest to the active volume of the detector. The Barium salt used to make conventional glass is chemically the same as Radium, and therefore very difficult to purify during the production of the glass. Various manufactures have developed recipes for low-background glasses, but the requirements of SuperNEMO have motivated this development to a new level of radio-purity. Photonis has provided preliminary samples of their new ultra-pure glass that have met R&D requirements.

3.1. Energy Resolution

Optimization of the energy resolution is the result of a high number of photo-electrons which reduces the statistical error $1/\sqrt{N_{pe}}$. This can be simplified into three experimental objectives which are described by formula (3).

$$\frac{N_{ph}}{E_e} \cdot \varepsilon_{col}^{light} \cdot \left(QE^{PMT} \cdot \varepsilon_{col}^{PMT}\right) = N_{pe} \tag{3}$$

N_{ph}/E_e is the number of photons per unit energy and is determined by the scintillator light output. $\varepsilon_{col}^{light}$ is the light collection efficiency and depends upon: scintillator geometry, transparency, reflector efficiency, optical coupling quality, etc. Intrinsic characteristics of the PMT include the quantum efficiency of the photo-cathode QE^{PMT}, and the cathode to first dynode collection efficiency ε_{col}^{PMT}. There has been a significant breakthrough in development new high quantum efficiency PMTs based on bi-alkali photocathodes by Hamamatsu and

Photonis. The SuperNEMO group is working very closely with PMT manufacturers on characterizing these new photo-detectors which have now a QE in the range of 35–43% at the peak wavelength (to be compared with ~25% QE for "conventional" photo-multipliers). Assuming that the energy resolution of the scintillator detector is mainly determined by the photon statistics we can express the resolution in terms of the number of collected photo-electrons (4).

$$\frac{\Delta E}{E} = \frac{FWHM}{E} = \frac{2.35\sigma}{E} = \frac{2.35}{\sqrt{N_{pe}}} \tag{4}$$

The scintillator must be a low Z material to minimize backscattering electrons and has to have a good timing resolution (a coincidence time resolution of $\sigma = 250$ ps at 1 MeV is required). It has to be cost effective and radio-pure. These requirements essentially rule out many popular non-organic scintillator, such as NaI(Tl), CsI(Tl), CaF_2(Eu) etc. which would otherwise provide a good energy resolution due to their high light output. The choice of reflective material is also limited to low density reflectors to reduce electron energy loss through the material.

4. Experimental Setup

The energy resolution measurement is carried out by exciting the scintillator under test with a flux of electrons of known energy and then analyzing the resulting distribution. The mono-chromatic source of electrons approximates the delta function and therefore any smearing of the distribution is due to the light collection of the scintillator and PMT under study. The test setup can be broken into three subcategories: the calorimeter block (scintillator + reflector + lightguide + PMT), the electron source, and the data acquisition (DAQ).

4.1. Calorimeter Block

Many different scintillator, reflector, and PMT combinations are being studied. Solid scintillator candidates include polystyrene (PST) based scintillators from ISM and JINR labs (1.5% PTP, 0.0175% POPOP) and polyvinyltoluene (PVT) based scintillators from Bicron (BC404, BC408) and Eljen (EJ204, EJ200) manufacturers. Liquid scintillators are toluene based and from CENBG, INR, ISM, and JINR labs (0.5% PPO, 0.0025% POPOP). Various specular and diffusive reflectors being tested include: Teflon, Kapton, Aluminized Mylar, and Enhanced Specular Reflector (ESR) from the Vikuiti and ReflechTech manufactures. The three PMT competitors are Hamamatsu, Photonis, and Electron Tubes Ltd. (ETL).

4.2. Electron Source

There are two methods used to obtain a mono-chromatic source of electrons. The first method is simplest to implement as one uses the K-shell 976 keV conversion electrons (CE) from a ^{207}Bi source. The drawback to this method is that the fitting function needs to incorporate the convolution of additional x-rays, gammas, L-shell and M-shell conversion electrons. The second method is more involved to set up, but in principle leaves a spectrum that can be easily fit with a Gaussian function. The β emission from a highly active ^{90}Sr source is passed through a magnetic field so that β's of a particular energy can be selected. For the energy resolution measurements, 1 MeV electrons are used.

4.3. Data Collecting and Analysis

Data acquisition is accomplished with a gated QDC (charge to digital converter). The PMT signal is split in two, half the signal is used for triggering of the electronics and generating the gate signal for the QDC, the other half of signal goes directly to the QDC after some passive delay to match the timing of the electronics. In the method of the ^{207}Bi source, three different data runs must be taken to obtain a pedestal, an energy spectrum of just the gammas (achieved by shielding out the electrons with 2 mm of Aluminum) and the energy spectrum of the gammas + CEs (conversion electrons). The Compton edges from the gamma distribution are sufficiently described by a modified Heaviside step-function. The free parameters of the gamma distribution are determined and then fixed while the gamma + CEs distribution is fit. The CEs are a sum of three Gaussian distributions from the K, L, and M shells.

5. Measurements

The calorimeter baseline design calls for 8" diameter PMTs, but as a check of physical limitations on achievable energy resolution, a detailed study of small (3") PMTs preceded. A resolution of 6.5% FWHM at 1 MeV was measured using Bicron BC404 scintillator, wrapped in Vikuiti ESR (Enhanced Specular Reflector), mounted on a 3" Hamamatsu Super-Bialkali type PMT (Figure 3). Using (4), this extrapolates to 3.8% at 3 MeV which is better than the goal of 4% stipulated by the R&D. This is an unprecedented result for plastic scintillators. Proving there are no physical limitations to reaching the 4% level, the challenge then becomes scaling up the PMT and scintillator size while maintaining resolution.

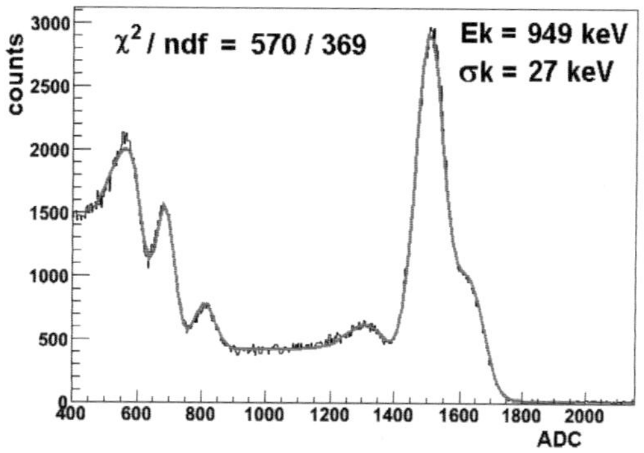

Figure 3. The fit to data (RED line) results in 6.5% FWHM at 1 MeV.

5.1. Light Collection Simulations with GEANT4

Extensive optical simulations were carried out in GEANT4 with all inputs being wavelength dependent, experimental measurements including: POPOP absorption and re-emission (Stokes Shifting), PMT QE, scintillator bulk absorption and emission, material indexes of refraction, and material reflectivities. The simulations revealed sensitive parameters of the setup. Polishing the side of the lightguide to give specular internal reflection as well as wrapping the lightguide with a specular reflector yielded a 2–3% improvement in the expected resolution. The simulations gave expected resolutions of 7.5% and 7.7% for the 8" and 11" PMTs with lightguide wrapped in ESR and 5 × 5 × 2 cm BC404 scintillator wrapped in ESR.

After optimizing the sensitive parameters in our experimental setup, the resolution measurements were 1–2% worse than the expectations from simulation. Current suspects awaiting investigation are photo-cathode QE uniformity and cathode to first dynode collection efficiency in the presence of Earth's natural magnetic field. Earth's magnetic field is known to influence large (>5") PMTs collection efficiency with non-negligible effects, and the effect increases with PMT diameter. These are both characteristics of the PMT which change from one PMT to another and are therefore difficult to simulate accurately.

5.2. Solid Scintillator Measurements

Large solid scintillator blocks are an ideal candidate for SuperNEMO because of low cost, high radio-purity, decreased number of channels, and the physical simplicity of the setup. Three possible variations under study are: small (<5") PMT with flat cathode window with scintillator coupled directly to cathode window (Figure 4A), large (>8") PMT with hemispherical cathode window with scintillator coupled to concave lightguide (Figure 4B), coupled to cathode window, and large (>8") PMT with hemispherical cathode window with concave scintillator coupled directly to cathode window (Figure 4C). Table 2 summarizes the best measurements for these configurations.

Table 2. Measurements with the Solid Scintillator Setup

Scintillator Dimensions and Type	PMT Diameter and Make	FWHM at 1 MeV
5 × 5 × 2 cm BC404	3" Hamamatsu SBA	6.5%
9 × 9 × 2 cm BC408	8" Hamamatsu SBA with Lightguide	10.1%
14 × 14 × 2 cm BC404	8" Electron Tubes Ltd. with Lightguide	9.2%
15 × 15 × 2 cm BC408	8" Hamamatsu SBA with Lightguide	10.3%
20 × 2 cm (hexagonal) BC408	8" Hamamatsu SBA with Lightguide	11.2%
∅ 20 × 2 cm PST	8" Photonis	7.5%
∅ 20 × 10 cm PST	8" Photonis	8.2%

5.3. Liquid Scintillator Measurements

In parallel an R&D program on liquid scintillator detectors is being carried out. The motivations for using liquid scintillator are the following: lower cost, no lightguide needed to couple to hemispherical PMT cathode, larger active volume increases gamma tagging efficiency for background rejection, good uniformity, and high radio-purity. The dominating drawback is the mechanical engineering of the containment structure and meeting safety requirements for an underground laboratory. Two main variations on the setup are under study: a semi-conical setup where the diameter of the liquid scintillator surface is larger than that of the PMT (Figure 4D), and a cylindrical setup where the diameter of the liquid scintillator surface matches that of the PMT (Figure 4E). Table 3 summarizes the best liquid scintillator measurements.

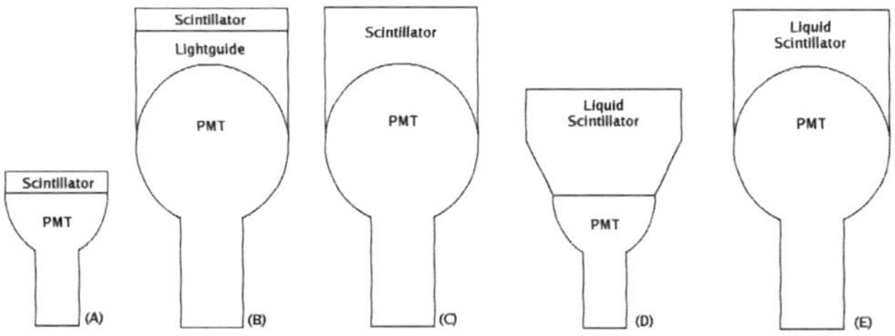

Figure 4. Configurations for the solid and liquid scintillator setup.

5.4. Liquid + Solid Hybrid Measurements

The mechanical engineering of the liquid scintillator containment is challenging. A thin film entrance window with low density and low Z must be used to minimize electron energy losses.

Table 3. Measurements with the Liquid Scintillator Setup

Scintillator Dimensions	PMT Diameter and Make	FWHM at 1 MeV
∅ 7.6 × 2 cm	3" Photonis	7.6%
∅ 7.6 × 10 cm	3" Photonis	8.0%
∅ 8.4 × 9.2 cm	5" Photonis	7.3%
∅ 20.3 × 20 cm	8" Photonis	11.3%
23 × 9.2 cm (hexagonal)	5" Electron Tubes Ltd.	10.8%

An alternative approach is to use a so-called active window where solid scintillator is used on the containment face. This approach utilizes the liquid scintillator as the lightguide and increases the active volume for gamma tagging efficiency. Table 4 summarizes the best hybrid measurements.

Table 4. Measurements with the Liquid + Solid Hybrid Setup

Liquid / Solid Scintillator Dimensions	PMT Diameter and Make	FWHM at 1 MeV
23 × 9.2 cm (hexagonal) / 5 × 5 × 2 cm	5" Electron Tubes Ltd.	12.3%
23 × 9.2 cm (hexagonal) / 23 × 2 cm (hexagonal)	5" Electron Tubes Ltd.	15.1%

5.5. Long Bar Scintillator Measurements

The detector "floor-space" requirement can drastically be reduced by implementing the long scintillator bar design. In this configuration, 2 meter scintillator bars span the tracker volume with a PMT coupled to each end of the bar. This configuration is also the cheapest because of the drastically reduced number of PMTs and the reduced floor-space required from an underground laboratory. With the timing from the two PMTs, an impact resolution of 1–2 cm (along the bar length) is achievable and this information is of additional use for background rejection. Moreover, due to a significantly reduced mass of PMT glass and their relatively remote locations from the detector fiducial volume, the bar design should have a much lower background from PMTs which is one of the main background sources of SuperNEMO. Table 5 summarizes the bar scintillator measurements so far.

A resolution of 7% at 1 MeV is probably impossible to reach with 2m bars. Thus the crucial question for feasibility of this design is whether a better background rejection and higher detection efficiency compensate a worse energy resolution. Rough estimates show that it might be a valid option if a resolution of 10–11% is achievable with the bars. Extensive physics simulations are under way to answer this question with certainty. In the meantime measurements are being carried out with high QE PMTs and optimized geometry to reach the best possible resolution with scintillator bars.

6. Summary and Future Plans

Exceptional resolutions of 6.5% at 1 MeV were measured for small PVT scintillators coupled to high QE PMTs. The SuperNEMO baseline design calls for large scintillator blocks (∅ 20 × 10 cm). Scintillators of this size read out through a lightguide showed an energy resolution of 9–10% at 1 MeV. Better results have been achieved by casting a large plastic

Table 5. Measurements with the Bar Scintillator Setup

Scintillator Dimensions and Type and Reflector	PMT Diameter and Make	FWHM at 1 MeV
200 × 10 × 1.25 cm / BC408 / Al. Mylar	3" Hamamatsu SBA	12.9%
200 × 10 × 1.25 cm / BC408 / Al. Mylar	3" Hamamatsu SBA with Lightguide	13.6%
200 × 10 × 1.25 cm / BC408 / ESR	3" Hamamatsu SBA	12.9%
200 × 10 × 1.25 cm / BC408 / ESR	5" Electron Tubes Ltd.	13.7%

scintillator directly on a hemispherical 8" PMT. With this configuration we have been able to reach the important milestone of 7–8% $1/\sqrt{E}$ MeV energy resolution for the baseline detector design. Consequently the R&D on solid scintillators will be focusing on cast scintillator solutions rather than lightguides to increase the light collection efficiency. The development program will also move away from the previous square-block designs and focus on more realistic hexagonal scintillator geometries. We note that there is room for further improvements by using a higher QE PMTs and more efficient scintillators.

Liquid scintillator provides an alternative while maintaining good resolution (7–8% at 1 MeV) and improving gamma tagging efficiency, but achieving the required resolution with large blocks as well as the engineering of the mechanical design and safety remain a challenge. The hybrid solution creates a more robust containment setup for the liquid, but achieving <7% is very challenging. Long scintillator bars design can potentially give a more efficient detector with more background rejection power. It will drastically reduce the number of PMTs and facilitate a more compact detector design. Measurements so far yield 12–13% at 1 MeV. Work is in progress to understand if this resolution can be improved to 10% and whether a worse energy resolution can be compensated by the above advantages of this detector configuration.

Last years have seen a significant progress in development of novel photo-detectors. PMTs with a QE of over 40% are now available. Using the latest achievements in PMT, reflector, and scintillator technology the SuperNEMO collaboration has demonstrated the feasibility of achieving the target energy resolution necessary to reach the sensitivity goal of the experiment. The remaining challenge is to demonstrate that the achieved energy resolution can be maintained at the mass production scale. SuperNEMO expects to make the final decision on the calorimeter design in mid-2009. The large scale construction will start in 2011 with the aim to reach the target sensitivity of $\langle m_{\beta\beta} \rangle$ = 50–100 meV by 2017.

References

[1] R. Arnold et al. Technical design and performance of the NEMO 3 detector. *Nucl. Instrum. Meth.*, A536:79–122, 2005.

[2] Ettore Majorana. Theory of the Symmetry of Electrons and Positrons. *Nuovo Cim.*, 14:171–184, 1937.

[3] W. H. Furry. On transition probabilities in double beta disintegration. *Phys. Rev.*, 56:1184–1193, 1939.

[4] Frank T. Avignone III, Steven R. Elliott, and Jonathan Engel. Double Beta Decay, Majorana Neutrinos, and Neutrino Mass. *arXiv*, nucl-ex:0708.1033, 2007.

[5] F. T. Avignone, G. S. King, and Yu. G. Zdesenko. Next generation double-beta decay experiments: Metrics for their evaluation. *New J. Phys.*, 7:6, 2005.

[6] T. Asch, H. Gemmeke, M. Kleifges, A. Kopmann, A. Menshikov, and D. Tcherniakhovski. Single photoelectron resolution for the calibration of photomultiplier systems. *Nuclear Science Symposium Conference Record, 2005 IEEE*, 2:887–890, 2005.

The ArDM project: A Liquid Argon TPC for Dark Matter Detection

V. Boccone[1], on behalf of the ArDM collaboration

[1] Physik-Institut der Universität Zürich, Winterthurerstrasse 190, CH-8057 Zürich, Switzerland

E-mail: boccone@cern.ch

Abstract. ArDM is a new-generation WIMP detector which will measure simultaneously light and charge from scintillation and ionization of liquid argon. Our goal is to construct, characterize and operate a 1 ton liquid argon underground detector. The project relies on the possibility to extract the electrons produced by ionization from the liquid into the gas phase of the detector, to amplify and read out with Large Electron Multipliers detectors. Argon VUV scintillation light has to be converted with wavelength shifters such as TetraPhenyl Butadiene in order to be detected by photomultipliers with bialkali photocathodes. We describe the status of the LEM based charge readout and light readout system R&D and the first light readout tests with warm and cold argon gas in the full size detector.

1. Introduction

Noble liquids such as argon or xenon are today two of the best options for large-size Dark Matter (DM) experiment such as WIMPs[1] searches because they serve as target and detecting media at the same time. WIMPs in fact, with particular regard to the lightest neutralino, are the most popular DM candidates within the Minimal Supersymmetric Standard Model (MSSM). Most of the DM models predicts that WIMPs interact almost only gravitationally and maybe weakly with matter thus forming cold relic halos around massive astronomical objects. For this reason DM is very difficult to detect directly nevertheless its evidence has been observed using astronomical observation [1].

Many groups in the world are focusing their attention on the direct detection. The detectors should have big target masses and excellent noise rejection capabilities because of the small cross section between DM and ordinary matter. Liquid noble gases (Ar, Xe, Kr, Ne) have a relatively low ionization energy and a good cross section for nuclear recoil processes, moreover they can act as active target because they not only have good scintillation properties but also a long electron lifetime. Argon, because of its availability and low cost, is the best one among those candidates and is very competitive for large detectors.

1.1. Interaction with liquid argon

The expected DM signal from a WIMP-argon head on collision is a single 10-100keV recoiling nucleus. The passage of ionizing particles in argon produces not only ionization but excitation

[1] Weak Interacting Massive Particles.

of atoms as well. Ionized and excited atoms form the excited molecular states Ar_2^+ and Ar_2^* respectively. Ar_2^+ eventually recombines with an electron, producing Ar_2^* in the molecular excited state which decays radiatively (see fig.1).

The lowest allowed radiative decays from the molecular excited states are the transitions from the singlet and the triplet states ($^1\Sigma_u^+$ and $^3\Sigma_u^+$) to the dissociative ground state ($^1\Sigma_g^+$) which consist of two independent atoms; the emission spectrum for those transitions has a narrow peak at (128 ± 10)nm in the Vacuum Ultra Violet (VUV) band [2]. Reabsorption cannot occur because the energy of the single atomic excited state is too high. Only impurities such as water, N_2 and CO_2 can eventually reabsorb VUV light and capture the electrons.

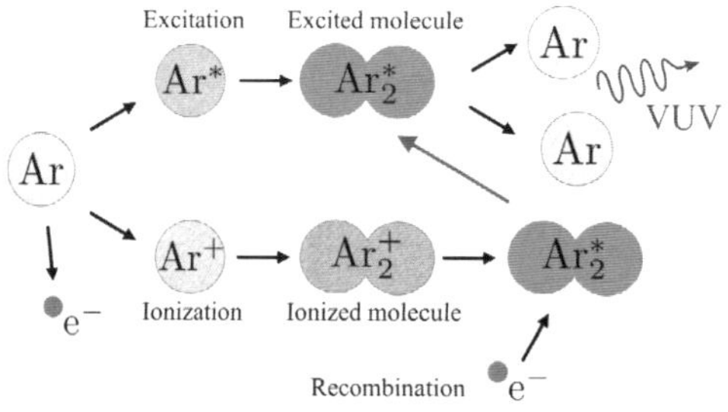

Figure 1. Scintillation and ionization in argon.

Those two transition have two different decay times, respectively τ_1 and τ_2, which do not depend on the type or on the energy of the incident particle; $^1\Sigma_u^+ \rightarrow ^1\Sigma_g^+$ is strongly allowed and has a short decay time while $^3\Sigma_u^+ \rightarrow ^1\Sigma_g^+$ is allowed only because of spin-orbit coupling and has a much longer life. In liquid argon (LAr) $\tau_1 = 5$ ns and $\tau_2 = 1.6$ μs.

The ratio between the populations of the two states $r_{ex} = \delta(^1\Sigma_u^+)/\delta(^3\Sigma_u^+)$ depends strongly on the ionization density of the track (see fig.2); for light projectiles such as electrons and γs $r_{ex} \approx 1/2$ while for αs and fission fragments is $r_{ex} \approx 4 \div 5$. The electron lifetime and the decay time of triplet state strongly depend on the on the purity of the liquid.

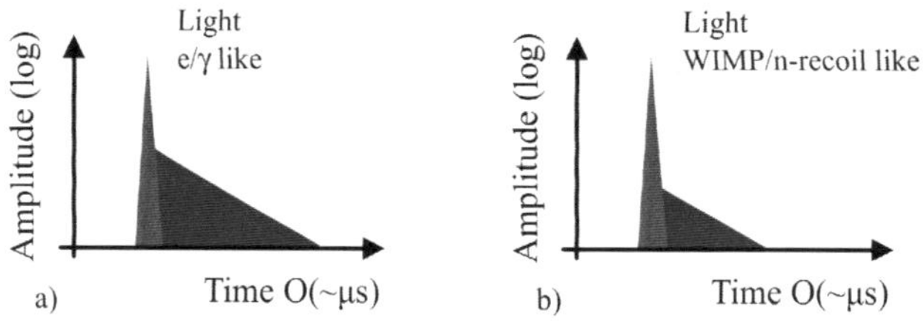

Figure 2. Argon scintillation event topologies.

2. The ArDM experiment

2.1. Detector description

The conceptual layout of the ArDM experiment [3] is illustrated in fig.3. An ionizing particle traveling through an active mass of 850 kg generate scintillation and ion pairs; a 500 kV Cockcroft Walton generator chain provides the electric field necessary to drift the electrons towards the gas phase on the top of the detector. The electrons are then extracted from the liquid to the gas phase (E_{extr} ~3 kV/cm) and accelerated towards the double stage Large Electron Multiplier (LEM)

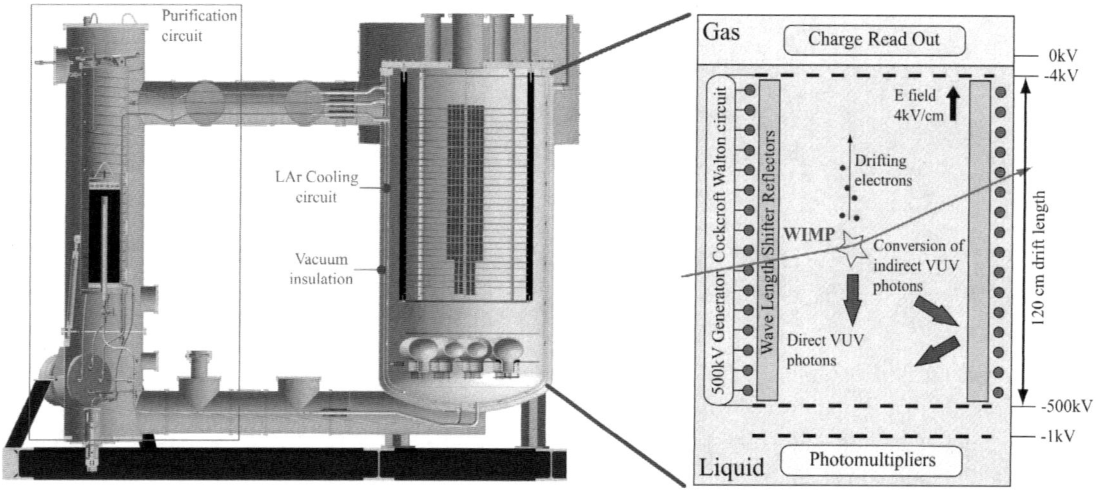

Figure 3. Conceptual layout of the ArDM experiment.

detectors which provides multiplication (gain $\approx 10^4$) and position reconstruction. Detection of light is performed by an array of 14 cryogenic photomultipliers (PMTs) located below the cathode. Most of the VUV scintillation light hits the side of the detector, therefore we decided to install wavelength shifter (WLS) coated diffusive reflectors which shift the light from 128 nm to 420 nm (λ_{peak} for bialkali photocathode) and diffuse the blue light in the detector. A WLS coating of the PMT surface is also foreseen to detect the direct light which is ~10% of the total light. We plan to achieve an average light collection efficiency of 2%. With a combined light/charge analysis we should reach an energy threshold of approximately 30 keV.

In order to reach this goal it is necessary to have a very high LAr purity and an excellent background rejection level. A good purity is obtained by recirculating the LAr through an active copper filter (purification circuit) with a specially developed cryogenic pump. Argon quality is monitored using two independent purity monitoring systems, which measure the electron drift length and the decay time of slow component (τ_2). The main vessel and the purification circuit are surrounded by an insulation vacuum; a LAr circuit embedded in the walls of the inner vessel allows to pre-cool the detector.

2.2. Background sources

Common background sources for this type of detectors are neutrons, electrons and γs. U and Th chains contaminating the detector components produce neutron and γ background, P isotopes (present in steel) have β activities. Muons induce neutrons and, if interacting with argon, produce the unstable long living ^{39}Ar isotope. LAr obtained from the liquefaction of air contains this isotope in a non negligible fraction; the WARP collaboration has reported an activity of 1 Bq/kg [4] which then makes ^{39}Ar one of the most problematic source of background for ArDM.

The rejection of this background is a complicated issue and involves more detailed studies which were the subject of simulations.

Since we measure the charge with its position, time distribution and the scintillation light with its time structure down to *ns* scale we are able to distinguish between electrons, γs and recoiling nuclei. An example of rejection studies can be seen in fig.4: it is possible to identify electron and γ events by looking at the "Charge/Light ratio" as a function of the total reconstructed energy (fig.4a) or if we want to use only the light information (for instance to reduce the trigger

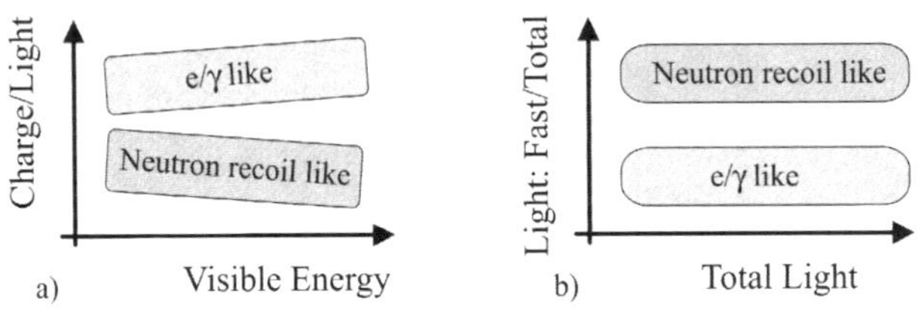

Figure 4. Event topologies.

rate) we can measure the fraction of light coming from the fast component as a function of the total light (fig.4b). This kind of analysis is fundamental to reject β radioactivity from ^{39}Ar.

WIMPs and neutrons lead to similar event topologies since they produce a recoiling argon nucleus; however the energy spectra and the scattering multiplicities are different. We can partly reject neutrons because more than 50% of them scatter more than once; moreover only 10% of them produce a single recoil event in the energy range between 30 keV and 100 keV.

We estimate with a MonteCarlo the total number of neutrons per year coming from the detector components; from simulation we expect ~1000 neutrons/year which will generate a WIMP-like recoil rate of ~50 fake WIMPs/year.

2.3. Charge readout system

The LEM is a scaled-up version of the Gaseous Electron Multiplier (GEM) ([5], [6] and [7]). We are developing a double stage LEM system composed of two LEM planes and a segmented anode plane; each LEM plane consists of a dual layer 1.5 mm thick printed circuit board (PCB) with 0.5 mm diameter holes with a pitch of 0.8 mm, made by precise machining (standard PCB technology). Holes in the last plane are grouped by strips similarly to the anode plane in order to provide a 2D segmented readout of ions and electrons. The third dimension along the cylinder is obtained by time projection imaging. A granularity of few millimeter will grant a sufficient vertex resolution for rejection of multiple scattered neutrons.

We have already achieved operation of a dual stage LEM system in dual phase argon, extracting electron from the liquid to gaseous phase with a small 10x10 cm^2 prototype in a dedicated setup (fig.5).

2.4. Light readout system

The light read out system must be able to detect and identify signals in a range from the single to few thousand photons. It is composed mainly of two parts, the reflectors and the detectors, which have been subject of dedicated R&D projects ([8] and [9]).

The internal surface of the active volume is covered by WLS reflector panels. We selected TetraPhenyl Butadiene[2] (TPB) as the best WLS for VUV to blue light conversion because of its light conversion efficiency and its diffusive reflection properties. We optimized the WLS surface density on different reflector materials in a small 1lt liquid argon setup in gas and liquid. A surface density of 1 mg/cm^2 was evaporated on 250 μm thin Tetratex[3] membrane. Each reflector (130 $\times$ 30 cm^2) was later mounted on 3M ESR[4] foils which provide mechanical support and shielding from the light coming from the nonactive volume. We prepared and installed in the detector 15 WLS reflector panels.

[2] 1,1,4,4-TetraPhenyl-1,3-Butadiene.
[3] High performance ePTFE membrane.
[4] 3M Vikuiti™Enhances Specular Reflection.

Journal of Physics: Conference Series **160** (2009) 012032 doi:10.1088/1742-6596/160/1/012032

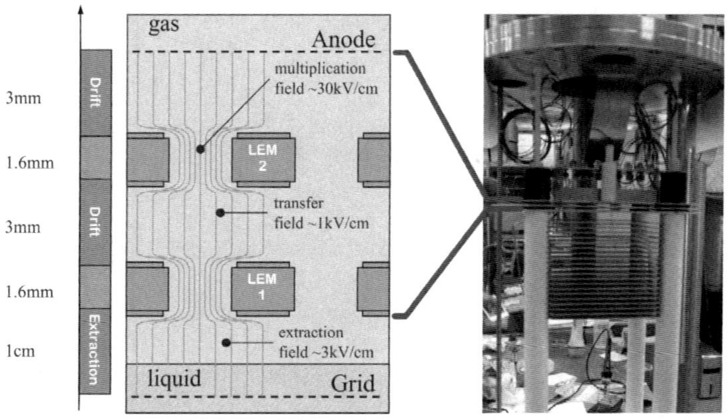

Figure 5. Left: Schematic of the 10×10 cm^2 dual LEM prototype layout. **Right**: Picture of the small R&D dual LEM.

The photon detector is composed of 14 special cryogenic PMTs with platinum underlay below the bialkali photocathode, which allows charge restoration at cryogenic temperatures. A low activity glass is desired to reduce the amount of self background. Each PMT is mounted on a 3 mm thick cryogenic PCB base which provides voltage supply, signal extraction and mechanical support. A detail of the prototype of the PCB base mounted on the photomultiplier can be seen in fig.6 (left).

Polyethylene collars positioned on the steel support frame keep the PMTs in the horizontal position at the right distance from the others. The PMTs are shielded from the -500 kV of the cathode with a grid placed at the same potential of the PMT photocathodes. The mechanical structure can be seen in fig.6 (right).

3. Test of light readout in argon gas

For the test in gas and liquid argon we installed the 15 light shifter reflector panels (full coverage) and 8 of the 14 cryogenic PMT modules. We have seven Hamamatsu R5912 PMTs (five with 14 dynodes and two with 10 dynodes) and one ETL 9357 KFLB; both types were studied at cryogenic temperatures [10] [9]. The PMT modules were tested in LAr (88K) with a blue LED ($\lambda_{pk} = 400$ nm) light before coating and assembly in the experiment. A brief list of the PMT

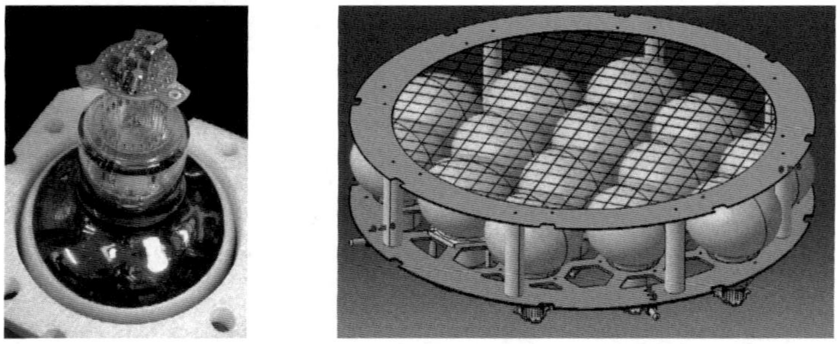

Figure 6. Left: Cryogenic PCB mounted on the photomultiplier. **Right**: Design of the light detector.

properties is listed in table 1. The E-Field protection grid (which will protect the PMT surfaces

Producer	Model	Quantity	Dynodes	Max. Gain (1500V)
Hamamatsu	R5912/02mod	x5	14	$\approx 1 \cdot 10^9$
Hamamatsu	R5912/01mod	x2	10	$\approx 1 \cdot 10^7$
ETL	ETL9357	x1	12	$\approx 1 \cdot 10^7$

Table 1. Types of photomultiplier used in the tests.

from the -500 kV potential of the cathode) and the electric field generator chain were installed, but not used for these tests. Ten PT-1000 temperature sensors, placed at different depth in the detector, allowed precise measurements of the inner temperature of the detector. An ^{241}Am α-source (40kbq) and two blue LEDs were installed on a movable magnetic actuator placed on the center of the main flange, since the LEM was not present. The α source has a very thin metallic window which led to a reduction of the α energy from 5.486 MeV to ~4.5 MeV. The test setup is schematized in fig.7 (left) while the picture of the assembled setup illuminated by UV light from inside is shown in fig.7 (right). The PMTs were tested in gaseous argon at room and cryogenic temperatures. The setup was first closed and evacuated down to 10^{-6} mbar. We then filled the dewar with gaseous argon[5] up to a pressure of 1.1 bar. Several measurement of gain, dark count rate and light yield as a function if the source position have been performed and repeated fig.9 (left); two PMTs were not working in gas because of sparking in the high voltage power supply circuit. Then the setup was evacuated down to 10^{-6}mbar and liquid argon was circulated in the cooling circuit for more than two days. We filled then the detector with gaseous argon up to a pressure of 0.4 bar, turned on the PMTs and let the temperature sensors inside the dewar stabilize over-night to 88K. Meanwhile a set of data was acquired each hour to study possible effects of gain drift and quantum efficiency (QE) change on the PMTs. The temperature variation as a function of time inside the detector is shown in fig.8.

[5] Ar60, from Carbagas, impurities ~ 1 ppb.

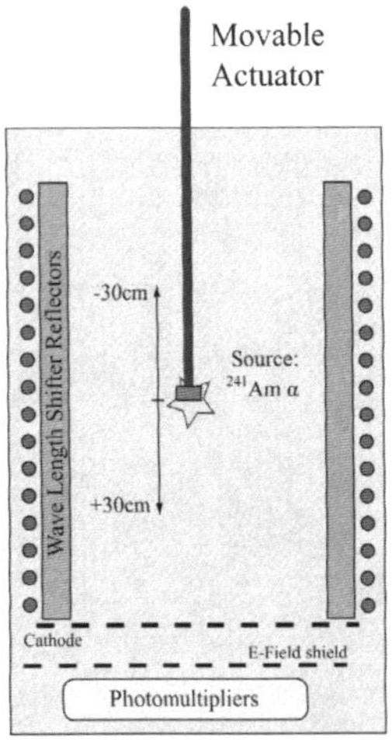

Figure 7. Left: Scheme of the setup used in the first tests. **Right**: Photograph of the assembled setup with internal UV light illumination.

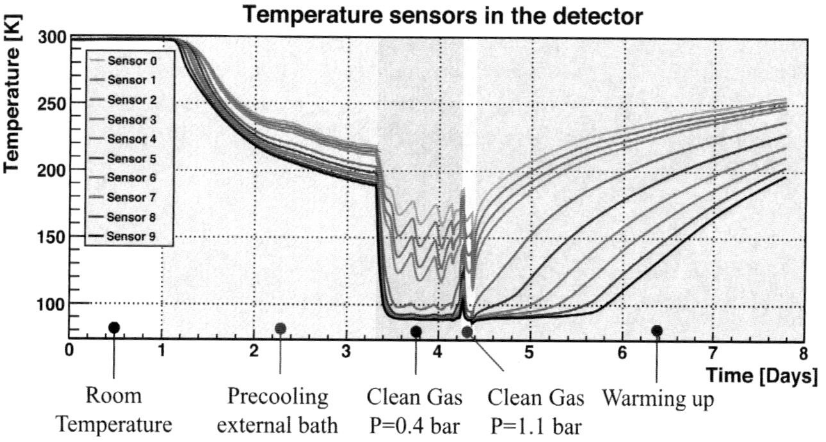

Figure 8. Time variation of the temperature inside the vessel during the light readout test period. Sensors are distributed along the detector height from top to bottom; sensor 9 is placed at the level of the PMT photocathodes.

At the beginning of day 4 temperature was stable at the photocathode level; we repeated the measurement of gain, dark count rate and light yield as a function of source position in cold argon gas at P=1.1 bar fig.9 (right). We measured a light yield increase of ~50% when passing from room temperature to argon gas at 88 K. This effect has not been yet fully understood, but is mainly related to the emission spectrum of the gaseous argon at those thermodynamic conditions. A detailed analysis will be performed with LAr for which we will be able to compare directly the light yield with the one we obtained in the small LAr cell.

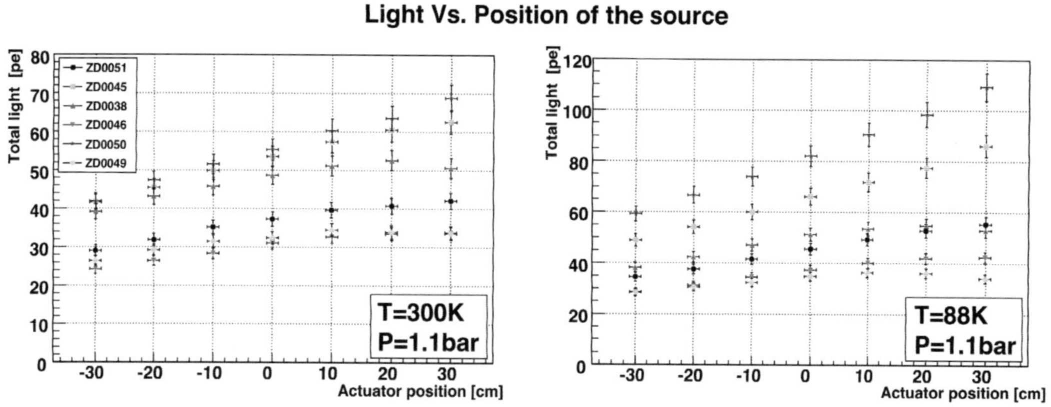

Figure 9. Light collected from the PMTs at different source positions in gaseous argon in the warm (left) and cold (right) condition.

4. Conclusion

In conclusion I report about the progresses done recently by the ArDM collaboration. Many R&D efforts have been invested in the last 3 years to develop a suitable charge and light readout which fulfills our requirements. ArDM is now in the commissioning phase; the light readout has been installed in its prototype version, the choice of the final PMTs will be done after the liquid argon test. The full size prototype of the LEM is in its design phase.

References

[1] D. Clowe et al., *Title: A direct empirical proof of the existence of dark matter*, [arXiv:astro-ph/0608407v1].

[2] T. D. Strickler and E. T. Arakawa, *Optical Emission from Argon Excited by Alpha Particles: Quenching Studies* J. Chem. Phys. **41**, (1964) 1783.

[3] A. Rubbia, *ArDM: A ton-scale liquid argon experiment for direct detection of dark matter in the universe*, J. Phys. Conf. Ser. **39**, (2006) 129.

[4] P. Benetti *et al.* [WARP Collaboration], *Measurement of the specific activity of Ar-39 in natural argon*, Nucl. Instrum. Meth. A **574**, (2007) 83.

[5] F. Sauli, *GEM: A new concept for electron amplification in gas detectors*, Nucl. Instrum. Meth. A **386**, (1997) 531.

[6] V. Peskov et al., *First attempts to combine capillary tubes with photocathodes*, Nucl. Instrum. Meth. A **433**, (1999) 492.

[7] R. Chechik, A. Breskin, C. Shalem and D. Mormann, *Thick GEM-like hole multipliers: Properties and possible applications*, Nucl. Instrum. Meth. A **535**, (2004) 303.

[8] C. Amsler et al., *Luminescence quenching of the triplet excimer state by air traces in gaseous argon*, JINST **3**, (2008) P02001.

[9] A. Bueno et al., *Characterization of large area photomultipliers and its application to dark matter search with noble liquid detectors*, JINST **3**, (2008) P01006.

[10] G.L. Raselli et al., *Characterization of ETL 9357FLA photomultiplier tubes for cryogenic temperature applications*, Nucl. Instrum. Meth. A **556**, (2006) 146.

Electromagnetic shower reconstruction with emulsion films in the OPERA experiment

Frédéric Juget
Institut de Physique, Université de Neuchâtel, rue A. L. Breguet 1, 2000 Neuchâtel, CH
E-mail: frederic.juget@unine.ch

Abstract. OPERA is a neutrino oscillation experiment designed to perform a ν_τ appearance search in the CNGS beam from CERN to Gran Sasso underground laboratory, 730 km from CERN. The identification of the τ lepton produced by a ν_τCC interaction is based on the use the Emulsion Cloud Chamber (ECC) which consists of a modular structure made of a sandwich of passive material plates interleaved with emulsion films. In addition of its very good tracking information capability, used to the detection of short-lived particles, the ECC can be used as a fine sampling electromagnetic calorimeter allowing the study of the $\tau \to e$ decay and the $\nu_\mu \to \nu_e$ oscillation channel. We will report the method developped to reconstruct the electromagnetic showers with ECC and the energy measurement in the 1-10 GeV range.

1. Introduction

OPERA [1] is a long-baseline neutrino oscillation appearance experiment designed to obtain an unambiguous signature of $\nu_\mu \to \nu_\tau$ oscillations in the parameter region indicated by atmospheric neutrino experiments [2, 3, 4]. The detector, located in the underground Gran Sasso Laboratories, plans to detect ν_τ's in the CERN to Gran Sasso (CNGS) ν_μ beam, which is optimised for ν_τ appearance [5]. It may also explore the $\nu_\mu \to \nu_e$ oscillation channel [6], improving the current $\sin^2 2\theta_{13}$ limit [7].

OPERA is an hybrid experiment with electronic detectors, iron magnets (Fig. 1) and Emulsion Cloud Chambers (ECC) (Fig. 2). The ECC combines in one cell the high tracking precision of nuclear emulsions (~ 1 μm) and the large target mass of the lead plates. The basic element, the "brick", has dimensions of $12.7\times10.2\times7.5$ cm^3; it is a sequence of 56 lead (1 mm thick) and emulsion plates. Each emulsion layer is composed of a pair of 44 μm thick emulsion films on either side of a 205 μm plastic base [8]. The bricks are arranged in "walls" separated by electronic detectors. The lead target has a total weight of $\sim$1.35 ton. Electronic detectors are used to identify the brick where the neutrino interaction occured and to track mainly muons escaping the brick.

Charged particles give a track segment in each emulsion film (*base-track*). The $\sim$1 μm granularity of the emulsions [9] ensures redundancy in the measurement of particle trajectories. If a τ is produced, its decay is detected by measuring the angle (kink) between the τ direction and the charged decay daughter. Furthermore, given its high granularity the OPERA ECC is well adapt for the identification of electromagnetic shower.

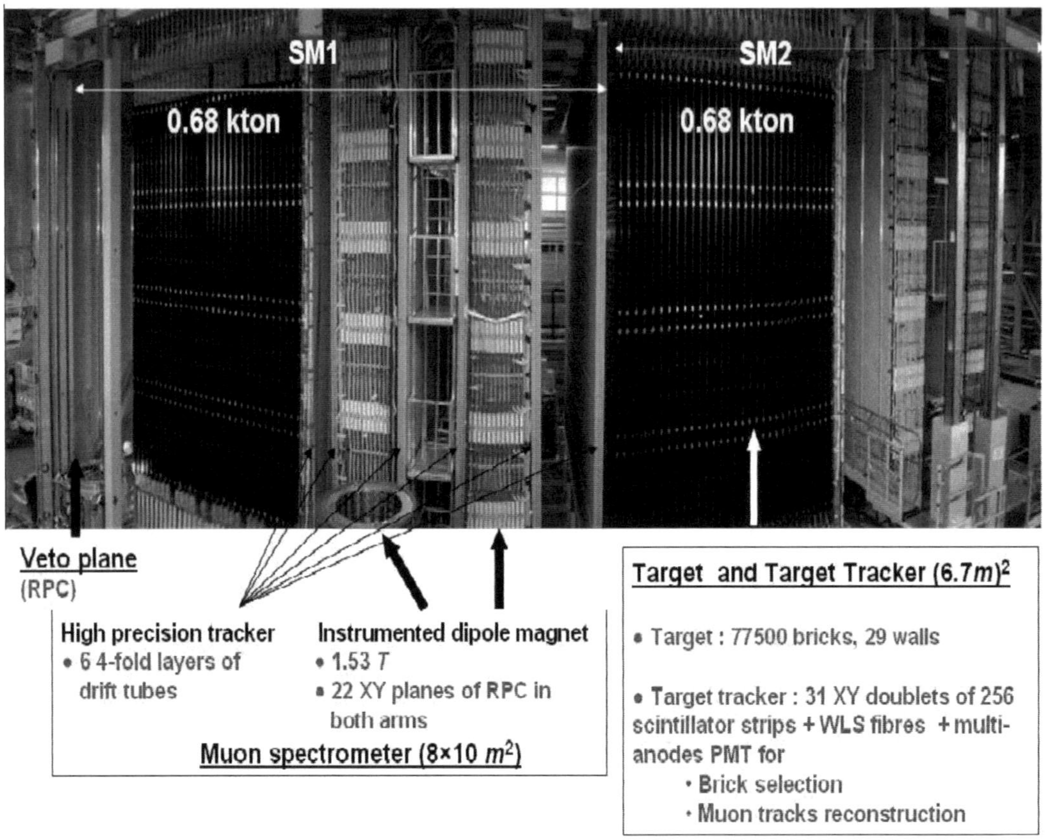

Figure 1. The OPERA detector.

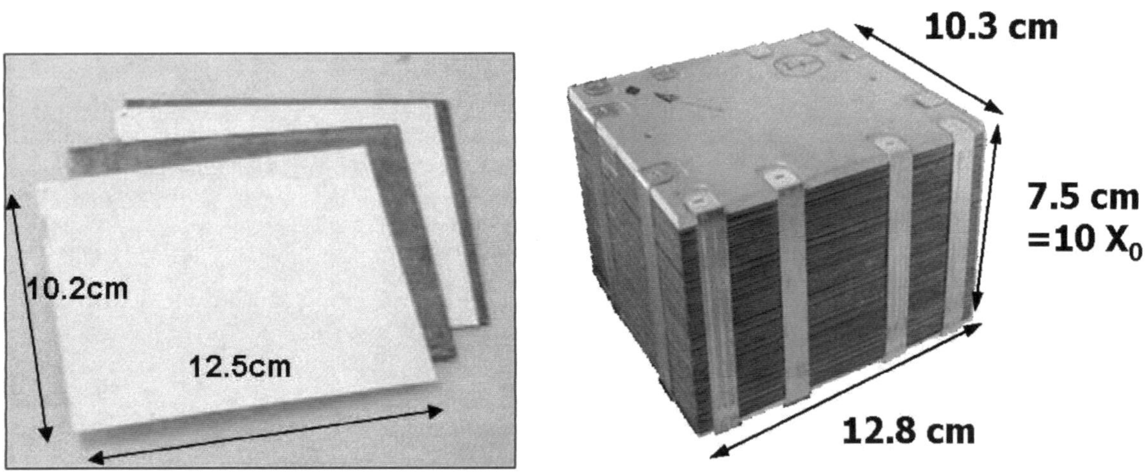

Figure 2. The OPERA Emulsion Cloud Chamber.

Here we will focus on the electromagnetic shower reconstruction, the electron/pion separation and the shower energy estimation achievable in an OPERA ECC. Several algorithms have been developed in order to fullfill the requirements of the OPERA analysis; minimize the misidentification of pion to electron and maximize the electron efficiency for an energy greater than 1 GeV.

2. The shower reconstruction

A good electron-pion separation may be obtained by studying the different behavior of these particles in passing through an OPERA brick. At the energies of interest (1 - 10 GeV) an electron loses energy essentially through bremsstrahlung processes while charged pions lose energy mainly via ionization. An electron quickly develops an electromagnetic shower in 10 X_0 of an ECC brick. The total number of base-tracks and the longitudinal and transverse profiles of the showers can be used for particle identification.

The algorithm for the shower reconstruction has already been described in [10]. Here we will briefly report the reconstruction method and give the new results obtained using of a most realistic background.

The principle of the reconstruction algorithm is to collect base-tracks inside a given cone using connection criteria on the position and the slope. The angular displacement $\delta\theta$ and the position displacement δr are optimize in order to keep background at around 5% of the total number of the collected base-tracks. In the present study we used as background scanned data, extracted from a brick placed inside the OPERA detector during the october 2007 run and exposed to cosmic rays for alignement. This kind of background is the one we will have in the OPERA experiment and is then more realistic than the previous one we used which was extracted from a brick exposed to test beam. As this new background was found to be lower, we were able to open up the connection criteria in the shower reconstruction algorithm. An adequate background rejection is achieved with $\delta\theta < 150\ mrad$ and $\delta r < 150\ \mu m$. In addition the maximum radius allowed for the cone is set to 800 μm.

A reconstructed showers electron at 6 GeV are shown in Fig. 3. It is worth to note that a not-showering event (as a passing-through pion) would be reconstructed by this algorithm, even if not accompanied by any shower.

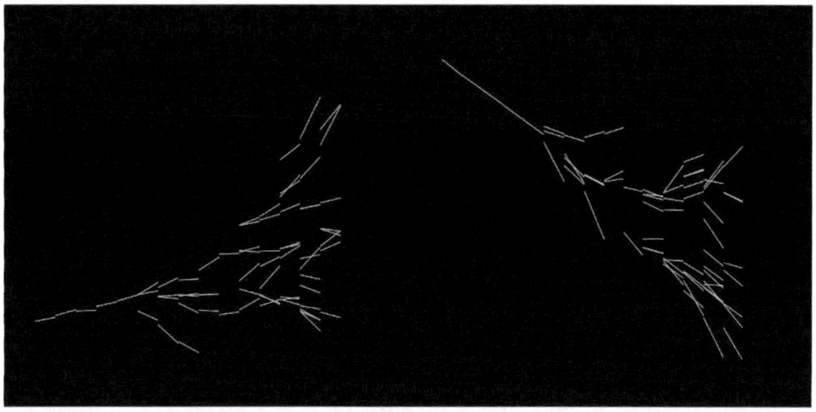

Figure 3. xz projection (left) and yz projection (right) of a reconstructed shower generated by a 6 GeV electron extracted from the brick exposed to an electron beam at DESY. 20 emulsions are used which corresponds to 19 lead layers (~3.3 X_0). Each green dash corresponds to a base-track attached to the reconstructed shower.

2.1. Neural network analysis

The classification in terms of particle identities is demanded to a Neural Network (NN) [11]. Then, each event which has been reconstructed by the shower algorithm will be used to feed a NN. As it is shown in Fig. 4, the longitudinale profile of electrons and pions is very different as well as the number of base-tracks in the reconstructed shower. These informations are used as input values to feed the NN. More details of the NN method can be found in [10].

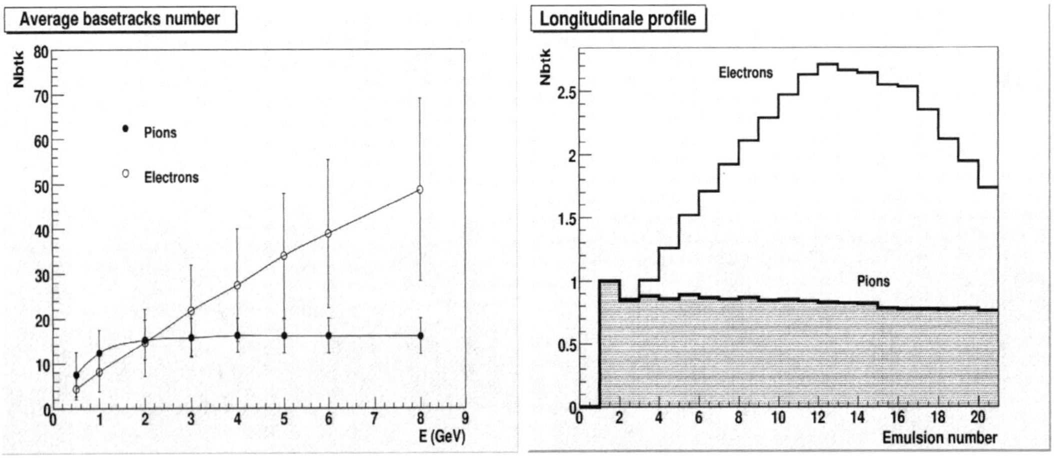

Figure 4. Left: Average number of base-tracks in reconstructed showers versus the energy for simulated pions and electrons. Right: Mean longitudinale profile for simulated electrons and pions at 4 GeV.

2.2. Results of the electron/pion separation

A good method for e/π separation should provide a high probability to correctly identify one type of particle and a small probability to misidentify the other type. To do that, we apply a threshold on the output neuron value in order to separate both particle type (see Fig. 5).
The electron efficiency, $\epsilon_{e\to e}$, and the pion contamination, $\eta_{\pi\to e}$, are defined as follow:

$$\epsilon_{e\to e} \equiv \frac{n_{e\to e}}{N_e} \quad \eta_{\pi\to e} \equiv \frac{n_{\pi\to e}}{N_\pi}$$

Where $n_{e\to e}$ and $n_{\pi\to e}$ correspond to the number of electrons, respectively pions, having an output value above the threshold. N_e and N_π are defined as: $N_e \equiv n_{e\to e} + n_{e\to\pi}$ and $N_\pi \equiv n_{\pi\to\pi} + n_{\pi\to e}$, they correspond to the total number of electrons respectively pions seen by the neural network. Using an equivalent definition we can also compute $\epsilon_{\pi\to\pi}$ and $\eta_{e\to\pi}$.
The threshold value will be chosen in order to have $\eta_{\pi\to e}$ as low as possible (< 1-2%) keeping $\epsilon_{e\to e}$ maximal (> 80-90%).

To train the NN, we used 18108 pions and 17988 electrons at randomized energy spectrum from 0.5 to 6 GeV. The validation sample, with different events from training sample, contains 1000 pions and 1000 electrons at each following energy 0.5, 1, 2, 3, 4, 5, 6, 8 GeV.
The results show a good discrimination for energy $\geq$ 1 GeV, indeed the pion contamination $\eta_{\pi\to e}$ is 1.0% at 1 GeV with an electron efficiency $\epsilon_{e\to e}$ around 80%. At higher energies, $\eta_{\pi\to e}$ diminishes below 1% and $\epsilon_{e\to e}$ is above 90%. For the lower energy, 0.5 GeV, the contamination remains high, 7-10%, with an efficiency around 65%. All the results are summarized in table 1.

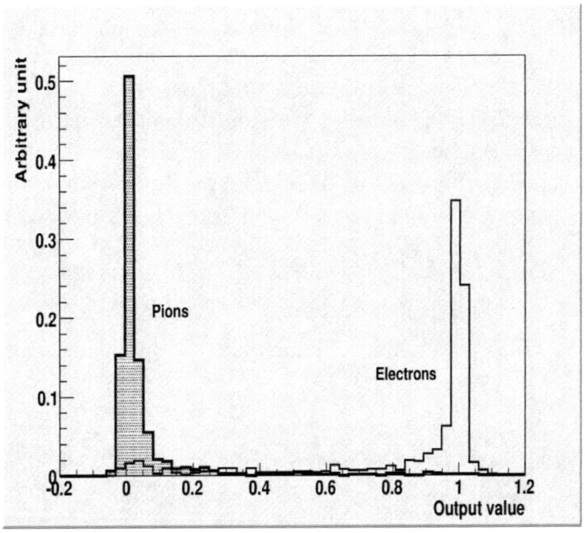

Figure 5. Output value given by the neural network for simulated pions and electrons at 2 GeV. Choosing an output value of 0.6, for instance, allows to well separate both particles. The small pion part above 0.6 will give the pion contamination.

E (GeV)	$\epsilon_{e\to e}$ %	$\eta_{\pi\to e}$ %	$n_{e\to e}/N_e$ %	$n_{\pi\to e}/N_\pi$ %	$\epsilon_{e\to e}$ %	$\eta_{\pi\to e}$ %
	20 sheets		30 sheets		50 sheets	
0.5	62.2 ±2.0	7.4±1.0	66.2±2.0	9.0±1.1	66.4±1.9	9.9±1.2
1	77.4± 1.4	1.0 ±0.3	79.9±1.4	1.0±0.3	79.8±1.4	1.0±0.3
2	91.3±0.9	0.2± 0.2	92.2±0.9	0.2±0.2	92.7±0.9	0.2±0.2
3	95.5±0.7	<0.3	96.6±0.6	<0.3	96.5±0.6	<0.3
4	96.2± 0.6	0.2 ±0.2	97.0±0.6	0.1±0.1	97.0±0.6	0.3±0.1
5	98.8 ±0.4	0.3±0.2	98.9±0.3	0.5±0.2	98.9±0.3	0.5±0.2
6	98.2± 0.4	0.2±0.2	98.2±0.4	0.4±0.2	98.0±0.5	0.4±0.2
8	98.8 ±0.4	0.4 ± 0.2	99.2±0.3	0.6±0.3	98.8±0.4	0.6±0.3

Table 1. Electron efficiency and pion contamination for the simulated data with added background extracted from real data using 20, 30, 50 sheets. The output NN value is fixed at 0.73, 0.7, 0.73 respectively. The upper limits (95% C.L.) for the 3 GeV case have been computed assuming a binomial distribution [12]. We computed an upper limits since none of the pion has been misidentified as an electron.

3. Electromagnetic shower energy estimation

An algorithm for the shower energy measurement has been developped and is currently under optimisation. Here we briefly present some preliminary results.

The algorithm is based on exactly the same NN structure as the one used for the e/π separation (section 2.2) except that as ouput we demand the NN to calculate the energy value (known from simulation during the training phase of the NN). As preliminary results we give the obtained resolution for simulated electrons reconstructed in 41 emulsions plates (Fig. 6):

$$(\frac{\sigma_E}{E})^2 = (\frac{49\%}{\sqrt{E}})^2 + (\frac{24\%}{\sqrt{E}})^2$$

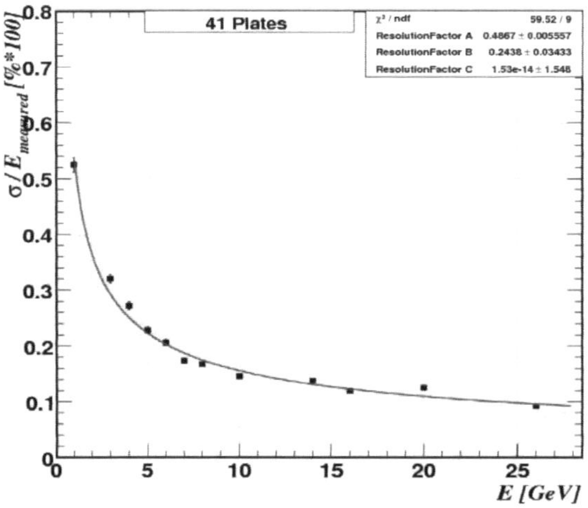

Figure 6. Energy resolution versus the energy for simulated data with background addition reconstructed in 41 emulsion plates.

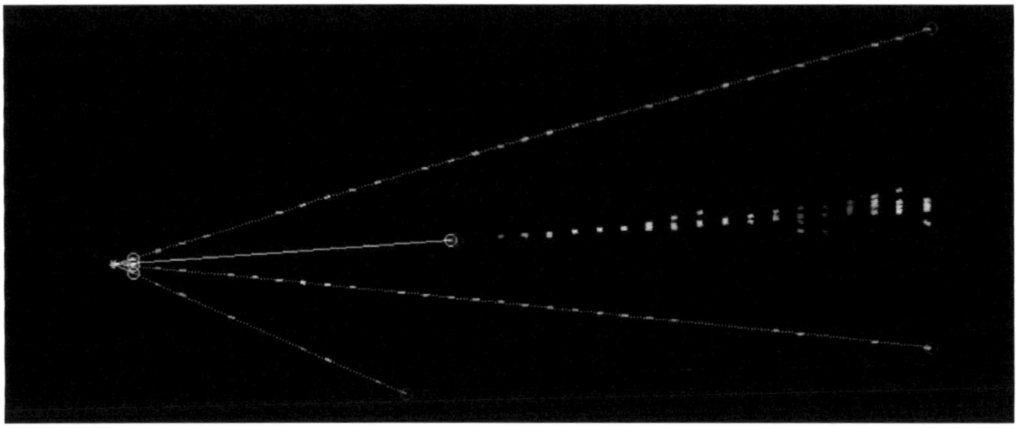

Figure 7. A first reconstructed $\nu_\mu CC$ event recorded during the october 2007 run in the OPERA detector. The 3 long straight tracks are direcltly attached to the primary vertex (the upper one is the μ). The colored shower in the middle is a gamma conversion coming from a π^0 decay and is also well attched to the primary vertex.

4. Conclusion

An algorithm for electron shower reconstruction have been developed and was used to study the e/π separation and the shower energy measurement in the OPERA brick. A method based on a Neural Network, using the electromagnetic shower properties was tested on simulated data with realistic background addition showing that in the OPERA case background, the pion contamination is below 1% for energies $\geq$ 2 GeV with an electron efficiency greater than 90%. At lower energy, around 1 GeV the pion contamination is $\sim$ 1 % for an efficiency close to 80%. Some preliminary results on the energy resolution gives $\frac{\sigma_E}{E} \sim \frac{50\%}{\sqrt{E}}$.

Although the presented results here are within the requirements for the OPERA kinematical analysis, it would be interesting to increase the efficiency and the energy resolution at low energy. Some improvement are still under investigation to improve the reconstruction at low energy (<1 GeV). This will allow for instance to diminish the background for both $\tau \to e$ and $\nu_\mu \to \nu_e$ analysis coming from π^0 decay in the hadronic shower.

References

[1] M. Guler et al., OPERA Proposal CERN/SPSC 2000-028
OPERA COLLABORATION, R. Acquafredda et al., *First events from the CNGS neutrino beam detected in the OPERA experiment, New J. Phys.* 2006 **8** 303
[2] MACRO Coll., M. Ambrosio et al., *Phys. Lett.* B 1998 **434** 451
[3] Soudan 2 Coll., W.W.M. Allison et al., *Phys. Lett.* B 1999 **449** 137
M. Sanchez et al., *Phys. Rev.* D 2003 **68** 113004
[4] SuperKamiokande Coll., Y. Fukuda et al., *Phys. Rev. Lett.* 1998 **81** 1562
Y. Ashie et al., *Phys. Rev. Lett.* 2004 **93** 101801
[5] G. Acquistapace et al., CERN 98-02 and INFN/AE-98/105 (1998); R. Bailey et al., Addendum to Report CERN 98-02, CERN-SL 99034 and INFN/AE-99/05 (1999); A.E. Ball et al., SL-Note 2000-063 (2000); *http://proj-cngs.web.cern.ch/proj-cngs/*
[6] M. Komatsu, P. Migliozzi and F. Terranova, *J. Phys.* G 2003 **29** 443
[7] CHOOZ Collaboration, M. Apollonio et al., *Phys. Lett.* B 1999 **466** 415
[8] T. Nakamura et al., *Nucl. Instr. Meth.* A 2006 **556** 80
[9] W.H. Barkas, Nuclear research emulsions, Academic Press, London (1963); C.F. Powell et al., The study of elementary particles by the photographic method, Pergamon Press, New York (1959).
[10] L. Arrabito et al., Electron/pion separation with an Emulsion Cloud Chamber by using a Neural network, *JINST 2* 2007 P02001
[11] *http://root.cern.ch/root/html/TMultiLayerPerceptron.html*
[12] G. D'Agostini, Probability and Measurement Uncertainty in Physics - a Bayesian Primer, arXiv:hep-ph/9512295

SciBar Detector for SciBooNE

H. Takei

Department of Physics, Tokyo Institute of Technology, 2-12-1 Ookayama Meguro Tokyo, Japan

E-mail: takei@fnal.gov

Abstract. SciBooNE is an experiment for measurements of neutrino-nucleus interaction cross-sections using the Booster Neutrino Beam at Fermilab. The SciBar detector in SciBooNE is a fully-active, finely segmented tracking and calorimetric detector. The SciBar detector can identify neutrino interactions by detecting charged particles. The design, performance, calibration and operation of SciBar are presented.

1. Introduction

SciBooNE [1] is an experiment for measurements of neutrino-nucleus interaction cross-sections. Precise knowledge of neutrino-nucleus interactions cross-sections is important for neutrino oscillation measurements to determine oscillation parameters. Figure 1 shows the ν_μ energy spectrum at SciBooNE. SciBooNE measures neutrino and anti-neutrino cross-sections at energies around 1 GeV. There are experimental data of low statistics of neutrino-nucleus and anti-neutrino-nucleus interaction cross-sections from previous experiments at energies below 1 GeV. SciBooNE helps the neutrino oscillation experiment T2K [2] since it uses the same neutrino energy region as SciBooNE. SciBooNE also helps MiniBooNE [3], which is on the axis of the same neutrino beam, as a near detector.

Charged current quasi-elastic (CCQE) scattering is the signal reaction in ν_μ disappearance oscillation measurement. The background comes from charged current single pion production (CC 1π) in which the pion is not observed; this happens because of pion absorption inside the nucleus or insufficient pion energy to be detected.

Neutral current pion production (NC $1\pi^0$) is the major ν_μ background of ν_e appearance measurement with a Cherenkov detector. Two γ's from a π^0 decay are detected as two Cherenkov rings. They can be mis-identified as a ν_e event if the opening angle between them is too small or they have highly asymmetric energies. SciBar can identify π^0's with higher momentum by looking at tracks of electro-magnetic showers. The measurement of these interactions enables oscillation experiments to know fractions of signals and backgrounds.

Neutrino-nucleus scattering gives us knowledge also about nucleon spin structure. The cross-section for neutral current elastic (NC elastic) scattering contains electro-magnetic form factors and axial form factor of the nucleon. Each form factor includes strange quark contribution. The strange quark axial form factor is proportional to the strange component of nucleon spin in the $Q^2 \to 0$ limit. Strange quark spin contribution to the nucleon spin can be determined from the NC elatic cross-sections as a function of Q^2. A previous experiment, E734 [5] at BNL in the 1980's, measured NC elastic cross-section with Q^2 down to 0.4 GeV^2. SciBooNE can measure cross-section with Q^2 downt to 0.2 GeV with more statistics.

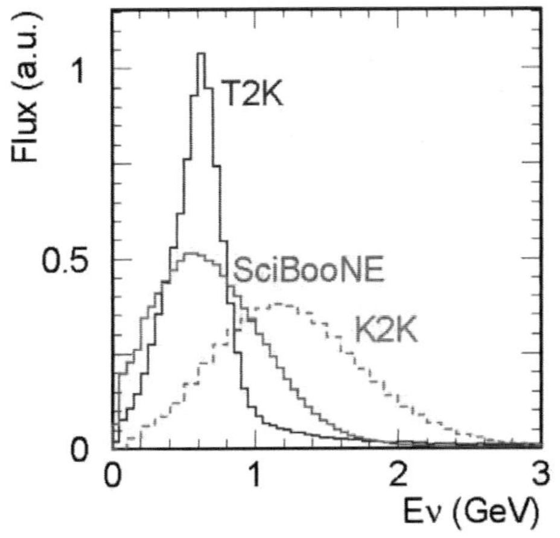

Figure 1. The ν_μ energy spectrum at K2K, T2K and SciBooNE. All curves are normalized by area.

SciBooNE uses the FNAL Booster Neutrino Beam (BNB) with a peak energy of 0.7 GeV. 8 GeV protons from the Booster hit a beryllium target surrounded by the magnetic focusing horn and produce mesons, primarily pions and kaons. The produced mesons are focused, and decay into neutrinos inside the 50 m decay volume. The polarity of the magnetic focusing horn can be changed to select the neutrino or anti-neutrino mode which enables us to study both neutrino and anti-neutrino cross-sections. SciBar, an electro-magnetic calorimeter (EC) [7], and a muon range detector (MRD) [8] are detectors for SciBooNE. They are located 100m downstream of the proton target.

The SciBar detector was used as a near detector in the neutrino oscillation experiment K2K [4] at KEK in Japan from 2003 to 2004. It was then moved from KEK to FNAL and re-assembled. We started commissioning with the neutrino beam in May 2007. Projected statistics are 1×10^{20} protons on target (POT) for both neutrino and anti-neutrino mode in one year. We completed collecting projected POT for neutrino data. We will complete data taking for anti-neutrino in August 2008.

2. SciBar Detector

The SciBar detector [6] is a fully-active neutrino detector with finely segmented structure. Carbon and hydrogen in the plastic scintillators are the target nuclei. Its volume is $3\times3\times1.7$ m^3 and weight is 15 tons. The SciBar detector consists of 14,336 plastic scintillator bars (Figure 2). Dimensions of each plastic scintillator bar are $1.3\times2.5\times300$ cm^3. The SciBar detector consists of 64 layers arranged perpendicular to the beam direction. Each layer has a vertical plane and a horizontal plane. One plane consists of 112 bars of plastic scintillator arranged vertically or horizontally. The scintillator bars are made of polystyrene, infused with PPO (1%) and POPOP (0.03 %) with 0.25 mm thickness TiO_2 coating. The emission wave length is 420 nm (at peak). One wave length shifting (WLS) fiber is inserted in each plastic scintillator bar. The diameter of the WLS fiber is 1.5 mm. The attenuation length of each WLS fiber is measured, and is about 350 cm on average. The scintillation light from the plastic scintillators is sent to a 64-channel multi-anode photo-multiplier tube (MAPMT) through the WLS fibers. In total, 224 of MAPMT (H8804 made by Hamamatsu Photonics K.K.) are used in the SciBar detector. Each has 64 anodes with 2×2 mm^2 area. The typical gain is 6×10^5 with a linear response up to $\sim$200 photoelectrons (p.e.); the gain uniformity over all channels is 20% in RMS. The fibers and the MAPMT are connected using a dedicated jig, called a "cookie" (Figure3) to align the fibers to the anodes. There is cross-talk mainly caused by light entering adjacent channels on the surface of the MAPMT. The average level of the cross-talk from an adjacent channel is measured to be 3.5%.

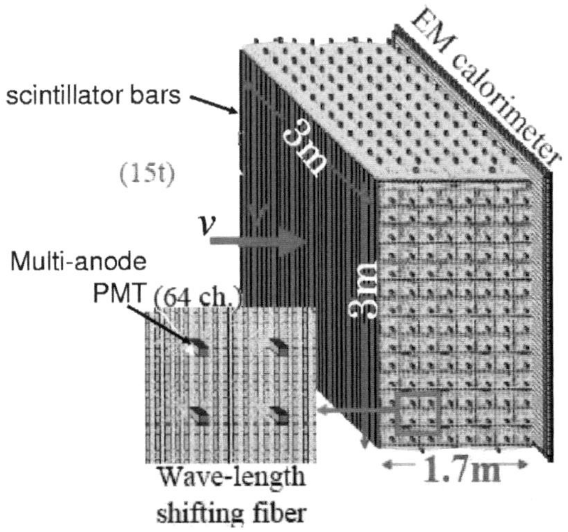

Figure 2. The SciBar detector consists of 14,336 of the combination of plastic scintillator bar and wave-length shifting (WLS) fiber. Volume of the detector is 3×3×1.7 m^3, weight is 15 tons.The Scintillator bars are arranged vertically and horizontally. Each WLS fiber is attached to one anode of a multi-anode PMT.

The gains of all the SciBar channels are monitored using a gain monitoring system which consists of LEDs, clear fibers, and light injection modules. The light from four LEDs is sent via clear fibers to each light injection module located between the plastic scintillators and every MAPMT. The light injection modules are designed to uniformly distribute light from the clear fibers to each WLS fiber. PIN-photo diodes are located near each LED to monitor the LED luminosity. The gain and pedestal data for each MAPMT channel are taken between every beam spill to monitor the detector stability.

Light yield and timing resolution of the SciBar detector are evaluated using cosmic-ray muons. Light yield is ~20 p.e./1.3cm for minimum ionizing particles at the detector edge close to the MAPMT. Timing resolution is ~1.6 ns. The number of dead channels of the SciBar detector is one out of 14,336.

Each MAPMT is mounted on a front-end board as shown in Figure 3. Each front-end board has two sets of Application Specific Integrated Circuit (ASIC), called "VA/TA". The VA is an ASIC with a preamplifier, a shaper and a multiplexer which serialize charge from 32 channels of the MAPMT to a single output. The TA is an ASIC which provides inclusive timing information for the 32 channels. The front-end boards process signals from the MAPMTs, and send them to VME-9U backend electronics boards. They process the charge and timing information from the MAPMT with 12-bit flash ADCs and 64-ch multi-hit TDCs.

Charged particles whose track length exceeds 8 cm inside the detector are reconstructed with 99% efficiency.

3. Neutrino Events in the SciBar Detector

Figures 4 and 5 show CCQE event candidates. The area of dots is propotional to the ADC count (i.e. energy deposit). Boxes represent TDC hit information by their colors. The track penetrating the MRD in Figure 4 is identified as muon (Figure 4). One of the tracks in Figure 5 is identified as muon which decayed inside the SciBar using the multi-hit TDC information. Existence of muon is the sign of charged current events. Difference in dE/dx between muons and protons can be seen. dE/dx distributions of muons and protons are shown in Figure 6. Protons are separated from muons with ~90% efficiency.

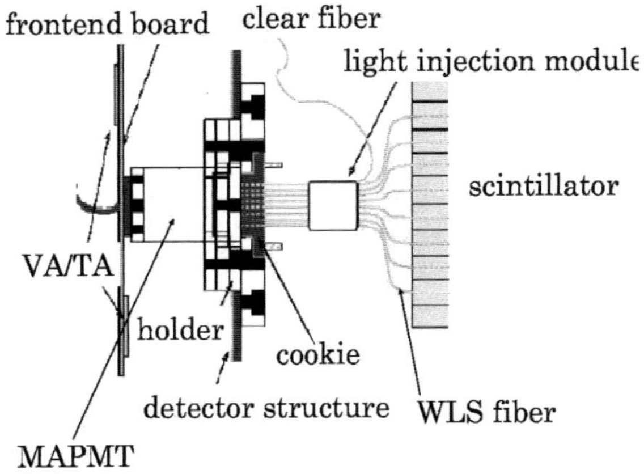

Figure 3. The WLS fibers are attached to the MAPMT by the cookie. The light injection module is located between the scintillators and the MAPMT. Each MAPMT is mounted on the front-end board.

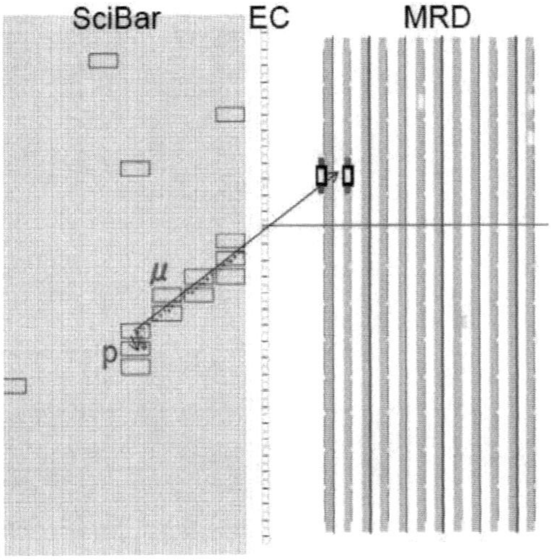

Figure 4. Event displays of CCQE candidate. the area of dots is propotional to ADC counts. Boxes represent TDC hit information. Tracks penetrating the MRD are identified as a muon.

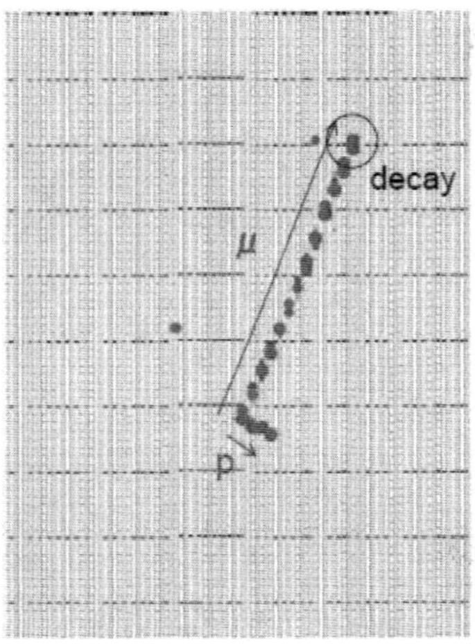

Figure 5. Event displays of charged current quasi-elastic candidate. The decay of the muon is tagged by multiple TDC hits.

CC 1π production can be categorized into three different groups: coherent interaction $\nu A \rightarrow \mu A\pi$, neutron scattering $\nu n \rightarrow \mu n\pi$ and proton scattering $\nu p \rightarrow \mu p\pi$. As shown in Figure 7, scattering on neutrons or coherent interactions appears as two charged tracks with common vertex in the SciBar, while scattering on proton appears as three charged tracks as in Figure 8. If one of the tracks is not long enough to be tracked in the proton interaction event, i.e., the proton track in Figure 8, the event can be separated from the neutron or coherent scattering events by looking at large energy deposit near the vertex.

There is no muon in the final state of neutral current events. Figure 9 shows a candidate of neutral current elastic scattering. The signature of this event is a single proton track. The proton is identified using dE/dx information. Figure 10 shows a candidate of neutral current

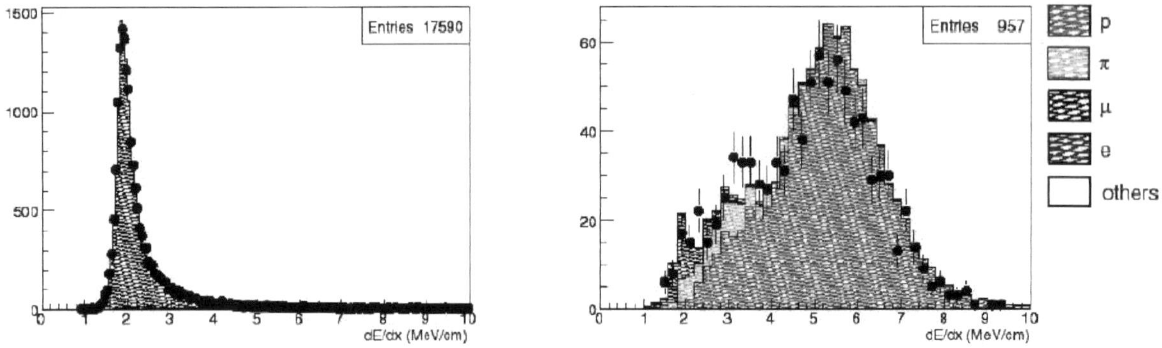

Figure 6. dE/dx distributions of muons (left plot) and protons (right plot). Protons are separated from muons with ~90% efficiency.

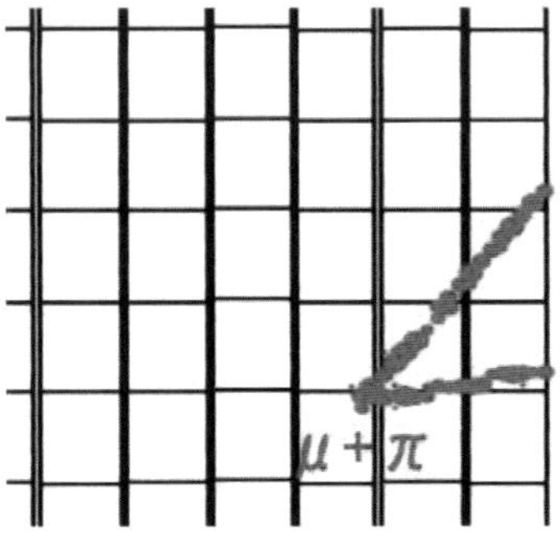

Figure 7. CC 1π coherent or $\nu n \rightarrow \mu n\pi$ candidate. Muon and pion are visible.

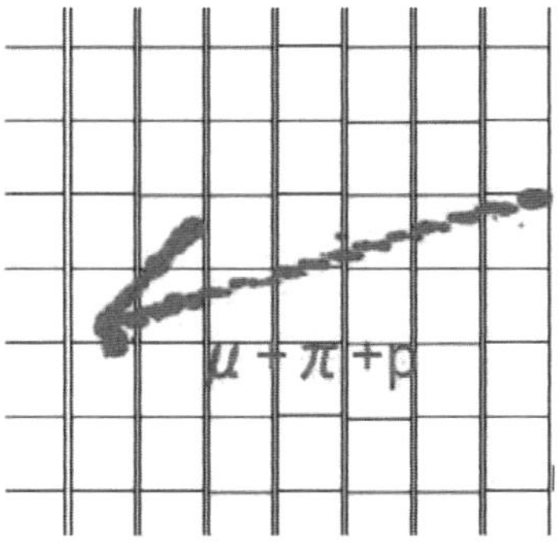

Figure 8. CC 1π $\nu p \rightarrow \mu p\pi$ candidate. Muon, pion and proton are visible.

π^0 production event. The 2 γ's from a π^0 decay produce electro-magnetic showers, which can be detected in SciBar. These features of SciBar enable us to make precise measurements of neutrino-nucleus interactions in the region of neutrino energy below 1 GeV. Analyses of these interactions are on going.

4. Event Rate

The number of CC event candidates is evaluated from tracks starting inside the SciBar every week during the data taking period. With the obtained number, one can monitor the SciBar status. Figure 11 shows number of CC event candidate normalized by POT. The flat shape for each mode and same values for two anti-neutrino mode periods shows that the SciBar took data stably since we started data taking. Neutrino event rate is ~4.5 times of anti-neutrino event rate because of difference between cross-sections of neutrino-nucleus and anti-neutrino-nucleus scattering, and production cross-sections of π^+ and π^- that are parent particles of neutrino and anti-neutrino. We started data taking in Jun. 2007, switched to neutrino mode in Oct. 2007. We completed taking data of projected POT with neutrino mode in April 2008, then switched back to anti-neutrino mode. We will complete taking data of anti-neutrino mode in August 2008.

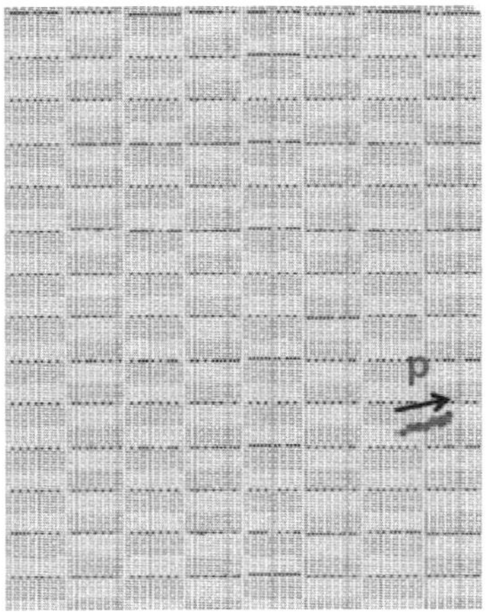

Figure 9. Candidate of neutral current elastic scattering. Signature of this event is single proton track. The proton is identified using dE/dx information.

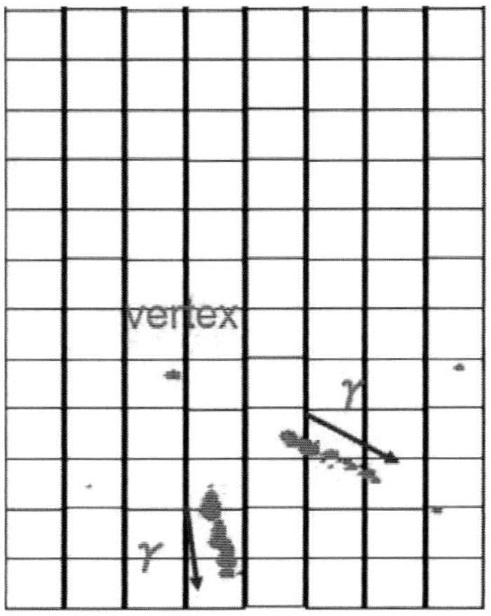

Figure 10. Candidate of neutral current π^0 production event. The 2 γ's from a π^0 decay produce electro-magnetic showers, which can be detected in the SciBar.

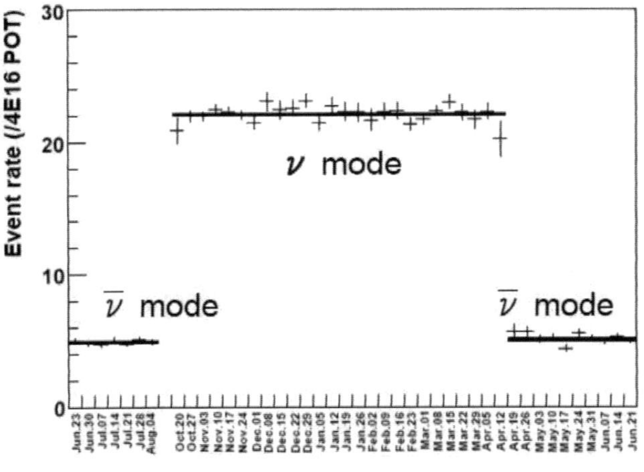

Figure 11. Number of CC event candidates normalized by POT. The flat shape for each mode and same values for two anti-neutrino mode periods shows that the SciBar took data stably since we started data taking.

Acknowledgments
The SciBooNE collaboration gratefully acknowledges support from various grants and contracts from the Department of Energy (U.S.), the National Science Foundation (U.S), the MEXT (Japan), the INFN (Italy) and the Spanish Ministry of Education and Science.

References
[1] A.A.Aguilar-Arevalo, *et al.* 2006 hep-ex/0601022.
[2] Y. Itow, *et al.* 2001 hep-ex/0106019.
[3] A.A.Aguilar-Arevalo, *et al.* 2007 Phys. Rev. Lett.98, 231801.
[4] The K2K Collaboration 2006 Phys. Rev. D 74, 072003.
[5] L.A.Ahrens *et al.* 1987 Phys. Rev. D35, 785.
[6] Nitta K *et al.* 2004 Nucl. Instrum. Meth. A535, p147.
[7] C. Giganti 2007 AIP Conf. Proc. 967, p301.
[8] J. Walding 2007 API Conf. Proc. 967, p289.

XIII International Conference on Calorimetry in High Energy Physics (CALOR 2008) IOP Publishing
Journal of Physics: Conference Series **160** (2009) 012035 doi:10.1088/1742-6596/160/1/012035

EC Detector at SciBooNE

Camillo Mariani
Columbia Physics Department, 538 West 120th Street, Mail Code 9324, New York, NY 10027
E-mail: camillo.mariani@nevis.columbia.edu

Abstract. SciBooNE is an experiment to measure neutrino and anti-neutrino cross-sections on the Booster Neutrino Beam at Fermilab. The EC is an extruded lead sheets and scintillating fibers "spaghetti calorimeter" to provide longitudinal containment and energy measurement for electrons and photons.

1. SciBooNE

The name SciBooNE[1] comes from the merging of SciBar and BooNE. SciBar[2] is a fine grained detector, while BooNE stands for Booster Neutrino Experiment. MiniBooNE[3] (Scintillating Bar Detector) is a neutrino oscillation experiment taking data on the Booster neutrino beam at Fermilab, designed to access the neutrino oscillation parameters claimed by LSND[4] (the collaboration has recently published data which exclude the LSND signal[5] for muon neutrinos and is now running in anti-neutrino mode). The idea of SciBooNE is to install the SciBar+EC (Electromagnetic Calorimeter) detector very close to the source on the Booster neutrino beam. SciBooNE can act as near detector of the MiniBooNE experiment and its very good performance, coupled with the high intensity of the beam, allow important measurements of the beam composition and of neutrino and antineutrino cross-sections. The neutrino beam obtained from the 8 GeV protons extracted from the Fermilab Booster is quite similar to the T2K[6] neutrino beam and the SciBooNE measurements of beam composition and neutrino cross sections will also be very useful for the T2K oscillation experiment. The SciBar and EC detectors were used in the K2K near detector at KEK. After disassembling they were shipped to Fermilab at the end of July 2006. SciBar and EC were assembled and partially commissioned in a self-contained structure in parallel with the excavation of the experimental enclosure, where it was moved in May 2007. The detector is complemented by a new muon range detector (MRD) made of iron and scintillator planes, that was built at Fermilab. The installation of SciBooNE on the neutrino beam line at Fermilab was completed in the middle of May 2007. After a period of commissioning of the detectors at the end of May 2007, data taking in anti-neutrino mode started in June 2007. The SciBooNE measurements will provide significantly improved inclusive and exclusive neutrino and anti-neutrino cross-sections in the energy region below 1 GeV in the near future.

2. Electromagnetic Calorimeter (EC)

The EC detector is an electromagnetic calorimeter installed just downstream of SciBar with respect to incoming neutrinos (Fig.1). SciBar acts as a fully active target for neutrino interactions.

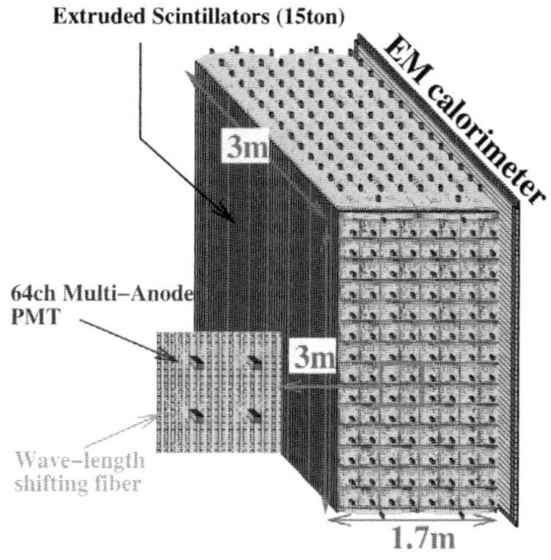

Figure 1. Diagram of SciBar and of the Electron Catcher (EC).

The main purpose of the EC is the longitudinal containment of the electromagnetic showers since the whole SciBar corresponds to only 4 radiation lengths. The EC provides 11 radiation lengths and has 85% energy containment at 3 GeV. It consists of one plane of 32 horizontal modules and one plane of 32 vertical modules. The two planes have both a cross sectional area of 2.7 m $\times$ 2.7 m. Each module is a sandwich of lead and scintillating fibers, built by piling up extruded sheets of grooved lead with scintillating fibers positioned in the grooves. A module consists of a stack of 21 lead sheets, 2,650 mm long, and 740 fibers of 1 mm diameter and 3,050 mm long (see Figure 2). The groove diameter is 1.1 mm and the sheet thickness is 1.9 mm. The sheet material is 99% lead with 1% antimony content to improve its mechanical properties. The stack is kept together by a welded steel case. An overall thickness non uniformity of less than 2% was achieved through the extrusion process. The fibers are manufactured by Kuraray (type SCSF81) and they have a 1 mm diameter polystyrene core, surrounded by a 30 μm thick acrylic cladding, with an emission maximum in the blue, around 420 nm. To improve the light collection uniformity an acrylic black paint is applied on the surface of the last 5 cm of fibers on each side. This has the effect of reducing the light coming from the cladding, which has a smaller attenuation length. In addition, to select the spectral component with a larger attenuation length a yellow filter (Kodak Wratten No. 3) is used. The attenuation length was about 500 cm when the modules were built in 1993 and was recently measured in SciBooNE to be 400 cm. The modules were originally produced for the CHORUS neutrino experiment at CERN[7]. At both ends of the readout cell (see Figure 3), fibers are grouped into two bundles with an hexagonal shape (22.2 mm apex to apex) and they are coupled to a plexiglas light guide, also with hexagonal cross section (24 mm apex to apex). The hexagonal shape and the length of the light guide were chosen to reduce non uniformities in the mixing of the light coming out of the individual fibers[8]. Each fiber bundle defines two independent readout cells of 42×42 mm^2 transverse cross section.

The light guides are coupled to 1-1/8 inch diameter photomultipliers, type R1355/SM from Hamamatsu, with a special green extended photocathode, of 25 mm useful diameter. The cathode material is Bialkali, with a quantum efficiency of 27% in the wavelength range 350-450 nm. The cathode is sensitive to wavelengths from 300 nm to 650 nm. A typical current amplification is 2.1×10^6 at the supply voltage of 1,600 V. The anode dark current is 10 nA. The PMT gain of each channel was measured before the installation. The non-linearity of the output signal vs input charge is 2% at 60 mA (corresponding to 600 photo electrons) at a gain 2×10^6. The pedestal width is approximately 0.7 photo electrons and the energy resolution was measured in a test beam as $14\%/\sqrt{E(\text{GeV})}$ for electrons [7]. From each PMT a differential signal is produced using the outputs of the cathode and the last dynode, read via multipolar

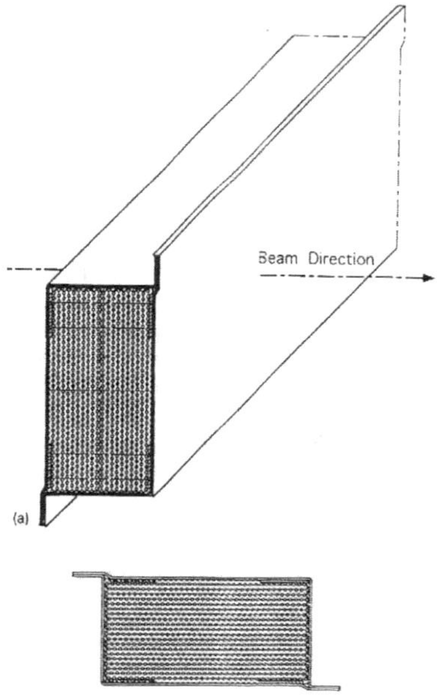

Figure 2. Cross section of one of the 64 electromagnetic calorimeter modules.

differential screened cables 100 m long.

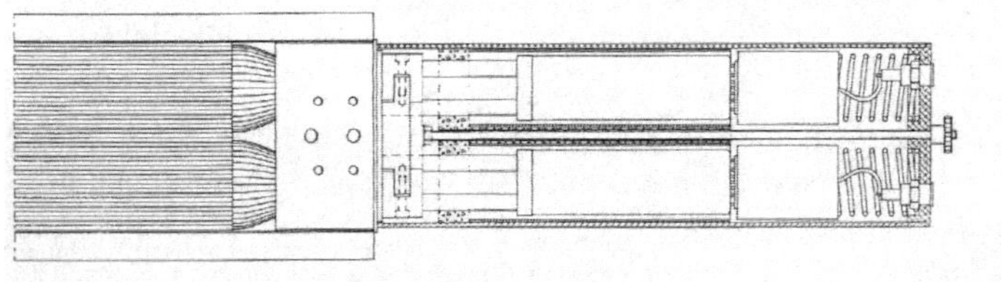

Figure 3. Schematic view of one readout side of a module.

2.1. EC HV and DAQ

At the front-end the 256 analog channels of the detector are digitized using 8 QDC VME (CAEN V792). These boards support 32 channels charge analog to digital conversion (12 bits) in 5.7 μs. Impedance matching cards (CAEN A992 custom modified) are used to convert the 110 Ω differential signals into 50 Ω single ended signals and to decouple the PMT's and the QDC grounds. The acquisition time for the EC DAQ is less than 1 ms. The EC DAQ can handle three types of trigger signal from the experiment: cosmic, pedestal and beam data. The HV system consists of two mainframes (CAEN SY527) each equipped with 9 HV distributor boards (CAEN A734N). The two CAEN mainframes are connected through a daisy chain to a PC that hosts the slow monitor and control system. The software uses the SMACS package[9] (see Figure 4), developed by INFN Roma for the CDF TOF system. The system performs cyclic read out of all the HV channels in less than 1 second to monitor the status of each voltage and current. Individual channel setting can be changed interactively and the status voltage, current and time plots are available for each channel through a tree browser interface. Fluctuations of the monitored voltage or current outside a user defined window around the set nominal value, trigger alarms. The windows were set to ± 5 V and ± 3 μA. All HV events are logged to a local database and the same tree browser allows navigation of all the HV channels where a led represents the status of each channel (green=ok, yellow=warning, red=alarm). The system has a server and a client part. The server is hosted on a pc that sits in the SciBooNE enclosure while the client runs in one of the control room PCs. The client receives the status of each channel from the server through a TCP/IP socket and it shows the operational status of the detector HV in real time. All alarms and warnings are logged and displayed on the client window, notifying the experts via emails and calling the shifter pager when appropriate.

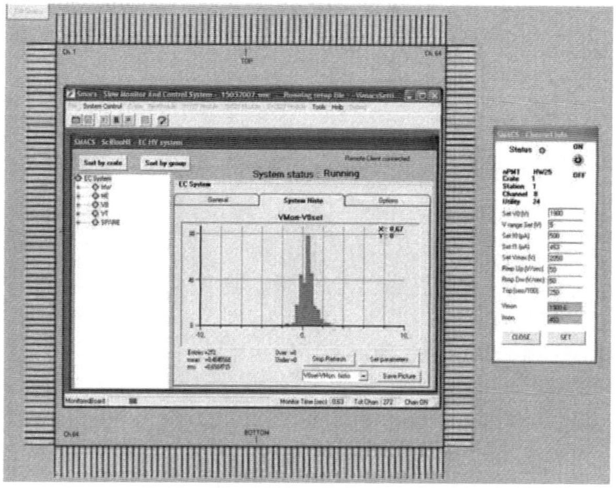

Figure 4. EC HV server display.

2.2. EC calibration and online monitoring

Before the transportation to the experiment enclosure, the detector was fully installed in an assembly area and partially commissioned with cosmic rays. During this commissioning the HV settings were tuned iteratively, achieving a spread of 10% in the pulse height distributions. The individual channel response was inter-calibrated by fitting the pulse heights with a Landau distribution convoluted with a gaussian to take into account resolution effects. Fig. 5 shows that the spread of the Landau most probable values after the HV tuning and applying the inter-calibration constants is better than 0.5%. Comparing the pulse height distribution for cosmics (most probable value and width) of the two PMTs in each readout cell, the number of photo-electron per MIP was measured to be 10.4 in average. The energy deposited in a read out cell is reconstructed from the geometrical average of the pulse heights of the two PMTs. The dependence on the longitudinal position along the cell due to the exponential attenuation of the light in the fibers factorizes out in the geometrical average. The energy deposited in a cell is given in Eq. 1.

$$signal = \sqrt{sig1 \times sig2} \times \sqrt{C_{inter-calib_1} \times C_{inter-calib_2}} \times absolute_{calib} \times C_{attenuation} \quad (1)$$

where sig_1 and sig_2 are the ADCs of the two PMTs, $C_{inter-calib_i}$ are the inter-calibration constants of the two channels, $C_{absolute}[GeV/ADC]$ is the absolute calibration constant and $C_{attenuation}$ is the correction for the light attenuation.

Cosmic rays taken in between the neutrino spills during normal data taking are used to calibrate the detector, producing sets of inter-calibration constants. Random triggered events (pedestal events) are used to compute the pedestal absolute values and widths every 100 events

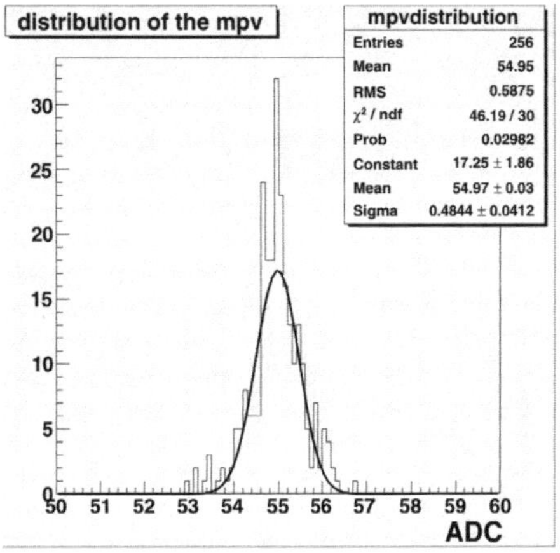

Figure 5. PMT's ADC response for the vertical and horizontal EC's planes after applying the inter-calibration factors.

(about 10 minutes). The response to cosmic ray muons is used to monitor the detector stability. In particular the shift crew monitor online the ADC hitmaps and the average deposited energy in each channel (see Fig. 6 for an example of the online monitor display).

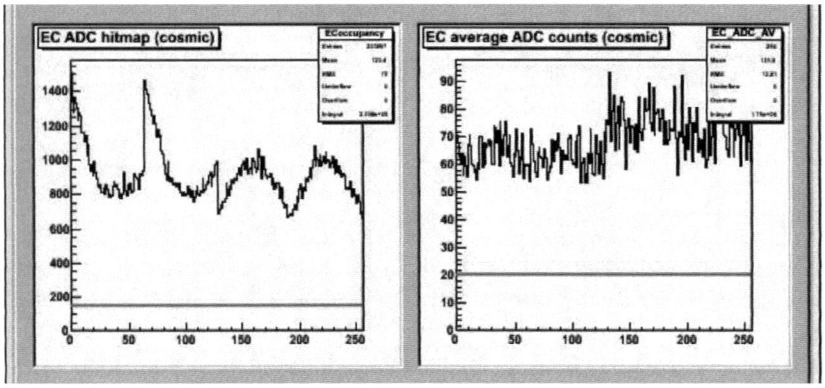

Figure 6. EC online monitor control.

2.3. EC MC simulation

In order to simulate the response of the EC detectors, the GEANT4[10] package is used to simulate the detector geometry and the tracking and interactions of particles. The energy loss of a particle in each individual EC sensitive fiber is simulated. The energy deposition in each cell is converted to a detector response value taking into account effects such as the light attenuation along the fibers, the poisson fluctuations of the number of photoelectrons, the PMT resolution and the pedestal fluctuations. The Bertini model in Geant4[10] is used to simulate the interactions of pions with the detector material. The input parameters of the detector simulation are derived from laboratory measurements and calibration data. The features of the simulation have been systematically compared and tuned with cosmic rays and neutrino data. The agreement between data and MC is shown in Figure 7. MC studies have been made of the energy measured for muons as a function of the true muon energy loss. The corresponding resolution is well fitted by $\sigma_E/E = (6.6 \pm 0.1)\%/\sqrt(E(GeV))$. This resolution for the dE/dx of a minimum ionizing particle corresponds to the pure sampling fluctuation. The resolution for electrons (and photons) will include also the shower fluctuations. In general the energy response to electrons of a calorimeter is different from the response to the same amount of energy released by muons. This effect is often called e/μ ratio and it is due to the fact that the physical mechanism of energy loss for electrons and muons is different. In our detector a smaller response to electron is expected since the photoelectric effect is larger in the (high-Z) lead absorber than in the (low-Z) active scintillator. Correspondingly, we expect $C_e/C_\mu > 1$ for the ratio of the absolute calibration constants for electrons and muons. Our result for the e/μ ratio is $C_e/C_\mu = 1.19 \pm 0.07$ and the energy resolution for electrons is $(12.8 \pm 0.1)\%/\sqrt{E(GeV)}$, compatible with the one measured in a test beam at CERN by the CHORUS experiment[7].

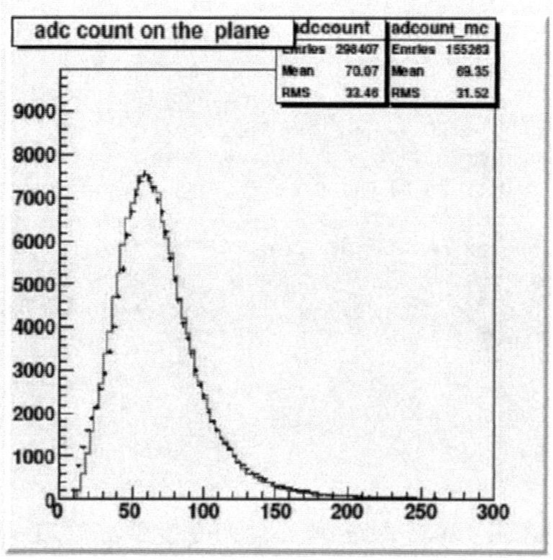

Figure 7. Data-MC agreement for ADC distribution for muons.

Acknowledgments

The EC detector has been commissioned and operated by the INFN Roma group. The author wish to thank in particular Dr. Lucio Ludovici for his support and encouragement. The SciBooNE collaboration gratefully acknowledge support from various funding agencies, grants and contracts and express gratitude to the Fermilab accelerator group for the excellent performance of the Booster Neutrino Beam.

References

[1] SciBooNE Collaboration. Bringing the SciBar detector to the Booster neutrino beam. *Fermilab-Proposal-0954, hep-ex/0601022v1*, 2006.
[2] H. Takei. The scibar detector at sciboone. 2008. Proceeding of this conference (Calor08), Pavia, Italy.
[3] MiniBooNE Collaboration. The miniboone experiment, hep-ex/0701040. 2006.
[4] LSND Collaboration. Neutrino oscillations with LSND. *Nucl. Phys. Proc. Suppl.*, 85:78–83, 2000.
[5] A. A. Aguilar-Arevalo et al. A search for electron neutrino appearance at the delta $m^2 \sim 1$ eV^2 scale. *Phys. Rev. Lett.*, 98:231801, 2007.
[6] T2K collaboration. *T2K letter of Intent.* http://jnusrv01.kek.jp/public/t2k/loi/loi.v2.030528.pdf.
[7] S. Buontempo et al. Construction and test of calorimeter modules for the CHORUS experiment. *Nucl. Instrument. Meth.*, A349:70, 1994.
[8] D. Acosta et al. Results of prototype studies for a spaghetti calorimeter. *Nucl. Instrum. Meth.*, A294:193, 1990.
[9] D. De Pedis et al. *SMACS: Slow Monitor And Control System*, 2006. http://www.top1.it/Smacs.
[10] S. Agostinelli et al. Geant4. *Nucl. Instrum. Meth.*, A506:250–303, 2002.

Journal of Physics: Conference Series 160 (2009) 012036 doi:10.1088/1742-6596/160/1/012036

The ANTARES underwater neutrino telescope

Oleg Kalekin on behalf of the ANTARES collaboration [1]

Erlangen Centre for Astroparticle Physics, Erwin-Rommel-Str. 1, 91058 Erlangen, Germany

E-mail: kalekin@physik.uni-erlangen.de

Abstract. The ANTARES collaboration has built an underwater neutrino telescope at 2500 m depth in the Mediterranean Sea, 40 km off the shore of Toulon. The experiment aims to detect high-energy cosmic neutrinos using a 3D array of 900 photomultipliers distributed along 12 lines. PMTs collect the Cherenkov light induced by charged particles in the water. Upward going muons and showers produced in neutrino interactions with the surrounding matter will allow the search and study of possible sources of extra-terrestrial neutrinos. The experiment delivers data since March 2006, when the first line was connected. Since January 2007 the ANTARES telescope consists of 5 lines. The completion of the detector has been made by the time of the Conference, on May 30, 2008.

1. Introduction

The ANTARES collaboration has been created in 1996. It consists of more than 200 physicists, engineers and technicians from France, Germany, Italy, The Netherlands, Romania, Russia and Spain. After an intensive research and development phase a proposal for a 0.1 km^2 detector [1] has been issued in 1999. Considering the environmental parameters in a deep sea and an existing shore infrastructure the site 40 km off La Seyne (French Riviera) at a location $42°50'N$ $6°10'E$ has been chosen for the deployment of the ANTARES detector.

2. Scientific motivation

The origin of cosmic rays (CR) was a puzzle for 100 years after their discovery by Victor Hess. Two recent observational breakthroughs have improved our understanding of CR's origin and spectral features.

Direct detection of CRs does not generally provide information about their origin because of CR's deflection on interstellar/intergalactic magnetic fields. Such a deflection is expected to be smaller for higher energy CRs. With increasing of energy a flux of CRs drops down rapidly and a huge detector area is needed for a significant detection of high energy CRs. Investigation of extended air showers (EASs) produced by CRs in the Earth's atmosphere is one of the most promising method to provide a reasonable detection area. 3000 km^2 large the Pierre Auger Observatory [2] combines 2 techniques of EAS detection - surface muon and fluorescence detectors. After a few years of observation the experiment has discovered anisotropy of the highest energy ($\sim 10^{20}$ eV) CRs and their correlation with positions of nearby active galactic nuclei (AGN) [3]. Thus an extragalactic origin of the ultra high energy CRs has been proved. Auger data do not however identify AGNs as the sources of cosmic rays unambiguously because

[1] antares.in2p3.fr

a distribution of AGNs correlates in turn with a mass distribution in metagalaxy and other sources can not be excluded.

Supernova remnants are the first candidates as galactic sources of CRs up to knee region ($\sim 10^{15}$ eV) in CR energy spectrum because of energy arguments [4] and existing mechanism of charge particle acceleration in shock waves in shells of SNRs (see e.g. [5]). The second observational breakthrough is the discovery of supernova remnant (SNR) RXJ1713.7-3946 in very high energy (VHE, 0.1-100TeV) gamma rays by the imaging atmospheric Cherenkov telescope H.E.S.S. [6]. The Galactic plane survey carried out by H.E.S.S. in 2005-2006 revealed a few more shell-type SNRs providing strong evidence for an acceleration of particles in SNRs up to energies above 100 TeV. If particles accelerated in shock waves are protons and atomic nuclei, then interactions of protons/nuclei with ambient matter lead to $\pi^0/\pi^\pm$ production and following production of photons via π^0 decay:

$$p/A + N \longrightarrow \pi^0\pi^\pm, \qquad \pi^0 \longrightarrow \gamma\gamma$$

Alternatively relativistic electrons can also produce γ-rays by non-thermal bremsstrahlung or by inverse Compton scattering. H.E.S.S. observations favour strongly the hadronic acceleration model but can not ruled out completely the leptonic one.

Neutrinos in cosmic hadron accelerators are produced in similar way as γ-radiation, via pion decay:

$$\pi^\pm \longrightarrow \nu_\mu\mu, \qquad \mu \longrightarrow \nu_\mu\nu_e e$$

Therefore detection of high energy (HE) neutrinos can be complementary method of studies of the CR's sources answering questions unsolved by other methods. Being definite signature of hadron acceleration, neutrinos travel like γ-rays through magnetic fields without deflection and point directly to their sources. In contrast to gammas neutrinos are not absorbed by intergalactic background radiation. It can increase a horizon of observable sources. Due to very small absorption by matter, neutrinos can also reveal new type of sources where dense matter obscures other kinds of radiation – messengers of acceleration processes.

3. ANTARES design

The small cross section of the neutrino interaction with matter is a disadvantage for the neutrino detection. Huge detector area is needed to observe cosmic sources of neutrinos. The water Cherenkov technique offers that for a reasonable efficiency/cost ratio. Such a detector consists of a 3 dimensional array of photodetectors placed in a transparent medium – water. They detect Cherenkov light radiated by charge leptons produced in charge-current neutrino interaction in the surrounding detector medium. Photomultipliers are used as photodetectors because of ability to detect low level signal (down to 1 photon) with a good time resolution ($\sim 1ns$) which is needed for efficient detection and track reconstruction.

The ANTARES detector is composed of 12 strings anchored on the seabed and pulled up by a buoy (see Fig. 1). A string is 480m long, it consists of an anchor and 25 storeys. The storey comprises 3 optical modules each one made of a 17" glass sphere housing a 10" HAMAMATSU photomultiplier tube, looking downwards at 45 degrees to the vertical. Fig. 2 displays the content of the ANTARES storey. Due to the flexibility of the lines, a redundant positioning system is incorporated on each line. It consists of compasses and tiltmeters measuring the roll, pitch and heading of each storey and 5 acoustic hydrophones spread over the line. For time calibration purposes, each line also houses 4 optical beacons equipped with blue LEDs, and 2 green laser beacons are placed at the bottom of 2 central lines. All the associated electronics, plus the readout computer and clock system, are contained in a titanium cylinder of a Local Control Module. The storeys of one line are chained on mechanical electro optical cable which carries the power, the 20 MHz common clock, and the digitised data on optical ethernet.

Figure 1. ANTARES telescope. Artistic view.

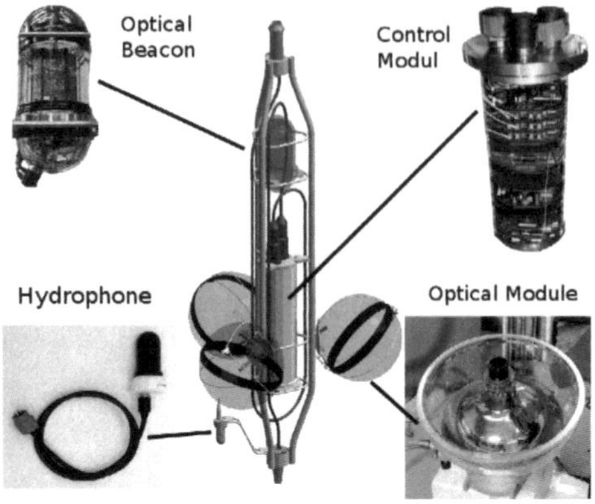

Figure 2. ANTARES storey with instruments on board.

4. ANTARES performance

The ANTARES telescope is able to detect all neutrino flavors. Electron and tau neutrinos can be detected as showers of secondary particles inside detector. For these two type of neutrinos and for muon neutrinos when the interaction vertex is also inside the telescope the energy resolution can achieve 30%. Such events are only very small part of detected neutrinos. The most part is muon neutrinos detected as muons traversing the detector. Beside this an accuracy of muon angle reconstruction is well below 1 degree in contrast to a few degrees angular resolution which can be achieved by reconstruction of showers. All this makes muon neutrinos the best candidates to study HE neutrino sources. Therefore reconstruction of main parameters – direction and energy are described in details only for muon neutrinos.

4.1. Angular resolution

Information about measured arrival time and position of each detected signal (hit) together with known Cherenkov angle ($\sim 42^\circ$) allows reconstruction of muon tracks and its precision increases with energy. Angular resolution of muon track reconstruction is shown as a red solid line in Fig. 3. The overall (neutrino) angular resolution is indicated by dashed blue line. It is better than 0.3° for energy above 10 TeV and achieves $\sim 0.15^\circ$ at 10^6 GeV. Below 10 TeV the reconstruction accuracy of neutrino arrival coordinates dominates by cinematic of neutrino interaction.

4.2. Energy resolution

Energy reconstruction of primary muon neutrinos is complicated. This task is divided into 2 subtasks: energy reconstruction of muon and having this information reconstruction of neutrino energy. They are described here separately.

Intensity of Cherenkov light for ultrarelativistic particles does not nearly depend on energy. Number of photons per unit of length:

$$\frac{dN}{dl} = 2\pi\alpha\left(1 - \frac{1}{1 - n^2\beta^2}\right)\int\frac{d\lambda}{\lambda^2}$$

Relativistic factor β^2 is only $10^{-4}\%$ lower than 1 already for 100 GeV muons in water. That means the energy reconstruction should not be possible using muon Cherenkov photons in the ANTARES detector. Cherenkov light energy loss however is at least 3-4 orders of magnitude

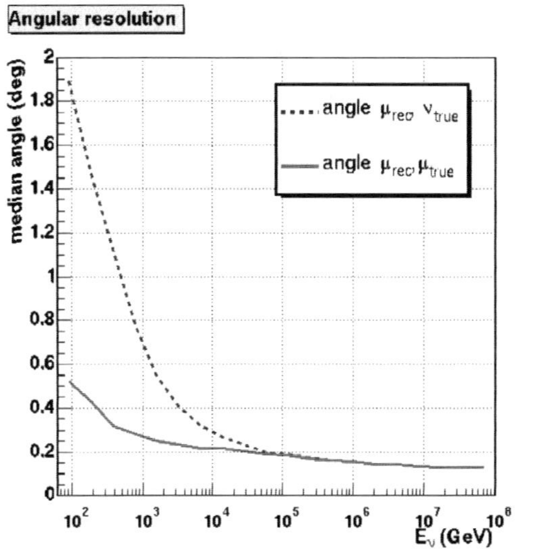

Figure 3. ANTARES angular resolution. Solid red curve – muon resolution, dashed blue curve – muon neutrino angular resolution.

lower other losses for muons with energies above 100 GeV. Pair production, bremsstrahlung and photo-nuclear interaction losses are roughly proportional to energy and start to dominate over weakly energy dependent ionisation loss above 600 GeV. Part of photons radiated by secondary particle in above mentioned reactions can be mixed up with Cherenkov photons from "primary" muon. Monte Carlo simulation confirms that picture and shows energy dependence of number of photons identified as Cherenkov ones [8]. Accuracy of the energy reconstruction by this method depends on energy of muon. It varies from a factor 3 for 10^3 GeV down to factor two for 10^6 GeV muons.

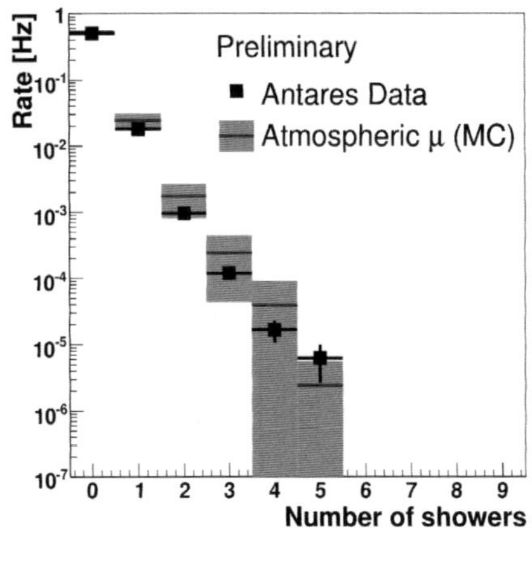

Figure 4. Distribution of the number of reconstructed atmospheric muon shower candidates for both data (points) and simulation (dotted line).

New method of the muon energy loss reconstruction has been proposed recently [9]. Above mentioned mechanisms of the muon energy loss in water are discreet. As a result electromagnetic (EM) showers are created. Number of showers along the track should be proportional to muon energy. Proposed algorithm permits to reconstruct EM showers. Good performance of the method has been validated by comparison shower counting in MC and ANTARES data. The shower multiplicity for down going muons is shown in Fig. 4. Further development of the method could improve a resolution of the muon energy reconstruction.

The neutrino energy can be reconstructed from muon energy. Generally the neutrino interaction vertex is outside of detector and muon energy loss between vertex and telescope is unknown. For higher neutrino energies, detected muons are produces at longer distances from the detector. Average ratio between the energy of the muon when it enters the ANTARES telescope and the neutrino energy drops from 0.8 for 100 GeV neutrinos down to 0.1 for 10^7 GeV neutrinos [8]. Energy resolution decreases correspondingly with increasing of energy. Inaccuracy of neutrino energy reconstruction is roughly factor five for 10 TeV muons.

4.3. Sensitivity

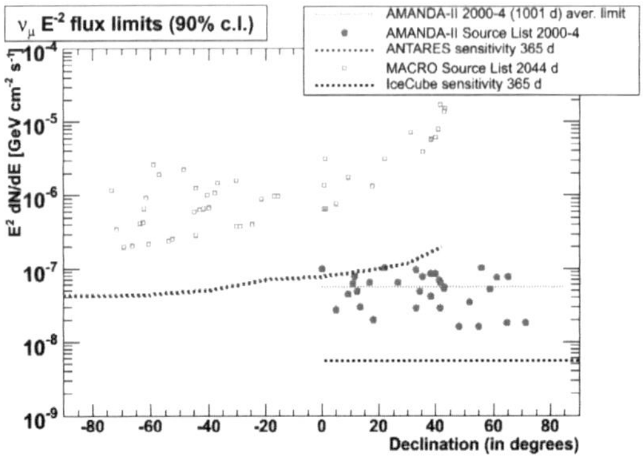

Figure 5. Sensitivity for point-like sources.

The ANTARES sensitivity to point-like sources depends on efficiency of neutrino detection and reconstruction. This efficiency depends on observational azimuth and zenith angles. The angles are changed with Earth's rotation. Sources with declination angles $\delta < -47°$ are always in the Antares field of view. Sources with $-47° < \delta < 43°$ are observable only part of time and sources with $\delta > 43°$ are above horizon and not visible. Combining all factors the sensitivity can be estimated as a function of δ of a potential source. This dependence is presented in Fig. 5. The 90% upper limit for muon neutrino flux in case of null signal after 1 year is $E^2\frac{dN}{dE} = 4 \times 10^8\ GeV cm^{-2}s^{-1}$ at $\delta = -90°$, and rises to $1.5 \times 10^7\ GeV cm^{-2}s^{-1}$ at $\delta = +40°$. Those limits are comparable to those obtained by Amanda–II for the northern sky [10]. ANTARES is able to observe the Galactic Centre ($\delta = -29°$) and nearby part of galactic plane where many VHE γ-ray sources have been detected. This part of sky is nonobservable by Amanda/IceCube.

5. First results

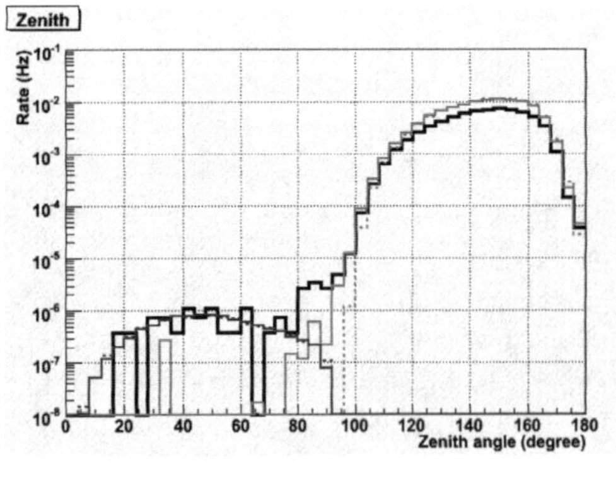

Figure 6. The zenith angle distribution of data taken in 2007. Black – observational data. Red – MC atmospheric muons, dashed – simulated, solid – reconstructed. Blue – MC atmospheric neutrinos.

Since the end of January 2007, the telescope consisted of 5 operating lines. With such a big detector, the good reconstruction quality is possible. During observations carried out in February-May 2007, $\sim 5 \times 10^6$ events have been collected. Zenith angle distribution of this data set is presented in Fig. 6. Results of the Monte Carlo simulations for the same telescope configuration are also drawn there. As expected, data are dominated by downward going muons (zenith angle $Z > 90°$). Events with $Z < 90°$ are neutrino candidates. Distributions of simulated atmospheric neutrinos and muons are in agreement with observational data. That validates good understanding of the detector and quality of the muon track reconstruction method.

Since December 2007 ten lines are in operation. First neutrino candidate obtained with 10-string detector is shown in Fig. 7. The Figure is the ANTARES event display. Plots represents all hits in each line in coordinates height–arrival time. Muon track has been reconstructed using hits of four lines as up-going with a zenith angle $\sim 52°$ from vertical.

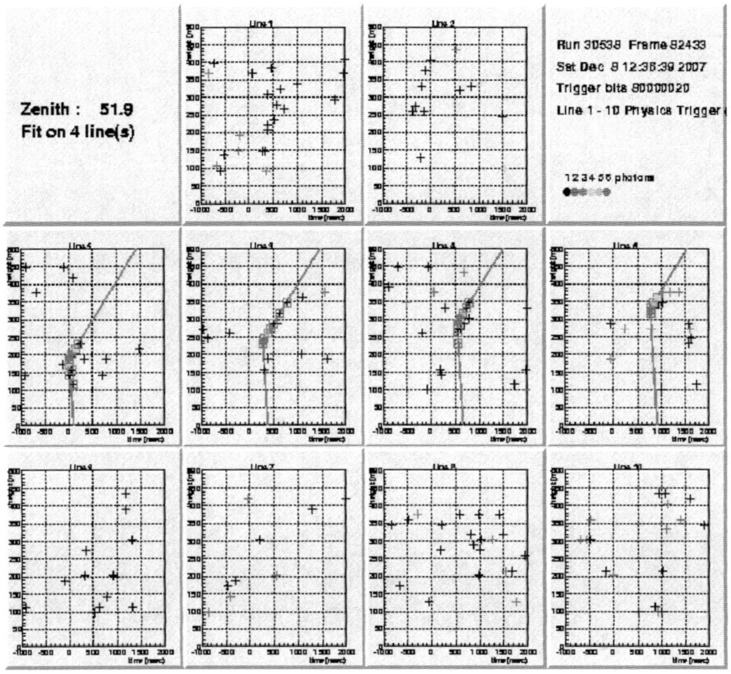

Figure 7. The first neutrino with 10 strings.

6. Conclusion

The ANTARES detector has been completed in May 2008. It is now the largest neutrino telescope in the northern hemisphere. Data taken with a half size detector in 2007 show, the performance of the telescope meets design specifications. During the next 5 years, the southern celestial sky will be observed with an unprecedent level of sensitivity for high energy neutrinos.

References

[1] ANTARES Collaboration, a deep sea telescope for high energy neutrinos *Preprint* arXiv:astro-ph/9907432v1

[2] Abraham J et al [Pierre Auger Collaboration] 2004 Properties and performance of the prototype instrument for the Pierre Auger Observatory *NIM* A *523* 50

[3] Abraham J et al [Pierre Auger Collaboration] 2008 Correlation of the highest-energy cosmic rays with the positions of nearby active galactic nuclei *Astropart Phys* **29** 188 0712.2843v2)

[4] Ginzburg V L and Syrovatski S I 1964 *The Origin of Cosmic Rays* (London and New York: Pergamon)

[5] Blandford R and Eichler D 1987 Particle Acceleration in Astrophysical Shocks: A Theory of Cosmic Ray Origin *Phys. Reports* **154** 1

[6] Aharonian F et al 2004 High energy particle acceleration in the shell of a supernova remnant *Nature* **432** 75 (Preprint arXiv:astro-ph/0411533v1)

[7] Aharonian F et al 2006 H.E.S.S. survey of the Inner Galaxy in very high-energy gamma-rays *Astrophys J* **636** 777 (Preprint arXiv:astro-ph/0510397v1)

[8] Zornoza J, 2005, Sensitivity to diffuse fluxes and energy spectrum reconstruction in the Antares neutrino telescope *PhD Thesis* can be found on http://antares.in2p3.fr/Publications/thesis/2005/Juande-Zornoza-phd.pdf

[9] Mangano S 2008 Algorithm for muon electromagnetic shower reconstruction *Nucl. Instrum. Meth.* A **588** 107

[10] Achterberg A et al [IceCube Collaboration] 2007 Five years of searches of astrophysical neutrinos with the AMANDA-II neutrino telescope *Phys. Rev.* D **75** 102001 (Preprint arXiv:astro-ph/0611063v1)

XIII International Conference on Calorimetry in High Energy Physics (CALOR 2008) IOP Publishing
Journal of Physics: Conference Series **160** (2009) 012037 doi:10.1088/1742-6596/160/1/012037

Atmospheric Calorimetry above 10^{19} ev: Shooting Lasers at the Pierre Auger Cosmic-Ray Observatory

Lawrence Wiencke for the Pierre Auger Collaboration
Department of Physics, Colorado School of Mines, 1532 Illinois St, Golden CO, 80401 USA
E-mail: lwiencke@mines.edu

Abstract. The Pierre Auger Cosmic-Ray Observatory uses the earth's atmosphere as a calorimeter to measure extensive air-showers created by particles of astrophysical origin. Some of these particles carry joules of energy. At these extreme energies, test beams are not available in the conventional sense. Yet understanding the energy response of the observatory is important. For example, the propagation distance of the highest energy cosmic-rays through the cosmic microwave background radiation (CMBR) is predicted to be strong function of energy. This paper will discuss recently reported results from the observatory and the use of calibrated pulsed UV laser "test-beams" that simulate the optical signatures of ultra-high energy cosmic rays. The status of the much larger 200,000 km^3 companion detector planned for the northern hemisphere will also be outlined.

1. Introduction

By the time this article appears in print, the base-line configuration of Pierre Auger Detector in Argentina (Auger South) will be completed. The tank commemorating completion of the surface detector was activated (Fig. 1), two weeks after this conference concluded.

Roughly the size of Provincia di Pavia, Auger South is the largest detector of cosmic rays above 5×10^{18}eV in the world. The instrument has been described in previous CALOR conferences [1] and elsewhere [2]. So this talk will focus instead on an aspect of this experiment that has not been presented in this conference series yet relates closely to its calorimetric theme. This is the use of atmospheric laser "test-beams" that simulate many of the optical properties of an extensive air-shower (EAS).

2. The Pierre Auger Southern Detector

A co-located surface detector (SD) and air-fluorescence detector (FD) comprise Auger South. The SD includes an array of more than 1600 water cherenkov detector stations [3] on a 1.5 km grid, 4 fluorescence "eyes", two test-beam laser facilities and an extensive collection of instrumentation ([4], [5], [6], [7]) dedicated to atmospheric monitoring ([8], [9], [10]) and photometric calibration [11].

The SD operates 24/7 to provide the statistical engine for Auger South. It is nearly fully efficient at 5×10^{18} eV. After correcting for zenith angle, the primary particle energy is estimated from the density of charged particles 1 km from the EAS core. Angular accuracy is a function of the number of tanks triggered. Essentially all events above the design threshold of $10^{19} eV$

Figure 1. June 13th 2008: Auger South staff celebrate deployment of the commemorative final surface detector tank. The design specification of 1600 operation tanks has been achieved.

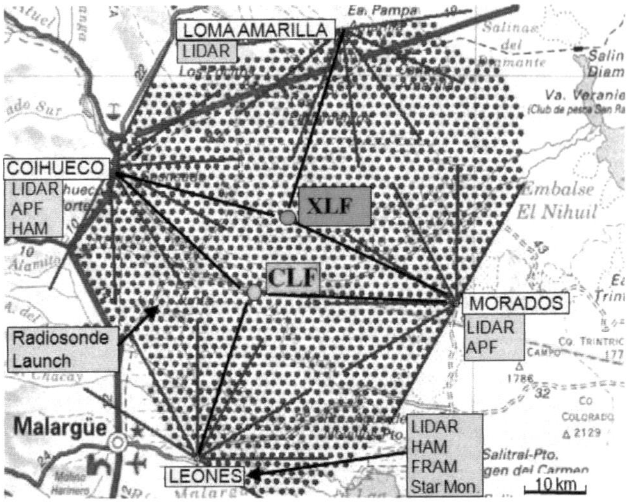

Figure 2. Auger South baseline configuration.

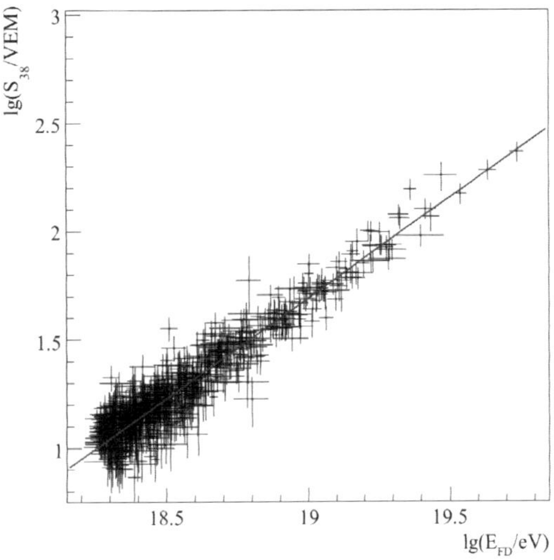

Figure 3. Correlation between energy as determined by the FD (horizontal axis) and by the SD for selected hybrid cosmic ray events.

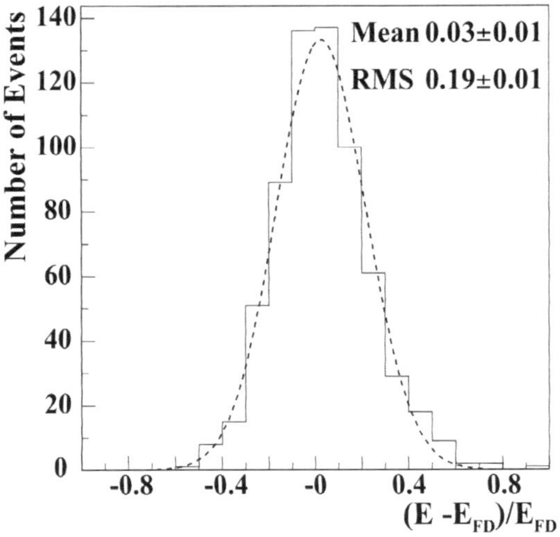

Figure 4. Energy correlation. E is the SD energy as calibrated by the FD using the fit in figure 3.

trigger 5 or more tanks. For EASs measured by at least 5 tanks, this accuracy is better than 1 degree [12].

The (FD) operates at night and records the longitudinal light profiles of EASs as they develop through the atmosphere. This calorimetric energy measurement does not rely on interaction models; the amount of fluorescence light emitted is essentially proportional to the energy deposited in the atmosphere.

About 10% of the observed EASs are observed by the FD and by the SD. This hybrid data sample currently exceeds 100,000 events. A subset, dubbed "golden hybrids", can be reconstructed using either the FD or the SD. FD energy measurements of 661 golden hybrids have been used to establish the energy scale (figure 3) of the high energy cosmic-ray spectrum measured by the SD [13]. The sigma of the correlation histogram (figure 4) is 19%. In terms of particle physics calorimetry, the level of consistency may seem suprising good; the SD is a sparse single-layer calorimeter. However, the large numbers of particles ($\geq 10^9$) comprising the typical EAS reduce measurement fluctuations. At present the systematic energy uncertainty is less than 25%. The three largest terms are fluorescence yield, detector photometric calibration, and atmospherics. Laser test beams play a critical role in the last two terms.

A motivation for reducing these systematics further is the strong dependence on energy of the propagation distance for protons through the CMBR. In the vicinity of the GZK suppression, a 25% decrease in energy scale from 8×10^{19} eV to 6×10^{19} eV corresponds to a 10-fold increase of observable volume.

3. Laser "Test Beams"

A variety of pulsed ultraviolet lasers (see table 1) are fired routinely into the atmosphere above the observatory. Their locations are shown in the map of figure 2.

Table 1. Lasers used at the Pierre Auger Observatory.

System	Type	Wavelength	Max. Energy	Width	Direction	Receiver
Roving	Nitrogen	337.1 nm	100 μJ	4 ns	vertical	FD
CLF	YAG	355 nm	7 mJ	4 ns	vert.+steered	FD
XLF	YAG	355 nm	7 mJ	7 ns	vert.+steered	FD
LIDARs(4)	YLF	351 nm	100 μJ	25 ns	steered	dedicated

When a pulsed UV laser is shot into the sky, the atmosphere scatters light out of the beam. The same FD detectors that record tracks produced by the passage of cosmic ray air-showers downward through the atmosphere can also record tracks produced by the passage of laser pulses upward through the atmosphere. But unlike cosmic-rays, the laser energy, absolute firing time (using GPS [14]), and direction can be specified and the pulses can repeated as necessary.

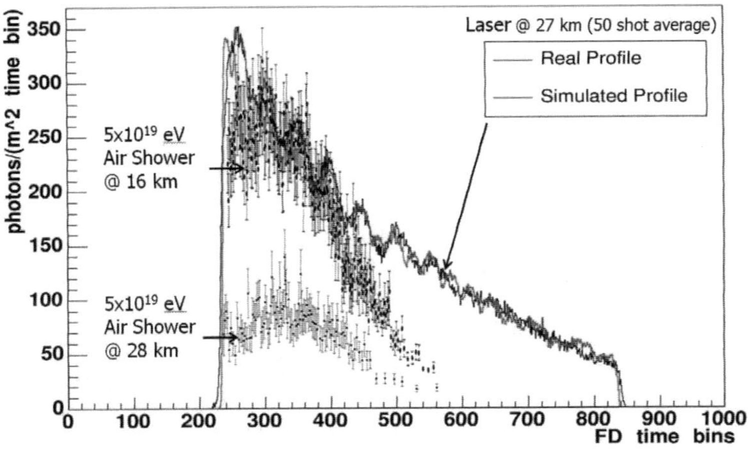

Figure 5. Comparison between vertical 7 mJ UV laser shots from the CLF and near-vertical cosmic ray air-showers observed by the FD. One time bin is 100 ns. The superimposed (downward going) cosmic ray profiles have been "flipped" so that the left edge of all profiles corresponds to the bottom of the FD field of view.

Superimposed profiles from vertical laser tracks and near vertical (downward) cosmic-rays are shown in figure 5 to illustrate the equivalence between laser energy and air-shower energy.

Under similar conditions of atmospheric clarity, a vertical laser fired from the central laser facility generates a track in the FD that is about twice as bright as a track from a near-vertical cosmic ray of 5×10^{19} eV seen at a similar distance of 28 km. The observed laser and cosmic ray optical profiles in this example are similar in that the brightest regions of both are near the bottom of the FD field of view. Their shapes differ because different interaction processes are involved. The laser profile is generated by atmosphere scattering which, in this example, largely traces the decrease in atmospheric density with height. By contrast, the cosmic ray profile viewed from the side is dominated by scintillation light emission along the electromagnetic component of the air-shower. However, once the light from either the laser beam or the air-shower leaves the interaction region, both its propagation through the atmosphere and its conversion to digitized traces in the FD follow the same process.

3.1. Roving Nitrogen Laser

A calibrated portable nitrogen laser system was configured (figure 6) to cross-check the photometric calibration of the FD. To reduce atmospheric effects the laser was fired vertically at distances of 3km from the FD eyes. At this distance, the atmospheric effects of scattering out of the beam (that would brighten the track) and subsequent attenuation from the laser beam to the detector (that would dim the track) tend to cancel [15]. The laser energy was measured by a calibrated energy probe and by reconstructing the tracks in the FD after the FD was calibrated using a "drum" light source placed over the telescope aperture. These independent energy measurements (see figure 7) were consistent with the 7% accuracy of the energy probe and the 10% accuracy of the FD photometric calibration.

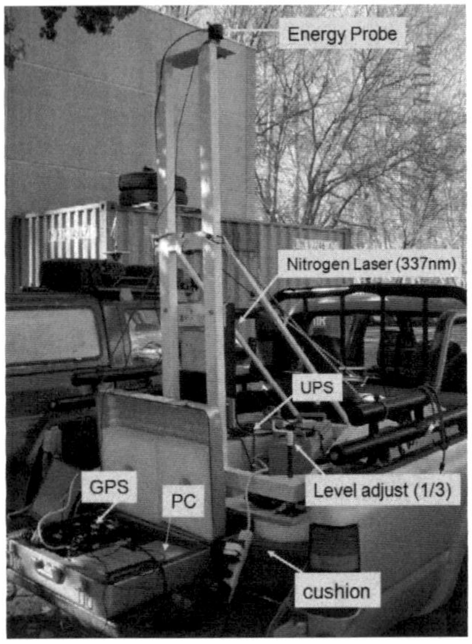

Figure 6. The roving nitrogen laser system is mounted on a small truck for deployment in the field.

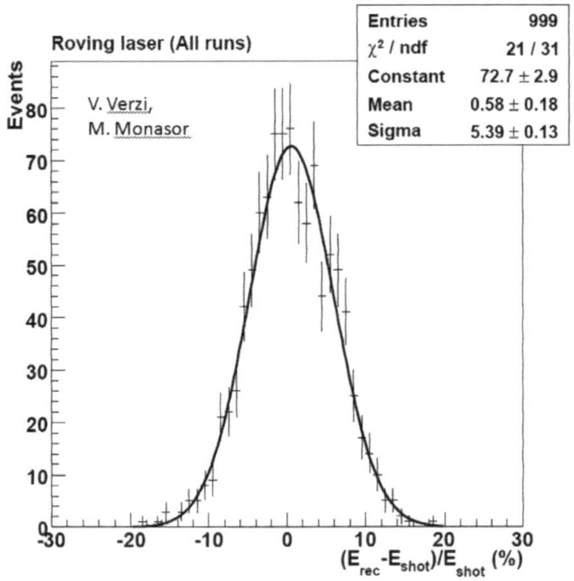

Figure 7. Cross-check of FD photometric calibration with the roving nitrogen laser system. E_{rec} is the energy reconstructed using the FD. E_{shot} is the energy measured by a probe at the laser.

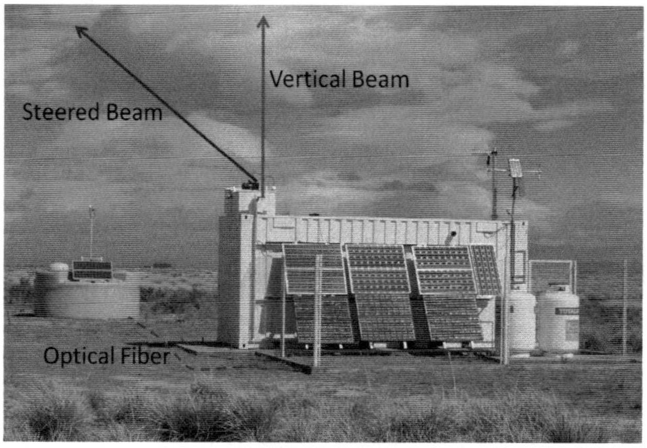

Figure 8. The Central Laser Facility at the Pierre Auger Southern Detector generates either a fixed-direction vertical or steered beam. A small fraction of light from each laser pulse is sent via optical fiber to the adjacent surface detector station. The facility is solar powered and is operated remotely from the central campus in Malargue.

3.2. Central Laser Facility (CLF) and Extreme Laser Facility (XLF)

Laser test beams from the CLF have been used extensively at the observatory [16]. Details of the CLF can be found in [17]. Its uses include FD commissioning (swapped cable finding), measuring and monitoring the offsets between the FD and SD clocks, testing the FD geometric reconstruction, hourly monitoring of the atmospheric aerosol vertical distribution and horizontal uniformity, and detection of clouds in the center of the observatory. The CLF will also supplement the LIDAR systems in a "shoot-the-shower" program. A preliminary cross-check of the photometric calibration of the FD has been performed with the CLF vertical beam (see figures 9 and 10). This cross-check includes the effect of atmospheric attenuation over the 27-30 km distance from the CLF to the FD eyes. At this point this cross-check is not robust because the atmospheric corrections applied were those determined by the CLF.

A second laser facility, dubbed the XLF, will provide an independent cross-check of FD photometric calibration including atmospheric effects. The XLF design is similar to that of the CLF with the addition of an automated system to measure the absolute beam energy and polarization each night of operation (see figure 11). The goal is to maintain the absolute energy calibration of the beam including polarization effects to an uncertainty of less than 5%.

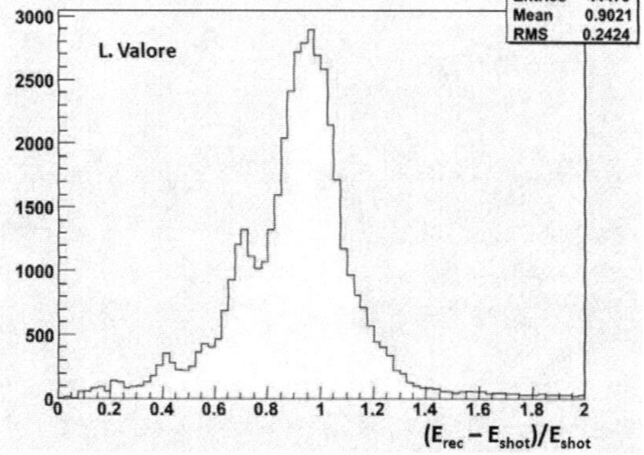

Figure 9. Reconstructed CLF energy using tracks observed in the Los Leones FD eye 27 km distant. This reconstruction assumed a clear atmosphere. Aerosol and cloud effects broaden the distribution.

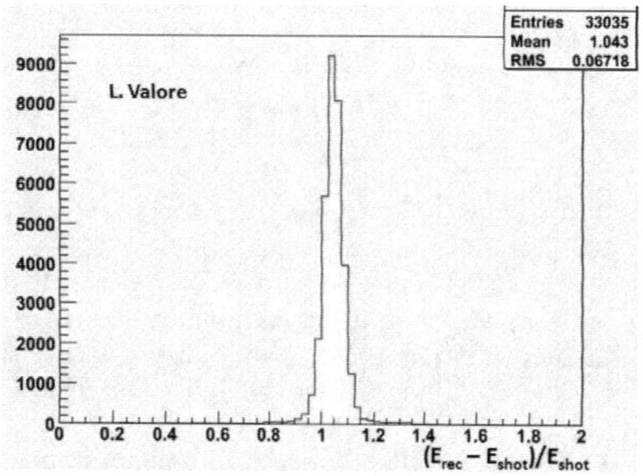

Figure 10. CLF laser data in previous figure, but reconstructed with the hourly atmospheric database corrections. Cloudy periods are also removed. Since the atmospheric corrections were obtained using CLF data, this cross-check is not robust.

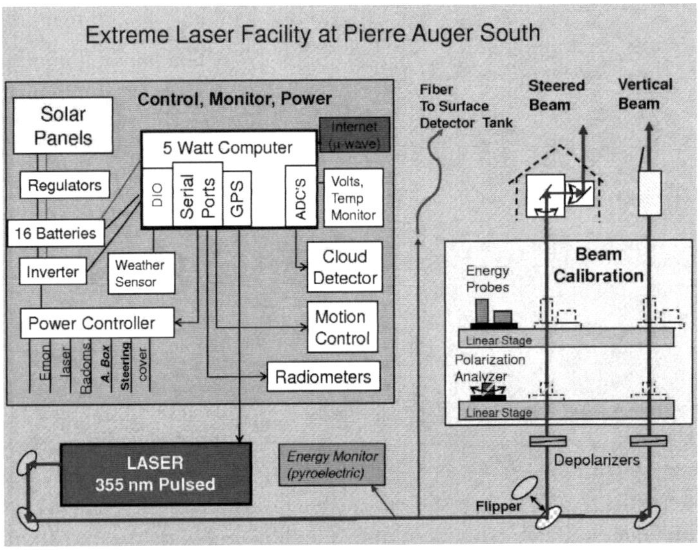

Figure 11. Functional diagram of the XLF The beam energy and polarization calibration will be measured remotely.

4. Auger North

The highest energy events recorded by Auger South are not distributed isotropically in the sky [18]. However, despite the large collecting area, the statistics from the south are limited to some 30 events per year above $10^{19.5}$ eV. To collect a sufficent sample to identify and study sources, a companion Auger North detector planned for Colorado [19] will be much larger than Auger South. The design features 4400 tanks covering 21,000 km^2 with $\sqrt{2}$ mile spacing. An FD of 4-6 stations will overlook most of the SD. An extensive R&D program is underway.

References

[1] B. Keilhauer [Pierre Auger Collaboration], Proc of XII International Conference on Calorimetry in High Energy Physics (CALOR06) Chicago [astro-ph/0608074].
[2] Pierre Auger Collaboration [J. Abraham et al.], 2004 NIM **A523** 50.
[3] I. Allekotte et al. 2008 NIM **A586** 409.
[4] S. Y. BenZvi [Pierre Auger Collaboration], (ICRC 2007)[arXiv:0706.1710].
[5] S.Y. BenZvi, B.M. Connolly, J.A.J. Matthews, M. Prouza, E.F. Visbal, and S. Westerhoff, 2007 Astroparticle Physics **28** 312 [arXiv:0704.0303].
[6] B. Keilhauer [Pierre Auger Collaboration],(ICRC 2005)**7** 123.
[7] S.Y. BenZvi et al. 2007 NIM **A574** 171.
[8] S.Y. BenZvi [Pierre Auger Collaboration], (ICRC 2007)[arXiv:0706.3236].
[9] M. Prouza [Pierre Auger Collaboration], (ICRC 2007)[arXiv:0706.1719].
[10] C. Bleve [Pierre Auger Collaboration], (ICRC 2007)[arXiv:0706.1491].
[11] R. Knapik [Pierre Auger Collaboration], (ICRC 2007)[arXiv:0708.1924].
[12] [Pierre Auger Collaboration](ICRC 2007)**7** p17.
[13] Pierre Auger Collaboration [J. Abraham et al.], 2008 PRL in press.
[14] J.D. Smith, S. Thomas, L.Wiencke (ICRC 2001)**HE1.** 825.
[15] L.Wiencke et al. 1999 NIM **A428** 593.
[16] L. Wiencke [Pierre Auger Collaboration], 2007 NIM **A572** 508.
[17] B. Fick, M. Malek, J.A.J. Matthews, R. Meyhandan, M. Mostafa, M. Roberts, P. Sommers and L Wiencke 2006 JINST **1** 11003.
[18] Pierre Auger Collaboration [J. Abraham et al.], 2007 Science **318** 939.
[19] D. Nitz [Pierre Auger Collaboration], (ICRC 2007)[arXiv:0706.3940].

Results of the Pierre Auger Observatory on High Energy Cosmic Rays

J. Lozano-Bahilo for the Pierre Auger Collaboration
Dpto. Física Teórica y del Cosmos, Universidad de Granada, Granada 18072, ES
E-mail: julio@ugr.es

Abstract. The Pierre Auger Observatory is a hybrid detector comprising water tank detectors covering an area of 3000 km^2 and fluorescence detectors located in 4 stations surrounding that area. The Pierre Auger Observatory has been in operation for the past few years while its components were being gradually assembled and deployed on the field. In spite of the limited acceptance the data which have been gathered have already delivered interesting results on High Energy Cosmic Rays. This paper reviews the most interesting results published by our collaboration so far, including limits on high energy photons and neutrino fluxes, the spectrum, the anisotropy of Ultra High Energy Cosmic Rays and their correlation with extragalactic objects.

1. Introduction

Cosmic rays provide the source of the most energetic particles known to scientists with energies reaching values far beyond those attainable at our largest colliders. Our knowledge on Ultra High Energy Cosmic Rays (UHECR) is still limited since the flux of cosmic rays decreases with energy and direct detection of sufficient numbers is not feasible. Only vast arrays of detectors capturing information of the Extensive Air Showers (EAS) produced by the primary cosmic ray particles can gather enough statistics over long periods of time. There are currently different models trying to explain the formation of those particles and basically all of them fall into two different categories : acceleration and *Top-Down* models. Acceleration mechanisms assume that those cosmic rays are accelerated within astrophysical objects of the appropriate size and magnetic field before coming to the earth [1]. In *Top-Down* models the cosmic rays are produced in decays of supermassive particles or topological defects. As we will see, our results tend to discard the latter possibility.

The Pierre Auger Observatory can be considered as the biggest calorimeter in the world since it makes use of an enormous portion of atmosphere as sensitive material. The particle showers produced by high energy cosmic rays interacting in air are detected with the help of two different techniques: surface water tanks recording Cerenkov light produced by the remnants of the showers at ground level and fluorescence detectors which track the evolution of the shower as it develops. These two independent detection techniques allow to extract optimum information of those uncommonly high energetic particles coming from outer space. There are currently more than 1600 high purity water tanks deployed which include 3 photomultiplier tubes that transform the Cerenkov light into a measurable electrical signal. The detectors are placed according to a triangular pattern with a distance of 1.5 km between them. This array of

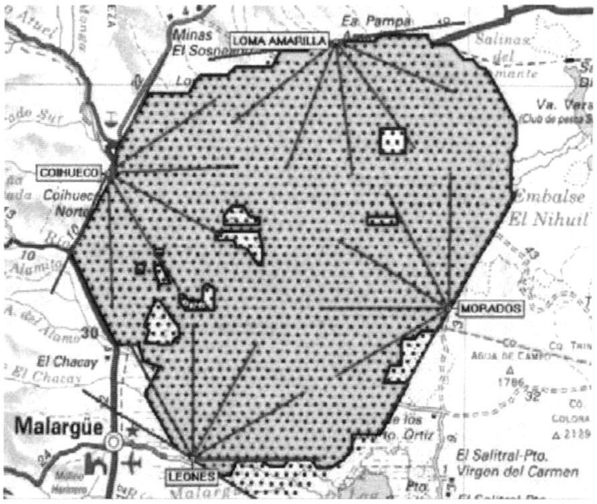

Figure 1. The observatory is located near Malargüe, in the province of Mendoza, Argentina. The dots represent the position of the water tanks. The four FD stations are named within yellow coloured rectangles. The azimuthal coverage of each telescope is represented with green lines

detectors is collectively known as the *Surface Detector (SD)* [6]. The *Fluorescence Detector (FD)* consists of 4 stations overlooking the area covered by the water tanks with 6 fluorescence telescopes each. Those telescopes contain mirrors covering 30^0 in azimuthal and zenithal angles which reflect the fluorescence light emitted by excited N_2 molecules onto an array of 22×20 photomultipliers. The layout of the tanks and the telescope stations is visible in Figure 1. The properties and performance of prototype instruments can be found in the following reference; [5]

1.1. Event Reconstruction

The position and the timing of the signals of the SD tanks allow to make a fit to the shower front of the cascade of particles which derives the zenithal and azimuthal angles of the incoming primary particle. Even for the lowest energy events which can be triggered by our detector (10^{18}eV) the angular resolution is smaller than a couple of degrees and as soon as the energy increases and more stations are involved in the reconstruction the resolution becomes roughly 1 degree or even less. The energy of the shower comes from another fit; this time it involves the Lateral Distribution Function (LDF) of the tank signals and their distances to the core of the shower. The fitted parameters are the core position and the expected signal value at 1000 meters from the core, S_{1000}. The choice of that distance guarantees the smallest fluctuations in the extracted signal parameter. The next step in order to compute the energy is to normalize the signal obtained at the measured zenithal angle to a given reference angle since there is a clear dependence of the signal with that angle. The chosen value is 38^0 which is the median of the zenithal distribution. The procedure to perform this normalization is called the Constant Intensity Cut (CIC) [4]. Assuming a nearly isotropic flux of cosmic rays beyond a given energy threshold it is easy to establish the normalization factors to estimate S_{38}. The last step to obtain the energy of the primary particle is the correlation of the value of S_{38} with the nearly calorimetric, model independent energy measurement of the FD detector. We use hybrid events where information of the shower is recorded in both detector systems, SD and FD. The dispersion of the SD estimates with respect to the FD energy values is about 20%.

The reconstruction of the arrival direction of the cosmic ray in the FD makes use of the position of the photomultipliers within the array and the timing of the signals recorded on those hit by reflected fluorescence photons. The angular fit does not precisely fix all the parameters and we need the timing of a tank from the SD detector in order to decrease the resolution from several degrees down to about 0.5^0. The energy estimate comes from a fit with a Gaiser-Hillas function [3] to the Longitudinal Shower Profile (LSP) obtained with the amount of signal collected on the PMTs. This measurement relies fully on the calorimetric properties of the atmosphere and not on hadronic models and is, therefore, quite accurate.

The hybrid approach of our detector provides obvious advantages. It allows a very precise determination of the energy and arrival direction of the cosmic primaries; we are able to calibrate the energies of the water tanks with the help of hybrid events and the timing of the tanks improves significantly the angular resolution of the FD. It is also easier to study systematic effects since both detectors provide totally independent measurements.

2. Energy spectrum

The spectrum of the cosmic rays follows approximately a power law with a negative index which is not completely uniform. When the energy of the primary reaches a value of $\simeq 4 \cdot 10^{18}$eV the index becomes smaller (*ankle*). The most common explanation is the onset of an extragalactic component. At even higher energies $E \simeq 6 \cdot 10^{19}$eV the GZK (Greisen-Zatsepin-Kuzmin) effect [2] should take a dominant role. It is the consequence of the ubiquitous presence of the Cosmic Microwave Background in outer space. The interaction of protons with the CMB at those energies produces pions whereas heavy nuclei disintegrate at even lower energies. Therefore an abrupt decrease of the flux of UHECR is expected unless their origin is considerably closer than the interaction length ($\lambda \sim 50$ - 100 Mpc) for these processes involving the CMB.

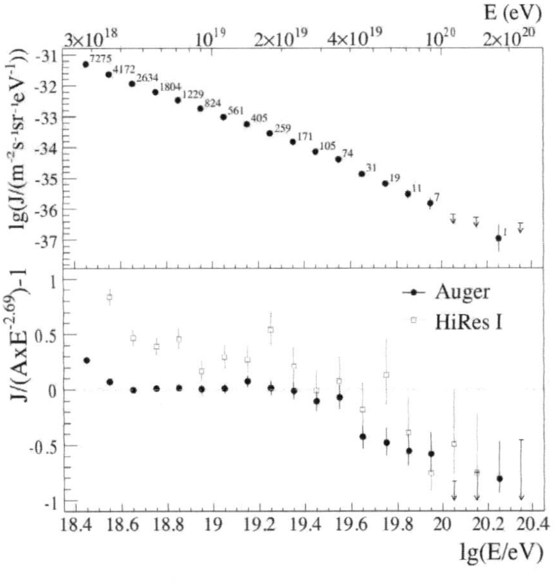

Figure 2. In the upper plot we present the flux and the number of events corresponding to each energy value. In the lower plot we show the spectrum for HiRes I and Auger normalized to a spectrum with a power index of −2.69.

Results presented by previous experiments were contradictory. AGASA, an array detector, obtained a flux which differed in the shape and the absolute scale from the flux measured by HiRes, a fluorescence telescope; the flux values measured by AGASA were higher and no evidence of the GZK-effect was noticeable [13] whereas HiRes observed a steepening of the curve at GZK cutoff energies [14]. The spectrum obtained by the Pierre Auger Observatory (see Figure 2) basically agrees with the recent results of HiRes establishing new evidence of the expected GZK effect [7].

3. Photons

Photons of the energy range studied in our experiment penetrate deeper in the atmosphere than protons or nuclei. It is easy to prove that the delay of the signals induced on the tanks by the particles of the shower front decreases with the height of the production point. Therefore, photon showers give higher delays on the stations as compared to proton or nuclei induced showers. This is related to the curvature of the shower front which can be extracted from a fit to the trigger times of the SD tanks. On the other hand, the time spread of the signals on a given tank corresponding to particles generated at lower heights is bigger than the spread for particles appearing at higher altitude. This, combined with the fact that showers induced by photons have a lower quantity of muons which are produced high in the atmosphere, help to establish selection criteria based on the risetime (time difference between 10% and 50% of integrated signal) of the ADC traces recorded in the water tanks.

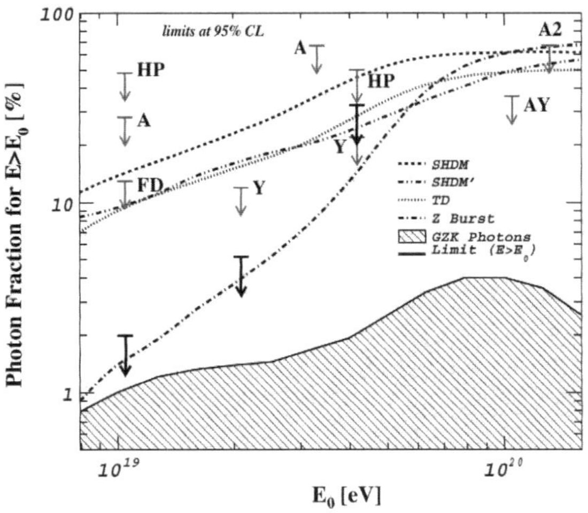

Figure 3. Photon fractions at 95% C.L. for different experiments and models; HP: Haverah Park, A1 and A2: AGASA, AY: AGASA-Yakutsk, Y: Yakutsk, FD: Auger hybrid and black arrows: Auger SD. The striped area represents the expectation for GZK photons.

The variables chosen for our selection cuts are the radius of curvature R and the expected risetime value at a distance of 1000 meters from the core of the shower $t_{1/2}(1000)$. The latter variable comes from a fit to the risetime values of the tanks corrected by their internal azimuthal angle and the shower zenith angle. Normalized (*pull*) variables involving the photon expectation values and the spread in the distributions are combined into a two-dimensional plot in order to perform a principal component analysis with a selected efficiency of 50%. More details can be found in reference [8].

No candidates were found after applying our selection criteria; thus, limits on the flux and the fraction of photons were calculated at 95% C.L. for different energy ranges. Figure 3 shows the

Journal of Physics: Conference Series **160** (2009) 012038 doi:10.1088/1742-6596/160/1/012038

fraction of photons obtained by our collaboration for three different energy values compared to the expectation curves of several models. Many of the Top-Down scenarios for HECR production are rejected with our results.

4. Ultra High Energy Neutrinos

The interaction probability of a neutrino with matter is very low and it can produce a shower anywhere along its path in the atmosphere. If we limit our study to showers which have already traversed a long distance within air, the electromagnetic component should be highly suppressed. This is the case of showers with big zenithal angles initiated by protons and heavy nuclei which are produced close to the point where the particle enters the earth atmosphere. On the other hand, neutrinos interacting deep in the atmosphere will produce a shower with a detectable electromagnetic component, especially in those tanks hit earlier by the shower front. The Pierre Auger Observatory is studying so called *horizontal* showers ($\theta > 60^0$) where the time spread of the *early* tanks signals is high according to the risetime and falltime ($50\% - 90\%$ of the signal) observables. This analysis is still in progress. Our collaboration has been also looking for τ neutrinos interacting with the crust of the earth or the mountains surrounding the SD area [11]. If they undergo charged-current interactions they can produce τ particles that may travel as far as 10 km before producing a shower which is nearly horizontal. The event selection for neutrino showers includes a cut on the time spread of signals within the FADC traces of the water tanks. It requires a certain elongation pattern of the triggered tanks by assigning a length and a width to the event and restricting its ratio. In addition, it is assumed that the event travels horizontally at a velocity which is close to the speed of light. The measured value is an average obtained for pairs of triggered stations. The RMS of their distribution also helps to discard non neutrino events.

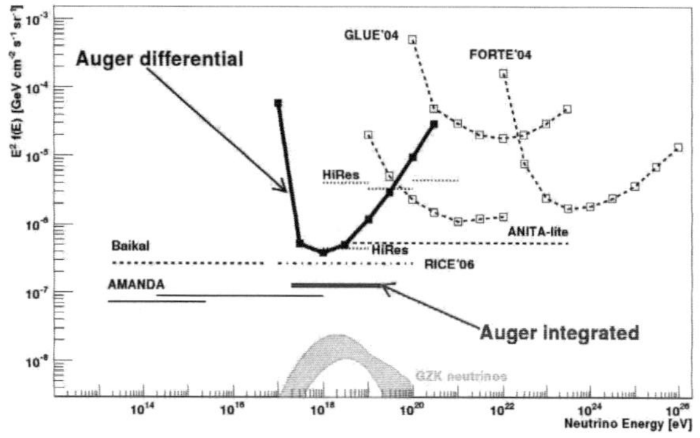

Figure 4. Limits on integrated and differential fluxes of high energy neutrinos for different experiments including those obtained by the Pierre Auger collaboration.

The selection criteria retain 80% of neutrino showers while no background event is kept. No candidates have been found until now, so we have established limits on the differential and integral neutrino fluxes. Those limits pose a big constraint on certain *Top-Down* models, even rendering some of them invalid. As can be seen in Figure 4 [18], our observatory may be capable of observing GZK neutrinos produced in the collisions of CRs with the CMB in the near future. The limit derived from our study on the diffusive neutrino flux in the energy range $2 \times 10^{17} < E < 2 \times 10^{19}$ eV is $E_\nu^2 dN_{\nu_\tau}/dE_\nu < 1.3 \times 10^{-7}$ GeV cm^{-2} s^{-1} sr^{-1}.

5. Large scale anisotropy of HECRs

The Galactic Centre (GC) harbors several objects which could be sources of ultra high energy cosmic rays and it is therefore an interesting region if we look for anisotropies in the HECR flux. It is even more attracting if we consider that previous experiments reported excesses in areas located close to it. In particular, AGASA observed a 4.5σ excess in the energy range $10^{18} - 10^{18.4}$eV in a 20^0 radius area close to the GC [15] [16] and SUGAR measured a 2.9σ flux excess in the range $10^{17.9} - 10^{18.5}$eV in a smaller region with a radius of 5.5^0 centered at a different point [17].

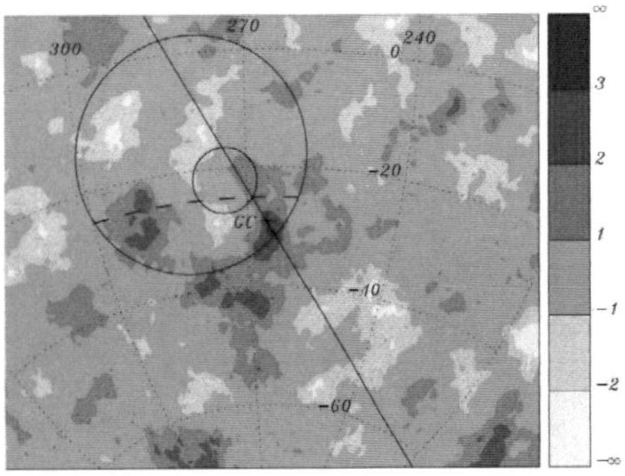

Figure 5. Map of overdensity significances around the galactic center. The regions where AGASA and SUGAR found excesses are also shown in circles.

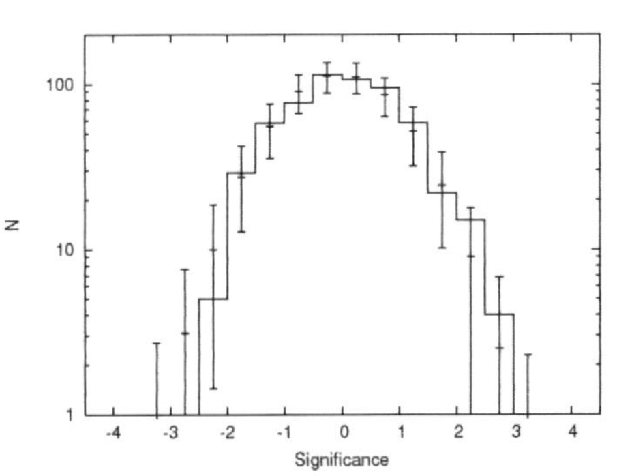

Figure 6. Histogram of overdensities (line) with superimposed isotropic expectations (average and 2σ bounds).

In our analysis [12] we considered the events with reconstructed energies in the range chosen by SUGAR. To search for anisotropy in the HECR flux, we compare our data to an estimate of the flux expected from an arrival direction distribution that is isotropic, but modified to account for the modulations of the exposure with right ascension as the array grew during construction. In Figure 5 we present the significances of the overdensities in circular windows of 5^0 radius. The distribution of observed overdensities is compared to the expectations for an isotropic flux in Figure 6. Our data is completely consistent with statistical fluctuations of an isotropic sky.

6. Anisotropy of UHECRs and their correlation with Active Galactic Nuclei (AGN)

The source of the UHECRs has been a topic of great interest since their discovery and according to acceleration models only certain objects with high magnetic fields covering big enough areas are capable of generating particles at such extraordinary energies. Active Galactic Nuclei are amongst those objects which could be held responsible for UHECR production. The quasars and active nuclei listed in the 12th edition of the Veron-Cetty and Veron catalogue [10] have been used in our correlation study. The catalogue, although thoroughly compiled to include all objects in the literature, can not be considered to be complete, especially near the galactic plane, nor an unbiased sample.

A first scan was conducted with a set of our data (*exploratory scan*) to obtain the values of the angular distance ψ, the maximum redshift z and the minimum energy E_{min} which minimized the chance probability for an isotropic flux. The selection parameters obtained with this *a priori* analysis were applied to future data according to a *running prescription*. This prescription, applied on an event by event basis, was established in order to confirm the anisotropy of the UHECRs. The procedure we followed avoids any penalty factors which would apply in *a posteriori* searches. More information on this can be found in references [19] and [9].

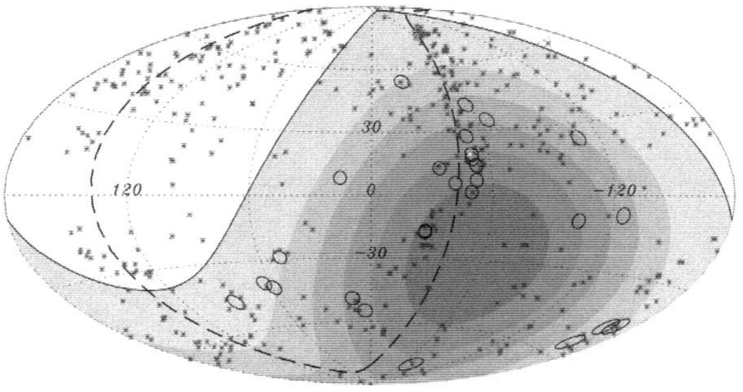

Figure 7. Aitoff projection of the celestial sphere in galactic coordinates. The positions of the AGNs with distances smaller than 71 Mpc and the 3.2^0 circles around the selected events are marked. Colors indicate equal exposure.

The analysis of the whole data set derived the following numbers for the maximum correlation significance: $E_{min} = 57$ EeV, $z_{max} = 0.017$ corresponding to distances $\lesssim 71 Mpc$ and $\psi = 3.2^0$. The angular distance is bigger than the resolution of our detector for cosmic rays with energies beyond 10^{19}eV (about 1^0 or lower) but we have to take into account the presence of magnetic fields in the trajectory of the rays. With the selected parameters we find that 20 out of 27 cosmic ray events correlate with an AGN where only 5.6 are expected for an isotropic flux. It is worth mentioning that 5 out of the seven which do not correlate are within 12^0 of the galactic plane where, on the one hand, the data of the VC-V catalogue is less reliable and on the other hand, the magnetic fields are stronger. Figure 7 shows the position of the correlated cosmic ray events and that of the AGNs within the obtained maximum distance.

In any case, the scientific claim of the Pierre Auger Collaboration is that the highest energy cosmic rays are not isotropic but in spite of the observed correlation we can not claim that the sources of those rays are, actually, AGNs. They could be the sources but they may be just tracers for the real sources since the spatial distribution of those objects agrees well with that of other extragalactic objects. It is important to notice that the calculated minimum energy and maximum distance lie within the GZK effect expectations. The fact that the angular distance is relatively small would favour that the highest energy rays are protons.

7. Conclusions

The Pierre Auger Observatory is an enormous air calorimeter which has been collecting data on UHECRs with a growing array of water tanks and fluorescence telescopes for several years and now is finishing its deployment. The data acquired so far has already allowed to improve our knowledge on several topics related to high energy cosmic rays. The spectrum of the cosmic rays above 10^{18} eV shows an abrupt decrease above 60 EeV confirming the expected GZK effect. Our collaboration has been able to establish very good limits on neutrino and photon fluxes which discard many *Top-Down* models. Additional data should provide evidence for photons and neutrinos produced in the interaction of protons and nuclei with the CMB. There is no evidence of anisotropy of HECR with energies in the range $10^{17.9} - 10^{18.5}$ eV around the galactic center, whereas UHECR rays with energies greater than 60 EeV show a clear anisotropic origin. Using a catalogue of extragalactic objects containing a compilation of AGN we have observed a correlation with our highest energy events,but this does not establish that AGN themselves are the sources. The significance of the correlation is maximized for distances and energies which agree with the existence of the GZK cut-off.

The successful completion of the Observatory is an important milestone which makes us confident that our laboratory will be able to continue producing interesting results in the future. Members of the collaboration are already working on the deployment of a test array in the vicinity of Lamar (Colorado) which will be the starting point for the northern hemisphere Observatory. It will cover the areas of the sky not reachable with the southern hemisphere Observatory and it will, obviously, increase our statistics. Thus, it is evident that there is a bright future ahead of us in the field of High Energy Cosmic Rays.

Acknowledgments

I would like to thank the organizers of the CALOR 2008 meeting held in Pavia for the wonderful atmosphere of the conference. It was a pleasure to present the results obtained by the Pierre Auger Observatory on behalf of all my colleagues. I wish to thank them for helpful discussions while preparing the presentation. I also want to acknowledge the technical and administrative staff of the Observatory for the enormous effort they have put into this project. The *Ministerio de Educación y Ciencia* of Spain is gratefully acknowledged for their financial support.

References

[1] A. M. Hillas, Ann. Rev. Anstron. Astrophys. 22, 425 (1984).
[2] K. Greisen, Phys. Rev. Lett. 16 (1966) 748; G. T. Zatsepin and V. A. Kuzmin, Pisma Zh. Eksp. Teor. Fiz. 4, 114 (1966).
[3] T. K. Gaisser and A. M. Hillas, Proc. of the 15th Int. Cosmic Ray Conf. 8 (1977).
[4] J. Hersil et al., Phys. Rev. Letters 6, 22 (1961).
[5] J. Abraham et al. [Pierre Auger Collaboration], Nuclear Inst. and Methods A523, 50 (2004).
[6] I. Allekotte et al. [Pierre Auger Collaboration], Nuclear Inst. and Methods A586, 409 (2008).
[7] M. Roth et al. [Pierre Auger Collaboration], Proc. of the 30th Int. Cosmic Ray Conf. (ICRC 2007), Merida (Mexico), July 2007, [arXiv:astroph/0706.2096].
[8] J. Abraham et al. [Pierre Auger Collaboration], Astroparticle Physics 29, 243 (2008).
[9] J. Abraham et al. [Pierre Auger Collaboration], Astroparticle Physics 29, 188 (2008).
[10] M.-P. Véron-Cetty and P. Véron, Astron. and Astrophys. 455, 733 (2006).
[11] J. Abraham et al. [Pierre Auger Collaboration], to appear in Physical Rev. Letters (2008) [arXiv:astroph/0712.1909].
[12] J. Abraham et al. [Pierre Auger Collaboration], Astroparticle Physics 27, 244 (2007).
[13] M. Takeda et al. [AGASA Collaboration], Astroparticle Physics 19, 447 (2003).
[14] R. Abassi et al. [HiRes Collaboration], Phys. Rev. Letters 100, 101101 (2008).
[15] N. Hayashida et al. [AGASA Collaboration], Astroparticle Phys. 10, 303 (1999).
[16] N. Hayashida et al. [AGASA Collaboration], Proc. of the 26th Int. Cosmic Ray Conf. (ICRC 1999), Salt Lake City, 3 [arXiv:astroph/9906056]; M. Teshima et al. [AGAS Collaboration], Proc. of the 27th Int. Cosmic Ray Conf. (ICRC 2001), Hamburg, 1, 337 (2001).
[17] J. A. Bellido, R. W. Clay, B. R. Dawson and Johnston-Hollitt, Astroparticle Phys. 15, 167 (2001).
[18] J. Abraham et al. [Pierre Auger Collaboration], Phys. Rev. Letters 100, 211101 (2008).
[19] J. Abraham et al. [Pierre Auger Collaboration], Science 318, 939 (2007).

Journal of Physics: Conference Series **160** (2009) 012039 doi:10.1088/1742-6596/160/1/012039

Performance of the PAMELA Si-W Imaging Calorimeter in Space

V Bonvicini[1], M Boezio[1], E Mocchiutti[1], A Vacchi[1], G Zampa[1], N Zampa[1], R Bellotti[2], A Bruno[2], F Cafagna[2], O Adriani[3], L Bonechi[3], M Bongi[3], S Bottai[3], D Fedele[3], P Papini[3], S Ricciarini[3], P Spillantini[3], E Taddei[3], E Vannuccini[3], G Castellini[4], M Ricci[5], G A Basilevskaja[6], A N Kvashnin[6], Y I Stozhkov[6], A M Galper[7], L Grishantseva[7], S V Koldashov[7], A Leonov[7], V V Mikhailov[7], S A Voronov[7], Y T Yurkin[7], V G Zverev[7], G Barbarino[8], G De Rosa[8], G Osteria[8], D Campana[8], M Casolino[9], M P De Pascale[9], V Malvezzi[9], L Marcelli[9], M Minori[9], P Picozza[9], R Sparvoli[9], E A Bogomolov[10], S Y Krutkov[10], G Vasilyev[10], W Menn[11], M Simon[11], P Carlson[12], P Hofverberg[12], S Orsi[12], M Pearce[12]

[1]INFN – Sezione di Trieste, Padriciano 99, I-34012 Trieste (Italy)

[2]INFN – Sezione di Bari and Physics Department of University of Bari, Via Amendola 173, I-70126 Bari (Italy)

[3]INFN – Sezione di Firenze and Physics Department of University of Florence, Via Sansone 1, I-50019 Sesto Fiorentino, Florence, Italy

[4]IFAC, Via Madonna del Piano 10, I-50019 Sesto Fiorentino, Florence, Italy

[5]INFN, Laboratori Nazionali di Frascati, Via Enrico Fermi 40, I-00044 Frascati, Italy

[6]Lebedev Physical Institute, Leninsky Prospekt 53, RU-119991 Moscow, Russia

[7]Moscow Engineering and Physics Institute, Kashirskoe Shosse 31, RU-11540 Moscow, Russia

[8]INFN – Sezione di Napoli and Physics Department of University of Naples "Federico II", Via Cintia, I-80126 Naples, Italy

[9]INFN – Sezione di Roma 2 "Tor Vergata" and Physics Department of University of Rome "Tor Vergata", Via della Ricerca Scientifica 1, I-00133 Rome, Italy

[10]Ioffe Physical Technical Institute, Polytekhnicheskaya 26, RU-194021 St. Petersburg, Russia

[11]Physics Department of Universität Siegen, D-57068 Siegen, Germany

[12] KTH, Department of Physics, Albanova University Centre, SE-10691 Stockholm, Sweden

Journal of Physics: Conference Series **160** (2009) 012039 doi:10.1088/1742-6596/160/1/012039

Correspoding author email: bonvicini@ts.infn.it

Abstract. The Payload for Antimatter-Matter Exploration and Light Nuclei Astrophysics (PAMELA), primarily designed to directly measure antiparticles (antiprotons and positrons) in the cosmic radiation, was launched successfully on June 15th, 2006, and, since then, it is in continuous data taking. The calorimeter of the PAMELA apparatus has been designed to identify antiprotons from an electron background and positrons from a background of protons with high efficiency and rejection power. It is a sampling silicon-tungsten imaging calorimeter, which comprises 44 single-sided silicon sensor planes (380 μm thick) interleaved with 22 plates of tungsten absorber (0.74 X_0 each). It is the first silicon-tungsten calorimeter to be launched in space. In this work we present the in-orbit performance of the calorimeter, including the measured identification capabilities. The calorimeter provides a proton rejection factor of $\sim 10^5$ while keeping a high efficiency in selecting electrons and positrons, thus fulfilling the identification power needed to reach the primary scientific objectives of PAMELA. We show also that, after almost two years of operation in space, the calorimeter is still performing nominally.

1. Introduction

PAMELA (a Payload for Antimatter Matter Exploration and Light-nuclei Astrophysics) is a satellite-borne experiment designed to study charged particles in the cosmic radiation, with a particular emphasis on antiparticles (antiprotons and positrons). PAMELA is installed inside a pressurized container mounted onboard the Russian Resurs DK-1 earth-observation satellite that was launched into space by a Soyuz-U rocket on June 15th, 2006 from the Baikonur cosmodrome (Kazakhstan) [1]. The satellite orbit is elliptical and quasi-polar, with an inclination of 70.4° and an altitude varying between 350 and 610 km. The minimum estimated lifetime of the mission is 3 years.

The primary scientific goal of the mission is the study of the antimatter component of the cosmic radiation, in order to:

- Search for evidence of dark matter particle annihilations;
- Search for antinuclei (in particular, antihelium);
- Test cosmic-ray propagation models through precise measurements of the antiparticles energy spectra.

Other important scientific goals include: the study of solar physics and solar modulation during the 24th solar minimum, and the study of the cosmic-ray electron energy spectrum up to several TeV, thereby allowing possible contributions from local sources to be identified [1].

Table 1 shows the design goals for PAMELA performance, presenting the cosmic-ray components and energy ranges over which PAMELA will give new results.

Table 1. Design goals for PAMELA performance

Cosmic-ray particle	Energy range
Antiprotons	80 MeV – 190 GeV
Positrons	50 MeV – 270 GeV
Electrons	50 MeV – 400 GeV
Protons	80 MeV – 700 GeV
Electrons + positrons	Up to 2 TeV
Light nuclei (up to Z=6)	100 MeV/n – 200 GeV/n
Antinuclei	Sensitivity 95% CL
Antihelium/Helium ratio	$O(10^{-7})$

2. The PAMELA apparatus

A schematic overview of the PAMELA apparatus is shown in figure 1. The apparatus is ~ 1.3 m high, has a total mass of 470 kg and an average power consumption of 355 W.

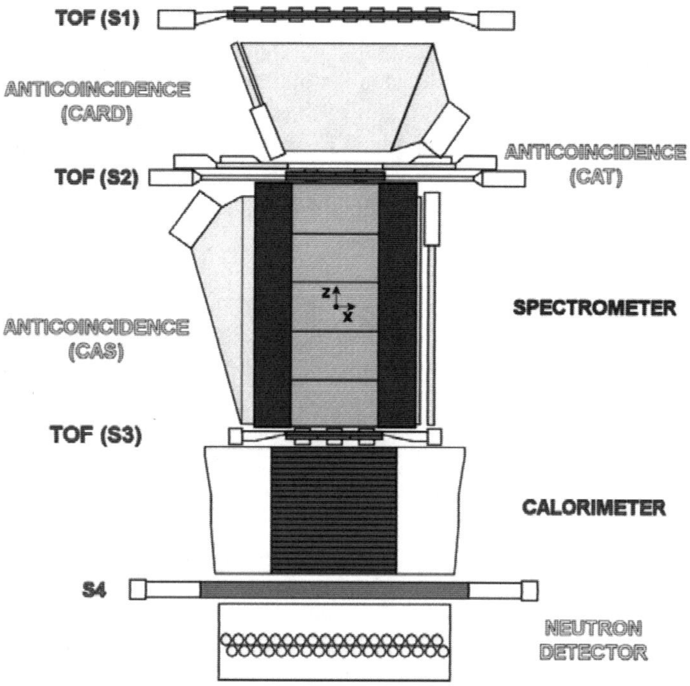

Figure 1. Schematic overview of the PAMELA instrument

The central part of the apparatus is a magnetic spectrometer, consisting of a permanent magnet and a silicon tracker [2]. The magnet provides a uniform magnetic field of 0.43 T along the *y* direction. The dimensions of the cavity of the magnet define the geometrical factor of the experiment to be 21.5 cm^2 sr. The tracking system consists of six planes of double-sided silicon microstrip detectors for precise reconstruction of the particle trajectory. The spectrometer measures the magnetic deflection of charged particles and, consequently, their momentum. The spatial resolution of the tracker in the bending view is observed to be, during flight, ~ 3 μm, which corresponds to a maximum detectable rigidity (MDR) of about 800 GV.

The electromagnetic calorimeter, to which the next sections of this paper are devoted, is installed below the magnetic spectrometer.

The Time-of-Flight (ToF) system of the experiment is based on three double layers of fast plastic scintillators (S1, S2 and S3), with alternate layers placed orthogonally to each other [3]. The ToF system provides a fast signal for triggering the data acquisition in the PAMELA subdetectors, and measures the time of flight and ionization energy losses of traversing particles.

The anticoincidence (AC) system consists of four plastic scintillators (CAS) surrounding the sides of the magnet, one (CAT) covering the top and four (CARD) enclosing the volume between the first two ToF planes [4]. The aim of the AC system is to identify false triggers and multiparticle events, generated by secondary particles produced in the apparatus.

Below the electromagnetic calorimeter, a single square plastic scintillator (S4) acts as a shower-tail catcher and is used to generate a high energy trigger signal for the underlying neutron detector (ND). The purpose of the neutron detector is to complement the electron-proton discrimination capabilities of the calorimeter by detecting the increased neutron production associated with hadronic showers in the calorimeter compared with electromagnetic ones. It consists of ^{3}He proportional counters surrounded by a polyethylene moderator enveloped by a thin cadmium layer [5].

3. The electromagnetic imaging calorimeter

3.1. Physics tasks of the PAMELA calorimeter

The main task of the calorimeter is to select positrons and antiprotons from the large background constitued by protons and electrons, respectively. Positrons have to be identified from a background of protons that is about 10^3 times the positrons component at 1 GeV/c, increasing to ~ 5 x 10^3 at 10 GeV/c. Antiprotons have to be selected from a background of electrons that decreases from ~ 5 x 10^3 times the antiproton component at 1 GeV/c to less than 10^2 times above 10 GeV/c. This means that PAMELA must be able to separate electrons from hadrons at a level better than 10^5. Much of this rejection power in PAMELA is provided by the calorimeter.

Besides the electron-hadron separation, the calorimeter must also directly measure the energy of electrons and positrons.

3.2. Calorimeter structure

The sampling imaging calorimeter comprises 44 single-sided silicon strip detector planes interleaved with 22 plates of tungsten absorber [6]. Each tungsten layer is 0.74 X_0 thick (2.6 mm) and it is sandwiched between two printed circuit boards, which house the silicon detectors as well as the front-end and digitizing electronics. Each silicon plane consists of 3x3, 380 μm thick, 8x8 cm^2 detectors, segmented into 32 strips with a pitch of 2.4 mm. The orientation of the strips for two consecutive silicon planes is shifted by 90°, thus providing 2-dimensional spatial information. Each of the 32 strips is wire-bonded to the corresponding strip on the other two detectors in the same row (or column), thereby forming 24 cm long readout strips. The total depth of the calorimeter is 16.3 X_0 (~ 0.6 λ_I). The high granularity of the calorimeter and the use of silicon strip detectors provide detailed information on the longitudinal and lateral profiles of particles' interactions as well as a measure of the deposited energy.

Figure 2 shows the calorimeter prior to integration with the other PAMELA detectors. The instrument is ~ 21 cm tall and the sesitive area of the silicon planes is 24 x 24 cm^2.

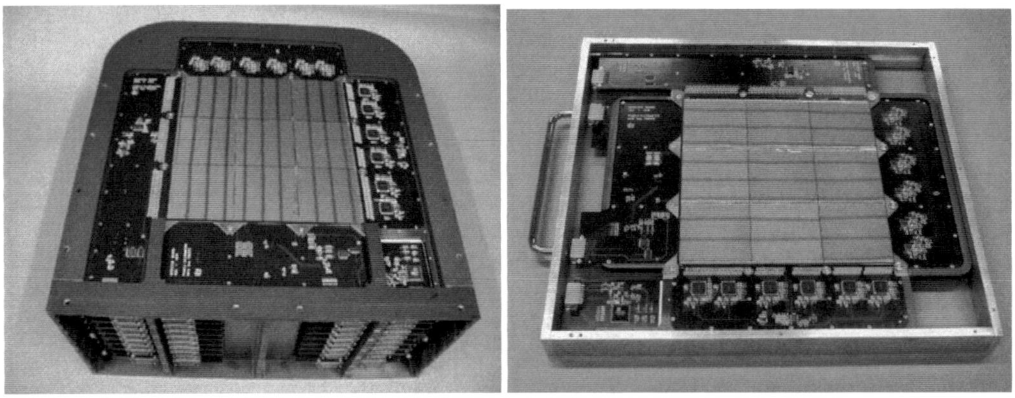

Figure 2. Left: The PAMELA Si-W imaging calorimeter with the topmost silicon plane visible. Right: Detail of a single calorimeter module (formed by 2 tungsten plates and 4 silicon planes) before being inserted into the calorimeter main frame.

3.3. Front-end and data acquisition electronics

The calorimeter front-end electronics is based on the CR1.4P full custom ASIC [7], which provides 16 channels featuring a charge-sensitive preamplifier, a CR-RC shaping amplifier, a track-and-hold circuit and an output multiplexer. An input-selectable calibration circuit and a self-trigger circuit are also implemented in the chip (see subsection 4.3).

A key feature of the CR1.4P is its wide dynamic range (~ 1400 Minimum Ionizing Particles, or about 7.1 pC for 380 μm thick silicon detectors). The large dynamic range is of crucial importance to correctly measure the interactions of high energy particles inside the calorimeter. Another important characteristics of the CR1.4P is the low noise slope as a function of the detector input capacitance, which allows the circuit to provide a good signal-to-noise ratio for 1-MIP signals (~ 9:1) while coping with the rather large (~ 180 pF) capacitance presented at its input by the 24 cm long calorimeter strips. Table 2 summarizes the main characteristics of the CR1.4P.

Six CR1.4P chips are used to read-out each of the 44 silicon planes (264 chips are used for the whole calorimeter). In each plane, the analog outputs of the front-end chips are multiplexed into a single 16-bit ADC (Analog Devices AD977A).

For redundancy purposes, the read-out is divided into 4 independent sections (exploiting the segmentation of the calorimeter), corresponding to x-even, y-even, x-odd and y-odd planes. Data from all 44 ADCs are processed by 4 read-out boards mounted on the front cover of the calorimeter. Four FPGA (Altera A54SX72), one per section, control all data acquisition and slow control processes and, in normal acquisition mode ("compress mode"), transmit the data to 4 DSPs (Analog Devices ADSP2187), which perform the tasks of first-level data analysis and compression and the calculations associated with the instrument on-line calibration (pedestal, baseline, noise rms, etc.). The calibration is performed once per orbit, at the equator crossing from the southern emisphere. After each calibration a special run is performed, in which both raw (non-compressed) and compressed data are transmitted ("full mode"). In this way it is possible to check the correct functioning of the compression algortihm. The FPGAs also monitor and check the status of the DSPs and, in case of failure of one of them, can bypass the malfunctioning DSP and transmit directly the raw data of that section to the main PAMELA DAQ ("raw mode").

Table 2. Main characteristics of the CR1.4P front-end circuit

Number of channels	16
Supply	±5 V
Total power consumption	97 mW
Max. output swing	7 V
Gain (sensitivity)	5 mV/MIP
Linear dynamic range	1400 MIP (7.1 pC)
Noise@C_{det} = 0 pF	2700 e^- rms
Noise slope	4.7 e^- / pF
Non-linearity	< 1% full range

4. In-flight performance of the calorimeter

After its launch from Baikonur on June 15th 2006, PAMELA was first switched-on on June 20th and, after a brief commissioning phase, the instrument has been in continous data taking mode since July 11th. At the time this paper is being written, more than 10 Tbyte of raw data have been downlinked to Earth and ~ 10^9 triggers have been recorded by PAMELA.

4.1. Basic performance

After two years of operation in space, no failures or loss of functionality have been observed in the calorimeter. All basic parameters (temperatures, pedestals, noise, gain) are stable and nominal.

For example, figure 3 displays the pedestal of one channel measured from May 2006 (i.e. before launch) to June 2007. This stable behaviour is common to all calorimeter channels.

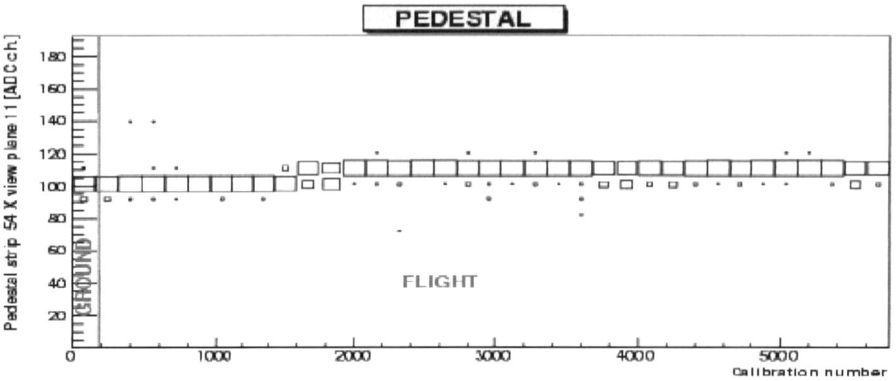

Figure 3. Pedestal of one strip of the calorimeter from May 2006 (prior to PAMELA launch) to July 2007 (one year after the launch) in ADC channels.

Figure 4 (left) shows the rms noise of another calorimeter channel measured in a time period spanning from July 2006 to January 2008, while figure 4 (right) shows, for the same channel and the same period, the position of the peak of the pulse height distribution. The stability in time of these parameters, hence of the signal-to-noise ratio, is clearly seen.

4.2. Physics performance

As mentioned in section 3, the imaging calorimeter of PAMELA has been designed to precisely reconstruct the longitudinal and lateral profiles of interactions and to measure the deposited energy. The longitudinal and lateral segmentation of the calorimeter, combined with the measurement of the particle energy loss in each strip, allows a very high identification (or rejection) power for electromagnetic showers. For example, figure 5 shows a 84 GeV/c antiproton interacting in the calorimeter and giving rise to a typical hadronic shower pattern, while figure 6 shows a 92 GeV/c positron giving an electromagnetic shower in the calorimeter.

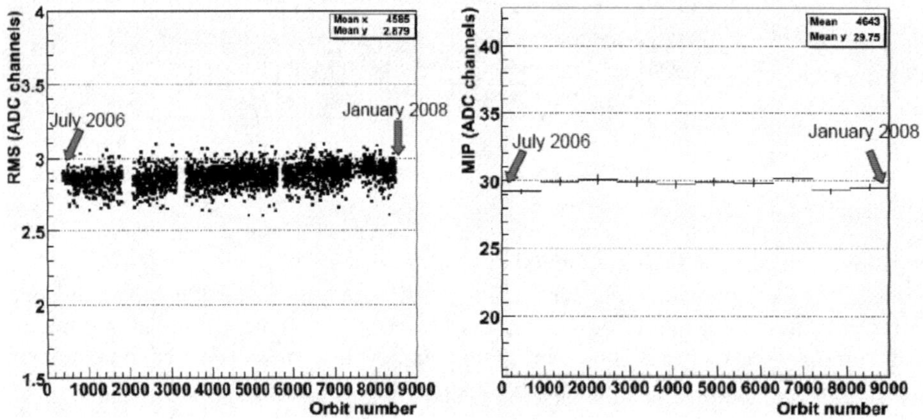

Figure 4. Left: rms noise (in ADC channels) of one calorimeter channel from the start of the mission to January 2008. Right: Peak of the pulse height distribution (in ADC channels) for the same channel and over the same time interval.

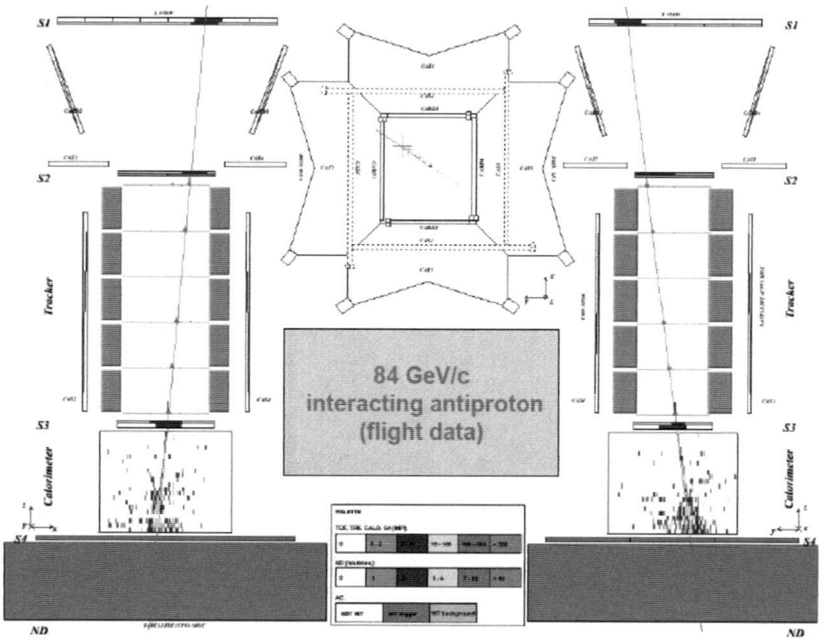

Figure 5. Event display of an interacting 84 GeV/c antiproton from flight data.

The e/h discrimination power of the calorimeter is obviously fundamental for the physics analysis, especially for the extraction of the positron signal from the proton backgound. An example of the calorimeter capabilities in the positron selection is illustrated in figures 7 and 8. Figure 7 shows a sample of flight data (positive and negative, with rigidity between 20 and 30 GV, as selected by the magnetic spectrometer) in which we have plotted the distribution of the ratio between the charge released along the track in the calorimeter (obtained by summing the signal in the hit strip and the signals in the left and right neighboring strips) and the total charge measured in the calorimeter. In these plots, the non-interacting antiprotons and protons are clearly seen, peaked at one. In the negative plot, the distribution due to the electrons is clearly seen (due to the paucity of the antiprotons), while in the positive plot the positron sample (expected to be peaked at an energy fraction corresponding to that of the electrons) is completely overwhelmed by the background of interacting protons. In figure 8 we have imposed several conditions: the request that the energy measured by the calorimeter matches the momentum measured by the magnetic spectrometer ("energy-momentum match") and various topological cuts related to the starting point of the shower and its longitudinal profile: the positron sample is now clearly visible in the data. It should be noted that the cuts mentioned above are just an example for the purpose of showing the calorimeter capabilities in e/h separation. In the positron data analysis, additional topological and energy-related conditions are used.

5. Conclusions

The satellite-borne experiment PAMELA has been studying cosmic rays for two years. Throughout this period, the Si-W imaging calorimeter of the apparatus has been performing nominally. In particular, all functional parameters (temperature, pedestals, noise, signal, gain) are nominal and stable. The physics performance are consistent with the expectations, with the simulations and test beam results. During two years of operation in space, no failures nor loss/degradation of performance have been observed and the calorimeter proved to be able to fulfill the requirements needed to achieve the scientific goals of PAMELA.

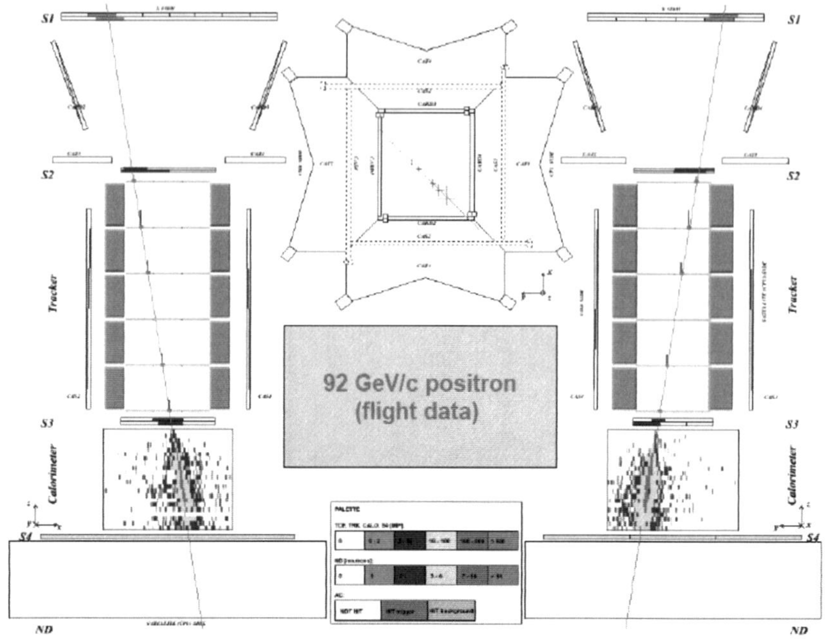

Figure 6. Event display of a 92 GeV/c positron from flight data. Compare the shower topology in the calorimeter with the hadronic shower in figure 5.

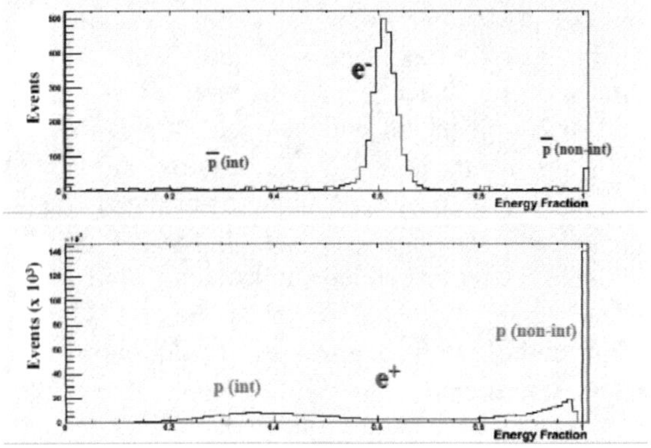

Figure 7. Plot of the energy fraction released along the track in the calorimeter for both negative (top) and positive (bottom) charge in the rigidity interval 20 ÷ 30 GV.

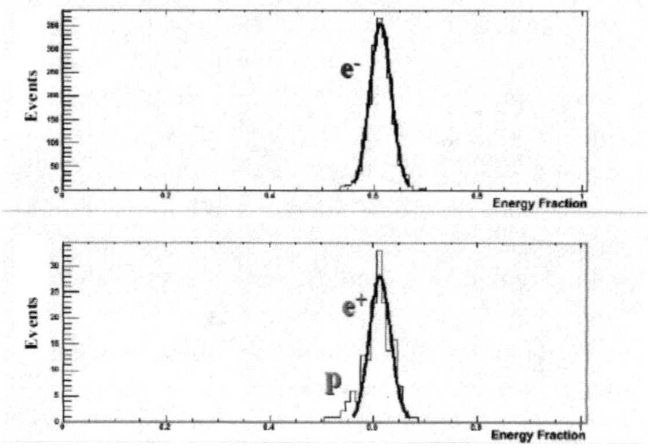

Figure 8. The distributions of figure 7 after applying an energy-momentum match condition and topological cuts related to the shower profile.

References

[1] P. Picozza et al., Astroparticle Physics 27 (2007) 296.
[2] O. Adriani et al., Nucl. Instr. and Meth. A 511 (2003) 72.
[3] G. Osteria et al., Nucl. Instr. and Meth. A 535 (2004) 152.
[4] S. Orsi et al., Adv. Space Res. 37 (2006) 1853.
[5] Y. Stozhkov et al., in: Proc. 19th European Cosmic Ray Symposium, Florence, 2004
Y. Stozhkov et al., Int. J. Modern Phys. A 20 (2004) 6745.
[6] M. Boezio et al., Nucl. Instr. and Meth. A 487 (2002) 407.
[7] V. Bonvicini, "The CR1.4P: an integrated front-end circuit for the imaging calorimeter of PAMELA", in: 5th Intern. Workshop on Front-end Electronics FEE 2003, Snowmass, 2003.

The Mini-Calorimeter on-board AGILE: the first year in space

M Marisaldi[1], C Labanti[1], F Fuschino[1], M Galli[2], A Argan[3], A Bulgarelli[1], G Di Cocco[1], F Gianotti[1], M Tavani[3], M Trifoglio[1] and A Trois[3]

[1]INAF-IASF Bologna, Via Gobetti 101, I-40129 Bologna, Italy
[2]ENEA Bologna, Via Martiri di Monte Sole 4, I-40129 Bologna, Italy
[3]INAF-IASF Roma, Via del Fosso del Cavaliere 100, I-00133 Roma, Italy

E-mail: marisaldi@iasfbo.inaf.it

Abstract. AGILE, the Italian space mission dedicated to gamma-ray and hard-X astrophysics, was successfully launched on 23rd April 2007 and is currently fully operative. The Mini-Calorimeter (MCAL) on-board the AGILE satellite is a scintillation detector made of 20 kg of segmented CsI(Tl) scintillator with photodiode readout with a total geometrical area of 1400 cm^2. MCAL can work both as a slave of the AGILE Silicon tracker and as an independent detector for gamma-ray bursts (GRB) detection in the 300 keV - 100 MeV energy range. Despite its limited thickness, due to weight constraints, MCAL has proven to successfully self-trigger GRBs at MeV energies providing photon-by-photon data with less than 2 μs time resolution and almost all-sky detection capabilities. The instrument design and characteristics, as well as the in-flight performance after one year of operation in space and the scientific results obtained so far are reviewed and discussed.

1. Introduction

AGILE[1, 2] is a small space mission of the Italian Space Agency (ASI) devoted to astrophysics in the gamma-ray energy range 30 MeV - 50 GeV, with a monitor in the X-ray band 18 keV - 60 keV. The AGILE Payload is composed of:

- a Tungsten-Silicon Tracker (ST) [3, 4], with a large field of view, good time resolution, sensitivity and angular resolution;
- a Silicon based X-ray detector, SuperAGILE (SA) [5], for imaging in the range 18 keV - 60 keV;
- a CsI(Tl) Mini-Calorimeter (MCAL) [6, 7] that detects gamma-rays or particle energy deposits between 300 keV and 100 MeV. ST and MCAL form the so called Gamma-Ray Imaging Detector (GRID) for observations in the energy range 30 MeV - 50 GeV;
- an anti-coincidence (AC) system [8], made with plastic scintillator layers, for the rejection of charged particles background;
- the Payload Data Handling Unit (PDHU) [9].

AGILE was successfully launched on April 23 2007 from Satish Dhawan Space Centre (India) on a PSLV-C8 rocket. The AGILE orbit is quasi-equatorial, with a 2.5 degrees inclination and an average altitude of 535 km.

Figure 1. Integrated MCAL during test benches, top view. The upper layer of the detection plane and two preamplifiers boards are clearly visible.

2. The Mini-Calorimeter instrument

2.1. Design

MCAL has been designed with stringent constraints, mainly concerning detector geometry, mass and power budget. The detector is designed to provide also information on particles produced in the tracker after gamma-ray interactions, so it must be located below the ST and have the same footprint. The weight constraints put limits on the detector thickness, and the power budget allocated constrained the overall architecture and front-end electronics.

MCAL is composed of 30 CsI(Tl) scintillator bars each one 15x23x375 mm^3 in size, arranged in two orthogonal layers, for a total thickness of 1.5 radiation lengths. In a bar the readout of the scintillation light is accomplished by two custom PIN Photodiodes (PD) coupled one at each small side of the bar. A permanent optical coupling between PDs and CsI is made by means of a clear siliconic resin; that medium has been chosen to realize an elastic bond between two components that exhibit a quite different thermal expansion coefficient. To maximize the light output and to keep the light attenuation inside the bar within an optimal range of values, the bars surfaces have been polished and the bars have been first wrapped with a reflective coating [10] and then by a thin adhesive layer. Each bar is then arranged inside a carbon fiber structure, about 1 mm thick, that provides rigidity and modularity to the detectors. For each bar the PDs signals are collected by means of low noise charge preamplifiers, and then conditioned in the Front End Electronics, (FEE). For each bar, the energy and the position of interaction of either a gamma-ray or ionizing particle can be evaluated combining the signals of the two PDs. The detection plane is hosted in the upper part of MCAL main frame; the preamplifiers are arranged in four boxes on each side of the detection plane and at its same level. Below the detection plane is placed the FEE board that has the same area of the whole detection plane. The overall mechanical envelope of MCAL constitutes the lower part of the whole AGILE payload. Figure 1 shows the integrated MCAL during test benches, prior to integration in the AGILE payload. The upper layer of the detection plane and two preamplifiers boards are clearly visible. Table 1 reports the main MCAL engineering characteristics and scientific performance.

2.2. Operative modes

MCAL works in two operative modes:

- in GRID mode a trigger issued by the ST starts the collection of all the detector signals

Table 1. MCAL engineering characteristics and performance

Detector property	Measured value
Active detector weight	20 kg
On-axis geometrical area	1400 cm^2
Number of independent detectors	30
Single detector dimensions	$15 \times 23 \times 375$ mm^3
Detector light output	21 ± 1 e^-/keV
Detector attenuation coefficient	0.028 ± 0.002 cm^{-1}
Total power consumption	< 6 W
Electronic noise / channel	< 1000 e^- RMS
Energy resolution	14% FWHM at 1.275 MeV
Position resolution	18 mm at 1.275 MeV
	7 mm at 11 MeV
Timing accuracy	2 μs
BURST mode energy range	0.33 - 110 MeV (near PDs) for each bar
GRID mode energy range	1 - 100 MeV (near PDs) for each bar
Effective area	$\sim$200 cm^2 at 0.4 MeV on axis
	$\sim$300 cm^2 at 1 MeV on axis
Field of view	$\sim 4\pi$ sr, non imaging

in order to determine the energy and position of particles converted in the tracker and interacting in MCAL;

- in BURST mode each bar behaves as an independent self triggering detector and generates a continuous stream of information of gamma events in the energy range 300 keV - 100 MeV. In the data handling system these data are used to detect impulsive variations in count rate. If a valid burst trigger is issued by the trigger logic, described in subsection 2.3, data are collected and stored in telemetry on a photon-by-photon basis. This means that for every photon the pulse height of both photodiodes of every triggered bar and an absolute time tag with 2 μs accuracy are available for analysis, so that energy and time binning are limited by counting statistics only.

Both operative modes can be active at the same time. Due to telemetry limitations BURST data are not sent on ground on a photon-by-photon basis unless a trigger for a transient is issued by a dedicated trigger logic. However BURST data are used to build a broad band energy spectrum (Scientific RateMeter, SRM) recorded and stored in telemetry every second. Scientific ratemeters provide information on the high energy gamma-ray background in space and its modulation through orbital phases. SRMs are organized in 11 bands for each of the two MCAL detection layers. An event's assignment to one of the energy bands is based on the most significant bit of the energy binary word. This assignment policy is straightforward from the computational point of view and also defines the energy bands. The first useful band includes events between 0.18 and 0.35 MeV, the second band between 0.35 and 0.7 MeV, the third is between 0.7 and 1.4 MeV and so on until the last band which includes events with energy higher than 180 MeV.

2.3. Trigger logic

In the PDHU a dedicated trigger logic[11] is implemented to detect gamma-ray bursts (GRBs) and other impulsive events. The trigger algorithm is based on the assumption that bursts are phenomena producing count rates above a threshold determined by the current background value. Since the burst signal is strongly energy and timescale dependant, the crucial first

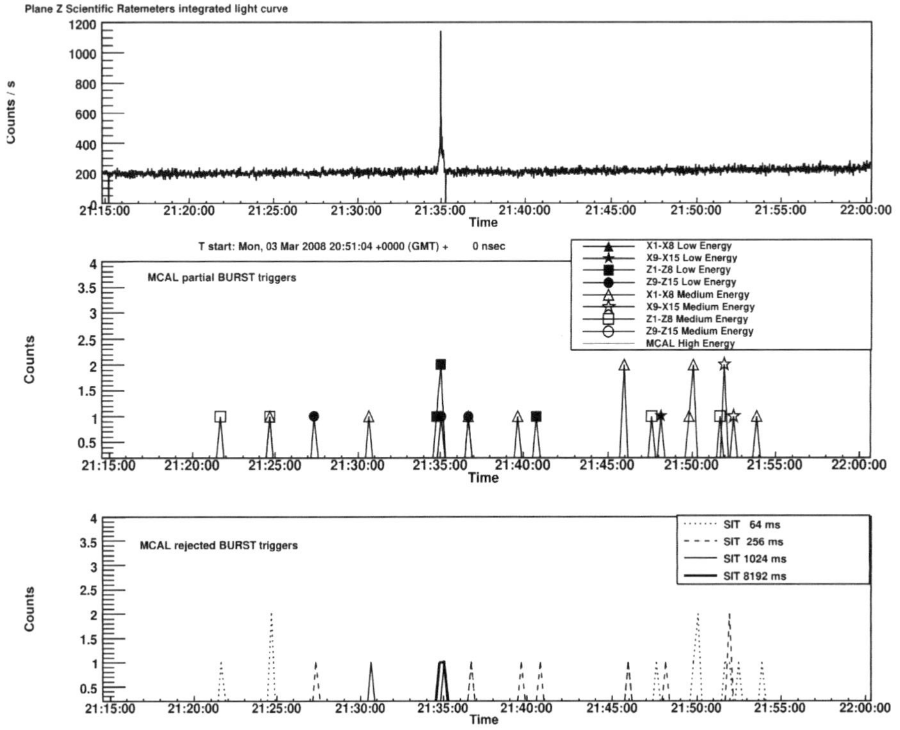

Figure 2. Trigger logic monitoring sequence. Data refer to the GRB triggered on 3 March 2008 at 21:34:37 UT. Top panel: plane Z integrated scientific ratemeters. Central panel: partial triggers ratemeter (from housekeeping data). Bottom panel: rejected trigger ratemeter (from housekeeping data).

task of the Burst Search (BS) software algorithm is the formation of several background ratemeters (RMs). These are evaluated integrating events on different time windows called Search Integration Times (SITs). SITs can be handled by Hardware (Sub-millisecond, 1 and 16 ms SITs) or by Software (64 , 256, 1024 and 8192 ms SITs). The RMs are generated depending on the detector, the events energy and position of interaction. For each of the 7 SITs, 9 RM are evaluated covering three ranges of energy: from 0.3 to 0.7 MeV (Low Energy, LE), from 0.7 to 1.4 MeV (Medium Energy, ME), and above 1.4 MeV (HE). The limits of the ranges are fully configurable. Events in the first two energy ranges contribute to generate different RMs depending on the place of interaction on MCAL; in this case MCAL is divided into four zones (two zones for each detection layer). The HE events contribute to a single RM, regardless the position of interaction. A partial trigger can be generated independently for every RM and for every SIT. After one or more partial triggers are issued, a validation step takes place by means of a programmable Look-Up Table (LUT) aimed at spurious trigger rejection. With current configuration a valid trigger is issued when at least two coincident partial triggers on different ratemeters are issued.

Figure 2 shows the partial and rejected counters for a real GRB triggered on 3rd March 2008. The trigger has been issued after the generation of two coincident partial triggers on the 8.192 s SIT, on the low-energy ratemeters of the bottom detection layer spatial zones.

3. In-flight performance

The scientific payload was switched on at the beginning of May 2007 and the commissioning phase lasted up to the end of June. During this phase the MCAL front-end configuration was tuned, the main task regarding the threshold setting. The average energy threshold was brought successfully from 650 keV, the on-board preset value for safety reasons, to about 330 keV, the expected final value. The threshold for each bar has been set independently in order to minimize the overall energy threshold but avoid triggers on electronic noise. A front-end configuration equivalent to the best one employed on ground was found to be stable and reliable. From early July to October 2007 the Science Verification Phase (SVP) took place, with activities mainly dedicated to in-flight calibration of the GRID and SuperAGILE instruments. During this period, for programmatic reasons, the MCAL on-board trigger logic was not activated. The trigger logic activation and tuning was carried out in November 2007, for time windows greater or equal than 64 ms, and finally switched on with the nominal configuration since the beginning of February 2008. During June 2008 another session dedicated to the MCAL trigger logic tuning has been successfully carried out, aimed at the configuration of the burst search algorithm on the 16 ms, 1 ms and sub-millisecond time windows.

The payload functional monitoring is carried out by means of housekeeping (HK) data. Several hundreds HK are collected and stored in a dedicated telemetry packet every 16 seconds. HK data include power supply voltages, input currents, reference voltages, basic ratemeters, configuration digital registers, detector and electronics temperature, and allow a complete monitoring of the health status of the payload and spacecraft's systems. The average detector temperature ranges between 4 and 8°C, with a typical variation of about 0.3 °C on the time scale of one orbit.

The in-flight background is monitored by means of the scientific ratemeters data, SRM, described in section 2.2. The background level above 330 keV has an average value of about 200 counts/s for each of the two detection layers, with a variation of about 20% along the orbit excluding the SAA passage. During the SAA passage SRM are acquired as well, but the effective background level is masked by the high increase in dead time due to the high count rate in the anti-coincidence panels that causes a significant depletion of SRM counts for the deepest passages.

4. Gamma-ray bursts detection with MCAL

4.1. Detection capabilities

AGILE has the study of GRBs among its main scientific targets. SuperAGILE, as well as MCAL, is equipped with an on-board trigger logic. Moreover, a localization algorithm provides GRB positions with few arcmin accuracy for events in the SuperAGILE field of view, allowing rapid dissemination of the coordinates ([12]). The GRID has a wide field of view that makes it a valuable instrument for GRB detection in the poorly explored 30 MeV-50 GeV energy band. A simultaneous GRB detection with GRID, MCAL and SuperAGILE would allow a spectral coverage over six orders of magnitude.

With respect to GRB detection, MCAL strength points are mainly the good effective area in the MeV range, the microsecond timing accuracy for triggered events and the flexible trigger logic. The weak point is the quite high energy threshold (330 keV). MCAL is so tailored to the detection of medium-bright GRBs with peak energy above few hundreds keV.

Several GRB detectors are currently active in space, each with its own specific characteristics. Apart from Swift-BAT ([13]), INTEGRAL-IBIS ([14]), SuperAGILE ([5]) and GLAST-GBM ([15]), all the other detectors have no or very limited imaging capabilities and rely on triangulation between different spacecrafts for GRB localization, by means of the 3rd Inter-

Planetary Network (IPN)[1]. Among the current IPN instruments, only three have spectroscopic capabilities at MeV energies, in an energy range partially overlapping with that of MCAL: Konus-Wind ([16]), Suzaku-WAM ([17]) and RHESSI ([18]). Among these, only the RHESSI spectrometer is capable of photon-by-photon data download. Also GLAST-GBM, supposed to join the IPN too, have both spectral capabilities in the MeV range and photon-by-photon data download for triggered events.

Despite MCAL has no (or very limited) directional capabilities it can act as a sentinel for GRB search in the GRID energy range, since a GRB detectable above 50 MeV is expected to be bright in the MCAL range. So, the fluence observed in the MCAL range provides a preliminary ranking of the GRB and a hint of the likelihood for a GRB to be detected in the GRID. On 14th May 2008 MCAL and SuperAGILE triggered GRB 080514B[19], which was also detected by the GRID above 50 MeV[20]. This event is the first GRB detected above 50 MeV after the five events detected during the nineties by the EGRET instrument on-board the Compton Gamma-Ray Observatory (CGRO), and is the subject of a dedicated paper in preparation.

4.2. GRB detections

Between 22nd June 2007 and 30th June 2008 MCAL detected 51 GRBs, with an average detection rate of about 1 GRB/week. Most of these detections have been independently confirmed by other instruments. Only 16 events have been localized, either by Swift, SuperAGILE or the IPN. Since most of the MCAL events have been also detected by other IPN instruments ([21]) the number of IPN localizations is expected to grow when the IPN catalogues will be available. The detection rate is in good agreement with the sensitivity estimations reported in [22].

Several GRBs have been detected at off-axis angles greater than 90°, with the record being the bright GRB071020, localized by Swift ([23]), detected at 166° off-axis, i.e. coming almost from a direction opposite to the AGILE pointing. Despite for $> 90°$ events it is difficult to provide reliable spectral information, due to the still incomplete modeling of the spacecraft shell with Monte Carlo simulations ([24, 25]), it demonstrates the MCAL all-sky detection capabilities.

Figure 3 shows the light curves for a GRB triggered on 3rd March 2008 at 21:34:37 UT. Both the light curve from SRM data for one of the detection layers, and the light curves in different energy bands obtained from photon-by-photon data are shown. The plot clearly points out the difference in the performance that can be achieved with SRM and photon-by-photon data, mainly concerning energy and time resolution. Before the permanent activation of the trigger logic in February 2008, 28 GRBs have been detected by scanning on-ground SRM data by means of a dedicated software task, but the scientific exploitation of these events is limited by the coarse time and energy binning. A detailed description of the first year of GRB detections with MCAL is reported in [26].

5. Conclusions

The MCAL design was limited by stringent constraints concerning geometry, weight and power consumption, but a clever architecture and a flexible trigger logic for transient events were implemented in order to exploit the instrument possibilities at best. MCAL main characteristics are its spectral capabilities in the MeV range and the photon-by-photon data download with microsecond timing accuracy for triggered events; very few instruments currently operating in space have these characteristics. The instrument in-flight performance are stable and nominal.

Between 22nd June 2007 and 30th June 2008 MCAL detected 51 GRBs, with an average detection rate of about 1 GRB/week. Moreover MCAL can work as a sentinel for the GRID, as actually happened for the detection of GRB 080514B. Since the end of June 2008 the trigger logic on the sub-millisecond, 1 ms and 16 ms time windows has been switched on and configured. The

[1] IPN web page: http://www.ssl.berkeley.edu/ipn3/

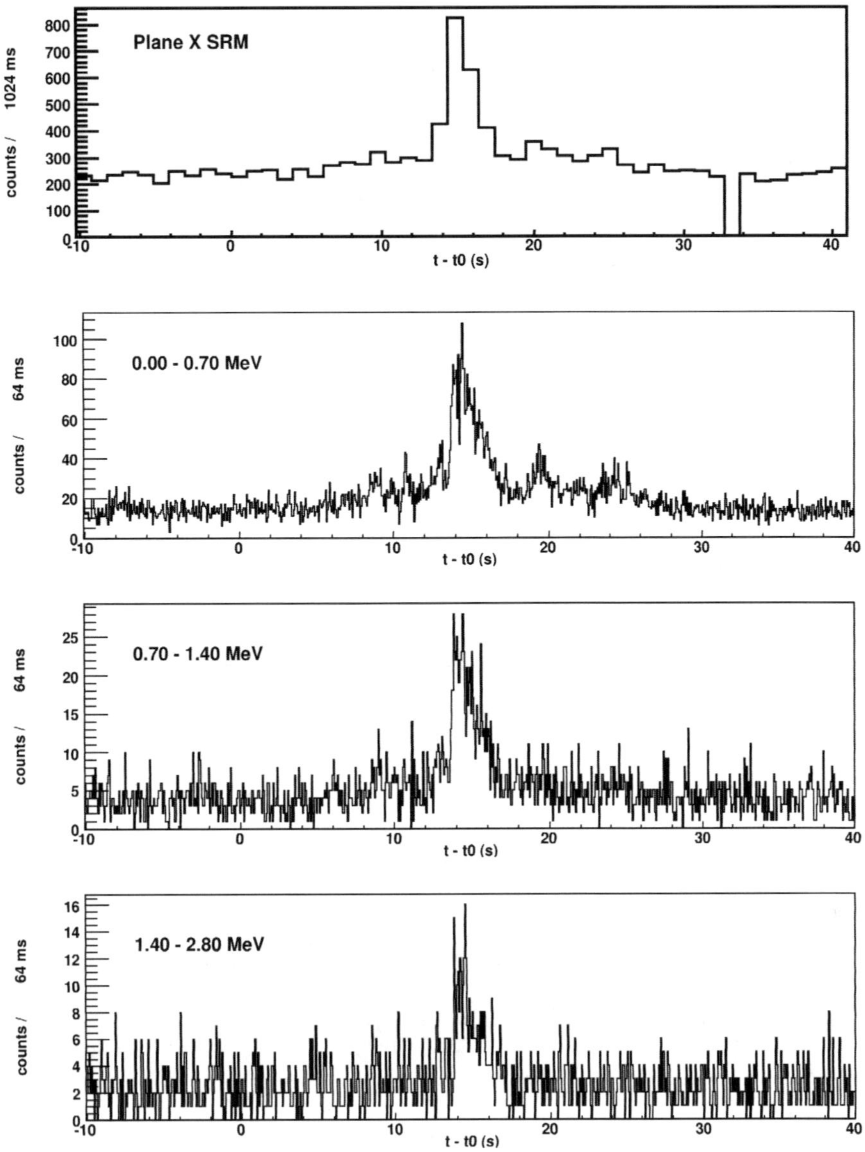

Figure 3. Light curves for the GRB triggered on 3rd March 2008 at 21:34:37 UT. Top panel: scientific ratemeters for the upper detection layer (plane X) with 1.024 s time bin. Other panels: light curves in three energy bands from photon-by-photon data with 64 ms time bin.

analysis of triggers on such short time-scales is in progress. Cross-calibration activity with the Konus-WIND instrument, which is mandatory to fully exploit the MCAL spectral capabilities, is in progress as well.

Acknowledgments

AGILE is a mission of the Italian Space Agency, with co-participation of INAF (Istituto Nazionale di Astrofisica) and INFN (Istituto Nazionale di Fisica Nucleare). The authors wish to thank all the AGILE team for their help and fruitful discussions. The authors wish to thank also the industrial partners, namely the AGILE people at Thales-Alenia Space Italia, Carlo Gavazzi Space and Telespazio, for their fundamental contribution to the construction, integration, testing and in-flight operation of MCAL. Kevin Hurley, Valentin Pal'shin and Kazutaka Yamaoka are greatly acknowledged for fruitful discussions and IPN support.

References

[1] Tavani M *et al.* 2008 *Nucl. Instr. and Meth. A* **588** 52–62
[2] Tavani M, Barbiellini G, Argan A *et al.* 2008 *submitted to Astronomy and Astrophysics*
[3] Barbiellini G *et al.* 2001 *GAMMA 2001: Gamma-Ray Astrophysics 2001, AIP Conference Proceedings Volume 587* pp 754–758
[4] Prest M *et al.* 2003 *Nucl. Instr. and Meth. A* **501** 280–287
[5] Feroci M, Costa E, Soffitta P *et al.* 2007 *Nucl. Instr. and Meth. A* **581** 728–754
[6] Labanti C *et al.* 2006 *proceedings of SPIE* **6266** 62663Q
[7] Labanti C, Marisaldi M, Fuschino F *et al.* 2008 *submitted to Nucl. Instr. and Meth. A*
[8] Perotti F *et al.* 2006 *Nucl. Inst. and Meth. A* **556** 228–236
[9] Argan A *et al.* 2004 *Conference Record of the IEEE Nuclear Science Symposium, October 16-22, 2004, Rome, Italy* pp 371–375
[10] Weber M, Stover C, Gilbert L, Nevitt T and Ouderkirk A 2000 *Science* **287** 2451–2456
[11] Fuschino F, Labanti C, Galli M, Marisaldi M, Bulgarelli A, Gianotti F, Trifoglio M, Argan A, Del Monte E, Donnarumma I, Feroci M, Lazzarotto F, Pacciani L, Tavani M and Trois A 2008 *Nucl. Instr. and Meth. A* **588** 17–21
[12] Del Monte E, Costa E, Di Persio G *et al.* 2007 *proc. of SciNeGHE 2007, Frascati Physics Series* **45** 201–208 arXiv: 0712.1548
[13] Barthelmy S D *et al.* 2000 *Proc. SPIE* **4140** 50
[14] Ubertini P, Lebrun F, Di Cocco G, Bazzano A *et al.* 2003 *Astronomy & Astrophysics* **411** L131–L139
[15] Meegan C, Bhat N, Connaughton V *et al.* 2007 *AIP Conf. Proc.* **921** 13
[16] Aptekar R L, Frederiks D D, Golenetskii *et al.* 1995 *Space Science Reviews* **71** 265–272
[17] Yamaoka K, Sugita S, Ohno M *et al.* 2006 *Proc. SPIE* **6266** 153
[18] Wigger C, Hajdas W, Smith D M *et al.* 2004 *Nucl. Phys. B Proc. Suppl.* **132** 331–334
[19] Rapisarda M *et al.* 2008 *GRB Coordinates Network* **7715**
[20] Giuliani A *et al.* 2008 *GRB Coordinates Network* **7716**
[21] Hurley K 2008 *private communication*
[22] Ghirlanda G, Galli M, Longo F *et al.* 2004 *AIP Conf. Proc.* **727** 704–707
[23] Holland S T, Barthelmy S D, Baumgartner W H *et al.* 2007 *GRB Coordinates Network* **6949**
[24] Longo F, Cocco V and Tavani M 2002 *Nucl. Instr. and Meth. A* **486** 610
[25] Cocco V, Longo F and Tavani M 2002 *Nucl. Instr. and Meth. A* **486** 623
[26] Marisaldi M *et al.* 2008 *submitted to Astronomy and Astrophysics*

The AMS-02 3D-imaging calorimeter: a tool for cosmic rays in space.

Corinne Goy LAPP, IN2P3/CNRS, Université de Savoie on behalf on the AMS calorimeter group.

Laboratoire d'Annecy-Le-Vieux de Physique des Particules, Chemin de Bellevue, F-74941 Annecy-Le-Vieux Cedex

E-mail: corinne.goy@lapp.in2p3.fr

Abstract. The AMS02 Electromagentic calorimeter (ECAL) is based on a lead/scintillating fiber sandwich. The flight model was tested and calibrated during Summer 2007 in a test beam at CERN, using 6 to 250 GeV electron and hadron beams. Preliminary results on the measurements of ECAL parameters and performances are reported in this paper.

1. Introduction

The AMS-02 experiment is a large acceptance spectrometer designed to operate on the International Space Station (ISS) for three years. The physics goals of the AMS experiment are:

- The measurement of the nature and fluxes of the cosmic rays up to the TeV range, including the identification of nuclei and light isotopes (^{3}He, B, C, ^{9}Be, ^{10}Be) and the galactic and extragalactic gamma rays spectra to a few 100 GeV.
- The search for an additional component originating from Dark Matter annihilations on top of the antiproton, positron or gamma spectra.
- The search for nuclear antimatter with an expected sensitivity of 10^{-9} for $\overline{\mathrm{He}}/\mathrm{He}$ and 10^{-7} for $\overline{\mathrm{C}}/\mathrm{C}$.

The experimental setup (Figure 1) consists of a 8 layers silicon tracker inserted in a superconducting magnet providing a 0.8 Tm bending power. A total of 4 layers of Time Of Flight scintillating system situated on the top and the bottom of the tracker provides the main trigger and gives the direction of the particles. A set of Veto-Counters surrounds the tracker to reject side tracks. A Transition Radiator Detector and the calorimeter allow for electron, positron and gamma identification, while the Nuclei species are identified by the RICH detector in conjunction with the Tracker. The total acceptance amounts to 0.5 m^2.sr.

2. The calorimeter

The aim of the electromagnetic calorimeter is to measure the energy of electromagnetic particles from a few GeV up to 1 TeV, to produce a stand-alone gamma trigger, to give a rejection against hadrons with respect to electrons and positrons of the order 10^{-4} combined with the tracker, and to identify gamma rays.

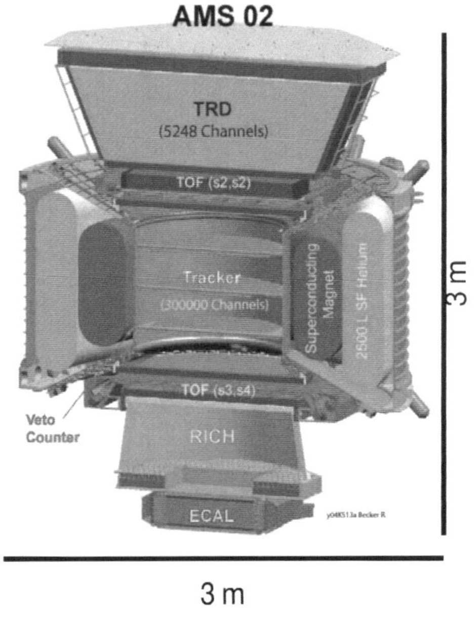

Figure 1. Schematic view of the AMS detector.

The calorimeter is an imaging calorimeter consisting of 9 modules of lead with embedded scintillating fibers corresponding in total to a thickness of 16.2 cm. Fibers are oriented along one or the other horizontal direction, alternatively for each module, as depicted in Figure 2, the incidence of the particles being vertical. The calorimeter has a sensitive surface of 648×648 mm^2.

Fibers are read at one end by 4-anodes photomultipliers (PMt :R7600-00-M4 Hamamatsu) giving an ultimate granularity of 72 anode-cells in the horizontal plane and 18 points longitudinally, each cell collecting the light on a surface of approximatively 9×9 mm^2. The four 30mm-long light guides, the PMt and the front-end electronics are inserted in magnetic shielding iron tubes which are maintained around the calorimeter by an aluminium structure.

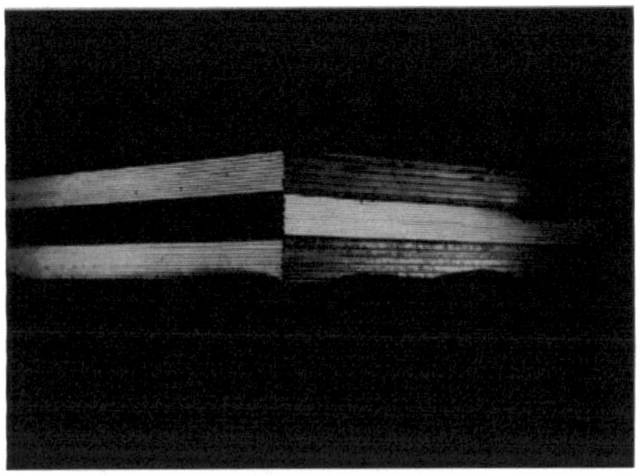

Figure 2. 3 modules of the calorimeter with enlightened fibers to illustrate the structure. There are 9 layers in total.

Figure 3. The calorimeter fully equipped with front-end electronics.

The flight model fully equipped with the frond-electronics is shown in Figure 3.

Dynamically, the expected signal from cosmic rays ranges from a few photoelectrons for Minimum ionizing Particles (non interacting protons) to about 10^5 for electromagnetic showers in the TeV range. To cope with this dynamic range, the front end electronics is designed with two gains and a ratio high to low of 33 on average.

3. Test-Beam setup

The flight model of the calorimeter was tested on a beam line of the CERN SPS accelerator ring in October 2006 and July 2007. Fully instrumented with the flight electronics, it was carefully wrapped in a plastic film to protect it from dust. It was fixed on a support structure which could move along X (horizontal) and Y (vertical) with respect to the beam direction (Z axis) and which could rotate around the vertical axis. With the test beam setup, 5 (resp. 4) layers were measuring the position and energy in X(Y). Electron beams of energies between 6 to 250 GeV and hadron beams with energy ranging from 20 to 100 GeV were used. Data were recorded essentially in the center of the calorimeter and scanning along the X and Y directions.

4. Slow Control

The calorimeter must be able to survive extreme temperatures without irreversible damages, the reference temperatures being -30 °C and + 50°C. Consequently, it should be able to operate for a wide range of temperature. During the test beam, thanks to sensors located close to the PMts' electronics on the aluminium back panel, the temperature was recorded in several points on each face and could then be extrapolated in each point. The pedestal dependance with respect to the temperature was studied and found compatible with the studies done in the laboratory. Figures 4 et 5 show respectively a map of the temperature and a typical pedestal dependance. With an average dependance of 0.45 ADC counts per degree, 3 channels out of 2592 have a 0 (negative) pedestal at -20°C.

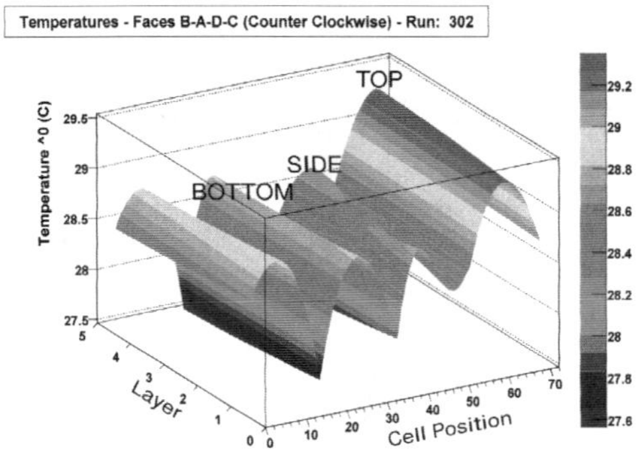

Figure 4. Map of the extrapolated temperatures on each face on the calorimeter.

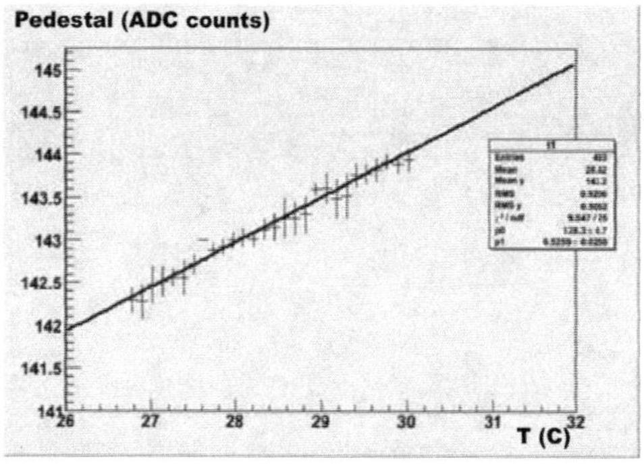

Figure 5. A typical dependence of the pedestal as a function of the temperature.

5. Energy reconstruction

Several corrections are applied to reconstruct the energy of electrons. They are detailed in the following subsections.

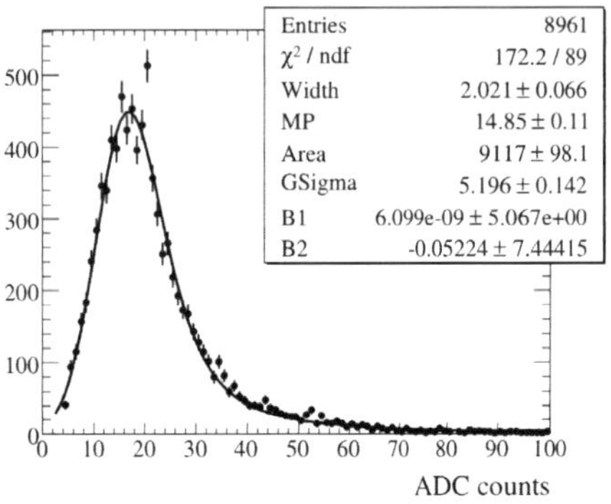

Figure 6. Example a MIP distribution fitted with a Landau convoluted by a Gaussian function + an exponential background,

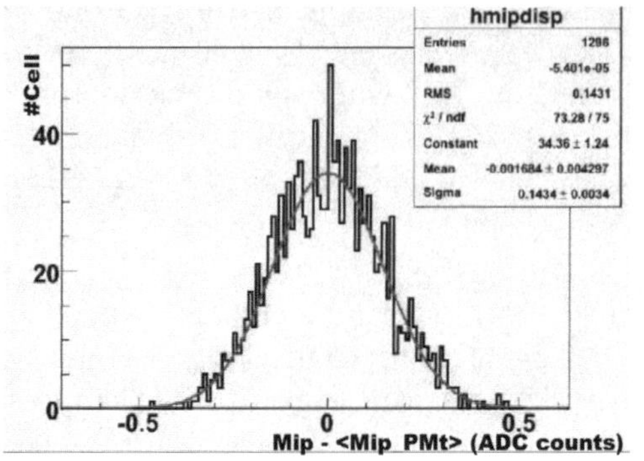

Figure 7. Dispersion of MIPs within PMts observed in test beam.

5.1. *Equalization*

The first set of corrections is meant to equalize the response of each anode and to treat the high to low gain ratio individually. The equalization is performed using the MIP signal obtained in each cell from a scan along the X and Y axis in the middle of the calorimeter by a 100 GeV hadron beam [1]. Each distribution is fitted by a Landau function convoluted by a Gaussian function plus an exponential background and an example is given in Figure 6. The MIPs were distributed with an average of 15 ADC counts and a sigma of 2.3 ADC counts. The dispersion within a PMt as shown in Figure 7, amounts to 15% in agreement with the specifications. Figure 8 displays an example of the correlation between high and low gain measurements. The distribution of gains measured in the test beam is shown in Figure 9.

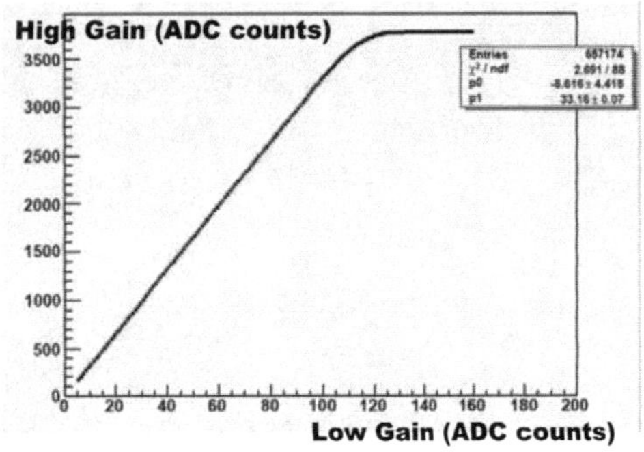

Figure 8. Relation between High to Low gain for a single anode.

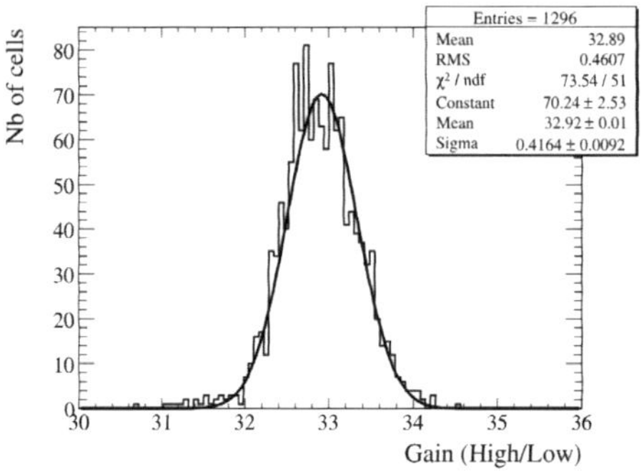

Figure 9. Distribution of the PMts' gains measured in test beam.

5.2. *Attenuation*

The signal is attenuated along the fibers and needs to be corrected depending upon the original position along the fibers. Scan along X and Y with electrons at 10 and 30 GeV and with hadrons

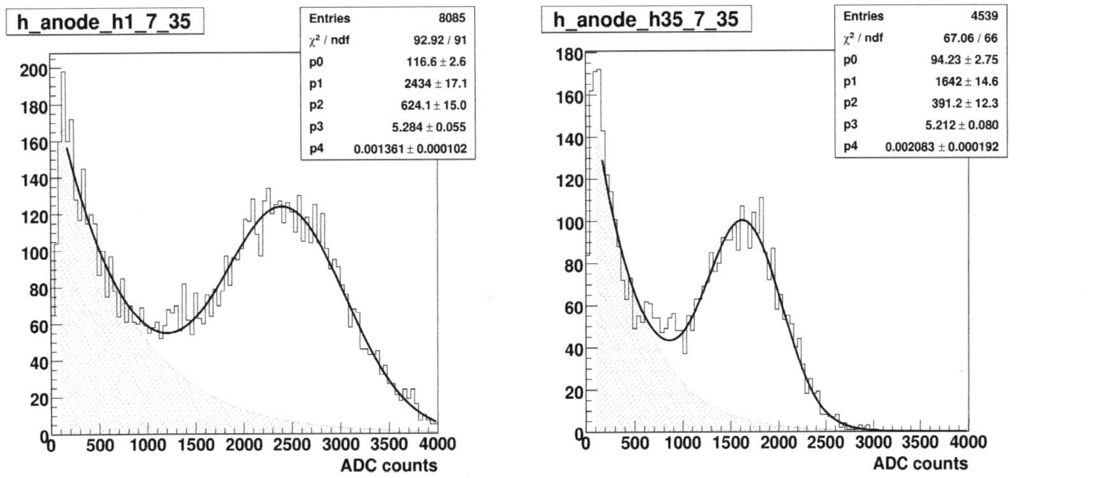

Figure 10. 10 GeV Electron signal measured close (left) and far (right) from the photomultiplier.

are used to measure the attenuation. An example of the signal of 10 GeV electrons recorded close and far from the PMt is shown in Figure 10. Similarly the MIP peak evolves along the fibers as shown in Figure 11. All attenuation curves are fitted simultaneously by the following function $f \times \exp(-x/\lambda_{fast}) + (1-f) \times \exp(-x/\lambda_{slow})$ letting free an individual amplitude per cell.

An example of the result of the fit is given in Figure 12. The correction is applied such that there is no correction in the center of the calorimeter: the effect of the correction on the energy measured in one layer is depicted in Figure 15.

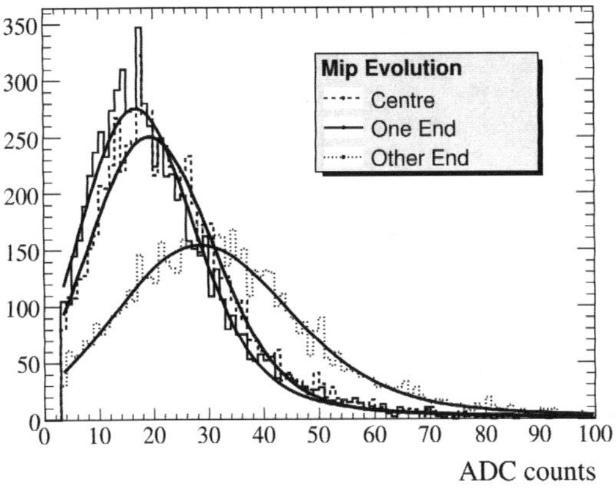

Figure 11. MIP signal measured at three different positions along the fiber.

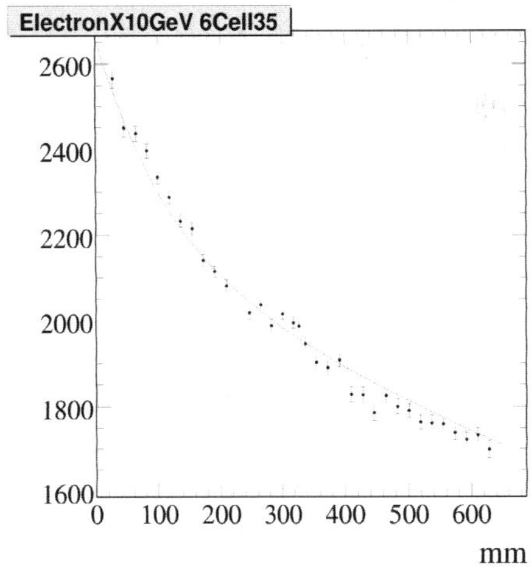

Figure 12. Example of fit of the attenuation.

5.3. Impact correction

The energy measurement is sensitive to the impact of the particle with respect to the boundaries of the cell as illustrated in Figure 13. To correct for this, a method adapted from [2] was used. It relies on the fact that the ratio S1/S3 of the energy in the cell with the maximum energy to the energy of the neighboring ones per layer and summed over all layers in one direction has a better sensitivity to the impact point (Figure 13). A simple dependance of this ratio to the energy can be observed as shown in Figure 14. The normalized inverse shape of this function is used to correct for the impact dependance and the result is given in Figure 16.

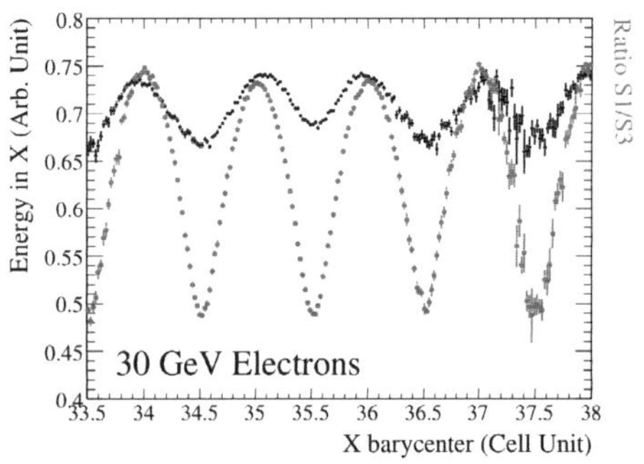

Figure 13. Sensitivity of the energy reconstruction to the impact position given by the barycenter (small black dots), Sensitivity of the ratio S1/S3 (See text) to the impact position (large red dots)

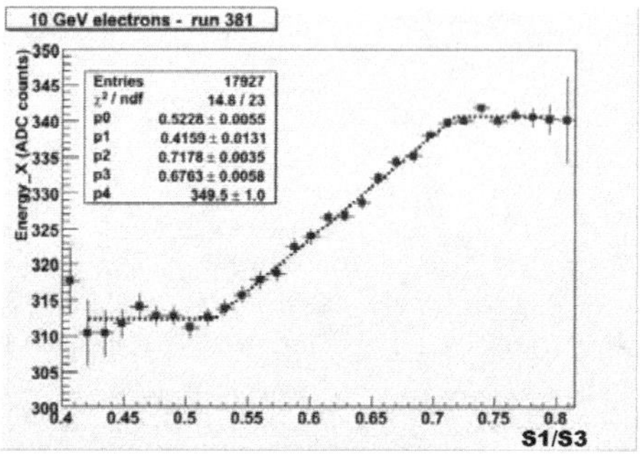

Figure 14. Dependance of the energy as a function of S1/S3 (See text).

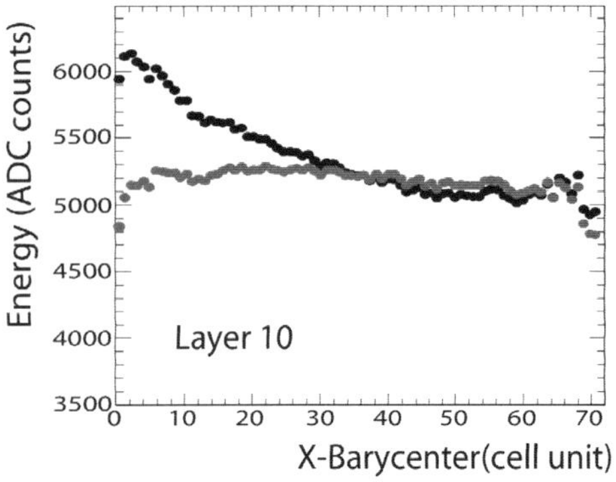

Figure 15. Energy measured in layer 10 before (black) and after (grey/red) correcting for the attenuation as a function of the X position.

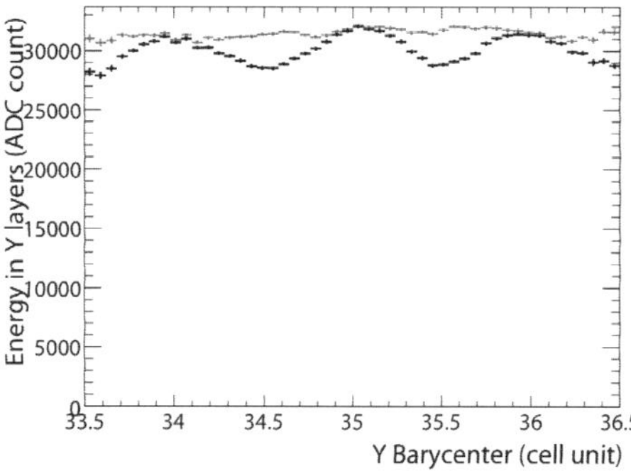

Figure 16. Energy before (black) and after (grey/red) impact correction.

5.4. Rear Leakage correction

The last correction will correct for the energy leakage. The relative leakage energy is approximated by a linear function of the fraction of energy measured in the latest layers of the calorimeter [3]. Figure 17 exemplifies this relation with 100 GeV electrons and in Figure 18, it can be seen that the linear relation scales with the beam energy. Therefore this dependance was taken into account using the energy divided by the longitudinal barycenter which is roughly proportional to the incident energy . Figure 19 shows the reconstructed energy before and after leakage correction.

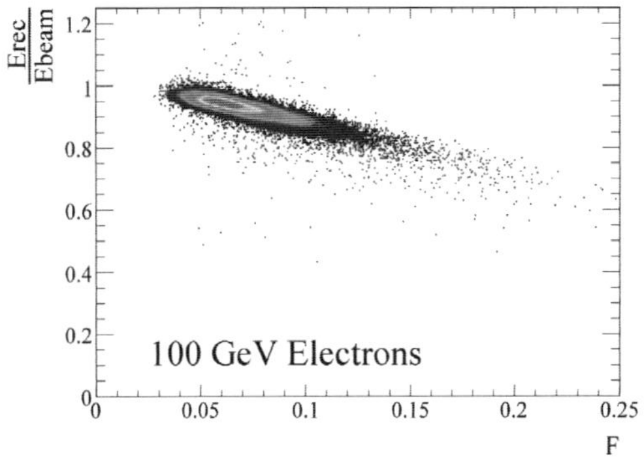

Figure 17. Example of the linear dependence of the relative leakage energy, given by the ratio of the reconstructed energy Erec, (including all preceding corrections) over the beam energy as a function of F, the fraction of energy in the last two layers for 100 GeV Electrons

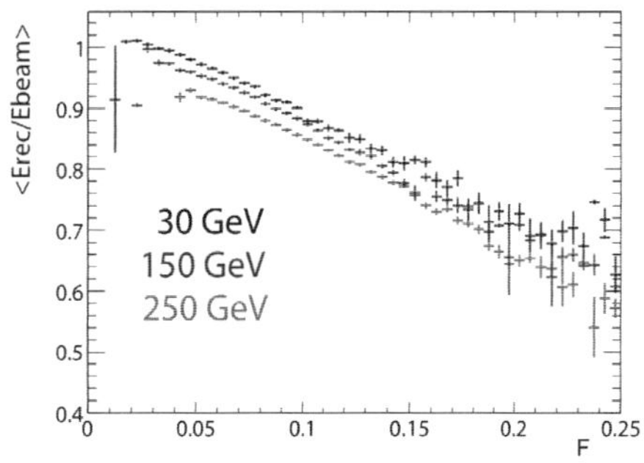

Figure 18. Average relative leakage energy as a function of the fraction of energy in the last two layers for 3 incident beam energies.

6. Performances

Preliminary results on the resolution and the linearity are shown in Figures 20 and 21. The resolution was fitted to be $\sigma/E = 10.3\%/\sqrt{E} \oplus 1.6\%$ while the linearity is kept within 2%. Using the same complete set of corrections, the linearity and the resolution are not degraded when the incidence angle of the beam is 15 degrees. Using the fit to the longitudinal profiles of all energies, displayed for 3 energies in Figures 22, the radiation length X_0 can be deduced: it was fitted to be 1.07(0.02) layer unit. Therefore, the calorimeter consists of 16.9 radiation lengths.

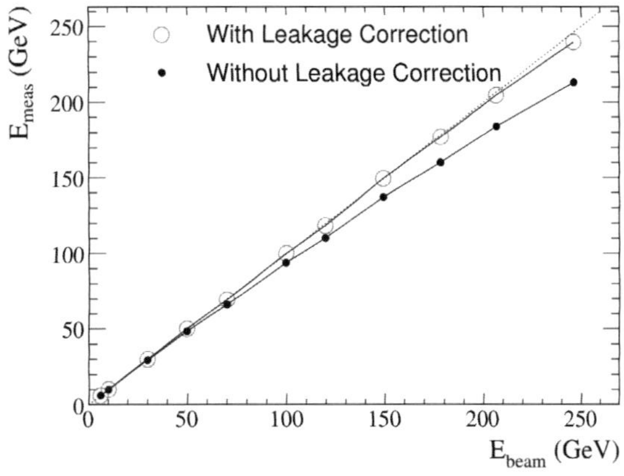

Figure 19. Energy before and after leakage correction in function of the beam energy.

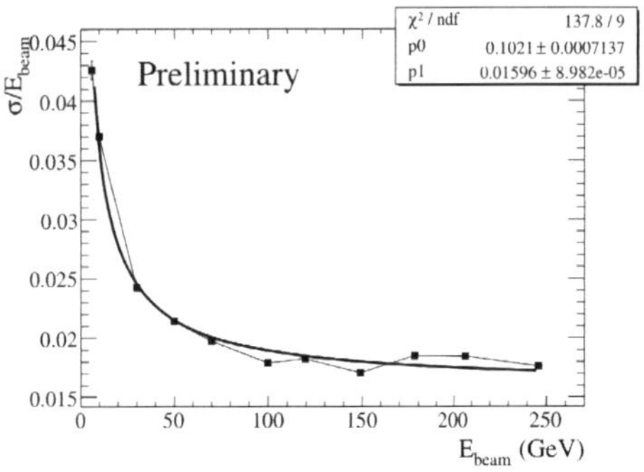

Figure 20. Preliminary results on the energy resolution.

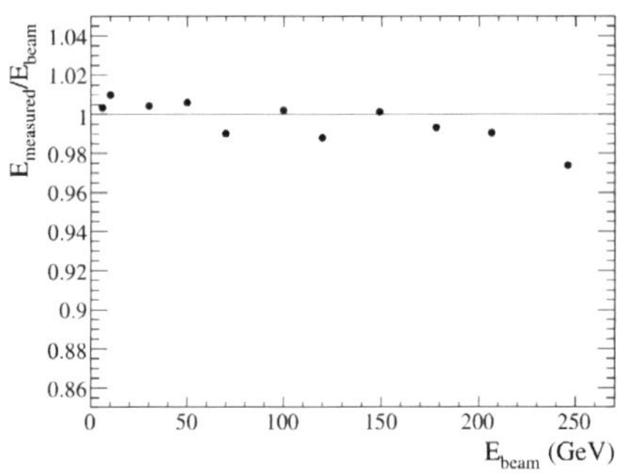

Figure 21. Preliminary result on the linearity.

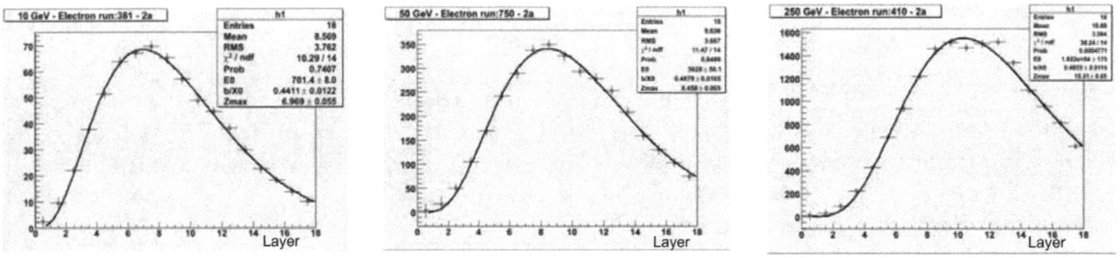

Figure 22. Longitudinal electromagnetic profiles fitted from left to right at the energies of 10, 50 and 250 GeV.

7. Conclusion

The beam tests of the flight model of the AMS calorimeter were successful. No channels were found dead or noisy. Preliminary results for the resolution and linearity are very encouraging: a constant term of 1.6% was obtained while the linearity is kept within 2% in the energy range from 6 to 250 GeV. Further refinements on the different corrections are expected to improve the constant term. The calorimeter was found to be 16.9 radiation lengths thick.

References

[1] AMS note 2008-01-01
[2] The L3 Collobration, *Nucl. Inst. Meth.* A **289**(1990) 35
L3 note 1712, (1995)
[3] Loic Girard. *PhD thesis*(2004) and reference therein.

Journal of Physics: Conference Series **160** (2009) 012042 doi:10.1088/1742-6596/160/1/012042

Calorimeters: key detectors for LHC physics

Egidio Longo

INFN Sezione di Roma and Università degli Studi di Roma La Sapienza, P.le A. Moro 2, 00185 Roma, Italy

E-mail: egidio.longo@roma1.infn.it

Abstract. The calorimeters designed for the LHC experiments are reviewed. Their expected performance is compared with the requirements set at the beginning of the project.

1. Introduction

Two decades after the launch of the projects, the LHC experiments are now almost completely installed and ready to take data at the start of LHC. Calorimeters are crucial detectors for this physics program. From the first edition to the present one, the CALOR Conference series were a privileged observatory to follow the design, development and construction of the LHC calorimeters. Today is probably the right moment for an assessment on the achievements of the different projects and on the perspectives of their exploitation.

2. Requirements for hadron collider calorimeters

Since the first concept of LHC, a hadron collider aiming to complete the exploration of the Standard Model, the physics goal was clearly set to the study of the origin of the electroweak symmetry breaking, whose mass scale is expected below 1 TeV. In this respect, a crucial benchmark is the discovery of the Higgs boson, whose small cross section demands high luminosity in the machine and high resolution in the detectors to fight backgrounds: in its search, calorimeters are needed for the detection of electrons, photons and jets. Later on, supersymmetry gained more and more favour as the simplest Standard Model extension providing natural mechanisms to avoid an implausible fine tuning of the parameters. For SUSY signals, missing energy is the key signature, enhancing once more the relevance of calorimetric measurements.

It is of some interest to look back at the CALOR91 edition in Capri, where for the first time a large fraction of the program was devoted to the calorimeters for hadron colliders. In the paper by D. Green [1] a list of requirements for calorimeters was given and discussed. The main items and their arguments are summarized here:

- hermiticity and speed were obvious requirements; the angular coverage was set to $\eta = 3$ for electrons and $\eta = 5$ for hadrons;
- trigger capability was a natural bonus of the fast and granular signals provided by the calorimeters;
- the containment of the upstream material was considered a must to many respects: a limit at less than $1X_0$ for the coil material and less than $0.1X_0$ for the more dangerous tracking material was set;

- the depth of the hadronic section was set to 10 absorption lengths by the requirement to reconstruct dijet masses up to 10 TeV;
- a transverse segmentation of 0.05-0.1 rad for the hadron section and 0.05 rad or better for the e.m. section was set to detect showers coming from isolated electrons and match them with their tracks;
- longitudinal segmentation was required for both sections, but the emphasis in the e.m. part was to give a handle, on an event by event basis, to correct for longitudinal inhomogeneities in the shower response, due to an expected severe damage by radiation;
- finally, the resolution: the e.m. one was clearly driven by the $H \to \gamma\gamma$ benchmark signal, but thanks to the $E^{-1/2}$ stochastic term, improving with energy, the key figure was identified in the constant term, with a better-than-1% goal; the hadronic stochastic term was set around 0.7%, with a strong requirement for compensation ($|e/h - 1|$ better than 30%) to provide the linearity needed to detect possible deviations of jet cross-sections with respect to QCD predictions: "linearity is more important than precision".

Even though some points of view changed with time, these requirements remain valid and can be used to benchmark the performance of the calorimeters as they have been built.

3. LHC experiments

The CERN Large Hadron Collider is coming into operation and should give first proton-proton collisions at $\sqrt{s} = 10$ TeV in the second half of 2008, moving on to the design energy of $\sqrt{s} = 14$ TeV in 2009. Four big experiments will take data at LHC: two of them, ATLAS and CMS, are general purpose detectors, aiming at discovering the Higgs boson and exploring the physics beyond the Standard Model; the other two are dedicated to specific physics: LHCb to the study of b physics and ALICE to heavy ion collisions.

General purpose experiments have to measure electrons, photons and jets: calorimeters play a key role not only to reconstruct with good precision invariant masses from two photons and two or four electrons in the Higgs decay chains, but also to identify complex decay chains (as the ones expected from the decay of supersymmetric particles) by reconstructing and measuring electrons, jets and missing transverse energy, as shown in figure 1.

Different choices in the magnetic field configuration lead to quite different general schemes for ATLAS and CMS:

- ATLAS [2] has an air muon spectrometer with three huge toroidal magnets. The tracker is inserted in a 2 T solenoid, while an highly segmented lead/liquid-Argon (LAr) e.m. calorimeter is placed behind it. Hadronic activity is measured by an iron/scintillating-tile sandwich in the barrel and LAr in the end-caps. In the very forward region ($3.2 < \eta < 4.9$) a tubular structure with very narrow LAr gaps integrates the e.m. section (LAR/Cu) with the hadron section (LAr/W).
- CMS [3] has a compact solenoidal magnet (4 T) containing the calorimeters and the silicon tracker, while muon chambers are embedded in the iron return yoke. The e.m. calorimeter is made by homogeneous $PbWO_4$ scintillating crystals, while the hadron calorimeter employs the scintillating tile technique everywhere but in the forward region ($3 < \eta < 5$), where an iron/quartz-fiber structure, sampling the Čerenkov light, is installed. In the barrel region, a tail catcher is placed behind the coil to detect the tails of the hadronic showers.

4. Triggering with calorimeters

As said before, calorimeters play a very important role in the trigger of the experiments, as they provide fast and granular signals able to identify the electromagnetic clusters, to detect the presence of jets, to tag both events with large transverse energy and those with high transverse

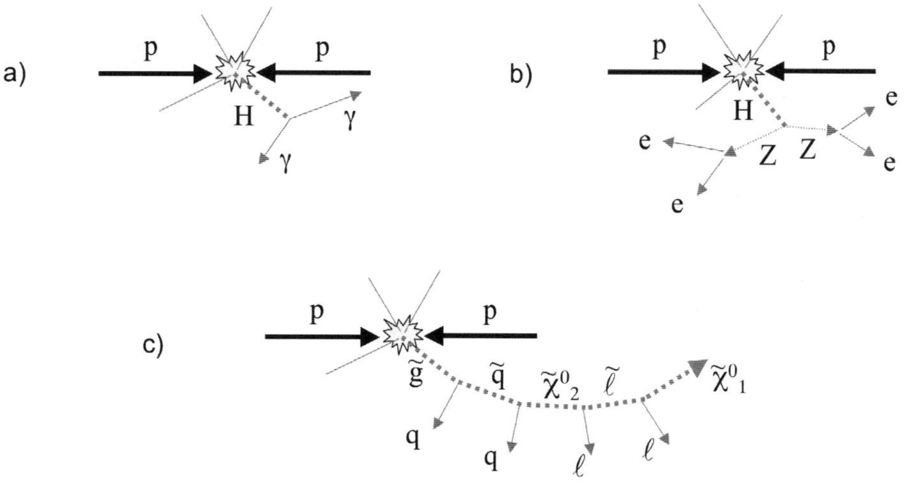

Figure 1. a) Higgs decay into two Photons. b) Higgs decay into four electrons. c) A typical SUSY particle decay chain: in this example a gluino decaying into a quark pair, a lepton pair and the lightest SUSY particle, appearing as missing transverse energy in the detector.

missing energy, and to determine the isolation for all the leptons and photons candidates. In practically all the cases, the calorimetric information is available already at the first level of trigger (L1), and can obviously be refined at higher trigger levels and in the off-line analysis.

An example is the CMS "regional calo trigger", which provides at L1 isolated electron or photon candidates not only by comparing the deposit in the electromagnetic trigger tower with the one in the hadronic towers (both with a granularity of $\Delta\phi\Delta\eta = 0.087 \times 0.087$), but also looking at the so called "fine grain profiles", namely dominos of 2×5 crystals in η, ϕ allowing to identify narrow shower profiles in η compatible with electromagnetic showers (being the spread in ϕ larger due to bremsstrahlung in high magnetic field), as shown in figure 2.

5. Electromagnetic calorimeters

5.1. Requirements for electromagnetic calorimeters

The requirements for the LHC electromagnetic calorimeters are direct consequences of the general requirements and experimental environment of LHC: they must provide large acceptance, good energy and position resolution for high energy em showers up to $|\eta| < 2.5$, they must be fast, compact and granular to cope with the expected high event rates, they have to be radiation tolerant, to provide a large dynamic range, from MIP to the TeV scale, they have to be linear and finally they must provide particle identification, in particular to separate electrons from jets and photons from π^0s.

Both calorimeters fulfill these requirements:

- the ATLAS LAr-Pb sampling calorimeter with the novel technique of the "accordion" shape of the layers gives a special emphasis on high granularity, both longitudinal and transverse. As a consequence it can provide a good γ/π^0 separation.
- the CMS calorimeter focuses more on an excellent energy resolution given by the choice of homogeneous $PbWO_4$ crystals as shower detectors, relaxing the requirements on γ/π^0 separation.

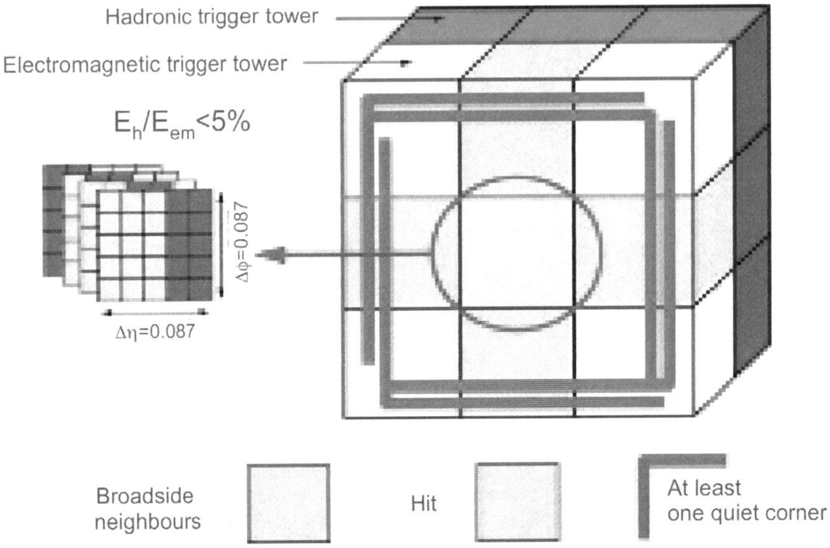

Figure 2. Schematic principles of the CMS "regional calo trigger" to define an isolated electron/photon candidate: first, a $E_{hadron}/E_{e.m.} < 5\%$ is required in the corresponding trigger towers; then the e.m. trigger tower is scanned by four "dominos" of 2×5 crystals in η, ϕ, looking for narrow shower profiles expected from an e.m. shower.

5.2. The ATLAS e.m. calorimeter

The Atlas electromagnetic calorimeter is made by layers of lead and liquid-argon, shaped in an accordion structure to allow an easy read-out of the longitudinal towers, as schematically shown in figure 3.

The length is 22 X_0 (47 cm) at $\eta = 0$, increasing with η. It is subdivided in 3 longitudinal layers with a pre-sampler just outside the superconducting coil. The first 4 X_0 are finely segmented in η by use of cathode strips, $\Delta\eta = 0.0031$ wide, to provide a good γ/π^0 rejection. The following 16 X_0 are read in towers of $\Delta\phi \times \Delta\eta = 0.0245 \times 0.025$ and measure the shower core. The last 2 X_0, with larger granularity, are to evaluate late showers. In total, there are 170000 readout channels.

One of the challenges of this design is the control of the mechanical uniformities. As an example, a 1% variation of the lead thickness would correspond to a 6 per mill change in the response. The measured spread on the absorber thickness has a σ of 9 μm around the average value of 2.2 mm. Other parameters to be controlled are the variation of the accordion slant angle and the sagging of the layer structure.

The resolution measured with an electron beam at $\eta = 0.687$ [4] can be represented as

$$\sigma(E)/E = (10.1 \pm 0.1)\%/\sqrt{E(\mathrm{GeV})} \oplus (0.17 \pm 0.04)\%$$

after noise subtraction. The linearity is within 1% from 15 to 180 GeV. The uniformity of the response was measured scanning in η the modules with 245.6 GeV monochromatic electrons, showing a variation within 0.44 % in the barrel. At this energy, the resolution goes from 0.7 % at $\eta = 0$ to 0.9 % at $\eta = 1.5$. From these figures a global constant term of 0.6 % can be evaluated.

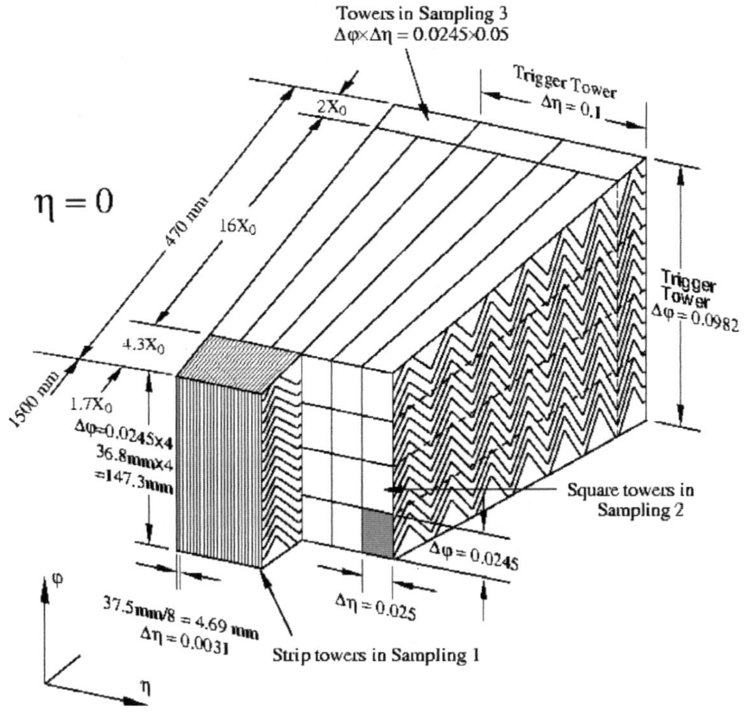

Figure 3. Detailed structure of ATLAS barrel e.m. calorimeter. See the text for the details.

5.3. The CMS e.m. calorimeter

The overall structure of the calorimeter is shown in figure 4.

The main features of $PbWO_4$ scintillating crystals are the extremely short radiation length and Molière radius (X_0 = 0.89 cm, R_M = 2.2 cm), allowing the realization of a compact calorimeter ($\Delta\phi \times \Delta\eta = 0.0245 \times 0.0245$ in the barrel), with fast light emission (80% in 25 ns) but a reduced light yield (100 photons per MeV). An intensive R&D program has been developed since the selection of these particular crystals to ensure the mass production of optically clear, fast and radiation resistent crystals [8]. Ionizing radiation produces a loss of light transmission without changes to the scintillation mechanism. The damage can be tracked and corrected for by monitoring the optical transparency with injected laser light. To ensure an adequate performance throughout the duration of the experiment, production crystals were required to exhibit radiation hardness properties quantified as an induced light attenuation length greater than approximately 3 times the crystal length when the damage is saturated.

The light is read in the barrel by Avalanche PhotoDiodes (two for each crystal), insensitive to the 4T magnetic field, especially developed by Hamamatsu for CMS (operated at gain 50) and Vacuum Photo-Triodes in the end-caps, insensitive to the expected high radiation level and able to operate in a magnetic field almost parallel to their axis.

The resolution of CMS ECAL can be parametrized as [5]

$$\sigma(E)/E = 2.8\%/\sqrt{E(\text{GeV})} \oplus 0.12/E(\text{GeV}) \oplus 0.3\%$$

for central impact electrons.

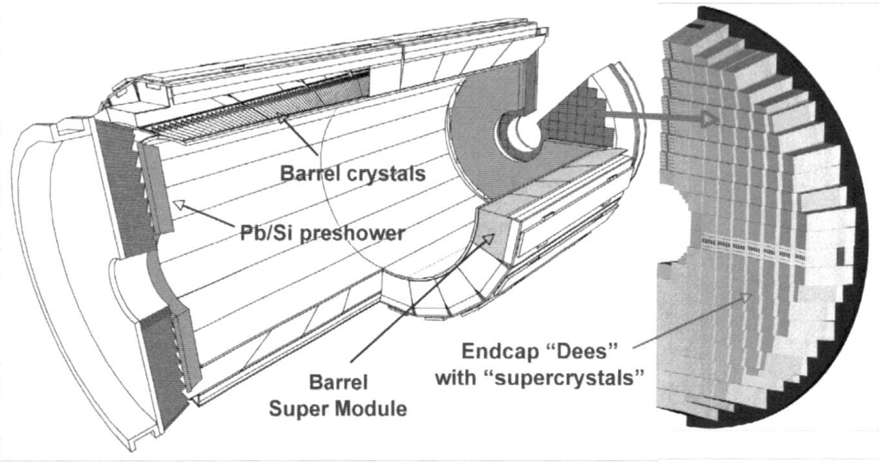

Figure 4. The structure of the CMS ECAL: Barrel crystals ($|\eta| < 1.48$) are grouped in 36 Supermodules containing 1700 crystals each, for a total of 61200 crystals. End-caps are divided into four "Dees" made by "supercrystals" (5×5 crystals each), for a total of 14648 crystals, covering $1.48 < |\eta| < 3$. A Pb/Si pre-shower is placed in front of the Endcaps.

5.4. Calibration of e.m. calorimeters

One of the main problems of a precision calorimeter is the calibration of the channels. The associated uncertainty directly contributes to the constant term, which tend to be dominant at high energies, where the stochastic term is naturally dumped. The conversion of the individual channel response to the incident particle energy requires several steps, each introducing some conversion factor. As an example, in the CMS energy reconstruction, first the digitized signals have to be multiplied by intercalibration constants and then a cluster algorithm selects the channels to be summed up. At this stage, an overall correction factor must be applied to the sum of energies, depending on the cluster algorithm and including containment corrections, but also the dependence of the response on the impact position, particle type and other possible parameters. Finally, an overall scale factor must provide the right absolute energy scale.

ATLAS will start data taking with channels intercalibrated at better than 1%, thanks to the mechanical uniformity in the modules construction and to an electronic calibration system. Checks of the calibration were performed on 10% of the modules in test beams. In the long term, precise *in situ* absolute calibrations will be provided by electron decays of W and Z: 50 k $Z \to e^+e^-$ events, corresponding to 0.1 fb^{-1} integrated luminosity will provide a global constant term better than 0.7%.

CMS calibrated one quarter of the calorimeter at an electron beam, reaching an intercalibration of 0.3%. All Supermodules were calibrated with cosmic rays prior to installation, with a precision better than 2%, as shown by comparison with test beam calibrations. *In situ* fast equalization at 1.5 to 2% can be reached exploiting the ϕ symmetry of the detector in the energy deposition of minimum-bias events. In the long term, intercalibration can be obtained by comparing the energy and the momentum of the electrons from W decays, while absolute calibration can be obtained by looking at the reconstructed invariant mass of $Z \to e^+e^-$ events. From W alone, 5 fb^{-1} are required to get better than 5 per mill intercalibrations in the center of the barrel, but Monte Carlo studies show that faster calibration is possible by π^0 mass reconstruction.

Journal of Physics: Conference Series **160** (2009) 012042 doi:10.1088/1742-6596/160/1/012042

5.5. Material budget in front of e.m. calorimeters

Material budget from the tracker and other services (including the coil for the ATLAS solenoidal magnet) has always been a big concern, often increasing with an improved definition of the central detector parameters. Because of the tracker material, electrons loose energy via bremsstrahlung, while photons convert in e^+e^- pairs. Moreover, the high solenoidal magnetic field bends electrons, so that the radiated energy is spread in ϕ. So, the effects of the material are different for photons and electrons, while the strategies for *in situ* calibrations rely on electrons but the results have to be applied to photons as well. The material of the tracker maximizes these effects, but in principle a detailed track reconstruction could help at least to identify more affected events. On the contrary, material immediately in front of the calorimeter can be considered as an additional sampling, less contributing to the energy spread in ϕ.

In CMS, the tracker material goes from 0.4 X_0 at $\eta = 0$ up to 1.4 X_0 (of which 0.4 X_0 is the material outside the sensitive area) around $\eta = 1.5$, then decreasing to 0.8 X_0 at $\eta = 2.5$. 50% of electrons undergo non negligible bremsstrahlung. Most of the energy spread can be recuperated by means of dynamic cluster algorithms, summing adjacent clusters in extended ϕ windows. In ATLAS, a constant amount of 2 X_0 is due to the coil everywhere, and the maximum amount of material, in the region of transition from barrel to endcap, reaches 6 X_0.

6. Hadron calorimeters

6.1. Role of hadron calorimeters at LHC

Hadron calorimeters are crucial in all the physics channels with jets, requiring good energy and position resolution for energies up to the TeV scale. This sets to 10 λ_{int} the minimum depth required. All the new particles searches demand hermeticity in jet measurement to provide good resolution in missing energy. Jets are made of an e.m. component as well, so good combined hadronic and e.m. resolution is required. In addition, forward calorimeters are also requested to tag forward jets in the "combined production channels" and must maintain a good jet energy measurement to preserve the missing E_T resolution.

6.2. ATLAS hadron calorimeters

The barrel ATLAS HCAL is a sandwich of iron and plastic scintillator tiles perpendicular to the beam axis, grouped in towers with granularity $\Delta\phi \times \Delta\eta = (0.1 - 0.2) \times 0.1$. Each tower is segmented in three longitudinal samplings, for a total of 9852 channels. Wavelength shifting fibers carry the light to photomultipliers (mounted on the back of the towers). The barrel covers $|\eta| < 1.7$ and surrounds both the EM barrel and the forward calorimeters, with a depth of 10 to 13 λ_{int}. The energy resolution for pions at $\eta = 0.35$ is

$$\sigma(E)/E = 56.4\%/\sqrt{(E(\text{GeV})} \oplus 5.5\%.$$

The End-cap HCAL is made of copper/Liquid argon layers mounted in two wheels for each side (behind the e.m. section), has a depth of 10 to 12.5 λ_{int} and a granularity $\Delta\phi \times \Delta\eta = (0.1 - 0.2) \times (0.1 - 0.2)$ in four longitudinal samplings, resulting in 5632 channels.

The forward calorimeter covers $3.2 < |\eta| < 4.9$ and integrates the e.m. and the hadronic section. Both have a tubular structure with very narrow LAr gaps. The absorber is copper for the 27.6 X_0 of the e.m. section and tungsten for the 7.48 λ_{int} of the hadronic section, divided into two samplings. There are 3524 channels in total.

6.3. CMS hadron calorimeters

CMS hadron barrel calorimeter is made by brass/plastic-scintillator tiles, parallel to the beam axis. Wave-length shifting fibers are coupled to clear fibers leading the light to pixel Hybrid

Journal of Physics: Conference Series **160** (2009) 012042 doi:10.1088/1742-6596/160/1/012042

Photo-Diodes mounted at the end-side of the barrel. There are 2593 towers without longitudinal segmentation. The HCAL is 7 λ_{int} deep at $\eta = 0$. The energy resolution is

$$\sigma(E)/E = 100\%/\sqrt{E(\text{GeV})} \oplus 8\%.$$

To improve the energy resolution and the linearity, an additional layer is placed outside the coil, adding 3 λ_{int} and improving by 10% the energy resolution for 300 GeV pions.

The forward region, $3 < |\eta| < 5$, is covered by a radiation resistant iron/quartz fiber structure, detecting the Čerenkov light emitted in the quartz. Fibers are of two different lengths, the longer ones starting from the front face, the shorter ones after 22 cm. With this arrangement, the energy deposited in the first section, corresponding to the e.m. showers, can be determined by subtraction of the signals from the two kinds of fibers. In total, there are 1728 towers, with a segmentation $\Delta\phi \times \Delta\eta = 0.175 \times 0.175$.

6.4. Missing E_T resolution

The differences in the hadron calorimeter resolutions reflect in the resolution on the sum of E_T, corresponding to the missing E_T resolution. This is $0.48/\sqrt{\sum E_T}$ for ATLAS and $0.97/\sqrt{\sum E_T}$ for CMS. However, both experiments show similar performance in the search for SUSY particles in inclusive channels with jets and missing E_T [6, 7]. The reason for that is that what really matters in these searches is the hermeticity of the detector (and, obviously, its correct description in the Monte Carlo simulations) rather than the missing E_T resolution.

7. Combined energy measurements

In the past years, a big effort was put in the test-beam measurements of the combined performances of the e.m. and the hadron calorimeters of both experiments. ATLAS launched this work since 1996, developing a "cell weighting" technique, where a different weight, varying with the energy, is applied separately to each e.m or hadronic cell. The result [9] is a linearity within ±2% from 20 to 300 GeV and a resolution represented by

$$\sigma(E)/E = 52\%/\sqrt{E(\text{GeV})} \oplus 1.6/E((\text{GeV}) \oplus 3\%.$$

The combined test of ECAL and HCAL in CMS is more recent, and some results are reported at this Conference [10]. The main result is that adopting an "optimized" sum of the energies, a linearity better than 2% from 5 to 300 GeV is obtained.

8. 1991 requirements revisited

If we confront the actual parameters of the calorimeters and their measured performance with the general requirements outlined in sect. 2, we see that most of them are fulfilled: hermeticity, angular coverage, speed and trigger capability are satisfied in both experiments. The same is true for the expected resolutions, which are even beyond expectation for CMS e.m. and ATLAS hadronic sections respectively. The transverse segmentation of the e.m. sections is a factor two below the 1991 figures. The hadronic transverse segmentation is not as good, but this is not considered a big deal. ATLAS also provides extended longitudinal segmentation in all sections, though the main purpose of this solution, the handle against radiation damage, is no more expected to be necessary because of the detector choices. The HCAL depth is above 10 λ_{int} for ATLAS only, though also CMS reaches this limit taking into account also the tail catcher.

None of the two experiments aimed at an intrinsic e.m./hadron compensation. However, both are hardly working to recuperate the required linearity by means of some kind of software compensation, based on a refined weighting of the different energy deposits, well performing on test beam data.

The only item where some concern remains, is then the material in front of both electromagnetic calorimeters. This material is much larger for ATLAS than for CMS, but due to the better nominal resolution, the effect could be important for the latter too, also because of the higher magnetic field, which is spreading the energy for electrons over a larger azimuthal region. The degradation of the resolution has been studied estensively in both experiments and tools for taking it into account have been developed. The effect is already included in all the simulations leading to the estimations of signal-to-noise ratios in all the main physics cases. However, everything relies for the moment on the Monte Carlo description of the inner material: in particular the different behaviour of photons, where we seek the best resolution, and electrons, which will be mainly used to get the best calibration of the response.

9. LHCb Calorimeters

The LHCb Collaboration [11] exploits the high production rate of B particles at small angle to study the CP-symmetry violation in the neutral B meson systems. The experiment is a single arm spectrometer, covering angles from 10 to 300 mrad.

The purpose of the calorimeters in LHCb is twofold: to provide electron/hadron separation at the trigger level and in off-line analysis; to measure the position and energy for electrons/photons and hadrons. The calorimetric section is placed at more than 12 meters from the interaction point and is organized in four layers:

- a scintillating pad detector (SPD), made of 6016 cells, whose light is transmitted trough WLS fibers to photomultiplier tubes;
- a preshower (PS) with a 2.5 X_0 lead sheet followed by a second scintillating pad with the same structure of SPD;
- an e.m. calorimeter (ECAL) made of a Pb/scintillator sandwich in the "shashlick" configuration, with WLS fibers read by photomultipliers; the lateral size is 7.8 $\times$ 6.3 m^2 with three tower granularities going from the center to the side;
- a hadron calorimeter (HCAL) made by 16 cm iron layers and 4 mm scintillating tiles, with a total depth of 5.6 λ_{int}; it has a surface of 8.4 $\times$ 6.8 m^2.

At the trigger level, the calorimeters alone can identify and measure electrons, photons and hadrons: the SPD tags the charged particles, the PS indentifies the electromagnetic nature of the shower, the two calorimeters provide the estimate of the transverse energy.

The resolution of the ECAL is

$$\sigma(E)/E = 10\%/\sqrt{E(\mathrm{GeV})} \oplus 1\%$$

up to 100 GeV. The ECAL will be calibrated *in situ* by energy flow, by comparing the energy of the electrons with their momentum measured in the spectrometer and from the π^0 mass reconstruction. The HCAL has been precalibrated by means of a Cs source and its response will be followed *in situ* by a LED system.

10. ALICE Calorimeters

ALICE Collaboration [12] is building a dedicated heavy-ion experiment for nucleus-nucleus interactions, to study strongly interacting matter at high densities where the formation of quark-gluon plasma is expected. To investigate the nature of this process, they have to detect and measure hadrons, electrons, muons and photons produced at LHC in the collisions of lead nuclei at 5.5 TeV/nucleon-pair.

One of the key detectors is the photon spectrometer, PHOS, acting as a quark-gluon plasma thermometer: the spectrum of the emitted photons is related to the temperature of the source.

To measure this spectrum, a large coverage is not required, so the solution adopted was a limited-coverage high-resolution calorimeter.

A wider angle electromagnetic calorimeter, EMCAL, has been recently approved and funded. It will be a projective Pb-scintillator sampling calorimeter, installed back to back with PHOS. The details of this project will be presented at this Conference [13].

10.1. The PHOS calorimeter

The purpose of PHOS is to measure photons, π^0 and η in the energy range from 0.5 to 100 GeV. It covers $|\eta| < 0.12$ and $\Phi < 100°$ with a granularity $\Delta\phi \times \Delta\eta = 0.004 \times 0.004$. It is made of 18000 $PbWO_4$ scintillating crystals, $22 \times 22 \times 180$ mm^3 each, positioned at 4.6 meters from the interaction point. The crystals are operated at -25 C° and read out by a single APD of the same type of the CMS ones. One out of five modules is already installed and the whole detector will be completed in 2010.

The resolution measured at electron test-beams can be parametrised as

$$\sigma(E)/E = 3.3\%/\sqrt{E(\text{GeV})} \oplus 0.18/E(\text{GeV}) \oplus 1.1\%$$

and is consistent with ALICE requirements.

11. Conclusions

The calorimeters for LHC experiments are almost ready to take data. Their design and construction gave a big contribution to the development of calorimetric techniques. In spite of the tigh requirements, some of them at the limit of the available technologies, these detectors passed extended tests demonstrating the ability to meet the design specification. In a few months, a new, unexplored physics domain will hopefully be reached and will try out their performance.

References

[1] Green D 1992 Physics requirements for LHC/SSC calorimeters *Proc. of the Second International Conference on Calorimetry in High Energy Physics, Capri 14-18 Oct 1991* 47-68 (Singapore: World Scientific)

[2] ATLAS Collaboration 1994 Technical proposal for a general-purpose p-p experiment at the Large Hadron Collider at CERN CERN-LHCC-94-43

[3] CMS Collaboration 1994 The Compact Muon Solenoid Technical Proposal CERN-LHCC-94-38

[4] Aharrouche M *et al* 2006 Energy linearity and resolution of the ATLAS electromagnetic barrel calorimeter in an electron test-beam *Nucl. Inst. Meth.* A **568** 601-23

[5] Adzic P *et al* 2007 Energy resolution of the barrel of the CMS electromagnetic calorimeter JINST 2 P04004 .

[6] A Airapetian *et al* (ATLAS Collaboration) 1998 ATLAS detector and physics performance technical design report Physics Performance Technical Design Report CERN-LHCC-98-13

[7] CMS Collaboration 2006 Physics Performance, Physical Technical Design Report Vol II *J. Phys.* G **34** 995 (CERN-LHCC-2006-021)

[8] Annenkov A A, Korzhik M V and Lecoq P 2002 Lead tungstate scintillation material *Nucl. Instrum. Meth.* A **490** 30

[9] Akhmadaliev A *et al* 2000 Response uniformity of the ATLAS liquid argon electromagnetic calorimeter *Nucl. Inst. Meth.* A **449** 461

[10] Yazgan E 2008 The CMS Barrel Calorimeter Response to Particle Beams from 2 to 350 GeV/c *these Proceedings*

[11] LHCb Collaboration 1998 LHCb Technical Proposal CERN/LHCC/98-4

[12] ALICE Collaboration 1995 A Large Ion Collider Experiment - Technical Proposal CERN/LHCC-95-71

[13] Ronchetti F 2008 The ALICE Electromagnetic Calorimeter Project *these Proceedings*

The ATLAS Liquid Argon Calorimeter: an overview

Henric Wilkens on behalf of the ATLAS LArg Collaboration
PH-Department, CERN, 1211 Geneva, Switzerland
E-mail: Henric.Wilkens@cern.ch

Abstract. The various cryostats with the ATLAS LArg calorimeter are installed in the ATLAS cavern since several years. Following this, an effort to install and commission the front end read-out electronics (infrastructure, crates, boards) has been ongoing and is converging, in time for LHC start. After the mechanical installation of the LArg calorimeter 99.9 % of the read-out channels were working, hence great care was taken to assure the same high level of quality after the installation of the read-out electronics. Following cautious procedures and with continuous testing-campaigns of the electronics at each step of the installation advancement, the result is a fully commissioned calorimeter with its readout and a small number of non-functional channels.

1. Introduction

An new energy frontier will soon open at CERN with the start-up of the Large Hadron Collider (LHC) providing proton onto proton collisions at nominal center of mass energy of 14 TeV. With a very high collision rate, it will deliver an integrated luminosity ranging from 10 to 100 fb^{-1} a year. These performances will allow, using the detectors located around the collision points to study the production of heavy particles, interactions resulting in the productions high transverse momenta particles, even with low production cross-sections. A prominent example is the possible discovery of the Higgs boson, which within the Standard Model provides the mechanism giving mass to the particles. A good energy resolution and a good linearity are needed to fully exploit the collisions.

2. Physics requirements for the ATLAS LArg Calorimeters

The performance requirements for the Electromagnetic (EM) calorimetry in ATLAS [2] has been driven by the potential production of the Higgs boson in LHC collisions. The two channels decay channels $\text{H} \rightarrow \gamma\gamma$ and $\text{H} \rightarrow 4\text{e}^{\pm}$ were used to benchmark the design choices in the EM calorimeter design. The accordion design provides a large acceptance and uniform response over the full azimuthal range with no gaps for services, whereas the liquid Argon active medium is intrinsically resistant to radiations. To address the large dynamic range requirement going from 30 MeV to 3 TeV, the radiation hard front end electronics implements three gains for each readout channel. The contribution to the energy resolution of the stochastic term for EM showers should be 10 $\%\sqrt{E}$, and the constant term around 0.7 %, which requires very precise mechanics, in terms of absorbers and electrod positions, as well as a calibration system for the readout system reaching an accuracy of 0.25 %. To address the precise measurement of the W-mass, an excellent linearity of the calorimeter response, of order 0.1 %, is needed. This is accomplished with a presampler, as well as 3 longitudinal segmentations of the calorimeter,

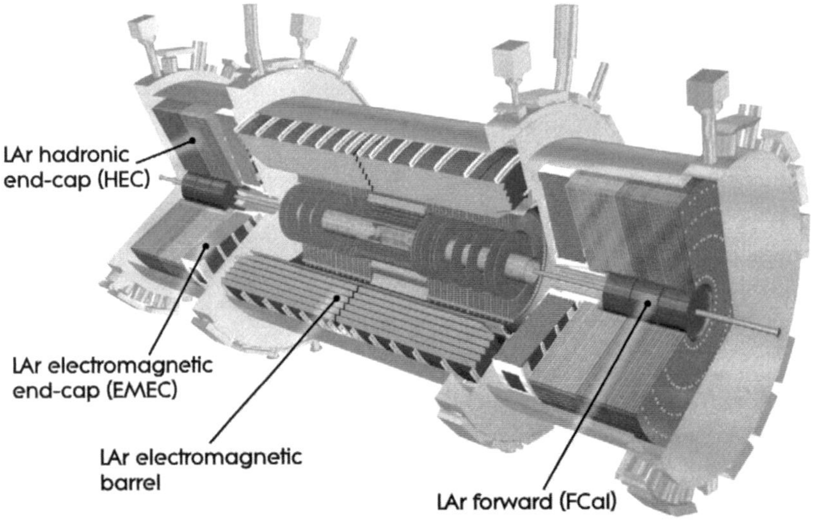

Figure 1. Schematic drawing of the ATLAS LArg calorimeter system. In the middle the barrel cryostat hosts two electromagnetic wheels, at each end the end cap cryostats host two concentric electromagnetic wheels, two hadronic wheels and thee forward calorimeter wheels.

which allow for correction for the upstream material using a layer weighting technique. The requirement for good particle identification, of electrons versus jets, and photons versus π_0, is achieved through a fine granularity of the detector, especially in the first sampling, allowing a fine position and angular resolution (50mrad/$\sqrt{E}$).

The potentiality that super-symmetric particles may be produced in LHC collisions invites for excellent missing energy determination. Therefore the ATLAS LArg calorimeters have an almost 4π acceptance, and sets the energy resolution criteria for the LArg hadronic and forward regions to 50 $\%/\sqrt{E}$ for the pseudo rapidity range $|\eta| < 3$ and 100 $\%/\sqrt{E}$ in the pseudo rapidity range $3 < |\eta| < 4.9$.

The very fine granularity and longitudinal segmentation of the calorimeter allow through layer weighting technique, when estimating the energy offline, to correct for the non compensating nature of the calorimeter. Finally the LHC 40MHz bunch frequency makes pile-up an important background. To minimize its contribution, and making use of the fast rise time of the ionisation signal, bipolar shaping is applied in the front end electronics.

3. The Atlas Liquid Argon Calorimeters

Three large aluminum cryostats host the LArg calorimeters as displayed in Figure 1. The three cryostats are surrounded by a barrel hadronic calorimeter build out of iron absorbers and scintillating tiles as active material, This calorimeter, TileCal, is described in a separate contribution at this conference [1]. The design and construction of the barrel and endcap EM calorimeter is precisely documented in the paper [3], and [4] respectively.

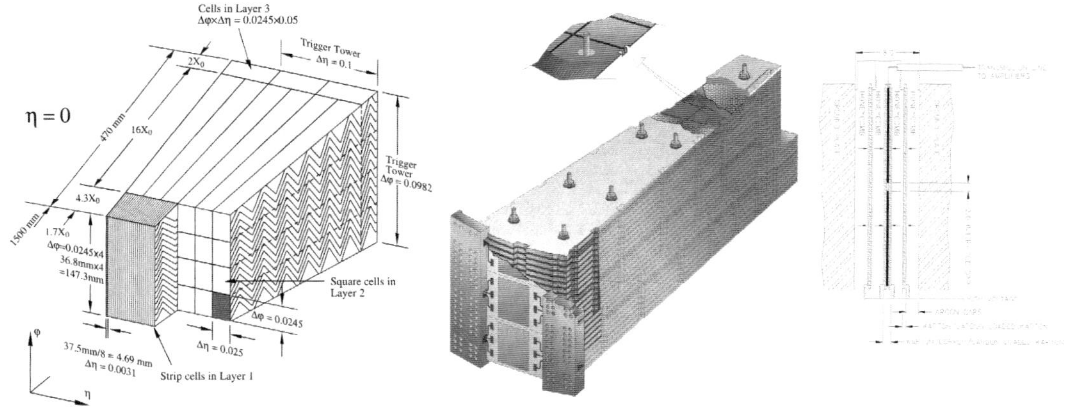

Figure 2. On the right side, schematic section of the EM barrel detector, displaying the granularity of the three samplings, in the middle a schematic view of a section of the HEC module. On the left side the signal gap structure in the HEC.

3.1. EM calorimeters

The Barrel EM calorimeter consists of two wheels of 1024 steel claded lead absorbers, interleaved with the same number of electrodes held in place by a 2.1 mm high honey comb structure, which defines the gap size. The electrode itself is build out copper etchings on polyimide, two outer conductive layers distribute the high voltage over the electrode outer surfaces, whereas a central layer, in between the two isolating foils collects the signals through capacitive coupling. When the nominal potential difference of 2000V between the electrodes and the absorbers is applied, the charges induced by ionisation in the argon drift in 450 ns towards the electrodes. The barrel calorimeter extends in pseudo-rapidity to 1.475, for a total thickness of 22 X_0. The readout electrodes are etched with readout pads defining three different sections in depth, and signals from adjacent electrodes are collected together, defining the geometry presented left in figure 2. The EM endcap detector extends the pseudo rapidity coverage from $1.375 < |\eta| < 3.2$, and is build similarly to the EM barrel out of accordion shaped absorbers and electrodes, assembled together to form two concentric wheels in each endcap cryostat, build out of 256 absorbers for the inner wheel, and 768 four the outer wheel. The presampling detectors are build by instrumenting a small gap of argon in front of the precision detectors, however the layout of the electrodes in different between barrel and endcap.

3.2. Hadronic End cap calorimeters (HEC)

The wheels of the hadronic section are build out of 32 modules. Each of them is a alternation of 25 mm thick copper absorbers (50 mm in the second wheel), and signal collection gaps. Such a gap, as can be seen in figure 2, is subdivided in 4 sub-gap, of 1.8 mm each, by three electrode of same nature as the EM electrodes. Two adjacent sub-gaps act as an electrostatic amplifier, allowing to operate at lower voltage than with a single gap. The signal is collected from the central electrode. The total depth of the hadronic section amounts 10 absorption length.

3.3. Forward calorimeters(FCAL)

The forward calorimeter wheels consist of a first copper followed by two tungsten matrices, traversed by cylindrical holes. The holes house the gaps and the electrodes which are build out of signal collecting tubes which in turn contain a copper (respectively tungsten) rod. The tube and the rod are kept in place by a fiber spiraled around them as illustrated in figure 3, the

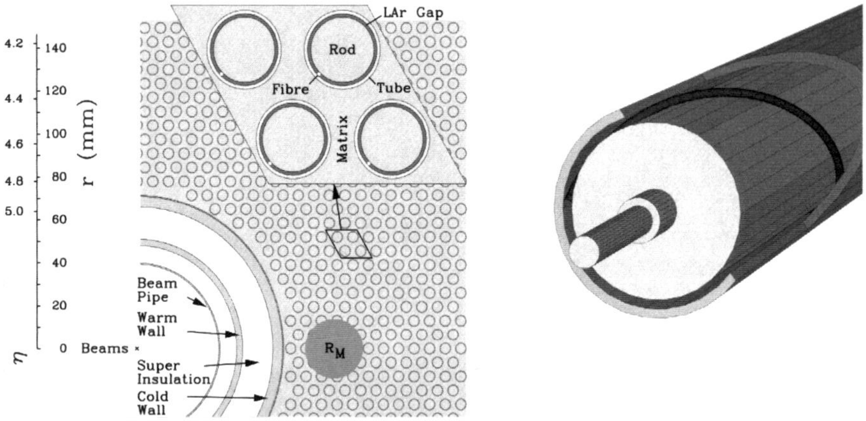

Figure 3. Left, schematic section of a fraction on the FCAL detector. Right, detail of the assembly of an electrode, with the signal collection tube, and the inserted rod, maintained by the plastic fiber.

fiber diameter defines the gap size of 0.25 mm, 0.375 mm and 0.5 mm respectively for the three wheels.

4. Installation and commissioning in the experimental hall

After construction of the detectors and cold testing at the surface, each of the cryostat have been lowered into the ATLAS experimental hall, in October 2004, December 2005 and April 2006 respectively for the barrel cryostat, and for each the two endcap cryostats.

4.1. cryogenics

All three cryostats have been connected to the cryogenic system and cooled down to the operating temperature of 87K. During the cool down, great care was taken to maintaining minimal temperature gradient across the instrument to prevent mechanical tress on the detector. Following its cool down each cryostat was filled with liquid Argon, via condensation of gaseous Argon, this minimizes the turbulences in the Argon bath and the risk to alter the high voltage distribution system. As the signal response is depending on the active medium temperature (2 %/K), the uniformity of the temperature inside the detector volume is important. The temperature of the detector is closely monitored through a large array of precision PT100 sensors, and the dispersion of the temperatures across the detector shows a satisfactory value of order 70 mK, and very stable over time. Also electronegative contamination can affect the quality of the calorimeter response. The contamination is monitored in each cryostat an found to be below 250 ppb in all three cryostats.

4.2. High voltage system

The high voltage system is build out of 157 commercial modules, in the ATLAS counting room, providing individual setting and monitoring of about 4700 channels. On the detector side, redundancy is achieved as different sides of electrodes are powered from different high voltage channels/module. During the commissioning phase, a number of problematic regions in the EMEC and presampler detectors could be cured by injecting a significant charge on the problematic channel. Other techniques have been implemented to address the remaining

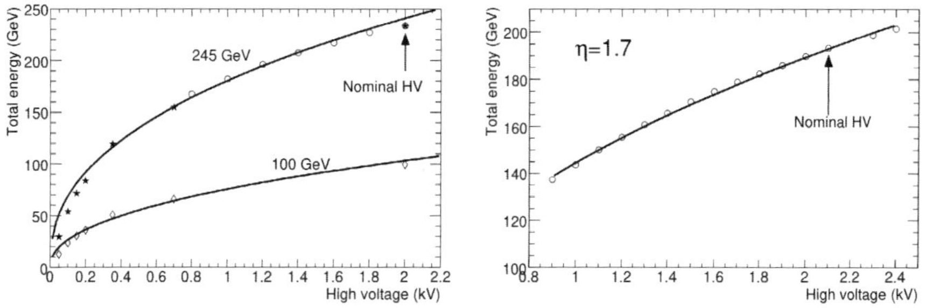

Figure 4. EM calorimeter response as function of the high voltage setting in test beams. Left for the barrel for 245 GeV electrons (circles), 100 GeV electrons (diamonds), and right for 193GeV electrons in a endcap module. Start show the results at nominal high voltage, scaled to the corresponding voltage value.

problems like powering individual sectors of an electrode from an individual channel for problems observed during construction. For remaining problematic region, some half gaps are not powered, which results in a signal reduction by a factor 2 that can be corrected for offline at the reconstruction level. Alternatively some half gaps can be powered from dedicated modules, allowing to flow a DC current through a problematic region, but keeping enough constant high voltage to operate the faulty sector, again this can be corrected for. Overall less than 1 % of the sectors are at reduced voltage, and all detector sectors are good for physics. Studies at test beam validate the corrections procedures for non nominal high voltage settings, see figure 4, showing the good agreement of data and the fitted $E = aV^b$ function.

4.3. Electronics

The LArg electronics, composed of the on detector electronics, so called front end, and off detector electronics, so called back end. It provides precision measurements of the energy in each readout channel for the triggered bunch crossings up to a rate of 75 kHz, and analogue inputs to the first level trigger system proportional with the energy in trigger towers of $\Delta\eta\times\Delta\phi = 0.1\times0.1$. A schematic description of the system is provided in figure 5.

The radiation hard front end electronics (10^{12} n/cm^2/year, 20 Gy/year), has been installed and extensively tested. It is located in 58 crates mounted on the cryostats. It consists of Calibration boards which inject precisely known pulses through high precision resistors onto the detector for EM and HEC, or later in the signal path for FCAL. The signals from the detector are collected on the 1524 Front End boards [6], after amplification and shaping, the signals are summed on one side to prepare the trigger inputs. These signals are further treated and time aligned in dedicated tower builder boards. On the other side after shaping through bipolar filters with 3 gains for each channel in ratios 1/10/100, signals are then stored in an analogue memories. Upon reception of a level 1 accept signal, signals are extracted from the memories, digitized on 12 bits analogue to digital converters and sent to the off detector system for further processing over 1.6 Gbits/s links. For each trigger received up to 32 time samples can digitized on the incomming signals. This is the readout mode used mainly for commissioning noise studies, and recording of cosmic signals, whereas 5 samples is used for physics.

Off detector the calorimeter data are processed in digital signal processors. These provide energy and time estimates through optimal filtering applied to the calorimeter raw signals, using preloaded coefficient determined from earlier electronic calibration runs. This has been demonstrated to operate without noticeable deadtime up to 90 kHz. The data is then further propagated to the ATLAS readout buffering system. The Analogue trigger input signals are

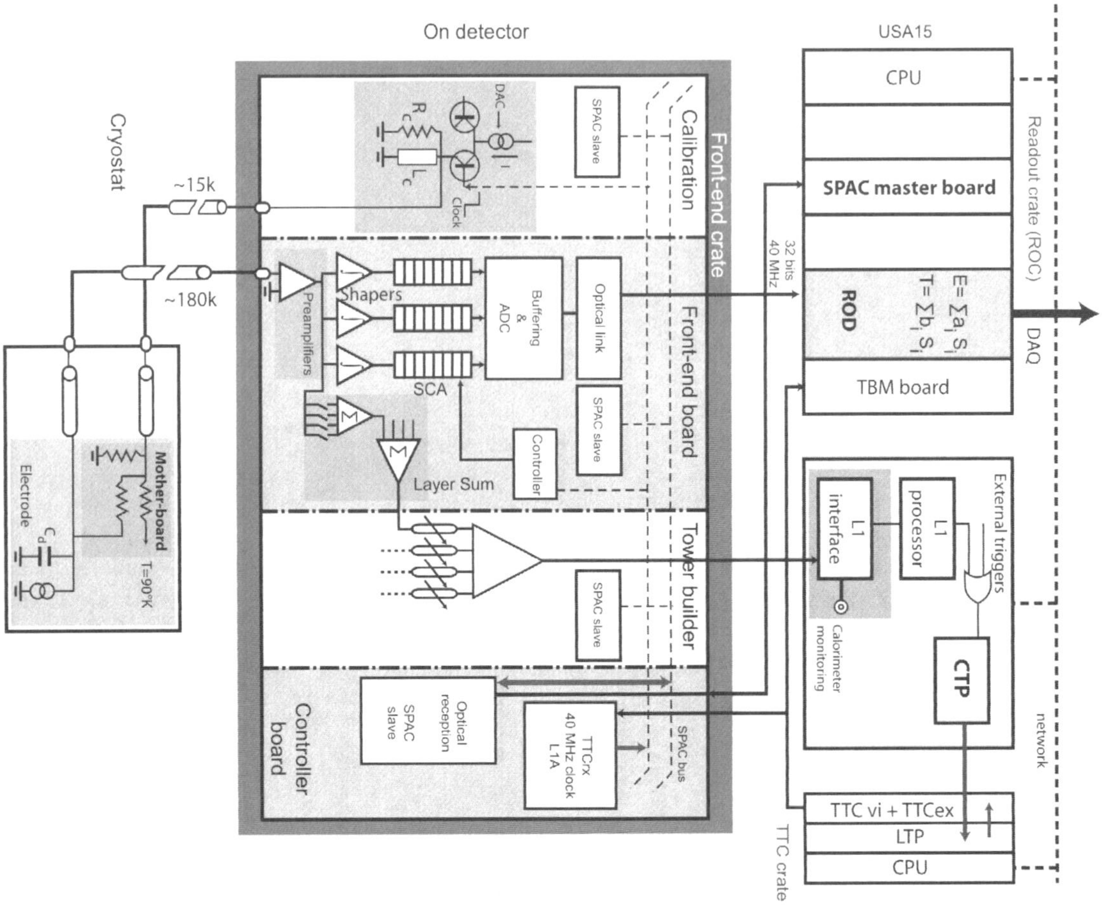

Figure 5. Block diagram of the LArg readout electronics, from left to right, the cold electronics, followed by the on detector electronics, and the off detector readout and trigger systems.

received off detector by a system allowing to adjust the gains before entering the Atlas trigger system. This system allows also to route signal from any trigger tower to a dedicated monitoring readout system and oscilloscopes for specific studies.

4.4. Commissioning

From the start of the electronics installation, the system has been used to study the behaviour of the detector, its readout and and calibration system. Typical data being collected are random triggered data, data collected when pulsing the calorimeter, either varying the pulse amplitude, or the phase of the pulse with respect to the system clock. This allows to study respectively the baseline response and the noise, the gain and linearity of the response, and finally the signal shapes for each channel. Carefully analyses allowed to identify and fix problems on the readout electronics, as wheel as to establish the scheme for the electronics calibration during ATLAS operations. From these studies it appears the calorimeter suffers only 0.015 % of isolated dead channels, and the number of calibration lines without signal, requiring dedicated calibration strategies is down to 0.05 %.

Dedicated runs pulsing the calorimeter and recording the trigger data at the monitoring readout in the trigger receiving system allowed to validate the integrity of the analogue signal path, to establish the gains by comparing the results with the signals obtained from the main readout. Also the relative timing of the trigger signal up to nano second precision has been established. A small number of mis-cablings could be identified and addressed in time.

Specific attention was spent on coherent noise in the calorimeter. The use of both the main and level 1 readout made it possible to identify a number of noise sources and their path into the cryostats. Additional filtering was added to some of the cryogenics services. A modification of the calorimeter grounding scheme at the level of the high voltage filtering system at their feed-through into the cryostat was needed. Also carefully routing and shielding the signal lines inbetween the cryostat feed-throughs and the front end electronics was important. Those actions addressed most of the coherent noise issues. In order to maintain the electrical isolation of the LArg cryostats with respect to other ATLAS systems, key for low noise in the detector, a dedicated system was used injecting a small current to the cryostats and verifying this current returns through the calorimeter ground connection. This system was crucial maintaining adequate grounding over the years of installation and cabling up of the detectors in the experimental hall.

The calorimeter is also tested using the signals induced by atmospheric muons, penetrating the earth and reaching the calorimeters. For this purpose a special trigger logic has been added to the Tile calorimeter electronics, and combined runs involving all of the available calorimeter system has been taken regularly starting from August 2006. Also a number of dedicated periods, aiming at integrating the different ATLAS detectors into a combined system, allowed to test operational strategies, procedures, developing appropriate monitoring. During these periods extensive time is devoted to collect cosmic muon data. Such data allows to validate the electronics calibration procedure, and to test the calorimeter response uniformity. Results from the Calibration runs taken in standalone mode and combined muon runs are presented a separate contribution in these proceedings [8].

5. Conclusions

At the time of this conference, the calorimeter installation has just been completed, and access to it is closed. Through careful testing at all stages of the installation of the detector, readout and services made it possible to deliver an instrument in very good condition: More than 99 % of the high voltage channels are at nominal high voltage, and all of the calorimeter is usable for physics. A very small amount of the signal lines, of order 0.015 % do not transmit signals to the readout, and 0.05 % of the calibration lines need refined strategies. A lot of experience has been accumulated over the commissioning period, resulting in available initial electronics calibration, readily tested on muon induced signals. Steadily moving towards a more and more routine operation of the calorimeter, the collaboration eagerly awaits first collisions from LHC later this year and the exciting physics to follow.

Acknowledgements

The design, construction, and commissioning of the ATLAS liquid argon calorimeter wouldn't have been possible without the excellent contributions of our technicians. We are indebted to their great implication into the construction work. The installation of the ATLAS LArg calorimeter in the cavern was under the responsibility of the ATLAS technical coordination, we are grateful for this very good collaboration.

References

[1] O. Solov´ Yanov, these proceedings.
[2] ATLAS Collaboration, "The ATLAS Experiment at the CERN Large Hadron Collider" accepted for publication in *JINST*

[3] A. Aubert et al, *NIM* A **558** , 2008 pp 388-418
[4] M Aleksa et al, *JINST* **3**, P06003, 2008
[5] A. Atamonov et al, *JINST* **3**, P02010, 2008
[6] N. Buchanan et al, *JINST* **3**, P03004, 2008
[7] J. Ban et al, *JINST* **2**, P06002, 2007
[8] C. Gabaldon Ruiz, these proceedings.

The Electromagnetic Calorimeter of CMS, Summary and Status

Werner Lustermann on the behalf of the CMS ECAL group
ETH Zurich, Institute for Particle Physics, ETH Hönggerberg, 8093 Zurich, Switzerland, c/o CERN, PH Department, 1211 Geneva 23

E-mail: werner.lustermann@cern.ch

Abstract. The construction of the lead tungstate crystal calorimeter for the CMS experiment at the Large Hadron Collider (LHC) is close to completion. The barrel part of the calorimeter composed of 61200 crystals is installed and operated inside CMS. Only 102 readout channels are problematic including the 21 entirely dead, corresponding to 0.17 % and 0.034 %, respectively. All 14648 end-cap crystals are mounted. The electronics installation and commissioning of one end-cap has finished and the second will be finished in few weeks. The construction of the pre-shower detectors installed in front of the end-caps is well advanced.

The many challenges of the design and construction imposed by a 4 Tesla magnetic field, radiation levels ranging from 100 krad up to several Mrad and a bunch crossing rate of 40 MHz were mastered by a huge effort in developing and testing appropriate crystals, photo-detectors and readout electronics carried out over the past $\sim$ 15 years.

Test beam results demonstrate that the energy resolution obtained is better than 0.5 % at high energies. All readout channels in the barrel are inter-calibrated to better than 2 % using cosmic muons.

1. Introduction

The CMS experiment [1] is a typical collider detector, designed to study 14 TeV proton-proton collisions at the LHC with a bunch crossing rate of 40 MHz and a design luminosity of $10^{34}\,\mathrm{cm}^{-2}\mathrm{s}^{-1}$. One of the primary goals is the discovery of the Higgs boson in the mass range suggested by precision electroweak measurements [2] of $m_{\mathrm{H}} \leq 140\,\mathrm{GeV/c}^2$. Depending on the mass of the Higgs boson the $\mathrm{H} \rightarrow \gamma\gamma$ and the $\mathrm{H} \rightarrow \mathrm{W}^+\mathrm{W}^-$ with a subsequent identification of the W and Z bosons through their leptonic decay channels are the most promising discovery channels. In both cases a precise energy measurement for photons and electrons is mandatory. This can be achieved with a crystal calorimeter [3] [4].

2. Construction of the Calorimeter

The crystal calorimeter is made of 61200 lead tungstate crystals for the barrel and 14648 for the end-caps (see fig. 1). A fine-grained pre-shower detector is added in front of the end-cap calorimeter. Major design constraints are imposed by the operation in a 4 Tesla magnetic field and a high radiation environment.

The barrel calorimeter is made of two rings of 18 so called super-modules. Each of them covers 20° in ϕ and a pseudo-rapidity range up to 1.48. A super-module comprises 1700 crystals, of $\sim 26\,X_0$, arranged in four modules. Sub-modules of 2×5 crystals arranged in glass fiber supports,

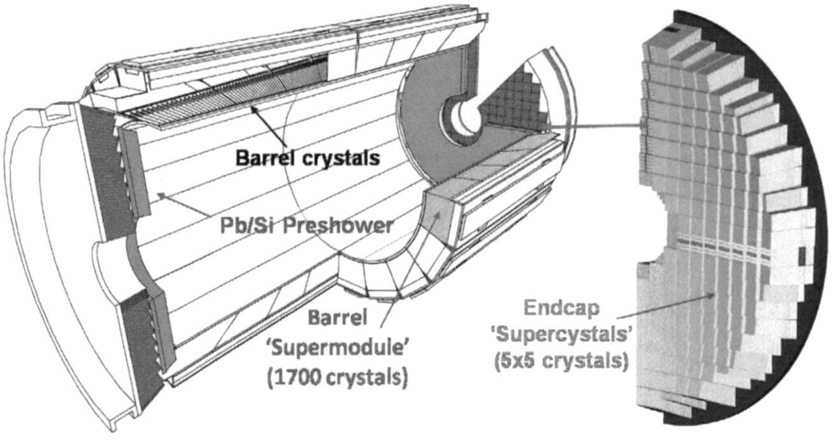

Figure 1. Schematic view of the CMS ECAL, including barrel, end-caps and pre-shower parts.

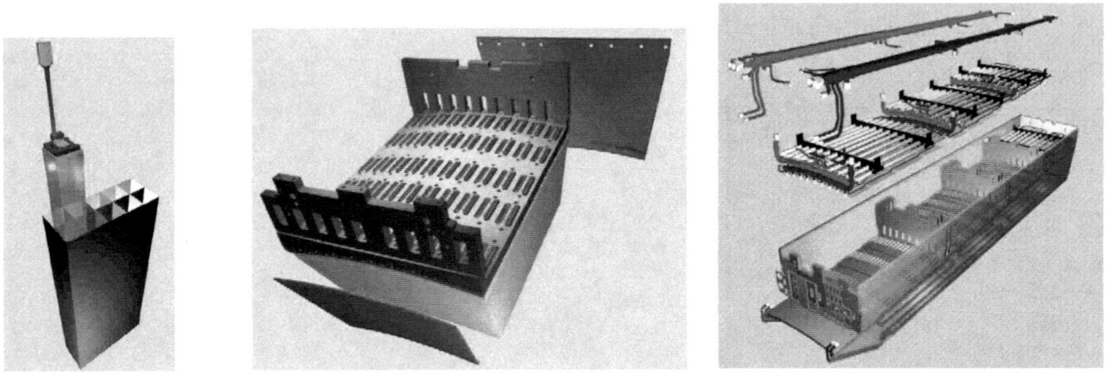

Figure 2. ECAL super-module assembly. Left : Alveolar structure during insertion of the first crystal; Centre : Complete module; Right : Expanded view of a super-module with cooling circuits.

the alveolar structures, together with their photo-detectors are the basic building blocks of the modules (see fig. 2). Hermeticity is achieved by an off-pointing geometry, obtained by a tilt of $\sim 3^\circ$ in η and ϕ of the crystals with respect to the interaction point.

The two end-caps consist of four so called "Dees" each made of 3662 crystals of identical shapes with $\sim 25\,X_0$ in length. Assemblies of 5×5 crystals together with their photo-detectors are arranged into alveolar structures forming so called super-crystals (SC) subsequently mounted onto one side of the Dees backplane (see fig. 3). The readout electronics and cooling circuits are attached to the opposite side. In total 156 SCs are mounted on a Dee. Eighteen of them are of special shape composed of less than 25 crystals in order to approximate the circular shape of the end-cap.

The pre-shower detector [5] is composed of two orthogonal layers of silicon micro-strip detectors placed behind (2 and 1) X_0 of lead absorber, respectively, covering a pseudo-rapidity range of $1.65 < \eta < 2.6$. It improves the π°/γ discrimination. The 4300, 63 mm long silicon sensors with a strip pitch of 1.9 mm are arranged into 496 ladders. They cover a total area of $16.6\,\text{m}^2$.

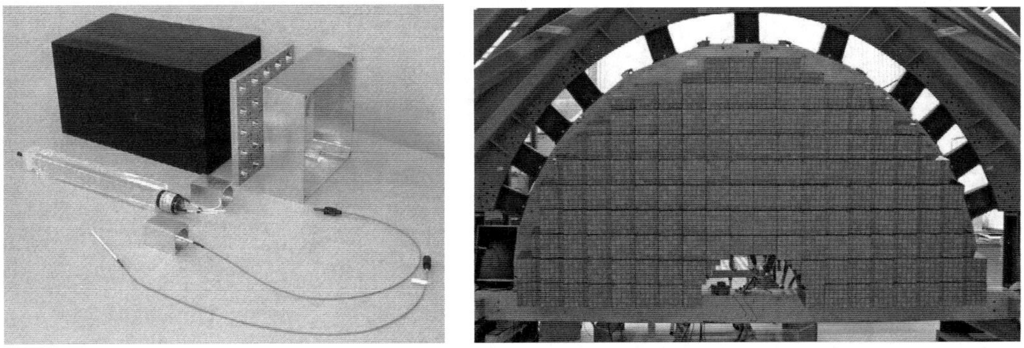

Figure 3. End cap construction: Left : Super crystal mechanics, crystal with VPT attached and monitoring fiber; Right : Dee backplane with all super-crystals mounted.

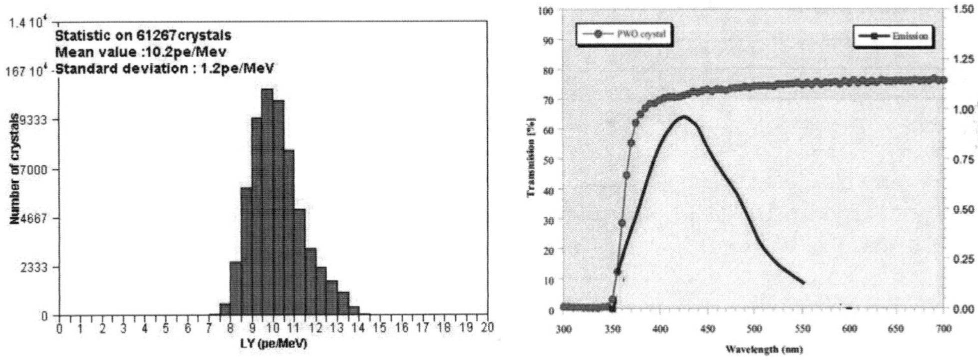

Figure 4. Light yield $PbWO_4$ (left). Transmission spectrum (red) and emmission spectrum (blue) of $PbWO_4$ crystals as a function of wavelength in nm.

3. Crystals

Lead Tungstate ($PbWO_4$) crystals were chosen for having a suitable combination of their properties. They have a short scintillation light decay time; in 25 ns $\sim$ 80% of the light is collected, a short radiation length of 0.89 cm and a small Molière radius of 2.19 cm. The light yield (LY) for 63000 crystals is shown in figure 4 [6]. Crystals with LY $<$ 8 p.e./MeV were rejected. The LY varies with $-2\,\%/^{\circ}$C requiring a very stable temperature. Figure 4 shows the emission and transmission spectrum of lead tungstate. The emission maximum is at 430 nm. The measured light transmission is $(69.2 \pm 2.3)\,\%$ [7]. The required radiation hardness of the crystals has been achieved by doping them with Y/Nb, optimizing the crystals growth and stoichiometric fine tuning. The only radiation damage is a reduced light transmission. Low dose rate irradiations at 0.15 Gy/h were performed up to total doses of 1.5 Gy. Crystals with LY loss larger than 6% were rejected.

4. Photo-detectors

Because of the low light yield of the $PbWO_4$ crystals photo-detectors with intrinsic gain are required. Standard photo-tubes do not work in the 4 T solenoid field.

In the barrel calorimeter Avalanche Photo-Diodes (APDs) with an active area of $(5 \times 5)\,\mathrm{mm}^2$ are used. They were specially developed by Hamamatsu Photonics in collaboration with CMS

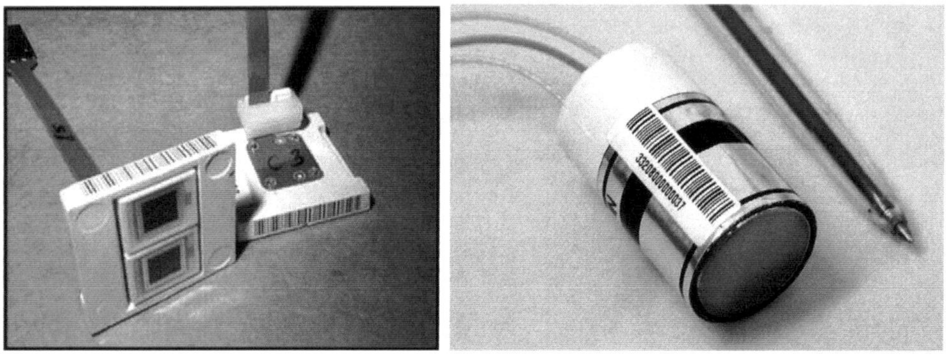

Figure 5. Capsules with two APDs and Kapton flex print (left), VPT (right)

(see fig. 5). The APDs are operated at gain 50 with a bias voltage of (340 – 440) V. They have a capacitance of 80 pF, a quantum efficiency of 75% at 430 nm and an excess noise factor of 2 [8]. The gain depends on the bias voltage and on the temperature with 3.1%/V and −2.4%/ °C, respectively.

The high radiation levels in the end-caps prohibit the use of APDs. Fortunately the angle between the photo-detectors and the magnetic field is limited to 24°. This allows the usage of Vacuum Photo Triodes (VPTs). The VPTs produced by RIE corporation in St. Petersburg have an active area of 280 mm^2 [9]. The quantum efficiency at 430 nm is 22%.

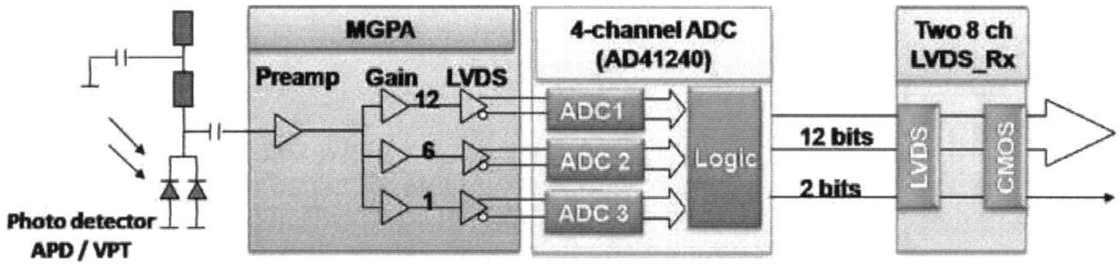

Figure 6. Schematic view of a single channel of the very front end electronics.

5. Readout electronics

The readout architecture of the ECAL is described in reference [10]. The on-detector electronics uses exclusively application specific integrated circuits. The Multi Gain Pre-Amplifier (MGPA) [11], amplifies and shapes the signal. In order to cope with the dynamic range of ~ 1.5 TeV, while maintaining the desired resolution of ~ 40 MeV three different gain ranges are implemented. The three outputs of the MGPA are digitized in parallel by a 4-channel, 40 MHz, 12-bit ADC, the AD41240 [12] and the data in the optimal gain range are selected (see fig. 6). Five identical channels are integrated in a very front end board together with output buffers. The digital data of 25 channels are stored in the so called front end (FE) card during the trigger latency. Every 25 ns the energy sum of 25 channels (barrel) or 5 channels (end-caps) is estimated and sent to the CMS trigger system using a 1 Gbit/s optical link. In case of a positive trigger decision the corresponding data are transferred using another Gbit optical link.

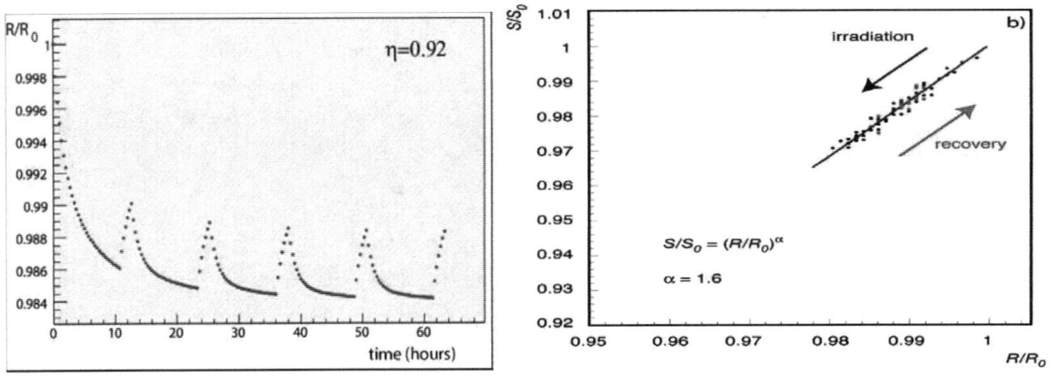

Figure 7. Simulation of the change of the crystal response to electrons due to irradiation (left). Measured correlation of the crystal response to 120 MeV electrons and to laser light (right).

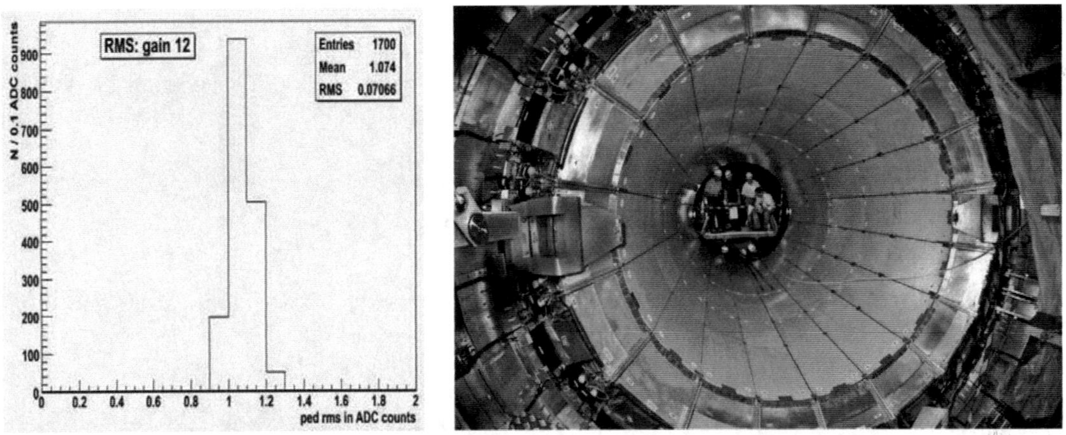

Figure 8. RMS of the pedestal in gain 12 for one super-module (left). All 36 barrel super-modules installed inside the CMS detector (right).

6. Light monitoring system

The calorimeter features a light monitoring system [13]. This is necessary because the light yield of the crystals varies with time due to radiation (see fig. 7). The light monitoring system injects light of two different wavelength, 440 nm and 700 nm into each crystal and measures the response. The correlation of the response of the crystals to laser light with respect to electrons was established during irradiation tests [14] (see fig. 7).

7. ECAL integration and installation

The integration of the ECAL electronics comprises the installation of almost 25000 printed circuit boards and about 8800 optical fibers [15]. Extensive testing at each installation step was mandatory. Therefore a dedicated ECAL electronics integration center was equipped with ECAL low and high voltage systems, cooling units, a detector control system, a laser system, dedicated special purpose readout systems and one final readout system. Originally designed for hosting three super-modules at a time finally up to six super-modules were accommodated simultaneously. The integration of the 36 barrel super-modules was completed within six months.

The performance of each readout channel was tested systematically during the installation. Figure 8 shows a typical distribution of the RMS of the pedestals of one complete super-module. The average noise of 1.1 ADC counts in gain 12 corresponds to $\sim 40\,\text{MeV}$. Beside the measurements of the pedestals we used the test-pulse injection system of the MGPA and the laser monitoring system for systematic checks during integration and commissioning. The installation of the 36 barrel super-modules inside the CMS detector was completed in July 2007 (see fig. 8). Out of the 61200 installed channels only 102 (0.17 %) have problems including the 21 (0.034 %) which are entirely dead.

Since August 2007 the infrastructure has been adapted for super-crystal mounting and electronics integration of the four end-cap Dees. Two of them are complete and the remaining two will follow by end of July. Installation and test sequences are the same as in the barrel.

8. ECAL energy resolution and inter-crystal calibration

The energy resolution of the ECAL barrel super-modules was studied in detail during a test-beam period in summer 2004 [16][17]. The super-module was fully equipped with final readout electronics and connected to the final configuration of services including low and high voltage supplies, cooling system and temperature monitoring. Figure 9 shows the energy dependence of the energy resolution for a central $(4 \times 4)\,\text{mm}^2$ impact of the beam into the crystal. Summing the energy in matrices of 3×3 crystals the obtained energy resolution for electrons of 100 GeV energy and higher is better than 0.5 %. A typical crystal has the following energy resolution for electrons entering the centre of the crystal:

$$\frac{\sigma_E}{E} = \frac{2.8\%}{\sqrt{E}} \oplus \frac{125\,\text{MeV}}{E} \oplus 0.3\% \tag{1}$$

As the energy is extracted by summing at least nine crystals, the relative calibration of the crystal readout channels is of primary importance. Main causes for different responses are the variation of the crystals light yield, the spread of the APD gains and the spread of the gains of the MGPAs.

While the final calibration is done in situ using data from π^0's, W and Z decays a good starting point is very desirable. Therefore nine of the 36 super-modules were calibrated using an electron beam [18], five with a beam energy of 90 GeV/c and four with a beam energy of 120 GeV/c. One of the super-modules was calibrated a second time one month after its first calibration. Figure 10 shows the distribution of the relative differences of the calibration constants. The measured

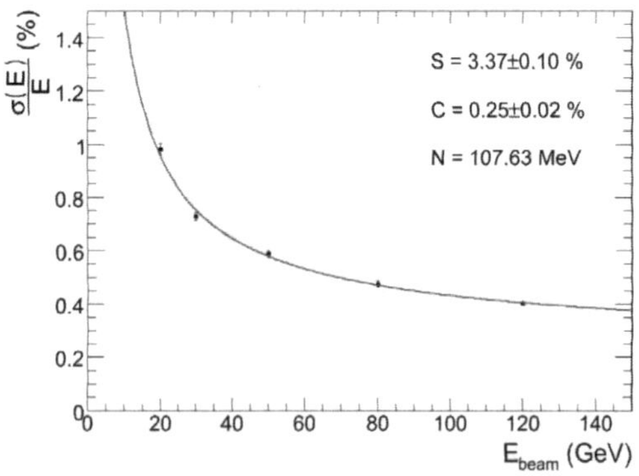

Figure 9. Energy resolution of a matrix of 3×3 crystals. The energy is reconstructed by summing the signals of the 3×3 crystals, for electrons entering the central $(4 \times 4)\,\text{mm}^2$ area.

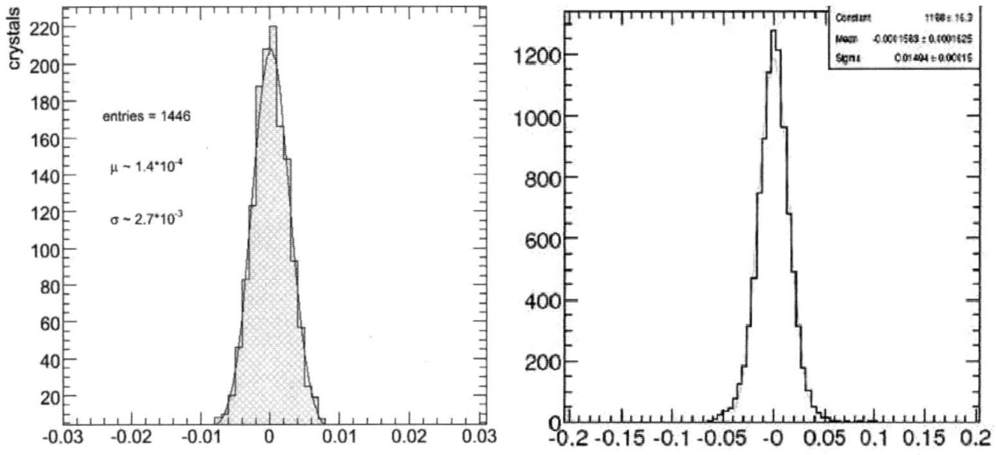

Figure 10. Relative difference of the calibration constants of two calibrations of one super-module (left). Relative difference of the calibration constants extracted from cosmic ray calibration and from electron beam calibration.

RMS of the spread is 0.27 % which is compatible with the statistical uncertainty obtained for the extraction of the calibration constants.

It was understood early enough that a calibration of all super-modules in an electron beam would not be possible. Therefore all 36 super-modules were calibrated with cosmic muon events. In order to assess the accuracy the relative difference of the calibration constants extracted from the cosmic muon calibration were compared with the results from the electron beam calibration. The spread depends on η ranging from 1.4 % to 2.2 % with an average of 1.5 % (see fig. 10) A similar inter-calibration of the four end-cap Dees was not possible, because of the tight schedule.

9. Summary and Outlook

CMS features a high resolution crystal calorimeter with an additional fine grained pre-shower detector in the end-cap region. The barrel part of the detector is installed inside CMS, commissioned and fully operational. It participates regularly to the CMS global data taking of cosmic muons. One of the two end-caps is ready for installation inside CMS and the second will follow soon. The pre-shower detector approaches completion. Systematic studies in an electron beam demonstrated an excellent energy resolution. Using cosmic muons and partially an electron beam a precise inter-calibration of all barrel crystals readout channels has been achieved. This will permit a fast understanding of the detectors in-situ calibration, mandatory for early physics results a the LHC.

10. Acknowledgment

First of all I would like to thank the organizers of the conference for the opportunity to present these results. It is also a great pleasure for me to thank the CMS ECAL group for their kind support in the preparation of the talk as well as of this report.

[1] "The Compact Muon Solenoid (CMS)", Technical Proposal, CERN/LHCC 94-38, LHCC/P1, 1994.
[2] The LEP Electroweak Working Group, "A Combination of Preliminary Electroweak Measurements and Constraints on the Standard Model", LEPEWWG/2003-01.

[3] G. L. Bayatian et al. "CMS The Electromagnetic Calorimeter Project-Technical Design Report", CERN/LHCC 97-33, 1997.
[4] G.L. Bayatian et al. "Detector Performance and Software, Physics Technical Design Report", Volume I, CERN/LHCC 2006-001
[5] E. Tournefier, "The Preshower Detector of CMS at LHC", NIM A461, pp. 355–360, 2001.
[6] E. Auffray et al. "Overview of the 63000 PWO Barrel Crystals for CMS ECAL production", IEEE Tans. in Nucl. Sci. vol 55, N03, June 2007,
[7] E. Auffray et al. "Status of the PWO Crystal production for the Electromagnetic Calorimeter of CMS and of its construction", Proceedings of the 7th International Conference on Inorganic Scintillators and Industrial Applications, SCINT2003, Valencia (Spain), 8-12 Sept. 2003, NIM A 537, pp 373-378, 2005
[8] Z. Antunovic et al. "Radiation Hard Avalanche Photo-diodes for the CMS Detector", 7th International Conference on Inorganic Scintillators and their Use in Scientific and Industrial Applications (SCINT 2003), Valencia, Spain, Sept. 2003, NIM A537, pp. 379-382, 2005.
[9] K.W. Bell et al. "The development of vacuum phototriodes for the CMS electromagnetic calorimeter", NIM A469, pp. 29–46, 2001.
[10] M. Hansen "The New Readout Architecture for the CMS ECAL", CERN-LHCC, pp. 78–82, 2003.
[11] M. Raymond et al. "The MGPA Electromagnetic Calorimeter Readout Chip for CMS", CERN-LHCC, pp. 83–87, 2003.
[12] G. Minderico et al. "A CMOS low power, quad channel, 12 bit, 40 MS/s pipelined ADC for applications in particle physics calorimetry", CERN-LHCC, pp. 83–87, 2003.
[13] J. Rander "The monitoring system for the CMS-ECAL", *Proceedings of the Eighth International Conference on Calorimetry in High Energy Physics*, Lisbon (Portugal), June 1999.
[14] A. Van Lysebetten and P. Verrecchia, "Performance and measurements of the light monitoring system for CMS-ECAL from 2002 test beam data", CMS NOTE, CMS RN 2004/001.
[15] W. Lustermann et al. "Installation and Test of the CMS Crystal Calorimeter Electronics", CERN-LHCC-2005-38.
[16] P. Adzic et al. "Energy resolution performance of the CMS electromagnetic calorimeter", CMS NOTE, 2006-140.
[17] P. Adzic et al. "Energy resolution of the barrel of the CMS Electromagnetic Calorimeter", CMS NOTE, 2006-148, JINST 2 P04004 (2007).
[18] P. Adzic et al. "Intercalibration of the barrel electromagnetic calorimeter of the CMS experiment at startup", CMS NOTE 2008/18.

The ALICE PHOS Calorimeter

Hisayuki Torii for the ALICE-PHOS Collaborations

Graduate School of Science, Hiroshima University, Kagamiyama 1-3-1, Higashihiroshima, Japan, 739-8526

E-mail: `htorii@hiroshima-u.ac.jp`

Abstract. The ALICE PHOton Spectrometer (PHOS) is designed to detect photons in nucleus-nucleus collisions at the Large Hadron Collider (LHC). The basic design and the measured performance with the first constructed module are shown. The installation schedule and the future construction plan are reported.

1. Introduction

The ALICE detector has been designed to study the strongly interacting matter created in nucleus-nucleus collisions at the Large Hadron Collider (LHC). Measurement of direct photons is probing the initial state of the interactions. In the heavy ion collisions program at LHC, it is very critical to measure thermal radiated photons, which is known to carry the temperature information of created hot medium, with very fine energy and position resolution.

The ALICE PHOton Spectrometer (PHOS) [1] consists of 17920 PWO crystals to each having size of $22 \times 22 \times 180mm$ and Avalanche Photo Diode (APD) for readout of scintillation light. The performance result of prototype modules for the PHOS calorimeter has been reported earlier[2]. First PHOS module consisting of 3584 crystals was constructed and tested by utilizing 2GeV$/c$ electron beam in the summer of 2006 and cosmic rays in 2007. The HV bias for APDs were calibrated in order to obtain equal gain for better trigger performance. Second PHOS module is under construction and installed for the first p+p collisions at LHC. Further research and development program for improved performance of PHOS as a electro-magnetic calorimeter are under studying.

In this paper, we will present the construction and installation status of the PHOS modules and performance results. In addition, physics potential with the PHOS module during the first physics run of LHC will be discussed.

2. PHOton Spectrometer (PHOS) in ALICE Experiment

Figure 1 depicts the ALICE setup consisting of two arms; a central arm inside the L3 magnet and a muon arm shown in the right part of the figure. The PHOS detector is placed below of the collision vertex at 460cm distance and covers 100deg in the azimuthal angle and pseudo-rapidity of between -0.12 and 0.12. It designed to measure photons in the heavy-ion collisions at LHC.

Several measurements of photons in heavy-ion collisions were reported at the several accerellator, AGS, SPS, and RHIC. One of the most interesting physics is the thermal photon measurement since it is expected to be probing the initial state of the strongly interacting matter, Quark Gluon Plasma (QGP), created in nucleus-nucleus colisions.

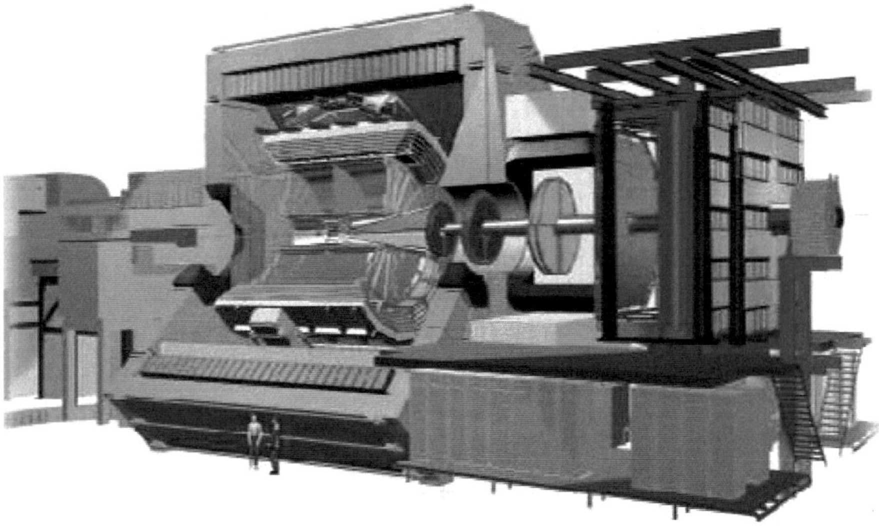

Figure 1. ALICE Setup

At the LHC energy in heavy-ion collisions, several theoretical predictions for thermal photon production [1][2][3]. Also see more compilation of general photon physics in heavy ion collisions at the LHC [4]. According to the predictions, the thermal photons are expected to be measurable in the energy range between 1 and 10GeV depending on the temperature of the medium and the strength of the background photons. One of the two major requirements on PHOS is to have good energy resolution down to the energy range of a couple of 100MeV. At the LHC energy, we expect that the heavy ion collisions produce about 8000 charged particles in a unit of rapidity. In order to survive the high multiplicity condition, another requirement is the segmentation size. The PHOS detector is specially designed to fulfill these both requirements by focussing the geometrical coverage at very small region. In addition, the high energy photon is one of good tools to study the QCD process. Especially, the jet measurement in the opposite direction of the photon is interesting physics process in the heavy-ion collisions in terms of the jet modification in hot medium.

All electro-magnetic calorimeters which are under construction at three LHC experiments; ATLAS, CMS, and ALICE are listed in Table 1. The rapidity and azimuthal coverage of PHOS is smaller than that of others, however PHOS is designed to fulfill the requirements for the thermal photon measurement; fine segmentation, good energy resolution and dynamic range to lower energy.

3. Basic Design

In this section, an overview of the PHOS structure is described. See ALICE Technical Design Report[5] for further detail.

3.1. $PbWO_4$ Crystal + Avalanche Photon Diode

The PHOS consists of combination of $PbWO_4$ (PWO) crystals and Avalanche Photon Diode (APD). In total, 17920 channels covers 100deg in azimuthal angle and ±0.12 in pseudo-rapidity. Total weight is 12.5ton.

The PWO crystal is known as its characteristics of the fast signal and the small Moliere radius of about 2cm. Each crystal of the PHOS detector has $22\times22\times180$mm dimention, which

Exp.	ATLAS		CMS		ALICE	
Name	LAr Barrel	LAr Endcap	ECAL(EB)	ECAL(EE)	**PHOS**	EMCal
Structure	Liquid Ar		PWO + APD		**PWO + APD**	Pb + APD
Coverage	$0<\|\eta\|<1.4$, 2π	$1.4<\|\eta\|<3.$ $2.\ 2\pi$	$0<\|\eta\|<1.5$, 2π	$1.5<\|\eta\|<3.$ $0.\ 2\pi$	**$0<\|\eta\|<0.12$, 0.6π**	$0<\|\eta\|<0.7$, 0.6π
Granularity $\Delta\eta \times \Delta\varphi$	0.003x0.100 0.025x0.025 0.025x0.050	0.025x0.100 0.025x0.025 0.025x0.050	0.0174x0.0174	0.0174x0.0174 to 0.05x0.05	**0.004x0.004**	0.0143x0.0143
Res.	$10\%/\sqrt{E} \oplus 0.5\%$	$10\%/\sqrt{E} \oplus 0.5\%$	$2.7\%/\sqrt{E} \oplus 0.55\%$	$5.7\%/\sqrt{E} \oplus 0.55\%$	$3.3\%/\sqrt{E} \oplus 1.1\%$	$7\%/\sqrt{E} \oplus 1.5\%$

Table 1. Comparison of Electro-Magnetic Calorimeter at ATLAS, CMS, and ALICE.

corresponds to $20X_0$ radiation length in the longitudinal direction. The crystals are kept at the temperature of -25deg, where the light yield is enhanced to be 230 photo-electron per MeV. The scintillation light has the wave length of 400nm to 500nm.

The APD of Hamamatsu S8664-55 type is employed for detecting the scintillation light from PWO. Its advantage is the high quontum efficiency of 60% to 80% depending on the light wave length. With the PWO crystal and the APD active surface of 5×5mm, 4.5 photo-electron per MeV is expected at the temperature of -25deg. Another advantage is its capability to be operational in magnet field. Each APD is mounted on one preamplifier board where the integration of charge is done.

3.2. Front-End Electronics and Trigger System

The front-end electronics (FEE) are designed to fit the compactness of the PHOS module. The research and development has started since 2004. Two 10-bit 10MHz sampling ADCs for two different dynamic ranges are implemented. The dynamic range in total is from 5MeV to 80GeV. The electronic noise is 615 electron, which is equivalent to 3.1MeV. One FEE board covers 32 APD reading and has 32 APD bias regulators.

The sum of 2x2 analog signals are processed through a fast shaper amplifier on the FEE board and transfered into Trigger Region Unit (TRU) for the purpose of triggering energy activities in PHOS. 12-bit sampling ADC are used for the digitization at the rate of 40MHz. The energy sum of overlapping 4x4 crystals are taken and transfered into Trigger OR (TOR) board with a programmable FPGA, where sophisticated decision for physics trigger can be performed for enhancing isolated photons or jet for example.

3.3. Control and Monitoring

The slow control and monitoring of the PHOS modules are implemented in an ALICE standard framework, Detector Control System (DCS). The six different parts, FEE, FEE cooling, PWO cooling, LV power supply, HV power supply, LED calibration system are controled and monitored by three computers. All are accesible from one computer together with the other detectors by a shift leader as for a global control in ALICE.

We also have a monitoring system of the data stream obtanied by the DAQ system. The same data stream is monitored through the High Level Trigger (HLT) system, where we check the quality of the data, export information for the online monitoring display, and perform compression of the raw data.

4. PHOS Performance

In this section, we will describe the measured performance with the constructed PHOS module consisting of 64×56 xtals in addtion to previous reports on 8x8 prototype [6], 16x16 prototype [8] and expected performance [7][9][10].

4.1. Test Beam in 2006

A construction of first PHOS module out of five modules was finished in the spring of 2006 and was tested at at a beam line. The test was done at T10 beam line of CERN-PS in the summer of 2006. As described in Fig. 2, several beam counters and electron identification gas counter are used in the beam line. The first PHOS module was put on a moving stage so that the beam can be irradiated into any positions of the PHOS module surface. Electrons with 2GeV/c momentum are irradiated into most of channels in order to calibrate the relative energy. During the beam time, the APD bias was successfully adjusted for equalized gain by accuracy of 4% . In addition, electrons with 1-5GeV/c and π^- with 5GeV/c were used to investigate the energy resolution and the shower shape.

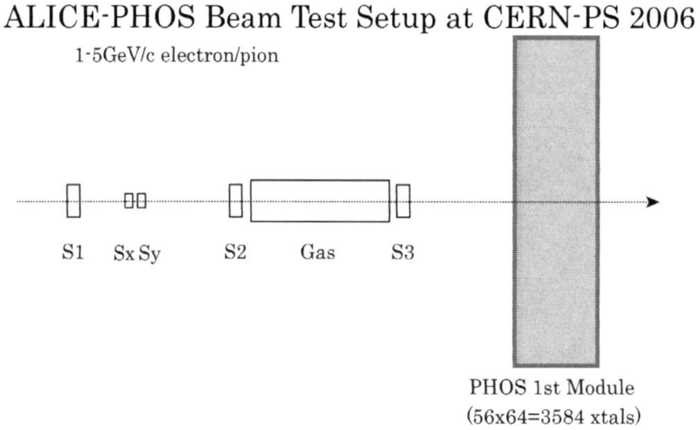

Figure 2. Test beam setup at T10 beam line of CERN-PS in the summer of 2006

Figure 3 shows the obtained energy resolution as a function of photon energy together with previous results from the prototypes [6][8] and a fitting result by following formula;

$$\frac{\sigma}{E(GeV)} = \sqrt{\frac{0.018}{E} \oplus \frac{0.033}{\sqrt{E}} \oplus 0.011.}$$

We concluded that the measured energy resolution of the first 64×56 module is consistent with previous results from prototypes and all can be described by a single formula. Also all results fulfill the requirements on the PHOS performance as shown in the figure.

4.2. Cosmic Ray in 2007

The first constructed PHOS module is also tested by cosmic rays during summer in 2007. The PHOS module was rotated so that the vertical cosmic-ray penetrates longer path (18cm) of

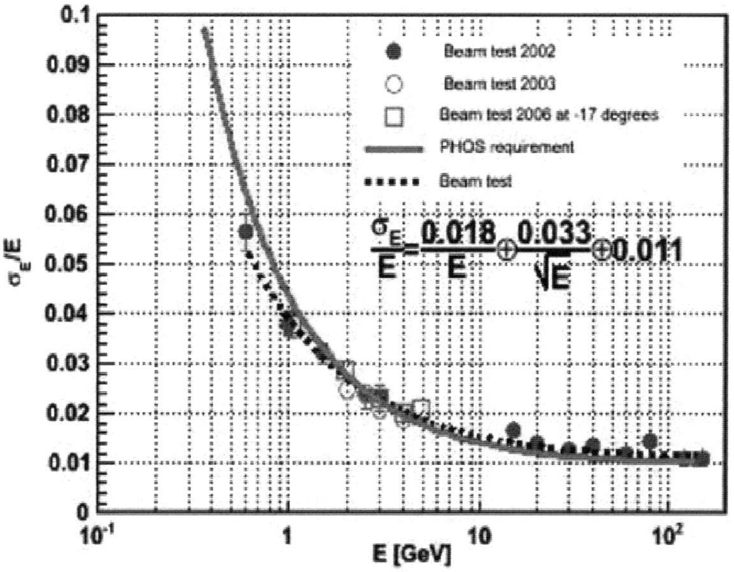

Figure 3. Energy resolution measured in 2006 for first PHOS module together with the measured resolution for prototypes [6][8]. All results can be described by a single fitting as shown in a dotted line.

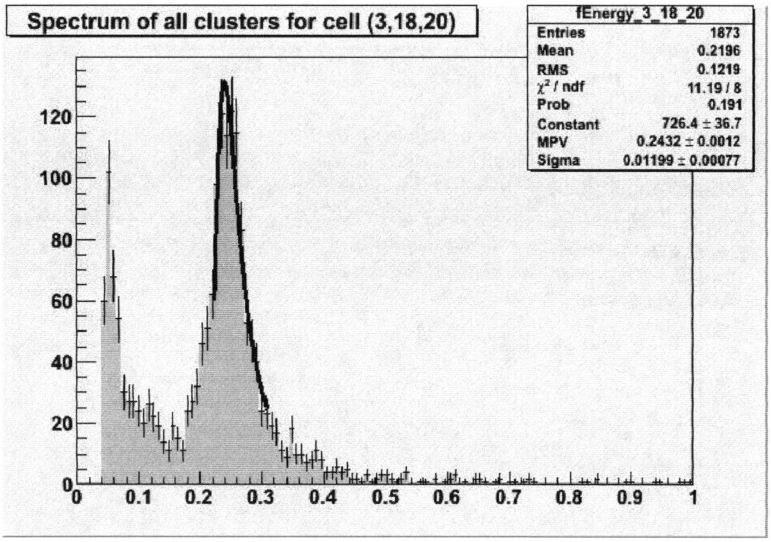

Figure 4. Cluster energy distribution by cosmic-rays in the summer of 2007 together with a fitting result.

xtals. By utilizing a prototype of the cooling system, the PWO crystals were cooled at the temperature of about -17deg. A scintillator of $40 \times 80 \times 2cm^3$ was set by 61cm distance from another scintillator of $50 \times 100 \times 4cm^3$. The bottom one was put just on the PHOS module and covered about 1107 (27x41) xtals. In order to cover all xtals, they needed to be moved on the PHOS surface several times. The analog signals from both were discriminated by thresholds of 300mV and used for starting the data taking. The trigger rates after coincidence was about 10Hz, which is consistent with an expectation.

Same APD bias as the test beam in 2006 was applied during this test, where we expect the gain of all channels were equalized by 4%. On the other hand, there was clearly seen that > 10% gain unbalance so that they were corrected in the analysis. Figure 4 shows the cluster energy distribution after the relative energy are calibrated. A clear peak at around 250MeV was observed together with an exponential background. These data were used for an initial calibration and also for quality check of all channels.

5. Installation Schedule

First p+p collisions at LHC is scheduled in 2008 summer at the center of mass energy of 900GeV and 10TeV. Secondly constructed PHOS module was succesfully constructed and installed on May 2008 and is operated at normal room temperature. First and third constructed modules are basically ready and they are under mechanical upgrade now to fulfill the air-tightness reuirement. They will be tested in the lab in 2008 with cosmic and electron beam and will be installed in ALICE during shutdown after first LHC beam and before the first Pb+Pb collisions, which is scheduled after second year of LHC.

The construction of fourth and fifth modules are scheduled after the installation of the other three modules. All five modules are scheduled to be ready before the full luminosity run of Pb+Pb collisions at LHC.

References

[1] P. Aurenche, *Nuclear Physics,*, A. 783, p. 395, 2006.
[2] S. Turbide *et. al.*, *Physics Review,*, C. 69, p. 014903, 2004. S. Turbide *et. al.*, *Physics Review,*, C. 72, p. 014906, 2005.
[3] F. Arleo, D. d'Enterria, D. Peressounko, *nucl-th* 0707 2357.
[4] F. Arleo *et. al.*, "Yellow Report on Photon Physics in Heavy Ion Collisions at the LHC", hep-ph/0311131
[5] ALICE Technical design report, *http://aliceinfo.cern.ch/Collaboration/Documents/TDR.*
[6] M. Ippolitov *et. al.*, *Nuclear Instrument and Methods in Physics,*, A. 486, p. 121-125, 2002.
[7] G. Conesa *et. al.*, *Nuclear Instrument and Methods in Physics,*, A. 537, p. 363-367, 2005.
[8] D.V. Aleksandrov *et. al.*, *Nuclear Instrument and Methods in Physics,*, A. 550, p. 169-184, 2005.
[9] ALICE Collaboration, *Journal of Physics G: Nuclear and Particle Physics,*, 30, p. 1517-1763, 2004, Physics Performance Report I.
[10] ALICE Collaboration, *Journal of Physics G: Nuclear and Particle Physics,*, 32, p. 1295-2040, 2006, PHysics Performance Report II.

LHCb Preshower(PS) and Scintillating Pad Detector (SPD): commissioning, calibration, and monitoring

Eduardo Picatoste Olloqui on behalf of the LHCb Collaboration

Universitat de Barcelona, Facultat de Física, Av Diagonal 647, Barcelona, 08028 Spain

E-mail: Eduardo.Picatoste.Olloqui@cern.ch

Abstract. The calorimeter system of the LHCb experiment comprises in front to the electromagnetic and hadronic calorimeters a double detector, aimed at tagging the electric charge and the electromagnetic nature of the calorimeter clusters for the first level of trigger. It consists of two planes of scintillating pads separated by a 2.5 radiation lengths lead sheet. The first scintillating plane is the SPD. The second plane and the lead form the Preshower. Their design and construction are recalled. Their performance and calibration strategies are discussed and illustrated by recent results.

1. Introduction

The LHCb experiment, one of the four particle physics detector experiments on the Large Hadron Collider accelerator at CERN, will perform a through study of CP-symmetry violation and rare decays of b-hadrons which should complete the understanding of quark flavor physics in the framework of the Standard Model, and may reveal assign of the physics beyond. Interest in CP violation comes not only from elementary particle physics but also from cosmology, in order to explain the dominance of matter over antimatter observed in our universe, which could be regarded as the largest CP violation effect ever seen.

The LHCb is a single arm spectrometer with a forward angular coverage from approximately 10 mrad to 300 mrad in the horizontal projection and 250 mrad in the vertical projection. The choice of the detector geometry is motivated by the fact that at high energies both the quarks from bb-pairs are predominantly produced at small angles to the beams, a feature exploited in the flavor tag (as described in the LHCb technical proposal [1]). The LHCb consists of a beam pipe, a magnet, the Vertex Locator (VELO), four stages of tracking system, two Ring Imaging Cherenkov, the calorimeters, the Muon System, and the Trigger [2] (**Figure 1**).

The LHCb trigger has two decision levels. Using custom made electronics, the first decision is made based on high transverse momentum hadrons or electrons found in the calorimeter system, or muons found in the muon system, at the bunch crossing rate of 40 MHz. All data from the detector are then read out at a rate of 1 MHz and sent to a CPU farm for further event reduction. For this purpose, all the detector information is available.

2. LHCb Calorimeter and SPD/PS detector role

The main purpose of the calorimeter system is the identification of hadrons, electrons and photons, and the measurement of their energies and positions. This information is the basis of the Level 0

Journal of Physics: Conference Series **160** (2009) 012046 doi:10.1088/1742-6596/160/1/012046

trigger, and therefore has to be provided with sufficient selectivity in a very short time. The set of constraints resulting from this functionality defines the general structure and the main characteristics of the calorimeter system and its associated electronics [1]. The ultimate performance for hadron and electron identification will be obtained at the offline analysis level. This stage, however, has little influence on the detector design.

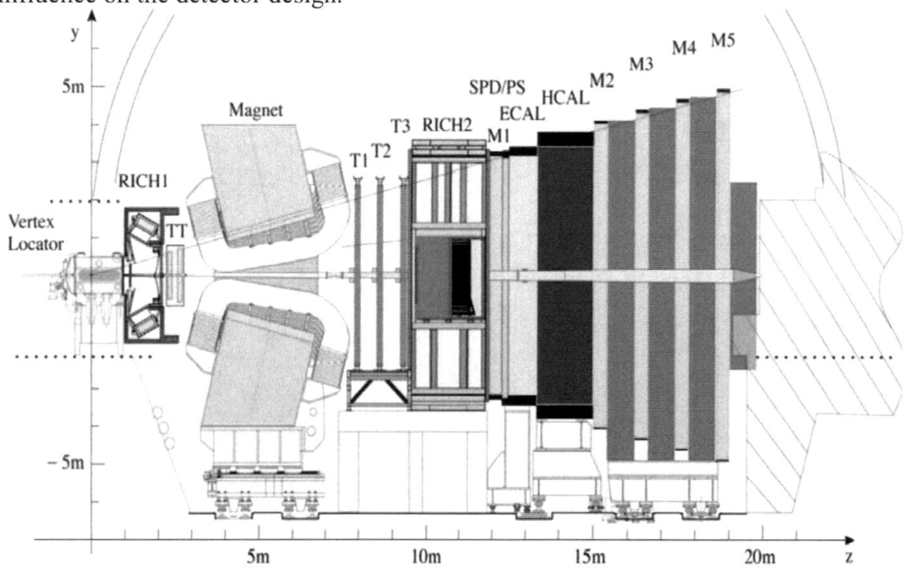

Figure 1. LHCb layout.

The Calorimeter is composed of an electromagnetic calorimeter (ECAL), followed by a hadron calorimeter (HCAL), and before both of them there is a double detector made by three layers, the Scintillator Pad Detector (SPD), a 2.5 radiation lengths lead wall, and the Preshower (PS) [3]. The SPD/PS system helps the calorimeter on the requirement of good background rejection and reasonable efficiency on the detection of photons with enough precision to enable the reconstruction of B-decay channels containing a prompt photon or π^0 and on the electron identification. The SPD (Scintillator Pad Detector) identifies charged particles, and allows electrons to be separated from photons. The PreShower detector identifies electromagnetic particles. The electro-magnetic calorimeter ECAL, of the shashlik type, measures the energy of electromagnetic showers. The hadronic calorimeter HCAL, made of iron with scintillator tiles, measures the energy of the hadrons.

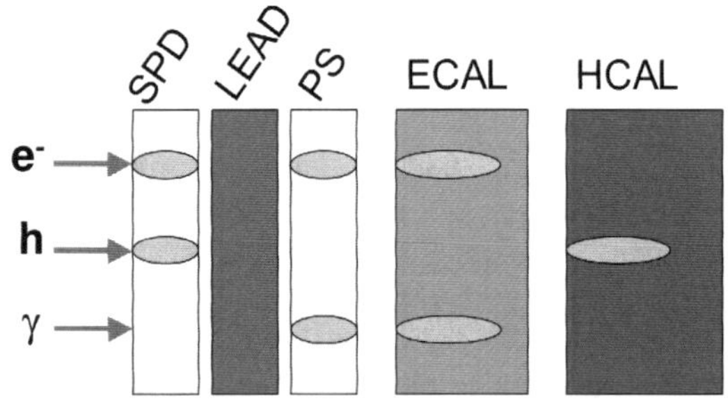

Figure 2. Signal deposited on the different parts of the calorimeter by an electron, a hadron, and a photon.

3. SPD/PS design

The SPD and PS are two planes of scintillator pads separated by a lead 12 mm thick sheet. A groove in each pad holds the helicoidal wave lift shifting (WLS) optical fiber which collects the scintillating light (**Figure 5**). The transmission of this light is done by long clear fibers to multi-anode photomultipliers tubes (MAPMT) that are located, along with the Very Front End (VFE) electronics in boxes above and below the detector in order to optimize the light yield at the MAPMTs. To handle the data for the first trigger level as quickly as possible, the signals are shaped directly at the VFEs. While on the PS the VFE just shapes and integrates the signal, the SPD VFE includes a discriminator telling whether a cell has been hit. The PS signal is digitised in the Front End (FE) electronics placed in crates on top of the calorimeter detectors after 20-30 m of cable (see **Figure 3**). The FE boards hold the data of each channel, sampled at 40 MHz, in digital pipelines waiting for the first level trigger decision. Next to the FEs, the SPD Control Boards (CB) includes the functionality necessary to configure the SPD VFEs and deliver the SPD multiplicity trigger.

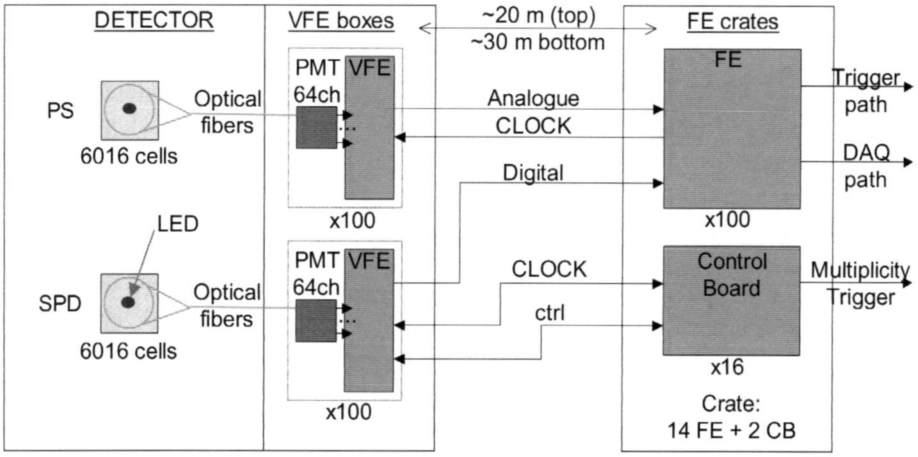

Figure 3. SPD/PS readout electronics scheme.

3.1. Cell geometry

The cell granularity of SPD/PS is projective as respect to ECAL. As displayed in **Figure 4**, the cells are about 4x4 cm^2, 6x6 cm^2, and 12x12 cm^2 for inner, middle, and outer parts, respectively. The exact size of the cells is proportional to their distance from the vertex in order to obtain a pointing geometry.

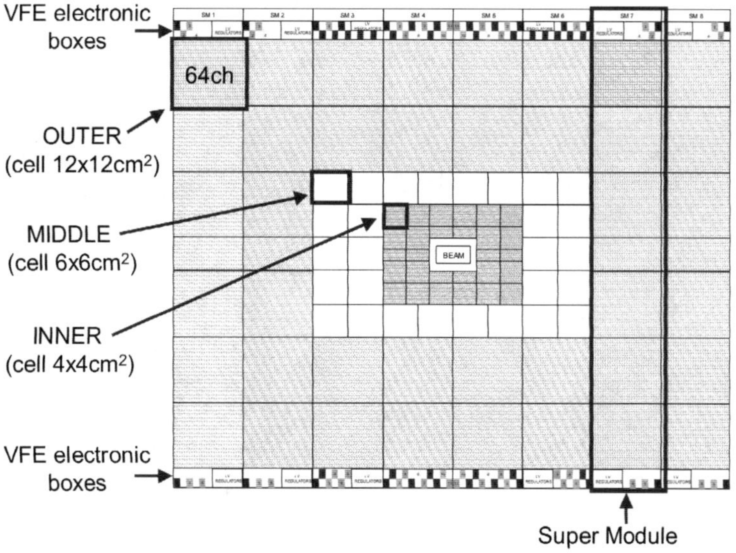

Figure 4. SPD/PS detector scheme.

Each square represents a group of 64 cells (a few are only 32) which corresponds to a readout electronics group of channels. Both SPD and PS planes are divided in 8 vertical parts called Super Modules (SM) which contain from 7 to 22 VFE units with its corresponding electronics positioned on the boxes over and below de detector cells.

3.2. Detector pads calibration

The detector structure provides an average of about 25 photoelectrons in response to a minimum ionizing particle (MIP) which is enough to allow to distinguish between electron and photon showers.

All the cells were calibrated before the installation using the final Super Modules and some calibrated MAPMTs. The SMs were placed horizontally and data was taken to obtain the number of photoelectrons per MIP on each cell. **Figure 6** shows an example of the measured number of photoelectrons of all the cells in a SM.

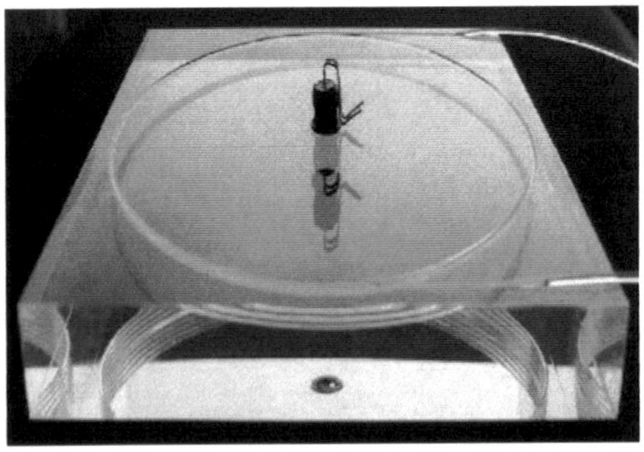

Figure 5. A SPD/PS detector cell. It can be observed the helicoidal WLS fiber and the LED.

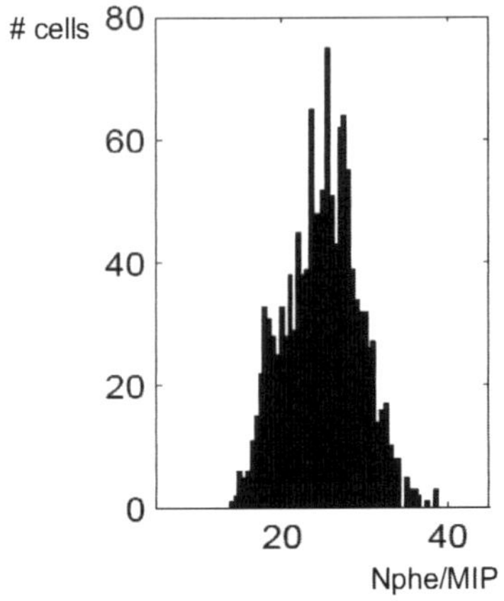

Figure 6. Number of photoelectrons per MIP on all the cells in one of the Super Modules of the SPD/PS.

4. Commissioning

4.1. Pedestals/offset and noise

Prior to any measurement with any of the SPD/PS detectors, it is necessary to perform an analysis of the noise and the pedestals (or threshold offset for the SPD) for all channels. The SPD/PS pedestals and noise have been extensively studied and the results are plotted in **Figure 7** and **Figure 8**. All the PS channel pedestals are below 300 ADC counts with a mean of about 140 ADC counts. And, whereas 1 MIP is approximately 10 ADC counts, the noise is essentially less than 1.2 ADC counts.

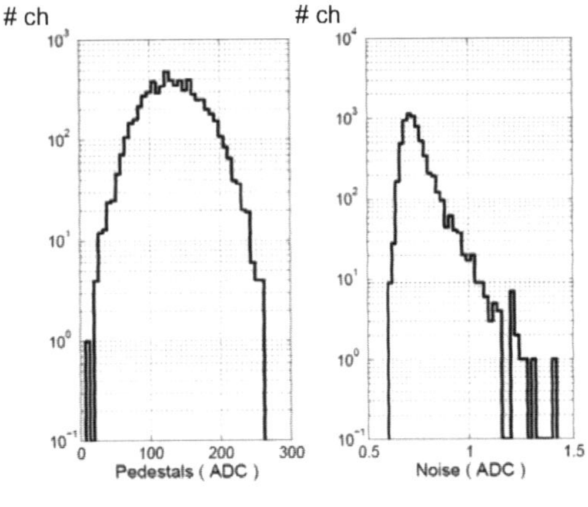

Figure 7. PS pedestals and noise (all 6016 channels).

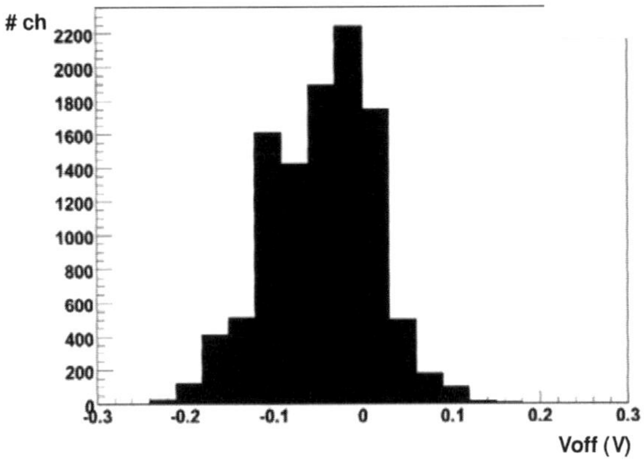

Figure 8. SPD offset values (all 6016 channels).

On the SPD side, the channel offset values are distributed around –70 mV with a standard deviation of about 70mV (as exposed in **Figure 8**). Although the noise on SPD channels is not represented here, it is confined to values lower than 3 mV, which is much less than the signal corresponding to 1 MIP (100 mV).

4.2. Time alignment

Since all signals between the VFE and the FE boards communicate through long cables (20-30 m) and the time requirements for the data capture are sharp (the detector clock is of 40 MHz), all the FE boards are equipped with time delay units to ensure a correct communication on each line.

For the PS electronics there are three phases (**Figure 9**) to be set so the data transmission has no errors. Additionally, these phases are correlated.

- The signal integration starting time (Φ_{VFE}). It depends essentially on the data cable length and integration clock.
- The sampling of the VFE signal by the ADC in the FE ($\Phi_{VFE\text{-}ADC}$).
- The sampling of the ADC signal by the FEPGA ($\Phi_{ADC\text{-}FE}$).

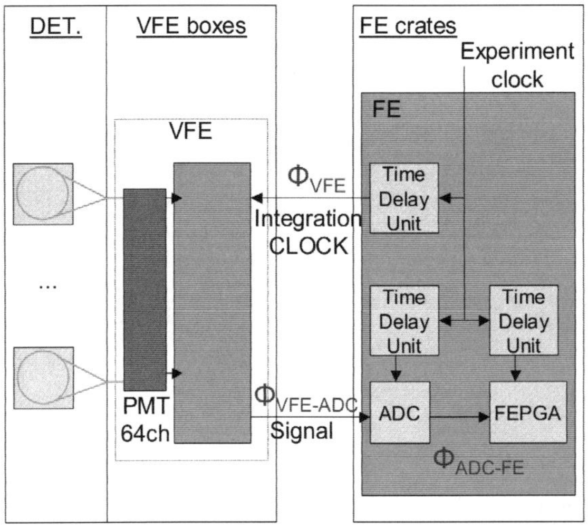

Figure 9. Phases to be set within PS boards.

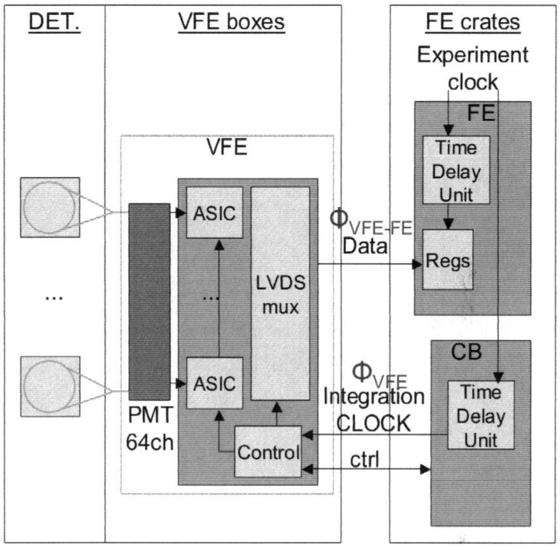

Figure 10. Phases to be set within SPD boards.

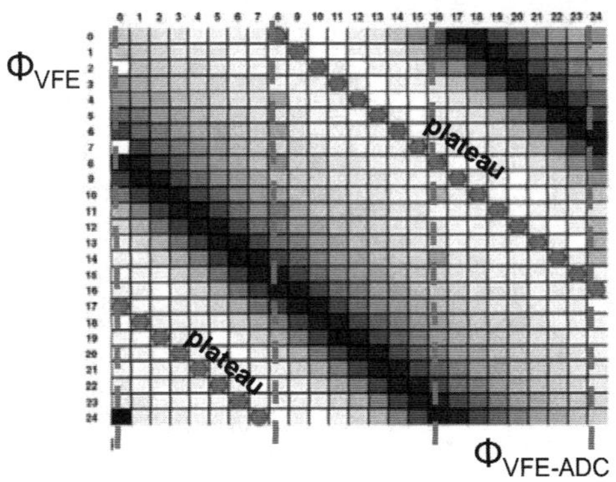

Figure 11. PS time alignment: sample VFE integration ($\Phi_{ADC\text{-}FE}$ fixed).

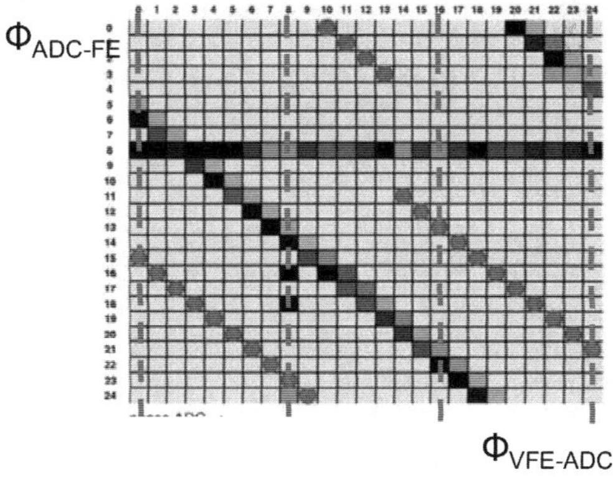

Figure 12. PS time alignment: re-sample in FEPGA (Φ_{VFE} fixed).

The alignment method followed for the PS phases comprises three steps:

- Φ_{VFE} is obtained scanning and searching for the maximum value from the VFE output.
- Fix a correct value of $\Phi_{ADC\text{-}FE}$, then, using LED signal, it can be scanned the allowed Φ_{VFE} - $\Phi_{VFE\text{-}ADC}$ difference (**Figure 11**).
- Taking a fixed Φ_{VFE}, it is possible to scan the allowed $\Phi_{ADC\text{-}FE}$ and $\Phi_{VFE\text{-}ADC}$ region from pedestal measurements (**Figure 12**).

In the case of the SPD electronics (**Figure 10**), the phases to be set within SPD boards are just the ones corresponding to the signal integration starting time (Φ_{VFE}) and the VFE data read at the FE ($\Phi_{VFE\text{-}FE}$), which are correlated as well. Unlike PS signal, the data sent by the SPD VFEs is only 1 or 0 for each channel. Fortunately, the VFE includes a special mode which allows different patterns to be sent and perform bit error rate tests. These can be done for all possible Φ_{VFE} and $\Phi_{VFE\text{-}FE}$. As for the PS, a 2D plot can be obtained but is not included in the present document for space saving.

4.3. LED measurements

All SPD/PS channels have taken measurements using the LED system. In **Figure 13** it is illustrated a 2D display of the PS detector showing measured LED peak signal in ADC counts above the pedestal value. Due to the fact that LED signals vary on a large scale, the LED system will be essentially dedicated to detect dead channels, check the detector stability, and help for the coarse time alignment.

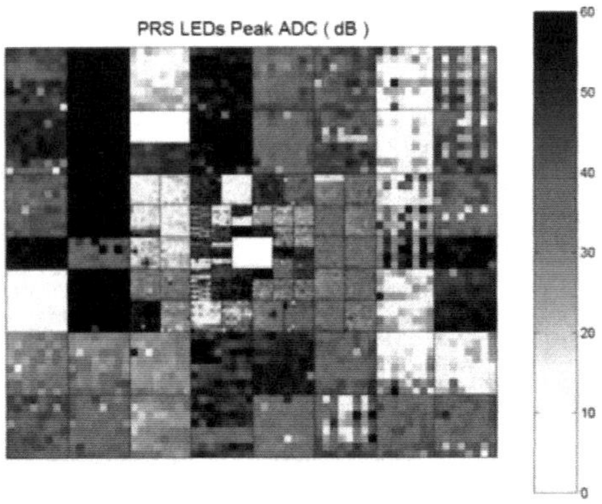

Figure 13. PS measures of LED light peak value in ADC counts above pedestal value.

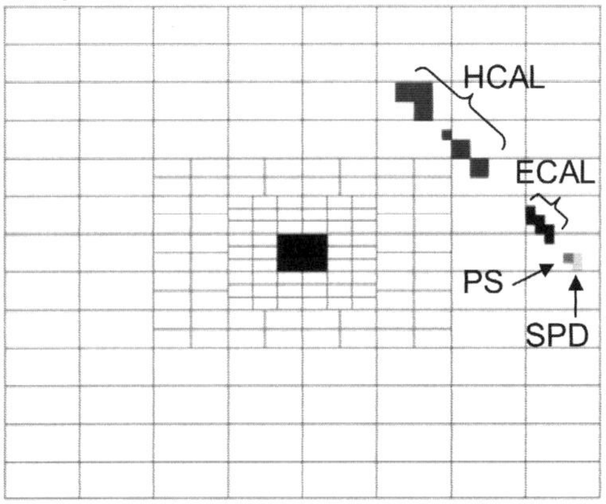

Figure 14. A cosmic event measured by all the calorimeter detectors (SPD, PS, ECAL, and HCAL).

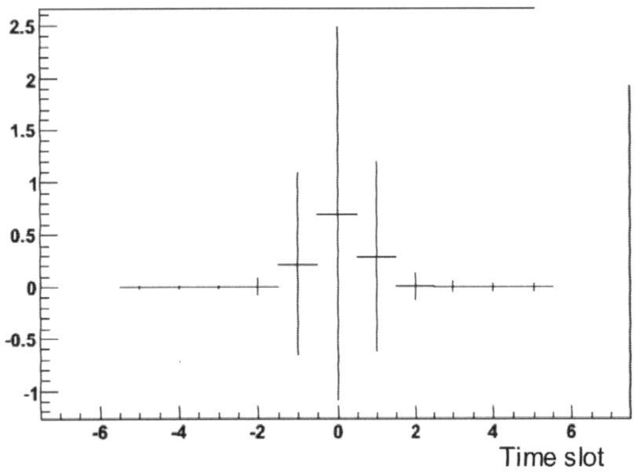

Figure 15. SPD coarse time adjustment example: accumulated signal during a cosmic run in consecutive time slots.

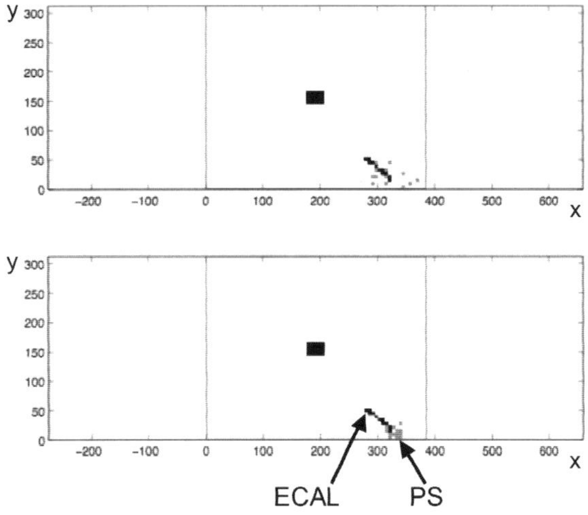

Figure 16. Example of a SPD/PS mapping problem. On top, ECAL signal and the scattered signal on PS are shown. On bottom, the mapping has been fixed.

4.4. Cosmics measurements

Data from cosmic events has been taken using all the calorimeter at an approximated rate of 10 Hz using the HCAL and ECAL coincidence as a trigger. These cosmic runs are the nearest approach to the normal data taking previous to the beam and include the full data and first level trigger paths. Even sometimes the runs are done together with other parts of the LHCb detector, such as the muon chambers and the trackers.

Cosmic runs are useful not only for checking the detector functionality, their data can also be analysed providing an initial time alignment between sub-detectors and to identify problems such as mapping issues like miscabling, mislabelling (**Figure 16**) and non-working channels.

The calorimeter presents a very useful feature which offers the possibility of storing +/-7 previous and next time slots (or experiment clock cycles). This way, the data taking strategy is simple; the events are triggered with ECAL and HCAL coincidence and include the data corresponding to the same time slot plus the 7 previous and 7 next time slots for time alignment.

After some runs, the pipelines in the FE electronics have been adjusted so most of the cosmic events are triggered in the same time slot as ECAL and HCAL for both SPD and PS. In **Figure 15** it is shown, as an example, that the accumulated signal on SPD in consecutive time slots is a peak centered in the same time slot as the trigger. At the present moment, further analyses are ongoing for fine time alignment (within one clock cycle) which would lead to a good starting point for the exact integration time when the first LHC beam will be available.

5. Calibration methods

Although SPD and PS are very similar in construction and electronics, their data output is not. While the PS digitizes the signal with 10 a bit ADC, the SPD provides a discriminator (i.e. 1 bit for each channel). This difference leads to dedicated calibration methods for each one. Along with the MAPMT gain values (set by their bias voltage), the SPD needs to adjust the threshold and measure the offset values whereas the PS, the gain corrective values and pedestals, which will be performed periodically with a frequency depending on the corresponding stability.

5.1. PS Calibration procedure

Taking advantage of the fact that the occupancy is relatively low (about the 10%), there is the possibility of running offline pedestal measurements on data.

On the FE boards, there are gain corrective factors for all channels to compensate for the non-uniformities of the detector and MAPMTs. A good way to find out the gain corrective factors with MIP particles involves the following steps: first select the expected gain (by setting the corresponding PMT bias voltage), then, identify the channel with the highest MIP value and tune the PMT bias voltage so 1 MIP corresponds to 10 ADC counts. Once this is done, adjust the other channels of the same MAPMT corrective factors so all of them present 10 ADC counts per MIP.

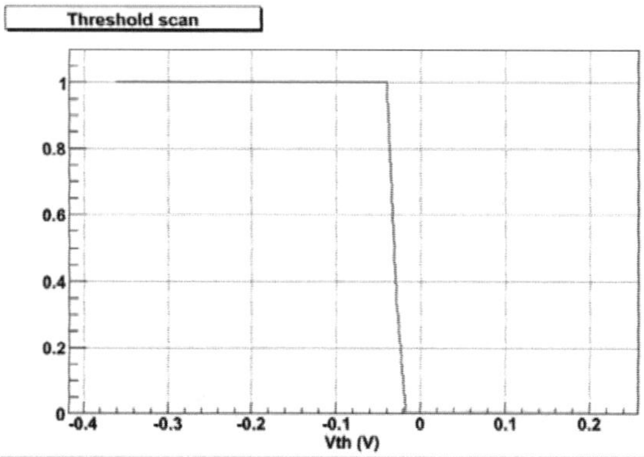

Figure 17. SPD threshold scan for one channel without signal. The 1 to 0 transition shows the offset value of the channel. Offset measurements require about 1000 triggers and 128 steps.

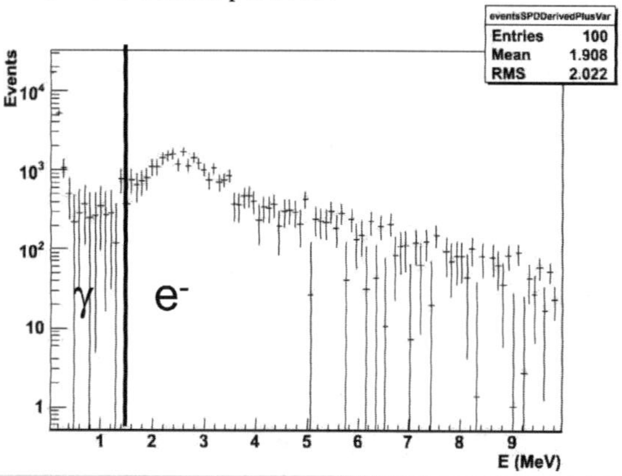

Figure 18. Energy deposited in a SPD cell derived from a threshold scan (from MC simulations). The expected optimum value to separate electrons for photons is 0.7 the energy of the MIP peak.

5.2. SPD calibration: threshold scan

SPD is a binary detector intended to distinguish between charged particles and photons. Each channel output is 0 or 1 depending whether the input signal is below or over a programmed threshold. Although a direct and accurate measurement of the signal is not feasible, there is the possibility to perform a threshold scan.

A threshold scan suitable for the offset measurements (or even LED periodic signals) is performed by, first, fixing a low threshold value, taking N events. Afterwards, it is repeated for increasing values of threshold. In the end, it can be represented the number of ones obtained divided by N for each threshold value. The transition from 1 to 0 is a measure of the VFE signal (**Figure 17**).

When a threshold scan is done using particles signal the output of the SPD on each step represents the number of events with an energy (proportional to the threshold) or lower. Therefore, the resulting accumulated histogram is differentiated in order to obtain the number of events with a deposited energy (**Figure 18**). The total estimated time is about 30 seconds per step by 20 threshold steps to be able to fit a curve of the MIP peak. Once the peak is fitted, the expected optimal threshold position to distinguish between photons and charged particles is 0.7 times the energy value of the peak.

6. Monitoring

A dedicated Monitoring farm will reconstruct events online at 5 Hz. Dedicated tools are being developed and will be useful to monitor:

- The PS trigger rate stability done by checking that the ADC MIP value is constant.
- The PS coefficient of energy for ECAL correction. It can be calibrated with electrons and refine the MIP calibration.
- The SPD efficiency on charged particles identification.
- The occupancy for checking the pedestal stability, the detector ageing, and noisy channels.
- The crosstalk between channels.
- Dead channels.

In addition there will be 10 Hz of LED triggers during normal data taking synchronized with empty bunches. LEDs will be flashed by regions and their data is aimed for the fast detection of dead and noisy channels.

7. Conclusions

The SPD/PS system, two planes of scintillator pads separated by a lead sheet, is designed for the tagging of the electric charge and the electromagnetic nature of the LHCb calorimeter clusters for the first trigger level. Both SPD and PS detectors are fully equipped and working at the present.

Runs have been made using cosmic data to test the detector functionality, electronics, data acquisition and trigger paths, and efficiencies. In addition, cosmic data has been used for time alignment between calorimeter sub-detectors.

SPD and PS calibration methods have been studied and require different approaches. PS needs MIP particles to adjust the gain corrective factors, while a threshold scan is essential in order to distinguish charged particles from photons at the SPD.

The tools for the monitoring the efficiency and correct operation of the SPD/PS have been defined and are under development.

References

[1] S. Amato et al. (LHCb collab.) 1998 *LHCb Technical Proposal* (CERN LHCC 1998-4)

[2] S. Amato et al. (LHCb collab.) 2003 *LHCb reoptimized Detector Design and Performance Technical Design Report* (CERN LHCC 2003-030)

[3] S. Amato et al. (LHCb collab.) 2003 *LHCb Calorimeters Technical Design Report* (CERN LHCC 2000-0036)

The LHCb electromagnetic calorimeter

Irina Machikhiliyan[1] for the LHC*b* calorimeter group

Institute for Theoretical and Experimental Physics,
SSC RF ITEP, Bolshaya Cheremushkinskaya 25, 117218 Moscow, Russia

E-mail: Irina.Machikhiliyan@cern.ch

Abstract. The design and construction characteristics of the Electromagnetic Calorimeter in the LHC*b* experiment are described. Current status of the subdetector is reported together with selected results of performance studies both with test beam and cosmic particles. Strategies for $ECAL$ monitoring and calibration are discussed.

1. Introduction

The LHC*b* [1] detector is a forward single-arm spectrometer aimed at studies of CP-symmetry violation and rare decays of b-quark in the LHC collider environment. The tasks of Electromagnetic Calorimeter ($ECAL$) in this experiment are the following:

- to provide high transverse momentum electron, photon or π^0 candidates to the LHC*b* trigger system [2] in order to pre-select b-containing events;
- to reconstruct precise kinematical parameters of neutral particles, for example to study B-meson decays containing prompt photon or neutral pion;
- to participate in the particle identification algorithms, particularly to ensure reliable $e^{\pm}$ / hadron separation.

$ECAL$ performance should comply with the following list of specifications [3] dictated by LHC*b* operating environment and physical goals:

- an energy resolution $\sigma_E/E(GeV)$ on the level of $10\%/\sqrt{E} \oplus 1\%$;
- a fast response time compatible with LHC bunch spacing 25 ns;
- a good reliability to operate for decades in a radiation hostile environment. Annual radiation dose can reach 2.5 kGy in the vicinity of the accelerator beam pipe (in the position of shower maximum);
- transverse segmentation should be small enough to separate significant fraction of γ-produced shower pairs from neutral pion decays and, at the general level, to minimize pile-up effects;
- for optimal detector and trigger performance the dynamic range should be adjusted according to the transverse energy (E_T) rule[2] with an upper limit 10÷12 GeV;

[1] Present address: European Organization for Nuclear Research CERN CH-1211 Geneve 23 Switzerland

[2] The exact formula is $E_{max} = 7+10/sin(\theta)$, where E_{max} is an upper limit of dynamic range in terms of deposited energies, $sin(\theta) = \sqrt{x^2+y^2}/\sqrt{x^2+y^2+z^2}$ and (x, y, z) are coordinates of the $ECAL$ cell with respect to the interaction point.

- Finally at the distance 12.5 m from the interaction point $ECAL$ lateral size is chosen to be 7.8 m x 6.3 m, excluding $\theta_{x,y} < 30$ $mrad$ central cut-out for the beam pipe.

2. Design overview

The general layout of the LHC*b* $ECAL$ is sketched in figure 1. Subdetector itself is realized as a rectangular wall constructed out of 3312 separate modules of square section. In purpose of the maintenance the wall is split vertically in two halves; each of them covers right or left hemisphere of $ECAL$ acceptance and is positioned on independently retractable platform.

The $ECAL$ is subdivided into three sections, Inner, Middle and Outer, comprising modules of the same size but different granularities. Namely, modules of the Inner / Middle / Outer section are subdivided into 9 / 4 / 1 readout cells or towers with transverse dimensions 4.04×4.04 / 6.06×6.06 / 12.12×12.12 cm^2 correspondingly.

$ECAL$ media structure employs the shashlik technology [4], when interleaving scintillator-absorber layers are pierced by optical fibers of light collection system. The light readout devices are photomultipliers (PMs). Their output signals are digitized and processed at 40 MHz rate by the dead-timeless Front-End electronics [5]. To control the behavior of the subdetector performance Monitoring System is foreseen.

2.1. The construction of the module

Construction of $ECAL$ module is illustrated in figure 2 with a module of the Inner section. The sampling stack comprises 66 lead plates and 67 scintillator (Sc, BASF-165H polystyrene based plastic doped with 2.5% p-terphenyl and 0.01% POPOP) planes, separated by thin (120μm) TYVEC paper sheets. Scintillator plane is constituted out of 1, 4 or 9 optically decoupled tiles thus ensuring required transverse segmentation of the module. Tile edges are chemically matted in order to improve efficiency and lateral uniformity of the light collection as well as to prevent tile-to-tile cross-talk.

Wavelength-shifting (WLS) fibers (KURARAY Y-11(250) MSJ, 1.2 mm diameter) of the light collection system run parallel to the beam axis and penetrate the entire module body. In order to improve the light collection efficiency and the lateral uniformity of the response, WLS fibers form U-shape loops at the front side of the module, so each fiber traverses the module twice. An additional clear fiber, penetrating each cell along the central axis, transports the light pulses of the Monitoring System from the front to the rear side of the module. Fibers serving the same cell are grouped together at the rear of the module; resulting bundle enters the PM

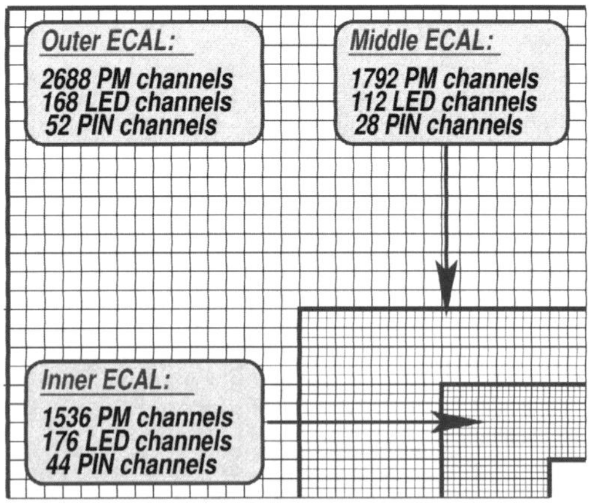

Figure 1. $ECAL$ layout (one quarter).

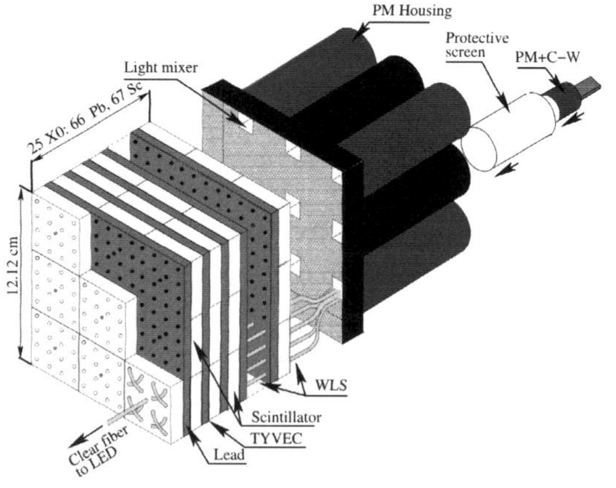

Figure 2. Construction of the Inner module.

Table 1. Selected parameters of *ECAL* modules.

	Inner	Middle	Outer
# of fibers per module (WLS+clear)	72+9	72+4	32+1
Average light yield, photoelectrons per GeV	3077	3516	2569
Response on MIP, cell-to-cell variation, %	8.0	5.3	6.7

compartment, where it is coupled with photomultiplier window via light mixer. At the front side the fiber loops are protected with a plastic cover, carrying an optical connector to inject the light of the Monitoring System. The entire module stack is held together by 100 μm stainless steel bands welded at the four sides of the module.

Effective depth of the module 42 cm is equivalent to 25 radiation lengths (X_0), the Moliere radius of *ECAL* media is 3.5 cm and the sampling ratio Pb:Sc is 2:4 (mm).

The energy resolution of the modules has been measured at the test beam at CERN [6]. Its value proved to be well within the design requirements, namely $\sigma_E/E(GeV)$ is $(8.2 \pm 0.4)\%/\sqrt{E} \oplus (0.87 \pm 0.07)\%$ and $(9.4 \pm 0.2)\%/\sqrt{E} \oplus (0.83 \pm 0.02)\%$ for the Inner and Outer modules correspondingly. A scan of module prototypes by means of 50 GeV electrons has shown that the non-uniformity of light collection over Sc tile is better than $\pm 1\%$. Cell-to-cell response on minimum ionizing particles (MIPs) varies within 8.0% over all modules accepted for the installation into the subdetector wall, as was measured during pre-installation data quality checks with the help of cosmic particles. The average light yield of the *ECAL* cell is found to be $\sim$ 3000 p.e. per GeV of the deposited energy. More details can be found in table 1.

2.2. The timing of PM signals

ECAL employs HAMAMATSU produced R7899-20 photomultipliers as photodetectors. Each of them is coupled with an individual power supply based on Cockcroft-Walton voltage multiplier (C-W) [7]. The duration of a typical pulse of PM anode current exceeds LHC bunch spacing (see figure 3), therefore output chains of the C-W contain a clipping line for cancellation of the

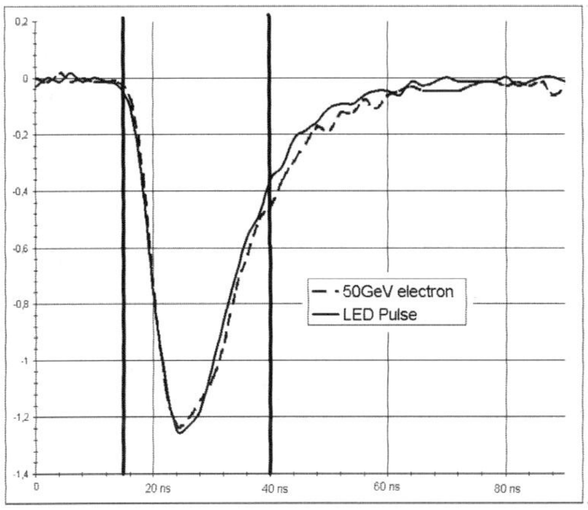

Figure 3. Typical shape of PM anode current pulse.

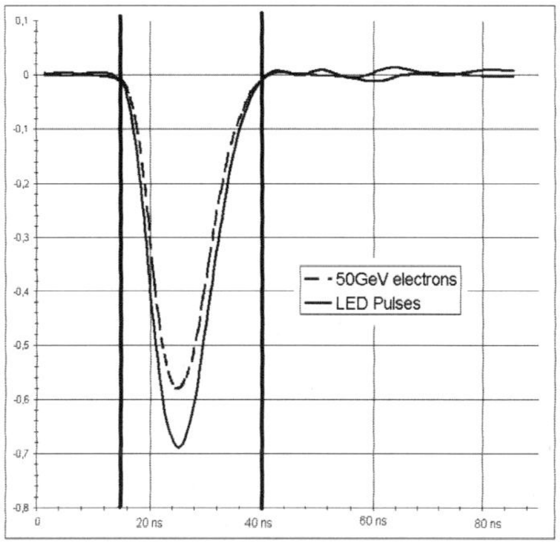

Figure 4. PM output pulse after the clipping chain.

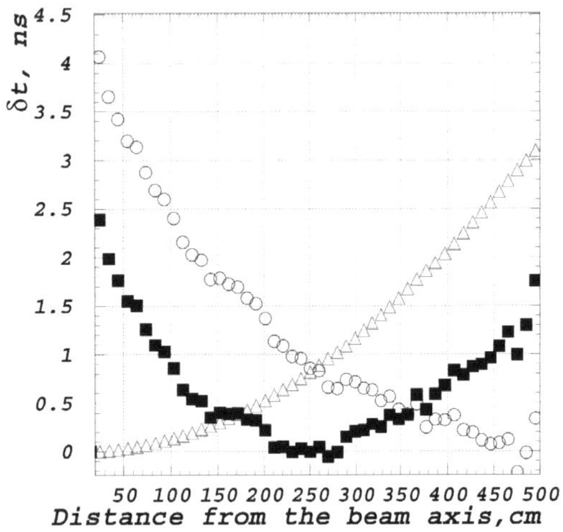

Figure 5. Relative contributions to PM signal timing: ○ - signal collection time, ▽ - time of flight. ■ marks resulting relative delay.

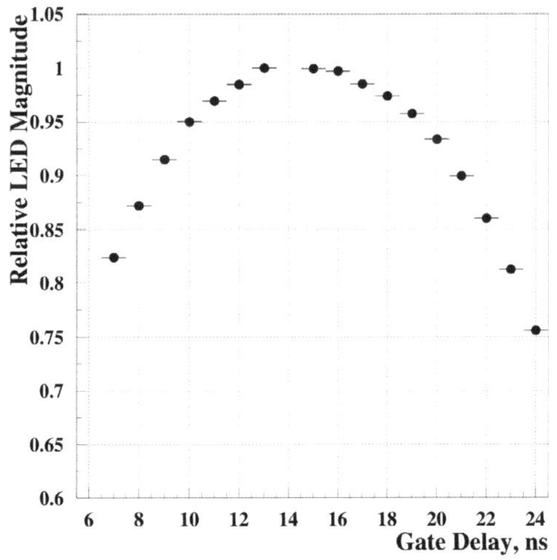

Figure 6. PM response to LED as function of relative ADC gate delay. PM gain setting is 1K.

signal tail and thus for minimization of the contribution to the next event. After the clipping the pulse duration becomes compatible with 25 ns as it is illustrated in figure 4.

Two major factors contribute to the PM signal timing. The first one is the time of flight of the particle, which rises from the beam axis to the periphery of $ECAL$ by ~3 ns as it shown in figure 5 by open triangles. The second factor originates from the large spread of PM gains depending on the coordinates of the cell with respect to the beam axis: from 1K to 20K from center to periphery to ensure the required dynamic range. As a consequence, the difference in signal collection time inside photomultiplier, which follows reverse square root dependence on high voltage supplied to it[3], amounts on average to 4 ns (open circles in the same figure). These two factors work in opposite directions and therefore compensate each other to a certain extend. Resulting relative time of arrival of PM signal at the Front-End electronics input varies in the range of 2.5 ns (full squares).

2.3. Front-End electronics

The readout system of $ECAL$ consists of 192 Front-End boards or FEBs, which reside in 14 crates on the top of the subdetector. Each FEB can read out up to 32 cells. Signal processing includes:

- integration of PM pulse by shaper integrator and posterior digitization by means of 12-bit 40 MHz two stage bipolar flash analog-to-digital converters (ADC) with dynamic range 80 pC. Fine synchronization of ADC gates and PM signal is done by programmable delay chips. Corresponding delay values (henceforth - FEB delays) can be varied in the 25 ns range with the step 1 ns. An example dependence of processed PM signal on FEB delay is shown in figure 6. The size of the top plateau, where the signal is processed in the optimal way, matches the expected dispersion in timing over all $ECAL$ cells;
- pedestal subtraction. The value of the pedestal is estimated as the minimal reading over two preceding bunch-crossings (BX). Such mechanism allows suppressing low-frequency noise;

[3] In its turn the gain of R7899-20 is related to the high voltage supplied as $G \sim HV^{\alpha}$ with $\alpha \sim 6$

- storage of the readings in memory, and, in parallel, conversion of them into the 8-bit values, re-calibrated to the transverse energies. These values are used in the trigger part of the Front-End board.

Trigger part of the FEB considers all 2x2 cells combinations (i.e. trivial clusters) and for each event selects a cluster candidate with the highest E_T to be sent to level-zero trigger [8]. The exchange by readings for border cells, necessary for cluster reconstruction, is done via backplanes of the crates. Each board also evaluates the full sum over all input channels useful for selection of energetic neutral pion candidates for the trigger.

Finally, if the event is accepted for further processing, the FEB formats corresponding data block and sends it to the DAQ.

2.4. Monitoring System

The general structure of LHC*b* *ECAL* Monitoring system is illustrated in figure 7. It employs super-bright red Light Emitting Diodes (LEDs) LUR3333H. One LED serves a group of readout cells: 16 in case of Middle/Outer and 9 in case of Inner section. LEDs reside beneath / on the top of the subdetector in several monitoring boxes, their light is delivered to modules front surface by means of long clear fibers. The fibers, transporting the light of the same LED, are grouped in a separated bundle, covered by protective opaque coat. The length of such bundles varies from 2 to 8 m introducing an additional dispersion into the timing of the LED signals up to 36 ns. To compensate this effect a controllable delay is foreseen for an individual LED fire pulse. The intensity of an individual LED is also adjustable and allows covering the substantial part of *ECAL* dynamic range.

The major tasks of the Monitoring System are to check the readout chain serviceability and to trace its stability (mostly dominated by the stability of PM gains). In order to fulfill it one should eliminate the effects, coming from the possible instability of the monitoring devices themselves or prove that these effects are negligible. To that purpose the stability of LEDs is monitored by means of silicon PIN photodiodes S1223-01 (HAMAMATSU) having negligible temperature variation coefficient. The PINs reside in the same monitoring boxes as the LEDs; the light between them is transported via short clear fibers, escaping each bundle. One PIN serves a group of two or four LEDs, its response is readout by FEBs of the same type as for PMs.

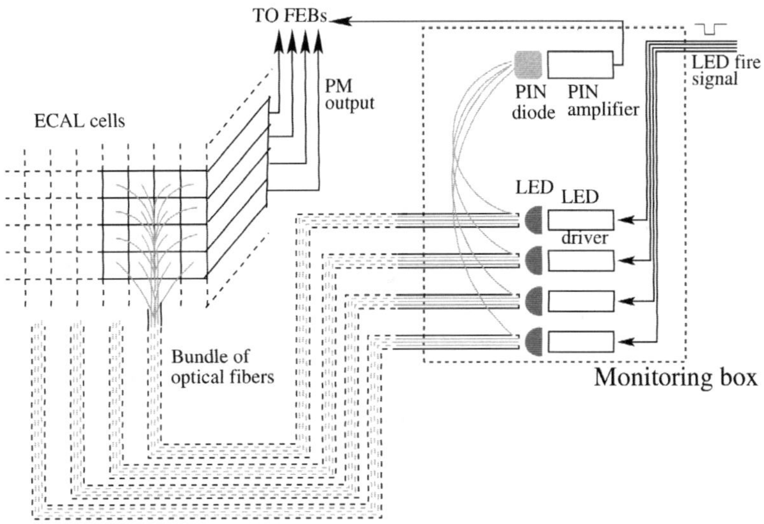

Figure 7. General structure of the *ECAL* Monitoring System.

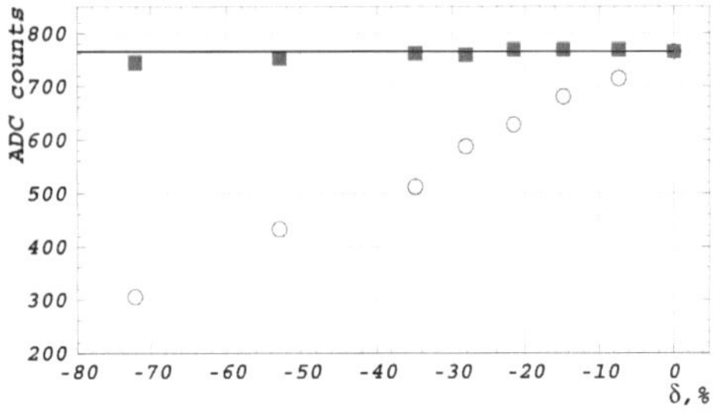

Figure 8. Correction of emulated LED drift. ○ - original PM readings, ■ - the same readings corrected according to the PIN.

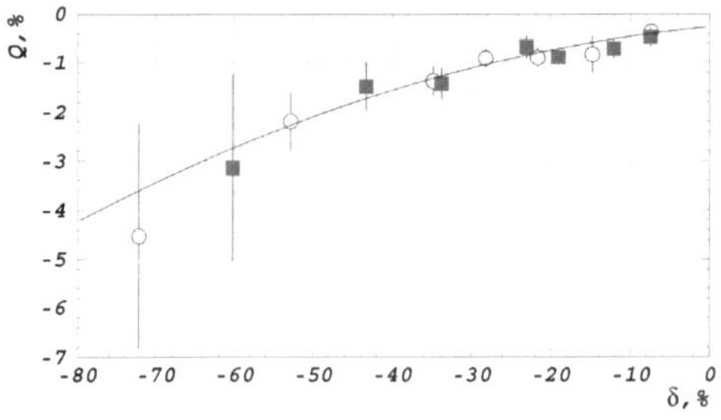

Figure 9. Precision of correction (by two LED / PIN couples). δ and Q are defined in the text of the article.

An example of the compensation procedure for possible LED instability is shown in figure 8. The LED magnitude was decreased in several steps down to 70% level with respect to the reference intensity to emulate the drift (X-axis, relative change in LED magnitude δ). The original PM readings (open circles) were corrected by the factor obtained from the PIN (full squares). It was found that the accuracy of such procedure Q is better than 1% up to 30% change in LED magnitude, as is shown in figure 9. Here the accuracy is estimated as $Q = (k_{PIN} \times I_{PM} - I_{PM}^{ref})/I_{PM}^{ref}$, where k_{PIN} is the correction coefficient, I_{PM} - PM readings at the current LED intensity and I_{PM}^{ref} are PM readings at reference LED intensity.

3. Monitoring and calibration of the subdetector response

Three basic conditions have to be met to ensure normal $ECAL$ operation:

(i) good timing of subdetector response on the level of individual cells;

(ii) periodical fine absolute calibration of the cell response;

(iii) monitoring of cell response in-between of absolute calibration runs. In case of a significant change the necessary correction factors for calibration constants will be provided according to the change of the response to the LEDs regarding also the information provided by the PIN system (so-called relative calibration).

3.1. Adjustment of subdetector timing

LHC*b* physical events. The first goal is to perform the synchronization of $ECAL$ readings with the LHC accelerator cycle in the way which ensures optimal processing of individual PM output pulse by readout ADCs. As it was shown in section 2.2, for the events under discussion the timing at uniform FEB delay settings is rather good; their remaining fine tuning, also allowing for possible additional spread introduced by the cabling, will be performed *in situ* when the accelerator starts to deliver interactions.

Monitoring (LED) events. As soon as the set of optimal FEB delays is fixed by the foregoing, the timing of the Monitoring System can be adjusted. To that purpose a tuning of individual LED fire pulse arrival times will be performed in the way which guarantees the appearance of all PM/PIN responses on LEDs flash synchronously with the same bunch-crossing and, preferably, with an optimal integration of the signals.

From technical point of view the time alignment will be performed by recording and further analysis of detector readings in several consecutive BX around the one under interest (so-called time-alignment events).

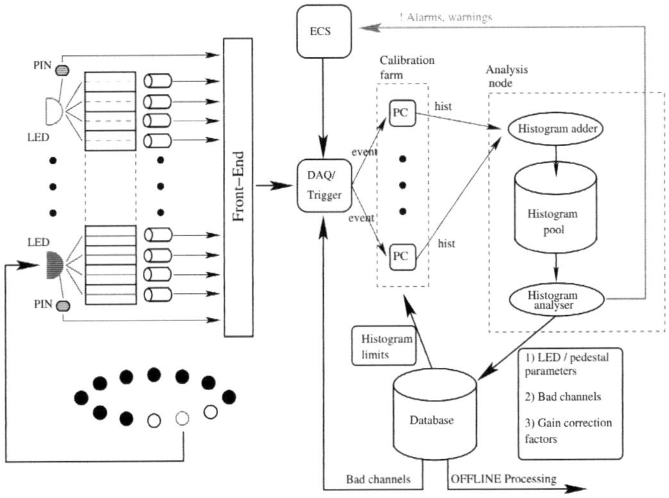

Figure 10. Monitoring with LEDs during data taking.

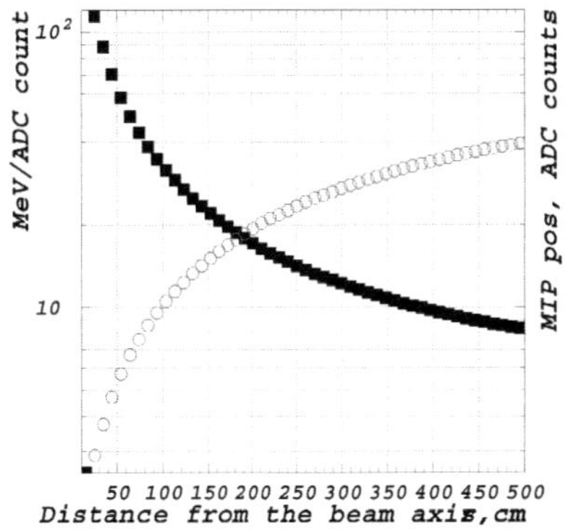

Figure 11. ○ - MIP position in terms of ADC counts. ■ - the factor to rescale ADC counts into MeVs of deposited energy.

3.2. Monitoring during data taking

LEDs will be fired in the midst of the group of several consecutive not-filled bunches with the frequency 50÷100 Hz (pre-scaled accelerator cycle). At the same time one LED over group served by the same PIN will be flashed at most. LED events will be read out and redirected to the calibration farm of few PCs (see figure 10), where dedicated software code will process them and produce spectra for individual cells containing pedestal and LED distributions as well as some summary histograms. Periodically spectra from all farm nodes will be retrieved via network to the analysis node, where the following sets of resulting parameters will be extracted:

- a list of malfunctioning cells;
- LED / pedestal distribution positions and widths;
- PM gain correction factors if any.

These sets will be stored in the database to be used afterwards in the offline analysis as well as to prepare new configuration sets for DAQ and Trigger system. In case of emergency alarms and warnings will be sent to the Experiment Control System.

3.3. Calibration algorithms

The dispersion in modules characteristics is less than 8% for all cells, the individual photomultiplier parameters are provided by the manufacturer, the reproducibility of PM gain setting over all C-W is on the level of 1.5% and the dispersion in ADC sensitivity values is better than 5%. All these conditions give a hope that trivial physical signal (like π^0 peak) will be seen in $ECAL$ as soon as LHC starts to deliver interactions. Nevertheless, to be on the safe side, additional methods of rough but fast calibration are now considered, allowing to achieve calibration accuracy 10÷15%[4]. These methods are based on such parameters as energy flow, occupancies, etc., but the details of their implementation strongly depend on LHC / LHC*b* start-up scenario (beam energy, magnetic field presence, etc). More detailed discussion of these methods can be found elsewhere [9].

Calibration with $\pi^0 \rightarrow \gamma\gamma$ decays. Neutral pions should be copiously produced in LHC environment and the noticeable part of their decays will be resolved by $ECAL$ as two clusters.

[4] Initial level, which allows seeing the π^0 signal

Therefore π^0 peak should be clearly seen in di-cluster invariant mass distribution as well as good signal-to-background ratio can be achieved using minimal set of cuts. The major obstacle for the method is 2.5 X_0-thick lead converter in the front of $ECAL$ between Scintillator Pad and Preshower (PS) detectors. Therefore $ECAL$ can not be calibrated in standalone mode: at least PS data have to be used either for full reconstruction of shower energy or for selection of decay photons which did not interact in the lead wall. Despite the fact that the last option means significant statistics loss, calibration with the help of neutral pion decay photons is considered as good starting point for the beginning of LHC operation.

Calibration with MIPs. The equivalent energy deposition of the MIP in $ECAL$ cell is 0.33 GeV (on condition of full containment). This value is small and, for the majority of $ECAL$ cells, ADC precision is too coarse to provide a reasonable accuracy of the calibration. The average response to a MIP in terms of ADC counts (see figure 11, open circles) changes from 3 counts in the Inner up to 40 counts in the very periphery of the Outer section. Therefore the best precision which could be achieved is 3% and, while PM gains are kept at nominal values, the MIP-based method might be somewhat useful to calibrate the very periphery of the Outer section.

Calibration with electrons. The major benefit of the calibration with a help of energy-to-momentum ratio for electrons is the high accuracy of the method (up to 0.5% according to simulations). However it requires precise information from tracker system and a pure sample of e^+/e^- candidates. Therefore this method is planned to be used later on, when the tracker chain of LHC*b* becomes fully operational.

4. Current $ECAL$ status

At present time all $ECAL$ sub-components are installed in the experimental hall and commissioned. Current activities cover the following set of tasks:

- study the behavior of $ECAL$ as whole system;
- development and debugging of software for monitoring, time alignment, calibration;
- participation in general LHCb commissioning on cosmic particles. $ECAL$ is involved in combined CALO trigger, which requires a coincidence between $ECAL$ and hadron calorimeter. To have MIP response as far from the pedestal as possible, $ECAL$ PM gains are set to 300K;

In general $ECAL$ is now ready to receive fist interactions.

Acknowledgments

Author is mostly grateful to Yu. Gilitskiy, D. Golubkov, V. Egorychev, I. Korolko and M. N. Minard for their comments and useful discussion. Author also would like to thank CALOR08 organizers for their hospitality during the conference.

References

[1] LHC*b* Collaboration 1998 LHC*b* technical proposal *Preprint* CERN/LHCC/98-4
[2] LHC*b* Collaboration 2003 LHC*b* technical design report: trigger system *Preprint* CERN/LHCC/2003-031
[3] LHC*b* Collaboration 2000 LHC*b* technical design report: calorimeters *Preprint* CERN/LHCC/2000-0036
[4] Atoian G S *et al* 1992 *Nucl. Instrum. Meth.* **A320** 144-54
[5] Machefert F 2005 Overview of the LHC*b* calorimeter electronics *CALORIMETRY IN PARTICLE PHYSICS, Proceedings of the Eleventh International Conference* ed G Cecchi *et al* (Singapore: World Scientific) pp 151-5
[6] Arefiev A *et al* 2008 Beam test results of the LHC*b* electromagnetic calorimeter *Preprint* LHC*b*-2007-149
[7] Avoni G *et al* 2006 *Nucl. Instrum. Meth.* **A560** 539-45
[8] Marconi U 2002 *Eur.Phys.J.direct* **C4S1** 31
[9] Korolko I, Obraztsov V, Popescu S and Yushchenko O 2000 On the possibility of *in situ* calibration of LHC*b* calorimeters *Preprint* LHC*b*-2000-051
Belyaev I and Voronchev K 2006 Energy flow calibration of LHC*b* $ECAL$ *Preprint* LHC*b*-2006-051

Journal of Physics: Conference Series **160** (2009) 012048 doi:10.1088/1742-6596/160/1/012048

Studies of the CMS Electromagnetic Calorimeter performance in the electron test beam

Roberta Arcidiacono (On behalf of the CMS ECAL Group)

Università degli Studi di Torino & INFN Torino, Via Pietro Giuria, 10125 Torino, I

E-mail: roberta.arcidiacono@cern.ch

Abstract. The Compact Muon Solenoid (CMS) is a general purpose detector for LHC. The electromagnetic calorimeter (ECAL), comprising about 76,000 $PbWO_4$ scintillating crystals, will allow a very accurate energy measurement of electrons and photons. The 36 supermodules (units of 1700 crystals) of the barrel have been successfully integrated inside the CMS solenoid. In 2006 nine supermodules with the final mechanics, cooling and electronics were exposed to electron test beams in order to perform precise inter-calibration and comprehensive studies of the detector response. A survey of results from these tests are here presented, with emphasis given to the energy resolution and linearity achieved.

1. Introduction

The CMS (Compact Muon Solenoid) [1] is one of the two multi-purpose experiments that will take data at the LHC proton-proton collider. The main physics goals of the CMS experiment are the discovery of the Higgs boson and the search for new physics phenomena, in particular the search for supersymmetric particles. For a mass lower than 150 GeV, the Higgs decay in two photons is the cleanest channel for the discovery. In this mass range, the very narrow Higgs signal will lie above an irreducible background: this led to the choice of a high resolution electromagnetic calorimeter.

2. The CMS electromagnetic calorimeter

The electromagnetic calorimeter (ECAL) [2] is an hermetic homogeneous calorimeter consisting of about 76,000 lead tungstate ($PbWO_4$) scintillating crystals. It covers the pseudo-rapidity (η) range from 0 to 3.0 by means of a barrel part ($0 < |\eta| < 1.48$) and two endcaps ($1.48 < |\eta| < 3$). ECAL is organized in 36 supermodules (SM), each containing 1700 crystals, in the barrel, and in 4 dees, each consisting of 3662 crystals, in the endcaps. The ECAL detector is mounted inside the CMS superconducting solenoid (4 Tesla). The target value of the energy resolution of the calorimeter at high energies is less than 0.5%.

The choice of lead tungstate [3], very dense material ($\rho = 8.28$ g/cm^3) with short radiation length (0.89 cm) and a small Moliere radius (2.2 cm), led to a very compact high granularity calorimeter design. The $PbWO_4$ produces fast signals (80% of light is emitted within the 25 ns LHC bunch crossing time) and is radiation resistant. Nevertheless this material presents major drawbacks: a low light yield (~100 photons/MeV, 0.2% with respect to NaI:Tl) which imposes a photodetector read-out with internal gain; a strong light yield dependence on temperature (-2%/°C around 18 °C), imposing a high precision cooling system to stabilize crystals temperature to $\pm$ 0.05°C; high refractive index (2.29 at peak emission wavelength = 420 nm) which makes the light extraction difficult.

Crystals in the barrel are read-out by Avalanche PhotoDiodes (APD), while in the endcaps the scintillating light is detected by Vacuum Photo Triodes (VPT).

The front-end electronics [4] and the trigger primitive generation are located on the detector with the advantage of minimizing external noise contribution and reducing the number of optical links to the off-detector read-out. Crystal signals are digitized sending the signal in parallel to the three amplifiers (gain 1, 6, 12) of the MGPA chip, choosing the highest not saturated gain, and sampled at 40 MHz. A time window of 10 samples is read-out for every L1 trigger.

In the following, test beam data results related to the ECAL barrel are described.

3. Test beam data taking in 2006

Two test beam campaigns, on the H4 and H2 beam lines of the CERN North Area, have been carried on during summer 2006.

In H4, where electrons in the 15-250 GeV range were delivered, detailed performance studies and precise inter-calibration of 9 SMs have been done. The final version of the read-out electronics, data acquisition and of all services have been tested. The supermodules were mounted on a rotating frame reproducing the same almost pointing geometry of CMS. Four planes of hodoscopes were used to reconstruct the trajectories of the electrons impinging on the crystals.

In H2, electrons/positrons in the 2-100 GeV range or pions were available for the combined test of the electromagnetic and hadronic calorimeters. ECAL response to low energy particles has been studied.

In parallel, each of the 36 ECAL barrel SMs has been tested for inter-calibration purpose with cosmic muons.

4. Performance studies

4.1. Detector response

The amplitude of the signal collected in each crystal is reconstructed using a digital filtering technique. A weighted linear combination of the time samples S_i, $A = \sum_i w_i S_i$, is performed, to minimize the contribution of electronics noise. The determination of the optimal set of weights makes use of a representation $f(t)$ of the time development of the signal pulse, as shown in figure 1.

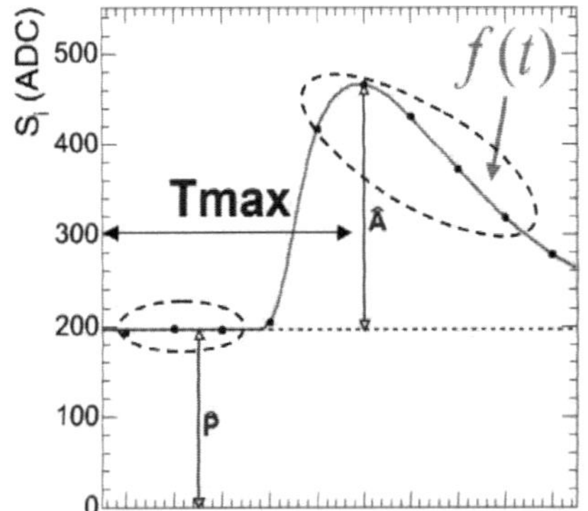

Figure 1. Signal amplitude reconstruction with digital filtering technique: time samples in the read-out window with superimposed the analytical representation $f(t)$ of the pulse shape time development.

Real data are affected by (high frequency) noise correlations between time samples, while any low frequency (e.g. pick-up) noise is taken away by pedestal measurements and subtraction. The best results for test beam data have been obtained with the "pedestal subtracting" or "3 + 5" weights scheme, where weights for samples on the rising edge of the signal pulse are set to zero, leaving 3 pedestal samples before the pulse and 5 samples during the pulse.

The measurement of the electronics noise is made by applying the amplitude reconstruction procedure to data taken with a random trigger. The distribution of such noise measurements for a large number of individual channels gives a mean width $\sigma_{1\times1}$ of 39 MeV with a rms dispersion of 3.3 MeV. The shower energy is measured by summing the energy deposits in matrices of either 3x3 or 5x5 crystals centered on the crystal with the largest energy deposit, with crystal channels inter-calibrated at gain 12 using 120 GeV electrons.

The noise distribution in the two clustering matrices has a width $\sigma_{3\times3} = 117$ MeV and $\sigma_{5\times5} = 195$ MeV respectively. This corresponds to about 3 and 5 times the single channel noise, as expected in absence of channel-to-channel correlated noise.

4.2. Position resolution

The distribution of the energy which is deposited by an electron in a NxN matrix depends on the electron impact point on the face of the central crystal. The impact position measured by the ECAL detector is reconstructed as a weighted average of the positions of the crystals in the matrix, $X = \sum_i w_i x_i / \sum_i w_i$. The weights $w_i = w_0 + ln(E_i/E_{tot})$, have a logarithmic dependence on the fraction of shower energy of the crystal to take into account the quasi-projective crystals geometry and the dependence of the shower density with the distance from the shower axis. The w_0 parameter controls the smallest energy fraction a crystal can have to be considered in the sum, and it has been selected to optimize the position resolution ($w_0 = 5$).

Dedicated large statistics samples collected at different energies on SM16 and SM6 are used for this study. The amplitude in the crystal is evaluated using the weight method with optimized weights calculated from a shape profile specific for each crystal. Inter-calibration coefficients calculated at 120 GeV are used.

The ECAL shower coordinates are compared to the coordinates measured by the hodoscope system positioned along the electron beam line (150 μs resolution), which are referred as "true".

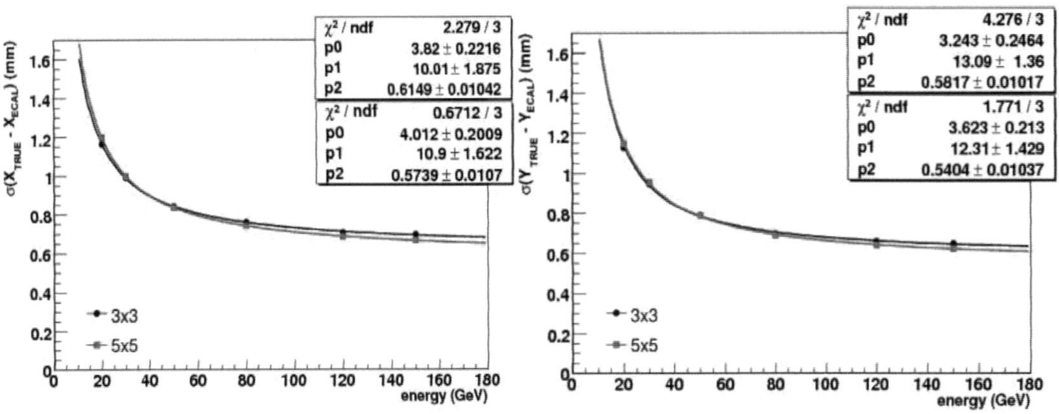

Figure 2. Position resolution in the X (left) and the Y (right) coordinate as a function of the beam energy. The crystals belonging to a 3x3 or a 5x5 matrix around the crystal with maximum energy deposition are considered as possibly contributing

Figure 2 shows the obtained resolution as a function of the beam energy for the X and Y coordinate (average of the 25 analyzed crystals of SM16). The hodoscope resolution has been subtracted. The plot shows results related to the 3x3 and 5x5 clustering algorithms, the second being slightly better.

4.3. Energy resolution

A uniform energy measurement response of ECAL is essential for precision physics. As already mentioned, the shower energy is reconstructed by summing signals in a 3x3 or 5x5 matrix centered on the crystal with the largest energy deposit. The variation of energy contained inside the clustering matrix, due to the electron impact position, leads to a degradation of the energy measurement. To achieve the desired energy resolution and uniformity of response when the incident electrons are distributed over the full crystal front face, a correction must be applied on event-by-event basis.

To study the calorimeter energy resolution decoupled from the shower containment correction, only electrons with central beam incidence, defined within an area of 4×4 mm^2 around the point of maximum shower containment, are considered. For such a small central area of incidence, there is no observable dependence on the incident position.

High statistics samples have been taken at 20, 30, 50, 80, 120, 180 and 250 GeV. The energy resolution versus energy is fitted by the following functional form:

$$\left(\frac{\sigma}{E}\right)^2 = \left(\frac{S}{\sqrt{E}}\right)^2 + \left(\frac{N}{E}\right)^2 + C^2 \qquad (E \text{ in } GeV)$$

where S is the stochastic term, N the noise and C the constant term. The fit is performed with the noise term fixed for each crystal at the value measured in the pedestal runs.

Figure 3 shows the relative energy resolution σ/E measured for a crystal array of 5x5 crystals in the 20-120 GeV energy range, where optimized weights have been computed for all the crystals of the matrix. The fit gives a stochastic term of 3.37%, a noise term of 0.107 GeV and a constant term of 0.25%, well in agreement with the designed values.

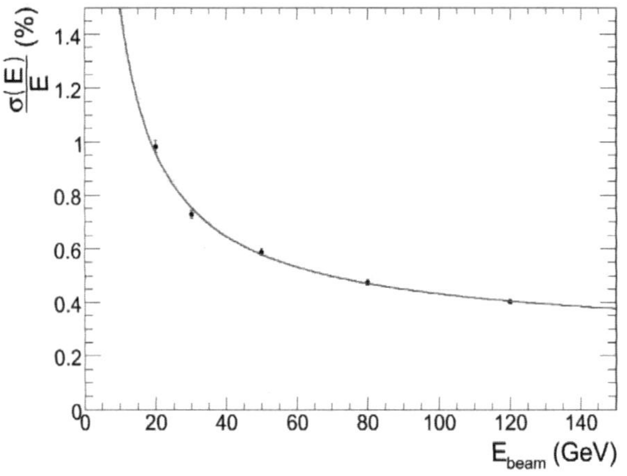

Figure 3. Energy resolution σ_E/E, as a function of the beam energy (average over 25 crystals of SM16), for central beam incidence. The energy is reconstructed using a 3x3 clustering matrix.

Compared to the results obtained for central incidence a slightly larger dispersion of the stochastic and constant terms is observed for uniform incidence. The resolution nevertheless remains better than 0.5% for electrons of energy greater than about 100 GeV.

4.4. Energy linearity

During the 2006 test campaign the H4 beam line has been equipped to be able to measure the incident electron energy with high precision. The precision achieved is $dE/E \sim 0.1\%$.

Special runs, where the beamline parameters (collimators, bends) have been carefully configured, are used to study at best the ECAL non-linearity as a function of the beam energy. Only very central electrons, impinging within an area of 2×2 mm^2 around the point of maximum shower containment, are considered for this study.

The reconstructed energy is obtained fitting the 5x5 array energy distribution (E25) with a Gaussian plus an exponential low energy tail (Crystal Ball function). Inter-calibration coefficients calculated at 120 GeV are applied.

The preliminary differential non-linearity (deviations from a linear fit of reconstructed energy versus beam energy normalized to the largest measured energy) is shown in figure 4. The beam and inter-calibration uncertainties have not been subtracted. The linearity of the Very Front-End (VFE) cards has also been measured in laboratory to be of the order of 0.1%.

The maximum deviation observed over the 20-180 GeV range (where gain switching occurs around 150 GeV) is of the order of 0.2%.

The linearity of ECAL response was also investigated at the ECAL-HCAL combined test beam in H2, using electrons (2-9 GeV) and positrons (9-100 GeV). The precision of the beam energy measurement in H2 was of the order of 0.5%.

The ratio E25peak/Ebeam, where E25peak is the E25 fitted energy peak, is used to quantify the deviation from linearity of every energy point.

Several corrections have been applied on ECAL energy measurements, related to the temperature variation during the data taking, the energy loss along the H2 beam line, the energy leakage of the crystals, the beam energy scale variation and beam energy spread.

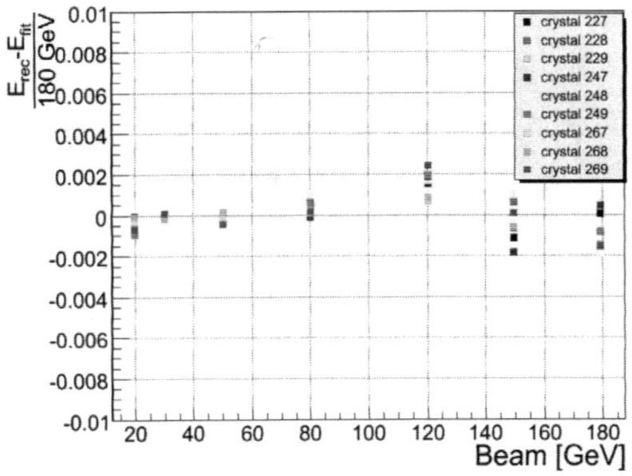

Figure 4. Preliminary differential non-linearity measurement in the 20-180 GeV range for 9 crystals - H4 beam line.

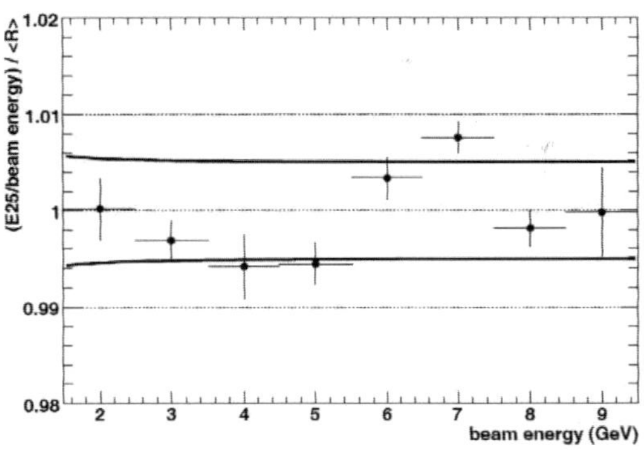

Figure 5. Profile of E25/Ebeam normalized to the weighted average $< R >$, for 16 crystals - H2 beam line.

In figure 5 the results for the low energy electrons events are shown, averaged over the 16 studied crystals distributed along the SM.

The plot displays, for every beam energy, the E25peak/Ebeam rescaled by the $< R >$ factor, which is a weighted average of the R_j=E25peak/Ebeam ratios for the various energies. The $< R >$ factor is directly related to the inter-calibration constant of the crystal; the rescaling of E25peak/Ebeam removes the contribution of the inter-calibration uncertainty from the linearity study. The two curves represent the beam energy scale uncertainty.

The RMS of the points, which is 0.51%, gives an indication that, in 2-9 GeV energy range, no deviation from linearity is observed at level of 0.5%.

4.5. Cosmic and Test-beam inter-calibration

All ECAL barrel SMs have been inter-calibrated with cosmics events in the cosmics test bench prepared in the CERN North Area, while, due to time constraints, only 9 SMs have been inter-calibrated with high energy electron beam.

The reference signal for the calibration with cosmic rays is provided by the energy released by minimum ionizing particles (about 250 MeV) crossing the ECAL crystals all along their length. The APD read-out gain is increased by a factor four with respect to the nominal working point, to enhance the signal to noise ratio to about 25. The trigger, given by the coincidence of a layer of plastic scintillators placed under the supermodule and of a plastic scintillator placed in the focal point, helps the selection of muons directed along the crystal axis.

About 5 million triggers, typically collected in seven days of data taking, ensures a statistical accuracy ranging from 1% to 2.5% according to the crystal position along η.

The inter-calibrations are obtained equalizing the responses of the different crystals to the muons. A detailed comparison with the coefficients from the inter-calibration with an electron beam shows an agreement over the single whole supermodule at 1.5% level (figure 6). The calibration with cosmics guarantees that the large majority of the ECAL channels are equalized to 1.5% already at the beginning of the data taking.

The high energy electron beam inter-calibration has been carried on in H4. For this study, each crystal of the SMs has been irradiated with more than 3,000 electrons. The beam has been set to the energies of 120 GeV and 90 GeV, allowing the study of systematic effects on the inter-calibration related to the energy. Two techniques, based on the energy deposited in each single crystal (S1) and on the energy deposited in a 5x5 matrix of crystals (S25), have been used. In the first case, the coefficients are calculated by equalizing the maximal response of each crystal to a reference value. In the second case, which will be used during the data taking, the

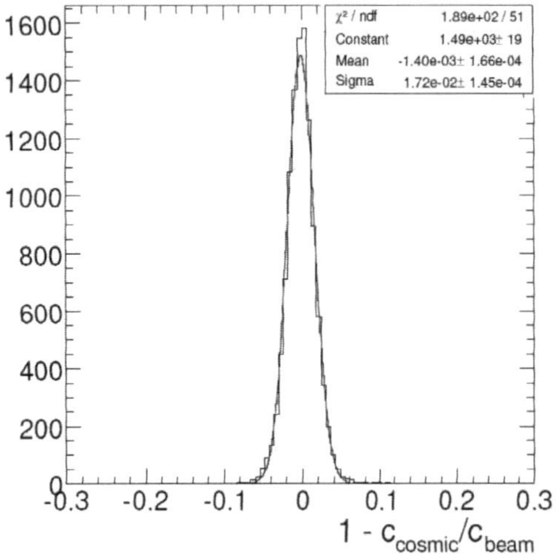

Figure 6. Distribution of the fractional difference between the inter-calibration coefficients measured at the test beam and the ones measured with cosmic rays (S1 method).

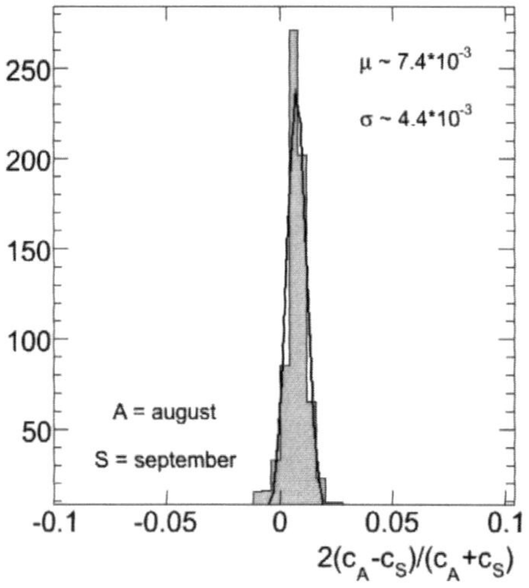

Figure 7. Relative difference between August and September inter-calibration coefficients (S25 method) for supermodule 22 that was under beam twice

energy deposited in each crystals cluster is compared to the one of the electron beam. The two methods are in agreement at a level of 0.3%.

The reproducibility of the inter-calibration measurements has been tested by performing the procedure twice on the same supermodule, with a time interval of one month. Figure 7 shows the distribution of the relative difference of the two measures for all the crystals of a supermodule. The width of the distribution is 0.27%, corresponding to an error on the coefficients of about 0.2%. The distribution of the relative difference between the measurements performed at 90 GeV and 120 GeV has been built. The inter-calibration coefficients are reproducible at a level of 0.3%.

5. Conclusion

The ECAL barrel has been pre-calibrated during the test beam campaign, installed and commissioned inside the CMS solenoid in the P5 experimental cavern. Many other extensive test beam studies have been carried on during summer 2006 on part of the ECAL barrel to evaluate the detector performance. Results look very promising and indicates that CMS ECAL will meet its ambitious design goal for energy resolution, uniformity of the response and noise level.

References

[1] 1994 *The Compact Muon Solenoid Technical Proposal* CERN/LHCC 94-38
[2] Lustermann W *The CMS crystal calorimeter* in this proceedings
[3] Annenkov A A, M V Korzhik and P Lecoq 2002 *Lead tungstate scintiallation material Nucl. Instrum. Meth.* A **490**
[4] Hansen M 2003 *The New Readout Architecture for the CMS ECAL* 78 CERN/LHCC 2003/55

Performance of the ATLAS Liquid Argon Barrel Calorimeter in the 2004 Combined Test Beam

Nicolas Kerschen (for the ATLAS test beam group)
University of Sheffield, Hicks Building, Hounsfield Rd, Sheffield S3 7RH, UK
E-mail: nicolas.kerschen@cern.ch

Abstract. The combined test beam (CTB) of a full barrel slice of the ATLAS detector in 2004 offered a unique opportunity to study the combined performance of the various sub-detectors down to energies as low as 1 GeV. In addition new analysis tools and techniques were developed that are already being used by the ATLAS offline reconstruction. This work presents studies related to the electromagnetic (EM) barrel calorimeter, both standalone and combined with the tracking and hadronic detectors. It also presents how these developments can be transferred to the ATLAS experiment. Here emphasis is given on studies of material effects on the electron energy reconstruction. The presence of extra material in front of the EM calorimeter leads to non-linear and non-uniform calorimetric energy response that requires detailed studies of sophisticated calibration methods. In this work we discuss these EM calibration methods and their application on the CTB data analysis. We also present final results on EM calorimeter uniformity, linearity and resolution for high energy beams (from 9 to 250 GeV), and studies of Monte-Carlo description of the calorimeter response at very low energies (from 1 to 9 GeV).

1. Introduction

ATLAS is a general purpose experiment mainly aimed at the discovery of the Higgs boson. From the collision point outwards it consists of a tracking detector, an electromagnetic and hadronic calorimeter and a muon spectrometer. The tracking detectors are operated within a magnetic field collinear to the beam axis. This configuration was reproduced in 2004 where a full barrel slice (from the innermost tracking detectors and magnetic field to the outermost muon spectrometer) was exposed to particles ranging from 1 to 350 GeV. The setup of the test beam was kept as close as technically possible to the ATLAS geometry. The distance between sub-detectors, the pointing geometry and the magnetic field orientation were preserved when permitted. The most important goals of this test beam campaign were the following:

- Test the detector performance with final or close to final electronics equipment, data acquisition (TDAQ) infrastructure and reconstruction software.
- Validate the description of the data by Monte-Carlo (MC) simulations down to energies of 1 GeV to prepare the simulation of the ATLAS data.
- Perform combined studies in a set-up very close to ATLAS (e.g. combined calorimetry, and ID-calorimetry).

In the following, after a description of the testbeam setup, we then focus on the energy reconstruction and calibration of electrons leading to the final results on linearity, uniformity and resolution (down to 9 GeV) for different material configurations. Subsequently we study the material dependence of the ratio of the energy deposited in the first sampling over the energy deposited in the second sampling as a possible way to correct for extra material in front of the calorimeter in ATLAS. The last sections cover the very low energy (1 to 9GeV) electron studies, the E/p measurements and the Bremsstrahlung recovery studies in magnetic field.

2. Description of the setup

Fig. 1 presents a schematic view of the setup which is described in detail in [1].

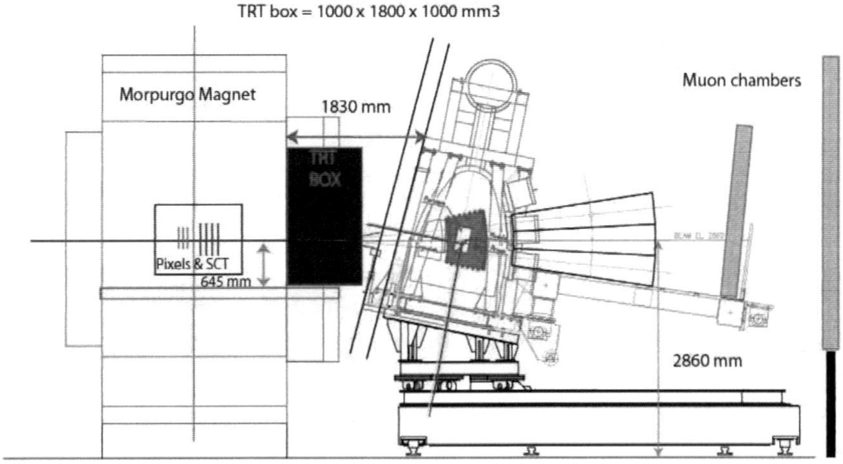

Figure 1. Schematic view of the H8 CTB setup, including the inner detector components and the LAr and Tile calorimeters.

The tracking part consists of a silicon pixel detector, a silicon strip detector (SCT) and a transition radiation tracker (TRT). Both pixel and SCT modules are installed in a magnet providing a field along z, as in the ATLAS detector. The TRT detector is not in the field (in ATLAS, the TRT detector is inside the solenoidal field). The orientation of the 1.4 T magnetic field is horizontal as in ATLAS, to get a vertical displacement. The deviation in the field for 10 GeV electrons is $\simeq$ 11 cm at the front of the electromagnetic calorimeter. In ATLAS, a 10 GeV electron produced at the vertex deviates by $\simeq$ 13 cm. The two calorimeters (electromagnetic and hadronic) are installed on a movable table which can rotate in η and translate in z and x. It is not possible to rotate in azimuth ϕ. The electromagnetic calorimeter modules were built for the ATLAS Combined Run, using lead sheets, electrodes, motherboards, connectors and cables left from the production of the 32 ATLAS barrel modules. The geometry of such a module has been extensively described in [5].

3. The energy reconstruction in the electromagnetic Calorimeter

The ionisation signal generated in the calorimeter is collected from the read-out electrodes and brought via cables to the front end electronics where it is amplified, shaped and sampled at a 40 MHz frequency. The samples are then stored in an analog pipeline until the trigger decision. The digitised samples are then transmitted by the calorimeter back-end electronics and the signal amplitude is reconstructed and converted to MeV. Once the cell energies are reconstructed, these cells are then summed to form a cluster over all three longitudinal compartments and the presampler of the EM calorimeter. A simple projective cone algorithm, optimised for the measurement of EM showers, was used. Starting from the cells of the middle layer, the most energetic cell is taken as a seed of the cluster and a fixed window of $(\Delta\eta\times\Delta\phi)$=0.075×0.075 (3×3 in middle cell unit) is opened around the seed. The choice of the cluster size is a compromise between electronic noise and shower containment.

The initial electron energy is found by performing a linear weighting of the longitudinal sampling energies which build up the cluster [2]. This MC based calibration procedure makes use of the longitudinal segmentation of the calorimeter and is described by the relation:

$$E_{\text{particle}} = \text{offset} + W_0 \times E_0 + \lambda \times E_{acc} + W_3 \times E_3 \quad (1)$$

where W_i and λ are the weighting factors that are to be determined. E_i is the measured energy in sampling i within the 3×3 cluster and $E_{acc} = E_1 + E_2 + E_3$. Each term of the calibration scheme is described as follows:

- offset $+ W_0 \times E_0$: this term parametrises the energy lost in the passive material before the presampler and between the presampler and the first layer of the EM calorimeter versus the energy deposited in the presampler. The purpose of the presampler is to measure the multiplicity of an early showering electromagnetic particle. The energy lost by pre-showering electrons is proportional to the presampler energy and is parametrised by the W_0 factor. The offset is due to residual energy loss through ionisation in the case of hard bremsstrahlung where the resulting photon does not interact in the presampler.
- $\lambda \times E_{acc}$: this term corrects for the energy lost outside of the cluster, it includes an implicit dependence of the sampling fraction with energy as shown in figure 2. This correction factor can be obtained by the ratio of the cluster energy by the total energy deposited in the calorimeter as can be seen in figure 2. Typically the leakage outside the cluster of about 5% of the energy deposited in the accordion of the EM calorimeter.
- $W_3 \times E_3$: a correlation exists between the deposited energy in the back sampling and the energy leaking from the back of the calorimeter. This factor can be obtained by parametrising the energy lost behind the calorimeter as a function of the energy deposited in the back layer of the calorimeter.

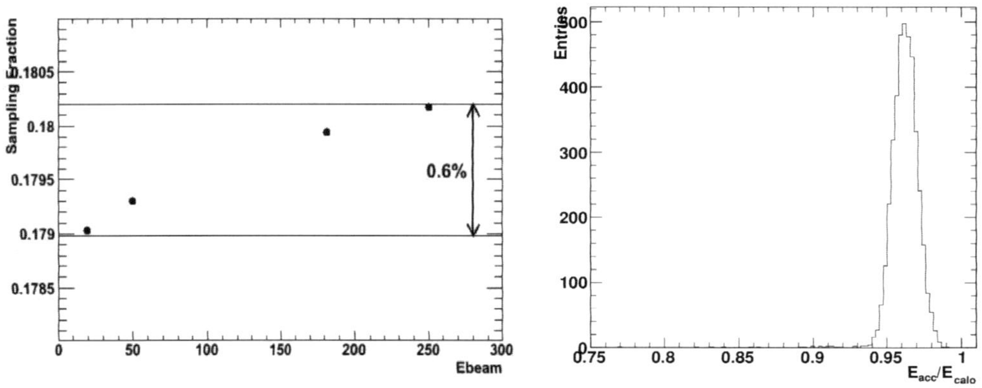

Figure 2. Sampling fraction versus the electron energy (left). Fraction of the energy measured within the cluster in the calorimeter (right)

4. Beam energy measurements

The electron momentum selection is based on a spectrometer consisting of two collimators and two triplets of bending magnets. The beam momentum is related to the bending angle and the integral of the magnet field along the beam path through the formula, $\Delta\theta = 0.3\frac{\int Bdl}{P}$, where P is the beam momentum in GeV/c, $\Delta\theta$ is the bending angle and $\int Bdl$ is the bending power. The bending power is a function of the current applied to the magnets, thus the beam momentum is deduced from these currents. The energy lost by electrons due to synchrotron radiation was estimated and subtracted. Also the shift of the mean beam momentum due to non-centered collimators was studied through a simulation of the beam line and corrected. There is an error on the absolute energy scale of up to 1.0 % coming from geometrical uncertainties. For uniformity and linearity measurements, high precision should be achieved on the relative beam energy measurements. Assuming that the beam conditions do not change from run to run, the only uncertainty on the relative beam energy comes from the uncertainty on the measurements of the currents in the bending magnets and on the uncertainty on the remnant field. In summary, a total precision of 0.2 % is achieved at low energy (20 GeV).

5. Comparison between Data and Monte-Carlo Simulation

The ATLAS calibration strategies rely heavily on the simulation of the experimental test-beam setup and the detector response. Thus it is important to study the level of agreement between data and MC prediction of the total energy distribution, the energies measured in each layer, their correlations, and the lateral development of the EM shower. Difference in the lateral and longitudinal shower profiles between data and MC may have significant effects in linearity due to the partial containment of the shower energy when a fixed cluster size is used. The longitudinal segmentation of the barrel calorimeter in four compartments allows for an evaluation of the simulation performance. The main goal of such an evaluation is to extract the level of agreement between the simulation of EM response and the response seen in the data. The level of agreement between data and MC in the reconstructed energy is at the level of 0.5%, consistent with the total expected systematic uncertainty discussed below.

The level of description of the EM shower development in the LAr Calorimeter is obscured by the presence of uncertainties associated with the accuracy of the detector description (thickness of the lead absorbers, the depth of the first sampling, the exact amount of material in front of the first sampling cables-electronics, the thickness of the cryostat and the amount of LAr in front of the presampler). The same uncertainties will be an issue during the actual ATLAS data taking, so their study in a controlled environment is important. All these uncertainties were extensively studied in the combined test beam and a cumulative systematic error in the level of agreement between data and MC from all effects of the order of 0.2-0.3% could be deduced.

6. Results on linearity and resolution

The energy linearity obtained after application of the longitudinal weights to the cluster sampling energies is summarised in Figure 3 for 4 different material configurations in front of the calorimeter cryostat: no material, 25, 50 and 75mm of Aluminium. From the figure a 0.5% non-linearity is observed. However a much smaller spread of the data-points at the level of 0.2% or less is seen for fixed beam energy. The variations from one energy point to another can be attributed to the systematics of the CTB setup itself. In particular, changes to beam conditions (collimator openings, beam-optics magnetic fields) seem to have large effects in the relative beam energy. In the case where the beam conditions are unchanged, as in the case of the 180GeV beam in Figure 3, a very small variation is observed.

The corresponding energy resolution for the 4 material configurations is presented in Figure 4. From the figure a good agreement between the data and MC points can be seen. The resolution worsens at the approximate rate of $0.5\%/\sqrt{E}$ per $30\% X_0$ increment of the material in front of the calorimeter.

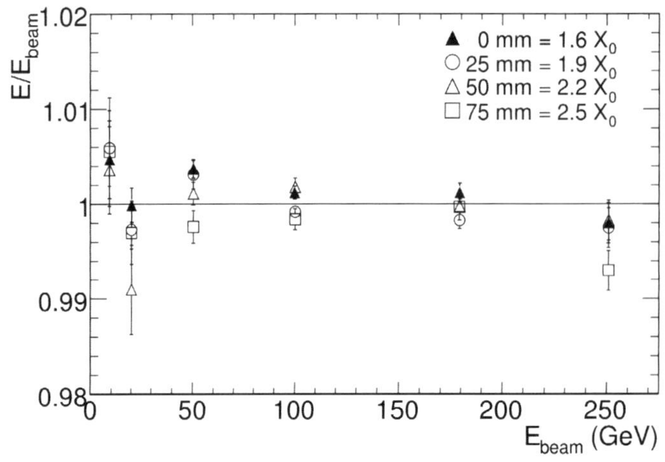

Figure 3. Energy Linearity after calibration for four different material configurations in front of the calorimeter: no material (black triangles), 25mm Al (red circles), 50mm Al (blue circles) and 75mm (black rectangles).

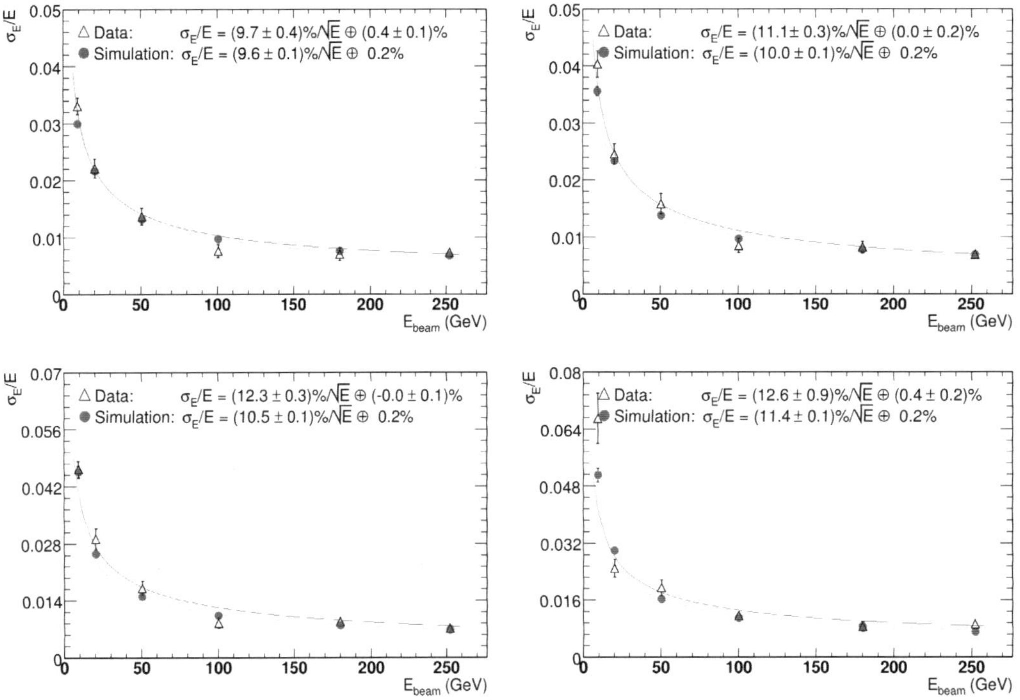

Figure 4. Energy resolution after calibration (using calibration hits) from Period 5 data and for 4 different material configurations in front of the calorimeter: no material (top left), 25mm Al (top right), 50mm Al (bottom left) and 75mm (bottom right).

7. Results on the uniformity of the calorimeter response

The uniformity of the calorimeter energy response depends on the impact point position within the cell. The cluster η position is determined from the barycenter of the energy deposition in the second sampling. The cluster size is a multiple of the size of a single cell. Consequently, the energy leakage outside the cluster is a function of the incoming particle impact coordinate within the cell. The measured energy is maximal when the impact point is in the center of the cell and drops by about 1% at the border of the cell.

The cluster energy also depends on the impact point within the cell along the ϕ direction due to the accordion geometry: the incident electrons traverse more passive material in the accordion fold than in the straight section due to the rounded shape of the folds and thus deposit more energy. There is an additional effect due to lateral leakage which is due to the limited cluster size (in ϕ), similar to the already mentioned modulation in η.

The High Voltage (HV) system of the electromagnetic calorimeter is divided in $\Delta\eta \times \Delta\phi$=0.2×0.2 regions (the so-called sectors). During the combined test beam the top side of one electrode had decreased HV, however the bottom part had the nominal HV of 2000V. Thus the energy measured in the cells belonging to the η region covered by this electrode has to be corrected.

Figure 5 gives the final result on the uniformity of the EM calorimeter. A uniformity level of 0.5% was observed.

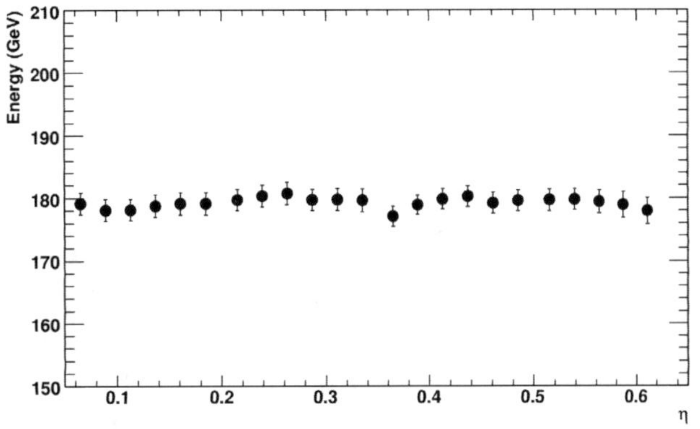

Figure 5. Uniformity of energy response of the EM calorimeter for 180 GeV electrons

8. Material effects and E1/E2

The ratio of the energy response in the strips and the middle sections of the EM calorimeter is sensitive to the material distribution in front of the calorimeter. This is demonstrated in figure 6, where the ratio is plotted for the runs where extra amounts of material (in steps of $0.3X_0$) were placed in front of the calorimeter. A significant shift in the mean and shape of these distributions is observed. The increase in the mean of the data distributions by 20-25% for $0.3X_0$ is well reproduced by the MC. In figure 6 the mean of these E1/E2 distributions are shown of as a function of the amount the extra material. This material dependence can be used to extract the calibration constants shown in Equation 1 from data instead of relying on the MC description. By unfolding the energy dependence of E1/E2 as well as the energy dependence of the weights, the direct dependence of E1/E2 on the material, and hence the weights can be derived. Application of such a strategy to the CTB data led to a determination of the "offset" and $W0$ constants with an accuracy better than 5%.

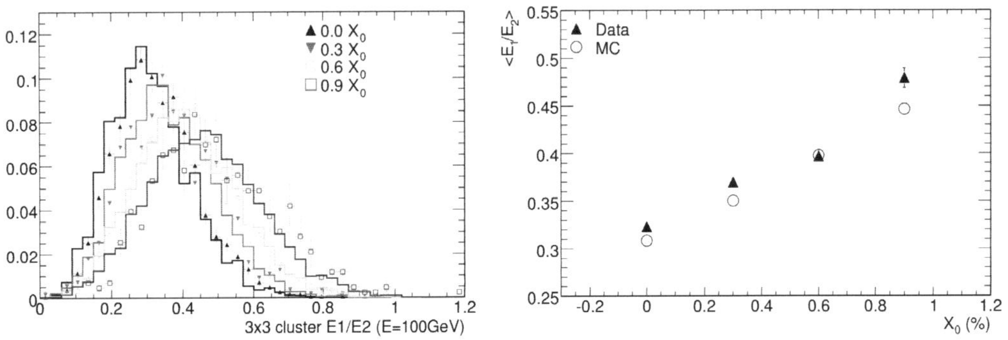

Figure 6. (left) Ratio of the energy response in the strips and the energy response in the middle sampling, E1/E2, for various thicknesses of material in front of the calorimeter cryostat and for electron energy of 100 GeV. Data are denoted by markers and MC by histograms. (right) Mean of E1/E2 distribution vs material thickness in front of the calorimeter cryostat.

9. Very low energy studies (1-9GeV)

During 2003 a very low energy (VLE) beamline was built as part of the H8 beam line. Particles from the HE beamline are diverted onto an additional target (T48), and a spectrometer consisting of 4 dipole magnets selects particles of the desired momentum. The precise momentum measurement is obtained from the deflection angle that is measured from the position of the beam spot in beam chambers placed before and after the magnets of the spectrometer. The estimated uncertainty on the beam energy comes from the uncertainty on the angle measurement due to the spatial resolution of the beam chambers ($\approx 0.2\,\text{mm}$) is 0.35 %. In addition, a systematic uncertainty of 1.5 % has to be taken into account due to the uncertainty of the alignment of the beam chambers and the exact position of the B8 magnet The calibration strategy for the very low energy is similar to that of the high energy electrons. One difference lies in the fact that the weights have a strong sampling fraction dependence that need to be taken into account. In Fig. 7 the comparison of the energy responses of the the four calorimeter compartments is shown for 5 GeV electrons. A detailed description of the calibration scheme for the very low energy electrons can be found here [6].

10. E/p measurements and Bremsstrahlung recovery

A number of studies on the combined performance of the inner detector and the electromagnetic calorimeter are ongoing. A quick overview of the main studies and their implications for ATLAS is given below:

- The energy distribution in the calorimeter and the p_T distribution in the inner detector can be parametrised and the convolution describes well the E/p distributions. A method has been developed to extract the energy scale for the EM calorimeter and the method can be used to compare the inner detector and LAr EM scales in ATLAS.
- A Bremsstrahlung recovery algorithm has been developed for ATLAS and tested successfully on the combined test beam. The implementation of the algorithm consists of dividing the (silicon detector) track into two parts, and refitting only the part close to the vertex together with the LAr cluster position as an ordinary hit.

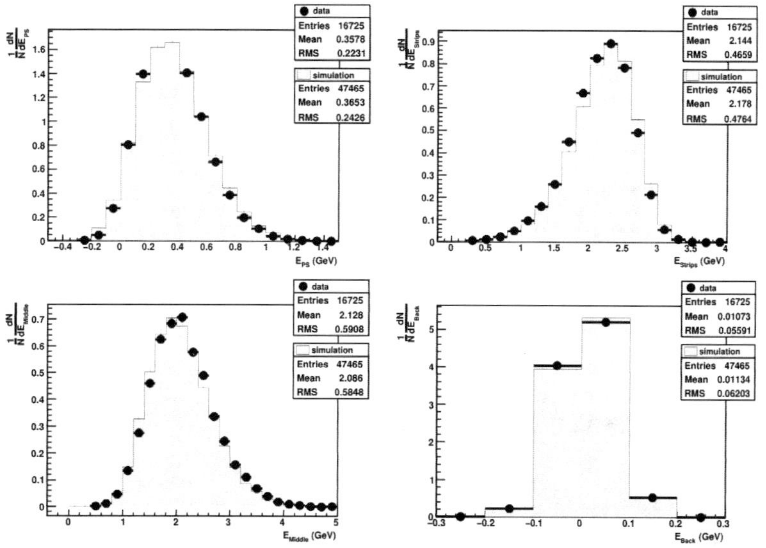

Figure 7. Layer energies comparing Monte-Carlo simulation (filled histograms) and data (black circles) for 5 GeV electrons

11. Conclusions

In this work final results on energy linearity, uniformity and resolution of the ATLAS barrel liquid argon electromagnetic calorimeter were presented. The calorimeter response was found to be linear at the level 0.5% limited by test beam systematics. The stochastic term of the resolution is around $10\%/\sqrt{E}$ and the unifomity is at the level of 0.5%. These results are consistent with earlier test beam studies ([3],[4]). Although the calibration presented here is MC based, the calibration constants can be tested with data. Observables like the E1/E2 ratio of response between the first and the second longitudinal layers of the calorimeter are very sensitive to the upstream material excesses or deficits. This ratio was used to extract the calibration constants directly from the test beam data. The calibration procedure presented here was also applied to very low energies (1-9GeV) where a very good agreement between data and MC was observed.

References

[1] B. Di Girolamo et al. "Beamline instrumentation in the 2004 combined ATLAS testbeam" *ATLAS-TECH-PUB-2005-001* 2005

[2] S. Paganis *et al.*, "ATLAS Liquid Argon barrel calorimeter performance in the 2004 combined test-beam", Nucl.Phys.Proc.Suppl.172:108-110,2007.

[3] M. Aharrouche *et al.*, "Energy linearity and resolution of the ATLAS electromagnetic barrel calorimeter in an electron test-beam", Nucl. Instrum. Meth. A **568** (2006) 601

[4] J. Colas *et al.*, "Response Uniformity of the ATLAS Liquid Argon Electromagnetic Calorimeter," Nucl. Instrum. Meth. A **582** (2007) 429

[5] B. Aubert *et al.*, "Construction, assembly and tests of the ATLAS electromagnetic barrel calorimeter," Nucl. Instrum. Meth. A **558** (2006) 388.

[6] M. Aharrouche *et al.*, "Electron Performance of the ATLAS Liquid Argon Calorimeter 11315 measured at the 2004 Combined Test Beam" submitted to Nucl. Inst. Meth.

XIII International Conference on Calorimetry in High Energy Physics (CALOR 2008) IOP Publishing
Journal of Physics: Conference Series **160** (2009) 012050 doi:10.1088/1742-6596/160/1/012050

Electronic Calibration of the ATLAS LAr Calorimeter and commissioning with Cosmic Muon Signals

Carolina Gabaldón
Department of Physics, Universidad Autónoma de Madrid, Madrid, Spain
on behalf of the ATLAS LAr Collaboration

E-mail: carolina.gabaldon.ruiz@cern.ch

Abstract. The three cryostats of the Liquid Argon calorimeter System have been gradually installed at their nominal position in the ATLAS cavern from end 2005 to beginning of 2008. After the cabling and installation, calibration runs are taken routinely, allowing to set-up and debug the electronics. Finally cosmic muon runs are also recorded regularly since august 2006 to test *in situ* the signal reconstruction procedure used to obtain the energy in each cell, as well as to check some of the detector performance. This paper presents the current status of these analyses.

1. Introduction

The ATLAS experiment is presently in its last stage of installation in the Point 1 of the Large Hadron Collider (LHC), the last CERN proton-proton collider, and will be complete and ready for the first collisions expected in the autumn 2008. One of the main ATLAS sub-detectors, the Liquid Argon (LAr) calorimeter System, composed of one barrel and two end-cap cryostats, has been gradually installed in its nominal position from the end of 2005 until the beginning of 2008. As a result, it was one of the first systems to start the commissioning phase with regular calibration data taking on its entire coverage. Before 2008, the muon spectrometer trigger was not available and a dedicated trigger set-up, based on the hadronic Tile Calorimeter that surrounds the three cryostats, was developed allowing to record several millions of cosmic muons in the LAr calorimeter system. The muon energy deposit in the calorimeters is generally small and therefore requires a very good understanding of the calibration and the signal reconstruction, especially in its ElectroMagnetic (EM) part. The current status of the cosmic muon analysis is presented, keeping in mind that this provides an unique occasion to test the $\sim$180,000 cells of the full system before the first proton-proton collision.

2. ATLAS LAr Calorimeter System

The ATLAS LAr Calorimeter System is composed of three sampling calorimeters with Liquid Argon (*LAr*) as sensitive material, covering a pseudorapidity region $|\eta| < 4.9$: the ElectroMagnetic Calorimeter in $|\eta| < 3.2$, the Hadronic EndCap Calorimeter (HEC) in $1.5 < |\eta| < 3.2$ and the Forward Calorimeter (FCal) in $3.1 < |\eta| < 4.9$.

The EM calorimeter is a device based on an accordion shape geometry consisting of lead absorbers and electrodes coupled to a fast electronic integration. This allows to combine the robust and very well tested liquid argon technology, with a perfect azimuthal hermeticity and a clean signal extraction in the harsh LHC environment. The EM calorimeter is divided in one barrel, covering $|\eta| < 1.475$ and two end-caps, $1.375 < |\eta| < 3.2$. In depth, three very granular compartments (S_1, S_2, S_3) allow a detailed measurement of the electron and photon EM shower lateral and longitudinal development (Figure 1). This is particularly usefull at LHC to separate efficiently electrons from jets and photons from π_0's. A presampler device is added in the most critical region for the amount of material upstream the EM calorimeter ($|\eta| < 1.8$) and allows to recover the electron/photon energy lost in the material in front (inner detector, cryostat and solenoid coil up to $|\eta| < 1.5$). As a result, the total number of read-out channels or cells is about 170,000, typically an order of magnitude above the calorimeters used in the Tevatron experiments. A smooth operation of all these cells together represents a major challenge needed for example to achieve the foreseen ATLAS Higgs discovery potential in the two photon Higgs decay channels.

The HEC uses the Cu-LAr sampling plate technology with "cold" preamps mounted on the calorimeter. It is designed to measure the jet energy escaping the EM calorimeter and therefore a coarser granularity, corresponding to the level one calorimeter trigger tower, is used. Finally, the FCal is a copper-tungsten/LAr detector which covers the region closest to the beam. It provides a hermetic calorimeter system up to $|\eta| < 4.9$ with more than 10 units of interaction length λ. The segmentation of the Liquid Argon System is recalled in Table 1. More details on its commissioning and installation can be found in Ref. [1].

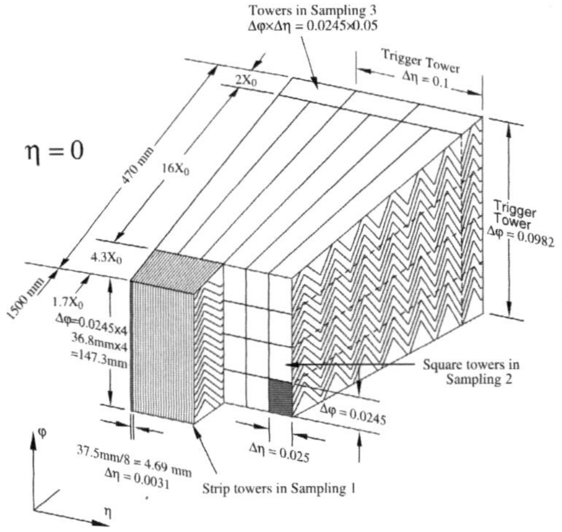

Figure 1. Sketch of the accordion and sampling structure of the EM Calorimeter.

Table 1. Covered region and number of read-out channels of the sub-detectors of the Liquid Argon System.

LAr detector	Channels	Region
EM calorimeter	173312	$\|\eta\| < 3.2$
HEC	5632	$1.5 < \|\eta\| < 3.2$
FCal	3524	$3.1 < \|\eta\| < 4.9$

3. Signal reconstruction with the EM Calorimeter

The LAr calorimeter signal is generated by the drift of the ionization electrons in the electric field provided by the High Voltage (HV) in the LAr gap. The ionization current has a triangular shape, where the peak is proportional to the energy deposited by the shower. The ionization signal is pre-amplified and then shaped by a $CR-RC^2$ bipolar filter at the end of the readout chain. The bipolar signal is sampled every 25 ns (the LHC bunch crossing period) and 5 samples are digitized and used in the signal reconstruction procedure. For special runs, like the ones with cosmic muons, more than 5 samples are digitized and recorded (typically 25 or 32). Figure 2 shows a comparison between the original triangular signal generated inside the LAr gap and the output signal after passing the readout electronics. The maximum has been normalized to 1. The bipolar shaper is designed such that the maximum of the triangular signal corresponds to the maximum of the shaped pulse. Hence, the maximum amplitude of the shaped pulse is proportional to the energy deposited by the ionizing particles in the readout cell.

Starting from the digitized samples, two relevant quantities are deduced with a digital filtering technique called Optimal Filtering [2]: the signal maximum amplitude (A_{max}) and the difference between the assumed and reconstructed time (Δt). The inputs to the method are: *i*) the covariance or autocorrelation matrix of the samples, which contains the noise information, *ii*) the predicted pulse shape (g^{phys}), and *iii*) its first derivative (dg^{phys}/dt). The outputs of the method are the weights a_i, b_i $i = 1,\ldots,n$ (n is the number of samples) which allows to reconstruct A_{max} and Δt as :

$$A_{max} = \sum_{i=1}^{n} a_i(S_i - p), \qquad \Delta t = \frac{\sum_{i=1}^{n} b_i(S_i - p)}{A_{max}} \tag{1}$$

where S_i $i = 1,\ldots,n$ are the 25 ns samples and p is the pedestal level of a given readout cell (both in ADC counts). Thanks to the autocorrelation matrix, the noise contributions to the signal are minimized. Two sources of noise are foreseen in the calorimeter during operation at LHC: the electronic noise and the pile-up noise. Since this article refers to cosmic muon data, only the first source of noise enters in the analysis.

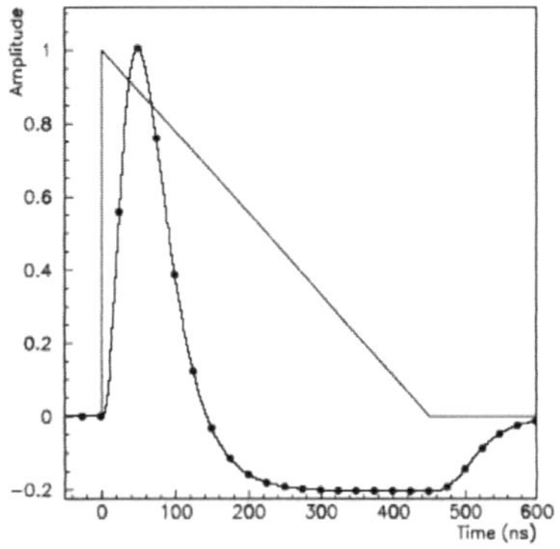

Figure 2. Ionization signal (a) and shaped signal at the end of the readout chain (b).

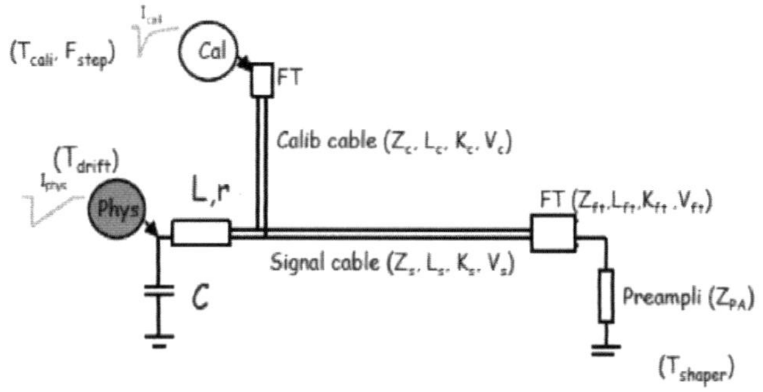

Figure 3. Schematic model of the readout chain of the LAr EM calorimeter cells.

To determine the Optimal Filtering Coefficients for each calorimeter cell the most difficult step is to predict the pulse shape of the physics signal. This can be done thanks to the calibration system, that sends a well-known exponential signal on the cold electronics which then follows the same readout path as the physics one (Figure 3). Two methods have been developed in the electromagnetic calorimeter. The first one, presently used to commission the detector [3], where the cell response to a calibration pulse shape is used to predict completely the physic pulse shape. And an alternative method [4], that uses an analytical model of the EM readout to predict the physic pulse shape per cell. In this last approach, all the circuits parameters are measured, except two, left free in order to match the measured calibration pulse response. A similar method is used in HEC [5], whereas FCal uses the physics pulse extracted from test beam data for the prediction [6].

Once the amplitude has been determined, the energy of each cell is then computed as:

$$E = F_{gain} \times \frac{1}{F_{I/E}} \times A_{max} \tag{2}$$

where F_{gain} is the electronic gain, converting the signal amplitude from ADC to current and $F_{I/E}$ is the absolute energy scale.

4. Commissioning of the ATLAS electromagnetic calorimeter with cosmic muons

The analysis of cosmic muon events is the only way to test the EM calorimeter *in situ* with physics signals before LHC collisions. The process of integrating more and more sub-systems has been ongoing since summer 2006. During the summer 2007, the situation for the EM barrel and one EM end-Cap liquid Argon calorimeters was stable and the cosmic data were taken nearly every weekend.

A dedicated trigger using only Tile calorimeter signals was configured to cope with the absence of the muon spectrometer trigger. Figure 4 shows the Tile towers (One trigger tower is the sum of all Tile cells in a region of $\Delta\eta \times \Delta\phi = 0.1 \times 0.1$) that were included in the trigger for the data taking.

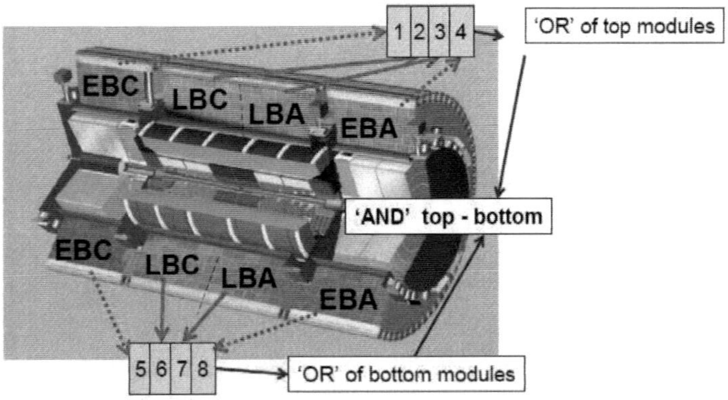

Figure 4. Tile trigger setup and logic of the top-bottom coincidence. Tile modules are labeled LB in the central part and EB in the end-cap part. A and C refer to $z > 0$ and $z < 0$ respectively.

Cosmic muons can be classified in two categories: the ones which radiate bremmsstralung photons and the ones behaving as Minimum Ionizing Particles (MIP). In the first case, high energy deposits can be selected ($E > 500$ MeV) to check the quality of the predicted physics signal. Note, that this is the only option for the EndCaps due to the absence of projective muons, i.e. muons pointing to the nominal Interaction Point [7]. In the second case, which corresponds to 99% of the events, a coarse check of global uniformity, both for (η, ϕ) spatial coordinates and for timing, is possible for the EM barrel calorimeter [8]. For these studies, projective or quasi-projective muons are required.

4.1. Studies with high energy muons

To compare the predicted physics pulses (g_i^{phys} normalized to one) with the data, the prediction must first be adjusted in time, to take into account the asynchronous muon arrival times, and in amplitude by multiplying g_i^{phys} by A_{max} computed with Equation (1). After this step, Figure 5 shows typical physics shapes for middle sampling (S2) in the EM barrel (left) and in the EM end-cap (right). The predictions agree nicely with the data in the raising and falling edges of the pulses. A similar comparison can be made with the other LAr detectors (Figure 6) that also shows a good agreement between the data and the predicted physic shape.

To perform a systematic quantitative comparison between data and pulse shape predictions over the full η coverage the following χ^2 can be computed:

$$\chi^2 = \sum_{i=1}^{n=5} \left(\frac{S_i - A_{max} * g_i^{phys}}{\sigma_{noise}} \right)^2 \tag{3}$$

where S_i is the amplitude of each sample i in ADC counts for data and σ_{noise} corresponds to the noise for a single sample in ADC counts.

Journal of Physics: Conference Series **160** (2009) 012050 doi:10.1088/1742-6596/160/1/012050

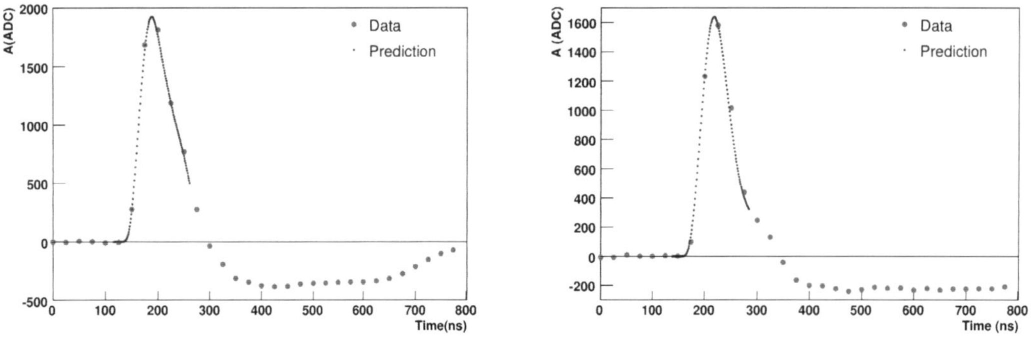

Figure 5. Typical cell response to high energy deposits in the EM barrel (left) and EM endcap (right).

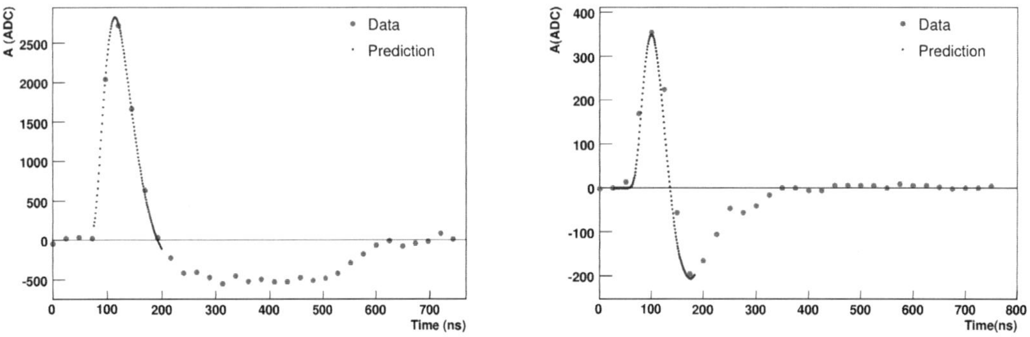

Figure 6. Example of cell response to cosmic signal in HEC (left) and FCal (right).

By construction, the χ^2 of Equation (3) depends on the square of the reconstructed amplitude A_{max} (the error on the predicted pulse shape can not be estimated) [9]. Therefore, an η − independant estimator, Q, of the prediction quality is build when dividing by ${A_{max}}^2$. Figure 7 shows this estimator as a function of the energy for the middle layer in the EM barrel (left) and in the EM end-cap (right). Its dependence with energy can be fitted by the following simple function :

$$Q = \frac{\chi^2}{n_{samples} \times {A_{max}}^2} = \frac{p_0}{E^2} + p_1 \qquad (4)$$

The first term, dominating at low energy, is due to the gaussian noise fluctuation for each sample. The second term, dominating at high energy, reflects the quality of the predicted shape as compared with the data. The results are only slightly better in the EM barrel compared to the EM end-cap. This is the first proof of the quality of an ATLAS-like signal reconstruction in the EM end-caps, despite its more challenging geometry: the LAr gap varying with η from 3 mm to 1 mm whereas it is 2 mm everywhere in the EM barrel. This is also illustrated in Figure 8, which shows the pulse shape prediction quality Q as a function of η. These results are obtained by applying a lower energy cut (1.5 GeV in S2), to minimize the noise contribution, and an upper cut (2500 ADC counts) to avoid high gain saturation. This assesses the coherence of the signal reconstruction quality, using 5 samples as foreseen in ATLAS, over the complete EM calorimeter coverage $0 < |\eta| < 3.2$.

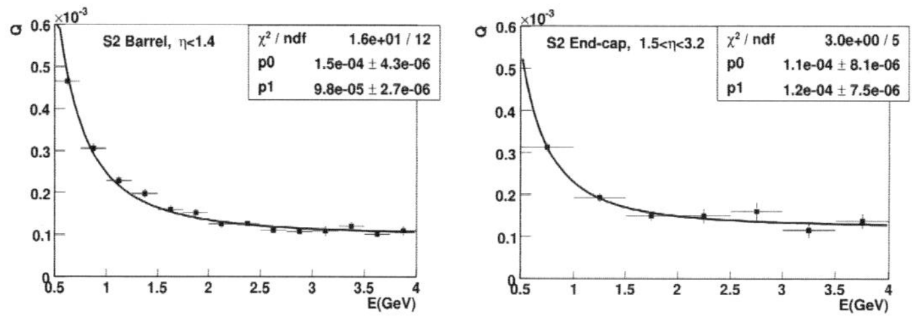

Figure 7. Estimator of the quality of the predicted physics pulse shape as a function of the energy in S2 cells of the EM barrel (left) and of the EM end-cap (right).

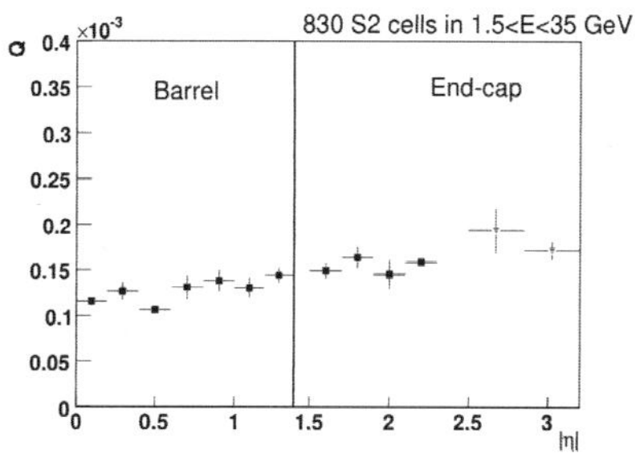

Figure 8. Estimator of the quality of the predicted physics pulse shape as a function of $|\eta|$ for S2.

4.2. Studies with MIP cosmic muons

In the previous section the quality of the standard signal reconstruction was tested with high energy muons. The following step is to extract a high quality sample of MIP cosmic muons to perform a global check of the EM calorimeter performance.

The two main difficulties in this analysis come from the small energy deposit of a MIP in the EM calorimeter (300 MeV in a S2 cell) and the non projectivity of cosmic muons. In the first case, the full noise reduction capability of the signal reconstruction method can be used (see section 3). As an example, Figure 9 shows that the signal to noise can be improved by a factor > 2 when considering 25 samples instead of 5. In order to extract a pure "projective" muon sample passing close to the nominal interaction point, a dedicated algorithm has been developed to reconstruct cosmic muon trajectories. The trajectory is extracted by fitting with a straight line the position of Tile cells in the top-bottom side (Figure 10). Only muons which pass within ± 30 cm of the interaction point in the $Y = 0$ plane are selected, representing $10,000$ events.

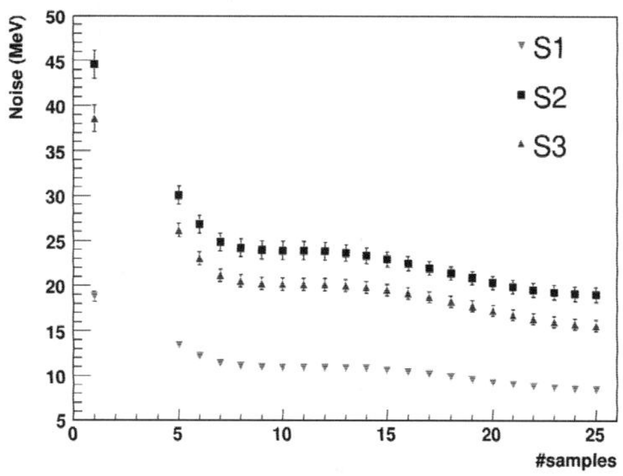

Figure 9. Noise as a function of the number of samples used by optimal filtering for EM calorimeter layers in high gain.

The first task to be performed with the LHC data is the timing of the calorimeter. Figure 11 shows the reconstructed EM time versus the reconstructed Tile time corrected for time of flight at $Y = 0$. A linear correlation between the EM cell time and the Tile time is observed. After fixing the slope to be unity in the linear fit, the error on the y-intercept decreases to just below 1 ns. The collection of this timing offsets allows for a first look at timing uniformity.

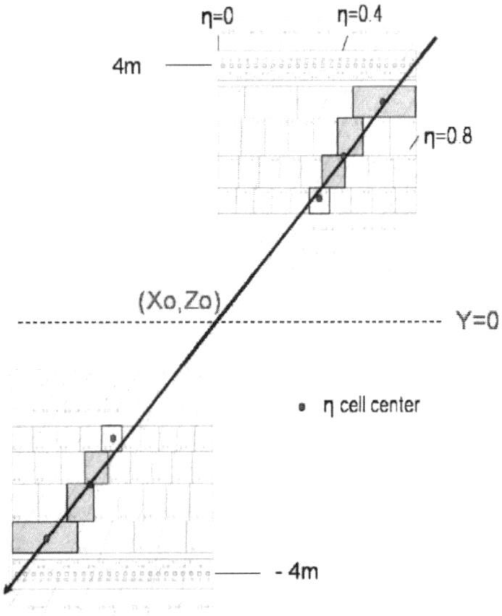

Figure 10. Schematic description of a reconstructed muon track. $Y = 0$ represents the horizontal plane with the interaction point (X_0, Z_0) in.

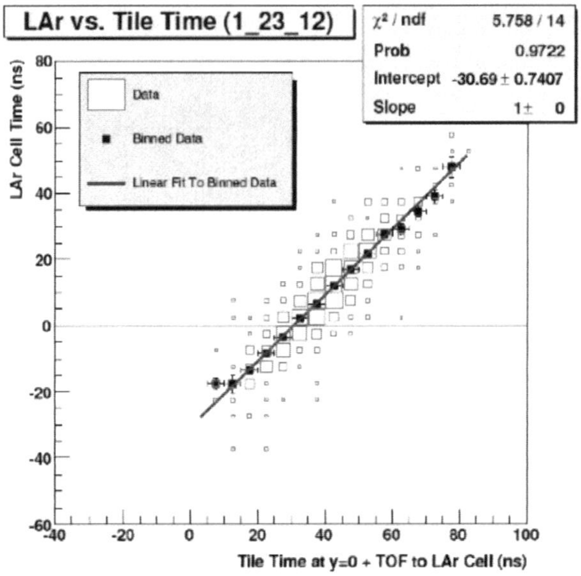

Figure 11. EM cell time vs. Tile Time at $Y = 0$ corrected for time of flight.

Cosmic muons also provide a good opportunity to study the calorimeter response, with more modules than in the test beam, before the start of the LHC. After a projective requirement their energy depositions should follow a Landau distribution. This is illustrated in Figure 12 which displays the measured energy distribution for clusters of $\Delta\eta \times \Delta\phi = 3 \times 3$ cells in the range $0.3 < |\eta| < 0.4$. The χ^2/NDF strongly supports the choice of a Landau convoluted with a Gaussian as the fit function which describes the data. The fitted value of the Gaussian width parameter, $\sigma_G = 60$ MeV, corresponds to the uncorrelated noise contribution of 9 cells ($20 \times \sqrt{9} = 60$ MeV) .

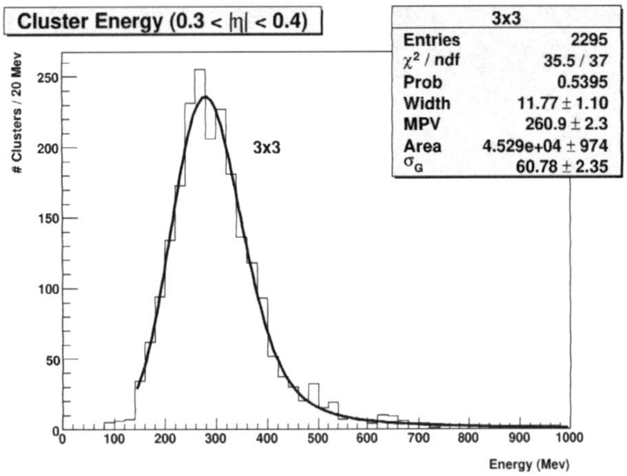

Figure 12. Measured 3×3 cluster energy distributions in the range $0.3 < |\eta| < 0.4$, using the $(|X_0|, |Z_0|) < (30$ cm, 30 cm$)$ projectivity selection criteria for data.

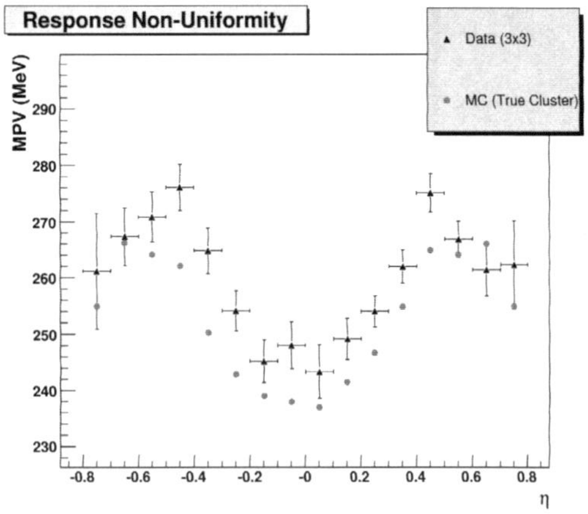

Figure 13. Most probably energy value vs η for 3×3 cluster (black triangle) and MC true clusters (pink circle).

Figure 13 presents the most probable value (MPV) as determined from the fits of the energy distributions binned in η intervals with a width of 0.1 for $|\eta| < 0.8$. The dependence of the MPV on η is compared to the true cluster values found in the MC analysis corresponding to the same event selection. The η dependence of the data is in good agreement with the MC simulation at the level of 3%. To probe the true detector non-uniformity for individual cells at the level of 1% would require approximately 160,000 projective events, which is out of reach before the LHC start.

Conclusions

At present, ATLAS is completely installed in its final position in the Point 1 cavern. The ATLAS Liquid Argon System was one of the first to begin the installation in 2006. These two years were an occasion to exercise the electronics calibration and assess the robustness of the signal reconstruction in all LAr detectors. Since October 2006, cosmic data have been collected nearly every weekend. This allows a full-scale test of the signal reconstruction chain which is now understood at 1% level. Despite the small and mostly non-projective signals, cosmic muons provide a first in situ test of LAr detectors. A first check of the overall energy scale has been shown to agree with simulation to within 3% for typical cosmic energy depositions.

All these results give confidence that the LAr system will be ready to take data in autumn 2008 when the first LHC pp collisions will occur.

Acknowledgments

The work presented here has been performed within the ATLAS LAr collaboration. This analysis would not have been possible without the dedicated effort of many people in our LAr detector group and the ATLAS collaboration over many years. I'm especially indebted to those who built, integrated and installed the LAr detectors in the ATLAS cavern, and those who operate the detector on a daily basis.

References

[1] H. Wilkens, these proceedings.

[2] W.E. Cleland et E.G. Stern, *Signal processing considerations for liquid ionization calorimeter in a high rate environment*, Nucl. Inst. Meth. **A 338** (1994) 467.

[3] D. Banfi, M. Delmastro and M. Fanti, *Cell response equalization of the ATLAS electromagnetic calorimeter without the direct knowledge of the ionization signals*, SN-ATLAS-2005-054, J. Inst **1** (2006) P08001.

[4] C. Collard *et al.*, *Prediction of signal amplitude and shape for the ATLAS electromagnetic calorimeter*, ATL-LARG-PUB-2007-010.

[5] C. Cojocaru, *Hadronic calibration of the ATLAS Liquid Argon End-Cap Calorimeter in the Pseudorapidity Region 1.6* $< \eta <$ *1.8 in Beam Tests*, NIM A531 (2004) 481-514.

[6] J.P. Archambault *et al.*, *Energy calibration of the ATLAS Liquid Argon Forward Calorimeter*, JINST 3 P02002.

[7] C. Gabaldon Ruiz, M. Kuna et al., *et al.*, *Signal reconstruction in the EM end-cap calorimeter and check with cosmic data in the region* $0 < \eta < 3.2$, ATLAS note ATL-LARG-PUB-2008-001.

[8] M. Cooke, P.S. Mangeard, M. Plamondon *et al.*, *In situ commissioning of the ATLAS electromagnetic calorimeter with cosmic muons*, ATLAS note ATL-LARG-PUB-2007-013.

[9] M. Delmastro, *Quality factor analysis and optimization of digital filtering signal reconstruction for liquid ionization calorimeters*, NIMA-D-08-00417.

Journal of Physics: Conference Series **160** (2009) 012051 doi:10.1088/1742-6596/160/1/012051

Calibration of the electromagnetic calorimeter of the CMS experiment

T. Tabarelli de Fatis [1]
Università di Milano Bicocca and I.N.F.N. Sezione di Milano Bicocca
Piazza della Scienza 3, 20126 Milano
E-mail: tommaso.tabarelli@mib.infn.it

Abstract. The calibration of the relative response of the individual channels, or intercalibration, and of the energy scale for electrons and photons are severe technical challenges for the operation of the electromagnetic calorimeter of the CMS experiment. The channel response uniformity within ECAL and stability in time will directly contribute to the overall energy resolution. Intercalibrations performed during the commissioning phase of the detector provide an acceptable performance at startup and a handle for validation of *in situ* calibration procedures based on the use of events collected during LHC operation. A laser monitoring system will be used to track response variations with time, as in the case of changes in crystal transparency caused by irradiation. Triggers, selections and reconstruction procedures of *in situ* calibration signals and their projected performance are reviewed. The interplay with reconstruction algorithms of electrons and photons is also discussed. Calibration procedures offer sufficient redundancy and flexibility to fast improve on startup conditions, intercalibrate the response of the individual channels towards a target precision of 0.5% and calibrate the energy scale of electrons and photons.

1. Introduction

The Compact Muon Solenoid (CMS) detector [1] is a general purpose detector installed at the new CERN proton collider (LHC). The Electromagnetic Calorimeter (ECAL) of CMS is a hermetic homogeneous calorimeter made of lead-tungstate ($PbWO_4$) crystals, equipped with avalanche photodiodes (APDs) in the central "barrel" part and vacuum phototriodes (VPT) in the end-caps for the light collection. The barrel calorimeter is organised into 36 supermodules, each containing 1,700 crystals arranged in four modules. It is closed at each end by an end-cap calorimeter consisting of two *dees*, each with 3662 crystals. The design of the calorimeter has been optimised for the detection of the Higgs boson through its electromagnetic decay ($H \rightarrow \gamma\gamma$). This requires an excellent energy resolution and fine granularity. The performance of the components of the calorimeter has been extensively tested with electron beams. The stochastic and electronic noise contributions to the energy resolution of ECAL have been demonstrated to match the design requirements [2, 3, 4].

An essential issue in CMS will be the channel response uniformity within ECAL, as this will contribute directly to the overall energy resolution. This uniformity is determined by the accuracy of the calibration of the relative response, or intercalibration, between different

[1] On behalf of the CMS Electromagnetic Calorimeter Group

channels across the detector. *In situ* calibrations with events collected during LHC operation are the main tool to achieve the target intercalibration precision of 0.5% required for detection of the $H \rightarrow \gamma\gamma$ decay. In addition, for optimal detector performance, the energy scale and linearity of the response to electromagnetic particles must be precisely calibrated. In CMS, electrons may lose energy through bremsstrahlung and photons may be converted into electron pairs in the Tracker material in front of ECAL. This will affect the energy measurement and resolution in ECAL and dedicated reconstruction algorithms have been developed to cope with the different cases [6]. These effects will also need to be calibrated *in situ*.

2. Intercalibration

The relative channel response spread affects the constant term of the energy resolution of ECAL in CMS with little dilution, as electromagnetic showers involve only a small number of crystals: unconverted photons deposit on average about 70% of their energy in a single crystal. The main sources of channel-to-channel response variation are the crystal light yield variation in the barrel, about 13% at construction, and the gain spread of the photodetectors in the endcaps, about 25%.

To reduce this spread and provide an acceptable performance of the detector at startup, different calibration procedures have been adopted during the construction and commissioning phase of ECAL. In particular, an intercalibration accuracy of 0.3% will be available at startup for nine ECAL barrel supermodules exposed to electron test beams. In addition, all the supermodules of the ECAL barrel have been commissioned in turn on a cosmic rays stand for a period of about one week. From these data, intercalibration coefficients with an accuracy better than 1.5% over most of the volume of a supermodule, raising to above 2% at the outer end (corresponding to $|\eta| \sim 1.5$) have been derived [7]. Due to tight construction constraints, the ECAL endcaps could not be intercalibrated with the same accuracy. Intercalibration coefficients at startup will be available with an accuracy of about 10-15%, derived from laboratory measurements of crystal light yield and photodetector response.

The pre-calibration accuracy of the ECAL barrel is adequate for LHC startup physics and for physics channels involving the measurements of electrons. For example, the impact of intercalibrations on the Z mass resolution in $Z \rightarrow ee$ decays is marginal for intercalibration accuracies better than 2%: other effects, including the impact of the electron reconstruction, also contribute to the mass resolution. On the other hand, the intercalibration precision in the endcaps should be quickly reduced to a few percent by means of *in situ* calibration procedures exploiting LHC data. Improvements in the intercalibration accuracy will be monitored from the width of known resonances, including the Z mass peak. *In situ* data must also be exploited to achieve the ultimate intercalibration precision over the entire detector.

With very first data, the invariance around the beam axis of the energy flow in minimum bias events (so-called ϕ-invariance) can be exploited to rapidly confirm, and possibly improve on, the startup intercalibrations within fixed pseudorapidity regions [8]. An external input, for example provided by intercalibrations performed prior to installation or by other physics events, must be used to intercalibrate these regions to one another. According to Monte Carlo studies, this method is expected to provide an accuracy of a few percent with of about 20×10^6 minimum bias events. This accuracy represents the systematic limit of the procedure assuming ignorance of the distribution of the material in front of ECAL, which is a source of ϕ-invariance violation. For fast calibration purposes at startup, a dedicated calibration trigger has been deployed: upon L1 minimum bias trigger, the High Level Trigger (HLT) module will unpack ECAL data, select and log only those crystals (about 20 per event on the average) above a tunable threshold. This will enable for logging data at 1 kHz rate, with minimum use of the bandwidth, and reach the systematic limit of the procedure within one day of data taking. Comparisons to pre-calibrations in the ECAL barrel, having a similar precision, will be used to validate the procedure. In the long term this procedure could be refined with corrections for tracker and geometry effects derived through comparisons to other calibration data.

A promising channel for intercalibrations exploits the π^0 mass constraint on the energy of the two photons from the $\pi^0 \rightarrow \gamma\gamma$ decay [9]. According to simulation, a clean sample of unconverted photons from π^0s decays can be selected by restricting the search of photon pairs within a few crystals (few degrees). Due to the magnetic field of CMS, converted photons of low energy are swept away or produces energy depositions well separated in ECAL. The use of unconverted photons reconstructed in 3×3 fixed arrays of crystals prevents the intercalibration procedure from being complicated by the effects of energy radiation in the tracker. Thus, this calibration method is only indirectly affected by tracker material, through an efficiency loss and a worsening of the signal to background ratio in the detector regions where the tracker is thicker. An iterative procedure is used to unfold the individual crystal constants. According to simulation, an intercalibration accuracy of 0.5% can be achieved in the barrel with about 2000 photons per crystal from π^0 decays, for a typical signal to background ratio of about two. This event sample can be collected in a few days of data taking, if π^0 can be logged at 1 kHz rate. To this end, a dedicated calibration trigger has been developed, which unpacks ECAL data only in restricted regions around L1 electromagnetic candidates in QCD events. In the HLT filter, π^0 candidates are selected upon simple shower-shape, kinematics and isolation cuts. Bandwidth constraints are satisfied by logging only a minimal information around the candidates and dropping all the rest of the event. Detailed studies for the endcaps, including the use of the pre-shower in front of ECAL, are ongoing.

In the medium and long term, when the tracker will be precisely aligned and understood, isolated electrons, mainly from W decays, can be used to derive intercalibration coefficients. In this procedure the energy measured in ECAL is required to match the momentum measured in the tracker [10]. Isolated electrons with negligible radiation in the tracker are selected by means of shower shape variables and tracker information. These electrons, similarly to unconverted photons, are well reconstructed in 5x5 fixed arrays of crystals. Thus the derivation of intercalibration constants is cleanly separated from reconstruction effects. The main drawback of this approach is a limited selection efficiency, of around 10%, in the regions at the end of the barrel and in the endcaps, due to the large amount of tracker material in front of ECAL. Yet, results of the full simulation shows that the target precision of 0.5% can be achieved after 5-10 fb^{-1}, depending on the pseudorapidity.

3. Time stability of the channel response

An important part of the calibration procedure is the time stability of the calorimeter response. During LHC cycles, the ECAL response will vary depending on irradiation conditions, which modify the transparency of the $PbWO_4$ crystals. These effects take place on a time scale of hours and cause transparency changes of a few percent at the design luminosity. The changes in crystal transparency will be monitored at 20 min intervals by means of laser light injected into each crystal through optical fibres. The capability of this system to allow correction for transparency changes was proved with test beam data [5].

4. Energy reconstruction of electrons and photons

The final goal of calibration is to achieve the best energy measurement for electrons and photons. Electromagnetic particles deposit their energy over several crystals and the energy estimate implies a sum over the corresponding channels. In addition to the intercalibration coefficients, a correction factor is also needed to relate the energy contained in these crystals to the energy of the incoming particle. This correction is particle, energy and position dependent: different reconstruction procedures are used for electrons, unconverted and converted photons. This is a consequence of the different interaction processes of these particles, of the significant variation of the thickness of tracker material as a function of the pseudorapidity, of the small continuous variation of the ECAL geometry along the length of the barrel and of the variation of the structure of ECAL between the barrel and the endcaps.

The best energy estimate for unconverted photons, reaching ECAL without interactions in the material in front of it, is obtained in by summing the energy deposited in fixed arrays of crystals, like 5×5 crystal matrices. In this case, containment corrections are precisely known from test beam measurements, where electrons were targeted on the calorimeter without any material in front of it.

For electrons and converted photons, more sophisticated "clustering" algorithms [6] are used to collect the energy radiated in the tracker material by bremsstrahlung or photon conversions and spread azimuthally by the intense magnetic field of CMS. Not all the energy is collected, due to algorithm inefficiencies, and part of the energy, lost in the tracker and swept by the magnetic field, never reaches ECAL. Thus, algorithmic corrections based on Monte Carlo simulations of the CMS detector must be applied. Due to unavoidable imperfections of the Monte Carlo model, in particular in the description of the material budget in front of the calorimeter, these corrections must be tested and tuned on data, using events collected during LHC operation.

$Z \to ee$ events are the primary data sample to set the energy scale and test the algorithmic corrections for electrons, by using the Z mass constraint [11]. This procedure is largely independent of the tracker performance, which could be not optimal at startup. While a clean sample of $Z \to ee$ can be selected, the event rate is insufficient for calibration of the individual channels. Still, these events are valuable to test and tune the η dependency of the algorithmic corrections. In addition, most methods of intercalibrations are local to a region of ECAL and $Z \to ee$ events are useful to intercalibrate these regions to one another.

Another sample under consideration is the $Z \to \mu\mu\gamma$ decay, where the photon comes from final state radiation. In these events, the Z mass constraint can be exploited to set the energy scale and tune the algorithmic corrections for converted photons. According to simulation, a large reduction of the major background, due to Drell-Yan events where one jet is mistaken for a photon, is possible by exploiting the $\mu - \gamma$ collinearity. An event rate of about 1 photon per crystal per 1 fb^{-1} is anticipated.

5. Summary and conclusions

Precise calibration of the ECAL detector in CMS is a severe technical challenge and several effects need to be carefully controlled to achieve the per mil precision. Calibration performed during the commissioning phase of the detector provide an acceptable performance of the ECAL barrel of CMS for startup physics and also a valuable handle for validation of in situ calibration procedures. The ultimate calibration accuracy, including precise intercalibrations and calibration of the electron and photon energy scales, will be achieved *in situ* by exploiting the redundancy of independent data sets.

References

[1] CMS Collaboration 1994 *The Compact Muon Solenoid Technical Proposal CERN/LHCC 94-38.*
[2] Auffray E et al. 1998 *Nucl. Instr. and Meth.* **A 412** 223; Alexeev G et al. 1997 *Nucl. Instr. and Meth.* **A 385** 425; Peigneux J P et al. 1996 *Nucl. Instr. and Meth.* **A 378** 410.
[3] Adzic P et al. 2006 *Eur. Phys. J.* **C44** 1-10.
[4] Adzic P et al. 2007 *J. Inst.* **2** P04004.
[5] CMS Collaboration 2008 *The CMS experiment at the CERN LHC*, Submitted to *J. Inst.*
[6] CMS Collaboration 2006 *The CMS Physics Technical Design Report, Vol. I CERN/LHCC 2006-001.*
[7] Adzic P et al. 2008 *CMS-NOTE/2008-018.*
[8] Futyan D and Seez Ch 2003 *J. Phys* **G29** 1229-1236.
[9] Litvin V 2007 *CMS-CR/2007-079* and *IEEE Nucl.Sci.Sym.* CR-N09-5
[10] Agostino L et al. 2007 *J. Phys.* **G33** N67-N84.
[11] Meridiani P and Paramatti R 2006 *CMS-NOTE/2006-039*

Journal of Physics: Conference Series **160** (2009) 012052 doi:10.1088/1742-6596/160/1/012052

Performance of CMS ECAL Preshower in 2007 test beam

Syue-Wei Li, Apollo Go, Chia-Ming Kuo
On behalf of CMS Preshower collaboration
National Central University, Taiwan,
No.300, Jhongda Rd., Jhongli City, Taoyuan County 32001, Taiwan (R.O.C.)

E-mail: Syue-Wei.Li@cern.ch

Abstract. The Preshower detector is part of the CMS Electromagnetic Calorimeter, located in the endcap regions, in front of the lead tungstate crystals. It consists of two orthogonal planes of silicon strip sensors interleaved with two planes of lead absorbers. A combined beam test of close-to-final prototypes of the Hadron calorimeter, the crystal calorimeter and the Preshower detector was performed in the summer of 2007. Calibrations were made using electron and pion data. Good signal/noise performance was obtained in both gains of measurements

1. Introduction

The CMS Preshower will be placed in front of endcaps of the electromagnetic calorimeter. The coverage of the Preshower is $1.653<|\eta|<2.6$. It consists of two lead absorbers (2 and 1 X_0 respectively) and two orthogonal planes of silicon strip sensors. The main purpose of the Preshower is to identify two closely spaced photons from π^0 decays (background of $H^0\rightarrow\gamma\gamma$) in order to separate these photons from single photons [1].
There is a great need to understand the overall performance of the calorimeters, before CMS completes the installation and goes into operation. An effort was made in summer 2007 to bring together close-to-final prototypes of the Preshower, Electromagnetic Calorimeter (ECAL) and Hadron Calorimeter (HCAL) in the H2 test area at CERN.

2. Experiment setup and the Preshower signal reconstruction

The setup including the 3 calorimeters HE (hadronic), EE (electromagnetic) and ES

Journal of Physics: Conference Series **160** (2009) 012052 doi:10.1088/1742-6596/160/1/012052

(Preshower) is shown in Fig. 1. For the Preshower, it was the first time we had a full read-out chain (on-detector, off-detector integrated into the main CMS DAQ system for the combined beam test), DQM (Date Quality Monitoring) and DCS (Detector Control System). It proved that a small scale ES can operate successfully in realistic conditions.

The Preshower is a sampling calorimeter and consists of two orthogonal layers of silicon strip sensors positioned behind two planes of lead absorbers (Fig.2). Each silicon sensor measures 63x63 mm^2, with an active area 61x61 mm^2 divided into 32 strips (1.9 mm pitch). The nominal thickness of the silicon is 320 μm. The thicknesses of two absorbers are nominally ~2 X_0 and ~1 X_0 [1].

Figure 1. The picture of test beam setup in H2.

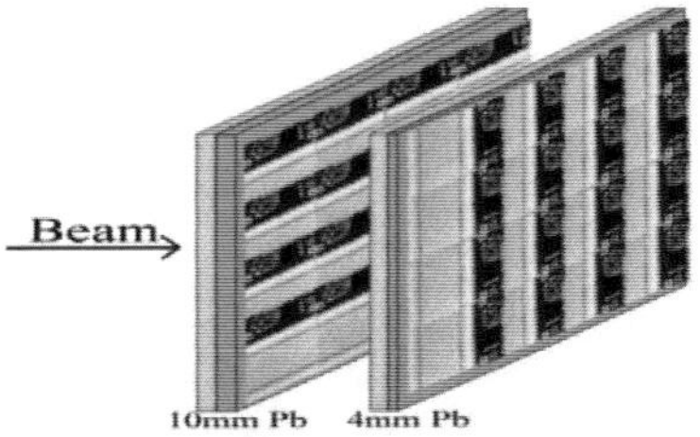

Figure 2. The setup of the ES in test beam.

The MIP is the energy unit in the Preshower and is defined as the deposited energy when a high energy pion/muon goes perpendicularly through the sensor. The Preshower absolute calibration can only be performed with MIPs.

Fig. 3 shows a typical pulse shape from a silicon strip. Three time samples are used to reconstruct the deposited energy in the silicon strip. The interval of each time sample is 25 ns. The total deposited energy in the silicon strip is given by $E = W_1*S_1+W_2*S_2+W_3*S_3$, where W_i is a weight function for the ith time sample and S_i is the signal in i^{th} time sample.

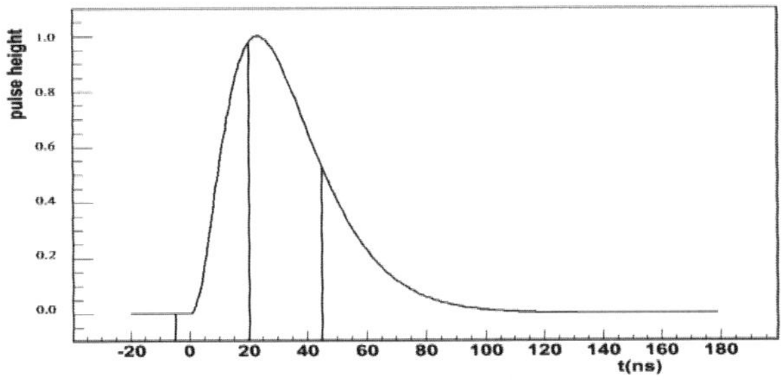

Figure 3. The typical pulse shape. The peak of the shape measures the total deposited energy in a silicon strip. Three time samples are used to reconstruct back to the total energy.

The standard operation of the Preshower is in "low gain (LG) mode". It has then a dynamic range (0~400 MIPs). The Preshower is also designed to operate in "high gain (HG) mode" with limited dynamic range (0~50 MIPs) in order to calibrate precisely the MIPs with good S/N ratio.

An electronics test signal can be injected and can be used to measure the pulse shape in both modes. Fig. 4 shows the pulse shapes in both gains. There is an obvious difference in the pulse shapes between HG and LG. It shows that MIPs need to be calibrated in HG and LG respectively, in order to obtain the corresponding weight functions.

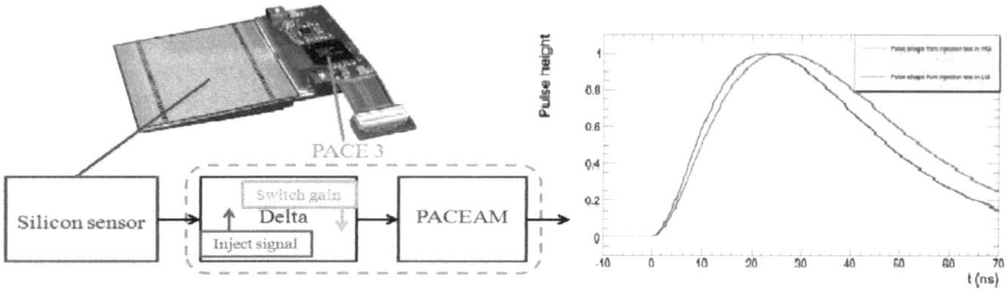

Figure 4. Left plot shows the workflow of the injection test. Right plot indicates the pulse shapes in HG (in red) and LG (in blue) respectively. The pulse shapes are normalized to 1.

3. Preshower MIP calibration

3.1. MIP calibration in HG

High energy (20 GeV/c) pion data was used to obtain the pulse shape, the weight functions and the MIPs, because the pulse shape from electronic charge injection may be different as the pulse shape observed with particle signals. Fig. 5 shows the image of the beam spot of pion data in the Preshower. The sensor in the 1st plane was used to obtain the pulse shape and the corresponding weight functions.

The data were divided in 1ns bins and fitted in each bin with a Landau distribution convoluted by a Gaussian to take the noise into account. The resulting pulse shape is shown in Fig. 7, and the corresponding weight functions were obtained using the method described in [2]. Fig. 8 shows the energy spectrum which is reconstructed using the weight functions. The signal to noise ratio is around 10.

The data obtained with the 2nd plane was used to cross check the pulse shape. The difference in the pulse shapes between these two sensors is negligible. It proves that the pulse shape is universal.

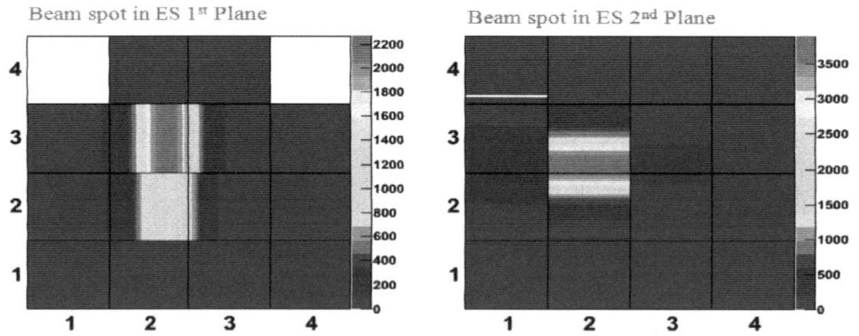

Figure 5. Beam spot in the 1st (left) plane and 2nd (right) plane of the Preshower.

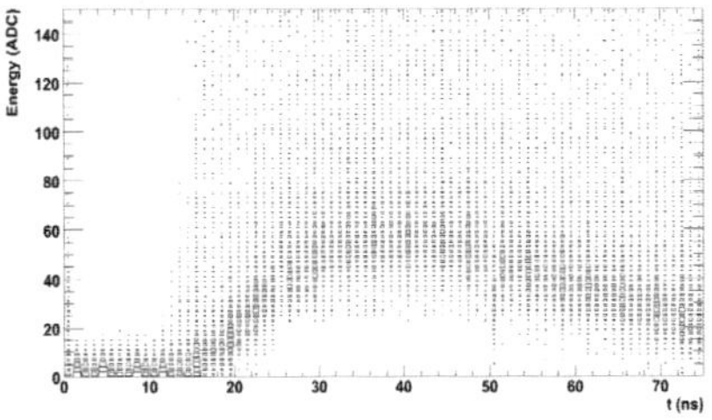

Figure 6. The energy distribution as function of t_{TDC}.

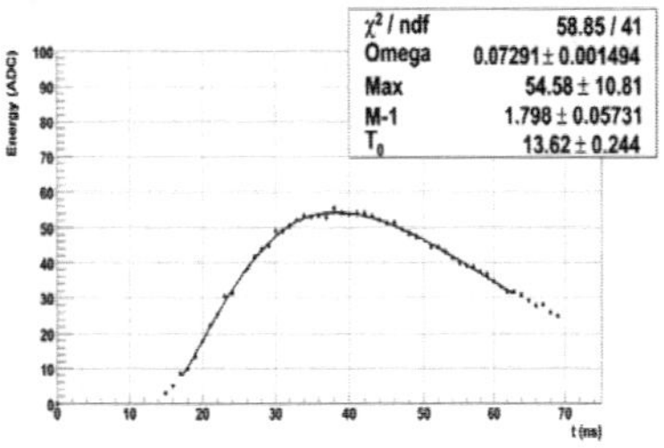

Figure 7. The pulse shape in HG.

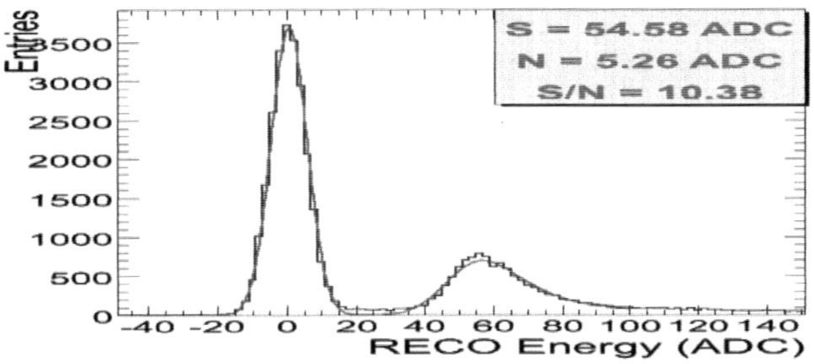

Figure 8. The reconstructed energy spectrum. The S/N ratio is around 10.

3.2. MIP calibration in LG

The MIP resolution in LG is not good enough to obtain the pulse shape directly, so electronics charge injection was used to deduce the pulse shape in LG (Fig. 9), and then to obtain the corresponding weight functions. Fig. 10 shows the MIP obtained from pion data.

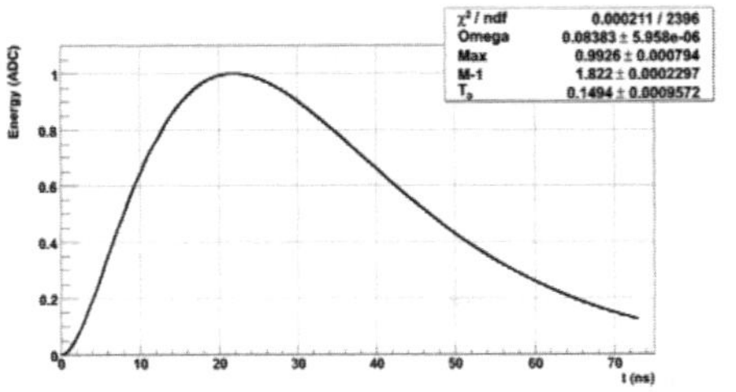

Figure 9. The deduced pulse shape in LG.

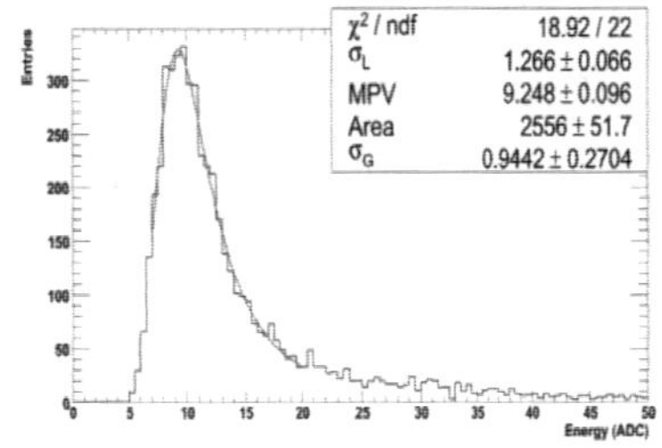

Figure 10. The MIP of the sensor. MIP is 9 ADC counts.

3.3. MIP cross check and additional correction

A cross check of the calibration in both HG and LG modes can be performed by measuring the energy deposited in the Preshower by electrons using data recorded with both gains. The distributions should be identical. The energy in the Preshower is defined as the sum of energies in a cluster containing the strip with highest deposit and 2 neighbors on each side. Fig. 11 shows the energy spectrum obtained in both gains with 10, 20, and 50 GeV electrons. The data of the two gains differ slightly, however in a consistent way.

Because the S/N (≈ 3) is marginal in LG, the MIP value is shifted by the effect from the noise.

This effect is shown in Fig. 12 which is a result of a simulation where the MIP signal was set to 9 ADC counts and the noise to 2.5 ADC counts. The Most Probable Value of the distribution shifts from 9 to ~ 11 ADC counts. Another effect is due to the fact that the pulse shape in LG was measured indirectly via charge injection.

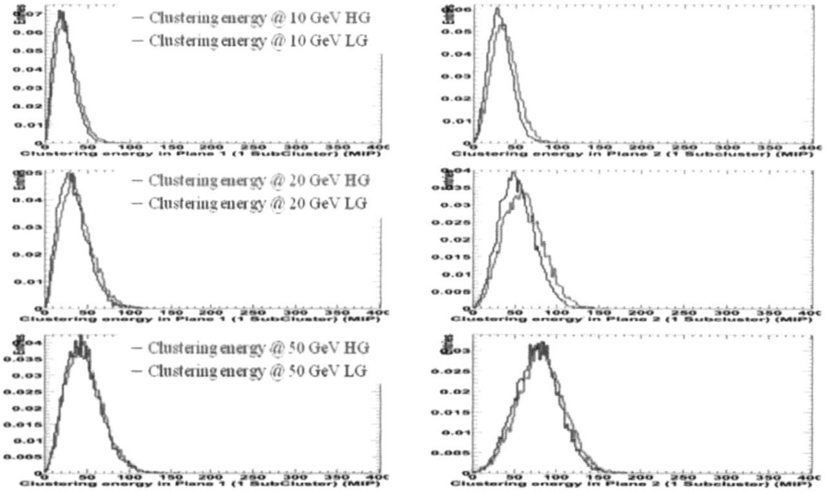

Figure 11. Cluster energy in MIPs from electron data at 10, 20, 50 GeV in the 1st (left) plane and 2nd (right) plane of the Preshower. Blue line indicates the results in LG and red line indicates the results in HG.

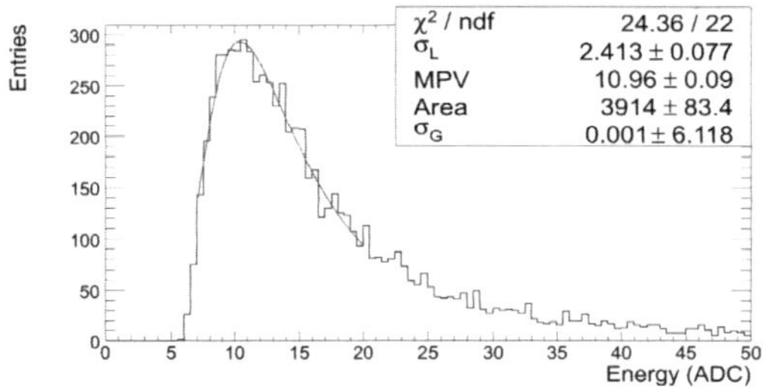

Figure 12. The fitting result from the simulation (see text).

By minimizing the χ^2 between the energy spectra in LG and HG, a global correction factor of 1.14 to be applied to the LG data was obtained. Fig. 13 shows the energy spectra after applying this correction factor to the reconstructed energy in LG. The distributions are in very good agreement for both planes and all energies. Fig. 14 shows the spectrum of the noise after applying correction and it implies that the S/N is around 3 in LG.

4. Preliminary result of combined EE and ES energy

The total deposited energy in the electromagnetic calorimeters is given by $E = E_{EE} + E_{ES}$, where E_{EE} is the energy measured by EE and $E_{ES} = \alpha(ES_1+0.7ES_2)$ is the energy measured by the Preshower. $ES_{1,2}$ are the deposited energies in the 1st and 2nd plane of ES respectively. The coefficient 0.7 results from an optimisation study done in earlier work [3]. The coefficient α is obtained from a linear regression fit of the correlation plot between E_{EE} and E_{ES}.

Fig. 15 shows value of α for different beam energies. The coefficient, α, is constant when the beam energy is above 20 GeV.

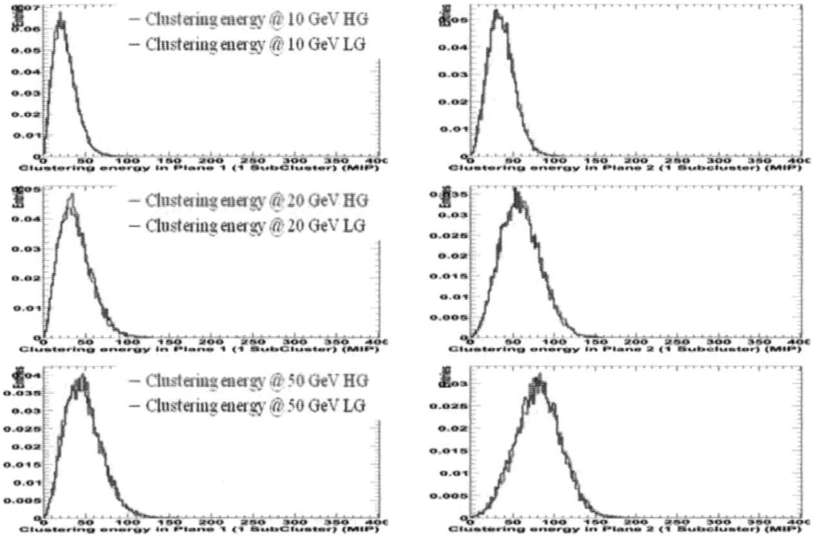

Figure 13. The energy spectrums after applying correction factor.

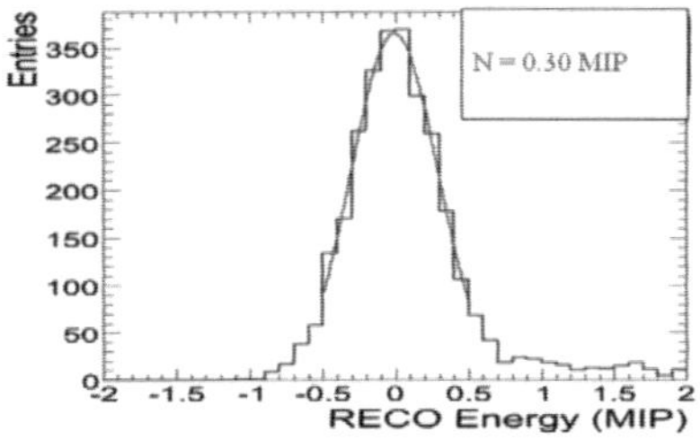

Figure 14. The spectrum of the noise in LG.

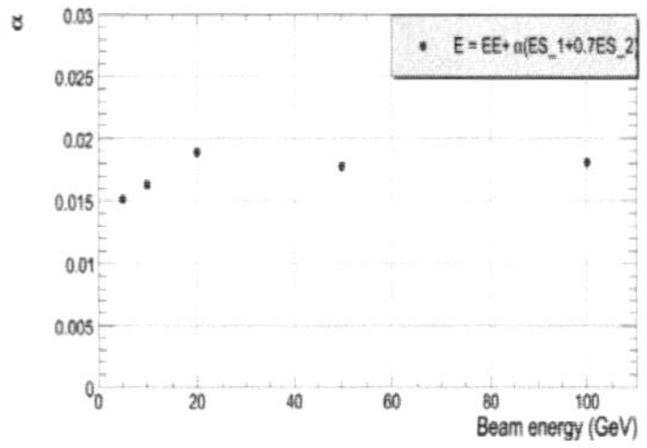

Figure 15. The Value of α as function of the beam energy.

5. Conclusion

For the first time a small scale prototype of the CMS Preshower operated successfully in realistic conditions and then took useful data for calibration. We have shown that a reliable calibration using minimum ionizing particles can be obtained, allowing to calibrate the detector in-situ during operation.

6. References

[1] CMS Physics TDR Vol. 1, CERN/LHCC/2006-001.

[2] I. Evangelou, *Nuclear Instruments and Methods in Physics Research Section A*, Volume **572**, Issue 2, p. 624-632.

[3] P. Aspell et al., CMS-NOTE 2000–001.

Journal of Physics: Conference Series **160** (2009) 012053 doi:10.1088/1742-6596/160/1/012053

The ATLAS Tile Calorimeter: Commissioning and Preparation for Collisions

O Solovyanov
on behalf of ATLAS Tile Community
Institute for High Energy Physics, Protvino 142281, Russia
E-mail: Oleg.Solovyanov@ihep.ru

Abstract. The ATLAS tile calorimeter has been assembled and commissioned in the underground cavern. Large data sets of cosmic rays and other calibration data have been acquired to test the stability of operation and validate the data acquistion system, detector control system and data quality monitoring.

Introduction

The ATLAS [1] TileCal[2] non-compensating sampling calorimeter, shown in fig.1, is constructed of steel plates (absorber) and scintillating plastic tiles (active material). It is designed as one barrel and two extended barrel parts. All the three sections have a cylindrical structure divided in ϕ by 64 wedge modules. The tile calorimeter covers the pseudorapidity $|\eta| < 1.7$.

Within modules, the light from the scintillating tiles, oriented perpendicular to the beam axis, is collected by PMTs via wavelength shifting fibers from both sides, as shown in fig.2. The module is divided into a number of cells corresponding to the projective geometry. Front-end electronics are located inside modules' girders and are organized in extractable *drawers*[3]. The total number of PMTs in entire calorimeter is about 10000. Each PMT signal is shaped and amplified with 2 gains, digitized by sampling ADCs, and the data are read out via optical fibres. Analog sums of projective trigger towers are available to the level-1 trigger.

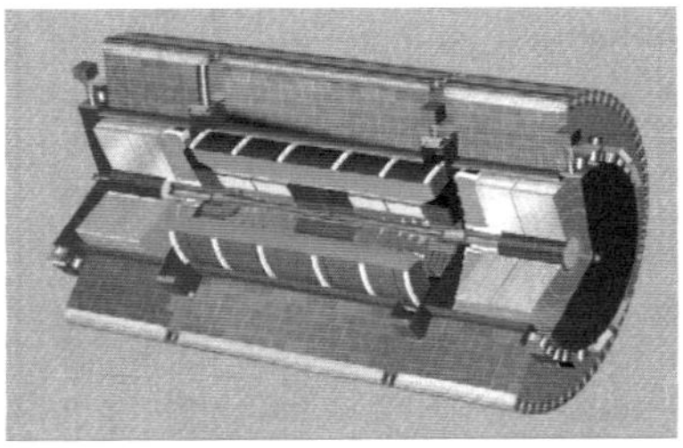

Figure 1. ATLAS calorimetry layout. Tile calorimeter is shown as outer cylinder.

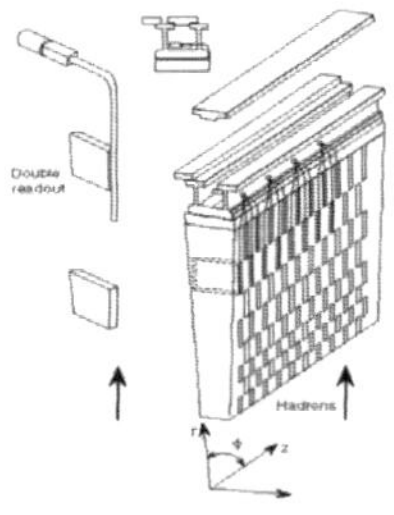

Figure 2. Tile cell geometry and readout.

Figure 3. Barrel calorimeter in place.

Figure 4. Insertion of the last extended barrel module.

1. Installation and services

By the end of year 2005 the barrel part of tile calorimeter was installed and positioned at the final place at Z=0 as presented in fig.3. During the year 2006, both extended barrels were assembled, see fig.4, and the services were connected, in parallel with standalone front-end electronics commissioning, as described in [4]. In 2007 and 2008 all low-voltage power supplies for the front-end electronics were fully installed, together with the refurbishment of the front-end electronics, to strengthen the power and signal distribution inside the drawers to avoid complete module readout failures.

The multitude of optical fibres for triggering and front-end configuration required a new set of tools and procedures to verify the connectivity and integrity of these services. Optical Time-Domain Reflectometers (OTDR) have been used to analyse the loss of signal in optical fibre by injecting short laser pulses into it and counting the back-scattering and reflection of photons as a function of time. This allows to locate any break or bad splice with the precision of a few centimeters. The precise fibre length is used to calculate correct trigger delays. A total of 12 fibres out of 512 of the barrel part had to be re-spliced (2%). These tests were performed for trigger, readout and laser calibration fibres.

2. Front-end electronics commissioning

Because of the lack of access to the front-end electronics after the closure of the detector for beam, it is very important to fully commission it beforehand. The following steps have been taken:

- in-situ certification with mobile readout unit MobiDICK[5]
- online commissioning with back-end electronics and short tests with MobiDAQ[6]
- offline commissioning with data written by ATLAS DAQ system

The certification with the MobiDICK system consists of detailed tests of all front-end electronic components of a single drawer and is used both before and after the installation to verify any additional failure caused during commissioning and to certify the repairs.

Tests with MobiDAQ station mimics all functions of the final read-out. It allows data taking with a set of 8 drawers at once. Besides electronic checks, it was used to develop commissioning software tools.

Finally, the ReadOut Driver (ROD) back-end system, which is capable of reading all tile calorimeter channels at once, was used to take various calibration runs, to be analyzed offline, and to develop the software to be used during ATLAS data taking.

To verify the performance of the electronics various kinds of data have been collected:

- pedestals - to measure the noise levels
- charge injection - to calibrate digitization electronics
- laser - to verify the photomultipliers performance
- cosmics - scintillating tiles, photomultipliers and readout combined

A rich set of automation tools was developed and successfully used to analyse the results of these tests, to present them, and to provide summary status of the commissioning, in a form of databases, plots and web interfaces. The summary status display, electronics performance history tables, and web interface for shifters proved to be invaluable for the commissioning process.

3. DAQ, DCS and monitoring systems

The data acquisition system (DAQ) for the tile calorimeter, based on standard ATLAS DAQ software, is routinely used to collect calibration data for the commissioning purposes and to record physics events, triggered by cosmic muons. The stability of the data acquisition system can be underlined by the 106 hours of continuous running over the Easter weekend, collecting cosmic rays data with the full tile calorimeter.

The detector control system (DCS) is fully operational and is used to control the low-voltage and high-voltage power supplies, logging all the essential control and environmental parameters to the database for later analysis. The detector safety system has been verified to turn-off detector electronics in case of major hazards like fire or cooling failure.

Data quality monitoring is proven to be crucial to acquire valid physics data. Tile calorimeter is being monitored at all levels, from channel-based quantities like pedestals to the physics objects, and results in form of histograms are checked in an automated way, as presented in fig.5.

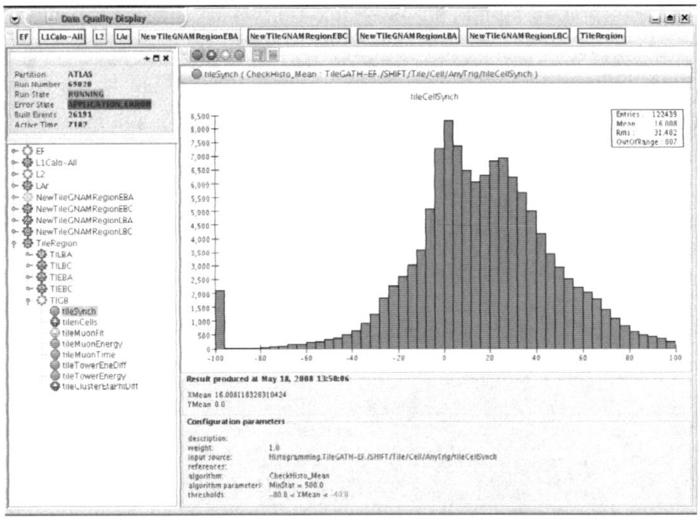

Figure 5. Data quality monitoring framework with automated histogram analysis. The status is organized as a color-coded tree on the left, and the histogram with the result of the automatic checks is presented on the right.

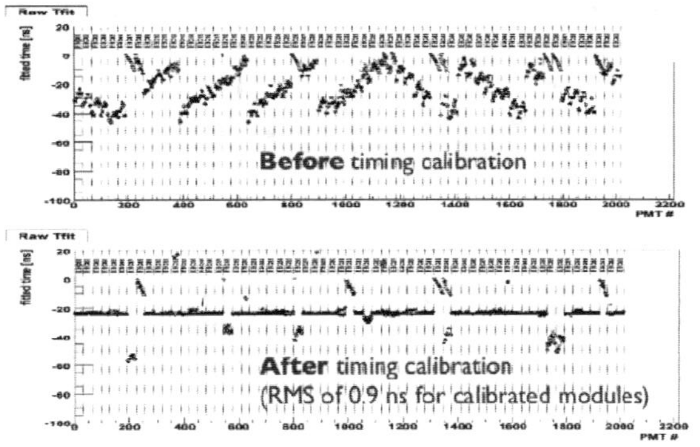

Figure 6. The effect of channel timing equalization with the laser system. Top plot shows the reconstructed time of the laser calibration pulse vs. channel number. The bottom plot shows the same distribution after the adjustment.

4. Calibration Systems

Tile calorimeter has a number of calibration systems at multiple levels of the readout chain, described in details in[7].

The charge injection system (CIS), designed to calibrate the response of readout electronics and to monitor the variations of the response it time, has performed very well to confirm the stability of the digitizer's ADC. During a 3 months period the calibration constants variation was less than 0.1%

The laser calibration system, which is designed to calibrate the responses of individual PMTs, and to monitor the short and long term PMT gain stability, has been installed and used for commissioning and timing studies. It also allows to establish the timing equalization of the signals from different channels, as shown in fig.6.

The cesium radioactive source calibration system[8, 9] that is conceived to determine the quality of optical response of the detector, is now fully installed and is used to perform the calorimeter cell response equalization. The source movement control software and data readout are performing well. First data from all the tile calorimeter modules have been collected, and the high voltage adjustment is in progress. The aim is to calibrate all 10000 channels with the precision better than 2% at electromagnetic scale.

As the tile calorimeter is providing the analog sums of the projective trigger towers to the level-1 trigger, a special set of joint calibration runs was taken to setup the energy scale and establish the correct timing of individual channels of the trigger system. Thanks to the charge injection system, it is also possible to simulate the physics kind of signal, like jets, to verify the performance of the level one calorimeter trigger system. The laser pulsed runs are used for the timing adjustments.

5. Integration

Starting from the end of the year 2006 a number of integration weeks was organized in order to move from stand-alone detector commissioning to the combined one. During these periods the various sub-detector groups were able to collect cosmic data with different trigger sources, perform calibrations, timing and trigger setup. These weeks were essential to train shifters, multiply experts, enhance monitoring tools and documentation, and to arrive well prepared for the first beams.

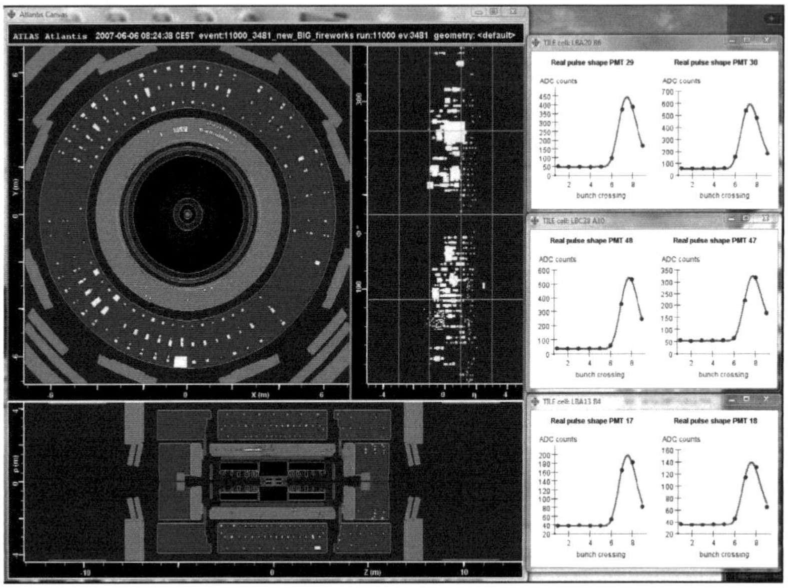

Figure 7. Event display of a large cosmic ray air shower, generating 120 GeV of $\not{E}_T$. On the right there are digitized samples from the ADC with the nominal pulse shape superimposed. The top vs. bottom signal time difference is about 16ns, that corresponds to the time of flight of a muon through the calorimeter.

6. Cosmic data taking

Before the circulation of the first beams in the LHC, the performance of the installed detector can be studied with the physics data from cosmic rays. During the 2005-2008 years we acquired a large number of data sets of such events, and used them for detector commissioning, monitoring tools development, systems integration and preparation for the physics data from the beam.

Apart from the calibration purposes[10], these events were also useful to develop a strategy, tools and methods to identify fake missing E_T from front-end electronics failures, large cosmic ray showers and hard bremsstralung events. The distribution of missing energy in these kind of events extends from the several hundred GeV, like in fig.7, up to a few TeV for very rare cases.

Methods to reject this background using calorimeter timing have been explored on simulated and real cosmic ray data. Further rejection is possible using the cuts on an electomagnetic fraction of energy deposited in calorimeters, to improve the quality of the data sample.

7. Conclusions

ATLAS tile calorimeter is fully installed and powered with all the necessary services. All front-end and back-end electronics have been thoroughly commissioned to arrive to the first collisions with minimal number of bad channels (<1%). DCS and DAQ systems are running in full swing. A number of software and hardware tools have been developed during commissioning and used to ensure good performance after the installation. A rich set of monitoring tools is ready to verify the performance and quality of the detector during data taking. All calibrations systems are installed and working to provide well-calibrated detector for the first beams. Tile calorimeter is fully integrated with other ATLAS subsystems, including L1 trigger. Large data sets of cosmics data have been acquired and being studied. Several methods to reject cosmic ray background events have been explored with the real data. An impressive number of experts and shifters are prepared for ATLAS running.

References

[1] Armstrong W W *et al.* (ATLAS) 1994 CERN-LHCC-94-43
[2] 1996 CERN-LHCC-96-42
[3] Teuscher R *et al.* 2004 *Proceedings of 11th International Conference on Calorimetry in High-Energy Physics (Calor 2004), Perugia, Italy, 28 Mar - 2 Apr 2004*
[4] Blumenschein U (ATLAS) 2007 *Nucl. Instrum. Meth.* **A572** 24–25
[5] Calvet D 2004 *Nuclear Instruments and Methods in Physics Research A* **518** 509–510
[6] Anderson K J *et al.* 2007 *JINST* **2** P07002
[7] Carvalho J 2006 *Calorimetry in High Energy Physics: XII* (*American Institute of Physics Conference Series* vol 867) ed Magill S R and Yoshida R pp 385–392
[8] Starchenko E *et al.* (ATLAS) 2002 *Nucl. Instrum. Meth.* **A494** 381–384
[9] Shalanda N *et al.* (ATLAS) 2003 *Nucl. Instrum. Meth.* **A508** 276–286
[10] Meirose B and Teuscher R 2007 Time of flight analysis using cosmic ray muons in the atlas tile calorimeter Tech. Rep. ATL-TILECAL-PUB-2008-004. ATL-COM-TILECAL-2007-015 CERN Geneva

The LHCb Hadron Calorimeter

Yu. Guz,
on behalf of the LHCb collaboration
Institute for High Energy Physics, Pobedy str, 1, 142281, Protvino, Russia
E-mail: Iouri.Gouz@cern.ch

Abstract. The Hadron Calorimeter of LHCb is a sampling iron-scintillator calorimeter of 5.6 λ_I thickness with structure arranged along the collider beam direction. The light readout is performed with WLS fibers running along the scintillator tile edges. The fast front end electronics allows to digitize events at 40 MHz without dead time. The self-calibration system based on ^{137}Cs radioactive source embedded into the calorimeter structure allows to obtain absolute calibration with precision of about 2It is also equipped with a LED monitoring system, which allows to follow the PMT response variations during physics data taking.

1. Introduction

The Hadron Calorimeter of LHCb is a part of the experiment's calorimetric system [1, 2] intended to provide the Level-0 trigger on presence of high transverse momentum hadrons, which is a signature of B-event. This imposes the following design requirements:

- it should be able to measure the particle's transverse momentum at 40 MHz bunch crossing rate;
- its elements should be radiation tolerant, the expected dose rate in the central area is ~50 krad/year ;
- moderate energy resolution is sufficient for triggering purposes.

HCAL is supposed to provide ~70% of Level-0 trigger output.

2. HCAL description

2.1. Structure and layout

The detailed description of the calorimeter system is given in [3]. For the LHCb HCAL the iron-scintillator sampling structure arranged parallel to the collider beam direction have been chosen. Such structure, first proposed in [4], was implemented in ATLAS TileCal [5]. The LHCb HCAL design [6] follows the same idea; the differences in structure and main parameters are caused by different design requirements and detector geometry.

The HCAL is placed at 13.33 m from the interaction point and has lateral dimensions of 8.4×6.8 m^2. Its instrumented depth is 122 cm, it has no longitudinal segmentation. The general view of HCAL is shown in Figure 1. It consists of two symmetrical halves, each of which is assembled out of 26 stacked horizontal modules and placed on a separate movable platform. Laterally it is segmented into cells of two sizes, 262×262 and $131\times131\,\text{mm}^2$ in the outer and inner sections, respectively. The two central modules of each half are 26 cm shorter than others,

Journal of Physics: Conference Series **160** (2009) 012054 doi:10.1088/1742-6596/160/1/012054

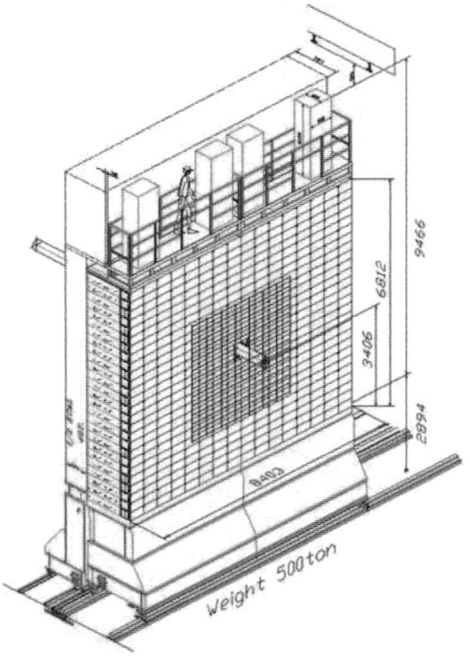

Figure 1. HCAL general view.

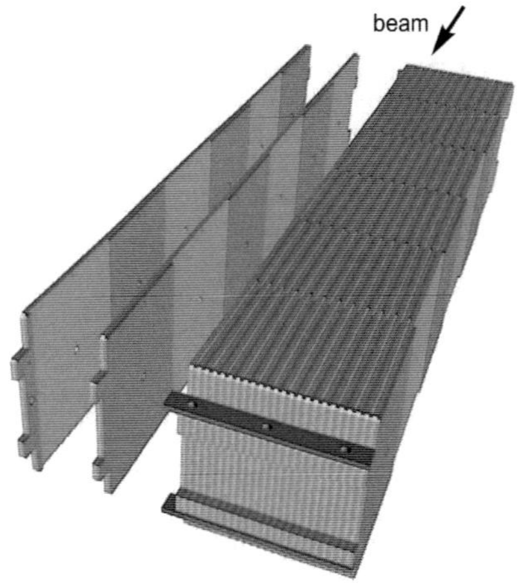

Figure 2. HCAL absorber structure.

forming a hole which lets the beam pipe pass through. In total HCAL contains 1488 cells, 608 outer and 880 inner.

One module is 262 mm high, it consists of self-supporting steel absorber structure and optical elements, scintillator plates and WLS fibers, embedded into it. The weight of one module is about 9.5 ton.

The absorber structure is periodical (Figure 2), one period is composed of two 1232 mm long, 6 mm thick master plates interlaced by six 4 mm thick spacer plates. The scintilator tiles are placed between spacers. This structure is repeated 208 times in each module. In longitudinal direction it presents six 202 mm rows with interchanging order of steel spacers and scintillator tiles. It was shown in [7] that, although the chosen depth does not provide full shower containment, it is sufficient for the trigger operation.

2.2. Light collection and readout

The active elements of the calorimeter are 3 mm thick scintillator tiles (Polystyrene + 1.75% Paraterphenyl + 0.05% POPOP). The tiles are wrapped by TYVEK®, in order to improve the light collection. In the outer section the tiles are 256×197 mm^2, while in the inner section, in order to arrange finer segmentation, the tiles are cut into two halves 127×197 mm^2.

The light readout is performed by WLS fibers (⌀1.2 mm Kuraray Y-11(250)MSJ) running along the tile edges at both sides of the module. In order to have optical contact between tile and fiber, the fiber is passed inside the TYVEK envelope.

The transverse segmentation is achieved by proper grouping of fibers into bundles. Each fiber collects light from tiles of three rows; the light from each full size tile in the outer section is read out by 2 fibers, the half size tiles in the inner section — by one fiber.

The WLS fibers' light attenuation length is about 3.6 m. In order to improve the light collection and longitudinal uniformity of response, one end of each fiber was cut by diamond mill and coated by aluminum, forming a mirror with reflectivity of about 85%. For long term

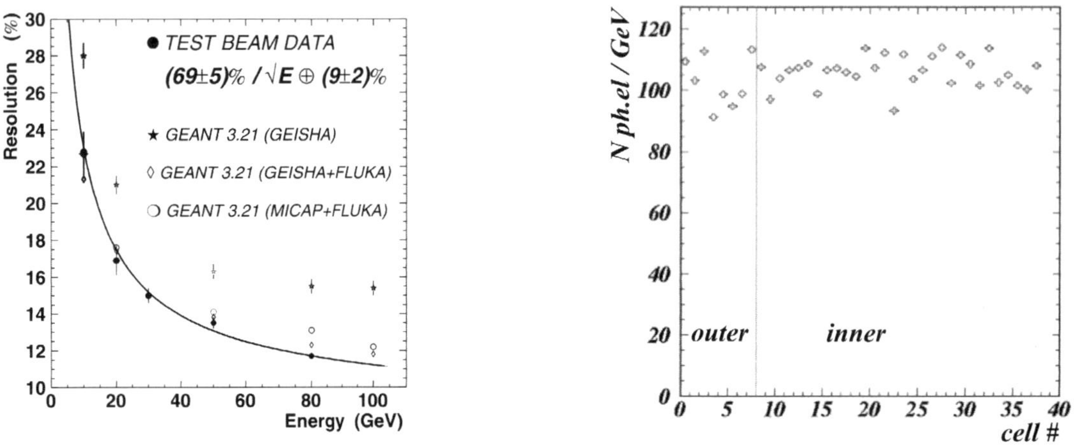

Figure 3. HCAL prototype energy resolution as a function of energy (left) and measured light yield in cells of a 40-channel HCAL module (right).

stability the mirror is protected with a layer of MgF_2.

Further improvement of the HCAL longitudinal uniformity was achieved by introducing artificial attenuation of light in the tile rows close to the photomultipliers. This was done simply by tuning the length of the optical contact between fiber and tile. Namely, the lengths of the optical contact were set to 197, 194, 189, 181, 169 and 149 mm for tile rows 1–6, respectively; the maximum light attenuation of $\sim$24% is applied in the row 6, which is closest to PMTs.

The photomultiplier used in the LHCb HCAL is HAMAMATSU R7899-20. This version of R7899 photomultiplier, specially designed for LHCb, differs from the basic version by its enhanced stability of gain with respect to variations of anode current. The High Voltage is supplied to each photomultiplier by an individually regulated Cocroft-Walton (CW) circuit installed on the photomultiplier [8]. The CW bases are therefore made of radiation hard components. The control voltages for CW bases are produced by DAC boards installed outside the radiation area. These boards communicate with the Experiment Control System via SPECS bus [9].

2.3. HCAL parameters from beam tests

The HCAL prototype beam test measurements [7, 10] are summarized in Figure 3. Its energy resolution is

$$\frac{\sigma_E}{E} = \frac{(69 \pm 5)\%}{\sqrt{E}} \oplus (9 \pm 2)\%, \tag{1}$$

the average light yield with HAMAMATSU R7899-20 is 105 photoelectrons per GeV.

2.4. Front end electronics

For the HCAL usage within the LHCb Level-0 trigger, it is essential that the signal duration is less than the time between two consecutive bunch crossings, 25 ns. The decay time of Kuraray Y-11 fiber $\tau_D = 7$ ns [11] leads to a longer signal; the clipping circuit shown in Figure 4 is used for shaping. The delay line of $\sim$6 ns necessary for this circuit is made of 1.15 m long 50 Ohm coaxiall cable, which is used also to supply the signal to the integrator board of the ^{137}Cs calibration system, which measures average PM current. The elements of the clipping circuit are also placed on the integrator board.

The signals from photomultipliers are processed by specially designed Front End Boards (FEB) [12, 13]. At the first stage the signal is digitized by dedicated dead timeless ADC. The

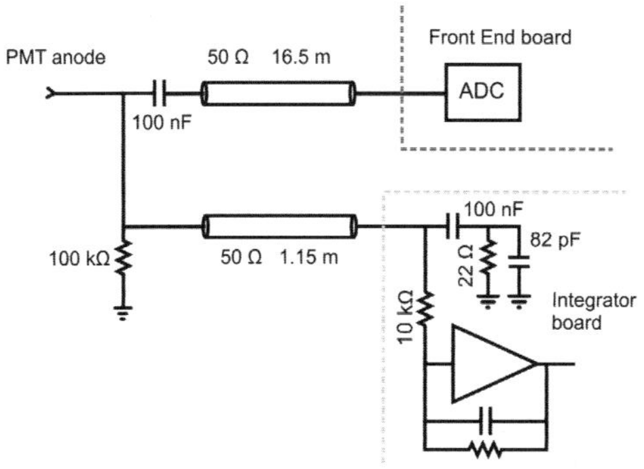

Figure 4. PMT signal shaping.

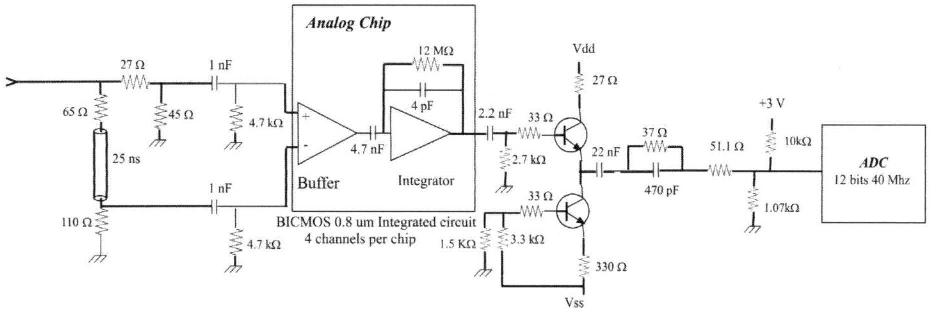

Figure 5. Front End Board input circuit.

principles of operation of the input circuit of such ADC are shown in Figure 5. The input signal is split into two parts, one of which is delayed by 25 ns. The delayed part is then subtracted from the direct one, and the difference is integrated, so that the output of the analog chip at any moment of time represents the integral of the input signal in preceding 25 ns. The output signal is then digitized by the 40 MHz 12-bit flash ADC (AD9042). The sensitivity of Front End Board is ~20 fC/ADC count.

After digitization Front End Boards perform pedestal subtraction. The pedestal value is calculated as a minimum of the two preceding ADC measurements. The pedestal subtraction eliminates also the low frequency noise which is present at the output of the integrator.

The information for the Level-0 trigger decision is also prepared in FEBs. It consists in detection 2x2 clusters with total energy deposition such that the corresponding transverse energy (E_T) exceeds certain limit. For that, the ADC value in each cell is multiplied by an individual calibration factor, such that the result is proportional to E_T.

In order to facilitate the calculation of the transverse energy in Front End Boards, the high voltages in the photomultipliers are supposed to be set such that the corresponding cell sensitivity (signal amplitude per unit energy deposition) is proportional to $\sin\Theta$. Namely, with 5 GeV trigger threshold, the PMTs' HV will be set as $E_{max} = 15$ GeV$/\sin\Theta$, where E_{max} is energy corresponding to maximum ADC count.

3. LED monitoring system

During physics data taking, the gain of the HCAL PMTs is supposed to be continuously monitored by LEDs [10]. Significant variations of gain will be recorded into the run conditions database and used in the offline analysis. The LEDs will be flashed during the series of empty LHC bunches foreseen for detector calibrations. For the HCAL monitoring, fast blue LEDs (WU-14-750BC) are used. The light of a LED installed in a HCAL module is distributed to all

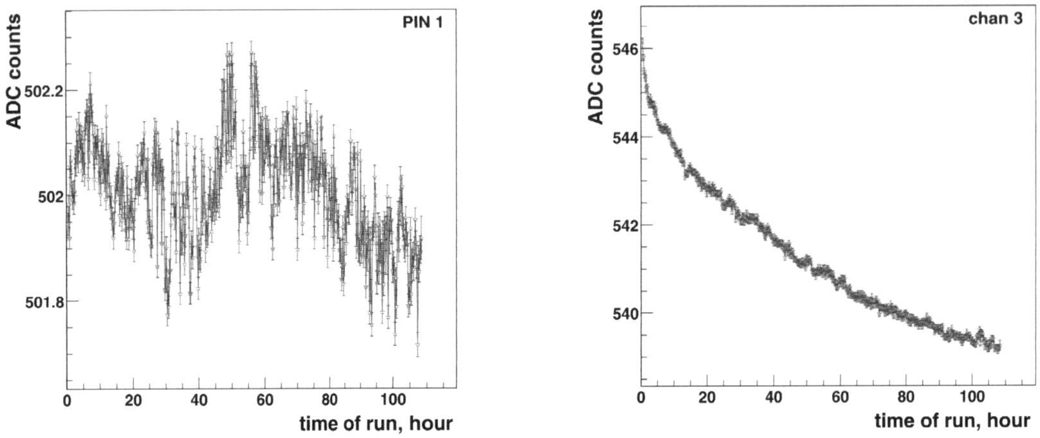

Figure 6. Long term stability of the LED flash intensity (left) and PMT response (right).

cells of this module through 2.4 m long clear fibers (⌀1.2 mm Kuraray CLEAR-PSMS40J). The flash intensity is set by control voltage supplied to the LED driver board. As the LED flash intensity itself can vary with time because of e.g. temperature variations, it is monitored by a PIN photodiode (Hamamatsu S1223-01) installed on the LED driver board next to the LED. Each HCAL module is equipped with two independent LED-PIN assemblies.

Results of a 108 hours long stability test made during the HCAL commissioning are shown in Figure 6. One can see that the LED flash intensity, according to the measurement with the PIN photodiode, is very stable (within 0.1%). In the same time, the PM signal amplitude shows slow drift by ~1.5%.

During the HCAL production and commissioning, the LED monitoring system was intensively used as a quality control and time alignment tool.

4. ^{137}Cs calibration system

The LHCb HCAL is situated in the downstream part of LHCb, behind the Electromagnetic Calorimeter, so it is difficult to calibrate it on physics events. HCAL will be calibrated using the ~10 mCi radioactive source ^{137}Cs moving through every cell. This method provides very detailed information about the calorimeter, it allows to measure the response of every individual scintillating tile.

The source integrated into the dumb-bell shaped aluminum capsule moves with flow of liquid inside the stainless steel pipes embedded into every HCAL module. The calibration system based on hydraulical movement of source was originally developed for the ATLAS TileCal [14]. The liquid used as moving medium in our case is pure demineralized water. Each HCAL module has an embedded pipe with inner diameter of 6 mm and wall thickness of 1 mm. The pipe is six-fold, it passes through the centres of each of the six tile rows. Its length is ~27 m (26 m in central modules). The pipes of all 26 modules of each HCAL half are connected together. Each half is then equipped with its own ^{137}Cs source and a hydraulic system for moving it.

A sketch view of one of the two hydraulic systems is shown in Figure 7. The source is driven by a pump; a system of valves determines the direction of water flow. Between calibration runs the source is housed in the container with 5 cm thick lead walls (called **garage**) which is inserted into the capsule pathway on top of the HCAL half. During the run the source passes sequentially through all 26 modules, with a speed of 20-30 cm/sec, then passes the same way in reverse direction and returns back to the garage.

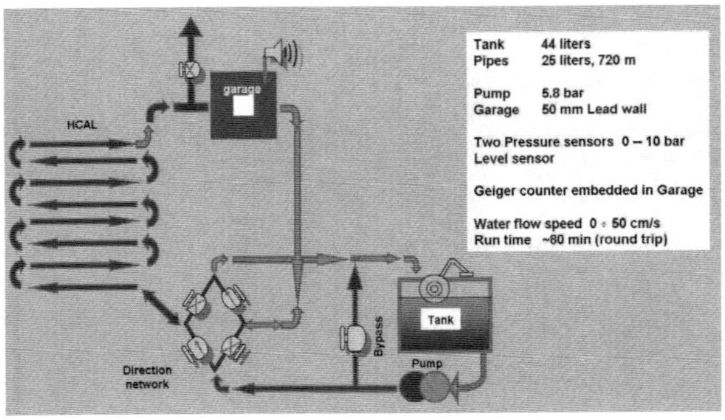

Figure 7. Sketch view of the HCAL hydraulic system.

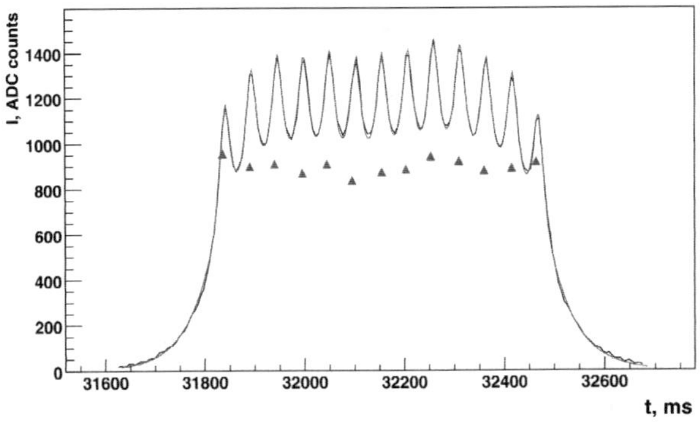

Figure 8. Anode current of a PMT as a function of run time, while the source passes through one tile row. The triangles represent the fitted response of every tile.

During the run, the anode current of photomultipliers is being measured every 3 ms and recorded to disk for the offline analysis. A typical picture of dependence of PMT anode current on time during the passage of the source through the 13 tiles of one tile row read out by this PMT is shown in Figure 8. The superimposed curve is a fit by a weighted sum of (empirically obtained) tile response functions placed at equal time intervals Δt:

$$I(t) = \sum_{i=1}^{N} c_i \cdot f(t - t_0 - i\Delta t), \tag{2}$$

where N is the number of tiles in this row read out by the PMT and c_i is a set of coefficients representing the response of each tile.

The ^{137}Cs calibration system was used as a quality control tool during the HCAL modules production. After the assembly the Cs run was performed in each module, aiming in the debugging of its optical system. Namely, it was required that the response of each tile should not deviate by more than 20% from the average over all tiles read out by this PMT. In case of finding defective optical elements, they were replaced and test repeated. The final result for one of the PMTs is shown in Figure 9. The resulting optical uniformity can be characterized by the average RMS of responses of tiles in groups read out by the same PMT which equals to 4.7%; their distribution is shown in Figure 10.

The precision of the ^{137}Cs calibration was studied at the beam tests [10]. Total of 47 outer and 30 inner cells were independently calibrated with ^{137}Cs and with 50 GeV π^- beam. The average ratio of sensitivities to the ^{137}Cs and to hadron showers was measured:

$$\frac{S_{Cs}}{S_h} = 41.07\ (20.88)\ \frac{\text{nA/mCi}}{\text{pC/GeV}} \tag{3}$$

for outer (inner) cells, respectively. The RMS of its distribution over tested cells, which represents the precision of the ^{137}Cs calibration, was 2 (3)% of average.

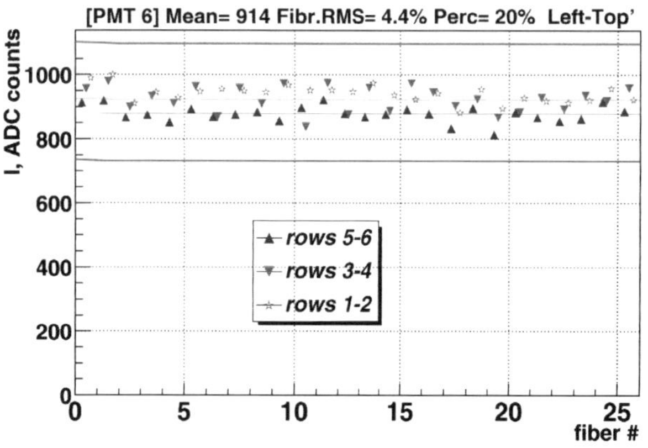

Figure 9. Tile responses for one HCAL cell.

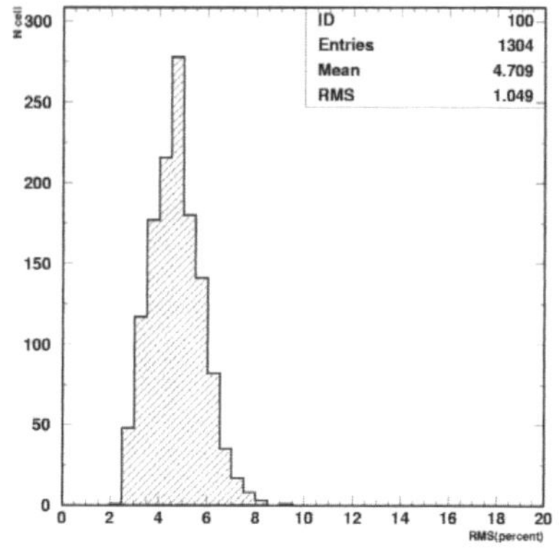

Figure 10. Distribution of RMS of tile responses.

The anode currents of PMTs are measured by sensitive integrating amplifiers, whose outputs are then multiplexed and digitized by a 12 bit ADC. The amplifiers have integration time of 1.5 ms and four selectable measurement ranges: 0.3, 1.5, 9 and 50 μA. The current measurements are read out through the SPECS bus independently on the main DAQ. Mechanically, the amplifiers and multiplexers are placed on 188 8-channel integrator boards, housing also the components of signal clipping circuitry. These boards are installed at the rear surface of HCAL, nearby the PMTs. They are read out by four control boxes, containing the ADC and SPECS interface. The control boxes are placed outside the radiation area.

The maximum anode currents in the HCAL PMTs during calibration at "nominal" HV settings, $E_{max} = 15$ GeV$/\sin\Theta$, are expected to vary in the range of 40–1200 nA, depending on the cell position. However the system for measuring currents will be used not only for the HCAL calibration: the anode currents of the PMTs will be continuously monitored during the physics data taking. The monitoring of the HCAL anode currents will give independent information on the relative luminosity and dose rates. The Monte-Carlo predictions for the anode currents in the HCAL PMTs corresponding to the "nominal" HV settings and "nominal" luminosity, $L = 2 \cdot 10^{32}\ cm^{-2}s^{-1}$, are shown in Figure 11. One can see that currents up to several tens of μA are expected in the central area of the detector. Therefore the two most sensitive measurement ranges of integrator boards, 0.3 and 1.5 μA, are intended for the ^{137}Cs calibration, while during data taking all the four ranges will be used.

5. Conclusion

The LHCb HCAL is an iron-scintillator sampling device with structure arranged parallel to the beam direction. The light is read out by WLS fibers to PMT. Its performance is adequate for providing Level-0 trigger for high-E_T hadrons. It is equipped with ^{137}Cs calibration system and LED monitoring system.

At present (July 2008) the HCAL is installed in the LHCb experimental area and is fully operational. Currently it is under intensive tests with LEDs; it is also used to provide trigger on cosmic events for LHCb. The first ^{137}Cs calibration run is scheduled for mid-July.

Acknowledgements

The author would like to express his thanks to all the members of the LHCb calorimeter group, and especially to Rustem Dzhelyadin and Marie-Noelle Minard for their great help in preparing this paper. The author would also like to express his gratitude to the organizers of the Conference for their warm reception and creative atmosphere.

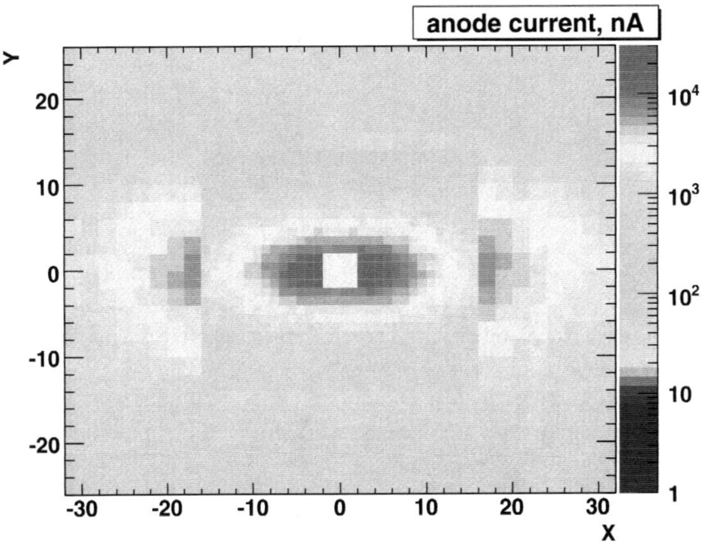

Figure 11. MC predictions for anode currents in HCAL PMTs at luminosity of $2 \cdot 10^{32}\, cm^{-2} s^{-1}$ and HV settings corresponding to E_T^{max}=15 GeV

References

[1] LHCb Technical Proposal 1998 *Preprint* CERN/LHCC 98-4

[2] LHCb Calorimeters Technical Design Report 2000 *Preprint* CERN/LHCC 2000-0036

[3] The LHCb Detector at LHC 2008 *accepted for publication in JINST*

[4] Gildemeister O, Nessi-Tedaldi F and Nessi M 1991 *Proc. 2nd Int. Conf. on Calorimetry in High Energy Physics, Capri, Oct 14–18, 1991* (World Scientific Pub) p 189

[5] ATLAS Tile Calorimeter Technical Design Report 1996 *Preprint* CERN/LHCC 96-42

[6] Dzhelyadin R 2002 The LHCb Hadron Calorimeter *Nucl. Instr. Meth.* A **494** 332

[7] Coca C et al 2000 The Hadron Calorimeter Prototype Beam Test Result *Preprint* LHCb note 2000-036;

[8] Gilitsky Yu, Golutvin A, Konoplyannikov A, Lefrancois J, Perret P, Schopper A, Soldatov A and Yakimchuk V 2007 *Nucl. Instr. Meth.* A **571** p 294

[9] Breton D and Charlet D 2003 SPECS: the Serial Protocol for the Experiment Control System of LHCb *Preprint* LHCb note 2003-004

[10] Gouz Yu, Dzhelyadin R, Konoplyannikov A, Matveev V, Novikov V and Soldatov M 2003 Design and integration of HV, LED monitoring and radioactive-source system for HCAL *Preprint* LHCb note 2003-147

[11] Brekhovskih V, Dzhelyadin R, Konoplyannikov A and Rykaline V 2000 The WLS Fiber Time Properties Study *Preprint* LHCb note 2000-39

[12] Konoplyannikov A 2000 Zero dead-time charge sensitive shaper for calorimeter signal processing 2000 *Preprint* LHCb Note 2000-041

[13] Hrisoho A, Lefrancois J, Machefert F, Tocut V and Truong K 2002 LHCb ECAL/HCAL front-end chip *Preprint* LHCb note 2002-002

[14] Starchenko E et al 2002 *Nucl. Instr. Meth.* A **494** 381

CMS HCAL Installation and Commissioning

Kerem Cankoçak [1], Pawel de Barbaro[2], Dima Vishnevskiy[2] and Yasar Onel [1]

[1] Department of Physics and Astronomy, The University of Iowa, Iowa City, IA 52242, USA
[2] Department of Physics and Astronomy, University of Rochester, Rochester, NY 14627, USA

E-mail: kerem.cankocak@cern.ch

E-mail: pawel.de.barbaro@cern.ch

E-mail: dma@cern.ch

E-mail: yonel@newton.physics.uiowa.edu

On behalf of CMS-HCAL Collaboration

Abstract.
The installation and commissioning of the Hadron Calorimeter system of the CMS detector is described and the performance of the various monitoring systems, the progress in the calibration work and the current plans for the HCAL calorimeter are summarized.

1. Introduction

The CMS hadron calorimeter system consists of a brass/scintillator sampling hadron calorimeter (HCAL) with coverage up to $|\eta| \leq 3.0$, followed by the iron/quartz-fiber Hadron Forward (HF) calorimeter with coverage $3.0 \leq |\eta| \leq 5.0$, comprising 9528 readout channels in total [1].

The Barrel Hadron calorimeter (HB) is 9 meters long, one meter thick and 6 meters in the outer diameter, consisting of two half barrels of 18 identical 20° wedges in ϕ, each made of flat brass alloy absorber plates with wavelength shifting fiber (WLS) readout, parallel to the z-axis in CMS coordinate frame. The two End Caps (HE) covering the geometrical region $1.305 < |\eta| < 3.0$, are also made of brass and scintillator, with a diameter of 0.8 to 6.0 m. and a thickness of 1.8 meters. HB and HE are inside the 4-tesla solenoid coil and have a $\Delta\eta \times \Delta\phi = 0.087 \times 0.087$ segmentation, except near $|\eta| = 3.0$, where the size of the segmentation is doubled. The two forward calorimeters (HF) are made of quarts fibers embedded in iron. Cherenkov light generated in the fibers is transmitted to the photo tubes (PMT). The central shower containment in the region $|\eta| < 1.26$ is improved with an array of scintillators located outside the magnet in the outer barrel hadronic calorimeter (HO). The inclusion of the HO layers extends the total depth of the calorimeter system to a minimum of $11\lambda_I$ for $|\eta| < 1.26$. The optical signal from the HCAL towers, except HF, is detected with hybrid photo diodes (HPD) mounted at the ends of the barrel.

Journal of Physics: Conference Series **160** (2009) 012055 doi:10.1088/1742-6596/160/1/012055

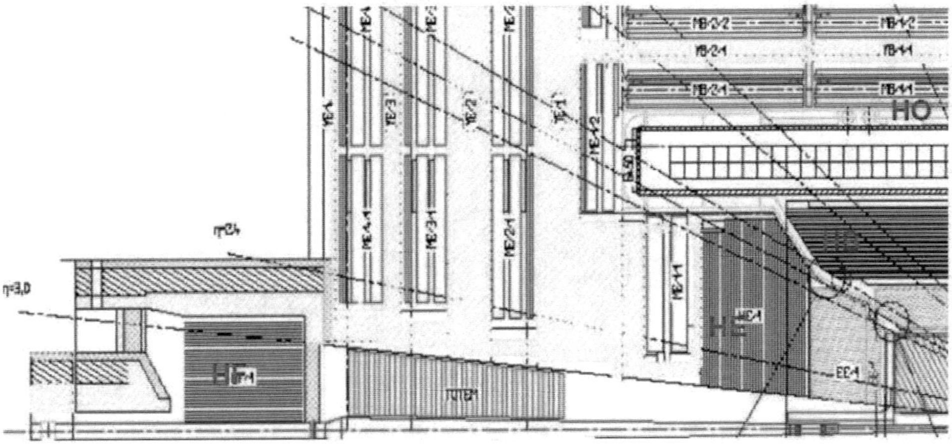

Figure 1. HCAL view in $r - z$ plane

1.1. HCAL Readout and data path

The optical signals from the subsystems are converted to the electrical signals using multichannel HPD's and PMT's. The analogue signal is converted to a digital signal by the charge-integrating ADC (QIE), which consists of four capacitors running one after another in 25ns time slices. The digital outputs of three QIE channels are combined together to create a 32-bit data word which is then transferred via the Gigabit Optical Link (GOL) at a rate of 40 MHz and transmitted to the counting house. In the counting house USC55, the data is received by the HCAL Trigger Readout (HTR) board, containing the Level-1 pipeline. The trigger primitives are sent to the Regional Calorimeter trigger (RCT) via Serial Link Board mezzanine cards [1].

2. HCAL Commissioning strategy

2.1. In the surface assembly hall (SX5)

Before lowering the CMS detector into the detector cavern, all sub-detector elements were commissioned using test beams and cosmic rays. Energy calibration of HCAL was done using test beam data and radioactive source measurements. The absolute energy scale of HCAL was established by measuring HCAL response to 50 GeV pions with data recorded in the H2 test beam in 2006 [2]. The relative (tower-to-tower) calibration of HCAL was performed using radioactive sources as part of commissioning the HCAL Barrel wedges in the CMS surface hall (SX5) in 2005 and 2006. Following the Magnet Test and Cosmic Challenge (MTCC) in summer 2006, all 15 heavy elements of the CMS detector have been lowered into the CMS cavern (Fig 2).

2.2. In the underground cavern (UX5)

The commissioning work in the underground cavern includes collecting data using using sub-detector DAQ systems (local runs) and central CMS DAQ and trigger systems (global runs). The global runs deploy larger and larger parts of detectors, as they are built and commissioned detector and then are included into simulations and data analysis. The main steps of the commissioning in the underground cavern are :

- commissioning using local runs (pedestals, LED);
- synchronization of various sub-detector systems (Muon, Calorimetry, Tracker);
- global commissioning runs (using cosmic ray triggers);

- preparing for 24 hours/7 days operation.

Figure 2. CMS cavern during the period 2006-2008

2.3. HCAL activities 2006-2008
Main HCAL activities between the period 2006-2008 were:

- 2005/2006: Installation and commissioning of hardware and readout electronics on the surface (SX5 assembly hall), initial testing of the HCAL readout with LED and Laser systems, relative calibration of all HCAL towers using radioactive source system and participation in MTCC data taking.
- 2007: Lowering of detector elements into underground hall, installation of readout and service cables between detector cavern and underground service cavern, establishing power up sequence, timing and calibration procedures.
- 2008: Full integration of HCAL with CMS system, monitoring of HCAL hardware (pedestals, LED), participation in CMS global runs.

2.4. Milestones for CMS commissioning
The main milestones for CMS in 2008 are:

- May-June, 2008 : Global runs with all CMS wheels/disks without Tracker, without the magnetic field
- July-August 2008 : Global runs with all CMS wheels/disks including Tracker, without the magnetic field
- late Summer 2008: Cosmic run at 3.8 Tesla, CRAFT (CMS closed and Field-on)

2.5. Commissioning with local runs
Data was taken for entire HB,HE,HO and HF subdetectors. Readout of all HCAL channels have been tested and verified that they have good optical connections. In particular, we monitored the stability of the pedestals and pedestal widths. LED runs were used to make sure that all the channels are functioning properly and to check the gain stability of HPDs and PMTs.

3. Global Runs
The basic goal of global runs is to allow the transition of CMS from a set of commissioned subsystems (HCAL, Muon, ECAL, Tracker) into a fully integrated detector (Fig. 3).

Muons traversing the HCAL Barrel tower are expected to deposit 1 to 2 GeV of energy, while the typical noise in HCAL (single channel, summed over 10 bunch crossings) is approximately 230 MeV. The first sample of cosmic ray muons in HCAL Barrel was recorded in September 2005 [3]. In September and October 2007 during global runs, HCAL collected the first cosmic ray muons in the underground cavern (Fig.4). In December 2007, over 0.5 million of cosmic ray muons events were recorded. This was the first global run which included. all 36 HCAL Barrel wedges in the readout. Finally, during the Cosmic Runs at Zero Tesla (CRUZET) in May, June and July 2008, a sample of over 40 million cosmic ray triggers has allowed us to apply stringent quality cuts on muons traversing HCAL.

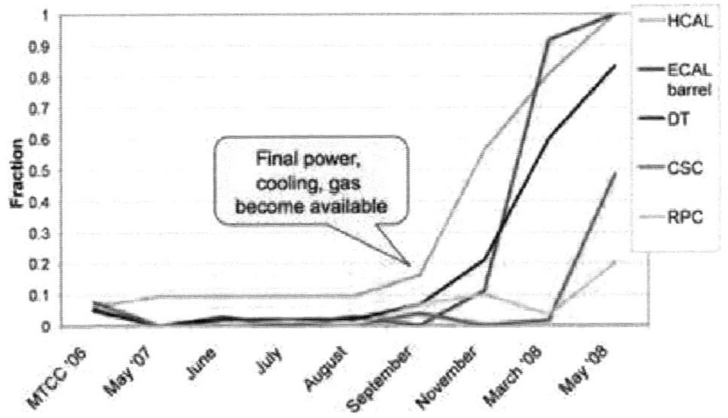

Figure 3. Fraction of Live Front Ends in CMS Global Runs

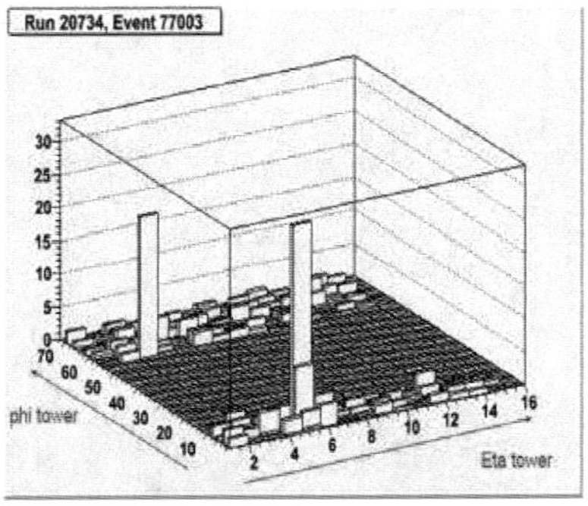

Figure 4. HCAL Barrel $\eta - \phi$ energy scatter plot in a Cosmic ray event triggered by muon track in Drift Tube (DT) during September 2007 Global Run.

3.1. Overview of Cosmic Run at Zero Tesla, May 5-14 2008, (CRUZET-1)

HCAL has collected over 23M good quality events with cosmic ray triggers. All Front End Devices were read out, including HCAL Barrel (HB) , Endcap(HE), Outer (HO) and Forward (HF). This is factor of 50 larger data set with respect to previous data sets (Sept 2007-March 2008).

For each Level-1 trigger (L1A) , HCAL readout acquires precision digi information from charge integrators (QIE cards) for ten 25 nanoseconds long time-slices or LHC bunch crossings (bx). It is crucial to verify that the signal from the cosmic ray muon which generates the trigger is registered well within HCAL readout window. In order to verify the synchronization of HCAL readout with respect to trigger, we looked at the timing of HCAL events selected by requiring that the sum of the signal in a single HCAL tower is above the noise threshold. Timing of the HCAL signal is determined on an event by event basis for each HCAL Barrel readout channel, requiring the ten time-slice (bunch crossing) signal sum to be at least 1 GeV above pedestal. Precision QIE digi information from four time-slices around the time slice with maximum signal is used to obtain energy weighted time average for each channel above energy threshold:

$$t_\mu = \frac{\sum_{i=E_{max-1}}^{i=E_{max+1}} \Big(E_i(ADC) - ped\Big) * i}{\sum_{i=E_{max-1}}^{i=E_{max+1}} \Big(E_i(ADC) - ped\Big)} \tag{1}$$

Fig. 5 shows the timing of the HCAL signals which are well timed in with respect to DT and RPC triggers, using data from the run 43434. Nevertheless, additional work is required for precise synchronization of triggers.

3.2. Muon DT and ECAL vs HCAL correlations

Using the cut on HCAL events mentioned above, we observed a good correlation between η and ϕ of DT tracks from cosmic ray muons and the position of corresponding signals in Hadron

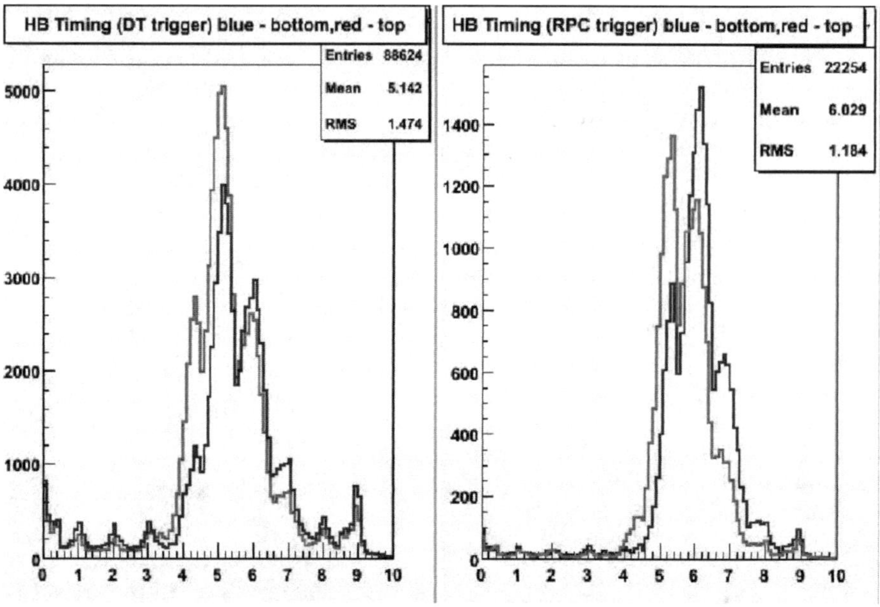

Figure 5. Timing of HB signal with respect to DT and RPC triggers from the Cruzet1 run 43434. Blue lines from HB-Bottom towers and red lines from HB-Top towers.

Barrel, and Outer, calorimeter towers. Fig. 6 show the correlation between the position of the DT track in ϕ and η, extrapolated into the center of HCAL Barrel towers. Similar correlations are obtained as well between ECAL and HCAL ϕ and η towers (Fig. 7). In these plots, the events were selected by requiring that track impact parameter in xy plane is less then 40 cm.

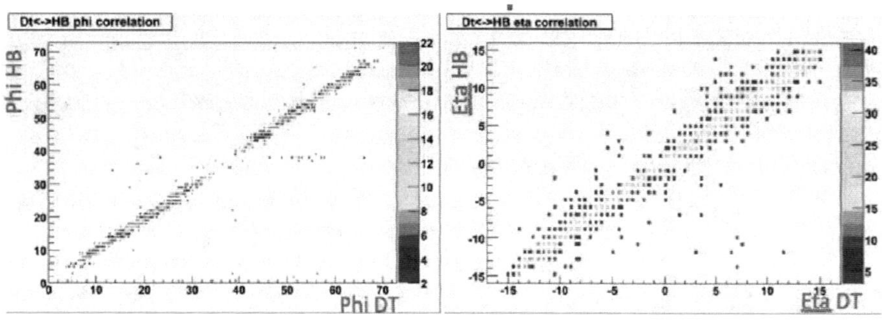

Figure 6. Correlations between DT track η and ϕ positions vs HCAL η (left) and ϕ (right) towers, during CRUZET-1

3.3. Selecting cosmic ray muons using Muon chamber (DT) tracks

To obtain an unbiased measurement of muon energy deposited in HCAL, we used DT tracks extrapolated into HCAL volume, to determine the block of HCAL $\eta - \phi$ towers crossed by a muon track (Fig 8). Cosmic ray muons which cross the detector with a large distance from the CMS interaction point, will traverse and deposit energy in multiple HCAL towers. For such events, one would need to sum up large number of towers, both in η (large impact parameter in z plane) and ϕ (large impact parameter in x-y plane) in order to fully contain muon energy

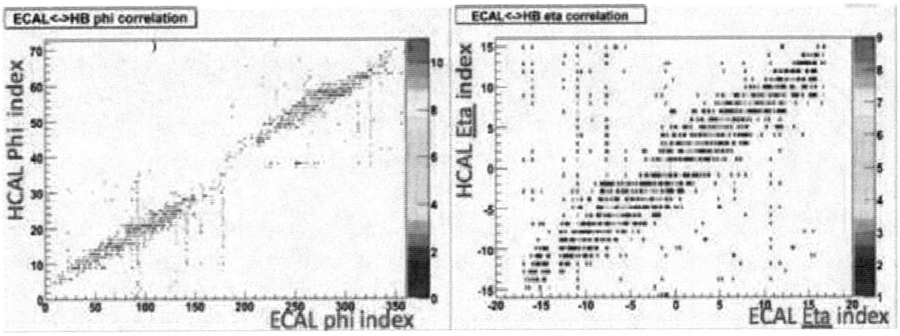

Figure 7. Correlations between ECAL η and ϕ positions vs HCAL η (left) and ϕ (right) towers, during CRUZET-1

deposit cluster. We require that muon track crosses only single HCAL 5 degree ϕ section. This requirement is imposed by checking position of the track at the inner and outer radius of HCAL. $\eta - \phi$ occupancy of the selected muons in HCAL are shown in Fig. 9.

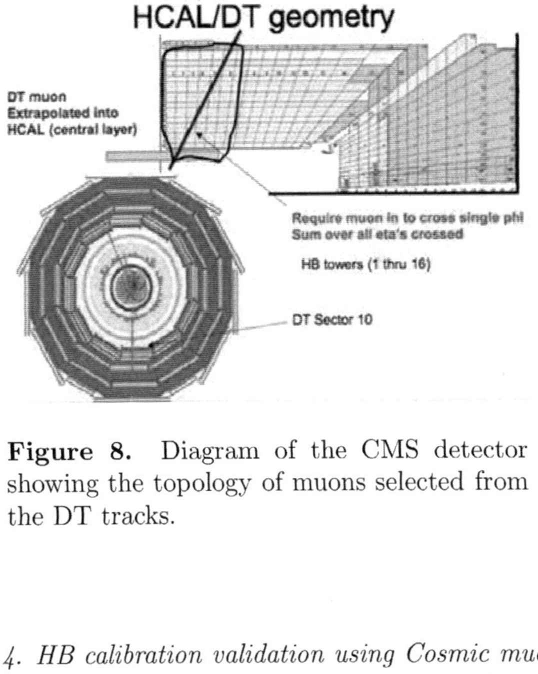

Figure 8. Diagram of the CMS detector showing the topology of muons selected from the DT tracks.

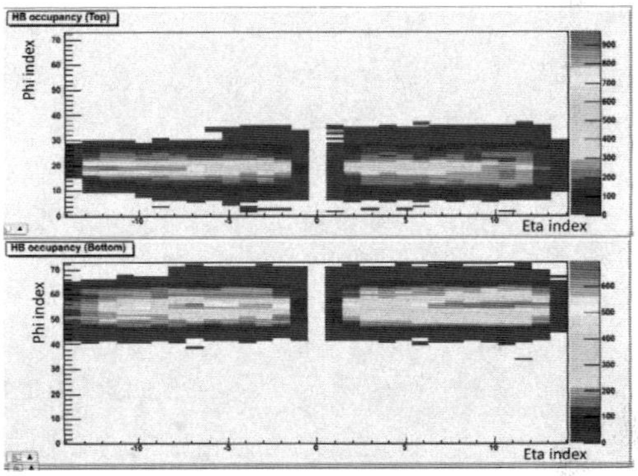

Figure 9. HB towers $\eta - \phi$ occupancy of selected muons from DT tracks. The muons enter HCAL though the top, negative half-barrel region and exit through the bottom, positive half-barrel region.

3.4. HB calibration validation using Cosmic muons

The source calibration was used to establish energy calibration from test beam wedges to the wedges assembled into Hadron Barrel calorimeter. The absolute energy scale and relative (tower-to-tower) calibration coefficients can be tested by comparing test beam and radioactive source results with the response of HCAL to cosmic ray muons [?].

Fig. 10 shows muon energy deposition in HCAL Barrel using Cruzet1 data set. The energy of muon is defined as:

$$E_{\mu}^{3\phi} = \sum_{\phi_0-1}^{\phi_0+1} \sum_{\eta_{min}-1}^{\eta_{max}+1} E(\phi_0, \eta)[GeV] * Cos(\theta) \tag{2}$$

Here, $E_{\mu}^{3\phi}$ is the HCAL energy summed over 3 ϕ towers and is obtained using HCAL Barrel $\eta - \phi$ tower clusters. The energy sum is performed over a range of towers crossed by DT track in η direction and expressed in GeV, after calibration coefficients have been applied. Three 5 degree ϕ sections in HB are included in the energy sum, one indicated by DT track (ϕ_0) and the two neighboring ones. All energies are corrected to correspond to normal incidence in the y-z plane.

The ratio of HCAL gains obtained from source calibration to uncalibrated response of HCAL to muons has of 4 to 5 percent RMS value. Energy deposited by cosmic ray muons in HCAL is consistent with measurements done in the test beam 2006 using 50 GeV/c pions [2]. One hour of running with cosmic ray triggers at 250 Hz rate yields approximately 1 million muon candidate events. Analysis of Cruzet1 data shows that using a 20 million sample of cosmic ray muon triggers, which can be collected in less than one day of running, one can validate HCAL calibration with precision better than 5 percent (Fig. 11). The accuracy of the validation process can be further increased by improving muon selection cuts.

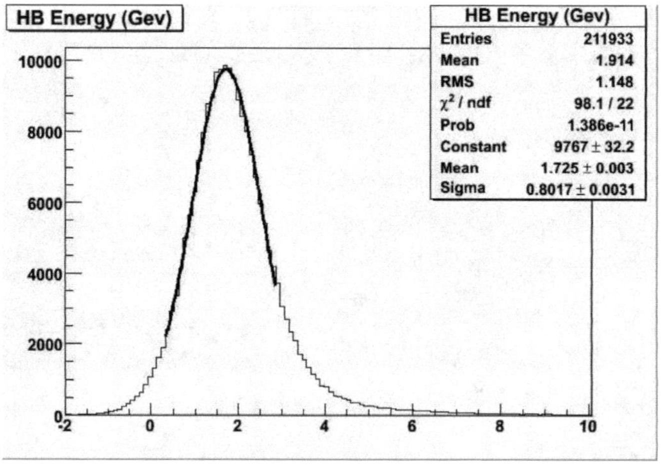

Figure 10. Muon energy deposition combined for the HCAL Barrel (HB).

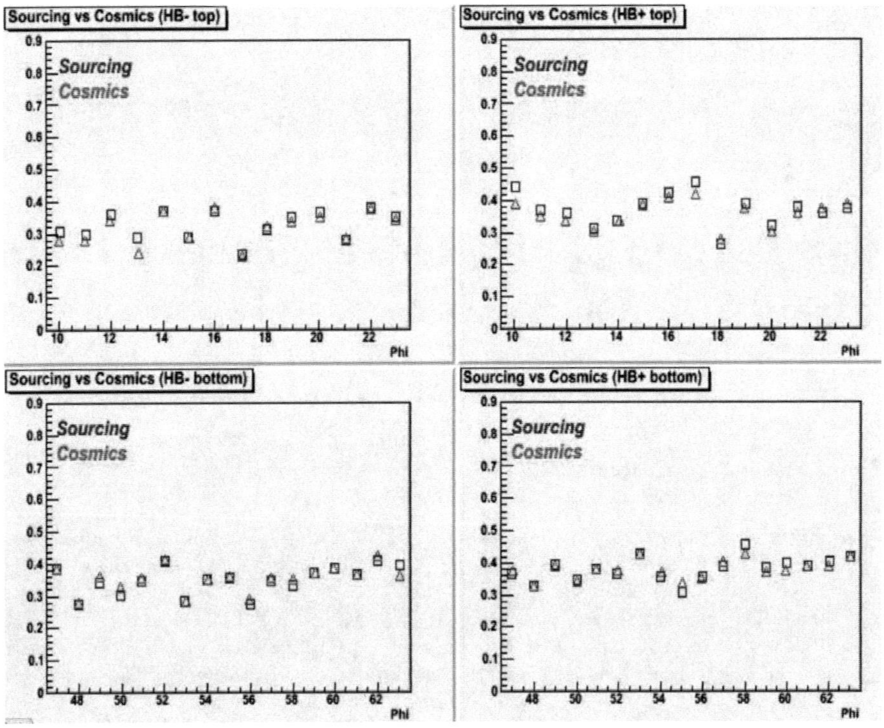

Figure 11. HCAL tower gains obtained from radioactive source calibration (blue squares) compared with the energy response of cosmic ray muons (red triangles).

4. Conclusions

- CMS Hadronic Calorimeter (HCAL) is fully installed and commissioned using local and Global runs,
- HCAL is timed in with respect to DT,RPC, CSC and ECAL during Global runs;
- DT/HCAL and ECAL/HCAL $\phi - \eta$ correlations are good;
- Relative timing of various triggers will be further adjusted;

References

[1] CMS Physics TDR, Volume I: CERN-LHCC-2006-001, 2 February 2006
[2] Design, performance and calibration of the CMS forward calorimeter wedges, G. Bayatian et al., Eur.Phys.J.C53:139-166,2008.
[3] The CMS Magnet Test and Cosmic Challenge (MTCC Phase I and II), CERN/LHCC 2007-011/G-129, 7 March 2007

Journal of Physics: Conference Series **160** (2009) 012056 doi:10.1088/1742-6596/160/1/012056

The CMS Barrel Calorimeter Response to Particle Beams from 2 to 350 GeV/c

Efe Yazgan for the CMS ECAL/HCAL Collaborations
Department of Physics, Texas Tech University, Lubbock, TX, USA
E-mail: efe.yazgan@cern.ch

Abstract. The response of the combined CMS barrel calorimeters to hadrons, electrons and muons over a range from 2 to 350 GeV/c has been measured. The analysis of the differences in calorimeter response to charged pions, kaons, protons and antiprotons and a discussion of the underlying phenomena are presented. Techniques to correct the signals from the considerably different electromagnetic (EB) and hadronic (HB) barrel calorimeters in reconstructing the energies of hadrons are also presented. Above 5 GeV/c, these corrections improve the energy resolution of the combined system where the stochastic term equals 84.7% and the constant term is 7.4%. The corrected mean response remains constant within 1.3% *rms*.

1. Introduction

The CMS calorimeters have distinct hadronic (HCAL) and electromagnetic (ECAL) systems. The central HCAL is made of brass and scintillators [1] while the ECAL comprises lead tungstate crystals ($PbWO_4$) [2]. The calorimeters are divided into the barrel (HB and EB) and the end-cap (HE, EE and pre-shower, ES) sections inside a cryostat of 5.9 m inner diameter, containing a superconducting solenoid coil providing a 4 T magnetic field. The HB design maximizes the number of interaction lengths (λ_I) inside the cryostat and is limited to $5.8\lambda_I$ at $\eta = 0$. The EB adds $\sim$1.1 λ_I. The outer hadron calorimeter (HO) was constructed to sample the energy leakage outside of cryostat [3]. There are also two very forward calorimeters (HF) made of iron and quartz fibers [4]. Figure 1 shows the calorimeters inside and around the solenoid coil.

This paper reports the responses of the barrel calorimeters to beam particles. The measurements were performed with production modules and front-end electronics as in the final CMS detector configuration. A special beam line was constructed to measure the calorimeter response down to 2 GeV/c. This was necessary since a large fraction of the particles reaching the CMS calorimeters at the LHC have energies below 20 GeV. More details on the CMS calorimeters can be found in [1, 2].

2. Test Beam Setup

The data were recorded during 2006 at the CERN H2 test beam. Figure 2 is a photograph of the moving platform that held two production HB wedges plus a production EB supermodule (SM) which was placed in front of the HB, and the HO behind the HB. The two-dimensional movement of the platform in the ϕ and η directions allowed the beam to be directed onto any tower of the calorimeter mimicking a particle trajectory from the interaction point.

Journal of Physics: Conference Series **160** (2009) 012056 doi:10.1088/1742-6596/160/1/012056

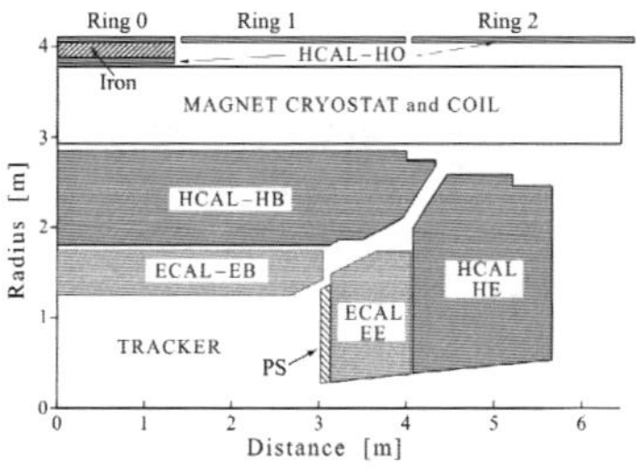

Figure 1. Location of the ECAL and the HCAL detectors (quarter slice-longitudinal cross section) in and around the CMS magnet.

Figure 2. The ECAL and the HCAL modules on a moving platform in the CERN H2 test beam area.

Temperature stability is critical for the ECAL as both the response of the crystals and the APDs change with temperature. The temperature was stabilized at 21°C by enclosing the EB SM (except in the beam direction) in 5 cm aluminum plates with cooling water pipes embedded in the plates and by wrapping the entire SM with a thermal blanket.

2.1. H2 Beam Line and Particle Identification

Figure 3 schematically depicts the CERN H2 beam line which is designed to operate in two distinct modes. In the high energy mode (15-350 GeV/c), particles are produced when 400 GeV/c protons from the Super Proton Synchrotron (SPS) strike a production target (T2) 590.9 m upstream of the calorimeters. The maximum usable momentum was 100 GeV/c for electrons and 350 GeV/c for hadrons. In the very low energy (VLE) mode ($\leq$9 GeV/c) , an additional target (T22) located 97.0 m upstream is used for particle production. As shown in Figure 3, a dog-leg configuration is utilized for the momentum selection of these low momentum particles.

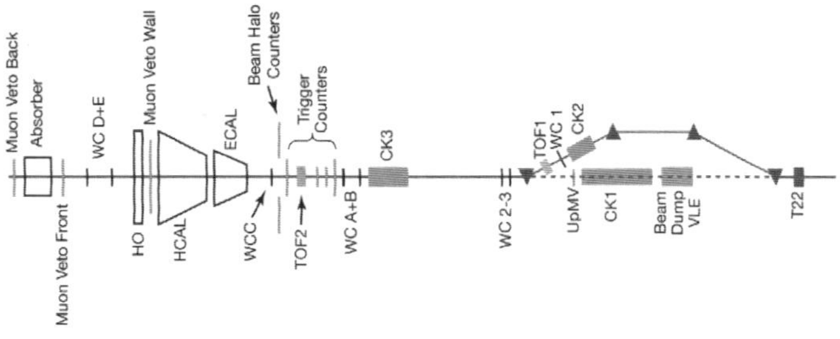

Figure 3. The CERN H2 beam line and the experimental setup.

In the VLE mode, two Cherenkov counters (CK2 and CK3), two time-of-flight counters (TOF1 and TOF2) and muon counters (Muon Veto Wall (MVW), Muon Veto Front (MVF) and Muon Veto Back (MVB)) were used to positively tag electrons, pions, kaons, protons, antiprotons and muons. CK2, which is filled with CO_2, was used to identify electrons in the VLE mode. CK3 was filled with Freon134a [5] and its pressure was set depending on the desired discrimination between electrons, pions, and kaons. At lower beam momenta, ($P_b \leq 3$ GeV/c), it was set to tag electrons. At higher momenta ($P_b > 4$ GeV/c), CK3 was used to separate

Journal of Physics: Conference Series **160** (2009) 012056 doi:10.1088/1742-6596/160/1/012056

pions from kaons and protons. Time-of-flight counters (TOF1 and TOF2) were separated by ~55 m. The time resolution obtained by this system was ~300 ps. Protons were well-separated from pions/kaons up to 7 GeV/c with the TOF system alone. Pions and kaons have $\pm 1\sigma$ TOF overlap at 5.6 GeV/c, while kaons and protons overlap at 9.5 GeV/c. Figures 4 and 5 display the identified particles in -3 and -8 GeV/c hadron beams. Energetic muons were tagged with MVF and MVB counters as well as the MVW counters. MVF and MVB were large (80×80 cm^2) scintillation counters and were placed well behind the calorimeters. To absorb the soft muon component in the beam, an 80-cm thick iron block was inserted in front of MVB. MVW consisted of 8 individual scintillation counters, each measuring 30 × 100 cm^2, placed closely behind the HB. In addition, six delay-line chambers, four scintillation counters (S1-S4) and four

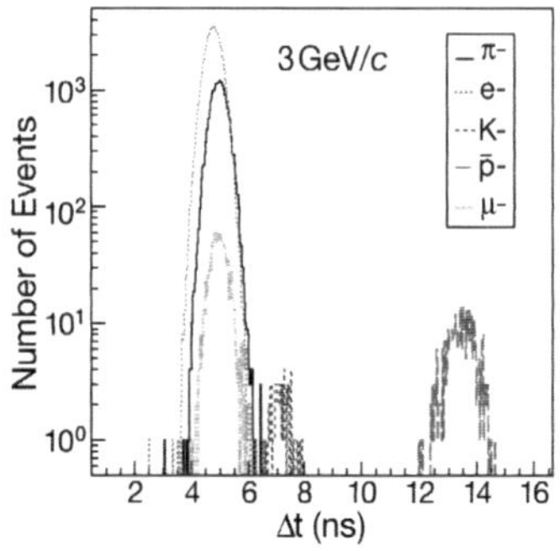

Figure 4. The distributions of the time of flight between TOF1 and TOF2 are shown for different particles.

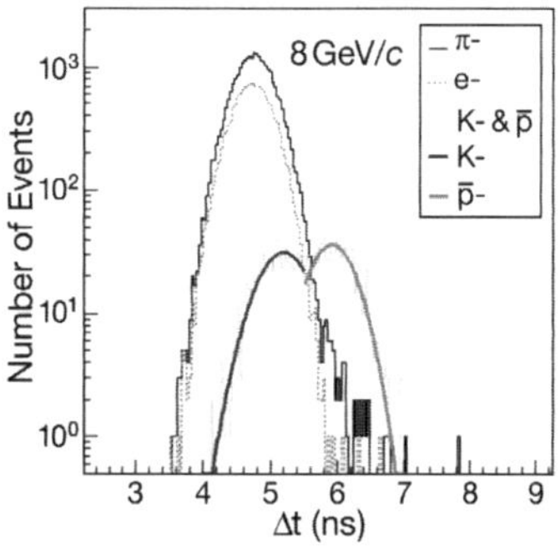

Figure 5. The same as Figure 4 but for a -8 GeV/c hadron beam. The solid blue and purple lines indicate fits to data.

scintillation beam halo counters (BH1-BH4) were used. The resolution afforded by the delay-line chambers was ~350 μm in both the x- and y-coordinates. The beam trigger typically consisted of the coincidence S1·S2·S4 which defined a 4×4 cm^2 area on the front face of the calorimeter. The S4 counter pulse height was used to eliminate multi-particle events off-line since it gave a distinct distribution for single and multi-particles in the beam (see Figure 6). BH counters, each measuring 30×100 cm^2 in size, were arranged such that the beam passed through a 7×7 cm^2 opening. These counters were positioned 17 cm downstream of the last trigger scintillator S4 and were effective in vetoing the beam halo and large-angle particles that originated from interactions in the beam line.

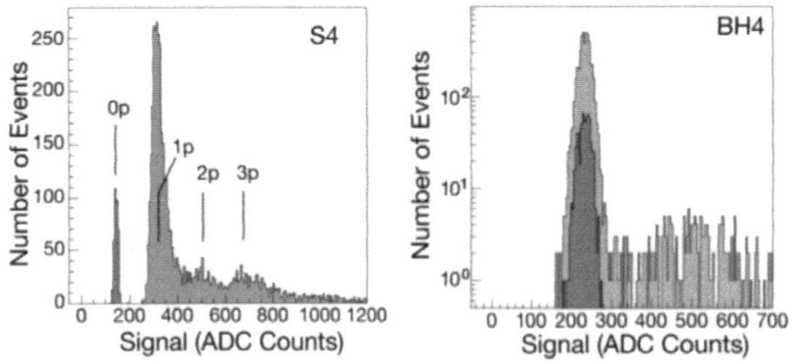

Figure 6. The signal distribution from the S4 trigger scintillator (left) for 50 GeV/c e^-s displays multi-particle events where up to three particles are discernable. The plot on the right shows the signal distribution of one of the four halo counters for 3 GeV/c negative pion beam. The red histograms indicate pedestal distributions.

In the high energy mode of the beam line where the data were taken with negative beams, there was no $\overline{p}$ contamination. If the beam line was configured for positive particles, at

350 GeV/c, the beam consisted almost purely of protons. At 20 and 30 GeV/c, the proton identification in the π^+ beam was possible when CK3 was pressurized to 1.7 bar of CO_2. The particle content depends on the momentum, P_b. At the higher end, pions dominate. At lower momenta, the beam consists mostly of e^-s. The beam consisted of 31% pions, 0.4% kaons, and 5.6% protons at +4 GeV/c, and the rest were e^+s. At +8 GeV/c, the beam contained 72% pions, 2% kaons and 7% protons, and the remaining were e^+s. In the negatively charged beam, the particle mixture was approximately the same but the $\overline{p}$ fraction was much reduced compared to that of the p in the positive beam. To enrich the hadron content of beam triggers at low energies, a S1·S2·S4·$\overline{\text{MVF}}$ trigger was employed.

3. Calibration of Calorimeters

Both the EB and HB calibrations were carried out with 50 GeV/c e^-s. The e^- beam was directed at the center of each HB tower. Similarly, the EB calibration data were collected by pointing the beam to a selected set of crystals that formed a tight grid pattern. For the EB, the signals from 7×7 crystals, and for the HB the signals from 3×3 towers were summed. In the case of the HO, the total energy was estimated by adding signals from 3×2 towers.

The response of each HB scintillator tile of each layer was also measured by using a 5-mCi Co^{60} moving wire radioactive source [6].

The HO modules were first calibrated by 150 GeV/c μ^- beam. A clear μ peak was observed in Ring 0 and Ring 2. In Ring 1 the μ peak was measurable but not as cleanly separated from pedestal. Next, the HO energy scale was determined by 300 GeV/c π^- beam impinging on η tower 4 of the HB. For this measurement, it was required that the energy in the EB be less than 1.2 GeV. The energy scale was determined by requiring the best energy resolution in HB+HO, as measured by *rms* width, for the 300 GeV/c π^- beam.

4. Combined Calorimeter (EB+HB) Response

Figure 7 shows the combined response of EB+HB to a variety of particles *vs.* momentum. At 5 GeV/c, the $\overline{p}$ response is ~70% of the e^- response. The responses to charged pions and protons are 62% and 47% of the e^- response at the same energy, respectively. At a given momentum, the available energy that is converted to a calorimeter signal varies by particle type. The available energy for protons is their kinetic energy. For $\overline{p}$, the available energy equals the kinetic energy plus twice the rest mass of proton. For pions and kaons, the available energy is their kinetic energy plus their mass. In Figure 8, the same data are plotted against the available energy. One expects roughly the same response characteristics for all hadrons, which is approximately observed in the data, but there are subtle differences which we discuss next.

The response to π^+ is systematically larger than the π^- response, increasing as the energy decreases. This is due to the characteristics of the charge exchange reactions, $\pi^+ + n \rightarrow \pi^0 + p$ (I) and $\pi^- + p \rightarrow \pi^0 + n$ (II). In these reactions, a large fraction of the pion energy is carried by the final-state π^0, that develops electromagnetic showers. The response to pions interacting this way is close to 1. Since the heavy nuclei contained in the calorimeter consists of ~50% more neutrons than protons, the relative effect of reaction (I) is larger than that of reaction (II), and therefore, the response to π^+ should be larger than the π^- response at lower energies where charge exchange is important. This difference, at 2 GeV/c, is ~10%.

The response to protons is systematically lower than that of negative pions. This effect, which is also observed at high energy, is a result of the fact that π^0 production is, on average, smaller in proton induced showers. This is a consequence of baryon number conservation, which favors the production of leading baryons, while pion induced reactions may have leading π^0s. This effect was observed in the HF calorimeter [7], where it caused a response difference $\geq$10%. Since the e/h values of EB and HB are smaller than for the HF, the effects are correspondingly smaller, but nevertheless significant.

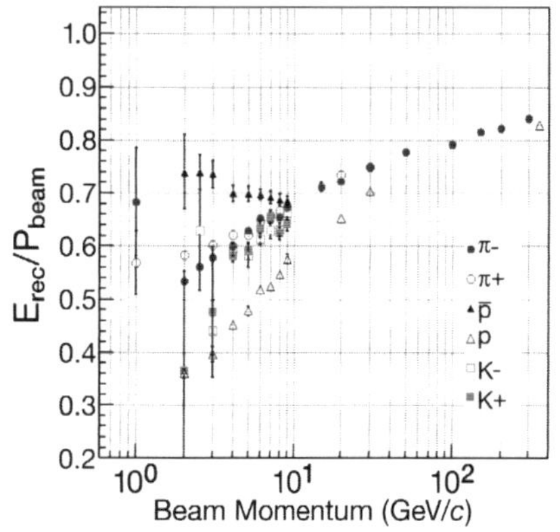

Figure 7. The response of the combined calorimeter system to six different particles *vs* beam momentum. Both the EB and HB are calibrated with 50 GeV/c e^-s.

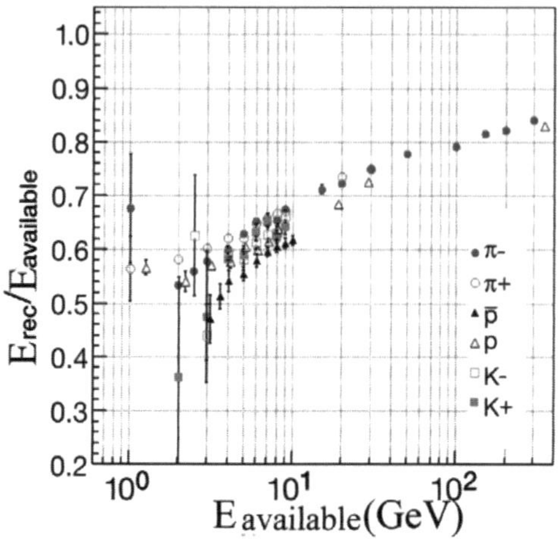

Figure 8. Same as in Figure 7 but the calorimeter response is plotted against the available energy.

Since the inelastic cross section for protons is larger than for pions, a larger fraction of the baryons start showering in the EB. It is found from 30 GeV/c π^+ and 30 GeV/c proton data that 41% of the pions penetrate the EB without starting a shower, versus 35% of the protons. The effective thickness of the EB is thus $1.05\lambda_I$ for protons and $0.89\lambda_I$ for pions.

The fraction of the beam energy deposited in the EB decreases from ~60% at 2 GeV/c to ~25% at 300 GeV/c. At the same incident momentum, protons deposit on average less energy than pions in the EB, while antiprotons deposit more. Antiprotons start their showers, on average, earlier than pions and therefore a larger fraction of the energy ends up in the EB. One would expect the same for proton induced showers. However, when a proton interacts in the EB, the interactions have limited energy transfered to secondary π^0s. This suppresses the proton signal in the EB, despite the fact that protons are more likely to start their showers in the EB compared to pions.

Pions deposit, on average a larger fraction of their energy in the HB. Since the e/h value of the HB is smaller than for the EB, the pions benefit more from the increased response to the non-electromagnetic shower components. The net result of the energy sharing is that for beam momenta less than 10 GeV/c, the total energy measured is smallest for antiprotons, then protons and then pions. This effect is observed in Figure 8.

Figure 9 shows the response of 150 GeV/c muons in the HB using 3×3 HB towers. Since the noise in a single tower of the HB is equivalent to ~0.2 GeV, this calorimeter system is superb in identifying single isolated muons. The HB trigger electronics is also designed to generate an isolated muon signal (bit) based on this capability. The mean energy deposited by a 150 GeV/c muon is 2.4 GeV.

5. Optimization of Energy Reconstruction

The e/h values are different for the EB and the HB, and thus corrections have to be applied to obtain the correct particle energy from the EB+HB+HO system. Figure 10 displays the measured EB *versus* HB energy for a number of pion beams. The events with near zero EB energy are due to the non-interacting π's in the EB. In a compensating calorimeter ($e/h = 1$), the events would normally lay about a straight line as indicated in Figure 10, and the line would intersect both axes at the beam energy. This is not the case for the EB+HB system, and we

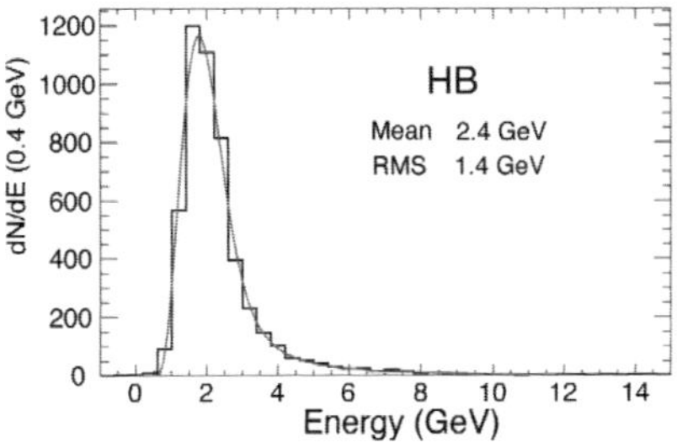

Figure 9. The HB response to 150 GeV/c μ^- from tower 4 ($\eta = 0.3$). The solid curve represents a fit using combined Gaussian and Landau distributions.

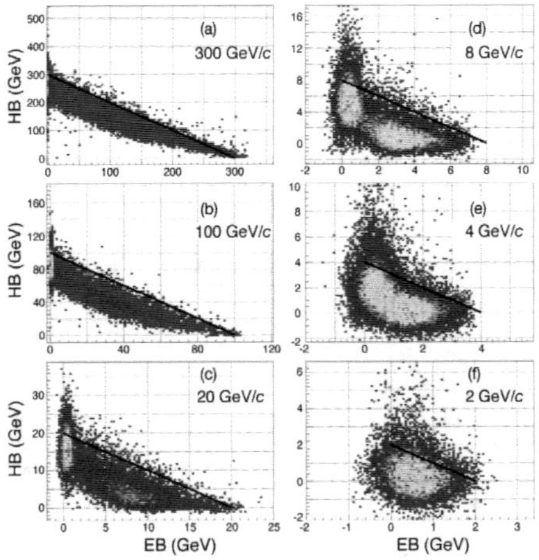

Figure 10. The "raw" EB *vs* "raw" HB energy scatter plot for incident π^- beam momenta of 300, 100, 20, 8, 4 and 2 GeV/c. The straight line indicates the expected behavior of an ideal calorimeter system.

choose to optimize the energy measurement in two major steps. The first correction is carried out for the energy response, and the second step involves correction of non-linear response as a function of the EB energy fraction. First, thresholds are applied to the EB and HB energy

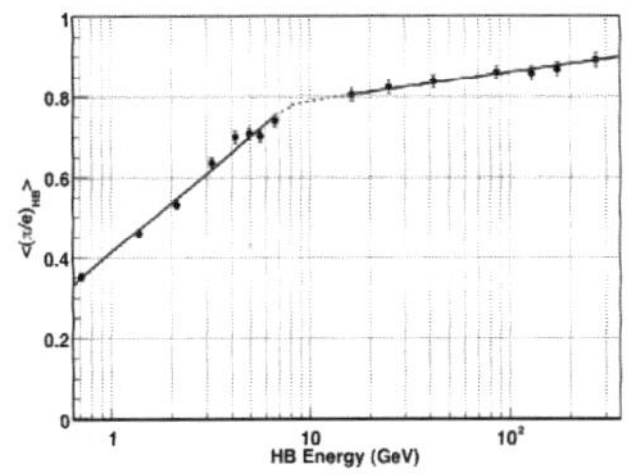

Figure 11. π/e *vs* E_{HB} for events interacting in the HB.

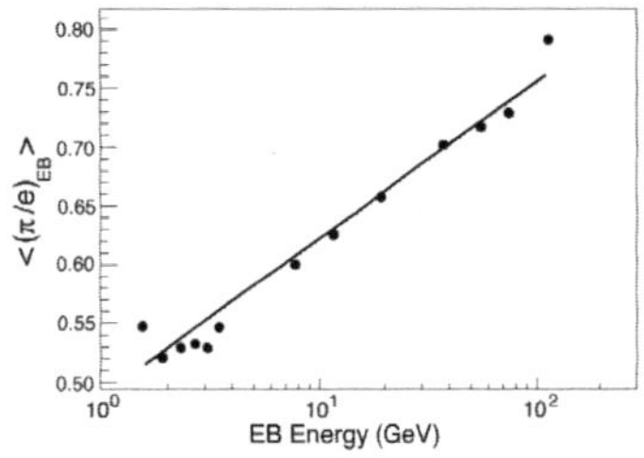

Figure 12. Measured $(\pi/e)_{\mathrm{EB}}$ *vs* E_{EB} after correcting the energies of pions that interacted in the EB.

clusters that are constructed from 7 × 7 EB crystals, 3 × 3 HB and 3 × 2 HO towers. If the energy in the cluster is less than the threshold, the cluster energy is set to zero. The thresholds are set at least 3σ away from the noise levels, and are 0.8, 1.0 and 2.0 GeV for the EB, HB and HO, respectively.

The next task is to parametrize the π/e for the HB as a function of the mean HB energy, using the events that deposit only minimum ionizing energy in the EB ($E_{\mathrm{EB}} < 1.2$ GeV). Figure 11 displays the π/e for the HB as a function of the log of the observed HB energy. The plot (in semi-log) shows two lines with a break point at about 8 GeV.

The π/e ratio is defined as $[1 + (e/h - 1)f_0]/(e/h)$. The electromagnetic fraction, f_0, is parametrized using Wigmans' function [8], where $f_0 = 0.11 \log P_b$ and P_b is the beam momentum. The fit above 8 GeV gives $e/h = 1.4$. Below ~8 GeV, π/e can be represented by the following function, $0.179 \log(E_{\mathrm{HB}}) + 0.413$. Further studies are underway to determine the physics behind the logarithmic function describing the low momentum data points.

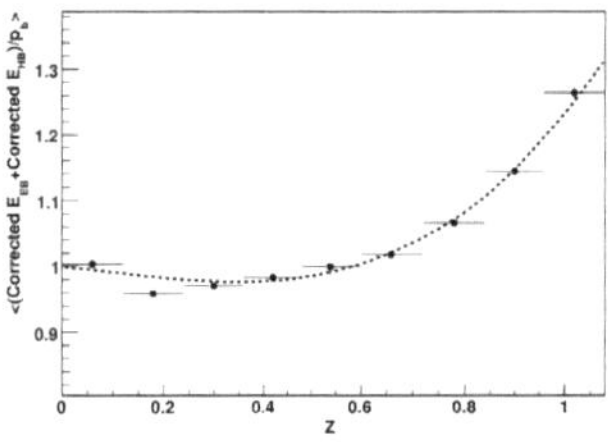

Figure 13. The π/e corrected response ratio for 100 GeV/c pions of the combined system as a function of the EB fraction.

Next, the mean π/e for the EB, $\langle(\pi/e)_{\mathrm{EB}}\rangle$, is estimated using the known beam momentum as a function of the observed mean EB energy, $\langle E_{\mathrm{EB}}\rangle$, for each beam momentum, $\langle(\pi/e)_{\mathrm{EB}}\rangle = \langle E_{\mathrm{EB}}\rangle/(P_b - E^*_{\mathrm{HB}})$ where E^*_{HB} is the event-by-event corrected HB energy, $E_{\mathrm{HB}}/(\pi/e)_{\mathrm{HB}}$. We calculate this ratio for events where the pion shower is shared between the EB and the HB with the requirement that $0.2P_b \leq E_{\mathrm{EB}} \leq 0.6P_b$. The mean π/e for EB as a function of the logarithm of the observed EB energy shows a linear behavior (see Figure 12). The function representing $\langle(\pi/e)_{\mathrm{EB}}\rangle$ is equal to $0.057 \log(E_{\mathrm{EB}}) + 0.49$. After correcting the EB energies event-by-event using the above function, $E^*_{\mathrm{EB}} = E_{\mathrm{EB}}/(\pi/e)_{\mathrm{EB}}$, we find that the π/e correction overestimates the energies for events with large EB energy fractions, $Z \equiv E_{\mathrm{EB}}/(E_{\mathrm{EB}} + E_{\mathrm{HB}}) > 70\%$ (see Figure 13). This is expected since these events correspond to the cases when a hadronic shower in the EB fluctuates largely to neutral particles.

The second step in the correction is to linearize by fitting the non-linear response to the correction function shown in Figure 13, $\langle(E^*_{\mathrm{EB}} + E^*_{\mathrm{HB}})/P_b\rangle = 0.412Z^3 - 0.096Z^2 - 0.084Z + 1.00$. This set of corrections has been determined to be insensitive to the beam momentum and 100 GeV/c data is a good representation for all other beam momentum data. Figure 14 displays the energy resolution and the response linearity of the combined EB+HB calorimeters for pions. The circles represent the "raw" and the squares represent the corrected data. The data in Figure 14 are compiled after having fit each energy response with a Gaussian down to 5 GeV/c. The energy resolution is parametrized as $\sigma/E = a/\sqrt{E} \oplus b$ where a is the stochastic and b is the constant term, and the terms are added in quadrature. The "raw" resolution of the EB+HB system is such that $a = 110.7\%$ and $b = 7.3\%$ as indicated by open circles within 5 to 300 GeV/c in Figure 14 (left). After applying the corrections, the energy resolution improves, as indicated by the solid red squares ($a = 84.7\%$ and $b = 7.4\%$). The corrected mean response remains constant within 1.3% *rms* as depicted in Figure 14 (right).

6. Summary and Conclusions

The CMS barrel calorimeter has been exposed to particle beams with momenta from 2 to 350 GeV/c. The particle identification detectors separated electrons, muons, pions, and protons over a substantial energy range. The response to different hadrons is studied and simple interesting regularities are observed. The ratio of negative charged to positive charged pion response, the ratio of negative pion to proton response and the ratio of pions to antiprotons are explored. The linearity and energy resolution for negative pions are optimized. The low energy response is explored where previously used parametrizations no longer fit the data well. The corrected data are linear within 1.3% rms for $p_b \geq 5$ GeV/c. The stochastic and the constant terms are 84.7% and 7.4%, respectively. The calorimeter remains non-compensating, so that a substantial

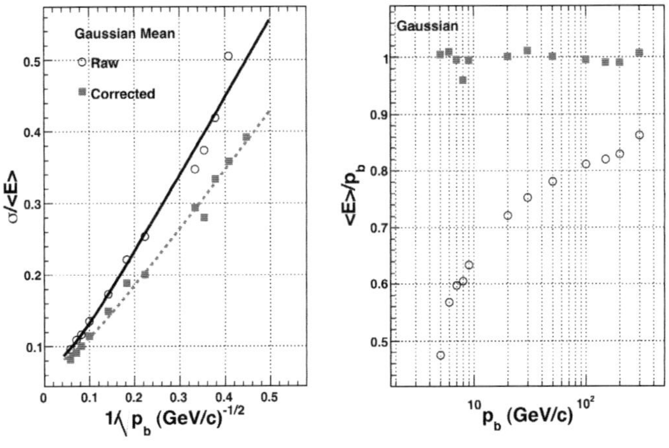

Figure 14. The energy resolution (left) and the response of the combined calorimeters (right) before (open circles) and after the corrections (squares).

deviation from $E^{-1/2}$ scaling is unavoidable. The correction method works for single isolated particles and the test beam environment. Direct application of the correction method to jets is difficult since jets are formed both from isolated as well as non-isolated objects. If the photons from π^0s in a jet can be separated from the charged hadrons, then the corrections could be applied on the charged hadrons and then the jet may be better reconstructed.

Acknowledgments
The results presented in this paper are partially based on the doctoral theses of J. Damgov [9], K. Gümüş [10] and E. Yazgan [11]. This project was carried out with financial support from CERN, Department of Atomic Energy and Department of Science and Technology of India, U.S. Department of Energy, U.S. National Science Foundation, RMKI-KFKI (Hungary, OTKA grant T 016823), Croatian Ministry of Science, Education and Sport (under grant No. 023-0982887-3064), French CNRS/Institut de Physique Nucleaire et de Physique des Particules, French Commissariat a l'Energie Atomique, Greek General Secretariat for Research and Technology, Italian Istituto Nazionale di Fisica Nucleare, Federal Agency for Science and Innovations of the Ministry for Education and Science of the Russian Federation, Federal Agency for Atomic Energy of the Russian Federation, Russian Academy of Sciences, Ministry of Science of Serbia, Swiss Funding Agencies, Scientific and Technical Research Council of Turkey (TÜBITAK), Turkish Atomic Energy Agency (TAEK), Bogazici University Research Fund (Grant no: 04B301), Science and Technology Facilities Council (UK).

References
[1] CMS Collaboration, *The Hadron Calorimeter Project Technical Design Report*, CERN/LHCC 97-31 (1997).
[2] CMS Collaboration, *The Electromagnetic Calorimeter Technical Design Report*, CERN/LHCC 97-33 (1997).
[3] B. S. Archaya *et al*, The CMS Outer Calorimeter, CMS NOTE-2006/127.
[4] S. Abdullin *et al*, Eur. Phys. J. C **53** (2008) 139.
[5] Freon134a is an ozone-friendly gas. Based on the measurements during the beam test, we find Freon 134a's refractive index to be 1.00065, which is also consistent with the estimates based on its molecular weight.
[6] M. Adams *et al* , *Nucl. Instrum. Methods* **A511** (2003) 311.
[7] N. Akchurin *et al* , *Nucl. Instrum. Methods* **A408** (1998) 380.
[8] R. Wigmans, *Nucl. Instrum. Methods* **A265** (1988) 273.
[9] J. Damgov, Ph. D. thesis, Institute for Nuclear Research and Nuclear Energy, Bulgarian Academy of Science, Sofia, Bulgaria (unpublished), 2008.
[10] K. Gümüş, Ph. D. thesis, Texas Tech University (unpublished), 2008.
[11] E. Yazgan, Ph.D. thesis, Middle East Technical University, Ankara, Turkey, Fermilab-thesis-2007-13.

ATLAS Tile Calorimeter performance for single particles in beam tests

T. Davidek on behalf of the ATLAS Tilecal system

Charles University in Prague, Institute of Particle and Nuclear Physics, V Holesovickach 2, 18000 Praha 8, Czech Republic

E-mail: Tomas.Davidek@mff.cuni.cz

Abstract. The modules of the ATLAS Tile hadronic calorimeter underwent extensive tests in the test beams at CERN. Studies were carried out with electrons, muons and hadrons ranging in energy from 10 GeV to 350 GeV. The Tilecal calibration systems and energy reconstruction algorithms were also studied in great details, the associated systematics has been evaluated. The updated calibration scheme led to improved linearity and uniformity of the response.

Electrons and muons were used to set and understand the electromagnetic (EM) scale and the uniformity of the calorimeter. The pion response shows the expected behaviour with energy. The performance of the real Tile calorimeter modules to pions in terms of linearity and resolution corresponds well to that of earlier Tilecal prototype modules, after accounting for the different lengths and segmentations of the calorimeters. The experimental results are also compared to MC simulation samples.

1. Introduction

The ATLAS experiment at the CERN Large Hadron Collider (LHC) is designed for precise measurements and fundamental discoveries in proton-proton collisions at the center-of-mass energy of 14 TeV. Calorimeters represent an indispensable part of the ATLAS detector system. The electromagnetic lead/liquid argon (LAr) calorimeter followed by the hadronic Tile calorimeter cover the central region of the ATLAS detector, other LAr based calorimeters span across the forward regions [1].

The Tile Calorimeter (Tilecal) is a hadron iron/plastic scintillator calorimeter. It consists of three cylinders – one central barrel (LB) and two extended barrels (EB) on its sides. Every cylinder is composed of 64 modules, each covering the azimuthal angle of $2\pi/64 \approx 0.1$ [2].

Tilecal modules are made of alternating layers of iron plates and scintillating tiles. The tiles are placed parallel to the incoming particles and radially staggered in depth. A schematic drawing is given in Fig. 1. The scintillating tiles are read-out by wave-length shifting (WLS) fibers on both sides of the module. These fibers deliver the light to photo-multipliers located in the outer iron structure that also houses the front-end electronics. Each cell is readout by 2 PMTs.

The calorimeter modules are segmented into three radial compartments. These compartments are further subdivided into cells of the dimension $\Delta\eta \times \Delta\phi = 0.1 \times 0.1$ (0.2×0.1 in the third radial compartment).

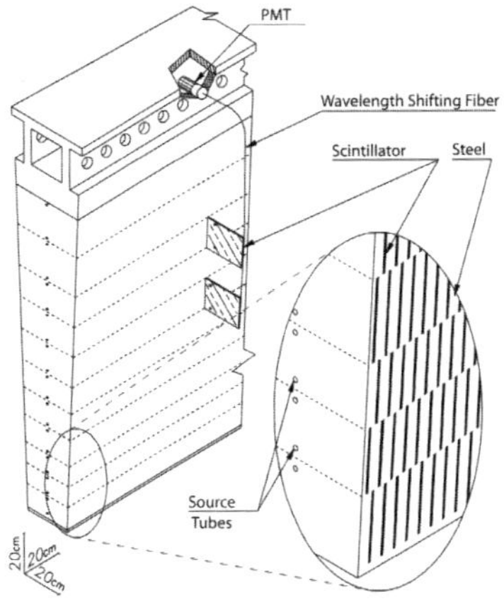

Figure 1. Mechanical structure of a TileCal module, showing the slots in the iron for scintillating tiles and the method of light collection by WLS fibers to PMTs. Radially, tiles are organized in eleven rows (tilerows) of different size.

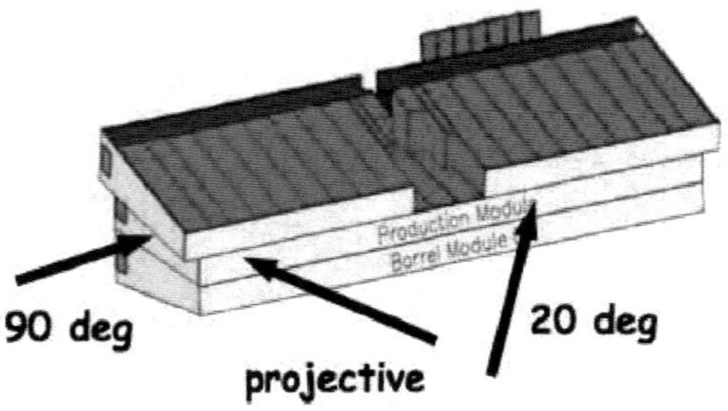

Figure 2. The typical setup for beam tests comprised the barrel module 0 in the bottom, final LB module in the middle and two final EB modules in the top. Also shown is the wall of muon counters downstream of the calorimeter stack, that assisted in tagging muons and/or calorimeter punchthrough.

1.1. Tilecal beam tests

The final Tilecal modules have been extensively tested with electron, muon and hadron beams, ranging in momentum from 10 to 350 GeV/c. Altogether, 11% of all modules were exposed to SPS beams in H8 beam-line at CERN during the years 2000 – 2003.

The modules under tests were placed on the scanning table, that could be positioned at various η, ϕ and z with respect to the incoming particle beams. A typical setup is drawn in Fig. 2. The stack consisted of the very first barrel module (so-called module 0) in the bottom, one LB module in the middle and either two EB modules or another LB module in the top.

The main goals of this testing program were to set and measure the energy-to-charge conversion factor, to explore the uniformity of the response and to determine the calorimeter performance at projective pseudorapidities.

In order to address these goals, data were taken in various configurations:

- Electrons impinging at the angle of 20° measured from the $\eta = 0$ direction were used to determine the EM scale as well as the uniformity of the cells in first radial compartment.
- Muons entering at ±90° were used to address the uniformity of the modules over all cells. Electron data were taken in this configuration too in order to determine the uniformity of the cells on the module edges. Beams impinging at ±90° were also used for special studies, e.g. hadron shower profiles [3], light-yield measurement etc.
- Runs at projective pseudorapidities were used to assess the detector performance in the "ATLAS"-like configuration. The attention was focused on the pion linearity and resolution, however the performance to electrons and muons was also explored.

2. Tilecal calibration systems

The Tilecal design includes four calibration and monitoring systems:

- Cesium (Cs) source system, that monitors the scintillators and PMTs responses
- minimum bias system for precise in-situ monitoring and (relative) luminosity measurement
- laser system, that monitors the PMT and front-end electronics

- charge injection system (CIS) to calibrate the front-end electronics

While Cs-source and minimum bias systems share the slow integrator readout (samples with approx. 100 Hz frequency), laser, CIS and physics data are readout through pulsed electronics chain that samples every 25 ns [4].

The Cs-source system and CIS are crucial for setting up the EM scale of the modules in the beam tests, therefore their main features are highlighted below.

2.1. Cesium source system

The Cs-source system is designed to measure the quality of the optical response of each calorimeter cell. Using this system, the response of every cell is equalized prior to the particle data taking. The ^{137}Cs radioactive source is passing at normal incidence through holes in every tile and absorber plate of the calorimeter, as indicated in Fig. 1.

Since the Cs-source movement through the calorimeter is very well controlled, one can reconstruct amplitudes of the individual tile responses (see Fig. 3). The gain of the cell is then determined as the average amplitude over all tiles in that cell.

2.2. Charge injection system

The pulse readout electronics for each PMT contains two analog scales, referred to as high- and low-gain. Both are digitized with 10-bit ADC. This bi-gain system allows to measure energies up to 1.6 TeV/cell, while it is still sensitive to small signals e.g. from muons.

CIS is designed to inject well-defined charge into both scales and to provide the respective conversion factors. Dedicated studies showed that a non-linearity up to 2% occurs in the low-gain due to the high-gain analog circuit saturation (see Fig. 4). The measured dependence is used offline for correcting data acquired via the pulsed electronics chain.

The uncertainty associated with the ADC performance and the calibration constant determination has also been evaluated [5]. It is below 0.7% for single particle response over the full energy range. The impact on jet energy measurement will be even smaller, since much more channels will be involved.

3. Data analysis

3.1. Event selection

The beams in are usually a mixture of electrons, muons and hadrons. In order to measure the response to a given particle type, a careful event selection has to be performed. Two kinds of selection criteria were combined – the beam and particle separation cuts.

The beam cuts require a MIP-like signal in upstream beam scintillators to avoid upstream showering and/or double particle events. The angular spread and impact point, measured by two x-y wire chambers, was also restricted in order to avoid halo particles that potentially have incorrect beam energy.

Particle selection criteria:

- Muons, that leave only small signal in the calorimeter, are selected by a cut on the total energy. The muon wall also assisted in the detection of high-energy muons that pass through the whole calorimeter.
- The electron/pion separation relies on the difference in size between the compact EM showers and rather large hadron showers. The average density is used for this purpose:

$$AvD = \frac{1}{N_{\text{cell}}} \sum_{i=1}^{N_{\text{cell}}} \frac{E_i}{V_i} \tag{1}$$

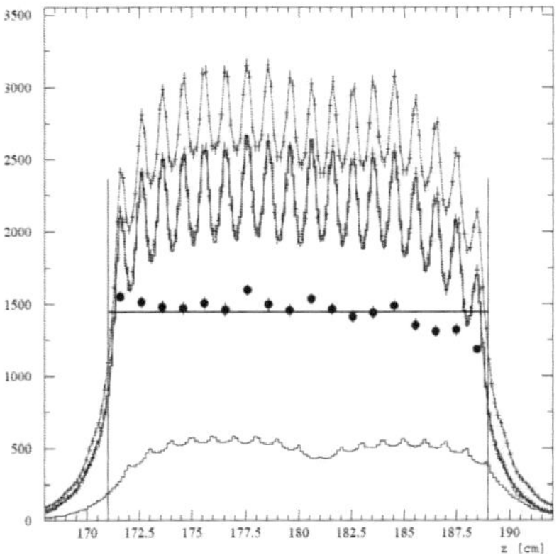

Figure 3. The Cs-source signal measured in one PMT as a function of the source position. The signal (middle curve) is obtained by subtracting the leakage estimate (bottom curve) from raw data (top curve). The individual tile amplitudes (full circles) are obtained by fitting the signal by a sum of pulse shape functions for individual tiles.

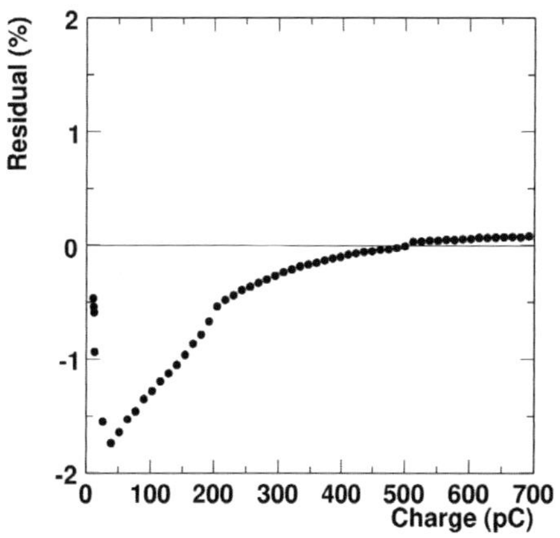

Figure 4. The non-linearity in low-gain as mapped with CIS. Horizontal axis indicate the true injected charge, the vertical axis shows relative difference between the measured and true charge.

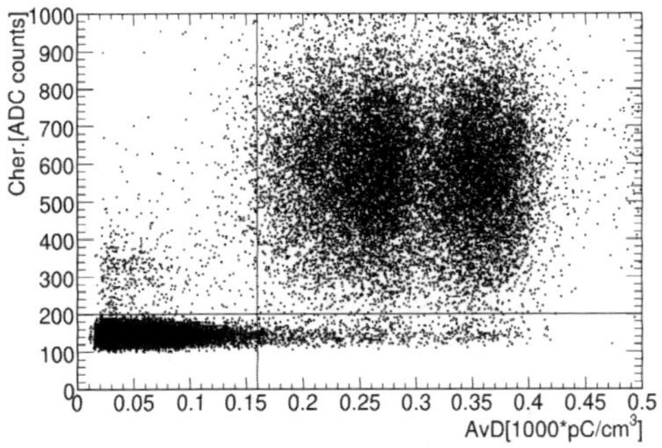

Figure 5. Example of electron/pion separation for $E_{\text{beam}} = 20$ GeV. The average density defined by Eq. (1) is used as a baseline, the Cerenkov counter signal is used to improve the selection. The two lines denote the applied cuts, the bottom left region corresponds to pions selected for further analysis.

where E_i, V_i denote the energy and volume of the i-th cell. For beam energies $E_{\text{beam}} \leq 20\ GeV$, the threshold Cerenkov counter is used to further improve the separation. An example is shown in Fig. 5.

- Apart from muons, electrons and pions, the positive-charged beams also contain a considerable fraction of protons. The pion:proton share is energy dependent and ranges approximately from 3:1 at 50 GeV up to 1:4 at 180 GeV. The threshold Cerenkov counter with appropriate settings is used for pion/proton separation.

3.2. Corrections applied to raw data

In all the analyses, the raw calorimeter data are corrected for dead channels. The energy is recovered by considering twice the signal from the second PMT reading the same cell in case the first PMT is malfunctioning.

The possible impact of the electron/pion separation procedure on the total pion response was also carefully studied. Since the event selection is based on the shower shapes, the pion events

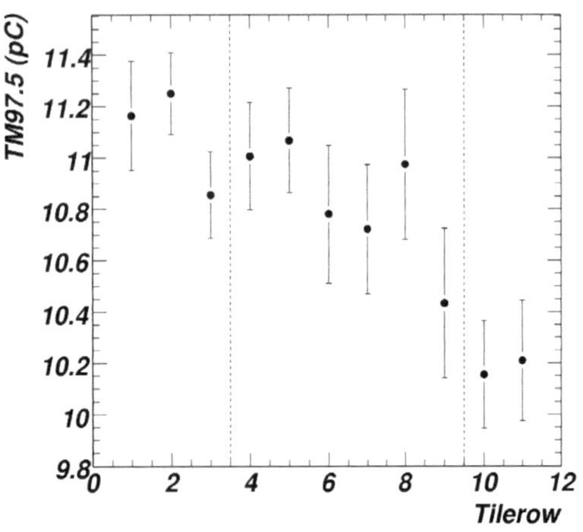

Figure 6. The muon response summed up over the whole LB module length in the 90° configuration. Mean value of muon signal truncated at 97.5% of the total number of entries (TM97.5) was adopted as the definition of the calorimeter response. Results are averaged over 5 LB modules for every tilerow, the error bars represent the corresponding RMS spreads. Vertical dotted lines indicate the individual radial compartments in the LB module.

with high EM content tend to be classified as electrons. This may lead to underestimation of the total pion response. The possible bias has been studied with MC and also with nominal pion beams, that contain only negligible fraction of electrons (as opposed to nominal electron beams, where the electron:pion ratio typically ranges from 1:1.5 to 2:1 depending on the energy and the exact beam settings). It was found that the above mentioned electron/hadron separation procedure causes an underestimation of the total pion response by less then 0.4% in electron runs, nevertheless we correct for this effect. The impact on the pion energy resolution was found negligible.

The real beam energy is calculated for every run using the data of the bending magnet and collimator settings in the beam-line. As a result, the real beam energies are slightly different for nominal pion and electron beams. The calculation also takes into account the synchrotron radiation of electrons, therefore electrons have lower energy than pions even in the same nominal electron beam.

Other two corrections are described in more detail in the following Sections 3.2.1, 3.2.2.

3.2.1. Particle/Cs correction High-energy muons entering the calorimeter modules at ±90° are used to study the uniformity of the individual cell response and to ensure that the EM scale is kept along all cells.

To check the EM scale in individual tilerows, the muon signal was summed up over the whole tilerow (i.e. approximately 6 m in LB and 3 m in EB modules). An example for LB modules is shown in Fig. 6, after the gains of all PMTs were equalized with Cs. The signal is clearly underestimated in the second and third radial compartment. This effect is due to trapezoidal shape of the tiles. The signal in a tile increases along the radial direction from the edge at lower radius to that at higher radius (approximately by $\sim 1\%/\mathrm{cm}$). Since Cs pipes run through the tiles close to their upper edge, the Cs signal is bigger than that induced by particles passing through the center of the tile, which in turn well represents the average signal over the tile surface.

The same result was obtained with electrons entering the tile-centers in 90° configuration and was further confirmed by special ^{137}Cs and ^{90}Sr source measurements.

This effect is corrected for by applying weights in individual radial compartments[1]. The weight in the first radial compartment is set to one, in order to preserve the EM scale as

[1] The individual tilerow corrections cannot be applied for beams entering at projective pseudorapidities, since every PMT collects the signal from several tilerows in the given cell.

Compartment	Barrel	Extended barrel
1	1.000	1.000
2	1.025 ± 0.002	1.009 ± 0.005
3	1.088 ± 0.005	1.055 ± 0.003

Table 1. The weights to correct for the particle/Cs difference, averaged over several analyzed modules. Statistical errors are shown as well.

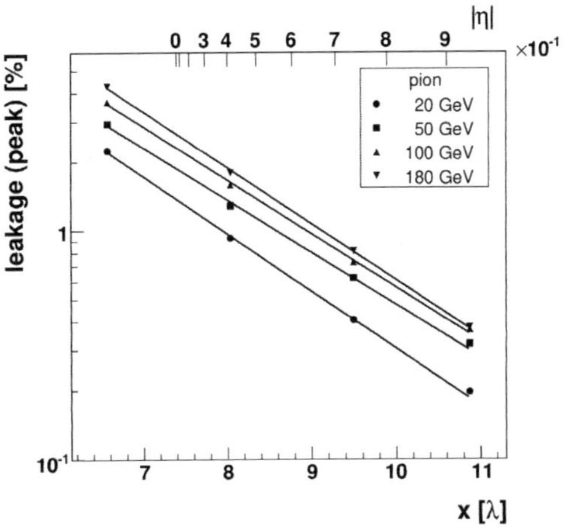

Figure 7. The longitudinal leakage correction of the pion response (Gaussian peak) as a function of calorimeter depth expressed in terms of interaction lengths [3].

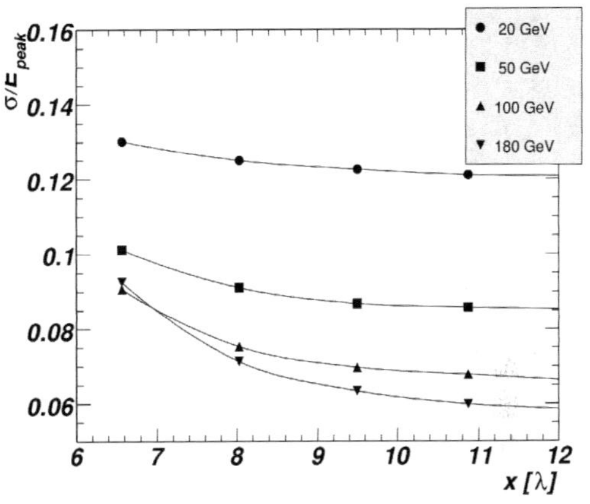

Figure 8. The energy resolution for pions as a function of the calorimeter depth expressed in terms of interaction lengths [3].

determined with electrons at 20°. The weights in the second and the third radial compartment are evaluated as the ratio of the mean muon responses in the respective tilerows.

The pattern observed in the EB modules is very similar to that shown in Fig. 6 for LB modules, nevertheless the weights differ due to different radial segmentation. Whereas the second and third radial compartment in LB modules is composed of six and two tilerows respectively, four tilerows constitute both second and third radial compartment in EB modules. The weights are averaged over all modules analyzed. Their values are listed in table 1.

3.2.2. Corrections for calorimeter length The pion response is corrected for the longitudinal leakage, that has been measured in the 90° configuration [3], see Fig. 7. The results presented in Section 3.3.2 are rescaled for an infinitely long calorimeter. This approach also enables a simple comparison to beam-tests results of earlier prototype Tilecal modules [2], that were radially longer – 9.30 λ at 20°, while final modules are 7.84 λ long at the same impact angle.

The energy resolution scales with the calorimeter length too. In order to compare the energy resolution for pions between the final and prototype calorimeter modules (see Section 3.3.2), the results from prototype modules were rescaled for the length of final modules. The measurements performed in 90° configuration were used to determine the scaling factor, see Fig. 8.

3.3. Results

The results reported in this paper come from recently re-calibrated data acquired during the beam test periods 2002 – 2003.

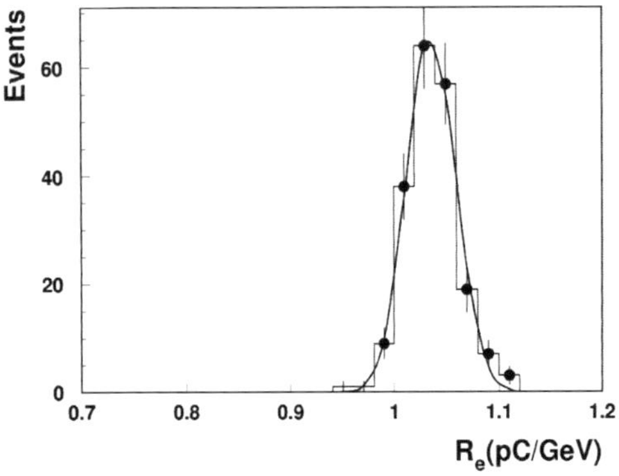

Figure 9. The cell response as measured for electrons at 20° of incidence. Each entry corresponds to one impact cell. The average EM scale factor 1.05 ± 0.003 pC/GeV is obtained, the cell uniformity shows an RMS spread of $(2.4 \pm 0.1)\,\%$.

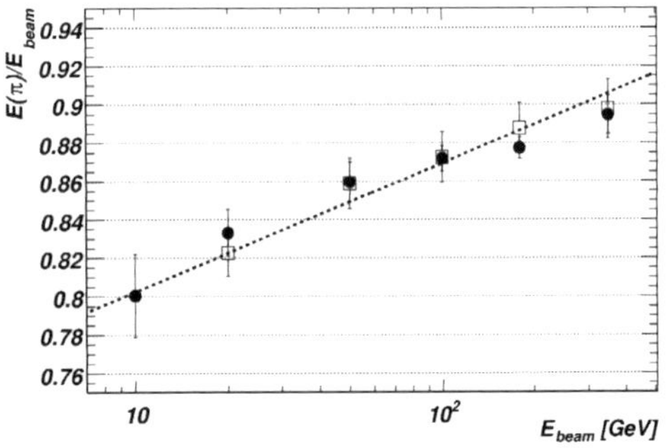

Figure 10. The relative pion response (black circles) as a function of the pion beam energy. Pions impinge the calorimeter modules at pseudorapidity $\eta = 0.35$. The pion response is corrected for the longitudinal leakage, i.e the results are scaled for an infinitely long calorimeter. Superimposed are the MC results (open squares), the dashed line corresponds to usual parameterization [6] with e/h = 1.36.

3.3.1. Electromagnetic scale The EM scale is measured with electron beams. As the calorimeter response to electrons slightly depends on the impact angle (due to sampling fraction frequency that also changes with incident angle), the EM scale is determined at a fixed angle in all modules. The angle of 20° was chosen for this purpose; it corresponds to pseudorapidity $\eta = 0.35$.

Electrons at the incidence angle of 20° were shot into all cells of the first radial compartment, entering the cell-center. The mean response measured over all such cells in all tested modules determines the EM scale factor, see Fig. 9. The uniformity of the response is fully explained by the tile-to-tile response variation (see also Fig. 3) as has been demonstrated by MC simulations.

3.3.2. Response to pions The important parameters of a hadronic calorimeter are the response linearity and the associated energy resolution to incoming hadrons.

The pion response[2] is shown in Fig. 10 relative to the beam energy of incident pions, after all corrections mentioned in Section 3.2 have been applied. The results demonstrate that Tilecal is a non-compensated calorimeter, the pion response is well described by the parameterization proposed by Fabjan [6] with e/h = 1.36 [7]. MC results (Geant4.8.3 using QGSP and Bertini-cascades models) match well the experimental data.

The energy resolution for pions is shown in Fig. 11. The data were fitted by the standard energy resolution formula

$$\frac{\sigma}{E} = \frac{A}{\sqrt{E_{\mathrm{beam}}}} \oplus C \tag{2}$$

The fit results in $A = (52.7 \pm 0.9)\,\%$, $C = (5.7 \pm 0.2)\,\%$. The results are in reasonable agreement with MC simulations (same version as for linearity studies).

[2] Defined as Gaussian mean fitted on the signal spectrum in the range from -2σ to $+2\sigma$ around this Gaussian mean value.

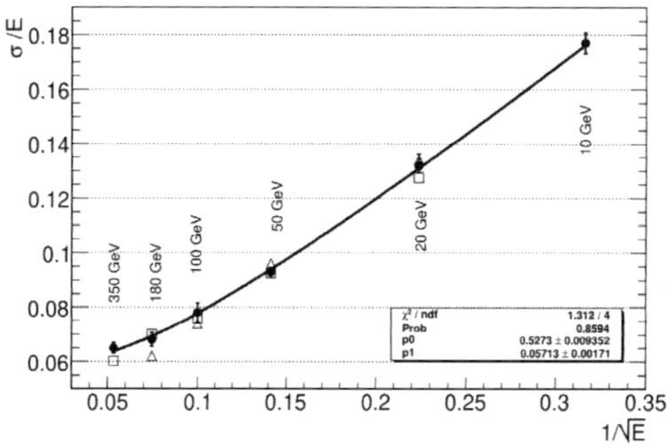

Figure 11. The energy resolution for pions incident at $\eta = 0.35$ (full circles), superimposed are MC results (open squares). Also shown are the earlier results from prototype modules (open triangle), that are rescaled to the length of the final calorimeter modules.

Earlier prototype data results[3] have been rescaled to the final Tilecal module length (see Section 3.2.2). A fair agreement is observed between the prototype and final Tilecal modules, as can also be appreciated from Fig. 11.

4. Conclusions

A big effort has been invested into understanding many details of the Tile calorimeter calibration and its performance for particles.

The calibration systems have been investigated in great details. The procedure to reconstruct the Cs-signal and to determine the gain of a cell has been improved as well as our understanding of the pulsed electronics calibration.

The EM scale was measured with electrons and the charge-to-energy conversion factor has been determined. The cell-to-cell uniformity as observed with electrons at 20° incidence has been explained by the variation in individual tile responses.

Muons provide a tool to check the EM scale in all cells of the calorimeter. The analysis of muon response resulted in additional particle/Cs correction that restores the EM scale in second and third radial compartment.

The response to pions has been carefully studied. The pion linearity curve, after correcting for all known effects, scales logarithmically with beam energy as expected. The results with final modules are compatible with recent MC simulations and also similar to results from earlier prototype modules. The energy resolution for pions corresponds well to that of prototype modules. It is also well reproduced by MC simulations.

The experience at the beam tests should allow to calibrate all Tilecal channels at the EM scale with a precision of $1-2\%$.

References

[1] ATLAS collaboration 1999 ATLAS Detector and Physics Performance Technical Design Report **1,2**, CERN/LHCC/99-14

[2] ATLAS/Tile Calorimeter collaboration 1996 Tile Calorimeter Technical Design Report, CERN/LHCC/96-42

[3] Hakobyan H Simonyan M Carli T and Henriques-Correia A M 2007 Measurement of Pion and Proton Longitudinal Shower Profiles up to 20 Nuclear Interaction Lengths with the ATLAS Tile Calorimeter, ATL-TILECAL-PUB-2007-008, CERN

[4] Anderson K et al 2005 *Nucl. Instr. and Meth. A* **551** p 469

[5] Anderson K Hurwitz M H Jen-La Plante I Pilcher J E and Teuscher R 2008 Performance and Calibration of the TileCal Fast Readout Using the Charge Injection System, ATL-TILECAL-INT-2008-002, CERN

[6] Fabjan C W Ludlam T 1982 *Ann. Rev. Nucl. Part. Sci.* **32** p 335

[7] Akhmadaliev S et al 2000 *Nucl. Instr. and Meth. A* **449** p 461

[3] It should be noted that the energy resolution parameters A, C quoted for prototype modules in Ref. [2] are better due to longer calorimeter. Moreover, the linear sum was used in the fit instead of the quadrature sum in formula (2).

Performance of the ATLAS forward calorimeter

Louise Heelan[1]

Department of Physics, Carleton University, Ottawa, ON K1S 5B6, Canada

E-mail: lheelan@physics.carleton.ca

Abstract. The ATLAS forward calorimeter is a liquid argon calorimeter based on a novel tubular electrode structure. This paper will focus on the performance of the calorimeter for electrons and hadrons, based on the analysis of beam test data taken in the H6 beamline at CERN in 2003. The ATLAS requirements on the forward calorimeter performance will be reviewed, the calorimeter design and construction will be discussed, and the performance for electron and hadrons will be described.

1. The ATLAS Forward Calorimeter

The Large Hadron Collider (LHC), located at CERN, is the highest energy collider built to date. It will collide together protons at a center of mass energy of 14 TeV at four points around the 27 km LHC ring. ATLAS is a detector located at one of these four collision sites. ATLAS is made up of several subsystems, used to measure the products of the proton-proton interactions. The main subsystems of ATLAS include the liquid argon and tile calorimeters, muon spectrometers, inner detectors, forward luminosity detectors, toroid and solenoid magnets.

The ATLAS detector is designed to have hermetic calorimetric coverage to provide a good measurement of jet energy and missing transverse energy. The calorimeteric coverage spans $|\eta|$ of 0 to 4.9. At the highest $|\eta|$ range ($3.1 < |\eta| < 4.9$) sits the liquid argon Forward Calorimeter (FCal). It is housed within the endcap cryostat, between the hadronic endcap calorimeter and the beam pipe. The location of the FCal within ATLAS is shown as one of the lightly shaded regions around the beam pipe in Figure 1.

Each FCal endcap is composed of three modules: FCal1, the electromagnetic calorimeter with copper as its main absorbing material, is closest to the interaction point. This is followed by FCal2 and FCal3, the hadronic calorimeters, using tungsten as their primary absorbing material. Each module has a longitudinal depth of approximately 45 cm, an inner radius of approximately 8 cm and an outer radius of approximately 45 cm. Detailed information about the design and construction of the FCal can be found in [1].

The LHC is designed to have a luminosity of 10^{34} cm^{-2} s^{-1} and a bunch crossing frequency of 40 MHz; resulting in an outward going flux of high energy particles. The high radiation and high frequency of collisions puts design constraints on each subdetector. For the liquid argon subsystem this implies a requirement of radiation hard material, and fast signal collection in the active medium. The FCal electrodes are composed of an inner anode rod surrounded by a cathode tube. The rod and tube are held at a potential difference of 250 V for FCal1 (375 V for FCal2, and 500 V for FCal3), and are filled with the active liquid argon medium. The liquid

[1] On behalf of the ATLAS Liquid Argon group.

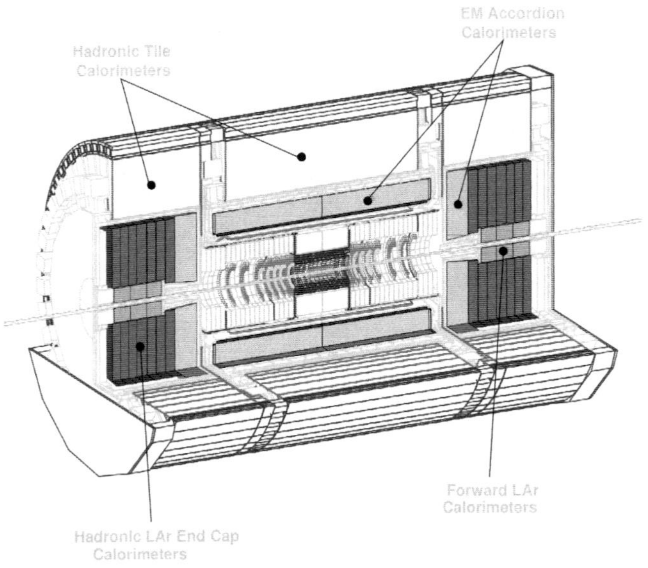

Figure 1. The ATLAS detector, with the four calorimeter technologies emphasized, including the forward calorimeter shown as one of the lightly shaded areas at the most forward regions.

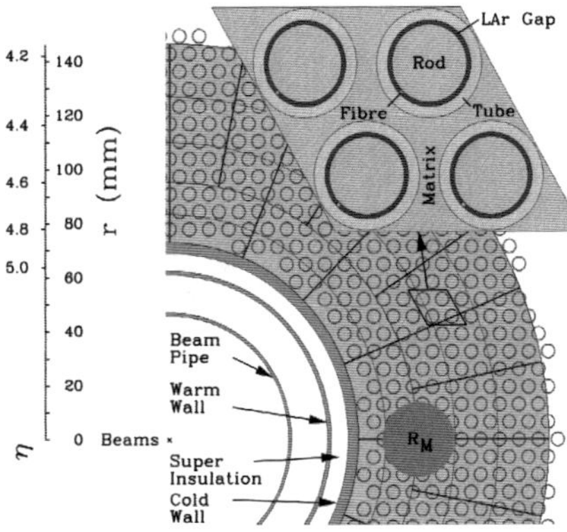

Figure 2. The front face of FCal1 as seen from the interaction point. The electrode rod, tube, and liquid argon gap are shown, along with electrode group structure. The Moliere radius (R_M) is superimposed on the electrodes and is larger than the electrode spacing.

argon gap is 269 μm, 376 μm, and 508 μm, in FCal1, FCal2, and FCal3, respectively. This small gap is necessary to provide fast signal collection, and reduce ion buildup that could alter the electric field across the gap.

The unique aspect of this calorimeter is that these electrodes run parallel to the beam pipe. Figure 2 shows the front face of FCal1 as viewed from the interaction point down the beam pipe. In this figure the rod and tube structure are clearly seen. In FCal1 four electrodes are grouped together to form an electrode group (in FCal2 and FCal3 this number is six and nine electrodes). Then typically four such electrode groups are summed together to form a readout channel; this is referred to as a cell in this paper.

2. Forward Calorimeter Test Beam Setup

A test beam was performed in 2003 with the CERN SPS H6 beamline to calibrate the FCal. The main goals of the test beam were to measure the intrinsic response of the calorimeter to electrons and hadrons, examine its performance under ATLAS-like conditions, and to understand the energy loss down the beam pipe and splashing across the beam hole. To achieve these goals, beams of electrons and hadrons with energies from 10 to 200 GeV were directed into one of the final FCal assemblies.

A schematic of the beamline is shown in Figure 3. The particles enter from the left, leaving the bend 9 magnet (B9). They traverse six beam profile chambers (BPCs), which provide tracking in the x and y directions. There are three scintillators (S1, S2, S3) that are used for coincidence triggering, and a veto wall, all to remove multiple scattered particles. The FCal sat in a cryostat filled with liquid argon. Downstream of the FCal were additional scintillators; a muon counter used to remove muons from the electron beam, and a tail catcher made of alternating steel and scintillator layers also used in the beam cleaning procedure.

A wide beam spot was used to ensure even sampling across the front face of the FCal. Five impact points on the front face of FCal1 were targeted; these are shown in Figure 4. For each

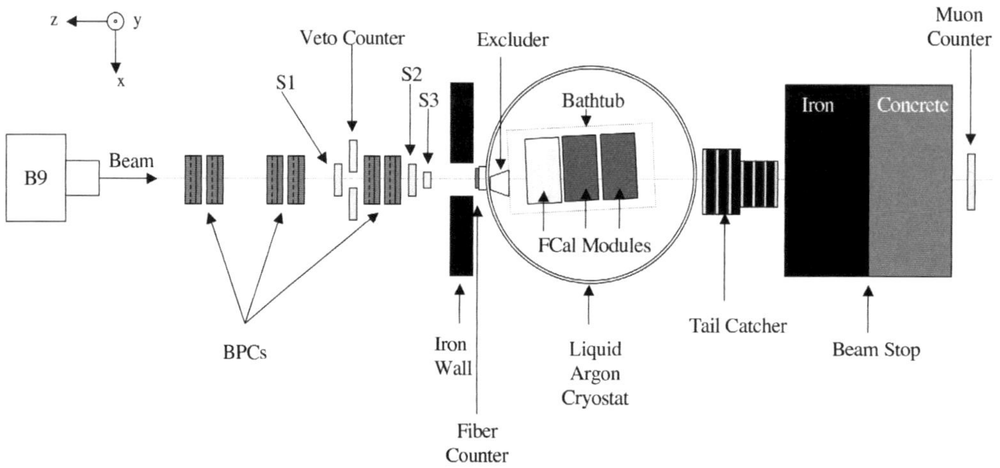

Figure 3. Schematic of the 2003 test beam setup used to calibrate the FCal.

impact point the FCal was rotated and translated such that the incoming particles hit the FCal as if they were coming from the ATLAS interaction point. The positions labelled 4H and 4L were chosen to fully contain particle showers. They differ in the amount of upstream material; position 4L had a minimal amount of upstream material to understand the intrinsic response of the FCal, whereas position 4H had additional material in front to simulate the ATLAS environment. The positions labelled 1, 2, and 3 were taken to study the detector response at the inner edge along the beam line.

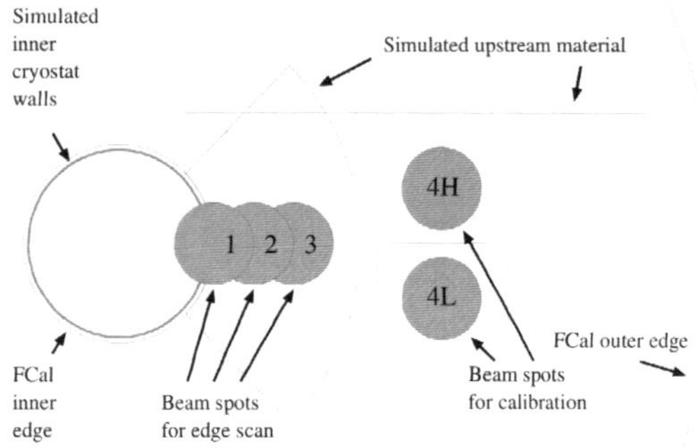

Figure 4. The five test beam impact points on the front face of the FCal. Positions 4L and 4H fully contain particle showers, with different amounts of upstream material. The positions closest to the beam pipe labelled 1, 2, and 3 are used for studies of energy loss at the inner edge.

3. Data Analysis of Impact Position 4L

This section describes the analysis of the electron and hadron data taken at the impact position labelled 4L. In this position there was minimal upstream material and particle showers were fully contained. This allows calculation of the intrinsic response of the FCal, which includes the energy resolution for electrons and hadrons, determination of the electromagnetic scale, and tests of different hadronic weighting schemes and clustering algorithms.

The pedestals and noise were calculated on a run-by-run and channel-by-channel basis using pedestal data taken during each run. The physics signal was reconstructed using the optimal filtering technique [2]. Seven samples were taken every 25 ns to reconstruct the pulse in each channel.

The electron and hadron data were clustered in the FCal using a cylindrical clustering algorithm. The cluster center was defined as the projection of reconstructed tracks using the BPCs. From the cluster center the energy of all cells within a fixed radius were summed. For the electron data analysis all cells within an 8 cm radius in FCal1 were used, and for the hadron data all cells within a 16 cm radius in FCal1, FCal2, and FCal3 were summed.

3.1. Electron Analysis of Impact Position 4L

Electron data were taken for eight beam energies between 10 and 200 GeV. A sample distribution for the 100 GeV data is shown in Figure 5. Similar distributions exist for the other seven energy points. The horizontal axis of Figure 5 is in units of ADC counts to set the electromagnetic scale. For each energy point the electron peak is best fit with a double Gaussian; it best describes the central peak and small high energy tail which is present due to the slight impact point dependence of the response. The primary background in the electron data sample was pion contamination. To properly account for the high energy tail of the pion distribution leaking under the low energy tail of the electron distribution, the pion background is modeled using pion data taken at the same energy, this modeling is shown on Figure 5 as the dashed line. The full fit to the double Gaussian and pion background is shown with the solid line. The mean and variance of this fit are used in the calculation of the linearity and energy resolution.

The electron linearity which is the mean reconstructed energy, in ADC counts, as a function of beam energy, is shown in Figure 6. The data are fit to a straight line, where the slope and intercept are free parameters. The best fit to the data yields a slope of 12.07 ± 0.7 (stat) ± 0.7 (sys) ADC/GeV, and an intercept of -12.27 ADC which is attributed to upstream energy losses, corresponding to a loss of approximately 1 GeV. Using the known properties of calorimetry one can calculate the electromagnetic scale for each module. This scale for each FCal module is predicted to be:

$$\alpha_1 = 12.0 \text{ ADC/GeV} \qquad \alpha_2 = 6.1 \text{ ADC/GeV} \qquad \alpha_2 = 5.4 \text{ ADC/GeV} \qquad (1)$$

The results presented in this paper agree with the prediction for FCal1, and a previous FCal test beam has confirmed the predicted electromagnetic scale of FCal2 relative to FCal1 [3].

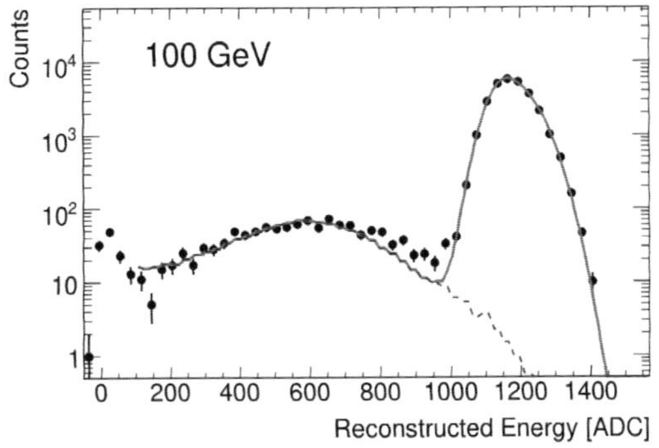

Figure 5. The double Gaussian fit and pion modeling to the 100 GeV electron data (plotted in ADCs). The electron peak is fit with a double Gaussian. The pion modeling is shown with dashed lines, and the full fit is shown with solid lines.

Studies were performed to quantify systematics arising from choice of binning, fitting technique, and cylinder cluster radial size. These sources were found to have a small impact on the overall linearity and energy resolution, contributing to an overall systematic of ±0.1-0.2%. The data is linear to within ±0.8% down to 10 GeV.

The intrinsic electron energy resolution can be found using the position 4L data. The electronic noise was not constant between runs due to a configuration problem, and so the noise is subtracted in quadrature energy by energy. This noise is calculated by clustering the same cells as those in the physics data, event-by-event but using the random data, and fitting the resulting energy distribution over all runs of a given energy point to a single Gaussian. On average the electronic noise clustered with the 8 cm cylinder cluster in FCal1 is 1 GeV. The noise-subtracted energy resolution is shown in Figure 6, where the data points at each energy are the solid points. These data points are fit to the energy resolution function:

$$\frac{\sigma_E}{E} = \sqrt{\left(\frac{a}{\sqrt{E}}\right)^2 + b^2} \tag{2}$$

where a is the stochastic term describing the statistical variations in the shower developement, and b is the constant term accounting for detector irregularities and miscalibrations, . The best fit values for these two parameters are a stochastic term $a = 28.5\%\cdot\mathrm{GeV}^{1/2}$ and a constant term of $b = 3.5\%$.

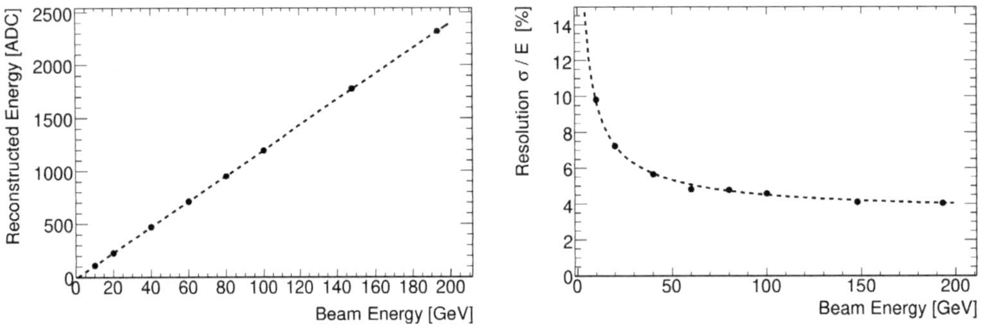

Figure 6. The left figure shows the electron linearity of the mean reconstructed energy versus beam energy. The data are the solid points and the dashed line shows the best fit line to the data, used to set the electromagnetic scale of FCal1. The figure on the right shows the electron noise-subtracted energy resolution at impact position 4L.

3.2. Hadron Analysis of Impact Position 4L

The FCal is a non-compensating calorimeter, as such it is necessary to apply a software correction to properly reconstruct the energy deposited by hadrons. This calibration has been performed with two techniques. The first is a longitudinal weighting scheme, in which there is a calibration constant for each module. These three constants are extracted by minimizing the energy resolution and requiring that the beam energy equals the mean reconstructed energy:

$$E = A_1\alpha_1 E_1 + A_2\alpha_2 E_2 + A_3\alpha_3 E_3 \tag{3}$$

where A_i (i = 1,2,3) are the longitudinal weights, referred to as flat weights, α_i are the electromagnetic scale factors as predicted from first principles (Equation 1), and E_i are the summed energies in FCal1, FCal2, and FCal3, respectively.

The second weighting scheme takes advantage of both the coarse longitudinal and fine transverse segmentation of the FCal, and is referred to as the radial weighting technique. The cluster center, as calculated using the projection of the BPC tracks, is taken as the starting point of this calibration. Going out with increasing radius from the cluster center the cells are split into radial slices. In this analysis 16 radial slices are taken at 1 cm each for each FCal module. The 16×3 radial weights are fit by minimizing the energy resolution at each point:

$$E = \sum_{i=1}^{3} \sum_{j=1}^{16} W_{ij} \times \sum_{k=1}^{n_{\text{cells}}} C_{ijk} \tag{4}$$

where i is the sum over the three FCal modules, j is the sum over the 16 radial slices, and k is the sum over all the cells in that radial slice. The W_{ij}'s are the 16 radial weights for each of the three FCal modules, and the C_{ijk} are the energy in the kth cell in the jth slice of the ith FCal module at the electromagnetic scale. The number of cells the jth radial slice of the ith FCal module is defined as n_{cells}.

The flat weights and radial weights extracted at the 200 GeV energy point are applied to all data to avoid energy dependent weights. Once applied one gets the energy resolution points shown in Figure 7. As in the electron analysis the noise has been subtracted in quadrature at each energy point. The average noise clustered in the 16 cm cylinder cluster using the flat weights is 5.5 GeV. These points are then fit to the noise subtracted energy resolution function, Equation 2. The fit results for the stochastic and constant terms for the two weighting techniques are shown in Table 1. The radial weights use more information from the calorimeter, and take advantage of the unique design of the FCal, hence provide a better performance than the flat weighting scheme.

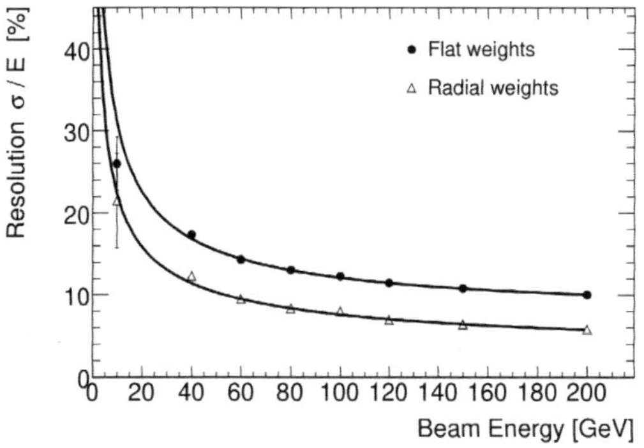

Figure 7. Hadronic energy resolution at impact position 4L. The 200 GeV flat weights and radial weights have been applied to the data. The noise-subtracted energy resolution function is shown as the solid line fit to the data of each weighting scheme.

Table 1. The stochastic and constant term fit results, as fit to the hadron data with the flat and radial weights applied. The noise has been subtracted in quadrature at each energy point.

Weighting Technique	Stochastic Term [%·GeV$^{1/2}$]	Constant Term [%]
Flat Weights	95.3 ± 0.6 (stat)	7.52 ± 0.06 (stat)
Radial Weights	70.0 ± 0.7 (stat)	2.97 ± 0.10 (stat)

The largest source of systematic uncertainty in the hadron analysis is the choice of hadronic weights applied to the data. The results presented here apply the flat and radial weights extracted with the 200 GeV hadron data, the highest energy data available in the test beam. In the flat weight analysis if instead the weights extracted at 100, 120, and 150 GeV are applied to the data one would find a variation of ±1.6%·GeV$^{1/2}$ in the stochastic term and ±0.4% in the constant term.

4. Ongoing Analyses

There are several ongoing analyses using the FCal test beam data. The results presented above show the response of the FCal at impact position 4L with a cylindrical clustering algorithm, and application of flat and radial hadronic weights to the hadron data. Analyses are underway to redo this analysis with alternative clustering algorithms that are used in ATLAS. Comparisons are being made to Monte Carlo simulations of the test beam with various physics models. There are four additional impact points that are currently under investigation (these were shown in Figure 4). A comparison of the 4H and 4L will show the effect of upstream material on the FCal performance. The inner edge analysis uses the data taken at impact positions 1, 2, and 3, to examine energy loss and splashing across the beam pipe.

Preliminary results of the test beam analysis using one of the default ATLAS clustering algorithms is presented below. This is followed by a first look at the energy loss in the vicinity of the beam pipe with data taken from impact positions 1, 2, and 3.

4.1. Topological Clustering

Topological clustering is one algorithm used by default in ATLAS to find significant sources of energy deposition in the liquid argon and tile calorimeters . The cylindrical clustering technique as described in this analysis cannot be used for the FCal in ATLAS because there is no tracking in front of the FCal necessary to set the cluster center. The topological clustering algorithm is based on the significance of a cell's energy divided by its noise, the significance is defined as σ. By default in the FCal, a cluster is seeded by a cell if the cell has a 4σ significance. A cluster is expanded upon by examining the significance of neighboring cells. In the default configuration, a neighbor cells must have a 2σ significance to expand the cluster. The last configurable option is the cell significance of the cells that are neighbors to an existing cluster. If the significance is greater than the set value (the default is 0σ), the cell is included in the cluster. The default topological clustering configuration is: seed/neighbor/cell = $4\sigma/2\sigma/0\sigma$

Using the default topological configuration and repeating the analysis as before on electrons in the 4L position, the best fit line yields an electromagnetic scale in FCal1 of 12.11 ADC/GeV, which agrees with the cylindrical clustering results of 12.07 $\pm$ 0.7 (stat) $\pm$ 0.7 (sys) ADC/GeV. The noise subtracted energy resolution has fits for the sampling and constant terms of 31.8%$\cdot$GeV$^{1/2}$ and 3.3%, respectively. These results are comparable to the results using the cylinder clustering, and confirm the use of the topological clustering algorithm for use in the FCal.

Studies are ongoing to optimize the topological clustering algorithm with both the electron and hadron data.

4.2. Inner Edge

The inner edge impact positions are labelled as 1, 2, and 3 in Figure 4. The primary goal of these data are to examine the energy loss down the beam line and energy splashing across the beam pipe. For these positions only 200 GeV electron and hadron data are available; in the future Monte Carlo simulations will be necessary to study this effect at a range of energies. Clustering is done with the topological clustering algorithm with the default seed/neighbor/cell configuration of $4\sigma/2\sigma/0\sigma$. The results presented below are shown at the electromagnetic scale.

The 200 GeV energy distribution for a given impact position is fit to a double Gaussian. The mean is calculated and used as the mean energy response of the FCal at that impact point. The average radial distance from the beam line is calculated using the BPC tracks for each impact point projected onto the front face of FCal1. Figure 8 (left) shows the mean energetic response of 200 GeV electrons as a function of the mean impact point for positions 1, 2, 3, and 4H (representing increasing radial distance from the beam line). As expected, closer to the beam line there is a decreased response. A similar figure for 200 GeV hadron data is shown in Figure 8 (right). The energy loss down the beam line is more severe for hadrons, as their showers tend to be larger than electron showers and more energy is lost down the beam line. Ongoing studies will examine how the energy is clustered on the opposing side of the beam pipe where it is known that energy is splashed.

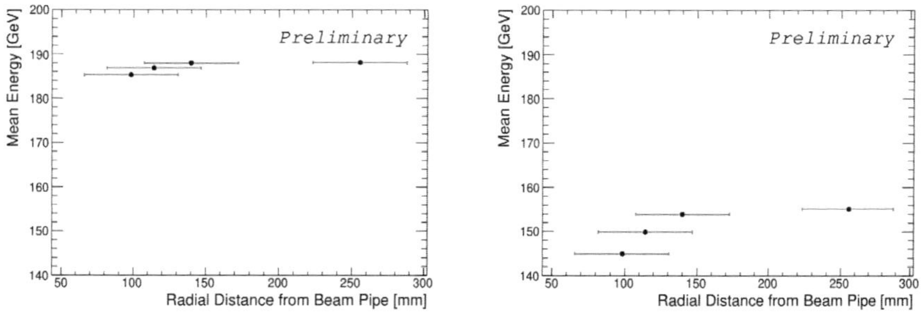

Figure 8. The mean reconstructed energy as a function of radial distance from the beam line for 200 GeV electrons (left) and hadrons (right). These four points represent data taken at impact positions 1, 2, 3, and 4H (with increasing radial distance from the beam pipe).

5. Summary

The performance of the ATLAS forward calorimeter has been studied using test beam data of electrons and hadrons at a range of energies from 10 to 200 GeV. Using a cylindrical clustering algorithm the intrinsic energy resolution of electrons is found to be $28.5\ \% \cdot \mathrm{GeV}^{1/2}/\sqrt{E} \oplus 3.48\%$. The electron data was used to confirm predictions of the electromagnetic scale for FCal1.

The test beam hadron data were used to investigage a hadronic weighting scheme in the FCal, making use of the unique geometrical arrangement of the electrodes. Using the flat weighting technique the hadron energy resolution is found to be $95.3\ \% \cdot \mathrm{GeV}^{1/2}/\sqrt{E} \oplus 7.5\%$. The energy resolution is improved by using the radial weights. Both weighting techniques result in an energy resolution that exceeds the ATLAS design requirements of $100\ \% \cdot \mathrm{GeV}^{1/2}/\sqrt{E} \oplus 10\%$.

Further information about this test beam setup and data analysis of the 4L position electron and hadron data can be found in [4]. Work is ongoing to investigate alternative clustering algorithms used in ATLAS that have the ability to suppress electronic and pileup noise, which will be a valuable tool for the FCal in its high flux environment. Additional work is being carried out with the test beam data and Monte Carlo simulations to study the effect of the presence of the beam line, and energy losses down the beam line. These analyses will be the subject of future publications.

Acknowledgments

This work has been supported by the ATLAS liquid argon collaboration. I would like to acknowledge those members of the forward calorimeter group. Also the CERN staff operating the SPS and the H6 beam line during the 2003 test beam.

References

[1] Artamonov A *et al.* 2008 *JINST* **3** P02010
[2] Cleland W E and Stern E G 1994 *Nucl. Inst. Meth.* **A 338** 467
[3] Armitage J C *et al.* 2007 *JINST* **2** P11001
[4] Archambault J P *et al.* 2008 *JINST* **3** P02002

Journal of Physics: Conference Series **160** (2009) 012059 doi:10.1088/1742-6596/160/1/012059

Performance of the combined zero degree calorimeter for CMS.

O A Grachov[1], M Murray, J Snyder, J Wood, V Zhukova
Departments of Physics and Astronomy, University of Kansas, Lawrence, KS, USA

A S Ayan, P Debbins, D F Ingram, E Norbeck, Y Onel
Department of Physics and Astronomy, University of Iowa, Iowa City, IA, USA

E Garcia
Department of Physics and Astronomy, University of Illinois at Chicago, Chicago, IL, USA

G Stephans
Laboratory for Nuclear Science, Massachusetts Institute of Technology, Cambridge, MA, USA

For CMS Collaboration

E-mail: grachov@ku.edu

Abstract. The combined zero degree calorimeter (ZDC) is a combination of sampling quartz/tungsten electromagnetic and hadronic calorimeters. Two identical combined calorimeters are located in the LHC tunnel at CERN at the straight section ~140 m on each side of the CMS interaction vertex and between the two beam pipes. They will detect very forward $|\eta| \geq 8.5$ photons and neutrons. ZDC information can be used for a variety of physics measurements as well as improving the collision centrality determination in heavy-ion collisions. Results are presented for ZDC performance studies with the CERN SPS H2 test beam.

1. Introduction

Two identical zero degree calorimeters (ZDCs) [1], will measure very forward neutrons and photons in heavy-ion and early p-p collisions and will cover pseudo rapidity region from 8.5 (a horizontal angular width of 580 μrad). The ZDC are located 140 m on each side of the CMS interaction vertex (IP5). They are between the two beam pipes inside the neutral particle absorber (TAN) downstream of the first beam dipole magnet at the straight section.

The number of spectator neutrons can be used for beam tuning and luminosity monitoring, and for determination of the impact parameter or centrality of interaction in heavy-ion collisions. With an additional shower maximum detector [2], the ZDC will measure of the reaction plane of A-A and p-A

[1] To whom any correspondence should be addressed.

collisions. The detector is useful for study of diffractive processes in p-p collisions [3] and also, as a neutron tagging detector for selection of γ-γ and γ-A collisions in ultra-peripheral heavy ion collisions [4]. The location of one ZDC calorimeter in LHC sector 5-6 is presented on figure 1.

Figure 1. Photo presents a part of LHC sector 5-6 at the location of one ZDC detector. The detector is located in a narrow space inside the neutral particle absorber (TAN) at 140 m from IP5.

2. Requirements

The calorimeter design was dictated by physics goals and by conditions at their location in the LHC tunnel. Specifically:

- The calorimeter should allow reconstruction of the energy of 2.75 TeV spectator neutrons and 50 GeV photons with a resolution of 10-15%;
- Radiation hardness consideration.

During the early p-p runs and design luminosity Pb-Pb runs, the expected average absorbed radiation doses will be approximately 180 MGy and 300 kGy, respectively, per data-taking year;

- The calorimeter should utilize fast detection technology.

Proton bunch crossing every 25 ns (40 MHz)..
The expected interaction rate for minimum bias event in Pb-Pb collisions is ~8000 event/sec ;

- Very compact device.

The calorimeter must fit in a space with dimensions: 96mm (W) x 607mm (H) x ~700mm (L)

3. Calorimeter design

A compact sampling quartz fiber calorimeter with W (tungsten)-radiators [1] meets all requirements. The design of each individual ZDC includes two independent sampling calorimeter sections: an electromagnetic section (EM section) and a hadronic section (HAD section). Each section consists of a tungsten plate/ quartz fiber ribbon stack.

The HAD section is longitudinally segmented into 4 readout segments (towers) of ~1.4 nuclear interaction length each. The electromagnetic section (19 radiation length or ~ 1 nuclear interaction length) segmented into 5 horizontal individual readout towers with transverse size of 15.56 mm. The ZDC EM and HAD section use a 90° or 45° – orientation geometry respectively (see figure 2 and figure 3).

Figure 2. 90° - orientation geometry for electromagnetic section .

Figure 3. 45° - orientation geometry for hadronic section.

Both sections have transverse dimensions of 79.3 mm and height of 100 mm. This small aperture is sufficient to capture all fragmentation neutrons in peripheral collisions. The layout and numbering scheme of ZDC detectors is presented on figure 4. The coordinate definitions are the same as the CMS coordinate system.

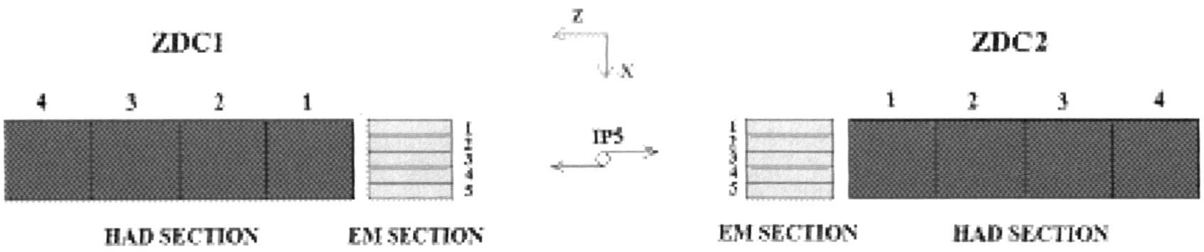

Figure 4. Layout and numbering scheme of ZDC detectors.

4. Experimental setup

The main goals of the test beam measurements were to study the energy resolution and linearity of the zero degree calorimeter. It also tested the electronic chain to be used with the CMS. Measurements were carried out in the SPS H2 beam at CERN in August 2006 and May 2007.

Journal of Physics: Conference Series **160** (2009) 012059 doi:10.1088/1742-6596/160/1/012059

A 400 GeV/c proton beam was extracted from the SPS accelerator and steered onto the primary target. Downstream from the primary target, a secondary beam was made with momenta from 10 to 350 GeV/c. The resulting beam then passed through a gas threshold Cherenkov counter, multi-wire proportional chambers and scintillating counters (see figure 5).
The trigger system required a coincidence between the scintillating counters and the appropriate signal in the Cherenkov counter. The normal beam spot exposed an area on the calorimeter of (2 x 2) cm^2.

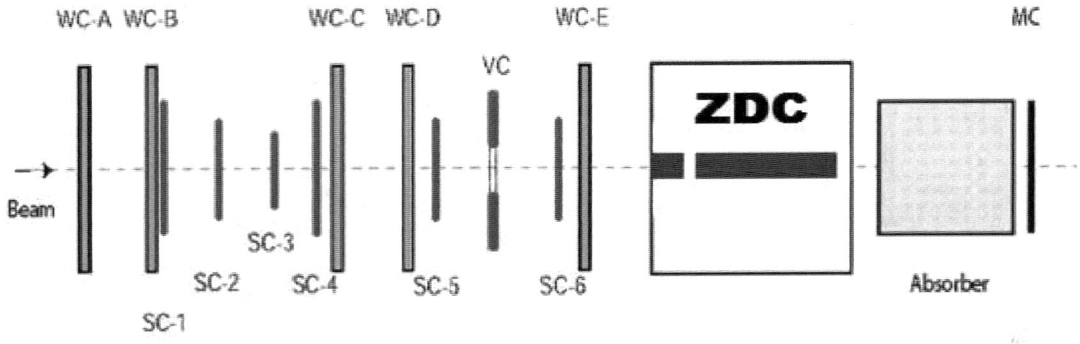

Figure 5. Test beam experimental setup.

The calorimeter was mounted on a table that could be moved in horizontal and vertical directions under remote control. This made it possible to position the beam anywhere on the front face of the calorimeter.

5. Read-out electronics

There are a total of 18 readout channels for the two ZDCs. Photomultiplier tubes (PMT) are used as photo detectors. The PMT is an 8 - stage Hamamatsu R7525 phototube with bi - alkali photocathode, with average quantum efficiency for Cherenkov light of approximately 10%.
The resistive chain high voltage base (ratio "B") is optimized to achieve a high gain and the maximum dynamic range of linearity.
For future work, the high-voltage for the PMTs will be provided by commercial CAEN NIM units. Five power supplies are mounted in a rack in NIM crate that resides in the counting room. These units will be controlled and monitored via HCAL slow controls.
The PMT's HV base has two connectors: a coaxial signal output connector and a SHV high voltage connector.
The PMT and base are housed in a shielding enclosure, so called PMT read-out box. A photograph of the EM section PMT read-out box is shown on figure 6.

Figure 6. PMT read-out box of the EM section

For data collection with CMS, signals from the PMT will be transmitted through a long (204 m, type C–50–11–1) coaxial cable to the patch panel mounted on the 9U VME crate located in the underground counting room (Electronic Rack S2E13). The electronic scheme of PMT signal transmission to patch panel and QIE is presented on figure 7. The patch panel splits each input signal into two output signals: S1, S2 with ratio S1 / S2 = 10/1. S1 provides the coaxial signal output connectors for the distribution of these signals to front - end modules contained QIE (Charge Integrator and Encoder) ADC chips for digitization. S2 is for monitoring and tuning of the interaction of beams and defining the real - time luminosity.
Trigger and front-end electronics run at 40 MHz, the maximum clock speed that will be used at the LHC.

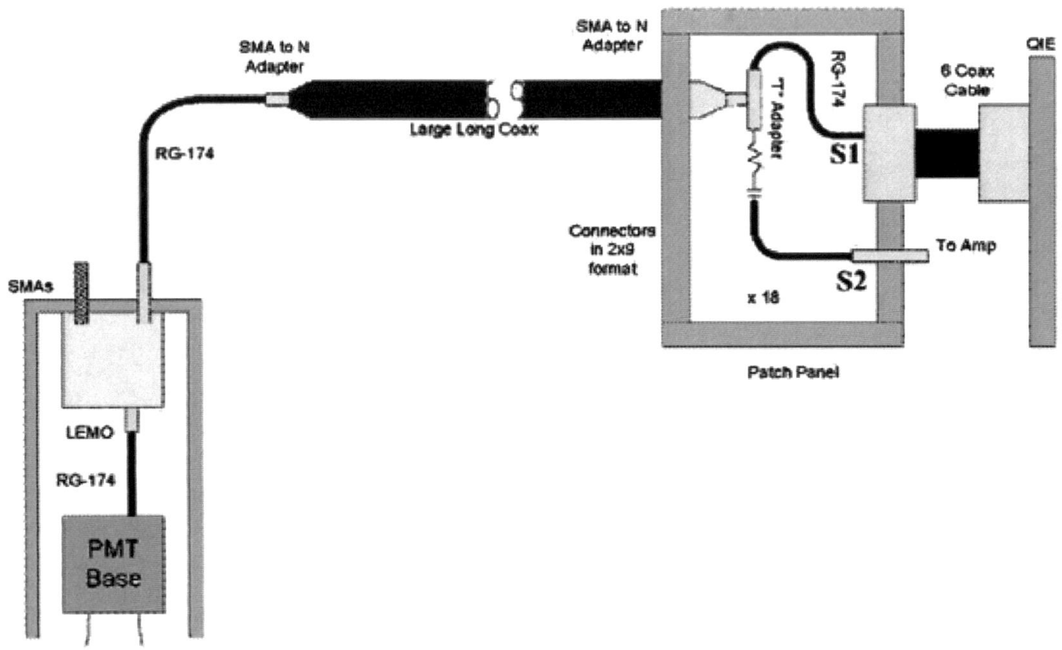

Figure 7. Electronic scheme of PMT signal transmission to patch panel and QIE

6. Response to positrons

The characteristics of the EM section were measured with positrons of seven different test beam energies (from 10 GeV to 150 GeV). To equalize response among towers the geometrical center of each tower (T1, T2, T3, T4 and T5) was irradiated by 50 GeV positron beams. The peak position obtained from a Gaussian fit of the amplitude distribution for each tower was used to determine the calibration coefficients. Energy deposition in the center tower T3 versus energy deposition in the neighboring towers T2 and T4 (figure 8 and figure 9) show approximately the same response, which shows that the towers were calibrated with a precision better than 10%.
The energy resolution for the different positron energies was obtained by Gaussian fits and was parameterized as $(\sigma/E)^2 = 0.49/E + 0.0064$. The GEANT4 simulations [3] reproduced well the EM section test beam data (figure 10). The linearity of the EM section response was defined as the ratio of the fitted mean energy to the beam energy. The calorimeter was found to be linear over a range from 10 GeV to 150 GeV to within 2%-3% (figure 11).

The analysis did not included corrections for dead material, leakage and the non-compensation nature of the calorimeter.

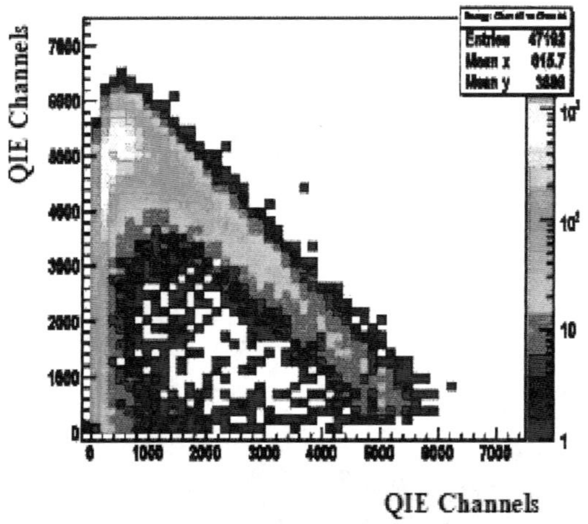

Figure 8 Energy deposition in tower T2 versus energy deposition in tower T3

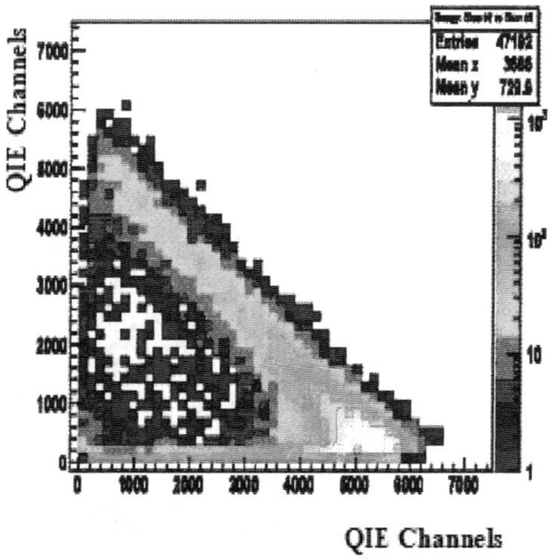

Figure 9 Energy deposition in tower T3 versus energy deposition in tower T4

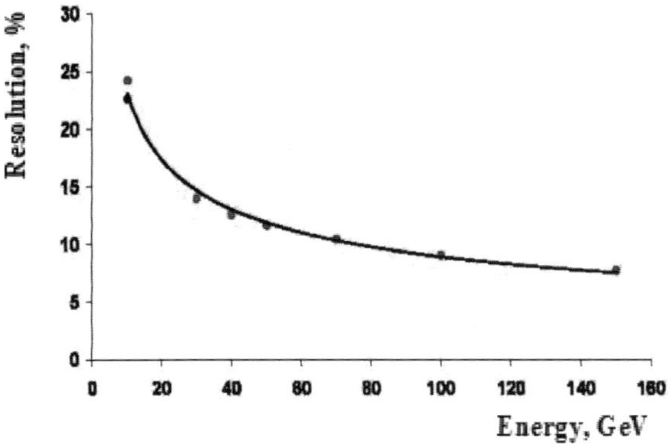

Figure 10. Positron energy resolution of EM1 and EM2 electromagnetic sections.

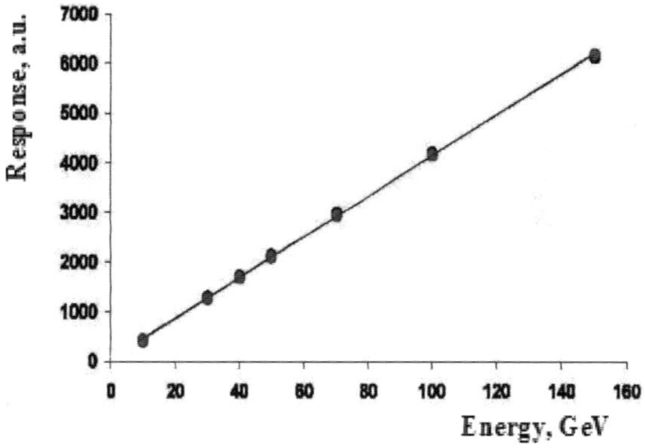

Figure 11 The linearity response to positrons of EM1 and EM2 electromagnetic sections.

7. Response to hadrons

Positive pions with energies of 150, 200, 300 and 350 GeV were used to measure the response of the combined zero degree calorimeter (EM + HAD system). The total depth of the combined system is ~7 hadronic interaction lengths (λ). The total energy was defined as the sum of the energy depositions in the EM and HAD sections. The energy dependent intercalibration parameter between the EM and HAD section was determined by minimizing the energy resolution for 350 GeV pions. Figure 12 presents energy deposition in the HAD1 section versus of energy deposition in the EM1 section for 350 GeV pions. The energy resolution was obtained by a Landau fit to be 21.5 % for 300 GeV pions and was parameterized as $\sigma/E = 138\%/\sqrt{E} + 13\%$ (see figure 13). An extrapolation to energy 2.75 TeV predicts a resolution of approximately 15% (GEANT4 - about 12% [5]). Measurements show good linearity of detector response to hadrons in the range of 100 to 350 GeV .

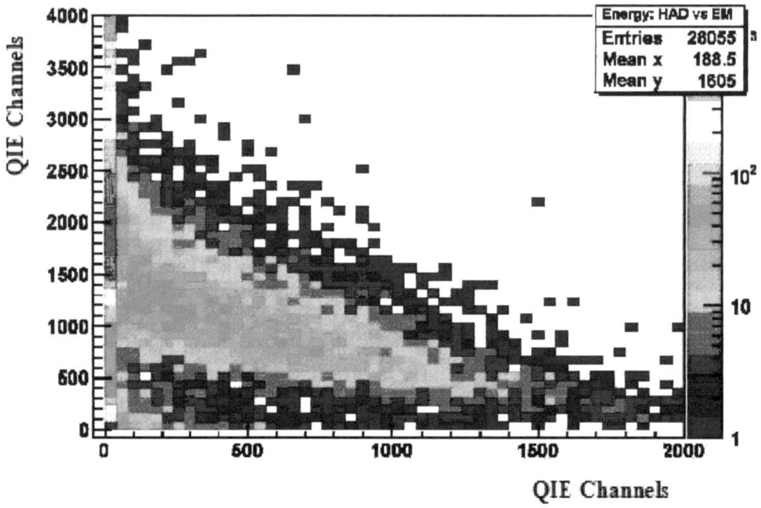

Figure 12. Energy deposition in HAD section versus energy deposition in EM section of ZDC1 for 350 GeV positive pions.

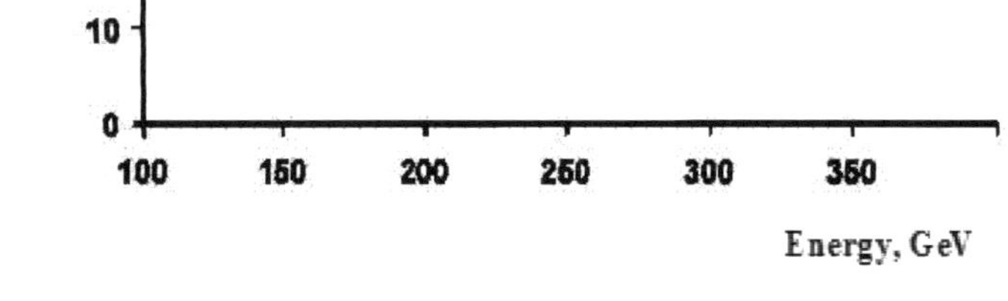

Figure 13. Positive pion energy resolution of the combined zero degree calorimeter.

8. Current status and summary

The installation of the ZDCs in the LHC sector 4-5 and sector 5-6 was finished in May 2008. A photograph of the ZDC installed in sector 4-5 is shown in figure 14. Both detectors were connected to read-out electronics and tested. Commissioning of detectors as a part of the CMS experiment and as a part of the LHC beam monitoring system has begun.

Figure 14. ZDC installed in sector 4-5 of the LHC.

8.1 Summary

- We have designed, constructed and tested the combined zero degree calorimeter;
- GEANT4 simulation reproduced well the test beam data;
- The calorimeter fulfils the technical, geometrical and physics requirements;
- The ZDC has sufficient energy resolution and linearity to meet our physics goals.

Acknowledgments

The detector studies were possible only because of the help of the CMS Collaboration and the SPS crew at CERN. It is our pleasure to acknowledge their important contribution.

References

[1] Grachov O A, Murray M J, Ayan A S, Debbins P, Norbeck E, Onel Y, d'Enterria D G 2006 *AIP Conf.Proc.867:258-265; nucl-ex/0608052*

[2] Grachov O A, Metzler B, Murray M, Snyder J, Stiles L, Wood J, Zhukova V, Beaumont W, Ochesanu S, Debbins P, Ingram D, Norbeck E, Onel Y, Garcia E 2007 *Int.J.Mod.Phys.E16:2137-2142; nucl-ex/0703001*

[3] Albrow M, et al Prospects for diffractive and forward physics at the LHC 2006 Preprint CERN-LHCC-2006-039, CERN-LHCC-G-124, CERN-CMS-NOTE-2007-002

[4] d'Enterria David G , et al 2007 *J.Phys.G34:2307-2455*

[5] Ayan A S, Debbins P, Norbeck E, Onel Y, Schmidt J, Grachov O A, Murray M 2006 GEANT4 Monte Carlo Simulation Studies for the Zero Degree Calorimeter for the CMS Preprint CMS IN-2006/028

Journal of Physics: Conference Series **160** (2009) 012060 doi:10.1088/1742-6596/160/1/012060

Commissioning and calibration of the Zero Degree Calorimeters for the ALICE experiment

N. De Marco[1], R. Arnaldi[1], E. Chiavassa[2], C. Cicalò[3], P. Cortese[4], A. De Falco[5], G. Dellacasa[4], A. Ferretti[2], M. Floris[5], M. Gagliardi[2], M. Gallio[2], R. Gemme[4], G. Luparello[2], A. Masoni[3], P. Mereu[1], A. Musso[1], C. Oppedisano[1], A. Piccotti[1], G. Puddu[5], E. Scomparin[1], S. Serci[5], E. Siddi[5], D. Stocco[2], G. Usai[5] and E. Vercellin[2] on behalf of the ALICE Collaboration

E-mail: demarco@to.infn.it

[1] INFN, Torino, Italy

[2] Università di Torino and INFN, Torino, Italy

[3] INFN, Cagliari, Italy

[4] Università del Piemonte Orientale and INFN, Alessandria, Italy

[5] Università di Cagliari and INFN, Cagliari, Italy

Abstract. The Zero Degree Calorimeters (ZDCs) for the ALICE experiment will measure the energy of the spectator nucleons in heavy ion collisions at the CERN LHC. Since all the spectator nucleons have the same energy, the calorimeter response is proportional to their number, providing a direct information on the centrality of the collision. Two sets of ZDCs are located at opposite sides with respect to the interaction point (IP), 116 m away from it. Each set consists of a neutron (ZN) calorimeter, placed between the two beam pipes, and a proton (ZP) calorimeter, positioned externally to the outgoing beam pipe. The ZDCs are spaghetti calorimeters, which detect the Cherenkov light produced by the shower particles in silica optical fibers embedded in a dense absorber. In summer 2007 the ZN and ZP calorimeters have been placed on a movable platform and then installed in the LHC tunnel. The results of the commissioning studies and in particular the solutions adopted to control the stability of the PMTs response will be shown: light injection with a laser diode and cosmic rays. The foreseen calibration with e.m. dissociation events in Pb-Pb collisions will also be discussed. Finally the first measurements carried out during the commissioning in the LHC tunnel will be presented.

1. Introduction

The ALICE experiment [1] at the LHC collider will study ultrarelativistic heavy ion collisions, where the formation of the Quark Gluon Plasma (QGP), a state of matter where quarks and gluons are deconfined, is expected. The Zero Degree Calorimeters (ZDC) will measure the energy carried by the non-interacting nucleons ("spectators") in order to select central nucleus-nucleus collisions, where the conditions for QGP formation are most favourable. Since the spectator nucleons will have approximately the same energy as the beam nucleons (2.7 TeV), the ALICE ZDCs will be used as multiplicity detectors. Two identical sets of calorimeters are located at opposite sides with respect to the beam interaction point (IP), 116 m away from it. Each set of detectors consists of a neutron (ZN) and a proton (ZP) Calorimeter. The ZN is placed at zero degrees with respect to the LHC axis, between the two beam pipes, while the ZP is positioned externally to the outgoing beam pipe. The spectator protons will be separated from the ion beams by means of the separator magnet of the LHC beam optics.

Table 1. *Synopsis of ZDCs parameters.*

Detector	ZN	ZP
Dimensions	$7.2 \cdot 7.2 \cdot 100$ cm^3	$22.8 \cdot 12 \cdot 150$ cm^3
Filling ratio	1/22	1/65
Absorber	W-alloy	brass
Density	17.6 g/cm^3	8.48 g/cm^3
Number of slabs	44	30
Slab thickness	1.6 mm	4 mm
Number of fibers	1936	1680
Fiber spacing	1.6 mm	4 mm
Fiber diameter (silica core)	365 μm	550 μm
Numerical aperture	0.22	0.22
Number of PMT	5	5
Type of PMT	Hamamatsu R329-02	Hamamatsu R329-02

2. ZDC description

The ZDCs are quartz-fiber spaghetti calorimeters [2, 3], with silica optical fibers as active material embedded in a dense absorber. Their principle of operation is based on the detection of Cherenkov light produced by the charged particles of the shower in the fibers. The response is very fast thanks to the intrinsic speed of the emission process, making such detectors suitable for triggering purposes. Moreover, the radiation tolerance of the quartz fibers ensures a high radiation resistance; in fact the ZDCs will operate in a highly radioactive environment, where the estimated dose is of the order of $\simeq 10^4$ Gy/day at a Pb–Pb luminosity of 10^{27} cm^{-2}s^{-1}. Finally, thanks to the dense passive material and to the detection of the Cherenkov light, a reduced transverse size of the detectable shower is achieved, allowing the construction of extremely compact devices. The technical specifications of the ZDCs are summarized in table 1. The

Figure 1. Photo of one set of ZDC calorimeters on the movable platform on the surface.

Figure 2. Photo of one set of ZDC calorimeters on the movable platform in the tunnel.

quartz fibers, hosted in the slab grooves, are placed at 0° with respect to the incident particle direction and come out from the rear face of the calorimeter, directly bringing the light to the photomultipliers. One out of two fibers is viewed by an individual photomultiplier (PMTc), while the remaining fibers are collected in bundles and sent to four different photomultipliers forming four independent towers [4]. The photomultiplier chosen is the Hamamatsu R329-02, with quantum efficiency of the order of 25% at a wavelength of 400 nm.

3. ZDC monitoring system and calibration

The ZN and ZP calorimeters have been tested in the H6 beam line of CERN SPS with hadron and electron/positron beams of various momenta (50–180 GeV/c). During the beam tests we inter-calibrated the individual PMTs and we verified the uniformity of the response as a function of the impact point of the particles on the front face of the calorimeters. For hadron beams we measured the light yield to be 0.9 and 1.1 photoelectrons per GeV and the energy resolution, $\sigma(E)/E = (257\pm3)\%\ GeV^{0.5}/\sqrt{E} \oplus (10.3\pm0.6)\%$ and $\sigma(E)/E = (237\pm2)\%\ GeV^{0.5}/\sqrt{E} \oplus (12.5\pm0.2)\%$, for ZN and ZP respectively. One of the two ZN has also been tested in the H8 beam line with an ^{115}In beam of 158 GeV/c per nucleon, in order to check the linearity of the response as a function of the number of the incoming nucleons. The performance of the detectors are summarized in [5, 6, 7, 8].

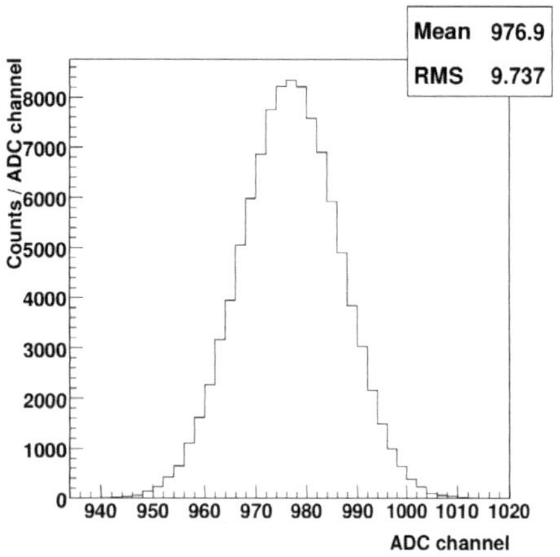

Figure 3. Typical PMT response to laser light.

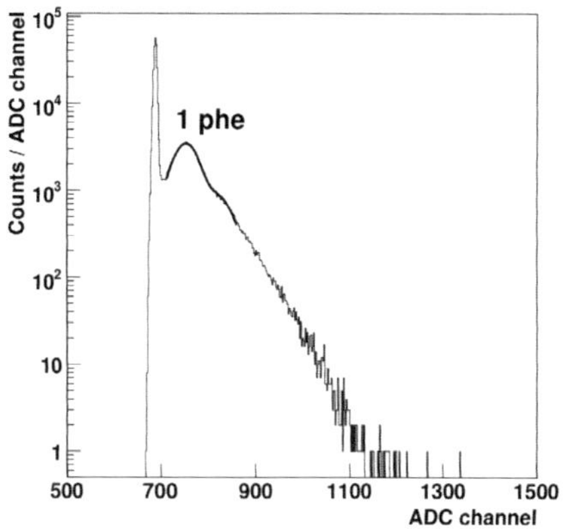

Figure 4. Single photoelectron signal obtained with a laser trigger.

3.1. ZDC monitoring system

The beam test phase is now finished. In summer 2007 two identical ZDC systems, each one composed of a ZN and a ZP calorimeters, have been assembled on the surface on a movable platform (Fig. 1) and then installed in the LHC tunnel (Fig. 2), at opposite sides with respect to the IP. The movable support structure will allow for lowering the detectors during injection to protect them from possible beam losses and to minimize the absorbed dose. The movement is controlled remotely from the ALICE DCS. Each ZDC system is equipped with a picosecond light pulser PLP-10 (405 nm, 70 ps) for the monitoring of the stability of the gain of the PMTs and the radiation damage of the fibers. The pulsed light source utilizes a laser diode head, which is positioned in a sealed box below the calorimeters. A few fibers in each calorimeter, longer than the normal ones, come out from the front face of the calorimeter, coupling the light source to the photomultipliers. A reference PMT, located in the sealed box, monitors the laser light stability. The laser is remotely controlled and a laser trigger will be used in order to monitor the PMT response variations with an accuracy at the level of 1%, as shown in Fig. 3. The PMT response depends on the PMT gain and on the light trasmission through the fibers, which both vary due to radiation damage.

In order to disentangle the PMT and fiber radiation damage two different systems are foreseen, which measure the absolute gain of the PMTs by means of the single photoelectron signal (Fig. 4). A cosmic ray trigger and a laser trigger, with an appropriate filter in front of the laser diode head, have been implemented. In fact the energy loss of cosmic rays (muons) in the calorimeters is negligible, producing in a small fraction of events only a few photoelectrons. For this purpose two scintillators are positioned respectively above and below each calorimeter. The cosmic ray trigger requires a coincidence between the signals coming from the two scintillators.

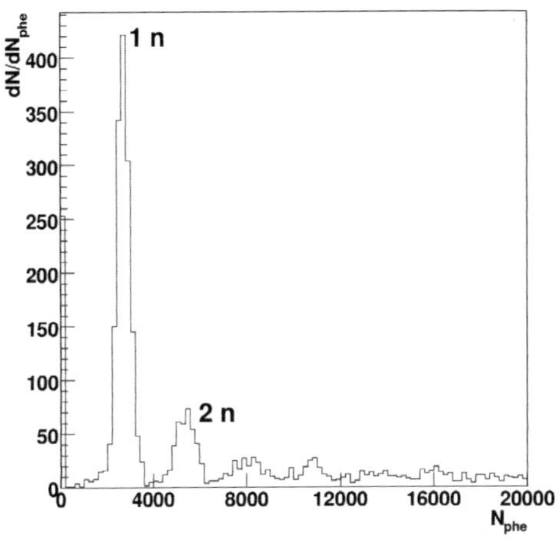

Figure 5. In-situ physical calibration. The expected light distribution in ZN in Pb-Pb e.m. dissociation events is shown. A clean separation of 1n-2n contribution is visible.

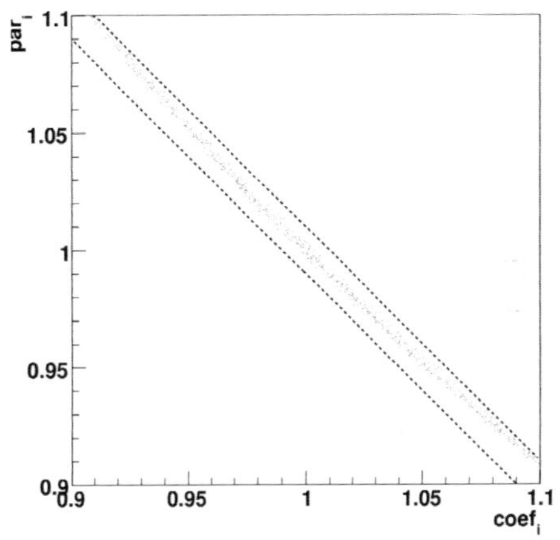

Figure 6. In situ tower inter-calibration. Parameters par_i obtained from the minimization vs input parameter $coef_i$ are shown. The dotted lines show the limits of ±1%.

3.2. ZDC in-situ physical calibration

During operation with ion beams it is foreseen that the ZN calorimeters will provide an estimate of the LHC luminosity, by measuring the rate of mutual electromagnetic dissociation of the colliding nuclei in the neutron channel. A prediction for neutron emission in Pb-Pb e.m. dissociation events at 2.76 ATeV is shown in Fig. 5. All emitted neutrons fall in the neutron calorimeter acceptance. Thanks to the good energy resolution (∼ 11% at 2.76 TeV), a clean separation of the single neutron contribution is achieved providing both physical calibration and stability monitoring.

3.3. ZDC in-situ tower inter-calibration

The information coming from the PMTc provides a complementary measurement of the shower energy, particularly useful for in-situ inter-calibration purposes. In order to verify this assumption 5000 Pb-Pb minimum bias events at 2.76 ATeV were generated with a fast simulation based on Hijing generator [9]. The light produced in each tower is multiplied by a factor $coef_i$ randomly chosen in the range from 0.9 to 1.1, in order to simulate a de-calibration of the four towers (10% at most). A minimization of the sum of the squares of the deviations of the light measured in PMTc (E_c) from the sum of the light in the four towers is performed with respect to a multiplicative parameter par_i:

$$\sum_{i=1}^{n}(E_c - \sum_{i=1}^{4} par_i coef_i E_i)^2 \tag{1}$$

where E_i is the light in the i-th tower PMT_i. The result of the minimization procedure, which is repeated 200 times, each time varying the four input factors $coef_i$, is shown in Fig. 6. The resulting values for par_i, which re-equalize the response of the four towers, are obtained with an accuracy of 1%, as shown by the dotted lines.

4. ZDC commissioning on the surface: gain measurements

During summer 2007 the gain of the PMTs of the four calorimeters have been checked at different supply voltages. The single photoelectron signal (Fig. 4), already measured in the laboratory, has been obtained triggering on cosmic rays and laser events. The position μ of the ADC peak corresponding to the single photoelectron signal is evaluated performing a fit to the sum of two gaussian functions:

$$f(x) = C_1 e^{\frac{-(x-\mu)^2}{2\sigma^2}} + C_2 e^{\frac{-(x-(ped+2(\mu-ped))^2}{2(2\sigma^2)}}. \tag{2}$$

where *ped* is the position of peak corresponding to the ADC pedestal, seen to the left of the spectrum. The two terms correspond to the detection of 1 and 2 photoelectrons respectively.

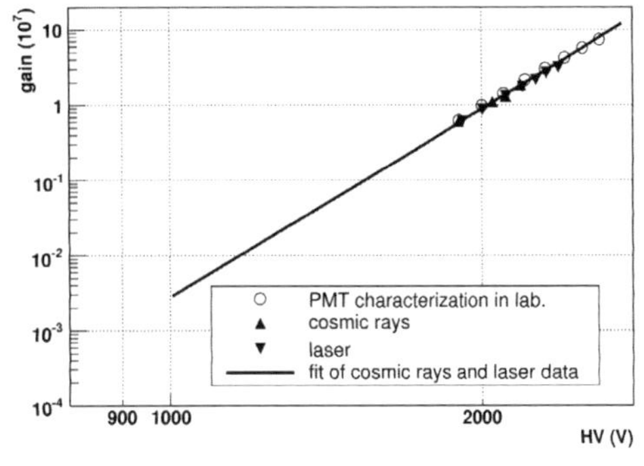

Figure 7. PMT gain as a function of the supply voltage; Cosmic rays (full triangle), laser (full triangle down) and previous PMT characterization (open circle) data are shown.

The gain is evaluated with the formula:

$$G = \frac{(\mu - ped) \cdot ADCgain}{At \cdot e}. \tag{3}$$

where At is the cable attenuation and e is the electron charge.

The gains, measured with the two methods, are plotted in Fig. 7 as a function of the PMT HV; the measurements have been fitted with a straight line on a log-log scale. A good agreement is achieved between this new set of measurements and the previous PMT characterization performed in laboratory.

In Pb-Pb collisions the PMTs will have to work at lower HV of the order of 1200–1300 V. Therefore in order to extend the gain measurements at low voltage we acquired laser events from HV $\approx$ 1200 V to HV $\approx$ 2300 V, using different filters, as shown in Fig. 8. Each time we changed the filter, we performed a junction measurement at the same HV, in order to be able to align the

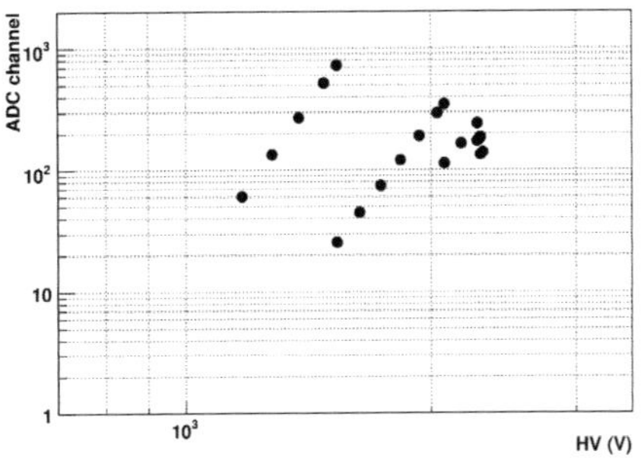

Figure 8. PMT response to laser light: ADC channel vs HV.

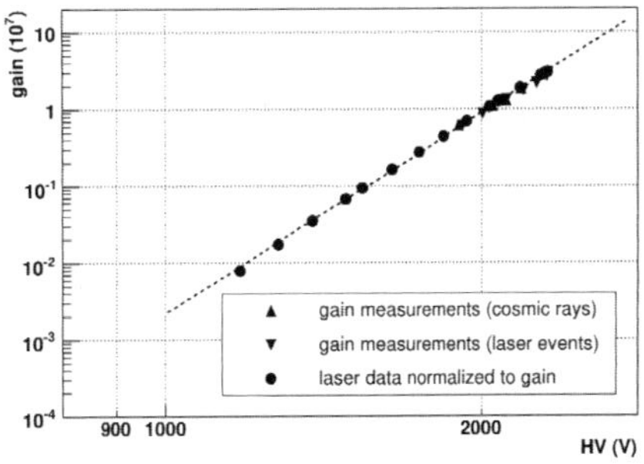

Figure 9. PMT response to laser light: gain vs HV.

points on a unique line on a log-log scale. In Fig. 9 the measurements performed with different filters have been rescaled and normalized to the gain previously measured at 2000 V. The PMT response to laser light, normalized to the gain at HV $\simeq$ 2000 V, is in reasonable agreement with the gain measurement extrapolation to low voltage.

5. ZDC commissioning in the tunnel: linearity measurements

The two ZDC systems, placed on their movable platform, are now installed in the LHC tunnel. Standalone runs in the framework of the general ALICE data acquisition system have been performed, flashing all the PMTs with the laser light. A preliminary check of the linearity of the entire system (PMT, bleeder and electronics in control room) has been done. The measurements have been carried out at two different laser light intensities in order to generate higher and lower PMT charge alternately, fixed at a ratio approximately 4:1. Different measurements of the ratio

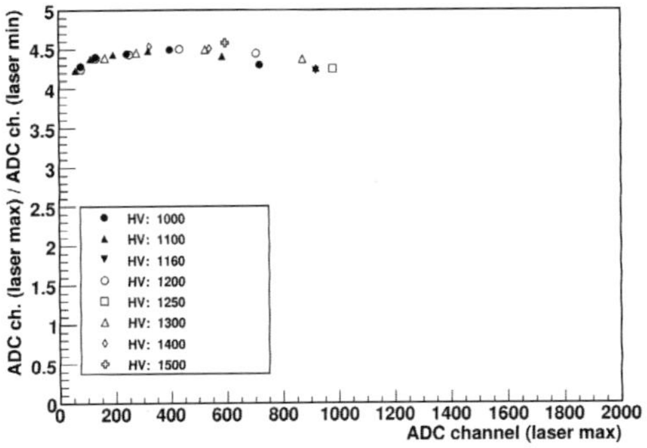

Figure 10. Linearity measurements for one of the ZN PMT: the ratio between higher and lower PMT charges is plotted as a function of the higher PMT charges for different HV of the PMT.

have been performed by inserting filters in front of the laser diode head, therefore varying the absolute intensity of the laser light. Preliminary results are shown in Fig. 10 for one of the ZN PMT, where the ratio between higher and lower PMT charges, measured by means of a QDC Caen V965, is plotted as a function of the higher PMT charges. Data have been collected at different HV of the PMTs, around the Pb-Pb working values. The data show a good linearity (within 6%) up to channel 1000 (200 pC). It was not possible to extend the measurements to higher charges due to the limitation of the input amplitude of the Linear Fan–In Fan–Out temporarily used ($\approx$1.4 V) in the ZDC electronic chain. We plan to replace these commercial modules with a custom Linear Fan–In Fan–Out with a higher maximum input amplitude, in order to extend the linearity measurements up to 800 pC, which corresponds to the full range of the QDC.

6. Conclusion

Two identical ZDC sets, each one composed of a ZN and a ZP calorimeters, are now installed at IP2 in the LHC tunnel. Each set is equipped with a laser light source for radiation damage monitoring of PMTs and fibers. A cosmic ray and a laser trigger will be used in order to measure the absolute gain of the PMTs, by means of the single photoelectron signal, in order to disentangle the PMT and fiber radiation damage.

Concerning the physical calibration, the single neutron signal from neutron emission in ion–ion e.m. dissociation events in the ZN calorimeter will provide a clean energy calibration. Moreover an inter–calibration method of the different towers, which use the complementary information coming from the PMTc, is discussed. The parameters which re–equalize the response of the towers are obtained with an accuracy of 1%.

Finally the results of the commissioning on the surface and in the tunnel, in particular the absolute gain of the PMTs and the linearity measurements, are presented. The absolute gain of the PMTs have been measured from HV $\approx$ 1200 V, the expected Pb–Pb working point, up to HV $\approx$ 2300 V. A good agreement is achieved between the cosmic rays, laser light measurements and the previous PMT characterization performed in laboratory. Linearity measurements of the entire system (detectors and electronics) have been performed. The data are consistent with linearity at the level of 6%.

7. Acknowledgment

We would like to gratefully acknowledge M. Arba, M. Tuveri and L. La Delfa from INFN Cagliari and G. Alfarone, R. Farano and O. Brunasso Cattarello from INFN Torino for their technical support.

References

[1] "ALICE Technical Proposal" *CERN/LHCC 95-71.*
[2] P. Gorodetzky et al.,"Quartz fiber calorimetry", *Nucl. Instr. Meth. A*, vol.361, pp.161-179, 1995.
[3] R. Arnaldi et al.,"The quartz-fiber Zero-Degree Calorimeter for the NA50 experiment at CERN SPS", *Nucl. Instr. Meth. A*, vol.411, pp.1-16, 1998.
[4] "Zero Degree Calorimeter Technical Design Report", *CERN/LHCC 99-5.*
[5] M. Gallio et al.,"The neutron zero degree calorimeter for the Alice experiment", *Proc. of XI International Conference on Calorimetry in Particle Physics, CALOR2004*, Perugia, Italy, March 29 - April 2, 2004.
[6] R. Arnaldi et al.,"The Neutron Zero Degree Calorimeter for the ALICE experiment", *Nucl. Instr. Meth. A*, vol.564, pp.235-242, 2006.
[7] R. Gemme et al.,"Test beam results on the Proton Zero Degree Calorimeter for the ALICE experiment", *Proc. of XII International Conference on Calorimetry in Particle Physics, CALOR2006*, Chicago, USA, June 5-9, 2006.
[8] N. De Marco et al.,"Performance of the Zero Degree Calorimeters for the Alice experiment", *IEEE Trans. on Nucl. Sc.*, vol. 54, no. 3, pp. 567-573, 2007.
[9] Xin-Nian Wang and Miklos Gyulassy, *Comput. Phys. Commun.* vol.83, pp. 307-344, 1994.

Journal of Physics: Conference Series 160 (2009) 012061 doi:10.1088/1742-6596/160/1/012061

Calorimetry triggering in ATLAS

Igonkina O.[32], Achenbach R.[15], Adragna P.[37], Aharrouche M.[24], Alexandre G.[11], Andrei V.[15], Anduaga X.[21], Aracena I.[41], Backlund S.[7], Baines J.[38], Barnett B.M.[38], Bauss B.[24], Bee C.[26], Behera P.[19], Bell P.[25], Bendel M.[24], Benslama K.[39], Berry T.[16], Bogaerts A.[7], Bohm C.[43], Bold T.[44,9], Booth J.R.A.[5], Bosman M.[4], Boyd J.[7], Bracinik J.[5], Brawn I.P.[38], Brelier B.[28], Brooks W.[47], Brunet S.[10], Bucci F.[11], Casadei D.[31], Casado P.[4], Cerri A.[7], Charlton D.G.[5], Childers J.T.[15], Collins N.J.[5], Conde Muino P.[20], Coura Torres R.[40], Cranmer K.[31], Curtis C.J.[5], Czyczula Z.[30], Dam M.[30], Damazio D.[3], Davis A.O.[38], De Santo A.[16], Degenhardt J.[36], Delsart P.-A.[28], Demers S.[41], Demirkoz B.[7], Di Mattia A.[23], Diaz M.[42], Djilkibaev R.[31], Dobson E.[34], Dova M.T.[21], Dufour M.-A.[27], Eckweiler S.[24], Ehrenfeld W.[14,10], Eifert T.[11], Eisenhandler E.[37], Ellis N.[7], Emeliyanov D.[38], Enoque Ferreira de Lima D.[40], Faulkner P.J.W.[5], Ferland J.[28], Flacher H.[7], Fleckner J.E.[15], Flowerdew M.[22], Fonseca-Martin T.[16], Fratina S.[36], Fhlisch F.[15], Gadomski S.[11], Gallacher M.P.[5], Garitaonandia Elejabarrieta H.[32], Gee C.N.P.[38], George S.[16], Gillman A.R.[38], Goncalo R.[16], Grabowska-Bold I.[9], Groll M.[24], Gringer C.[24], Hadley D.R.[5], Haller J.[14,10], Hamilton A.[11], Hanke P.[15], Hauser R.[23], Hellman S.[43], Hidvgi A.[43], Hillier S.J.[5], Hryn'ova T.[7], Idarraga J.[28], Johansen M.[43], Johns K.[2], Kalinowski A.[39], Khoriauli G.[39], Kirk J.[38], Klous S.[32], Kluge E.-E.[15], Koeneke K.[10], Konoplich R.[31], Konstantinidis N.[45], Kwee R.[17], Landon M.[37], LeCompte T.[1], Ledroit F.[13], Lei X.[2], Lendermann V.[15], Lilley J.N.[5], Losada M.[6], Maettig S.[14,10], Mahboubi K.[15], Mahout G.[5], Maltrana D.[42], Marino C.[18], Masik J.[25], Meier K.[15], Middleton R.P.[38], Mincer A.[31], Moa T.[43], Monticelli F.[21], Moreno D.[6], Morris J.D.[37], Mller F.[15], Navarro G.A.[6], Negri A.[35], Nemethy P.[31], Neusiedl A.[24], Oltmann B.[24], Olvito D.[36], Osuna C.[4], Padilla C.[4], Panes B.[42], Parodi F.[12], Perera V.J.O.[38], Perez E.[4], Perez Reale V.[8], Petersen B.[7], Pinzon G.[6], Potter C.[27], Prieur D.P.F.[38], Prokishin F.[47], Qian W.[38], Quinonez F.[42], Rajagopalan S.[3], Reinsch A.[33], Rieke S.[24], Riu I.[4], Robertson S.[27], Rodriguez D.[6], Rogriquez Y.[6], Rhr F.[15], Saavedra A.[46], Sankey D.P.C.[38], Santamarina C.[27], Santamarina Rios C.[27], Scannicchio D.[35], Schiavi C.[12], Schmitt K.[15], Schultz-Coulon H.-C.[15], Schfer U.[24], Segura E.[4], Silverstein D.[41], Silverstein S.[43], Sivoklokov S.[29], Sjlin J.[43], Staley R.J.[5], Stamen R.[15], Stelzer J.[10], Stockton M.C.[5], Straessner A.[11], Strom D.[33], Sushkov S.[4], Sutton M.[45], Tamsett M.[16], Tan C.L.A.[5], Tapprogge S.[24], Thomas J.P.[5], Thompson P.D.[5], Torrence E.[33], Tripiana M.[21], Urquijo P.[11], Urrejola P.[42],

Vachon B.[27], Vercesi V.[35], Vorwerk V.[4], Wang M.[39], Watkins P.M.[5], Watson A.[5], Weber P.[15], Weidberg T.[34], Werner P.[7], Wessels M.[15], Wheeler-Ellis S.[44], Whiteson D.[44], Wiedenmann W.[48], Wielers M.[38], Wildt M.[24], Winklmeier F.[7], Wu X.[11], Xella S.[30], Zhao L.[31], Zobernig H.[48], de Seixas J.M.[40], dos Anjos A.[48], Åsman B.[43], Özcan E.[45]

[1] Argonne National Laboratory, Argonne, Illinois
[2] University of Arizona, Tucson, Arizona
[3] Brookhaven National Laboratory (BNL), Upton, New York
[4] Institut de Fsica d'Altes Energies (IFAE), Universitat Autnoma de Barcelona, Bellaterra (Barcelona)
[5] School of Physics and Astronomy, The University of Birmingham, Birmingham
[6] Universidad Antonio Narino, Bogot, Colmbia
[7] European Laboratory for Particle Physics (CERN), Geneva
[8] Nevis Laboratory, Columbia University, Irvington, New York
[9] Faculty of Physics and Applied Computer Science, AGH University of Science and Technology, Cracow
[10] Deutsches Elektronen-Synchrotron (DESY), Hamburg and Zeuthen
[11] Section de Physique, Universit de Genve, Geneva
[12] Dipartimento di Fisica dell' Universit di Genova e I.N.F.N., Genova
[13] Laboratoire de Physique Subatomique et de Cosmologie de Grenoble (LPSC), IN2P3-CNRS-Universit Joseph Fourier, Grenoble
[14] University of Hamburg, Germany
[15] Kirchhoff Institut fr Physik, Universitt Heidelberg, Heidelberg
[16] Department of Physics, Royal Holloway and Bedford New College, Egham
[17] Institut fr Physik, Humboldt-Universitt, Berlin
[18] Indiana University, Bloomington, Indiana
[19] Iowa State University, Ames, Iowa
[20] Laboratorio de Instrumentaao e Fisica Experimental, Lisboa
[21] Universidad Nacional de La Plata, La Plata
[22] University of Liverpool, Liverpool
[23] Michigan State University, Department of Physics and Astronomy, East Lansing, Michigan
[24] Institut fr Physik, Universitt Mainz, Mainz
[25] School of Physics and Astronomy, University of Manchester, Manchester
[26] Centre de Physique des Particules de Marseille, IN2P3-CNRS, Marseille
[27] Department of Physics, McGill University, Montreal
[28] University of Montreal, Montreal
[29] Moscow State University, Moscow
[30] Niels Bohr Institute, University of Copenhagen, Copenhagen
[31] Department of Physics, New York University, New York
[32] Nikhef National Institute for Subatomic Physics, Amsterdam
[33] University of Oregon, Eugene, Oregon
[34] Department of Physics, Oxford University, Oxford
[35] Universit di Pavia, Dipartimento di Fisica Nucleare e Teorica, Pavia
[36] Department of Physics, University of Pennsylvania, Philadelphia, Pennsylvania
[37] Physics Department, Queen Mary, University of London, London
[38] STFC Rutherford Appleton Laboratory, Harwell Science and Innovation Campus, Didcot, Oxon
[39] University of Regina, Regina
[40] Universidade Federal do Rio de Janeiro, COPPE/EE/IF, Rio de Janeiro
[41] Stanford Linear Accelerator Center (SLAC), Stanford
[42] Pontificia Universidad Catlica, Santiago
[43] Fysikum, Stockholm University, Stockholm
[44] Department of Physics and Astronomy, University of California, Irvine, California
[45] Department of Physics and Astronomy, University College London, London
[46] University of Sydney, Sydney
[47] Universidad tecnica Federico Santa Maria, Valparaso
[48] Department of Physics, University of Wisconsin, Madison, Wisconsin

E-mail: Olga.Igonkina@cern.ch

Abstract. The ATLAS experiment is preparing for data taking at 14 TeV collision energy. A rich discovery physics program is being prepared in addition to the detailed study of Standard Model processes which will be produced in abundance. The ATLAS multi-level trigger system is designed to accept one event in $2 \cdot 10^5$ to enable the selection of rare and unusual physics events. The ATLAS calorimeter system is a precise instrument, which includes liquid Argon electro-magnetic and hadronic components as well as a scintillator-tile hadronic calorimeter. All these components are used in the various levels of the trigger system. A wide physics coverage is ensured by inclusively selecting events with candidate electrons, photons, taus, jets or those with large missing transverse energy. The commissioning of the trigger system is being performed with cosmic ray events and by replaying simulated Monte Carlo events through the trigger and data acquisition system.

1. Introduction

The ATLAS detector is a multipurpose spectrometer aiming at Standard Model precision measurements as well as discovery of physics beyond Standard Model. Operating at a center of mass energy of $\sqrt{s} \sim 14$ TeV and with a bunch crossing rate of 40 MHz, the ATLAS experiment will register unprecedentedly high number of events, most of which will contain several inelastic $p-p$ collisions. The fraction of events of interest is, however, small, as can be seen from fig. 1. The cross-section of Z and W bosons production is expected to be six orders of magnitude smaller than the rate of inelastic $p-p$ interactions, and the predictions for Higgs or Supersymmetry production are an additional three order of magnitude smaller. From the other side each recorded ATLAS event takes about 1.5MB of disk space and the storage of even small fraction of produced events requires significant capacity. To record only potentially interesting events, ATLAS is preparing a trigger system with a rejection factor of 10^5. This is 3 level system with first hardware level (L1) and with software second (L2) and third (EF) levels of *high level trigger* (HLT). The output rate is tuned to be 200Hz. Following detector design philosophy, the ATLAS trigger is a multipurpose system, oriented to select basic objects with high transverse momentum (p_T) or events with significant missing transverse energy ($\mathrm{xE_T}$). The basic objects include electrons, photons, muons, jets and taus.

All triggers except the ones aiming at registering muons and events with B mesons, are calorimeter based. At L1, these triggers use only calorimeter, electro-magnetic and hadronic, while at higher levels the tracking information is added. For the $\mathrm{xE_T}$ trigger also data from muon system are used at L2 and at EF. About 70% of the total bandwidth is allocated to calorimetric triggers.

2. ATLAS calorimeter and trigger system

The details of the ATLAS detector and of its calorimeters can be found elsewhere [?]. The high granularity liquid Argon (LAr) electro-magnetic (EM) sampling calorimeter covers the pseudo-rapidity range $|\eta| < 3.2$. The hadronic calorimetry (HAD) in the range $|\eta| < 1.7$ is a scintillator-tile calorimeter, which is separated into a large barrel and two smaller extended barrel cylinders, one on either side of the central barrel. In the end-caps ($|\eta| > 1.5$), LAr technology is also used for the hadronic calorimeters, matching the outer $|\eta|$ limits of end-cap electro-magnetic calorimeters. LAr forward calorimeters provide both electro-magnetic and hadronic energy measurements and extend to $|\eta| = 4.9$. The EM calorimeter is divided in four samplings around the z-axis and the hadronic part has three samplings. The total thickness of the EM calorimeter is > 22 radiation lengths in the barrel and > 24 radiation lengths in

Journal of Physics: Conference Series **160** (2009) 012061 doi:10.1088/1742-6596/160/1/012061

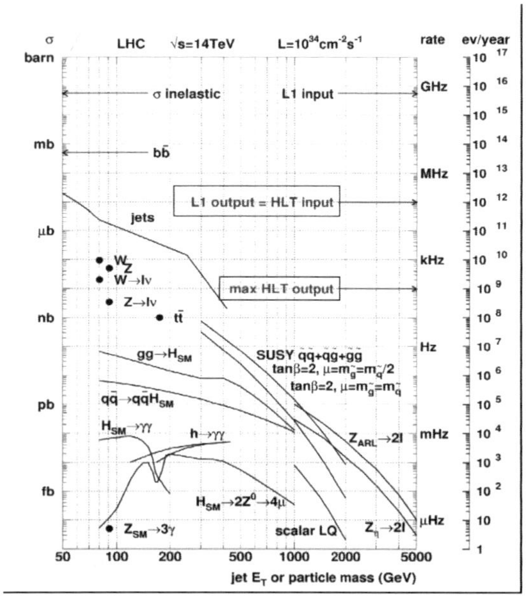

Figure 1. The expected cross-section (left axis) and corresponding event rate at 10^{34}cm^{-2} s^{-1}instant luminosity for different processes. Also shown L1 and HLT output rates.

end-caps. The calorimeter has 9.7 interaction lengths in the barrel and 10 in the end-caps. In total there are about 175 thousand readout channels in the EM and over 15 thousand in the HAD calorimeters.

The calorimeter based trigger system covers the complete acceptance provided by the detector. The electron, photon and tau trigger cover a pseudo-rapidity region $|\eta| < 2.5$, while jet and $\mathrm{xE_T}$ triggers use all barrel, end-cap and forward regions.

2.1. ATLAS L1 trigger

The first level trigger is a hardware based system that reduces the rate to 75 MHz within a fixed latency of 2.5 μm. The L1 calorimeter trigger has a dedicated readout, where the energy deposition is summed in about 7000 trigger towers of $\eta \times \phi = 0.1 \times 0.1$ which are connected to FPGAs. The trigger towers are larger in the forward region. The electro-magnetic and hadronic parts of calorimeter are analyzed separately. All local maxima are considered as a trigger. The L1 trigger for electrons, photons (EM) and taus is based on 4×4 towers, while the jet trigger uses blocks of 2 × 2 towers and consider either 2×2, 3 × 3 or 4×4 blocks. L1 electron and photon triggers require a large energy deposition in the two central EM towers and limit energy deposition in the EM and HAD isolation rings as well as in HAD central 2×2 area. L1 tau trigger requires significant energy deposition in central EM and HAD areas and limits energy in EM isolation ring. L1 jet trigger cuts only on the total energy deposition of the candidate. All variables are calculated in a transverse plane. The energy summation algorithm produces also sums of E_T, E_x and E_y energy deposits over all trigger towers, which are used to construct the total E_T and missing E_T L1 triggers. Two calibrations are used at L1 - one, particle based, is applied to electron and photon candidates, while jet-based calibration is applied for all other types. The jet based calibration re-weights EM and HAD parts with an average ratio of charged and neutral pions in jets to estimate the original transverse energy of the jet.

The trigger configuration allows to set up to 16 electron/photon, 8 tau, 8 jet and 8 $\mathrm{xE_T}$ thresholds. The L1 trigger menu allows 256 combinations of these thresholds (*items*). In addition to physics items there are items for detector monitoring, calibration and to collect control samples. Not all of selected events, e.g. for calibration, can be recorded, therefore *prescale factors* are applied. The L1 configuration allows to use different prescale factors for the items.

2.2. ATLAS L2 trigger

The L2 is a software trigger. The full granularity data from any sub-detectors are available within *regions of interest (RoI)*. Usage of RoI allows to reduce the data volume at L2 nodes by about 50 times. Only fast rejection algorithms run at L2. Among them are a simple clusterization of calorimeter information and a track reconstruction algorithm. The properties of the cluster and their correlation with the number of tracks found are tested against various hypotheses. The L2 rejection of about 40 is achieved within an average event processing time of 40 ms. About one third of the time is spent on data preparation (retrieving from the pile-up buffer, decoding). Therefore it is important to avoid repetition of this operation, in case the same region of interest is requested by different trigger items. This is achieved with a *cashing* procedure, where unpacked data are stored in memory and are reused if a second call is issued. Also the results of clusterization and track reconstruction are cashed and re-used to test different hypothesis, e.g. for electron items with different p_T thresholds.

Most of electron/photon and tau cluster shape parameters are calculated using the 2nd EM sampling, while the total transverse energy deposition of taus, jets and $\mathrm{xE_T}$ is calculated using all 3 samplings of EM and 4 samplings of HAD calorimeter. Tracks are reconstructed and matched to the cluster for electrons, photons and taus. The L2 jet calibration takes the weight of HAD part as a logarithmic function of energy and as a function of the pseudo-rapidity. The accuracy of jet E_T reconstruction is shown on fig. 2 for different pseudo-rapidity regions. The tau trigger uses jet calibration at all trigger levels.

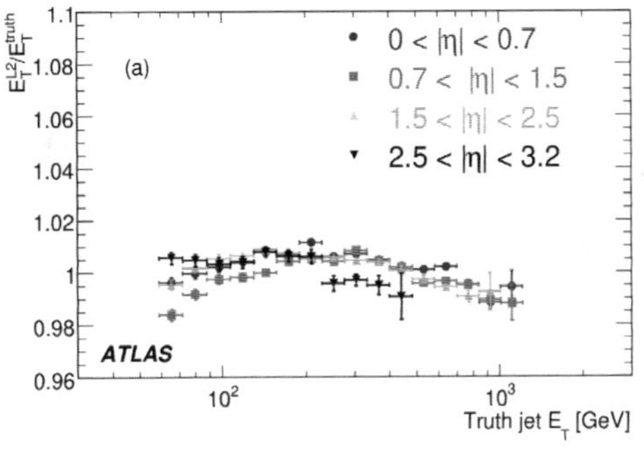

Figure 2. Jet energy scale for the L2 jets as a function of the truth jet E_T for four different bins in η..

2.3. ATLAS EF trigger

The third level trigger or *Event Filter* runs on a farm of PCs separated from L2 PCs. The complete event is built and available for analysis, however, only the data within RoIs accepted by L2 are analyzed due to time considerations. The algorithms executed are similar to the ones used for final offline reconstruction, but unlike offline reconstruction, the alignment of the various detector parts and calibration of the calorimeter are not final and are not updated during the run. The mean event processing time is about 4s and the output rate is tuned to 200Hz. The output rate is limited by the capacity of the ATLAS storage system, and although short spikes in the rate are not a problem, the average rate should not exceed design value. The data storage speed is about 300 MB/s.

More refined calibration and more advanced cuts are applied at this level. The selection cuts are optimized to provide highest possible efficiency for physics channels at the affordable rate. Different tightness cuts are introduced, such that one can use looser cuts together with prescale factor to collect control sample (e.g. for the measurement of efficiency) while tighter

Journal of Physics: Conference Series **160** (2009) 012061 doi:10.1088/1742-6596/160/1/012061

cuts without any prescale are used to collect signal sample. There are L2 and EF prescale factors additional to L1 factors, in case the same L1 item is used for several HLT selections.

2.4. Trigger menu

The trigger menu is a list of items and of the corresponding prescale factors for each trigger level. It is foreseen that several menus will be deployed, one for the start-up period of LHC and the others for the different instant luminosity periods of 10^{31}, 10^{32} and 10^{33} cm^{-2} s^{-1}. The start up menu will mostly include soft L1 triggers at first, slowly migrating to more advanced HLT selection and variety of items. The 10^{31} menu is considered as a basic one for the first physics run. It will include :

- items for physics,
- items for commissioning, calibration, alignment,
- back-up triggers to be used for physics if the actual trigger rate of the chosen items will be significantly different from the expectations,
- items for the trigger efficiency measurements,
- HLT pass-through items for the debugging/monitoring of the software triggers,
- minimum bias and random triggers.

The physics goals for the first physics run at $10^{31}cm^{-2}$ $s^{-1}$are illustrated by tab. 1, where subset of unprescaled physics items are given together with expected rate at L1 and HLT and physics motivation. The complete menu for the electron trigger for 10^{31} run is shown in tab. 2. The expected rates are given for 14 TeV center-of-mass energy.

Table 1. Tentative unprescaled trigger items for 14 TeV LHC run with instant luminosity of $10^{31}cm^{-2}$ s^{-1}.

Signature	L1 rate (Hz)	HLT rate (Hz)	Comments
Minimum bias	Up to 10000	10	Pre-scaled trigger item
e10	5000	21	$b, c \to e, W, Z$, Drell-Yan, $t\bar{t}$
2e5	6500	6	Drell-Yan, J/ϕ, Υ, Z
γ20	370	6	Direct photons, γ -jet balance
2γ15	100	< 1	Photon pairs
mu10	360	19	$W, Z, t\bar{t}$
2mu4	70	3	B-physics, Drell-Yan, J/ϕ, Υ, Z
mu4 + $J/\phi(\mu\mu)$	1800	< 1	B-physics
j120	9	9	QCD and other high-pT jet final states
4j23	8	5	Multi-jet final states
tau20i + xE30	5000	10	$W, t\bar{t}$
tau20i + e10	130	1	$Z \to \tau\tau$
tau20i + mu6	20	3	$Z \to \tau\tau$

3. Tests of the trigger system

Various tests of the system are performed in preparation for data taking. This includes *cosmics runs* where cosmic rays are triggered with either L1 muon or L1 calorimeter and *technical runs* where L1 output is replaced with simulated data. Also, a variety of Monte Carlo tests are performed to predict change in the signal efficiencies and in the trigger rates due to imperfect detector conditions.

Table 2. Summary of triggers for the first physics run assuming a luminosity of $10^{31}\text{cm}^{-2}\ \text{s}^{-1}$. For each signature rates and the motivation for this trigger are given. The prescale factors are given for L1 and for HLT.

Signature	Level-1 Pre-scale	Level-1 Rate [kHz]	Event Filter Sel-ection	Event Filter Pre-scale	Event Filter Rate [Hz]	Motivation
e5	60	0.7	medium	1	4.8 ± 0.2	$J/\Psi \to ee$, $Y \to ee$, Drell-Yan
2e5	1	6.5	medium	1	6	$J/\psi \to ee$, $Y \to ee$, Drell-Yan
Jpsiee	1	6.5	medium	1	1	$J/\psi \to ee$, $Y \to ee$
e10	1	5.0	medium	1	21	$e^{\pm}$ from b,c decays, E/p studies
e10_xe30	1	0.2	medium	1	0.3 ± 0.3	access low p_T-range for $W \to e\nu$
2e10	1	0.5	loose	1	0.4 ± 0.2	$Z \to e^+e^-$
Zee	1	0.5	loose	1	< 0.1	$Z \to e^+e^-$
2e12i_L33	1	0.5	tight	1	< 0.1	trigger for $10^{33}\text{cm}^{-2}\ \text{s}^{-1}$
e15_xe20	1	0.2	loose	1	1.0 ± 0.4	access low p_T-range for $W \to e\nu$
e20_passL2	1	0.3	loose	200	< 0.1	check L2/EF performance
e20_passEF	1	0.3		125	0.1	check L2EF performance
em20_passEF	1	0.3		750	0.5 ± 0.1	check HLT performance
em20i_passEF	1	0.1		300	0.5 ± 0.1	check L1 isolation
e22i_L33	1	0.1	tight	1	1.2 ± 0.1	trigger for $10^{33}\text{cm}^{-2}\ \text{s}^{-1}$
em105_passHLT	1	1		1	1.0 ± 0.1	New physics, check for possible

3.1. Cosmics run

Cosmic rays interact with media of the ATLAS detector and leave hits in muon system as well as non-negligible energy deposits in the calorimeter. Although the cosmic ray traverses detector vertically and rarely crosses the interaction point, still the data are very useful to test the calorimeter readout and the L1 calorimeter response. The signals up to tens of GeV are registered in the calorimeter. The first objective of such test is to verify time and energy of the signals seen by L1 trigger which has its own dedicated readout versus signals recorded with the main calorimeter readout. Fig. 3 shows the energy deposition in the LAr and in the hadronic calorimeters versus energy reconstructed by L1. One can also see the saturation of the L1 readout in hadronic calorimeter at about 120 GeV. Another important purpose of cosmic test is to identify hot and dead calorimeter cells which might increase the trigger rate.

The calorimetric signals from the cosmics runs are also send through the L2 and EF trigger. As many triggers require tracks from the interaction point or a significant calorimetric energy deposition, no hypothesis testing is applied during the cosmics run, and most triggers are configured in *pass-through* mode, where each L1 candidate is accepted. Such exercise allows to test the data acquisition system (DAQ) and data quality monitoring procedures.

In addition to check of the trigger and DAQ performance, such tests involve large number of experts and allow to test and to prepare procedures for the operation of the detector by the shift crew and experts.

3.2. Technical runs

During the technical runs the L1 output is replaced with the simulated data. Several goals are achieved during such runs:

- test of the data acquisition system and HLT with the realistic input rate,

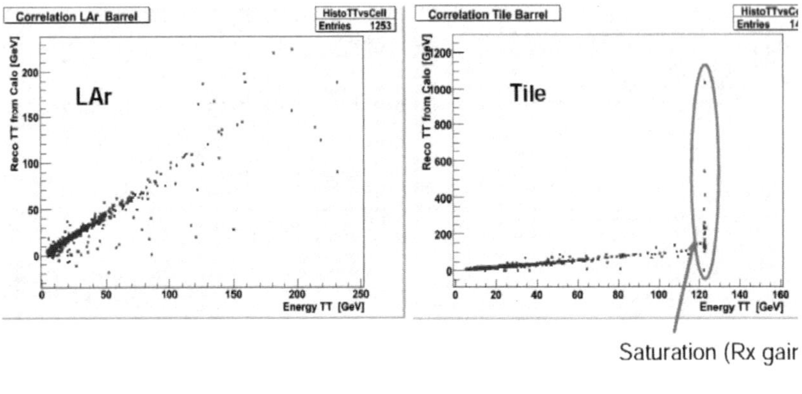

Figure 3. Energy deposition in the LAr (left) and Tile (right) barrel areas seen by detector and trigger readout during cosmics runs.

- CPU time used by the HLT algorithms as well as time required for the data preparation,
- robustness of the software triggers such as memory leaks, errors, ability to deal with incomplete data,
- stability of the performance during the long runs (usually, events are recycled many times during such long tests),
- validation of the output stream,
- ability of the data quality monitoring system to catch and report problems in timely manner,
- flexibility of tools, database access and time for initialization of the run.

The complete 10^{31} trigger menu with about 200 trigger items was successfully tested during a technical run using minimum bias events corresponding to the average detector occupancy as well as $t\bar{t}$ or events with black holes, taken as an example of the extreme occupancy.

3.3. Checks of robustness of the system

It is expected that the performance of the trigger system can be affected by a number of effects. In particular, presence of pile-up interactions or extra material in the detector can compromise the efficiency of the signal reconstruction. The misalignment of the inner-detector and the mis-calibration of calorimeter, the presence of hot and dead channels can significantly degrade the performance. Several ways are foreseen and being applied to predict behavior of the trigger and to tune algorithms such that they are less affected by a non-ideal detector response. It includes the simulation of different detector effects on Monte Carlo and the preparation of analysis software to extract efficiency from the data.

4. Estimated trigger performance

The trigger performance was studied in details in preparation for the first physics run. A large variety of physics subjects was considered, among them, searches for Higgs, Supersymmetry and a number of exotic states. The realistic description of the detector was simulated, including mis-alignment effects, imperfect response of the calorimeter readout and presence of pile-up interactions. The results could be found elsewhere [?]. In fig. 4,5,6 we give as an example, the efficiency of the main triggers of each type for the first physics run: e22i for electrons, tau16i for taus decaying hadronically, j255 for jets. Fig. 7 shows the resolution of xE_T reconstruction at different levels.

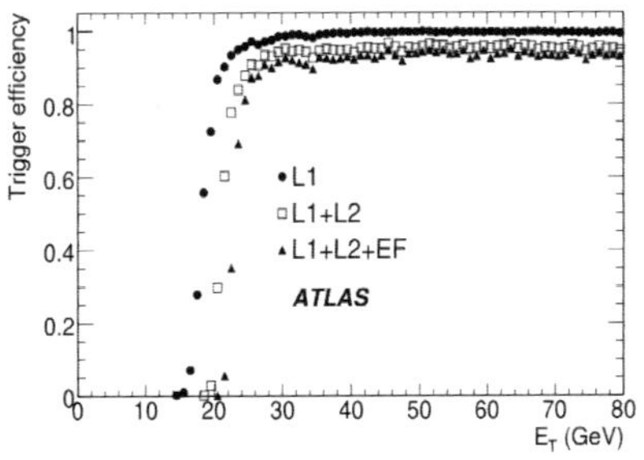

Figure 4. L1, L2 and EF efficiencies of e22i trigger item for single electrons as function of electron p_T.

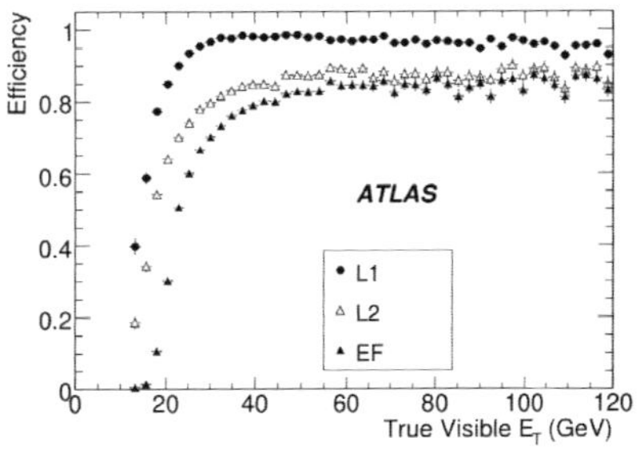

Figure 5. L1, L2 and EF efficiencies of tau16i item for the tau from W and Z bosons decaying hadronically as function of the P_T of the hadronic decay products of tau.

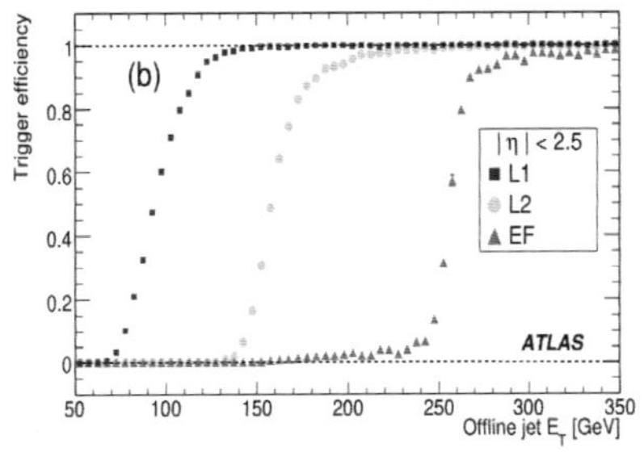

Figure 6. L1,L2 and EF efficiency of jet255 item as function of the offline jet transverse energy. The thresholds applied are 70/150/255 GeV for L1/L2/EF, respectively.

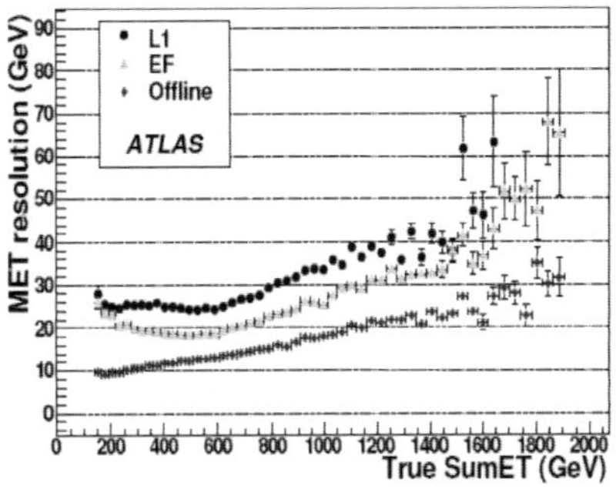

Figure 7. The resolution of $\mathrm{xE_T}$ as function of true sum of E_T of the event objects.

5. Conclusions

The ATLAS trigger is an advanced multipurpose system oriented to select rare events of interest out of large sample of inelastic *pp* events with a rejection factor of 10^5 and an output rate of 200Hz. The calorimeter is one of the main components of the system. The calorimeter based triggers take about 70% of the total output bandwidth. The ATLAS trigger consists of 3 levels: one hardware-based and two software based. All parts are installed and are being tested in preparation for the start-up of the LHC. Several types of tests are performed, among others tests of the complete system with cosmics rays and test of the HLT and DAQ with Monte Carlo events replacing L1 output.

The preparation of the trigger menu is in advanced stage with selections prepared for physics events as well as for commissioning and calibration of the detector. A set of menus are available for different luminosity periods 10^{31}, 10^{32} and $10^{33}\mathrm{cm}^{-2}\ \mathrm{s}^{-1}$as well as for initial period of trigger commissioning.

[1] ATLAS Collaboration, 1999, ATLAS detector and physics performance. Technical Design report, CERN/LHCC 99-14/15
[2] ATLAS Collaboration, 2008, The ATLAS book, paper in preparation

Journal of Physics: Conference Series **160** (2009) 012062 doi:10.1088/1742-6596/160/1/012062

CMS Electromagnetic Trigger commissioning and first operation experiences

Pascal Paganini on behalf of the CMS collaboration
LLR Ecole Polytechnique, route de Saclay 91128 Palaiseau, France
E-mail: pascal.paganini@cern.ch

Abstract. The CMS electromagnetic calorimeter (ECAL) is a high-resolution calorimeter made of 75848 lead tungstate crystals and optimized for the discovery of the Higgs boson in its two photon decay mode. In view of the high raw event rate at the Large Hadron Collider, the ECAL Trigger will play a major role. This paper reviews the strategy and the tests completed to ensure that the ECAL Trigger reaches the required specifications. The results from the commissioning and the first experiences with cosmic ray data are presented.

1. Introduction

The Compact Muon Solenoid (CMS) [1] is a general-purpose detector that operates at the Large Hadron Collider (LHC). This proton-proton collider will reach an energy of 14 TeV in the centre of mass and a high luminosity of 10^{34} cm^{-2} s^{-1} with a bunch-crossing rate of 40 MHz. The CMS trigger system has the formidable task of reducing the input data rate by a factor of $O(4.10^5)$. This reduction is made in 2 stages. The CMS data acquisition system is designed to accept an input rate of 100 kHz, with event size of the order of 1.5 MB. In the first stage, the trigger system uses a custom Level-1 [2] processor to select these 100 kHz of events from the input bunch crossing rate. This selection is based on coarsely segmented data from the calorimeters (electromagnetic and hadronic) and the muon detector systems. During the 3.2 μs latency of the Level-1 trigger, the event data are stored in front-end pipelines. The second stage of the event rate reduction, the High Level Trigger, is made in a farm of standard commercial processors that achieve an output rate of $O(10^2)$ Hz which will be written to permanent storage.

The CMS electromagnetic calorimeter (ECAL) is one of the 3 subdetectors involved in the Level-1 trigger decision. It is a high-resolution calorimeter made of 75848 lead tungstate ($PbWO_4$) crystals and optimized for the discovery of the Higgs boson in its two photon decay mode. Avalanche photodiodes (APDs) are used as photodetectors in the barrel and vacuum phototriodes (VPTs) in the endcaps. The ECAL barrel crystals are slightly tapered with front-face dimensions of $2.2 \times 2.2\ cm^2$ and a crystal length of 23 cm (25.8 X_0). The barrel covers a range $|\eta| < 1.48$ and consists of 36 supermodules each containing 1700 crystals. The 2 end-caps extend the coverage up to $|\eta| < 3$. A trigger tower in the ECAL barrel is made up of 5×5 $PbWO_4$ crystals, following the HCAL segmentation $\Delta\eta \times \Delta\phi = 0.087 \times 0.087$, whereas in the end-cap, the size of the trigger tower varies with the pseudorapidity in order to follow approximately a projective geometry.

This paper is organized as follow: in the next section, the calorimeter trigger system is described with an emphasis on the ECAL trigger primitives. The section 3 reviews the different

Journal of Physics: Conference Series **160** (2009) 012062 doi:10.1088/1742-6596/160/1/012062

steps of ECAL trigger commissioning. Finally, the last section is devoted to the first operation experiences with cosmic ray events.

2. The Level-1 calorimeter trigger

The Level-1 calorimeter trigger is able to define up to 4 isolated or non-isolated electromagnetic objects (electron or photon), 4 tau-jets, 4 central or forward jets, the total transverse energy and the missing transverse energy.

2.1. Algorithm of the electron/photon Level-1 trigger

The electron/photon Level-1 trigger algorithm [2] displayed on figure 1 uses a 3×3 trigger towers sliding window. The transverse energy of an electron/photon candidate is defined by the transverse energy deposit in the central tower of the sliding window summed with the largest deposit in one of its 4 neighbor towers adjacent by side (see figure 1). Electromagnetic showers being characterized by a compact lateral extension, only candidates with a central tower containing 2 adjacent strips (5 crystals in ϕ are called a strip) with a significant fraction of the tower energy (typically 90%) are kept. This criterion characterized by 1 bit is called the Fine Grain (FG) veto bit. Moreover, the associated HCAL energy contribution is required to be below a threshold (typically H over E $<$ 5%). Non-isolated electron/photon candidates require passing the previous criteria (FG and H over E). In addition, the isolated candidates must have a quiet neighborhood characterized by at least five adjacent trigger towers among the 8 nearest ones with their transverse energy below a threshold.

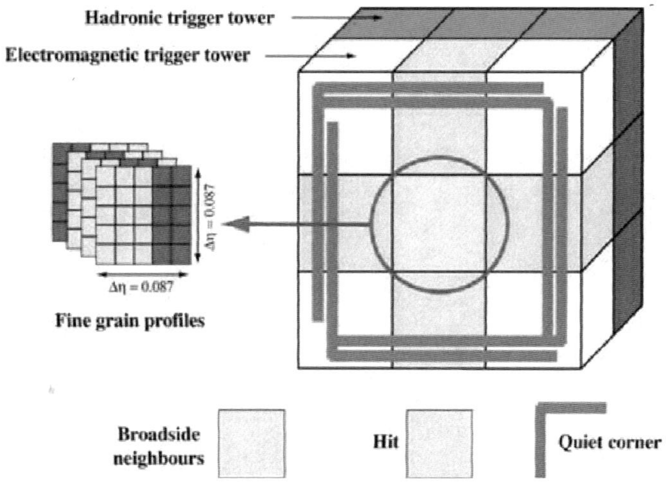

Figure 1. The L1 trigger algorithm of calorimeters

2.2. The ECAL trigger primitives

The trigger primitives of the electromagnetic calorimeter are basic quantities allowing the identification of electromagnetic showers by the Level-1 calorimeter trigger. They are computed for each bunch crossing and for all trigger towers. The trigger primitives are generated by two different sub-systems located on the On-Detector electronics and the Off-Detector electronics. The On-Detector electronics is composed of radiation resistant circuits located just behind the $PbWO_4$ crystals and the trigger primitives are generated by the Front-End (FE) boards. The trigger primitives are then completed by a dedicated 9U VME board called Trigger Concentrator Card (TCC) [3] belonging to the Off-Detector sub-systems located in the electronics cavern. The two sub-systems are connected by about 70 m serial optical links working at 800 Mbits/s. 3072 Front-End boards and 108 TCC boards are necessary to manage the trigger flow of the whole calorimeter.

The ECAL trigger primitive algorithm is displayed on figure 2 and 3. The 5 digital signals from the crystals of a strip are first linearized (see figure 2). The linearization process takes into account the gain used in the readout electronics and corrects the ADC counts by the calibration coefficients. A multiplicative factor is applied to get linearized ADC counts proportional to the transverse energy deposit in the crystal. The 5 linearized signals of the crystal are then summed up to produce the strip signal. An amplitude filter is applied to measure the amplitude of the strip pulse. The filter is based on linear weighted sums where the weights take into account the expected shape of the signal and subtract dynamically a possible residual pedestal. The output of the amplitude filter is then filtered by a peak finder stage that keeps only the maximum as a measure of the transverse energy contained in the strip.

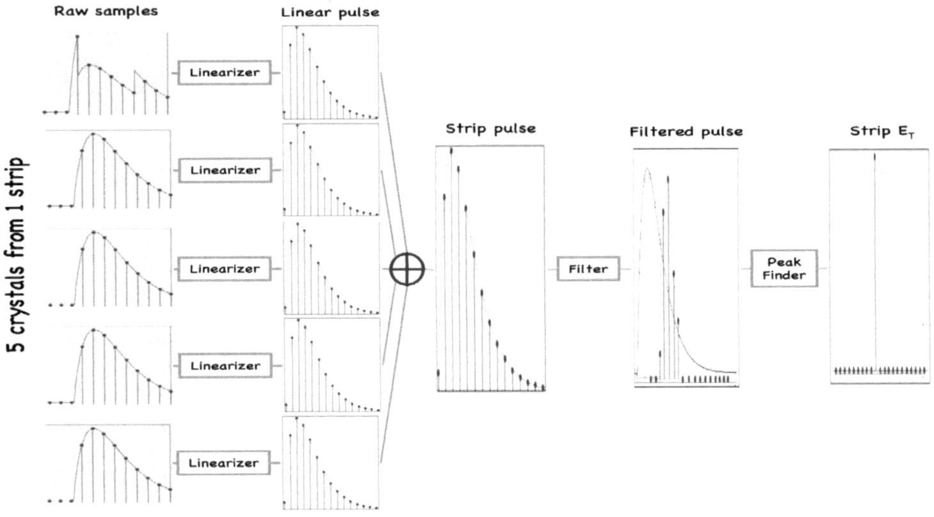

Figure 2. The ECAL trigger primitive algorithm: from the crystals to the strip.

The signals of the 5 strips of a tower are finally summed up to provide a measurement of the total transverse energy of the trigger tower (see figure 3). The dynamic range of the transverse energy must then be reduced from 10 bits to 8 bits. Look-up tables are used for this purpose. The look-up tables are chosen so that the loss of resolution due to the compression follows approximately the intrinsic ECAL resolution. In parallel, the signals of the strips are combined in a Fine Grain filter producing 1 bit, the fine grain veto bit, indicating the transverse extent of the electromagnetic energy deposit.

In summary, the ECAL trigger primitives are made of the total transverse energy of the trigger tower uncoded on 8 bits, accompanied by the fine grain bit. They are computed for all trigger towers and for each bunch crossing.

2.3. The Level-1 ECAL trigger path

The trigger primitives from both electromagnetic and hadronic calorimeters are sent to the Regional Calorimeter Trigger (RCT). One RCT crate covers a region of the calorimeters of 40°

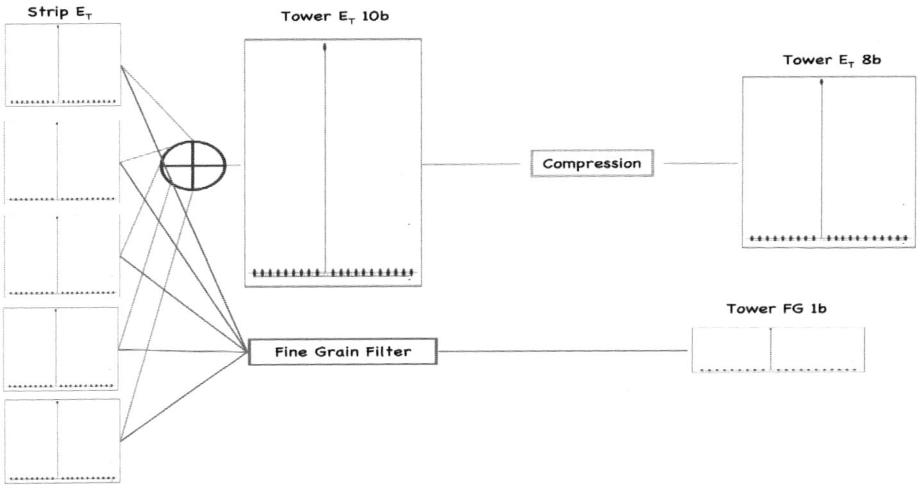

Figure 3. The ECAL trigger primitive algorithm: from the strips to the tower

in ϕ (equivalent to 2 ECAL supermodules) and up to $|\eta| < 5$. RCT applies the algorithm described in section 2.1 to produce isolated and non-isolated electron/photon candidates. The 4 candidates of each region with the highest energy are sent to the Global Calorimeter Trigger (GCT) that sorts out all candidates according to their energy. Only the 4 most energetic are sent to the Global Trigger. The decision to issue a Level-1 accept signal is thus based on these candidates for what concerns the ECAL trigger.

3. Commissioning of the electromagnetic calorimeter trigger

The commissioning of the electromagnetic calorimeter trigger is divided in two parts: the commissioning of the electronics components, boards and links, and the commissioning of the ECAL trigger primitives.

3.1. Commissioning of the electronics cards

There are 3 kinds of ECAL electronics cards involved in the ECAL trigger:

- the Front-End (FE) boards performing the pre-trigger primitives calculation per trigger towers,
- the Trigger Concentrator Cards (TCC) finalizing the trigger primitives calculation for a whole supermodule (barrel) or 20° sector in ϕ (endcaps),
- the Synchro Link Boards (SLB), mezzanine cards plugged on the TCC, in charge of the synchronization (time alignment) of the trigger primitives.

For the barrel, it represents a total of 2448 FE boards, 36 TCC boards and 324 SLB cards. All cards were first tested after the production thanks to dedicated test-benches located in the different labs in charge of the production. See for example [3] and [4] for more details. In a second stage, when the boards were assembled, integration tests were performed. The FE boards were tested during the assembling of the supermodules in the integration center at CERN. For these tests, the FE boards were configured to send a word representing its 'Id'. Prototypes of the TCC, configured in 'spy mode' (filling of input buffers allowing the capture of the data) were used to receive and check the word content with respect to the expectation. The TCC and SLB boards were tested during the assembling of the ECAL Off-detector crates in the electronics integration center at CERN. For this purpose, a tester module able to emulate the front-end was used: patterns were loaded in its memory and sent after serialization to the TCC and SLB. The data received by the TCC and SLB were then captured and analyzed quasi on-line. Once the supermodules and the Off-detector crates had been installed in the experimental and service caverns, the previous integration tests were reproduced.

3.2. Commissioning of the links between ECAL and the Regional Calorimeter Trigger

The previous tests involve only ECAL components. However, the testing of the links between ECAL and the Regional Calorimeter Trigger (RCT) is also important. These tests, also known as pattern tests, use the TCC boards as a pattern generator. Indeed, patterns were loaded (though VME) in dedicated memories of the TCC. In this mode, the TCC doesn't process the data from its optical inputs but the ones from its memory. The patterns were chosen to allow to activate an individual output channel per clock. On the RCT side, the data were received and compared on-line to the expectation. These simple tests revealed among other things few errors in the cabling. It is foreseen to apply this procedure regularly to check the quality of the signals between ECAL and RCT, in particular the stability of the phase of the signals.

3.3. Commissioning of the trigger primitives

The commissioning of the trigger primitives constitutes a severe test of the ECAL trigger. It is based on the comparison to emulated trigger primitives. The emulator is software that reproduces at the bit level the hardware response. It is therefore used to monitor the hardware and was initially tuned and checked with test beam data. All the operations described in section 2.2 are performed by the emulator taking into account the number of bits used for each calculation. The principle of the test with the emulator is the following: for each trigger tower, there are 2 streams of data sent to the DAQ when a level-1 accept signal is issued: the 10 consecutive ADC samples of the 25 individual crystals and the trigger primitive. The emulator uses the crystal data to reproduce the trigger primitives pipeline. However from the 10 samples of crystal data, only 5 trigger primitives can be computed (the amplitude filter needs 5 samples). The trigger primitive stored in the event is then compared to the 5 emulated trigger primitives. If none of them matches the real data, it indicates a problem in the hardware. Moreover, if the data are properly timed-in with respect to the level-1 trigger signal, the same emulated trigger primitive should match the data for all trigger towers. The procedure previously described has been regularly applied during runs and was found to be very useful to properly time-in the ECAL trigger with the detector.

4. First operation experiences with cosmic ray events

Since May 2007, the CMS collaboration organizes regularly global runs lasting several days. These global runs constitute a coherent exercise of CMS data taking in preparation for LHC collisions. The level of complexity is increased for each new global run, more and more subsystems becoming involved. In 2007, the electromagnetic calorimeter participated several times in global runs. However, only the readout of ECAL data was tested. In 2008, for the first time, ECAL participated in the global runs as a component of the trigger system trying to trigger on cosmic ray signals deposits in the crystals. However, a muon crossing longitudinally a $PbWO_4$ crystal loses about 250 MeV as a minimum ionizing particle. Since the noise level per channel is equivalent to about 40 MeV, it is too large with respect to mips signal to provide an efficient trigger. Therefore, a special setting of ECAL is used to increase the signal to noise ratio: the bias voltage applied to the APD is increased to multiply the nominal APD gain by a factor 4, the noise level being then equivalent to 10 MeV. In addition, a threshold is applied at the ECAL trigger tower level: trigger primitives below about 200 MeV are zeroed. The mips signal being mainly contained in a single trigger tower, this threshold decreases the noise contribution due to the second tower used by the Level-1 calorimeter trigger algorithm (see 2.1). Finally to keep the trigger rate under control, the double electron stream is used requiring a coincidence between the top and bottom sectors of ECAL.

Many events have been taken with the configuration 'mip trigger' described before, where the whole barrel was readout and the 8 supermodules located in the top-bottom region were involved in the trigger decision. Given the small amplitude of the mips signal, any problematic channel can pollute the signal and may pass the very low threshold applied. ECAL trigger is thus a perfect tool to reveal noisy channels. A systematic scan of all supermodules has been performed and about 2% of the ECAL trigger towers have been declared as noisy for mips trigger conditions. However, as soon as the threshold will be increased for beam collision (the typical trigger threshold will be of the order of 20 GeV) only very few towers will remain classified as noisy. The measurement of the noise rate as function of the threshold for all towers of the whole barrel is still an on-going activity.

During the global run of May 08, more than 23 million events were recorded by ECAL. ECAL mips trigger and usual muon subdetectors triggers were active. An analysis searching for cosmic events, requiring at least 1 crystal with an energy greater than 135 MeV or 2 adjacent crystals with each containing more that 45 MeV has shown that more than 2.2% of the events passing the cosmic selection were triggered by ECAL, the muon detector trigger bits (mainly from Drift Tubes) being also fired in a large fraction of these events. An example of such an event is shown in figure 4. The tracks in the muon chambers are clearly seen in both top and bottom sectors. The energy deposit in the top cluster of ECAL is 15 GeV shared by 17 crystals, so clearly not a muon at minimum ionization but rather an energy loss by bremsstrahlung. The energy deposit of the bottom cluster is 1.15 GeV. Both top-bottom clusters and muon tracks are compatible with a muon crossing the CMS detector.

5. conclusion

The commissioning of the ECAL barrel trigger has been presented. The ECAL trigger has been successfully commissioned with all the other subsystems of the CMS trigger chain, starting from the front-end of ECAL up to the Global Trigger system. It has been operated successfully and a large number of cosmic ray events have been recorded during the global runs.

Acknowledgments

The author would like to thank his ECAL-DAQ colleagues and the members of the 'ECAL Prompt Analysis Feedback Group' for their support and help during commissioning activities.

[1] CMS Collaboration, the Compact Muon Solenoid, Technical Proposal, *CERN/LHCC 94-38*, 1994

[2] The CMS collaboration, "The TriDAS Project Technical Design Report Vol. 1", *CERN/LHCC 2000-038*, 2000

[3] P. Paganini, "Tests of the Boards Generating the CMS ECAL Trigger Primitives: from the On-Detector Electronics to the Off-Detector Electronics System", *CMS-CR 2004-066*, 2004

[4] N. Almeida et al, "Calorimeter Trigger Synchronization in CMS, Implementation and Test System", *CMS-CR 2004-068*, 2004

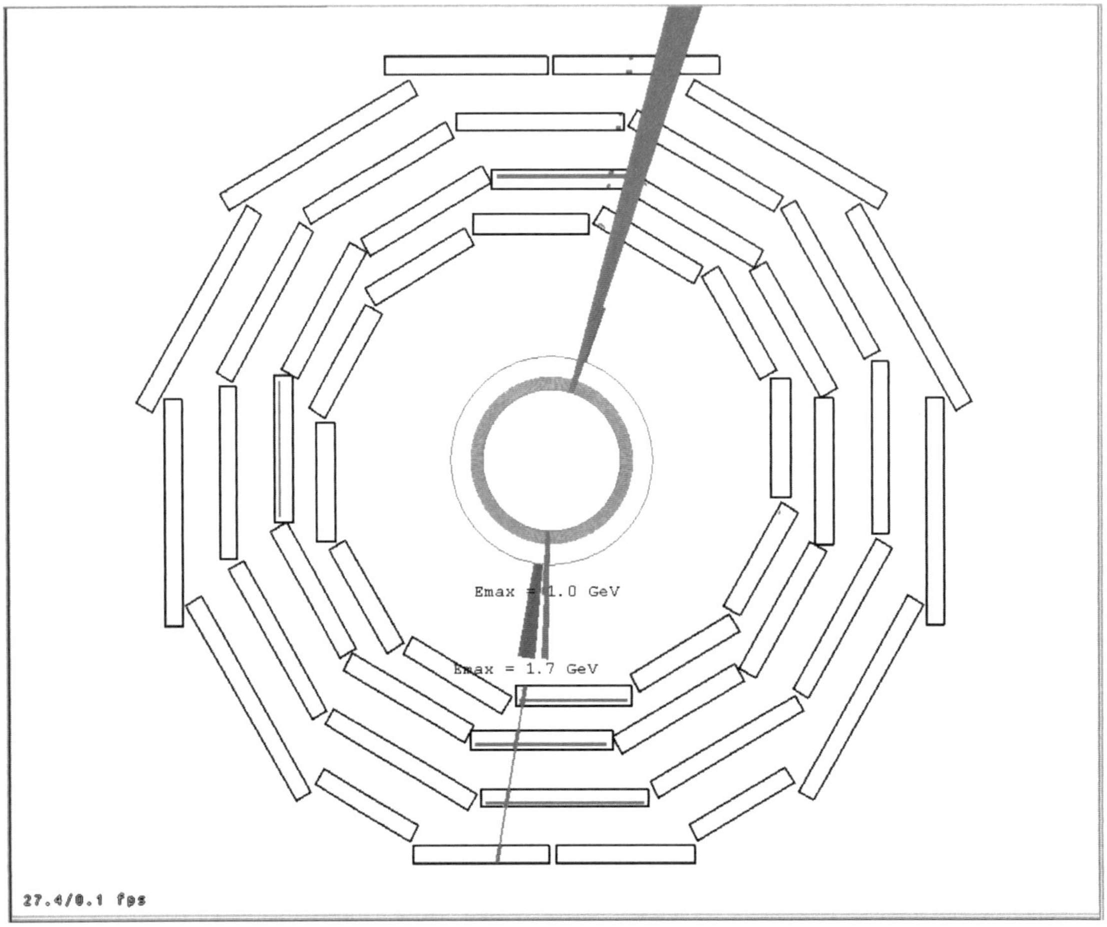

Figure 4. A cosmic ray event triggered by ECAL detector and seen in muon chambers and ECAL.

Triggering with the LHCb calorimeters

Régis Lefèvre on behalf of the LHCb Collaboration

Laboratoire de Physique Corpusculaire, Université Blaise Pascal - IN2P3/CNRS, 24 avenue des Landais, 63177 Aubière Cedex, France

E-mail: `lefevre@clermont.in2p3.fr`

Abstract. The LHCb experiment at the LHC has been conceived to pursue high precision studies of CP violation and rare phenomena in b hadron decays. The online selection is crucial in LHCb and relies on the calorimeters to trigger on high transverse energy electrons, photons, π^0 and hadrons. In this purpose a dedicated electronic has been realized. The calorimeter trigger system has been commissioned and is used to trigger on cosmic muons before beams start circulating in the LHC. When the LHC will start, it will also provide a very useful interaction trigger.

1. Introduction

The LHCb experiment [1] is dedicated to extensive, high precision studies of CP violation and rare decays in b-flavoured hadrons. As shown in figure 1, the $b\bar{b}$ pair production at the LHC is forward peak so LHCb has been designed as a single-arm forward spectrometer to reduce the cost. Figure 2 shows the layout of the experiment which covers polar angles from 10 to 300 (250) mrad in the bending (non-bending) plane. While the Vertex Locator and the tracking system (TT, T1-T3) provide very good vertexing and tracking capabilities, excellent particle identification is achieved thanks to two ring imaging Cherenkov detectors (RICH1 and RICH2), to the calorimeters (SPD/PS, ECAL and HCAL) and to five muon stations (M1-M5).

LHCb will benefit of the large $b\bar{b}$ cross section at the LHC ($\sim 500\ \mu$b). It will be operated at luminosities limited to few 10^{32} cm^{-2}s^{-1} in order to maximize the probability of single interaction per crossing, facilitating the trigger and the reconstruction. At such luminosities the $b\bar{b}$ production rate is around 100 kHz but, with small branching ratios, interesting B decays in the acceptance typically correspond to rates smaller than few Hz and a powerful trigger system is required to select them.

2. Trigger Strategy

The LHCb trigger strategy relies on a large output bandwidth of 2 kHz achievable thanks to a relatively small event size ($\sim$ 35 kbytes). 200 Hz will be dedicated to exclusive selections of B candidates in specific final states for the core physic program but also for tagging efficiencies studies using self-tagged final states. The remaining 1800 Hz will be divided among three inclusive samples: a single muon sample for trigger efficiencies and data mining, an high mass di-muon sample for proper time resolution studies as well as alignment and momentum calibrations, and an inclusive D^* sample to measure particle identification performances and for charm physics.

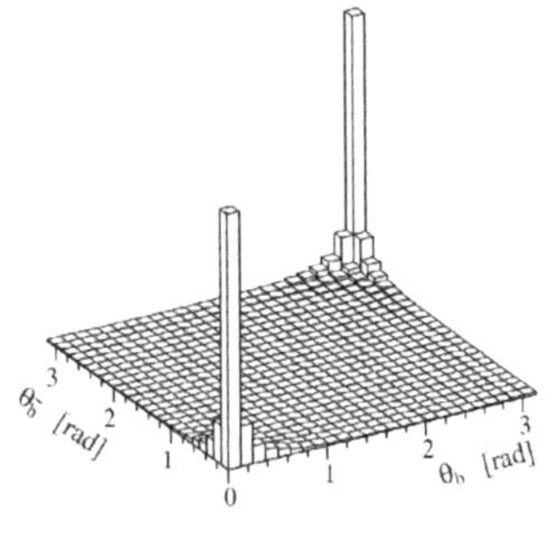

Figure 1. Polar angles of the b and $\bar{b}$ hadrons in $b\bar{b}$ events produced in proton-proton collisions at $\sqrt{s}$ = 14 TeV.

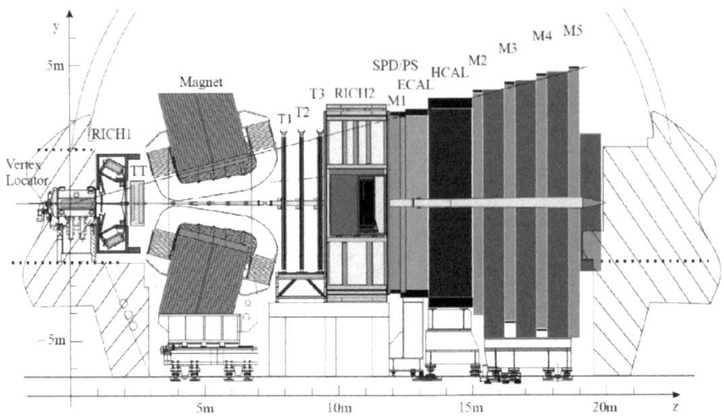

Figure 2. Layout of the LHCb spectrometer.

The LHCb trigger fully exploits the topology of B decays characterised by significant transverse momentum due to the high b quark mass and by long lifetime. To reduce the rate from the 40 MHz input of the LHC clock to the 2 kHz output to the data acquisition (DAQ), the trigger is divided in two levels. The first level, called Level 0, is implemented in custom electronic boards working in a fully synchronous architecture. It has a fixed latency of 4 μs and an output rate of 1 MHz. The second level, called High Level Trigger (HLT), uses a farm of about 2000 central processing units.

The Level 0 consists of four subsystems: the pile-up system, part of the Vertex Locator and used to veto events with more than one interaction and events with very high charged multiplicity; the calorimeter trigger which provides high transverse energy (E_T) electron, photon, π^0 and hadron candidates; the muon system which provides high transverse momentum muon and di-muon candidates; and the Level 0 Decision Unit (L0DU) which takes the decision combining the data of the three over subsystems.

The HLT first reduces the rate to something like 30 kHz using the tracking information to confirm the Level 0 candidate, eventually adding an impact parameter cut. Only then inclusive and exclusive selections are built using the full event reconstruction. Even if the calorimeters also participate to the HLT, their most critical contribution is to the Level 0 for which a dedicated electronic had to be designed, built and commissioned.

3. LHCb Calorimeters

The LHCb calorimeter system [2] is made of four consecutive semi-projective detectors shown on figure 3. The scintillating pad detector (SPD) and the preshower (PS) are two 15 mm thick scintillator planes separated by a 2.5 radiation length lead converter. The light is collected by wavelength-shifting (WLS) fibres inserted and glued in round grooves made in the square pads. The WLS fibres are connected at both ends to clear fibres that transmit the signal to multi-anode photomultipliers. The electromagnetic calorimeter (ECAL) has a shashlik sampling structure with alternating scintillating tiles and lead plates. WLS fibres penetrate the lead/scintillator stack through holes and are read out by photomultipliers. The ECAL is made of about 6000 cells with a transverse size varying from 4×4 cm^2 close to the beam pipe, to 6×6 and 12×12 cm^2 at larger polar angles, corresponding to one to three Moliere radius. The cell granularity is the same in the SPD and PS for a combined use in $\gamma/e/\pi$ separation. The ECAL accounts for 25 radiation lengths and 1.1 interaction lengths. The hadronic calorimeter (HCAL) consists of thin

Figure 3. Pictures of the calorimeters in LHCb cavern. The transverse dimensions of the calorimeters are about 8 m horizontally and 6.5 m vertically.

iron plates interspaced with scintillating tiles arranged along the beam axis. WLS fibres run along the edges of the scintillating tiles and collect the signal to photomultipliers. The HCAL has about 1500 readout cells with a transverse size of 13×13 and 26×26 cm^2 in the inner and outer parts respectively. The HCAL depth is 5.6 interaction lengths.

The purpose of the LHCb calorimeter system is twofold. It gives a fast identification of high E_T electrons, photons, π^0 and hadrons making a crucial contribution to the Level 0 trigger as already mentioned. Offline and in the HLT, it significantly contributes to the electron identification and provides precise measurements of prompt photons, π^0 and η.

4. Calorimeter trigger system

The calorimeter trigger [3] relies on the E_T measurements in ECAL and HCAL clusters made of 2×2 cells. The SPD and the preshower are used to identify respectively the charged and the electromagnetic nature of the cluster. Five types of candidates are then built: electron, photon, local π^0, global π^0, and hadron. The local π^0 candidate corresponds to two photons on the same front-end board and is obtained as the E_T sum of the board (4×8 cells). The global π^0 candidate corresponds to two photons on neighbouring boards and is obtained by adding the highest E_T clusters of two vertically adjacent boards. The highest E_T candidate of each type is selected and send to the L0DU where different thresholds, specific to each type of candidate, are applied to eventually trigger a positive decision. The calorimeter system also provides the total SPD hit multiplicity and the total E_T, two global inputs used to veto respectively very busy events and muon halo background.

Figure 4 shows the global architecture of the calorimeter trigger system. While the selection boards and the L0DU are located in a barrack behind a concrete wall, the calorimeter crates are on a platform on top of the calorimeters in a high radiation area. Thus, anti-fuse and flash FPGAs are used, registers are protected by triple voting, and the data contain parity bits. The total processing time from the input of the front-end board to the input of the L0DU, not counting the optical transmission from the calorimeter platform to the barrack, is around 1 μs.

Figure 5 shows the data flow in the calorimeter crates. In addition to the PS, ECAL and HCAL front-end boards, the crates contain boards dedicated to the trigger processing, namely the SPD multiplicity boards and the Trigger Validation Boards (TVB). Each crate also contains a Calorimeter Read Out Controller board (CROC) dedicated to the interface with the Experiment Control System (ECS) [4] and the DAQ. A specific backplane permits the necessary point to point transmissions between boards within the same crate via the backplane itself or in different

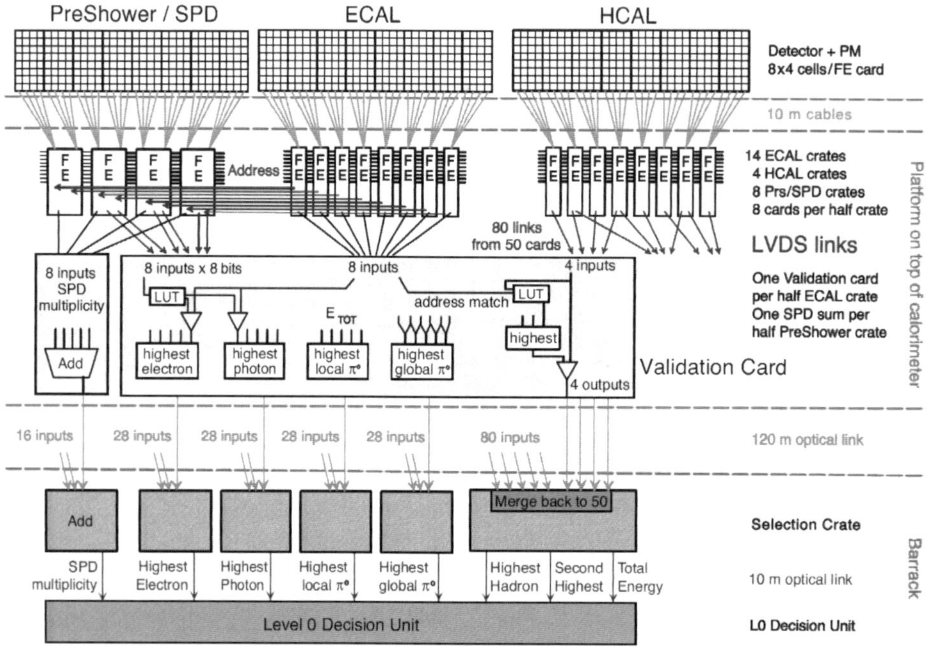

Figure 4. Architecture of the calorimeter trigger system.

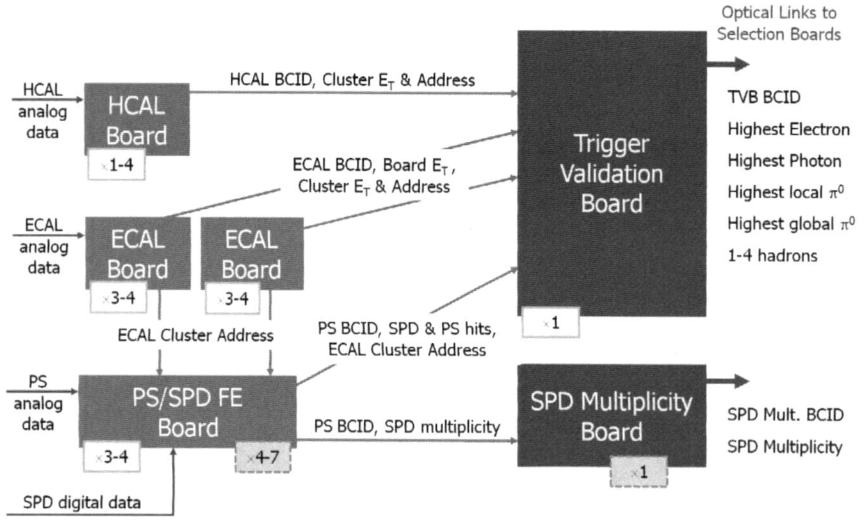

Figure 5. Trigger path data flow in the calorimeter crates. One has three or four PS/SPD front-end boards (64 SPD and 64 PS channels each), six or eight ECAL boards (32 channels each), and one to four HCAL boards (32 channels each) for each TVB. There are 28 TVB in total. One has four or seven PS/SPD boards for each SPD multiplicity board and 16 SPD multiplicity boards in total.

crates via LVDS links. It also permits the transmission of the bottom, left, and bottom-left cells respectively to the bottom, left, and bottom-left neighbour boards in order to build the 2×2 clusters at the edge of several boards.

While ECAL and HCAL front-end boards directly receive the signal from the photomultipliers, the data from the PS and the SPD are first integrated on very front-end boards. In the case of the SPD, the very front-end ASIC also corrects the data for spill-over and makes a 1-bit digitization. ECAL and HCAL front-end boards evaluate all the 2×2 clusters. Each board identifies its highest E_T cluster and sends it to the TVB. Each ECAL board also computes the board E_T summing up its 32 cells and sends it to the TVB. It also sends the address of its highest E_T cluster to the PS/SPD front-end board which sends the corresponding SPD and PS hits to the TVB. In addition, the PS/SPD board computes the number of SPD hits and sends the sum to the SPD multiplicity board.

The TVB uses the SPD and PS hits to identify electron and photon clusters within the ECAL clusters. The highest E_T electron and photon are then sent to the selection boards. The local π^0 and global π^0 candidates are respectively obtained as the highest ECAL board E_T and as the highest sum of two clusters coming from vertically adjacent ECAL boards. For each HCAL candidate, matching ECAL clusters are looked for comparing the addresses. The E_T of the matched ECAL cluster with the highest E_T is then added to the HCAL candidate in order to improve the hadron energy estimate. All the hadron candidates are then sent to the selection boards. The SPD multiplicity board sums the SPD multiplicity over the PS/SPD boards and sends it to the selection boards.

The selection boards find the highest E_T candidate of each type and sum the SPD multiplicity over the SPD multiplicity boards. They also compute the total E_T summing over the E_T of the hadron candidates from the TVB after removing duplicates related to the fact that, in order to search for matched ECAL clusters, over the fifty HCAL front-end boards thirty send their candidates to two TVB. The selection boards send their outputs to the L0DU which centralizes the Level 0 information and takes the decision. Another important part for the trigger system is the TFC (Timing and Fast Control) [5] which broadcasts the clock as well as the triggers to the front-end boards.

5. Expected performances

The Level 0 thresholds have been tuned to maximize the trigger efficiency on signal events passing the offline selection using few mainstream modes such as $B_s \to D_s^- (K^+K^-\pi^-) K^+$, $B^0 \to J/\psi (\mu^+\mu^-) K_s (\pi^+\pi^-)$, $B^0 \to J/\psi (e^+e^-) K_s (\pi^+\pi^-)$ or $B^0 \to K^* (K^+\pi^-) \gamma$. The 1 MHz output bandwidth of the Level 0 is dominated by the hadronic trigger with an allocated bandwidth of 700 kHz while the inclusive bandwidths for the electromagnetic (electron, photon

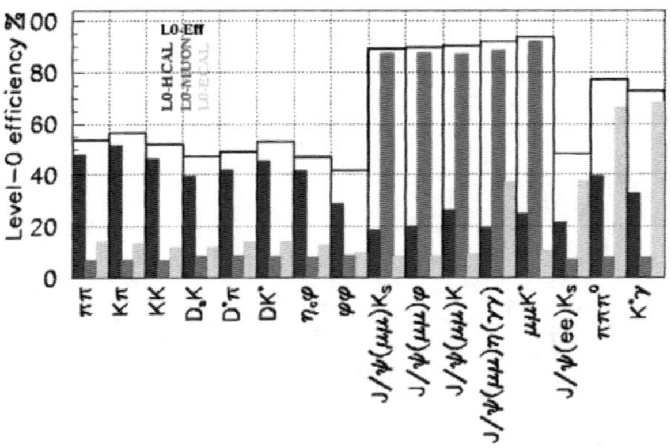

Figure 6. Level 0 efficiencies for some typical B decays. For each mode, the overall efficiency is reported as well as the inclusive efficiencies of the hadronic, muonic and electromagnetic triggers.

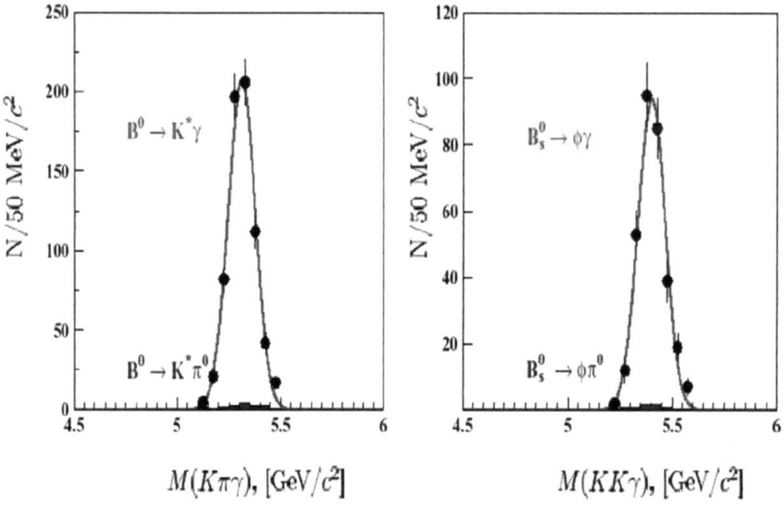

Figure 7. Reconstructed mass of the B meson in $B^0 \to K^*\gamma$ events (left) and in $B_s \to \phi\gamma$ events (right). The small background contributions respectively from $B^0 \to K^*\pi^0$ and $B_s \to \phi\pi^0$ decays are indicated.

and π^0) and muonic triggers are respectively 280 kHz and 160 kHz. Figure 6 shows some typical Level 0 efficiencies which are around 50%, 70% and 90% respectively for hadronic, electromagnetic and muonic B decays.

Essential to the first level trigger, the calorimeters also have an important contribution to the offline reconstruction providing for instance precise measurements of photons and π^0's [6]. Figure 7 illustrates the LHCb physics potential with prompt photons on radiative B decays. It shows the reconstructed mass of the B^0 and B_s respectively in $B^0 \to K^*\gamma$ and $B_s \to \phi\gamma$ events. The expected resolution on the B meson mass is 65 MeV/c^2. In the $B^0 \to K^*\gamma$ ($B_s \to \phi\gamma$) mode, the expected yield for 2 fb^{-1} is $35 \cdot 10^3$ ($9 \cdot 10^3$) selected signal events with a background over signal ratio lower than 0.7 (2.9) at 90% of confidence level [7].

As for photons, the π^0 reconstruction is based on ECAL energy deposits clusterised around local maxima [8]. Low transverse momenta π^0's are mostly reconstructed as a *resolved* pair of well separated photons. At higher momenta, the distance between the two photons at the entrance of the calorimeters becomes of the order of the size of one ECAL pad and energetic π^0's often lead to a single cluster. In order to identify those so called *merged* π^0, an iterative procedure is used to build two virtual interleaved sub-clusters sharing the cell energies according to the expected transverse shape of photon showers. Figure 8 shows the reconstructed mass of resolved and merged π^0 candidates in $B^0 \to \pi^+\pi^-\pi^0$ events. At the π^0 mass, true π^0's represent about one fourth of the resolved π^0 candidates which suffer from a large combinatorial background and about one half of the merged π^0 candidates. Resolved π^0's are reconstructed with a mass resolution of around 10 MeV/c^2 while a core mass resolution of about 15 MeV/c^2 is obtained for merged π^0's. Because of their intrinsic large transverse momentum, merged π^0's are of major importance for the reconstruction of π^0's from B decays. In $B^0 \to \pi^+\pi^-\pi^0$ events, the π^0's from the B^0 meson represent only around one eighth of the true resolved π^0's but more than half of the true merged π^0's. In those events, the average reconstruction efficiency of the π^0 from the B^0 decay is 53% and the expected resolution on the B^0 mass is about 80 MeV/c^2. Those performances lead to an expected precision of 10° in the determination of the CKM angle α for 2 fb^{-1} [9] through a time dependent Dalitz analysis of $B^0 \to \rho\pi$ decays [10].

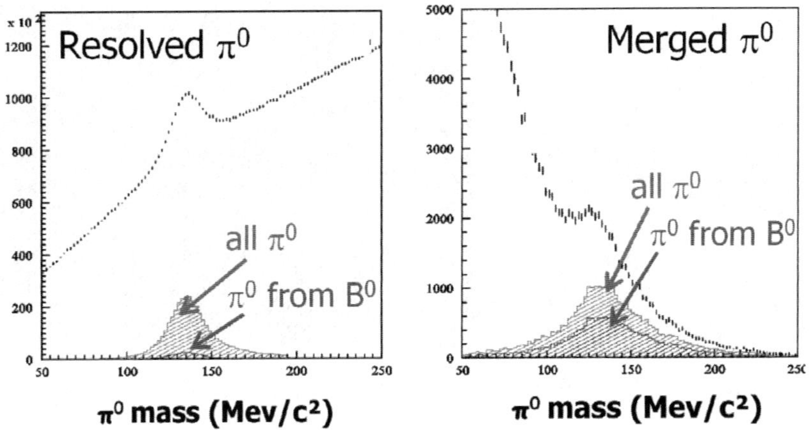

Figure 8. Reconstructed mass of resolved (left) and merged (right) π^0 candidates in $B^0 \to \pi^+\pi^-\pi^0$ events.

6. Commissioning

The commissioning of the calorimeter trigger system started in early 2007. The first in situ test of the full trigger chain was done in May 2007 as follows. First the TFC was configured by ECS to send calibration pulses used to drive the calorimeter LEDs. The front-end boards were reading out the detector and computing the trigger data at 40 MHz. Sequentially, the TVB, the selection boards and the L0DU were receiving their inputs from the previous step, making their computations and sending their outputs to the next step. The LED calibration pulses were associated to high E_T candidates in the L0DU which then sent positive trigger decisions to the TFC. Finally, the TFC broadcast the trigger to the front-end boards which sent the data, via the CROC, to the DAQ for monitoring in the control room and persistent storage. The test was fully successful and gave great confidence in the full trigger chain.

Since December 2007, the calorimeters provide a cosmic trigger to the experiment. This trigger is based on a coincidence between ECAL and HCAL with a special high voltage setting used to detect minimum ionizing particles by increasing the photomultiplier gains. This trigger has a high purity, close to 90%, and a rate of about 10 Hz. It is very useful to experience real data taking, to check cabling and mappings, and for timing and alignment studies. Figures 9 and 10 show two examples of cosmic events selected by this trigger. In the first picture, one can see, in the plan perpendicular to the beam axis, the cosmic track reconstructed in the calorimeters with hits in the SPD, the PS, the ECAL and the HCAL. In the second picture, one has, in the plane containing the beam and vertical axes, the hits left in ECAL and HCAL but also in the muon stations M2 to M5.

7. Triggers at the LHC start-up

Three kinds of trigger are foreseen for the very first collision data: a zero bias trigger, an interaction trigger and an inclusive muon trigger. The zero bias trigger will be based on the LHC bunch structure and will have to be heavily prescaled relatively early: asking for colliding bunches in LHCb will lead to rates of 214 kHz and 765 kHz respectively in the "43" and "156" LHC schemes envisaged for 2008.

This prescaled random trigger will be complemented by an interaction trigger based on some minimal activity in the calorimeters. Few solutions are being investigated and for the moment the best variable seems to be the E_T of the hadron candidate. For instance, for a cut at 1 GeV,

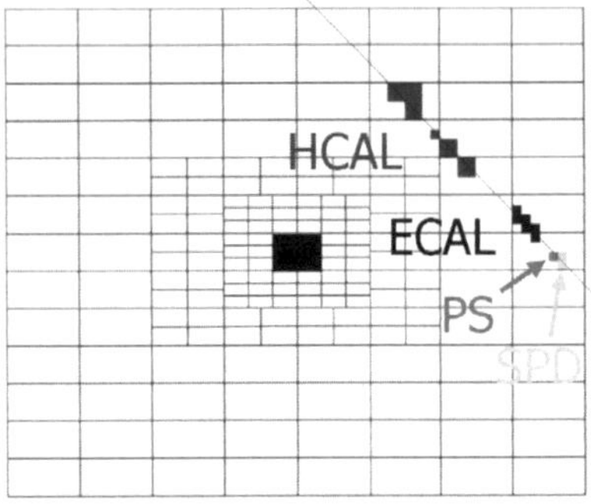

Figure 9. Hits left by a cosmic muon in the calorimeters.

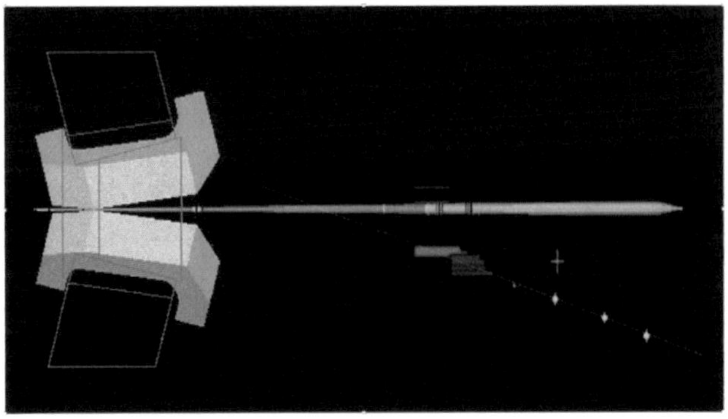

Figure 10. Hits left by a cosmic muon in ECAL and HCAL as well as in the muon stations M2 to M5.

the noise level would be very low, around 1 Hz, but the efficiencies would already be rather good: 79% for hard interactions and 97% for inclusive B production.

8. Conclusion

In the LHCb experiment, the trigger is a key point and heavily relies on the calorimeters. More than 80% of the first level trigger output bandwidth will be dedicated to high E_T electron, photon, π^0 and hadron candidates identified by the calorimeter trigger system. This system has been commissioned and is now routinely used to trigger on cosmic tracks. Thanks to their low electronic noise, the calorimeters will also provide a pure and efficient interaction trigger when the LHC will start. All expected triggers from the calorimeter system will be ready to be registered in pass through mode from the start of the LHC in order to get to the standard configuration as soon as possible.

References

[1] Antunes Nobrega R *et al.* (LHCb Collaboration) 2003 LHCb Reoptimized Detector Design and Performance Technical Design Report *CERN-LHCC-2003-030*

[2] Amato S *et al.* (LHCb Collaboration) 2000 LHCb Calorimeters Technical Design Report *CERN-LHCC-2000-036*

[3] Antunes Nobrega R *et al.* (LHCb Collaboration) 2003 LHCb Trigger System Technical Design Report *CERN-LHCC-2003-031*

[4] Gaspar C, Jacobsson R, Jost B, Morlini S, Neufeld N, Vannerem P and Franek B 2004 An integrated experiment control system, architecture, and benefits: the LHCb approach *IEEE Trans. Nucl. Sci.* **51** 513

[5] Guzik Z , Jacobsson R, Jost B 2004 Driving the LHCb front-end readout *IEEE Trans. Nucl. Sci.* **51** 508

[6] Deschamps O, Machefert F, Schune M H, Pakhlova G and Belyaev I 2003 Photon and neutral pion reconstruction *LHCB-2003-091*

[7] Shchutska L, Golutvin A and Belyaev I 2007 Study of radiative penguin decays $B^0 \to K^{*0}\gamma$ and $B_s^0 \to \phi\gamma$ at LHCb *CERN-LHCB-2007-030*

[8] Breton V, Brun N and Perret P 2001 A clustering algorithm for the LHCb electromagnetic calorimeter using a cellular automaton *LHCB-2001-123*

[9] Deschamps O, Monteil S, Perret P, Robert A, Machefert F, Robbe P and Schune M H 2007 The CKM angle α at LHCb *CERN-LHCB-2007-046*

[10] Snyder A E and Quinn H R 1993 Measuring CP asymmetry in $B \to \rho\pi$ decays without ambiguities *Phys. Rev.* D **48** 2139

Journal of Physics: Conference Series **160** (2009) 012064 doi:10.1088/1742-6596/160/1/012064

The CALICE Test Beam Programme

F Salvatore[1]
Royal Holloway University of London,
Physics Department, Egham Hill, Egham, Surrey, TW20 0EX, UK

E-mail: p.salvatore@rhul.ac.uk

Abstract. A very challenging test beam programme is being undertaken by the CALICE collaboration as part of a major R&D directed towards the design of an ILC calorimeter. This design has to be optimized for both performance and cost, where particle flow (PFA) calorimetry and software compensation are the main aim of the studies. This paper will concentrate on describing the experimental set-ups for the 2006, 2007 and 2008 test beams that have been carried out by the CALICE collaboration at CERN and FNAL.

1. Introduction

The CALICE collaboration is involved in a major programme of R&D into calorimetry for the International Linear Collider (ILC)[1]. The aim of the project is to compare the performance of different technologies for electromagnetic and hadronic calorimeters in terms of ILC requirements in a common framework. The main direction of the collaboration R&D is to study particle flow (PFA) calorimetry[2], software compensation and individual particle reconstruction, and therefore the studies are concentrating on fine granularity calorimeters with a high degree of longitudinal segmentation. These studies include comparison of test beam data with simulation models to measure their degree of agreement, the technical issues of building a detector optimized for PFA calorimetry, and development of algorithms for software compensation and particle flow reconstruction. For this purpose, a very intense test beam programme is being undertaken for extensive tests of calorimeter prototypes.

2. The 2006 test beam set-up at CERN

In August and October 2006, the CALICE Silicon Tungsten Electromagnetic calorimeter (SiW-ECAL)[3] and Analogue Hadronic calorimeter (AHCAL)[4] prototypes were tested in the H6B[5] experimental area at the CERN SPS. A detailed description of the SiW-ECAL prototype is given in reference [3]. It consists of 3 sets of 10 layers of tungsten of 1.4, 2.8 and 4.2 mm thickness respectively, for a total of 24 radiation lengths at normal incidence. Thirty layers of silicon PIN diode pads interleaved between the tungsten plates are used to sample the shower development. Each silicon pad is 1x1cm^2 and the sensors are made on 4 inch wafers in units of 6x6 pads. The layers consist of a 3x2 array of wafers, corresponding to 18 pads horizontally and 12 pads vertically, leading to a total of 6480 pads for the whole ECAL prototype (see figure 1). The readout of the SiW-ECAL is through a custom on-detector ASIC and VME readout boards[3].

The AHCAL prototype[4] is a sampling calorimeter with 38 layers of steel absorber sheets, instrumented with scintillator tiles which are read out using SiPMs[6]. The tiles are of varying size,

[1] p.salvatore@rhul.ac.uk

with the highest granularity central region using 3x3cm^3 tiles, increasing to 12x12cm^2 for the outmost tiles, for a total of number of ~8,000 channels (see figure 2). The readout from the SiPM is through a custom on-detector board and the AHCAL uses the same VME readout boards as the SiW-ECAL. The total interaction length of the AHCAL prototype is 4.5λ.
The hadronic calorimeter is complemented by a Tail Catcher and Muon Tracker (TCMT) detector[7], consisting of 96cm of iron instrumented with 16 layers of 0.5x5cm^2 scintillator strips, which tags the shower leakage and detect muons (see figure 3). It has a total of around 300 channels and the scintillator strips use the same SiPM readout and have the same downstream readout electronics as the AHCAL.

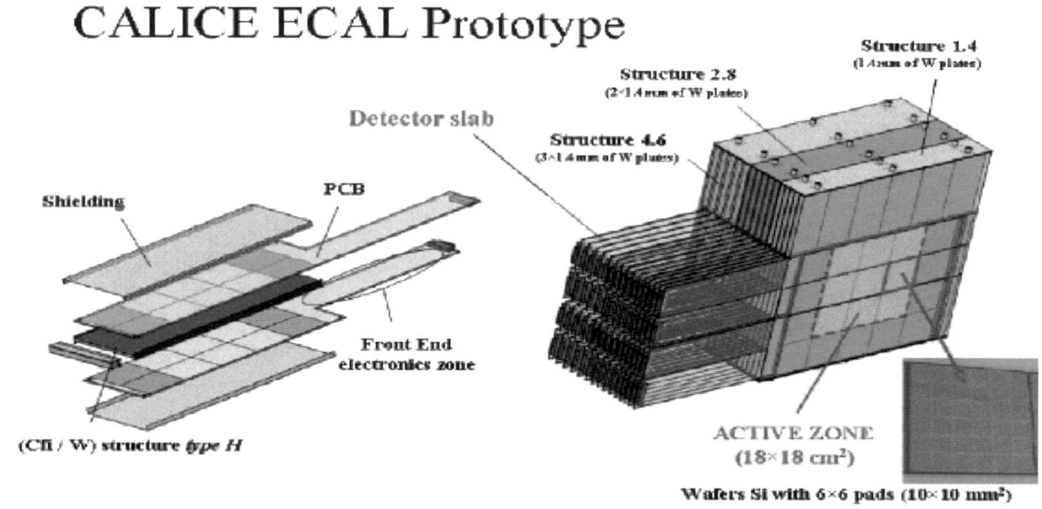

Figure 1. The CALICE SiW-ECAL prototype.

The beam line installation at CERN included locally provided beam detectors (multi wired proportional chambers – MWPC) and custom made scintillation detectors for the experimental trigger. A sketch of the experimental setup is shown in figure 4.

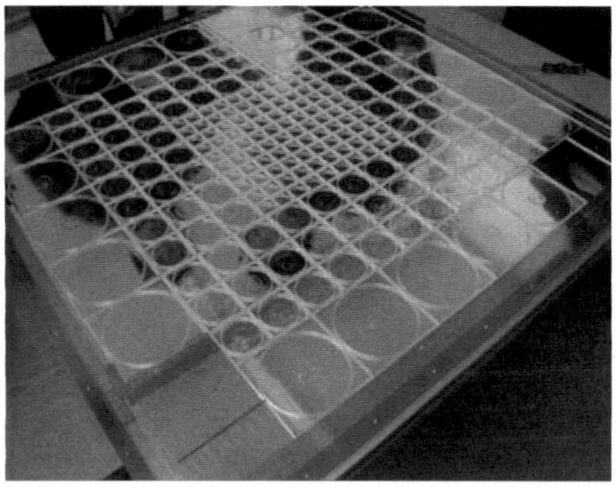

Figure 2. One scintillator layer of the CALICE AHCAL prototype

Figure 3. The CALICE TCMT prototype.

Journal of Physics: Conference Series **160** (2009) 012064 doi:10.1088/1742-6596/160/1/012064

All the prototypes have been tested on the CERN SPS beam line during two separate test beam periods, in August and October 2006. The three prototypes were installed in the H6B experimental line, as detailed in figure 5.

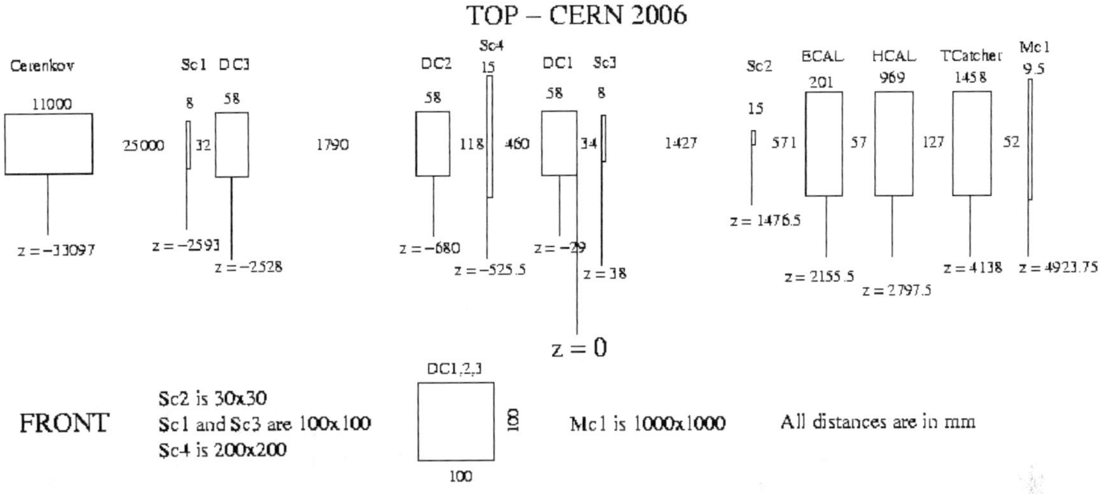

Figure 4. Sketch of the 2006 CERN test beam set-up.

In the August period, data have been taken with the SiW-ECAL and AHCAL, and also with only the SiW-ECAL on the beam line. Electron and pion beams have been used, with energies between 6 and 50 GeV for electrons and 30 to 80 GeV for pions. Since the SiW-ECAL was mounted on a platform that allowed it to be rotated, data have been taken for 4 different SiW-ECAL angles in the ECAL+AHCAL period (0°, 10°, 20°, 30°), and for 5 different angles in the ECAL alone period (0°, 10°, 20°, 30°, 45°).

Figure 5. The CALICE SiW-ECAL, AHCAL and TCMT prototypes in the H6B experimental area at the SPS at CERN during the 2006 test beam .

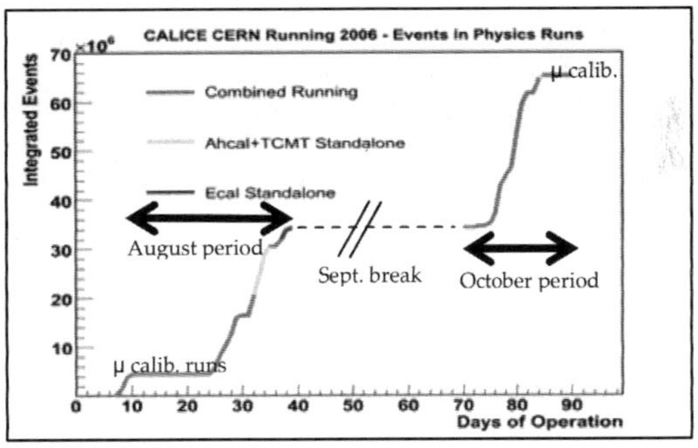

Figure 6. Summary of the data taken by the CALICE collaboration at the 2006 CERN test beam.

A total of ~11 million triggers were collected in the first beam period only. In October, the full system consisting of SiW-ECAL, AHCAL and TCMT was exposed to the SPS beam, using $e^{\pm}$ beams from 6 to 45 GeV and $\pi^{\pm}$ beams from 6 to 60 GeV. A total of ~6 million triggers were collected. For calibration, an additional 70 million muon events were recorded in both beam periods, for a total of ~6 TB of data (beam+calibration) events collected on disk. A summary of the total data recorded at the

2006 test beam is shown in figure 6. The analysis of both the electron and pion data in the SiW-ECAL and AHCAL is well under way and is detailed in various contributions in these proceedings (see [8-12]).

3. The 2007 test beam at CERN

Between June and August 2007 the Calice collaboration has successfully commissioned and operated the full chain of calorimeter prototypes in the H6B experimental area at the CERN SPS (figure 7).

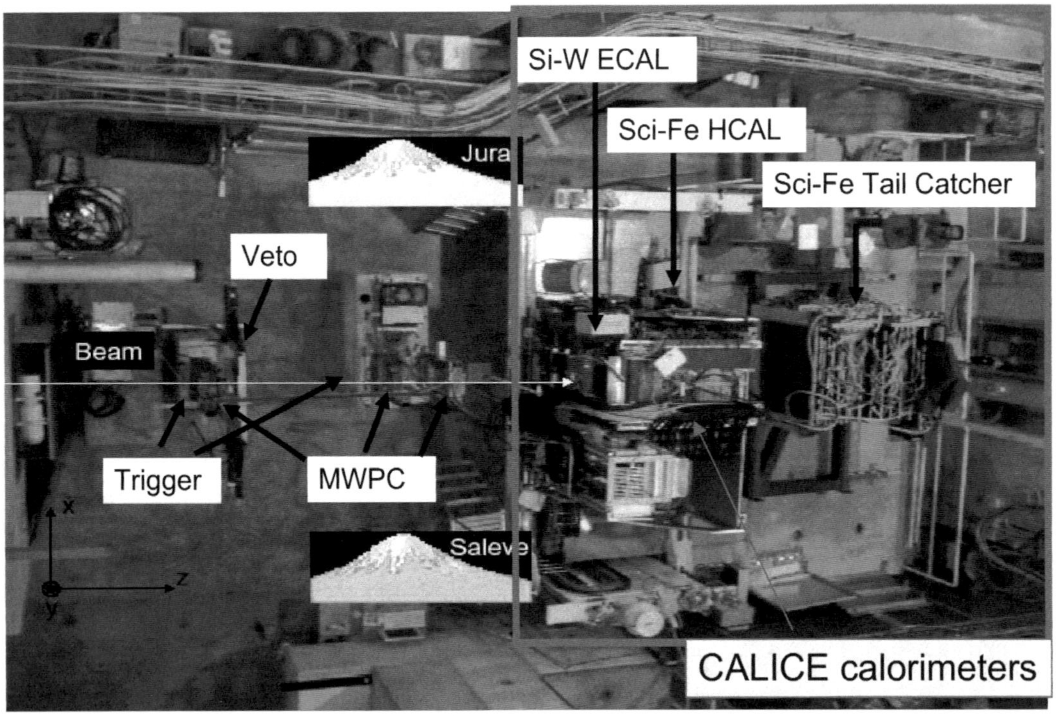

Figure 7. The installation of the CALICE SiW-ECAL, AHCAL and TCMT at the 2007 CERN test beam.

In 2007, the ECAL was equipped with 30 sensitive layers of silicon pads, corresponding to a total of 54 PCBs. The total number of readout channels was 9072, corresponding to 216 channels/PCB in the central part of the detector and 108 channels/PCB in the bottom part.

A total of 38 fully commissioned modules of the AHCAL were installed on the beam line; 30 modules with fine granularity (216 scintillator tiles) and 8 modules with coarse granularity (141 tiles) were present. Each tile is readout by a silicon photo-multiplier (SiPM), for a total of 7608 readout channels. The TCMT was completely installed with all 16 active layers fully instrumented and a total of 320 readout channels.

The trigger to the experiment is provided by the coincidence of two 10x10cm^2 scintillator plates with photo-multiplier readout. In addition a coincidence with a muon wall downstream of the detector may be used either for muon rejection or as a muon trigger during calibration. The analogue readout of an additional 20x20cm^2 scintillator plate serves as veto for events with double particles or showers initiated in the material upstream of the detector. In order to tag the halo of the beam, an additional 100x100cm^2 scintillator with a 20x20cm^2 hole has been employed as an outer veto. All triggers are digitized and recorded event by event by the VME-based data acquisition (DAQ), and can be used offline for data selection. A threshold Cherenkov counter filled with helium gas has been used to discriminate electrons from pions, in the range 6-20 GeV. The same detector has also been used with

nitrogen gas in order to discriminate pions and protons in the range 30-80 GeV. The gas pressure in the 11m long Cherenkov vessel needs to be adjusted depending on the beam energy. With optimal settings, efficiencies of 90% are obtained, going to 30 % with increasing energy. The discriminated Cherenkov signal is recorded as a trigger bit. For particle tracking, three sets of delay multi-wire proportional chambers provided by CERN have been included in the CALICE DAQ. Three pairs of x and y planes with two wires each are read out for each event by a TDC implemented in the DAQ. The spatial resolution of the tracking system is better than 200μm.
The system, with more than 16000 channels and an acquisition rate capability of 120 Hz (see figure 8), is a compact HEP experiment in itself.

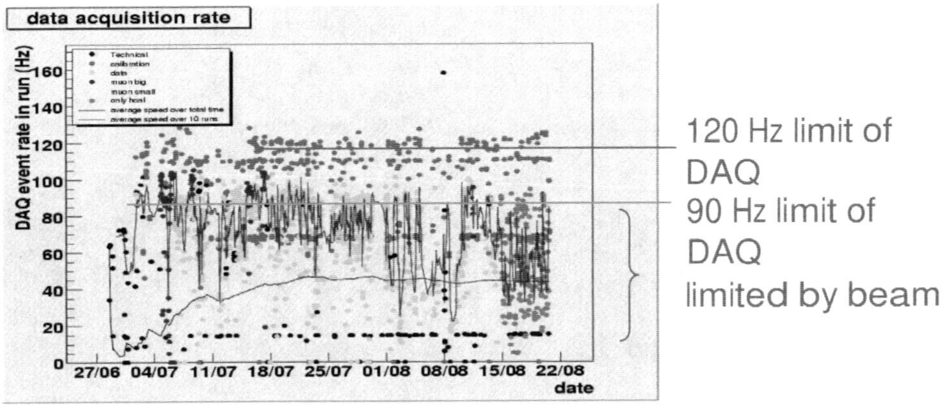

Figure 8. Data acquisition rate for the 2007 CALICE test beam at CERN. The average acquisition rate (black line) is ~50 Hz.

The performance of all beam-line detectors as well as that of the 3 calorimeter prototypes has been monitored online during data taking. A special fast analysis tool has been developed to access in real time the relevant beam and detector qualities. Several checks have been possible to monitor the response of the calorimeters at different energies. The preliminary response for the SiW-ECAL and AHCAL (with no calibration) can bee seen in figure 9 and 10.
The programme for the test beam has been very intense and has been completely fulfilled at the end of the 7 weeks of data taking. The collaboration has collected more than 200 million events (see figure 11 and 12), completing the muon calibration of all components, the electromagnetic program of both ECAL and AHCAL and hadronic program for the combined detector at four different incident angles of the beam. Both 'minus' (e^-/π^-) and 'plus' (e^+/p^+/proton) beam events have been recorded. A full scan of the calorimeters' front faces has been performed, as detailed in figure 13.
An important part of this year's test beam has been the irradiation of a test SiW-ECAL PCB with embedded electronics, to evaluate a second generation prototype of electronics for the ILC. The irradiation has been performed using 70 GeV and 90 GeV electron beams, and doing a complete position scan of the four chips present in the test board, as shown in figure 14. The test PCB has been inserted in the SiW-ECAL structure at the point of the shower maximum.
The most ambitious part of the test beam programme has been the rotation of the SiW-ECAL and AHCAL prototypes, with subsequent re-staggering of the active parts of the calorimeters. The success of this part of the programme has been possible thanks to the movable stage on which the SiW-ECAL and AHCAL have been installed.

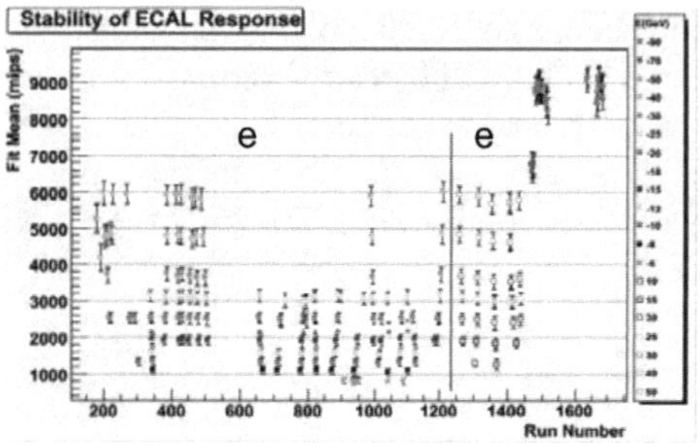

Figure 9. Response of the CALICE SiW-ECAL to electron beams, obtained analysing the data taken at the 2007 CERN test beam using the on-line monitoring programme.

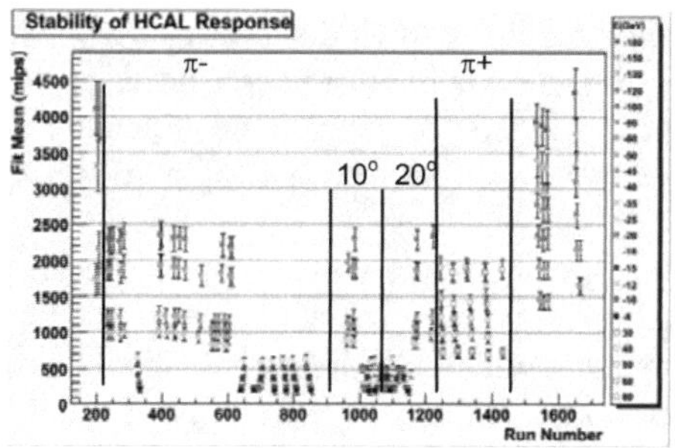

Figure 10. Response of the CALICE AHCAL to pion beams, obtained analysing the data taken at the 2007 CERN test beam using the on-line monitoring programme.

This 16 tonnes structure, designed and built at Desy, allowed for the X and Y movement of all the calorimeters. Moreover, the SiW-ECAL and AHCAL have been mounted on a steel platform that could be rotated to a maximum of 30 degrees with respect to the direction of the beam. Data have been collected with the SiW-ECAL and AHCAL rotated by 10°, 20° and 30° with respect to the normal beam incidence direction (see figure 15).

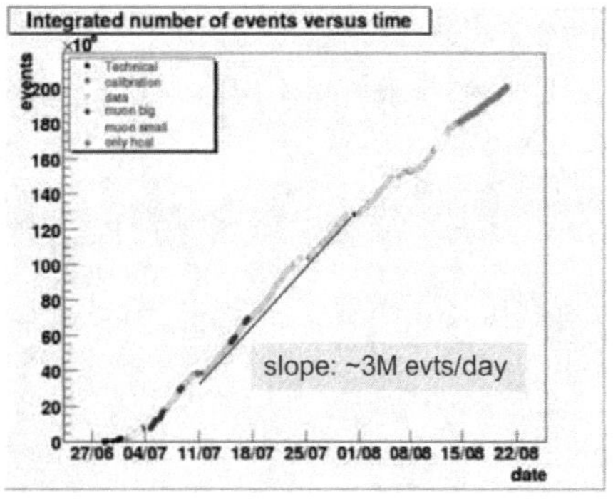

Figure 11. Total integrated luminosity collected at the 2007 CALICE test beam at CERN.

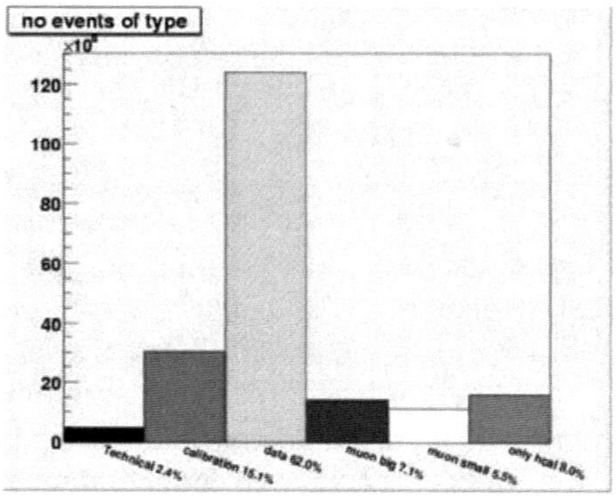

Figure 12. Data types collected during the 2007 CALICE test beam at CERN: 62% of the collected data are e/π/p interactions in the calorimeters.

All data collected during the test beam where immediately processed and reconstructed using the Grid tools. The recorded runs were available from the Desy dcache to the whole collaboration within hours of being collected at CERN.

4. The 2008 test beam at FNAL

In April 2008, the full chain of calorimeter prototypes has been installed on the MTBF test beam line[13] at Fermi National Accelerator Laboratory (FNAL) in Batavia (IL, USA), see figure 16. The prototypes will be tested using electron and pion beams from 6 to 60 GeV, in order to get a data sample that could later be compared to the one collected at CERN. A full programme of integration of different calorimeter prototypes, like the SiW-ECAL with a Digital HCAL (DHCAL) prototype[14] or a Scintillator ECAL prototype[15] with the AHCAL, will be performed, in order to optimize different choices of electromagnetic and hadronic calorimeters for the ILC.

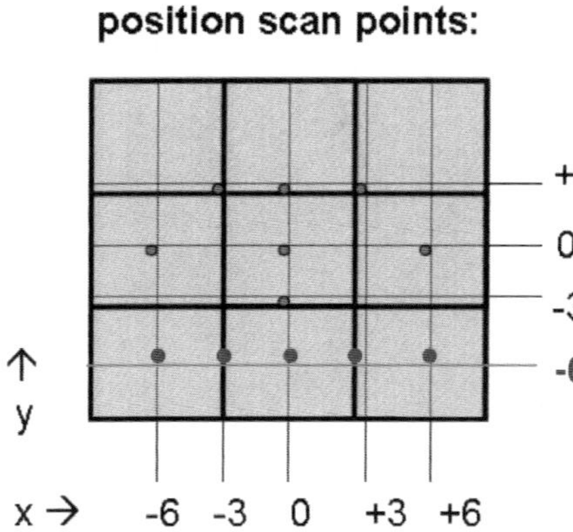

Figure 13. Summary of the position scanning on the front face of the SiW-ECAL performed at the 2007 CALICE test beam at CERN.

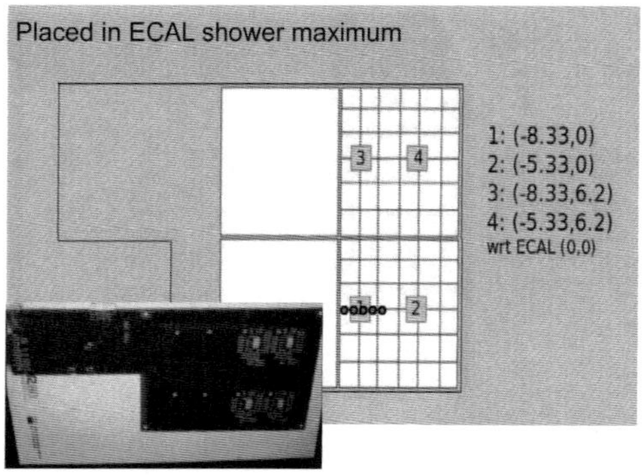

Figure 14. Sketch of the scanning of the SiW-ECAL PCB with embedded electronics performed at the 2007 CALICE test beam at CERN.

5. Conclusions

During 2006 and 2007 the CALICE collaboration has performed extremely successful test beams at the H6B experimental area at CERN. More than 260 million events have been collected during both tests and all the data are available on the Grid.

Figure 15. Set-up of the SiW-ECAL and AHCAL after the 30° rotation of the movable platform.

The analysis of the data is well under way and preliminary results have been shown during this conference and are summarized in these proceedings. The next phase of beam tests of the CALICE calorimeter prototypes is started in April 2008, when the detectors have been installed at the MTBF test beam area at FNAL. One of the aims of this year's tests will be to expose different combination of calorimeter prototypes to electron and pion beams (e.g. SiW-ECAL with DHCAL or Sci-ECAL with AHCAL), in order to optimize the choice of calorimeters for a future Linear Collider detector.

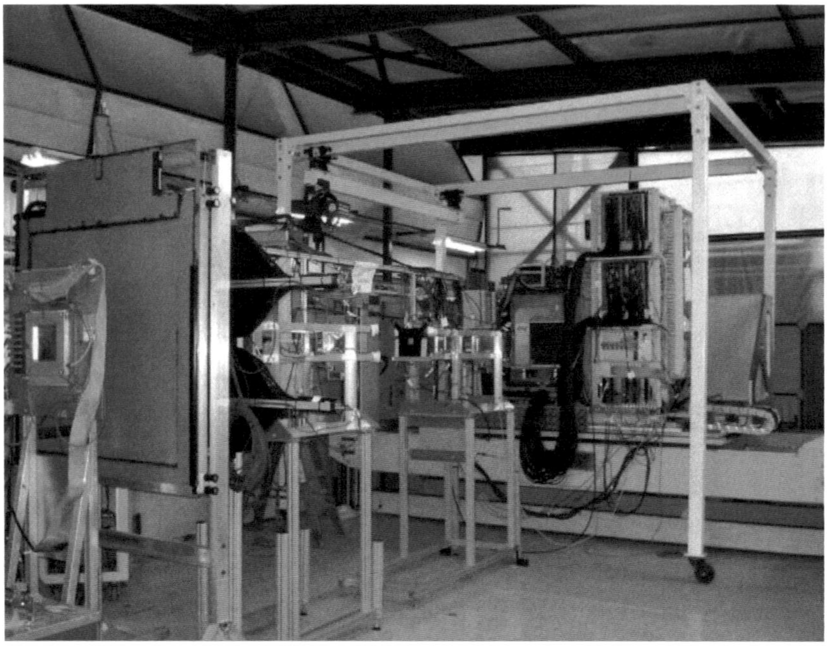

Figure 16. Set-up of the SiW-ECAL and AHCAL at the MTBF beam line at FNAL (April 2008).

6. References

[1] Benke T *et al (ed)*, "Reference Design Report, Volume 4: Detectors", available at http://lcdev.kek.jp/RDR
[2] Morgunov V L, "Calorimetry design with energy-flow concept", in Proc. of "X International Conference on Calorimetry in High Energy Physics", Pasadena, CA, 2002
[3] Repond J *et al*, "Design and Commissioning of the Physics Prototype of a SiW Electromagnetic Calorimeter for the International Linear Collider", *submitted to JINST*
[4] Eigen G, "The Calice scintillator HCAL Test Beam Prototype", in Proc. of "XII International Conference on Calorimetry in High Energy Physics", Chicago, IL, 2006
[5] Description of the CERN H6 area available at: http://ab-div-atb-ea.web.cern.ch/ab-div-atb-ea/BeamsAndAreas
[6] Buzhan P *et al*, *Nucl. Inst. Meth.* A **504** (2003) 48
[7] Chakraborty D, "The Tail-Cathcer/Muon Tracker for the CALICE Test Beam", in Proc. of the "2005 International Linear Collider Workshop (LCWS 2005)", Stanford, CA, 2005
[8] Boumediene D, these proceedings
[9] Cornat R, these proceedings
[10] Bartsch V, these proceedings
[11] Garutti E, these proceedings
[12] Lucaci-Timoce A, these proceedings
[13] Description of the FNAL MTBF area available at: http://www-ppd.fnal.gov/MTBF-w/
[14] Repond J, these proceedings
[15] Jeans D, these proceedings

Journal of Physics: Conference Series **160** (2009) 012065 doi:10.1088/1742-6596/160/1/012065

Response of the CALICE Si-W ECAL Physics Prototype to Electrons

Djamel BOUMEDIENE on behalf of the CALICE Collaboration
Laboratoire de Physique Corpusculaire, CNRS/IN^{2}P^3, Univ. B. Pascal
24 Av. des Landais, F-63177 Aubière Cedex - France
E-mail: Djamel.Boumediene@cern.ch

Abstract. The CALICE [1] Si-W electromagnetic calorimeter [2] has been tested with low energy electron beams at DESY (1 to 6 GeV) in May 2006, as well as higher energy electrons (6 to 45 GeV) at CERN in August and October 2006. The large amount of high quality data allows the characterisation the ECAL physics performance.

1. Introduction

In the frame of Particle Flow Analysis and in a high particle multiplicity environment, the jet measurements require efficient single particle reconstructions. A way of achieving it is by building high granularity calorimeters as proposed for ILC [3] by the CALICE collaboration.

The ECAL prototype developed by the collaboration is a highly segmented, sampling Si-W calorimeter. It has 30 layers, each with 216 1×1 cm^2 detection pads distributed on six wafers leading to a total of 6480 channels. Around the pads in each wafer, a non-active region of 1.8 mm width was used for a grounded guard ring structure.

The prototype was tested during two test beam campaigns at DESY and CERN. Four drift chambers were used to monitor the beam. Čerenkov detectors were also available for e/π discrimination. The large sample of collected data was used to characterise the ECAL physics performance in terms of energy resolution, linearity of the response, spatial resolution as well as the longitudinal and radial development of the electromagnetic showers. The spatial homogeneity of the detector was also addressed.

Some comparisons are performed between data and Monte Carlo simulations based on Geant4 [4] [5].

2. Electron selection

The total energy recorded in the ECAL, E_{raw}, is defined as follows :

$$E_{\text{raw}} = \sum_{i=0}^{i=9} E_i + 2\sum_{i=10}^{i=19} E_i + 3\sum_{i=20}^{i=29} E_i,$$

where the three stacks are weighted in the ratio 1:2:3 according to the tungsten thickness in each of them.

The selection of single electrons is performed in three steps :

Journal of Physics: Conference Series **160** (2009) 012065 doi:10.1088/1742-6596/160/1/012065

- an energy window, scaled to the beam energy, is applied on E_{raw} :

$$125 < \frac{E_{\text{raw}}(\text{MIP})}{E_{\text{beam}}(\text{GeV})} < 375$$

  This cut rejects the low energy peaks due to muons and pions ;
- rejection of pions is obtained by requesting higher then threshold signals from the Čerenkov counter ;
- the beam halo which contains low energy electrons is rejected on a run-per-run basis.

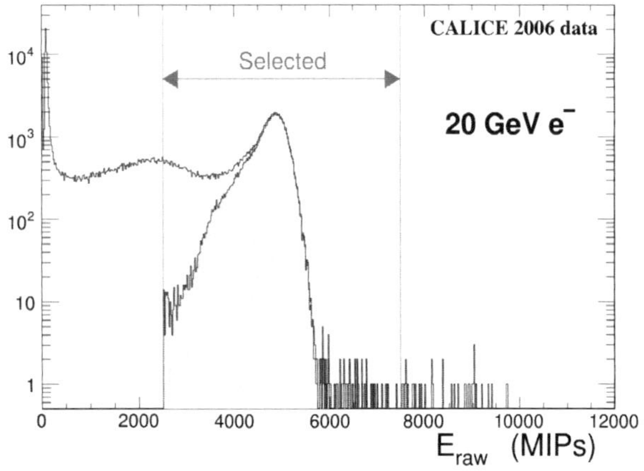

Figure 1. Distribution of E_{raw} for a 20 GeV beam. The E_{raw} selection window and the shaded area obtained by demanding a signal from the Čerenkov counter are shown. The low energy peak is due to MIP like particles (muons and pions).

3. Interwafer gap effects

The guard-rings represent a non-active area of 2 mm width leading to energy loss when particle traverse them. They represent the dominant source of lateral inhomogeneity. This non-uniformity is studied as a function of the position of the shower barycentre. The shower barycentre is given by the mean position of the hits weighted by their energy :

$$(\overline{x}, \overline{y}) = \left(\sum_i E_i x_i, \sum_i E_i y_i\right) / \sum_i E_i$$

The effect of the guard rings on the uniformity of the ECAL energy response is illustrated in figure 2. The ECAL energy response, $f(x, y) = E_{\text{raw}}/E_{\text{beam}}$, is modeled by a Gaussian function :

$$f(\overline{x}, \overline{y}) = \left(1 - a_x e^{-\frac{(\overline{x}-x_{gap})^2}{2\sigma_x^2}}\right)\left(1 - a_y e^{-\frac{(\overline{y}-y_{gap})^2}{2\sigma_y^2}}\right)$$

$f(\overline{x}, \overline{y})$ is fitted to the data, as displayed on figure 3, to extract the values of its parameters, a_x, a_y, σ_x and σ_y.

The energy loss in the interwafer gaps can be corrected by applying a $1/f$ correction factor to the reconstructed energy of each shower. Since this method relies on calorimetric informations only, it can be applied on photons as well as on electrons.

As shown on figure 4, the correction improves the detector uniformity to the percent level. It also improves the shape of the energy distribution, which becomes more Gaussian as shown on figure 5.

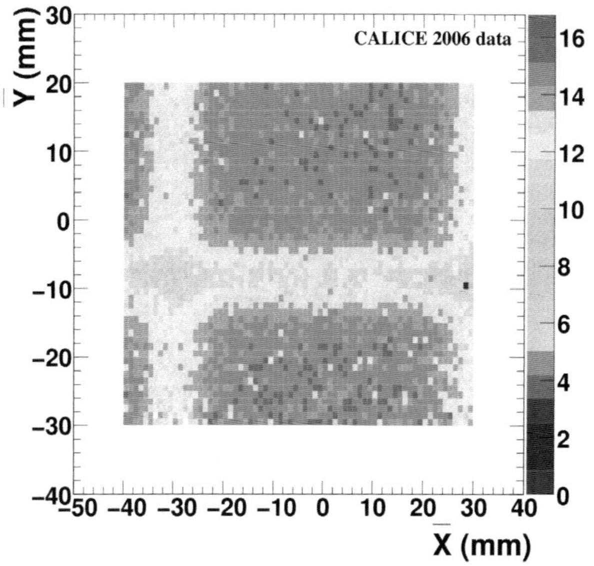

Figure 2. Mean values of E_{raw} for a 15 GeV e^- beam as function of the barycentre coordinates.The square pattern corresponds to the effect of the guard rings on the reconstructed energy.

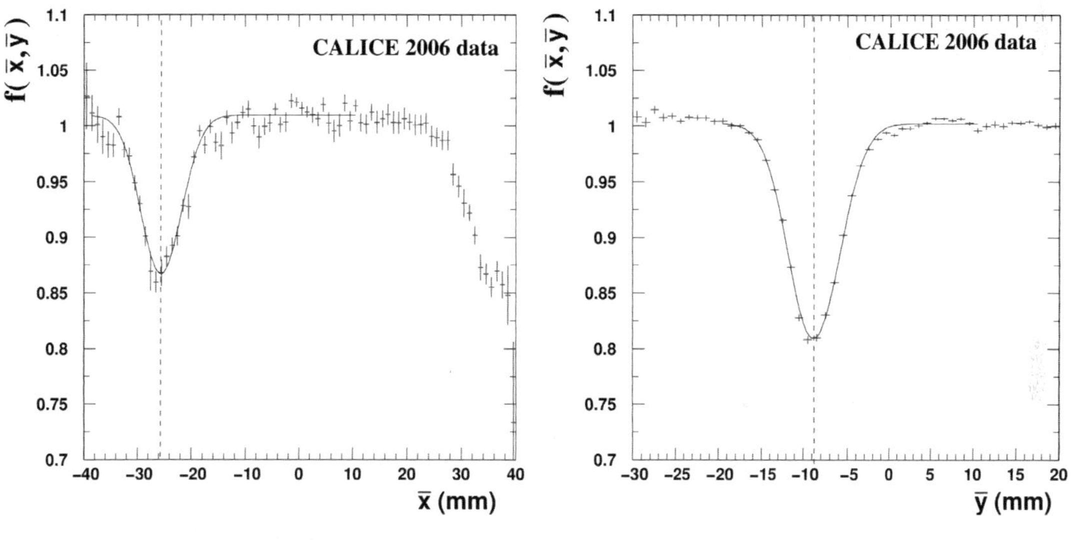

Figure 3. $f(\overline{x}, \overline{y})$ function of the shower barycentre coordinates, for a combined sample of 10, 15 and 20 GeV electrons. To characterise the x (y) response, the events were requested to be outside the interwafer gap in y (x).

4. Performance studies

4.1. Spatial and angular resolution

The spatial and angular resolution of the ECAL are studied with the DESY data at normal incidence. The shower direction and position at the ECAL front face are constructed on an event-by-event basis using a linear two-parameter chi-square fit to the shower barycentre positions in each layer for the x and y coordinates separately. The correlation matrix is determined from simulations for each beam energy.

The fit results are compared with the position and angle measured by the tracking system. The expected e^- position and direction at ECAL front face is obtained from a linear fit of the drift chambers. Sources of systematic uncertainties as residual misalignment, material modeling,

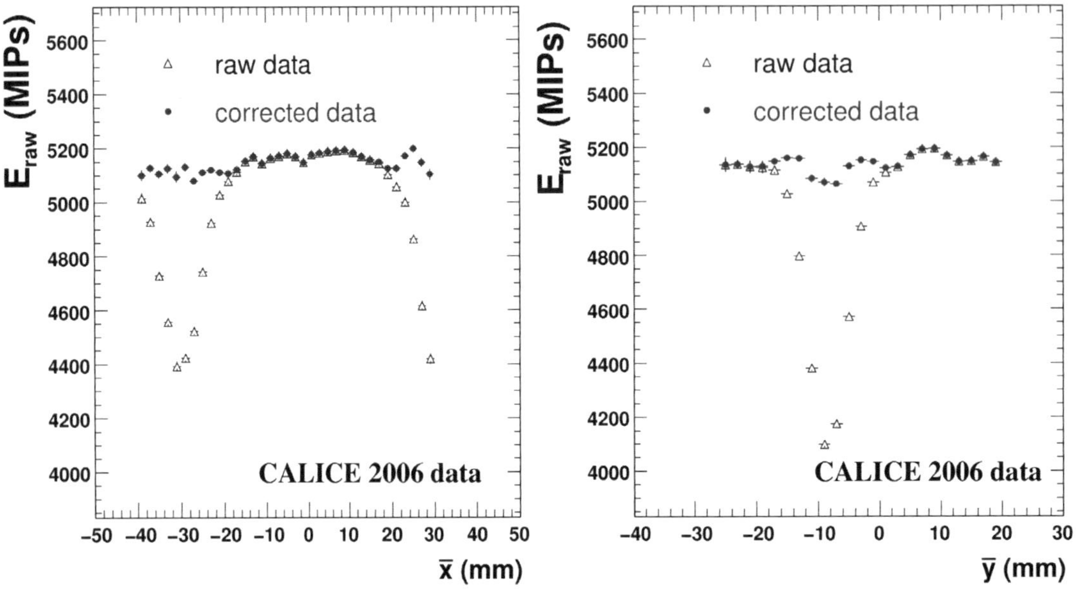

Figure 4. Mean E_{raw} function of the shower barycentre coordinates for 20 GeV electrons, before (black triangles) and after the corrections (blue circles) were applied on E_{raw}.

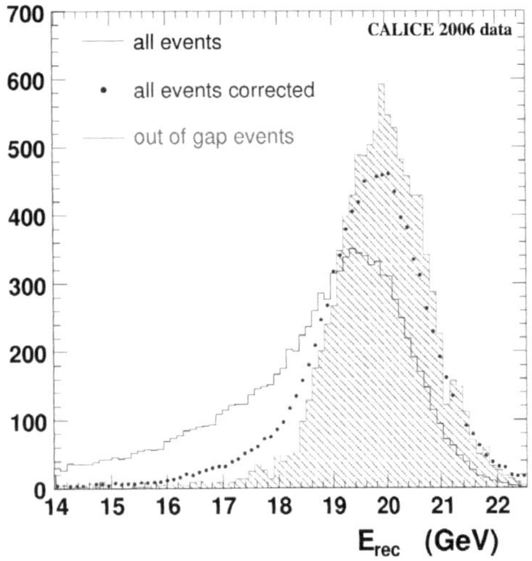

Figure 5. Energy distribution for 20 GeV electrons in case of events outside the interwafer gaps (red histogram), all events without corrections (black histogram) and all events with corrections (blue circles). The histograms are normalised to the same number of entries.

and background rate are estimated for the extrapolation to the ECAL front face.

The ECAL resolution, deconvoluted from the tracking errors, is displayed in figure 6.

4.2. Selection of showers well contained in the ECAL

A fiducial volume selects the showers well contained in the ECAL module and in the wafers. A first cut is applied to the distance between the barycentre of the electron shower and the borders of the ECAL in order to minimise the effect of the lateral leakage, while a second cut is applied on the distance between the electron track and the wafer borders in order to disentangle the effect of the guard rings. A summary of the selected electron and positron data is shown in table 1.

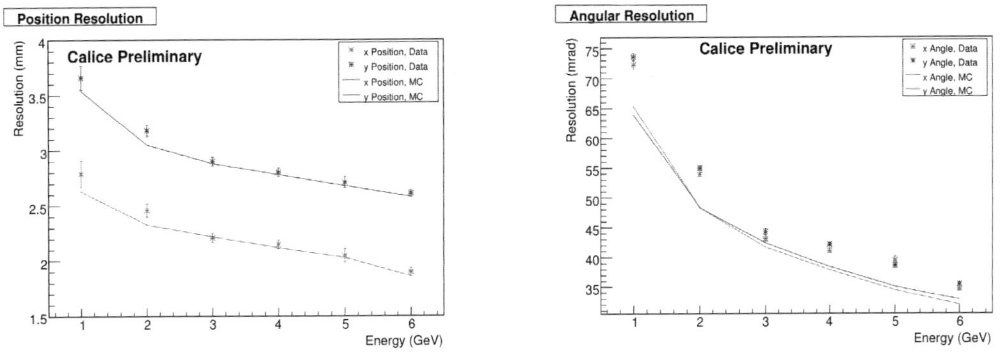

Figure 6. Resolutions in position (left) and angle (right) as a function of the beam energy. The data are shown as points with error bars. The expectation from simulation is shown by the continuous line.

Table 1. Number of selected electron events used to characterise the response of the calorimeter.

Beam energy (GeV)	Particle	Sample size (kevts)
6	e^+, e^-	10.6
10	e^+, e^-	55.9
12	e^+, e^-	32.1
15	e^+, e^-	60.4
20	e^+, e^-	76.9
30	e^+, e^-	43
40	e^-	27
45	e^-	129.3

4.3. Linearity and energy resolution

The total energy response of the calorimeter is calculated as

$$E_{\text{rec}} = \sum_{i=0}^{i=29} w_i E_i$$

where w_i is the sampling fraction of the layer i. This weight takes into account the tungsten thickness as well as the contribution of the passive materials of the calorimeter : PCB, aluminum, glue, etc. The distribution of E_{rec} for 20 GeV electrons is shown on figure 7. A Gaussian fit is performed in the range $[-\sigma, +2\sigma]$ in order to extract the mean energy response and the dispersion. The sensitivity to the pion background and the influence of the inter-wafer gaps are reduced by the asymmetry of the range.

The error on the beam energy, ΔE_{beam}, is taken into account and evaluated using the dispersion of the mean value of E_{rec}, observed on different runs at the same energy :

$$\frac{\Delta E_{\text{beam}}}{E_{\text{beam}}} = \frac{0.12}{E_{\text{beam}}(\text{GeV})} \oplus 0.1\%$$

The linearity of the response with respect to the beam energy is described by :

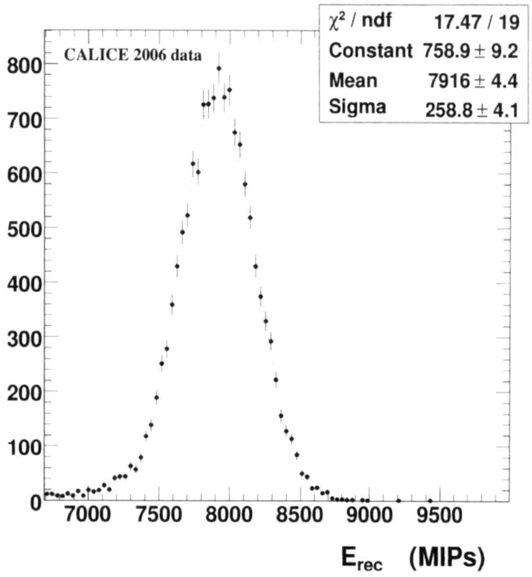

Figure 7. Gaussian parametrisation of E_{rec} for a 30 GeV electron run; The range of the fit is $[-\sigma, +2\sigma]$.

$$E_{\mathrm{mean}} = \beta E_{\mathrm{beam}} - \alpha,$$

where α and β are extracted from a fit to the data as shown on figure 8. The response of ECAL is linear at percent level as deduced from the residuals, displayed on figure 9, and the measured energy can be defined as $E_{\mathrm{meas}} = (E_{\mathrm{mean}} + \alpha)/\beta$.

The relative energy resolution $\Delta E_{\mathrm{meas}}/E_{\mathrm{meas}}$ can be parameterised by a quadrature sum of stochastic and constant term (figure 10) :

$$\frac{\Delta E_{\mathrm{meas}}}{E_{\mathrm{meas}}} = \frac{16.69 \pm 0.13}{\sqrt{E(\mathrm{GeV})}} \oplus (1.09 \pm 0.06)\%$$

from which the intrinsic momentum spread of the beam was subtracted.

4.4. Shower Development

Only events outside the interwafer gaps were used to characterise the longitudinal development of the shower. The mean energy distribution is well fitted by the standard parametrisation, $\gamma(t) = ct^{\alpha}exp(-\beta t)$, where t is the calorimeter depth, c is an overall normalisation, α and β are constants (figure 11). The position of the shower maximum grows logarithmically with the beam energy (figure 12).

An important issue in the development of a calorimeter is to achieve the smallest possible effective Molière radius ($\mathrm{R_M}$), in order to provide the best shower separation. It requires the use of an absorber with a small intrinsic Molière radius, but also the minimisation of the gaps between the absorber layers. $\mathrm{R_M}$ is often quoted as the radius of 90% (95%) of shower containment. With this definition our measurement values are 20 (28) mm. The results for the various energies studied are summarised in figure 13.

The geometry of the ECAL prototype, with 2.2 mm thick interlayer gaps leads to a $\mathrm{R_M}$ which is expected to be larger by a factor of about 2 with respect to the Molière radius of solid tungsten. The results from the test beam studies are therefore in agreement with expectations. R&D effort towards the use of Si pads with integrated readout is under way and will hopefully lead to a significant decrease of the interlayer gap and therefore of the $\mathrm{R_M}$ for ECAL.

Journal of Physics: Conference Series **160** (2009) 012065 doi:10.1088/1742-6596/160/1/012065

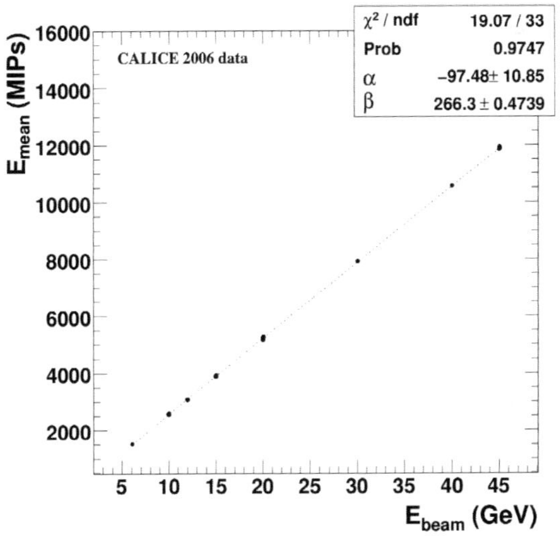

Figure 8. Linearity of the energy response of ECAL with respect to the beam energy. All the runs around a nominal energy of the beam were combined in one entry for clarity of the plot.

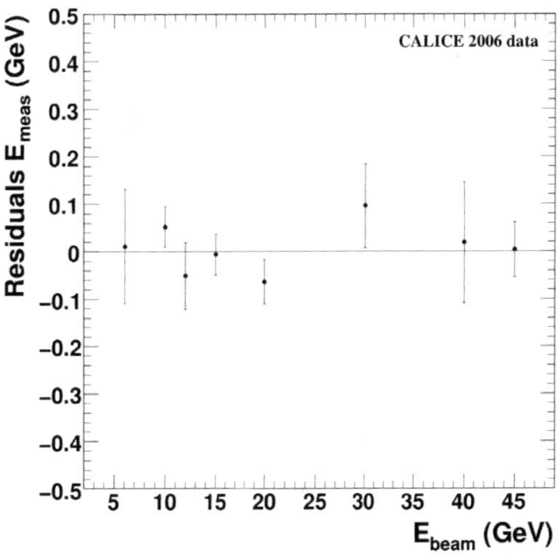

Figure 9. Residuals of the energy response of ECAL with respect to the linear function.

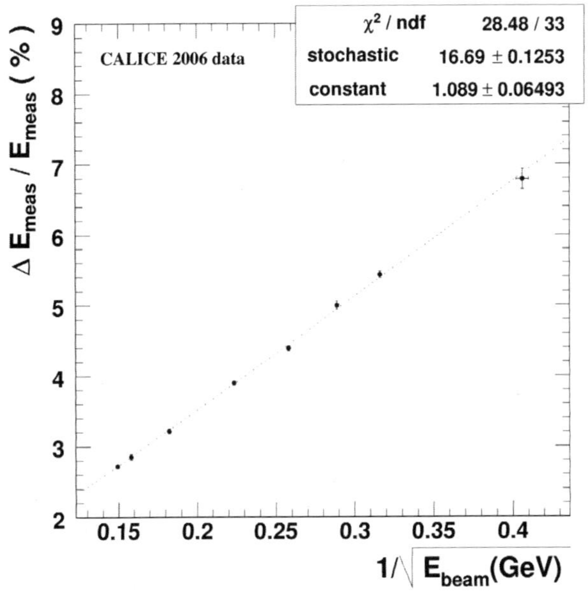

Figure 10. Relative energy resolution ($\Delta E/E_{meas}$) function of the beam energy. For clarity sake, the 35 runs available were combined into 8 different beam energy points. For the parametrisation of the energy resolution each run was however treated individually.

5. Conclusion

The Si-W ECAL prototype operated successfully during 2006 in its first testbeam at DESY and at CERN. The response of the prototype was measured using the recorded electrons and its physics performances were presented.

Additional data were collected at CERN in summer 2007 and at Fermilab in summer 2008. Their analysis is ongoing.

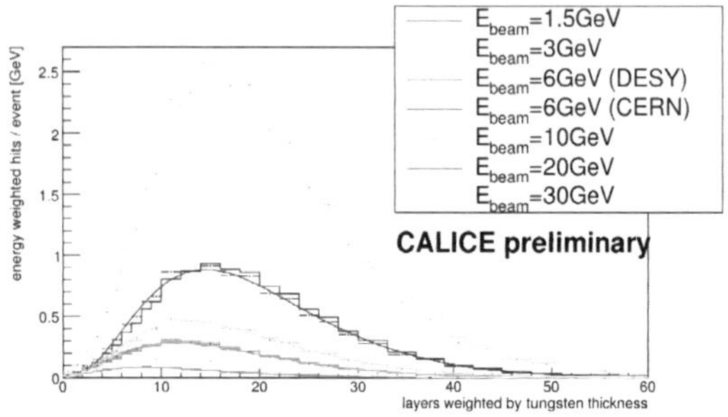

Figure 11. Longitudinal shower profile for the data (points) and Monte Carlo simulation (histogram). The smooth curve is the used parametrisation of the shower profile.

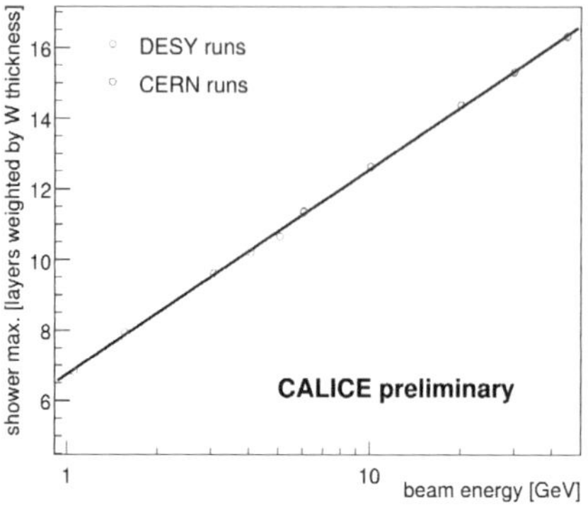

Figure 12. Evolution of the shower maximum as a function of the beam energy.

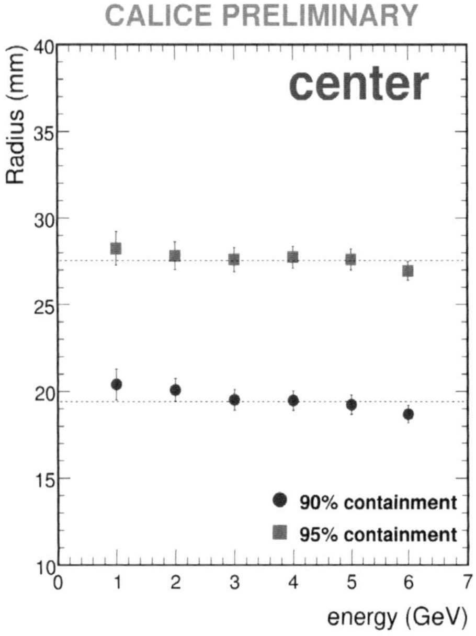

Figure 13. Radii for 90% and 95% containment of e^- as function of the beam energy.

References

[1] The CALICE collaboration : calorimeter studies for a future linear collider, 2002. `http://www.ppd.clrc.ac.uk/AnnReps/2004/exp/Calice-317.pdf`
[2] J. Repond *et al*, "Design and Commissioning of the Physics Prototype of a Si-W Electromagnetic Calorimeter for the International Linear Collider", to be submitted to NIM.
[3] T. Behnke *et al* (ed.), Reference Design Report "Volume 4 :Detectors", available at `http://lcdev.kek.jp/DR`.
[4] "Mokka Geant4 Application for Linear Collider Detectors" described at `http://polzope.in2p3.fr:8081/MOKKA`
[5] S. Agostinelli *et al*, "Geant4 – A simulation Toolkit", NIM A 506 (2003) 250-303.

Journal of Physics: Conference Series **160** (2009) 012066 doi:10.1088/1742-6596/160/1/012066

Tests of a Digital Hadron Calorimeter

José Repond

Argonne National Laboratory, 9700 S. Cass Avenue, Argonne, IL 60439, U.S.A.

repond@hep.anl.gov

Abstract. We present the concept of a Digital Hadron Calorimeter with Resistive Plate Chambers as active elements for use in a detector optimized for the application of Particle Flow Algorithms to the measurement of hadronic jets. We report on tests of a small scale prototype calorimeter using muons, positrons and pions from the Fermilab test beam. The results are compared to Monte Carlo simulations of the set-up and provide a first validation of the concept of a Digital Hadron Calorimeter. We conclude with a progress report on the construction of a larger prototype section.

1. The concept of a digital hadron calorimeter

We develop a Digital Hadron Calorimeter (DHCAL) optimized for the application of Particle Flow Algorithms (PFAs) [1] to the measurement of hadronic jets. PFAs require new types of calorimeters with very fine granularity of the readout in order to identify and measure energy deposits from charged and neutral particles. The DHCAL concept trades the high resolution on a small number of cells (towers) in a traditional calorimeter with the low (one-bit or digital) resolution on a large number of cells. This is a novel concept which needs to be validated through detailed tests with prototype calorimeters and subsequent comparison with Monte Carlo simulations.

Our approach uses Glass Resistive Plate Chambers (RPCs) with a single gas gap of 1.2 mm as active elements. The chambers are operated in avalanche mode with a negligible probability for streamers. The area of the individual readout pads was chosen to be squares of 1 x 1 cm^2.

We have taken a staged approach to the development of a DHCAL. In a first step we developed two competing designs of RPCs, optimized for the use in a calorimeter. In a second step we developed an electronic readout system capable of dealing with a large number of channels at an affordable cost. Both together were then assembled into a small scale prototype calorimeter and extensively tested with cosmic rays and in particle beams. Finally, we initiated the construction of a 1 m^3 prototype section large enough to contain most of the energy of hadronic showers.

Our collaboration consists of physicists and engineers from Argonne National Laboratory, Boston University, Fermi National Accelerator Laboratory and University of Iowa. Our project is an integral part of the program of the CALICE collaboration [2], which develops fine grained calorimeters for future lepton colliders.

2. Resistive Plate Chambers

We developed two competing designs of RPCs. The traditional design utilizes two glass plates (each with a thickness of 1.1 mm). The more 'exotic' design features only one glass plate while the readout

Journal of Physics: Conference Series **160** (2009) 012066 doi:10.1088/1742-6596/160/1/012066

pads provide the anode and close the gas volume. In both cases a gas gap with a thickness of 1.2 mm is maintained with the help of fishing lines spaced every 5 cm.

A number of prototype chambers based on either design have been tested with single and multiple readout pads, and with an analog (14-bit) or digital readout (1-bit) system. As an example of a test with cosmic rays and the analog readout, Fig. 1 shows the measurement of the signal charge as a function of the distance to the pad hit, identified as the pad with the maximum charge. Note that the cosmic rays are evenly distributed over the area of the pad and that no attempt was made to correct for this effect.

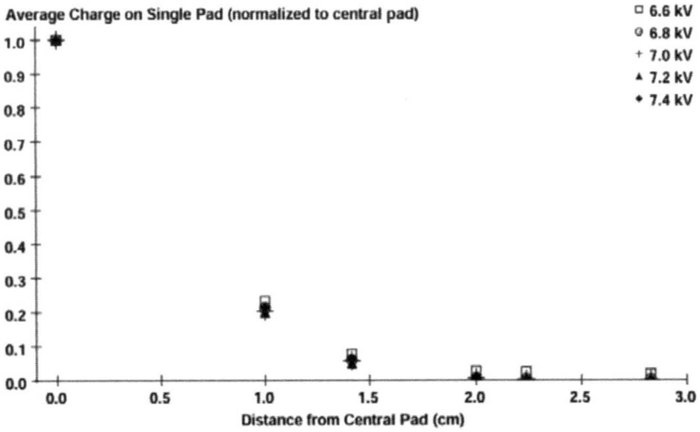

Figure 1. Average charge (normalized to the charge of the pad hit) as a function of the distance to the pad hit. The different symbols correspond to measurements with various high voltage settings.

Based on measurements such as these, we concluded that glass-RPCs constitute an ideal choice for fine grained hadron calorimetry. Due to the large variation in signal charge when RPCs are operated in avalanche mode, we also concluded that the possible additional information gained with a multi-bit readout system does not justify the increased complexity and cost of such a system. More details concerning the tests with the analog readout can be obtained from [3].

3. The electronic readout system

The electronic readout system consists of several subsystems:

a) The *DCAL chip* [4] reads out 64 channels either in (externally) triggered or in triggerless mode. A threshold common to all channels can be set within the signal range of 10 to 700 fC. The output of the chip consists of a time stamp with a 100 ns resolution and the hit pattern of the 64 channels.

b) The *pad- and front-end boards* contain the 1 x 1 cm^2 readout pads on one side and the DCAL chips on the other. The two boards are glued together with dots of conductive epoxy to provide the signal path from the pads to the DCAL chip.

c) The *data concentrator boards* are located on one edge of the front-end boards, read out multiple DCAL chips and transfer the data to the back-end.

d) The *data collector* is a VME-based readout system with a connection to the data acquisition computer via a PCI-bridge.

e) The *timing and trigger module* provides the necessary timing signals and distributes the trigger via the data collectors to the front-end.

All individual sub-systems were prototyped, followed by the assembly, commissioning and extensive testing of the entire system. Up to rates of about 1kHz, the system operates dead-time and error free. The noise rate (with no high-voltage on the RPCs) was measured to be less than $4 \cdot 10^{-5}$ Hz/pad, even for threshold settings at a few fC, close to the noise floor. As an example of the performance of the system, Fig. 2 shows the response to a charge of 10 pC injected into channel 54. It is seen, that for thresholds above the noise floor (around 18 DAC counts) channel 54 always records hits, as expected. The neighbouring channels register hits only at threshold values below 25 counts. This corresponds to a cross talk of less than 0.3 %. Additional details concerning the readout system can be obtained from [5].

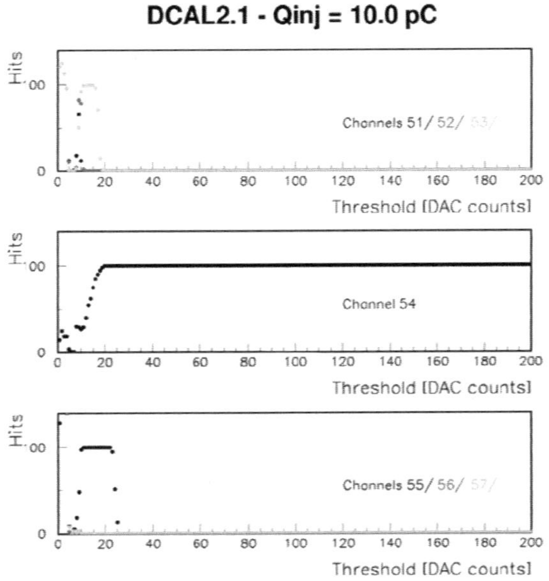

Figure 2. Measurement of electronic cross talk. The response of neighbouring channels is shown as a function of threshold setting, when a signal with a charge of 10 pC is injected onto channel 54.

4. Calibration of a Digital Hadron Calorimeter

With a digital hadron calorimeter the energy of an incoming hadron, E_{hadron}, is reconstructed from the number of hits associated with that particle. Ignoring effects of high-density sub-clusters, which might require non-linear corrections, Eqn. 1 provides a general formula for reconstructing E_{hadron}

$$E_{hadron} = \alpha_{samp}(\Sigma_i H_i) \cdot \Sigma_i(H_i - B_i)/(\varepsilon_0^{MIP}/\varepsilon_i^{MIP} \cdot \mu_0^{MIP}/\mu_i^{MIP}), \quad (1)$$

where

i is an index running over all pads associated with an incoming particle,
H_i is set to 1 (0) depending on whether a hit (or no hit) has been recorded in pad I,
B_i are the expected number of background hits (from accidental discharges, electronic noise or cosmic rays) for pad i,
ε_0^{MIP} is the average efficiency to record at least one hit when a layer is traversed by a minimum ionizing particle (MIP),
ε_i^{MIP} is the efficiency for pad i to fire when traversed by a MIP (typically uniform within an RPC),
μ_0^{MIP} is the average number of pads firing when a layer is traversed by a MIP,
μ_i^{MIP} is the average number of pads firing when pad i is traversed by a MIP (typically uniform within an RPC), and
$\alpha_{samp}(\Sigma_i H_i)$ is a sampling term, possibly depending on the total number of hits.

The sampling term can be regarded as independent of specific running conditions and so to be time independent. It needs to be determined from test beam data and Monte Carlo simulations.

The chamber's calibration constants (B_i, ε_i^{MIP}, μ_i^{MIP}) depend on both operational (such as high voltage and threshold settings) and environmental conditions (such as temperature, humidity and atmospheric pressure). Below we study the dependence on the operational parameters, assuming that the environmental effects are small in comparison.

5. Monte Carlo simulations

The test set-up was simulated with a package based on the GEANT4 code. The detailed simulation of the response of RPCs was reproduced with a standalone program. The latter generated for each charged particle traversing the gas gap in the GEANT4 simulation a signal charge according to the measured charge distribution for operation with a given high voltage [3]. In a second step this charge is distributed over the readout pads assuming an exponential decrease with a slope **a** as a function of distance [3,6]. Pads with a charge above a given threshold **T** are assumed to have recorded a hit. The parameters **a** and **T** were tuned to reproduce the measured MIP detection efficiency and pad multiplicity of the muon test beam data (see below). Since RPCs do not generate multiple avalanches within short distances, an additional parameter $\mathbf{d_{cut}}$ is introduced to describe this effect. The latter was tuned using the hit distribution of the 8 GeV positron data.

6. The Vertical Slice Test (VST)

As an intermediate step in the development of a DHCAL a small prototype calorimeter was constructed and assembled. This calorimeter was labelled Vertical Slice Test (VST), since it required the design, prototyping and commissioning of the entire readout chain. The VST involved up to 10 RPCs (including one chamber based on the one-glass design), each with an area of 20 x 20 cm^2. The area of the chambers was chosen for maximum exposure in a test beam, but the system was conceived such that an expansion to planes of 1 x 1 m^2 is relatively straightforward. The RPCs were interleaved with absorber plates consisting of 16 mm Steel and 4 mm Copper plates. The latter were required for the cooling of the front-end ASICs.

The VST prototype was tested extensively with Cosmic Rays and at the Fermilab testbeam [7]. The latter provided broad band muons, protons at 120 GeV and secondary beams at 1, 2, 4, 8, and 16 GeV/c. Figure 3 shows a photograph of the set-up in the Fermilab test beam.

At the time of the test beam (July 2007) the data acquisition had not been fully debugged, logging errors at about the 1% level. In the meantime the system has been fully debugged and operation is virtually error free.

Figure 3. Photograph of the Vertical Slice Test prototype calorimeter in the Fermilab test beam.

6.1 Measurement of noise rates

The noise rates B_i were measured using the self-trigger mode of the DCAL chips. For the default high voltage setting of 6.3 kV, Fig. 4 shows the noise rate as a function of threshold for six individual chambers. As expected the rate is seen to decrease with increasing threshold. At the default threshold value of 110 DAC counts the rate measured 0.1 – 0.2 Hz/cm^2. The noise hits were seen to be coming predominantly from pads located above the fishing lines. For a $50 \cdot 10^6$ channel calorimeter this rate corresponds to 1 – 2 hits within a 200 ns gate. For additional details see [8].

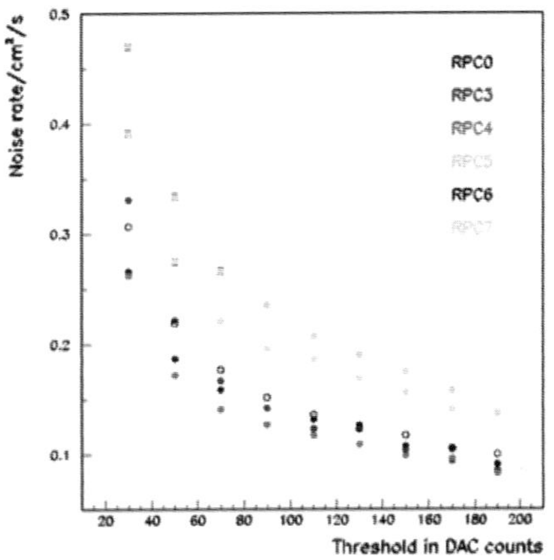

Figure 4. Measurement of the noise rate for six individual RPCs as a function of the threshold setting. The chambers were operated at 6.3 kV.

6.2 Calibration with muons

The primary purpose for exposing the VST to muons was to establish the calibration constants, i.e. the MIP detection efficiencies ε_i^{MIP} and pad multiplicities μ_i^{MIP}, as a function of high voltage and threshold. The calibration runs were taken with the primary 120 GeV proton beam and a 9 foot (~3m) iron beam blocker to produce muons and filter out the non-muon component. Figure 5 shows various views/projections of two selected muon events. Figure 5.a shows the event display of a clean muon track, where in every layer a single pad fired. Depending on the location of the avalanche with respect to the pad boundaries up to four pads might fire. The event display of Fig. 5.b shows a muon track entering the calorimeter with a slight inclination, both horizontally and vertically. As a result some layers show more than one hit in both projections.

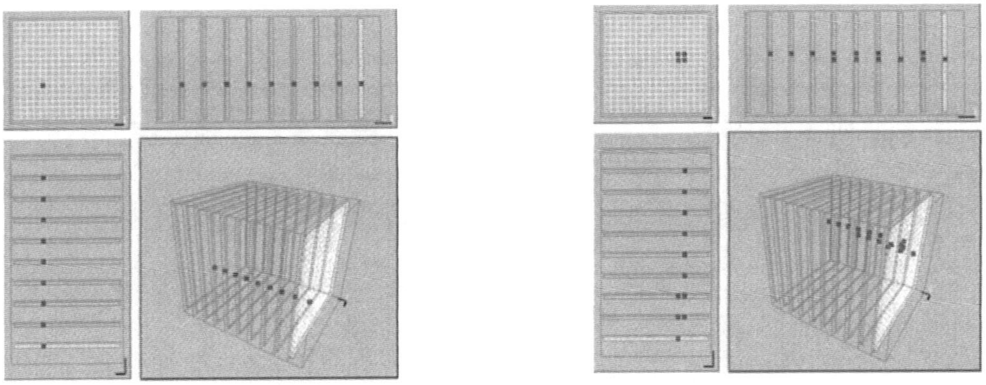

Figure 5. Event displays of single muon tracks a) with one hit per layer, and b) entering the calorimeter at a slight angle and therefore leading to multiple hits in some layers.

The calibration constants were measured for a selection of high voltage (remaining in the avalanche regime of the RPCs) and threshold settings. As an example Fig. 6 shows the MIP detection efficiency as a function of threshold for four different operating voltages. The MIP detection efficiencies, with values between 85% and 97%, decrease with increasing thresholds, as expected. Based on measurements such as these we chose a high voltage of 6.3 kV and a threshold of 110 DAC counts as default for the subsequent measurements with positrons and pions. For more details on the muon calibration see [8].

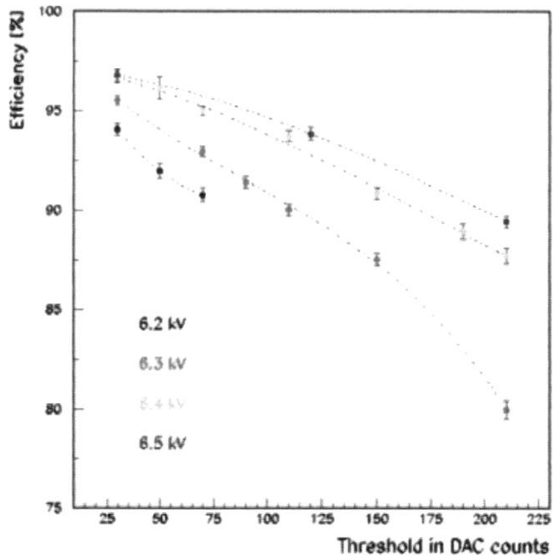

Figure 6. MIP detection efficiency as function of threshold for selected high voltage settings. The efficiencies are averages over four different chambers. The dashed lines are fits to higher order polynomials and only serve to guide the eye.

6.3 Response to positrons

Positrons were produced with a secondary target in the 120 GeV proton beam line and selected by requiring a signal in one of the Čerenkov counters [7] upstream of the test set-up. Figure 7 shows the number of hits summed over six layers (corresponding to about 7 X_0) for five different beam momenta. The distributions are fit to Gaussian functions, which adequately describe the experimental data.

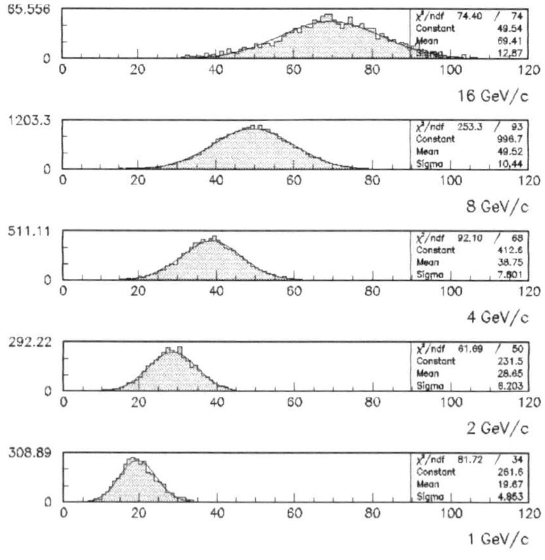

Figure 7. Number of hits in a stack of six chambers for 1, 2, 4, 8, 16 GeV positrons.

The mean number of pads and the width of the fitted distributions as function of beam momentum are shown in Fig. 8. Due to the large leakage out the back of the stack the response is highly non-linear. This behaviour is adequately reproduced by the Monte Carlo simulation. The simulation of an infinite stack still displays some non-linearity of the response, mostly due to the high density of electromagnetic showers and the finite size of the readout pads. The width of the distributions is also well reproduced. Note, however, that the results shown here are not corrected for the observed non-linearity. If such a correction is applied, the resolutions degrade to 50 – 150 % /√E, again mostly due to leakage.

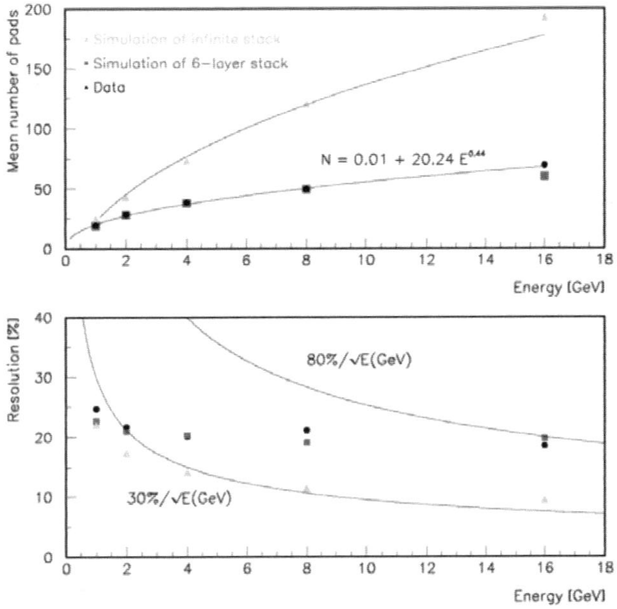

Figure 8. Mean number of pads and width of the distributions as obtained by Gaussian fits to the data in Fig.7 (black points). The results of a simulation of the 6-layer stack (red circles) and of an infinite stack (green triangles) are also shown.

6.4 Response to pions

Pions were produced by a secondary target in the 120 GeV proton beam line and selected with a veto on the .OR. of two Čerenkov counters [7] upstream of the test set-up. The pion offline event selection requested a small number of hits in the first layer (to reduce the number of pions interacting upstream of the detector) and a minimum of 5 hits in the second layer (to enhance the number of events with early hadron showers). Figure 9 shows the number of pads summed over six layers (corresponding to approximately 0.7 λ_I). The response is again adequately described by a Gaussian function. Note, however, that despite the event selection the small size of the calorimeter leads to large leakage. The simulation (not shown) is able to reproduce the response at 16 GeV/c without further tuning of the parameters **a**, **T**, and $\mathbf{d_{cut}}$. The average number of pads at 8 and 4 GeV/c is larger in the data compared to the Monte Carlo. This is most likely due to a remaining contamination from positrons in the data, enhanced by the multiplicity cut on the second layer.

7. Construction of the 1 m^3 prototype section

Based on the success of the Vertical Slice Test, we initiated the construction of a larger prototype section with the approximate dimensions of 1 m^3. This larger calorimeter will be used to measure

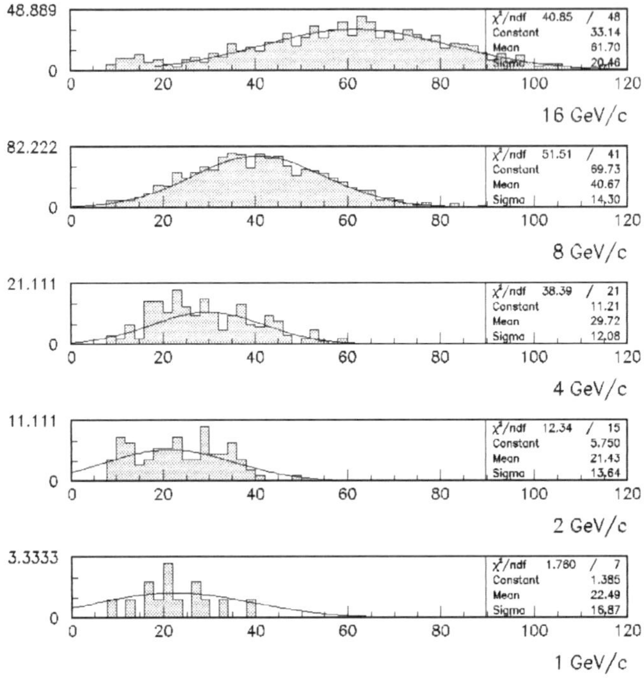

Figure 9. Number of hits for 1, 2, 4, 8, and 16 GeV/c pions. The distributions have been fitted to Gaussian functions.

hadron showers in great detail (for comparison with both simulation and similar measurements with a scintillator based hadron calorimeter [9]) and to gain further experience with the construction and operation of an RPC-based detector.

The prototype section will consist of 40 layers each with an area of approximately 1 x 1 m^2. Thus, with 1 x 1 cm^2 readout pads the calorimeter will count of the order of 400,000 channels. The readout resolution of an individual channel will again be a single bit. The planes will be divided into three chambers, each with an area of 32 x 96 cm^2. The chambers will be covered with two readout boards (32 x 48 cm^2) with connections to the back-end of the readout system on both sides of the detector.

Most preparatory work for the construction of the prototype section has been completed. Prototypes of full size RPCs have been constructed and are now undergoing detailed tests with cosmic rays. A few minor modifications still need to be implemented into the DCAL chip. However, these do not necessitate an additional prototype run. The pad- and front-end boards have been redesigned to eliminate the high cost associated with blind vias. The new boards tested successfully and showed no increase in noise levels. The remainder of the readout system will be identical to the one used for the Vertical Slice Test (see above).

Our plans foresee the construction of ten layers by early 2009, followed by the construction of the remaining layers later in the year. Funding for the construction of the first ten layers has been secured.

8. Further developments

In addition to the construction of the prototype section, our group initiated further R&D on both the RPCs and the electronic readout system towards a viable calorimeter for a future lepton collider. We are investigating ways of reducing the overall thickness of the chambers. Here the one-glass design looks particularly promising. Different methods of protecting the front-end electronics from accidental discharges are being explored.

Concerning the electronic readout system, several avenues are being pursued: power pulsing (to reduce power consumption and avoid active cooling), higher multiplexing (to reduce both the cost and complexity of the system), reliability of operation (necessary for electronics which will be buried inside the detector), etc…

Furthermore, ways to integrate the gas and high and low voltage distributions into wedge shaped modules is being investigated. Due to the adverse environmental impact of the major component (Tetrafluoroethane) of the chamber gas, a system for recycling and purifying the chamber gas is being discussed.

9. Summary

We have developed a concept of a digital hadron calorimeter with RPCs as active elements. The concept was tested with a small scale prototype in the Fermilab test beam. This so-called Vertical Slice Test involved the entire readout chain as conceived for a larger system. The tests were very successful and constitute a first validation of the novel concept of a digital hadron calorimeter. The group is in the process of building a larger prototype section with the approximate dimensions of 1 m^3. The latter will be tested in the Fermilab test beam starting in 2009.

The author would like to express his deep gratitude to the organizers of the conference. The excellent scientific program of the conference was greatly enhanced by a wonderful venue and superb food and drinks.

References

[1] 'R&D for a PFA Calorimeter', J.C. Brient, these proceedings.
[2] http://llr.in2p3.fr/activites/physique/flc/calice.html
[3] 'Resistive Plate Chambers for Hadron Calorimetry: Tests with Analog Readout', G.Drake et al., Nucl. Instr. Meth **A578**, 88 (2007).
[4] 'DCal: A Custom Integrated Circuit for Calorimetry at the Linear Collider', J.Hoff et al., Proceedings of the 2005 IEEE Nuclear Science Symposium, Puerto Rico (2005).
[5] 'A New Readout System for "Digital Hadron Calorimetry" for the International Linear Collider', J. Butler et al., Proceedings of the 2007 IEEE Nuclear Science Symposium, Honolulu, Hawaii (2007).
[6] 'RPC Investigations Using Finely Spaced 2-D Strip Readout', D. Underwood et al., Proceedings of the 2007 IEEE Nuclear Science Symposium, Honolulu, Hawaii (2007).
[7] http://www-ppd.fnal.gov/MTBF-w
[8] 'Calibration of a Digital Hadron Calorimeter with Muons', B.Burak et al., 2008 JINST **3** P05001.
[9] 'CALICE Scintillator HCAL – Electromagnetic and Hadronic Shower Analysis', E. Garutti, these proceedings.

Journal of Physics: Conference Series **160** (2009) 012067 doi:10.1088/1742-6596/160/1/012067

Semiconductor sensors for the CALICE SiW EMC and study of the cross-talk between guard rings and pixels in the CALICE SiW prototype

Rémi CORNAT, on behalf of the CALICE collaboration
Laboratoire Leprince-Ringuet
CNRS/IN2P3-Ecole Polytechnique (Palaiseau, France)
E-mail: `remi.cornat@in2p3.fr`

Abstract. The CALICE collaboration is studying high granularity and compact calorimeters for the future International Linear Collider. The electromagnetic calorimeter is made up of tungsten as absorber and silicon detectors implementing PIN diodes arrays. Calibration tests of a first prototype are describe first. Data taken in test beam environment have shown an unexpected behavior of the sensors found to be related to a crosstalk. This crosstalk is suspected to generate the so called "square events" seen in the test beam data where all bordering pixels of the sensor are illuminated. An analytic model is proposed to describe the crosstalk couplings within the sensors. Analytic results are compared with electrical simulation with the aim to understand forthcoming measurements on real hardware. A new design of guard rings is then discussed according to its impact on square events.

1. Introduction

The Calorimeter for the Linear Collider Experiment (CALICE) collaboration [1] is involved in the study and the development of prototypes of calorimeters targeting the needs of the future International Linear Collider experiments.

With the aim to be compliant with Particle Flow Algorithm method, the electromagnetic calorimeter is made to provide a high granularity and density [2]. A prototype intended to test Physics performance has been made up of tungsten as absorber and silicon as sensitive material to implement 6×6 pixels sensors. In a first part, the prototype design is described together with calibration and commissioning results highlighting on uniformity and stability of this 10000 channels detector [3]. In a second part of this article, the silicon sensors will be discussed about the crosstalk point of view. Indeed, the analysis of the data taken during test beam have pointed out an unexpected behaviour of the sensors. It consists in seeing all the bordering pixels to be illuminated. Three kind of models are proposed, compared and discussed to explain this phenomenon. A new way to design the sensors is then proposed being compliant with the constraint of the low cost of the overall sensors production.

2. The prototype of the electromagnetic calorimeter

Designing a sampling calorimeter, the absorber was chosen to allow to separate particles within a jet as well as to build a compact apparatus. Tungsten has appeared to be a good candidate for the absorber with both a small Molière radius of 9 mm and a small radiation length of 3.5 mm. The sensors are burned on high resistivity 500 μm thick silicon wafers thanks to microelectronics manufacturing technologies. It supports compactness and granularity as the pixel size is about 1 cm × 1 cm, close to the Molère radius. The prototype has 24 X_0 split on 30 absorber layers interleaved with sensitive layers. It ensure that 99.5% of the energy for 5 GeV electron showers is enclosed within the detector.

2.1. *Structure and instrumentation of the detector*

A good energy resolution is achieved for a wide range variating the absorber plates thickness according to three areas of 10 layers each which are respectively made up of tungsten plates of 1.4 mm, 2.8 mm and 4.2 mm.

The sensitive layers are all the same and are made up of modules of nine sensors providing a detection area of 18 cm×18 cm (pads are 1 cm^2 wide) together with the readout electronics. The electromagnetic calorimeter thus includes 9720 channels. The figure 1 shows a schematics of the architecture of the detector with its 3 independent alveolar structures made up of carbon fiber and epoxy supporting the tungsten plates. The figure 2 a picture of the detector on test beam with the sensitive layers slided into the alveola.

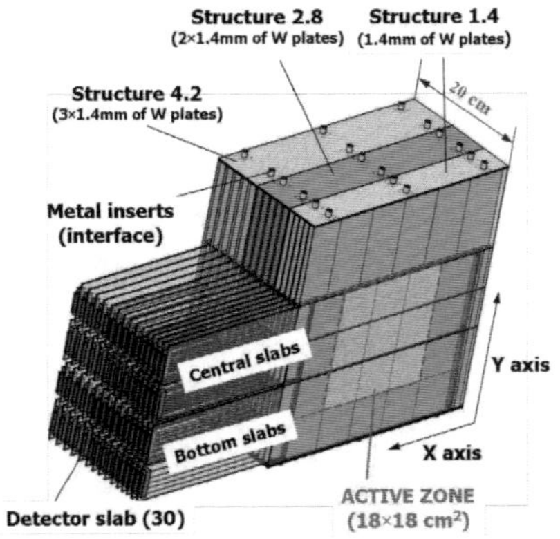

Figure 1. 3D view of the prototype.

Figure 2. Picture of the prototype taken at DESY in 2006.

2.2. *Calibration with cosmic rays*

The modules have been calibrated thanks cosmic rays right after their assembly. The assembly process require a lot of different steps, in particular the gluing of the sensors on the printed circuit board where are located the FLC_PHY3 front-end chips designed at LAL [4]. The chips are set to have a dynamic range from 0.1 MIP to 600 MIP on 13 bits (it could be increaded to 15 bits).

As shown on figure 3 and 4 the MIP calibration (precision of about 5%) as well as the noise level are mesured for each channel. The signal-to-noise ratio is 9.1±0.2 with a standard deviation of 1.0±0.1. This result is close to the final goal to have a S/N ratio of 10 and is therefore encouraging for further test in beam conditions. With only 0.1% dead channels, modules assembled from September 2004 do not show performance degradation over the time.

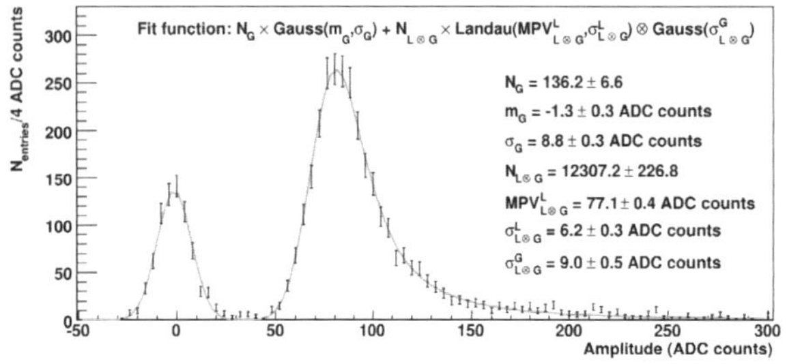

Figure 3. Example of the cosmic-ray muon calibration of a half module showing the signal recorded by the corresponding readout chip. The first peak, centered on zero, is due to the electronics noise (pedestal) and is fitted with a Gaussian function. The second peak, centered on 82 ADC counts, corresponds to the MIP signal, and is fitted with a Landau convolved with a Gaussian distribution.

Figure 4. Picture of the cosmic rays test bench setup.

3. Performance in Beam conditions

Many test beam campaigns at DESY and CERN during year 2006 have allowed to quantify the detector performance as well as to validate the simulations models. A total of 300 million of event have been recorded for an energy range going from 1 to 6 GeV at DESY and 6 to 50 GeV at CERN. A run was structured to start with 500 pedestal events, 500 events with internal charge injection at the level of the chips and few tens of thousands data events. A part of the pedestal and calibration events have been taken without beam thanks to a random trigger.

Hardware performance have also been characterized and the main aspects are summarized in the following.

3.1. Pedestal and MIP

The pedestal is defined as the mean value of the signal without beam. The first 500 events of a run are taken. The figure 5 shows the pedestals of one front-end board as a function of time. The shift is due to a bad decoupling of the power supply due to a bad isolation of the PCB traces resulting in changes in the front-end chips working points. It will be corrected in the next version of the front-end electronics. For the data analysis the pedestal are corrected on an event by event basis to compensate the shifts. As shown on figure 6, once corrected, the residual pedestal with respect to the channel is -0.03 ± 0.01 with a standard deviation of 1.05 ± 0.01 ADC counts (a MIP is 46 ADC counts).

The MIP calibration of the ECAL prototype have been performed for each channel thanks to the energy distribution of muon events as shown in figure 7 with a residual pedestal of 0.2% MIP. A fit with a convolution of a Landau and a Gaussian gives the MIP with a statistical uncertainty of 0.5% and a systematic uncertainty of 0.4%. The uniformity over the detector is shown on figure 8. The colors correspond to various sensors batches and MIP values between 44 and 48 ADC counts according to the batch.

3.2. Noise

The noise is evaluated using the same runs as for the pedestals. It is extracted from the standard deviation of the ADC counts distribution on a channel per channel basis. Noise is found to be

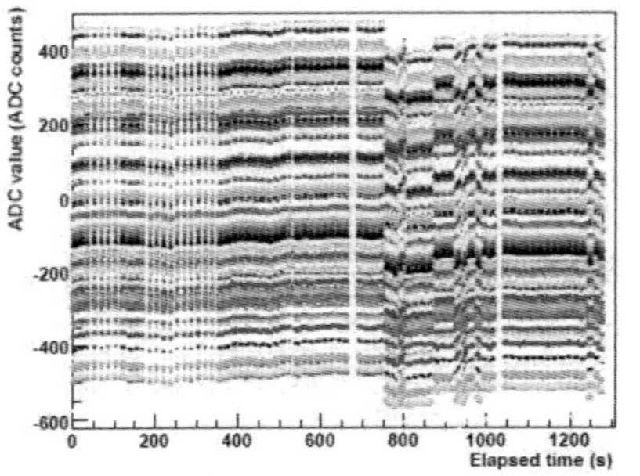

Figure 5. Pedestals as a function of time for 216 channels (muon run taken at CERN in 2006).

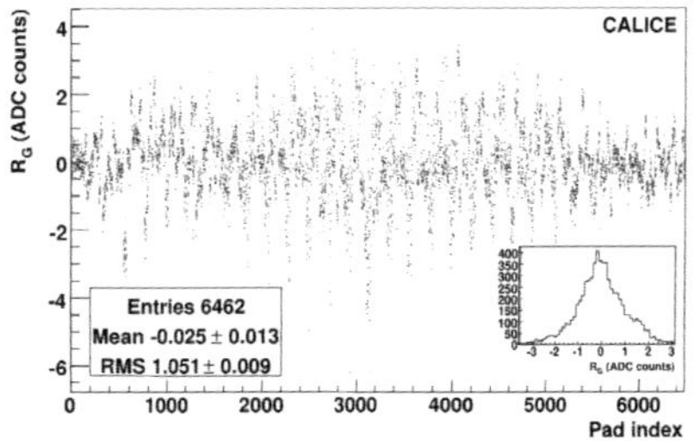

Figure 6. Uniformity of the residual pedestals R_G as a function of the pad index, for a 45 GeV electron run taken at CERN in August 2006.

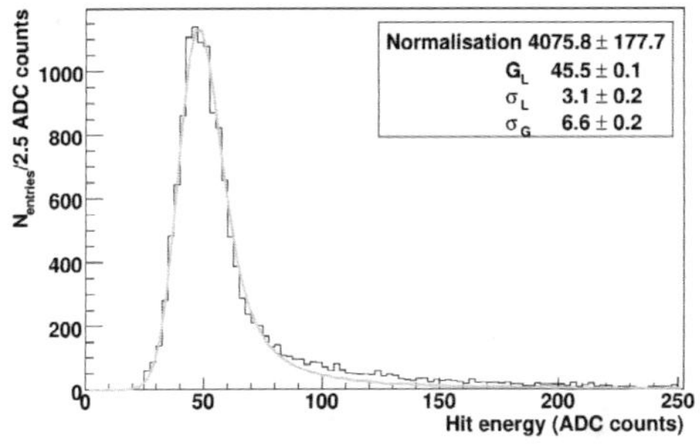

Figure 7. Energy distribution of hits in muon events for a particular channel. The fit function is a convolution of a Gaussian and a Landau. Normalisation, G_L and σ_L refers to the constant, most probable value and width of the Landau function, and σ_G to the width of the Gaussian function.

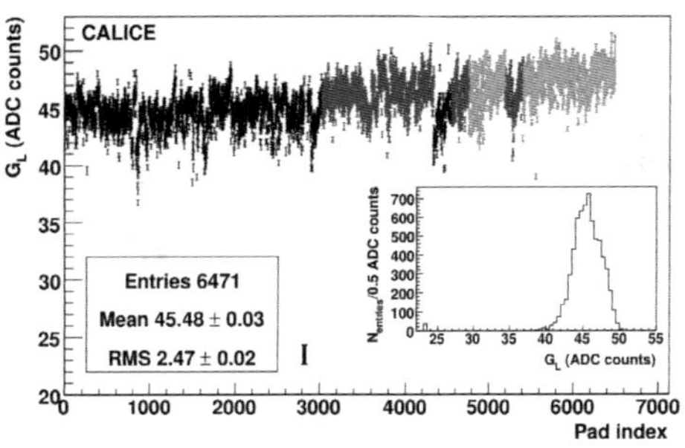

Figure 8. Equivalent ADC value G_L for a MIP according to the channels number (muon events). Colors correspond to channels respectively from different sensor batches. The group of 36 channels at 23.5 ADC counts corresponds to a non-fully depleted module.

at the level of 5.919±0.004 ADC counts with a standard deviation of 0.345±0.003 ADC counts. The channel to channel dependence is about 9% of the mean noise but run to run variations are less than 1%.

The pedestal shift effect has an influence on the noise due to the resulting offset with respect to the initial pedestal correction. The noise appears to be larger than expected. The pedestal shift occurs at the level of the PCB, thus the apparent noise is correlated between all channels. An off-line correction procedure have been defined. It is base on successive iterations conditioned by the RMS deviation from the mean pedestal over the channels until the effective noise is compatible with the mean expected noise. The remaining correlation is -0.0017 ± 0.0003 thus validating the correction procedure done event by event.

For sensors exposed to a high level of signal, the signal induced pedestal shift results in an increase of the apparent noise for all the 36 channels of the sensor. This effect is not systematic according to the PCB or to the time. It is identified by a strong correlation between the signal and the pedestal shift of all the channels of the corresponding sensor. It can be interpreted as coupling by a common impedance on the bias voltage line. It can hopefully be corrected (similar iterative method at the scope of the sensor) and the remaining mean correlation is 0.0058±0.0001 to be compared to 0.019±0.001 before correction performed on a event by event basis.

3.3. Stability

Stability in time can be checked comparing the calibration constants for various test periods. The figure 9 shows a correlation coefficient of 93.2% between the calibration constants for two datasets taken with a 3 months delay. The mean difference for all channels is found to be -0.76 ± 0.01 with a spread of 0.67±0.01 ADC counts. At a front-end chip level, a stable and reliable behaviour is confirmed.

Similar comparisons of datasets do not show clear correlations as function of time, temperature or beam energy for the residual pedestal as well as for the noise.

3.4. Conclusion on the hardware performance

Targeting a signal over the noise ratio of 10, the detector expressed an actual S/N ratio of 7.63±0.01 during test beam operations. This 10 thousand channels detector has shown a good behaviour in time and a satisfactory uniformity with 1.7±0.1% MIP variation over the channels (0.4% with muon runs) and 1.1±0.1% MIP variation over the runs. Building technologies and assembly methods allows to provide a reliable detector to achieve the granularity and compactness requirements of a particle data flow approach for an ILC experiment.

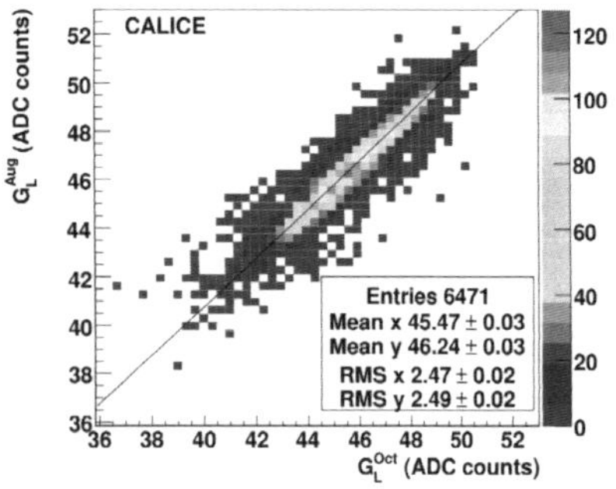

Figure 9. Correlation of the calibration constant between August and October runs showing a good time stability.

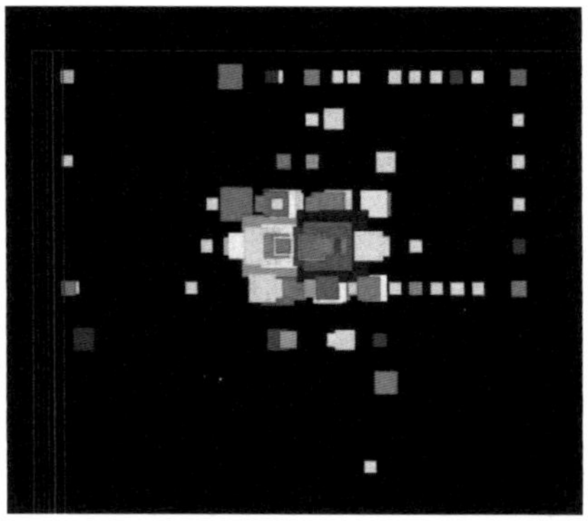

Figure 10. Event display including a "square event".

4. The silicon sensors

The sensors are made from float zone silicon wafers providing a resistivity greater than 5 kΩ·cm. It is necessary to ensure a full depletion with a reasonable bias voltage, and to obtain low bulk leakage currents. A wafer thickness of 525 mm was chosen, to obtain a signal-to-noise ratio of about 10 at the end of the whole readout electronics chain. A minimum ionising particle (MIP) produces about 80 electronhole pairs per mm of silicon, hence 42,000 electrons are obtained for the 525 mm thickness, giving an allowable noise range for the readout electronics of up to 4000 electrons.

Each wafer is used to make a sensor, consisting of a matrix of 6 6 PIN diodes (pixels) of 1 cm^2 surrounded by a guard ring structure [5], as shown in figures 11 and 12. In order to keep the price and the rate of rejected processed sensors low, the manufacturing process must be as simple as possible, with a minimum number of steps during processing. The final detector will need about 3000m^2 of such detectors. This explains the stringent requirements about the low cost.

Figure 11. Picture of a sensor (pixel side). The pixel size is 1 cm^2.

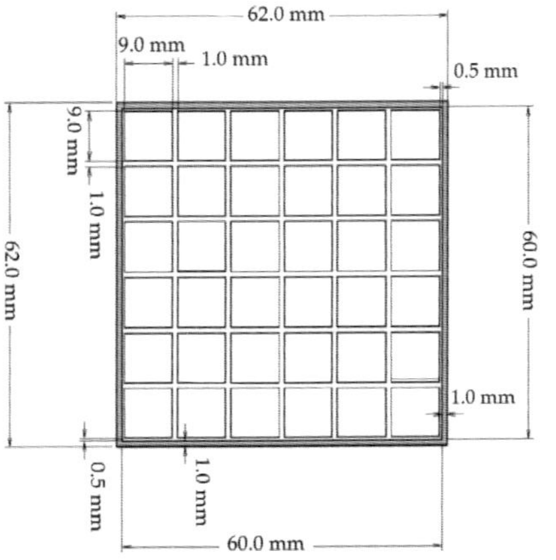

Figure 12. Details of the matrix dimensions in a module (cathode side). The area shaded in grey corresponds to the guard ring.

The edge termination is made using the guard rings (various number according to the manufacturer). The guard rings are made in the same way as for the pixels. The major advantage is that it does not require any extra step in the fabrication process of modules, helping to reduce the overall cost.

Before gluing, the modules are characterized one-by-one by measuring the leakage current as a function of bias voltage (I-V curve), the full depletion voltage (extracted from C-V curve) and the stability in time (leakage current versus time at a nominal bias). The test bench was used to characterize 550 modules, with an overall yield of 54%.

5. Study of the crosstalk

A characterization step in test beam conditions have shown an unexpected behaviour of the sensors (see figure 10). A look at the data reveals illuminated pixels arranged in a square shape corresponding to the silicon wafer border where the floating guardrings structure is located.

A crosstalk phenomenon between this guardring structure and the nearest pixels to this structure (pixels at the border of the wafer) is suspected to generate the square shape called "square event". The understanding of square event is crucial for the detector development. Prior to direct measurements on real sensors, a simulation step can help to understand the couplings between pixels and guardrings. Simulation models can then be used to evaluate various guardrings designs without the need to manufacture and test expensive real sensors samples.

6. Wafer structure and crosstalk models

From the point of view of crosstalk, the wafer can be seen as a capacitance network. This network is defined by the internal structure of the wafer and its geometry in particular the capacitances within the matrix of square pixels and the guardrings.

To overcome signal propagation along the guardring, the idea is to "cut" the guardrings which took the shape of a dashed line instead of a continuous one. In such topology, the internal capacitances which produce crosstalk are put into series and are divided lowering the crosstalk.

Many design parameters can vary : number of guardrings, distances, segment length, *etc.* Keeping in mind future measurements on small relatively cheap devices including 3×3 pixels, models and simulation runs have been limited to this low number of pixels. Nevertheless, it is sufficient to evaluate the crosstalk coupling. In future developments of the sensors, the number of guardrings should be limited to 4 to avoid a too large dead area around the sensor. This is also the case in this study. Three different topologies have been simulated, the options are described below:

- 4 continuous guardrings.
- 4 segmented guardrings with segment length equivalent to the pixel size.
- 4 segmented guardrings with segment length lower than the pixel size (factor 3).

The figure 13 shows a matrix of 3×3 pixels and the nearest guardring with internal coupling impedances for a given topology (here, the segment length is equal to the pixel width).

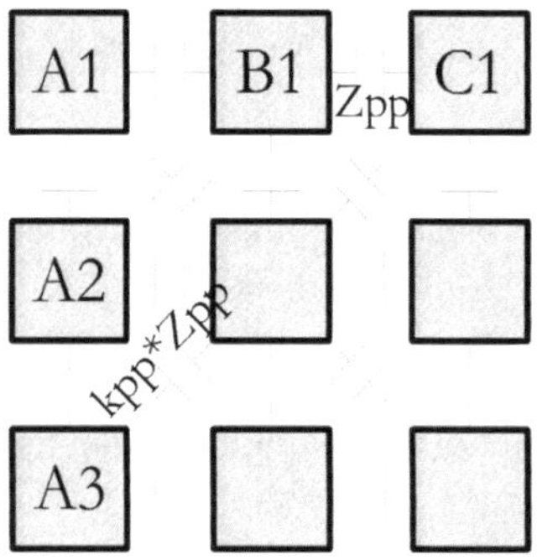

Figure 13. Wafer model with internal coupling impedances only, to be added to the external coupling model for electrical simulation of the complete sensor.

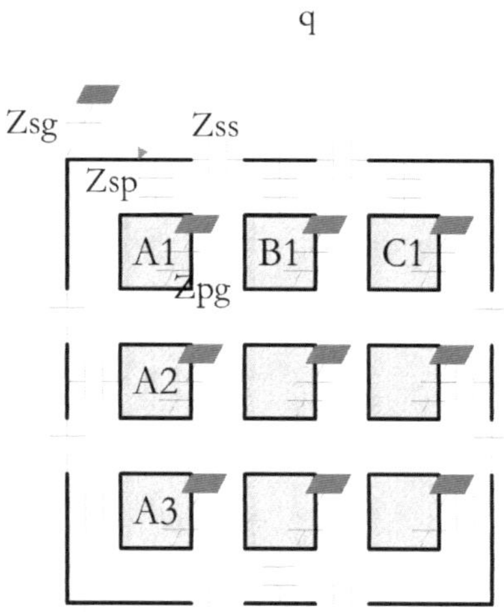

Figure 14. Wafer external coupling due to the guardring for the analytic model. Only one guardring is drown in the case of segments length similar to the pixel dimensions.

With the point of view of square events, a signal is induced on the guardrings by a high energy particle. Charges are collected as for a pixel by the guardrings. The signal can propagates itself through the guardrings or through the pixel matrix due to internal crosstalk. In the followings it is assumed that these two kind of propagation modes are independent.

6.1. Analytic model of the crosstalk

This model can be fitted to many geometrical configurations without the need to rewrite an electrical simulation netlist and can then help to evaluate various design options prior to manufacturing and complementary to electrical simulations. It models the crosstalk related to the guardrings only.

The whole measurement chain is modeled in order to take into account the effects of the electronics (bandpass filter).

The figure 14 shows the coupling within the sensor taken into account for the analytic model. Capacitances to ground and coupling of bordering pixels to the guardring are modeled.

The transfer function $H(s)$ can be computed from the equivalent quadripole transfer matrix T :

$$H(s) = \frac{\det(T)}{T_{2,2}}$$

where

$$T = T_{elec} \times T_{wafer,p} \times T_{inj}$$

With T_{elec} the transfer matrix of the measurement electronics (amplifier), $T_{wafer,p}$ the transfer matrix of the device under test (wafer or copper-epoxy model) associated to pixel number p and T_{inj} the transfer matrix of the signal injection electronics. With T_{pix} the transfer matrix of the pixel, T_{seg} the transfer matrix of the segment and T_{ss} the transfer matrix of the gap between two segments, the transfer matrix of the whole measurement chain can be written as :

$$T = T_{elec} \times T_{pix} \times (T_{seg} \times T_{ss})^{N_{seg}} \times T_{seg} \times T_{inj}$$

N_{seg} is the number of segment along the signal path following the guardring. This approach neglects the crosstalk between pixels themselves and is limited to the 3 first pixel in a column close to the guardrings. A Decrease of the signal amplitude of respectively 21 dB and 75 dB are expected for the second pixel in the cases of 1 cm and 3 mm segment length.

Table 1. Analytic model response for options 1, 2 and 3 showing a decrease in the signal amplitude (dB) for pixel (A2 and A3) located far from the injection point in the case of segmented guardrings

Pixel	Continuous	1 cm	3 mm
A1 (ref)	0	0	-14.5
A2	-6	-21.6	-75
A3	0	-34.6	-138

6.2. Electrical simulation

The capacitances network can be described as an input netlist to an electrical simulator such as SPICE. The internal crosstalk coupling between pixels and guardring can then be simulated with less limitations than the analytical model (as far as the coupling impedances are put into the simulation netlist). The table 2 sums up the simulation results. The bold numbers corresponds to pixels simulated thanks to the analytic model.

Similar results are found compared to the analytic model except for the 3 mm option. In this case the internal crosstalk (coupling shown on figure 13) seems to be dominant.

6.3. Printed circuit models

As a first step for measurements, some models of the sensors have been made in copper and epoxy (printed circuit boards). Various topologies of guardrings are then measured (figures 15, 16, 17). Measurements are compared to theoretic models and electrical simulation. The measurement electronics is made up of a waveform generator injecting signal through a capacitor, the device

Journal of Physics: Conference Series **160** (2009) 012067 doi:10.1088/1742-6596/160/1/012067

Table 2. Electrical simulation results (3×3 matrixof pixels) for options 1, 2 and 3 of the guardrings design. the signal amplitude in db is given with respect to a reference signal taken on the closest pixel from the injection point in the case of continuous guardring.

Continuous			1 cm			3 mm		
0	-4.4	0	**0**	-18.3	-35.6	**-11.5**	-36.0	-49.8
-4.4	-14.1	-4.4	**-18.3**	-34.2	-51.2	**-36.0**	-49.0	-71.1
0	-4.4	0	**-35.6**	-51.2	-64.3	**-49.8**	-71.1	-85.7

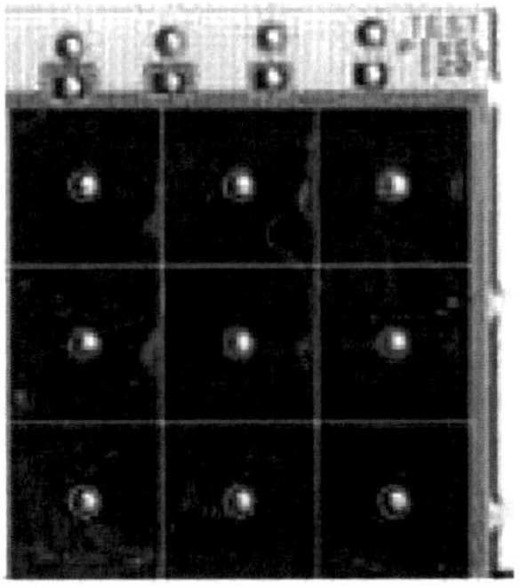

Figure 15. Picture of a test PCB.

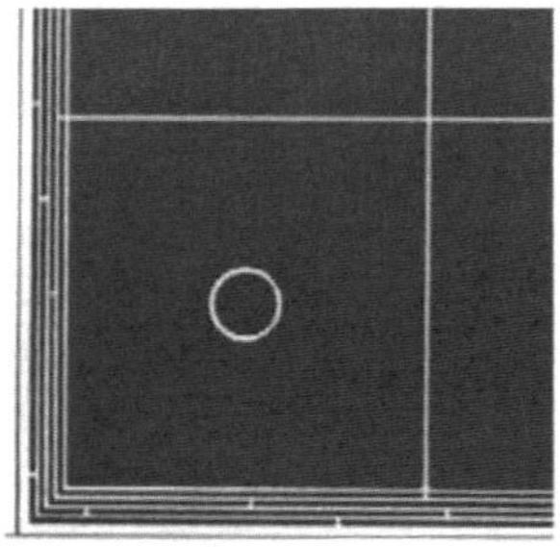

Figure 16. Layout for the 1 cm long segments.

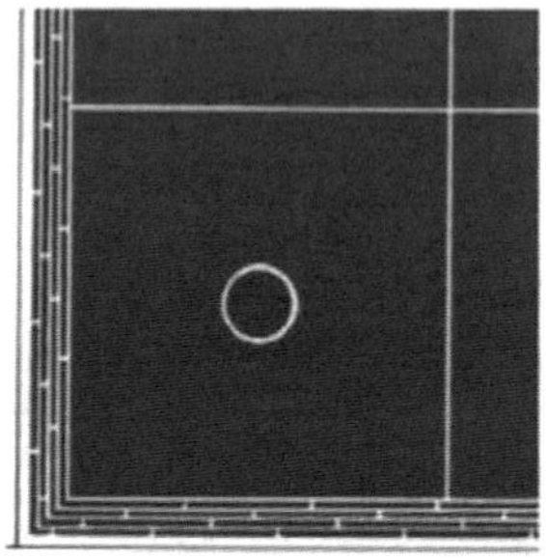

Figure 17. Layout for the 1 cm long segments.

under test (wafer or hardware model) and a amplifier to output reasonably high signal to a scope (modeled in the analytic model).

The measurements on copper hardware model have been performed at LPC (Laboratoire de Physique Corpusculaire), Clermont-Ferrand with the contribution of F. Morisseau and M. Benyamna.

The figure 18 shows the analytic model response (curves) for the various design options (Continuous: red, 1 cm: blue, 3 mm: green) and measurements (marks) made on the copper-epoxy hardware model. The measurements errors are comprised within the mark thickness. The measurements fit well with the predicted results (continuous guardring data have been used to fit some model parameters). Slight differences are due to effects not taken into account and the coarse method used to fit some model parameters.

The measurements below 50 dB are imprecise due to the noise level near 55 dB. The test bench should be upgraded for the next measurements steps.

For continuous guardrings, models and measurements are in accordance (20% precision). The analytic model of guardring crosstalk explains the measurments as well as electrical simulation including internal crosstalk.

For 1 cm segemented guardrings, this is still the case but for A2 analytic response is close to the internal crosstalk level. Nevertheless, measurements shows an higher level close to the sum of this two effects. It is confirmed by electrical simulation. For A3, the dominant effect is still the crosstalk related to the guardrings.

For 3 mm segmented guardrings, the full electical model explains the measurements and its response is close to electrical model of pure internal crosstalk while the analytic model gives extremely low level of guardring crosstalk (A3 not seen). Segment length smaller than the pixel size are then expected to make the guardring crosstalk as secondary effect compared to pixel to pixel crosstalk.

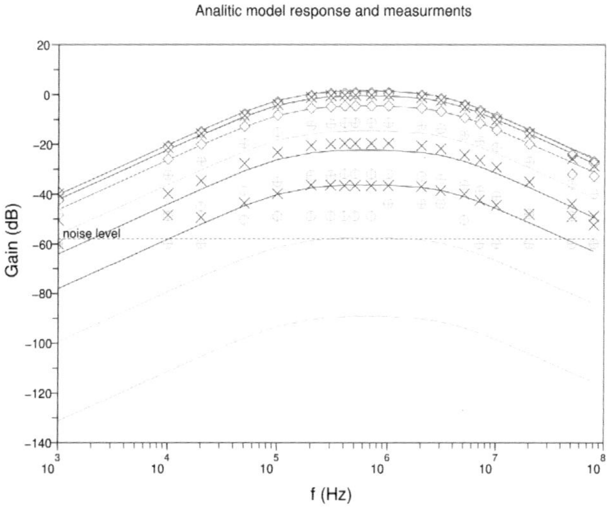

Figure 18. Model response (lines) and measurements (marks) made on hardware model: signal amplitude in dB for the 3 first pixels in a column close to the guardring structure according to the frequency of incoming signal injected on the guardring.

7. Conclusion

A ten thousands channels prototype was build and successfully tested achieving compactness and granularity requirements for a detector compatible with an ILC. The data accumulated from various test beam campaigns at CERN and DESY have allowed to characterize the detector. The signal over the noise ratio was found to be 7.6. This value is close to the goal of 10 and is therefore encouraging for the next development step. With the cost issue in mind, further work is needed about the sensors in order to understand and fix the guard rings related crosstalk. The use of advanced design techniques do not comply with the requirements on the overall cost which has to remain low (less than 10 euros/cm^2). Floating guard rings are used as they are manufactured with the same processing steps as for the other parts of the sensors. A segmented layout of the rings have been investigated thanks to analytical and electrical models as well as with measurements on small printed circuits. Results are showing that the crosstalk decreases by about 20 dB on the second pixel from the injection point when using segmented floating guardrings. Measurements on real sensors should confirm that point.

References

[1] The CALICE collaboration `http://polywww.in2p3.fr/activites/physique/flc/calice.html`.

[2] Brient J-C 2004 Improving the jet reconstruction with the particle flow method: an introduction *11th International Conference on Calorimetry in High-Energy Physics* (Calor 2004),Perugia, Italy.

[3] Respond J *et al.* Design and electronics commissioning of the physics prototype of a Si-W electromagnetic calorimeter for the international linear collider *Journal of Instrumentation* .

[4] Fleury J *et al.* 2007 SKIROC : A front-end chip to read out the imaging silicon-tungsten calorimeter for ILC, IEEE-NSS, Honolulu, HI, USA **3** 1847.

[5] Evensen L *et al.* 1993 Guard ring design for high voltage operation of silicon detectors *Nuclear Instruments and Methods in Physics Research Section A* **337** 44.

Effects of temperature dependence of the signals from Lead Tungstate

Gabriella Gaudio

Università degli Studi di Pavia and INFN Sezione di Pavia, Pavia, Italy

E-mail: gabriella.gaudio@pv.infn.it

Abstract. In recent papers [1] it has been demonstrated that Lead Tungstate signals contain both scintillation and Cherenkov components. In order to further assess and evaluate the relative contribution of the Cherenkov component, we performed measurements at different temperatures, ranging from $13°C$ to $45°C$ being only the scintillation component affected by the temperature change. Over this temperature range, the total light yield was measured to decrease by a factor of 2, while the relative Cherenkov contribution to the signals increased by the same factor. We also studied the decay time of the scintillation process and observed it to decrease as well.

1. Introduction

The results described in the following come from beam tests performed in 2007 as part of the DREAM project. It is well known that one of the main limitation to hadronic calorimetry is due to different response to electromagnetic and non-electromagnetic shower components, which led to non linearity and poor hadronic resolution. A possible solution to overcome this limitation is to measure the electromagnetic fraction (f_{em}) on event-by-event basis, for example, by separating the scintillation and the Cherenkov contributions, being the latter sensible only to the electromagnetic part of the shower. The capability of such a separation using different media [2] or crystals have been already proved. The measurement of the temperature dependence, of course, is not a technique to analyze data in an experiment, but allows us to evaluate the amount of Cherenkov contribution with respect to the scintillation one. More detail on the analysis can be found in [3]. Section 2 describes the experimental setup, Sections 3 and 4 are dedicated to the experimental results for both charge and time information analysis, respectively, while conclusions are drawn in Section 5.

2. Experimental setup

The beam tests took place in the H4 beam line of the Super Proton Synchrotron at CERN. We used a $PbWO_4$ crystal[1] with a length of 18 cm and a cross-section of $2.2 \times 2.2\ cm^2$. The transverse dimension, relevant for our measurements, corresponds to 2.5 radiation lengths. For the measurement described in this paper we used a 50 GeV electron beam. The light produced by particles traversing this crystal was read out by two PMTs, called in the following left (L) and right (R), located at opposite ends and coupled to the crystal by means of silicone cookies. This

[1] The crystal was provided by the Alice collaboration

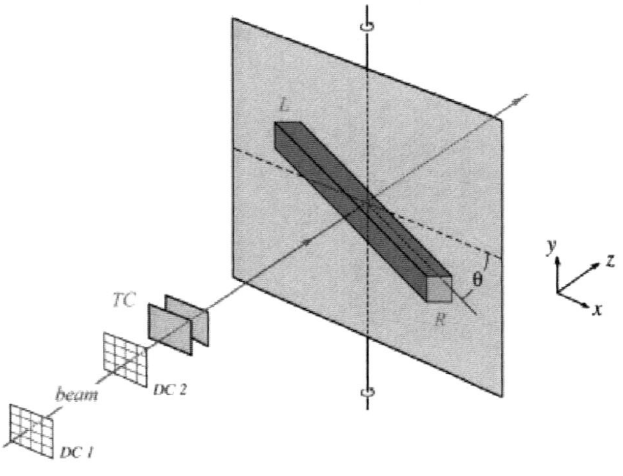

Figure 1. Sketch of the experimental setup. When $\theta > 0$ the PMT R is located downstream, while for $\theta < 0$ PMT L is downstream, as in the case shown here

crystal was positioned on a rotating platform in order to perform angular scans, as shown in Fig. 1. The angle increases when the crystal is rotated such that the crystal axis LR approaches the direction of the traveling beam particles. The signal coming from the PMT are read with both an high sampling rate (2.5 GHz) oscilloscope and a 12-bit ADC module ($100fC/counts$), in order to study the time structure and the charge distribution, respectively. Typical signal can be seen in Fig. 2 and Fig. 3.

2.1. Temperature measurement and control

The temperature of the crystal was changed by means of a thermoelectric system, based on the Peltier effect, remotely controlled by a computer, which allows to achieve a temperature stability of about $\pm 0.3°C$ within a single run (100000 events acquired in about 5 minutes) and $\pm 1°C$ within a complete angular scan, lasting about 3-5 hours, as shown in Fig. 4. In both case only oscillation around the mean value can be observed. Therefore we can use in the following the

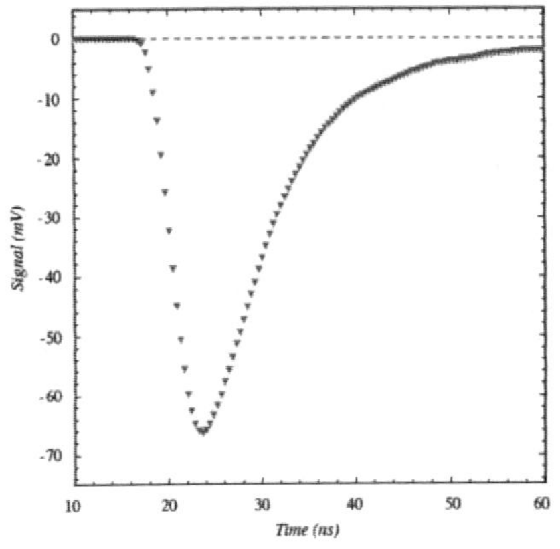

Figure 2. Typical time structure of the signal.

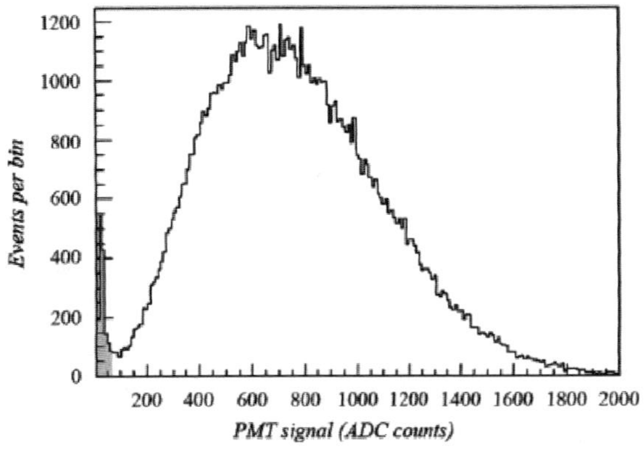

Figure 3. Typical ADC distribution for a PMT signal.

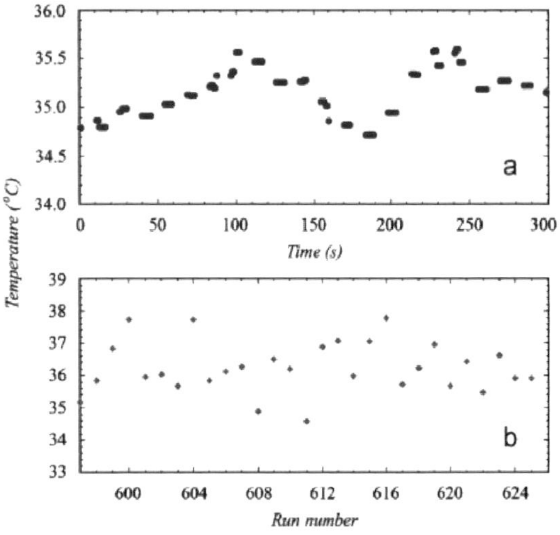

Figure 4. Typical temperature trend distribution within a run (top) and within a complete angular scan (bottom).

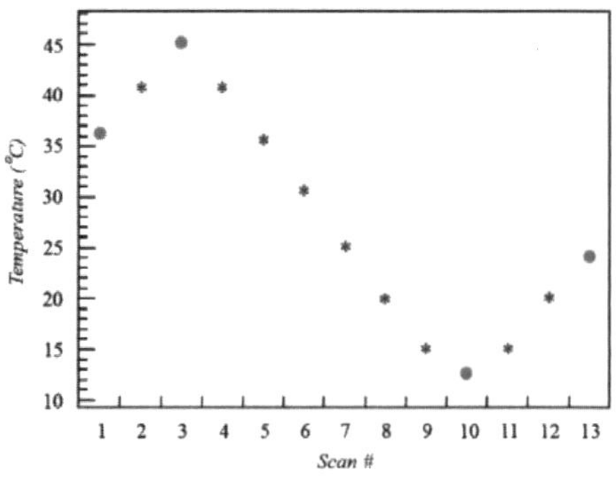

Figure 5. Temperature cycle performed on the crystal. The red full circles represent the measurement for which we performed a complete angular scan.

average temperature measured in each run. The dataset relevant for the work described in this paper consists of 13 angular scans performed at different temperatures, ranging from $13°C$ to $45°C$. Almost all angular scans consist of quick scan at relevant angles, i.e. 0°, ±25° and ±30°, while 4 scans where performed in step of 5°, between −60° and +60°, shown as red full circles in Fig. 5.

3. Charge distribution analysis

The ADC charge distribution gives information on the energy deposit of electrons traversing the crystal. The signal shows a pedestal, the electromagnetic shower deposit and a MIP peak, probably due to beam contamination. The pedestal subtraction is performed using the mean value obtained from distributions of dedicated pedestal run taken for each data run.

3.1. Systematic studies

The contribution of the MIP peak has been evaluated to be constant on all the runs at about 1%, therefore we decided not to perform any selection to eliminate it. We also considered different parameterization of the ADC signal, and due to only 5% variation in peak/mean ratio we decide to use the mean of the distribution as parameter to characterize the ADC distribution. Moreover we studied the presence of long tails, which could bias the mean value, and it only appear in less then 0.5% of the events. A selection on beam position, as determined by Beam Chambers positioned just upstream our setup, is applied to reduce events in which the beam is not hiting the crystal.

3.2. Results on ADC signal analysis

From observation of the ADC signal at different temperature it can be deduced that the light yield of the $PbWO_4$ decreases as temperature grows. This effect is due to the decrease in the amount of scintillation light produced by the crystal as can be demonstrated by Fig. 6. This picture shows the average signals measured in PMT L (a) and PMT R (b) for rotation angle of the crystal of 0° and ±30° as a function of the temperature. The fact that the light yield at ±30° is larger than at 0° is due to the longer path traveled by electrons in the rotated crystal;

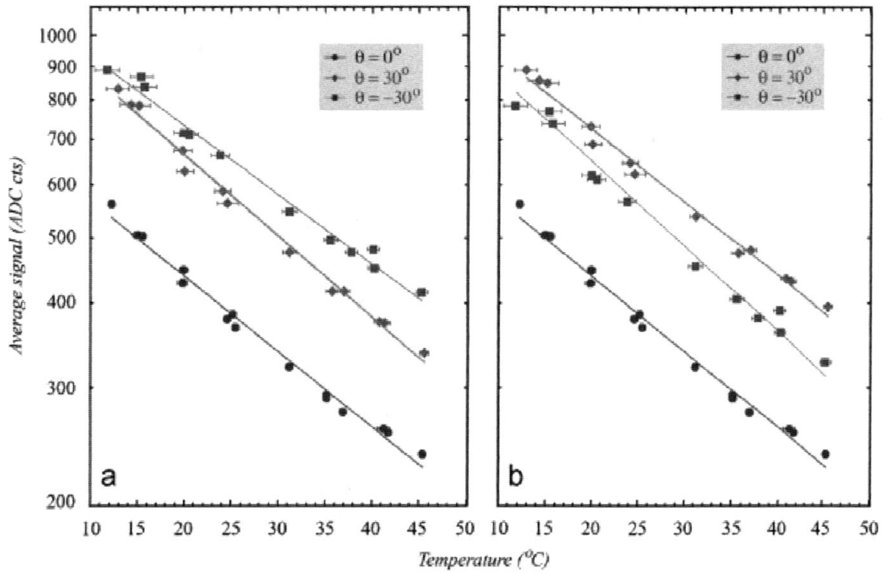

Figure 6. Average signal of left (a) and right (b) photomultipliers as function of the temperature at angles 0° and ±30°. The lines represent the results of least-squares exponential fits to the experimental data. Error bars are due to statistical uncertainties

Table 1. Temperature dependence of the light yield measured in the two PMTs reading out the $PbWO_4$ crystal, for three different orientation.

Angle θ	Slope PMT L ($\%/°C$)	Slope PMT R ($\%/°C$)
$-30°$	2.61 ± 0.02	2.99 ± 0.02
$0°$	2.81 ± 0.02	2.80 ± 0.02
$+30°$	2.95 ± 0.02	2.66 ± 0.02

the difference observed between the light yields measured at $+30°$ and $-30°$ is the result of different contributions of Cherenkov light to the signals. In fact the Cherenkov light is emitted at fixed angle, and therefore is collected by the PMT positioned downstream at proper rotation angle. The results indicate in all cases an exponential decrease in the light yield as a function of temperature, but the slope of the fitted exponential clearly depends on the angle. The temperature dependence of the light yield, expressed in terms of this coefficient, is summarized in Table 1. It turns out that the decrease in the light yield for increasing temperatures is considerably steeper when the signals contain no (or very little) Cherenkov light, i.e., at 30° for PMT L and at $-30°$ for PMT R. In those cases, the light yield decreases by 2.97%, compared with 2.64% for angles at which the Cherenkov content of the signals is largest. For 0°, a value of 2.80% was measured, indicating that there was at least some Cherenkov light contributing to the signals.This angular dependent pattern of the slope coefficients is consistent for the two PMTs, i.e., the measured slopes for the two PMTs at opposite angles are the same within experimental errors.

Due to different characteristic of isotropy of both scintillation and Cherenkov, being the latter emitted only at defined angles, it is possible to define a quantity, called anisotropy, ξ,

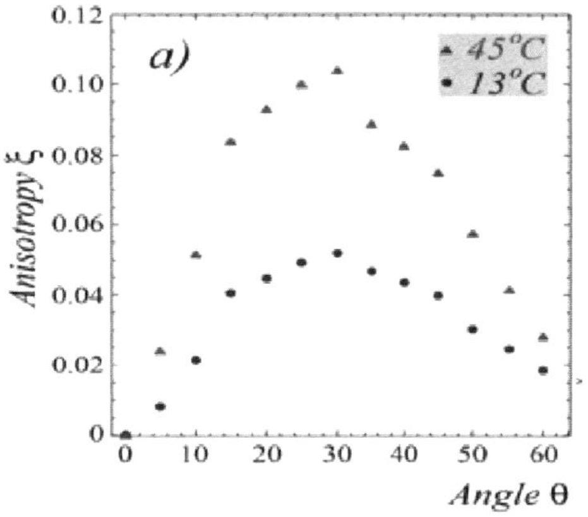

Figure 7. The response anisotropy ξ as a function of the angle of incidence of the beam, measured at the temperature of $13°C$ to $45°C$

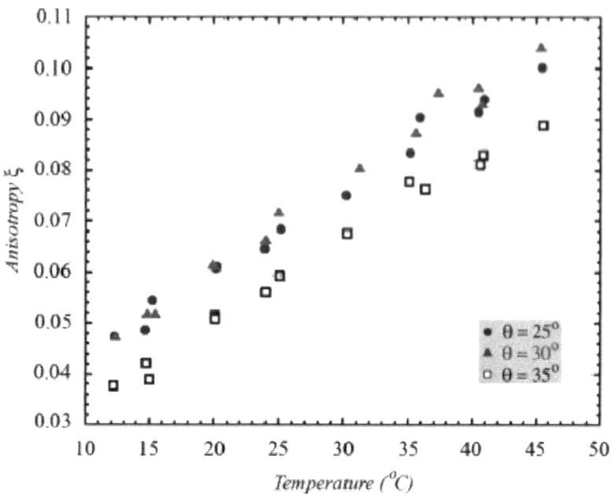

Figure 8. The response anisotropy ξ as a function of the temperature at angles 25°, 30°and 35°.

which depends on the angular position and which allows to measure the fraction of Cherenkov component in the signal:

$$\xi(\theta) = \left| \frac{R_\theta - R_{-\theta} - L_\theta + L_{-\theta}}{R_\theta + R_{-\theta} + L_\theta - L_{-\theta}} \right|$$

A non-zero value of ξ means presence of non-isotropic component, i.e. Cherenkov signal. Since we normalized signal at 0°, by definition the anisotropy is null at this angle. Data show that the maximum of the anisotropy is seen at Cherenkov angle, as demonstrated by Fig. 7. Moreover, as discussed before, since at higher temperature the Cherenkov fraction is higher, due to reduction in scintillation light produced, the anisotropy increases as well as function of temperature, as shown in Fig. 8.

4. Time structure analysis

The Cherenkov and scintillation signals can be distinguished also on a timing base, since the former is prompt, while the latter has usually a larger decay time. As a consequence the Cherenkov contributes only to the leading edge of the signal, while the trailing edge is only due to scintillation light. Many characteristic features of the time structure of the signal can be derived by Fig. 9. The top plots show the time structure of the signal for the PMT R[2] at $\pm 30°$ at $13°C$ (left) and $45°C$ (right), respectively. The red line corresponds at an an angle of $+30°$, for which the PMT R is downstream and collects almost all the Cherenkov light. It can be noted that the amplitude of the red curve is larger and steeper in the leading edge. The trailing edge, on the contrary, shows the same shape and amplitude for both angles, due to isotropy emission of the scintillation light. As a confirmation of the light yield analysis described in Section 3.2 a difference in amplitude for data at $13°C$ (left) and $45°C$ can be observed. The difference of the signals at $\pm 30°$ is shown in the bottom plots of Fig. 9. Both left and right bottom plots show the same amplitude, further confirming the independence of the Cherenkov signal from the temperature.

[2] Analogous plot can be produced for PMT L, with swapped angles

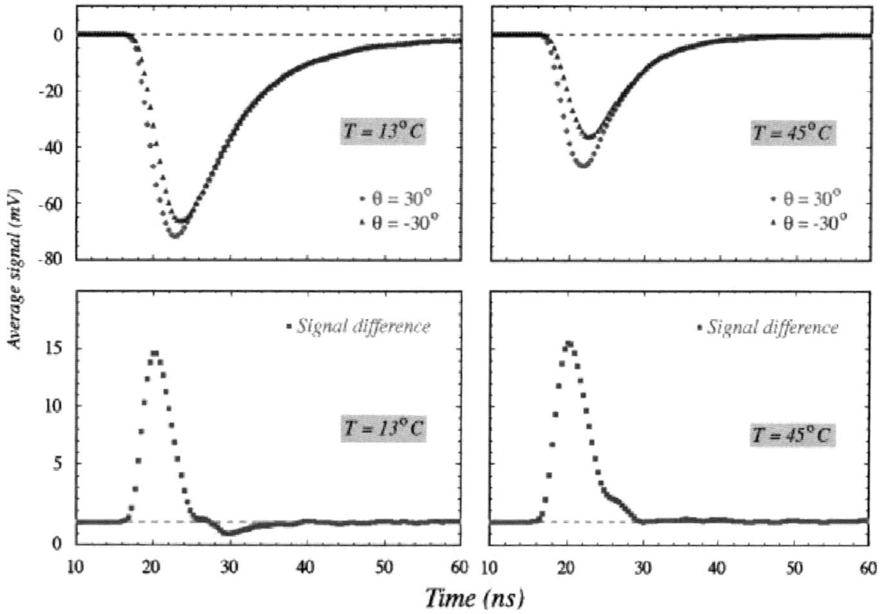

Figure 9. Average time structure of the signal for the PMT R at angles ±30° (top) and difference of the two signals (bottom) at 13°C (left) and 45°C (right)

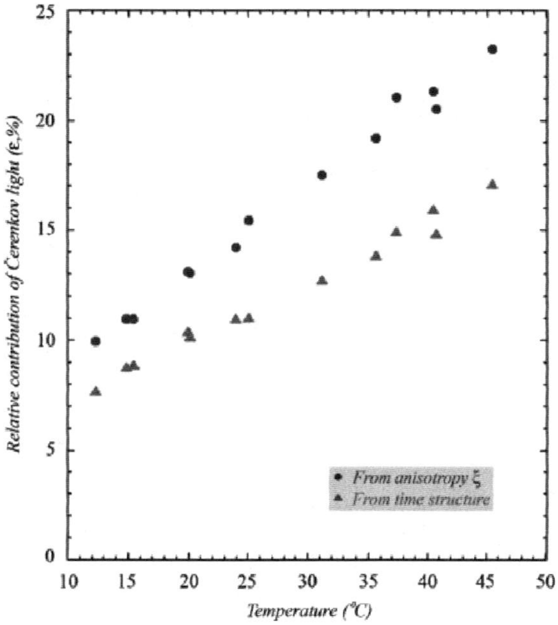

Figure 10. Temperature dependence of the Cherenkov/scintillation signal ratio measured in one of the PMTs. This ratio is derived from the response anisotropy, or from the relative contribution of the prompt component to the signals (averaged over the two PMTs) for Cherenkov angle.

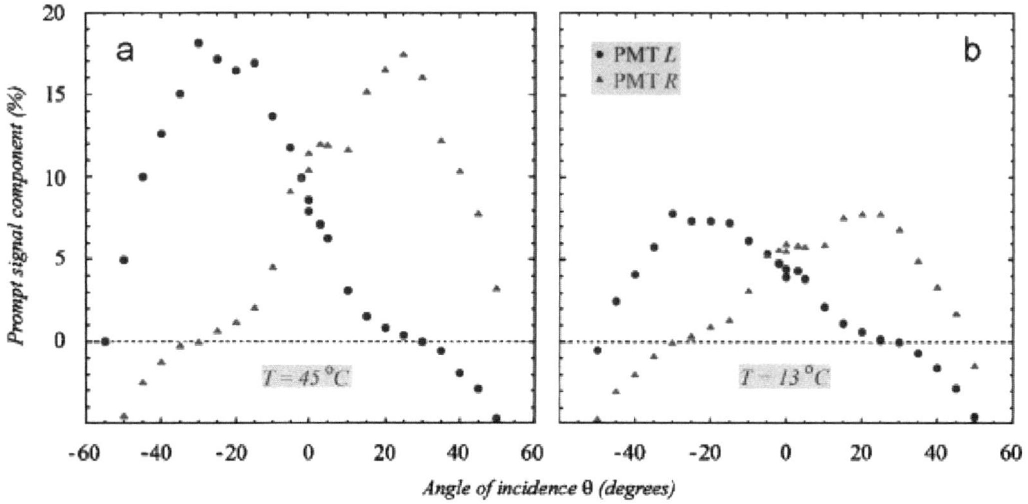

Figure 11. Fraction of the total signal represented by the prompt component as function of the angle of incidence measured at 45°C (left) and 13°C (right). The curves are shown for both PMT L (blue full circle) and PMT R (red triangle).

4.1. Results on time structure analysis

The fraction of Cherenkov component present in the signal can be evaluated by integrating the difference of the signal obtained at ±30° (the curves in bottom plots of Fig. 9), normalized with respect of the total signal obtained at angle opposite to the Cherenkov emission angle. The relative contribution of Cherenkov and scintillation signal thus obtained in shown as function of the temperature in Fig. 10. For comparison the analogous quantity evaluated from anisotropy is reported in the same figure. In both cases an increase by a factor of two in about 30°C of temperature variation is observed, the absolute values is slightly different, anyway, due to the different definitions and normalization points used in the two analysis. If the fraction of Cherenkov component is evaluated as described before but for every angle with respect to the signal at angle opposite to the Cherenkov one, which is mostly a pure scintillation signal, plots as those shown in Fig. 11 can be obtained. The maximum of the curves, which correspond to the larger fraction of Cherenkov component in the signal, is reached at +30° for PMT R and −30° for PMT L, as expected. By definition, the curves reach a null value at angle opposite to the Cherenkov one (−30° for PMT R and +30° for PMT L).

4.2. Scintillation decay time study

As discussed before, the trailing edge of the time structure of the signal is dominated by scintillation light, which is sensible to the temperature variation. It can be observed (Fig. 12) that at higher temperature the trailing edge become steeper. A study of the decay time constant as function of the temperature has been performed and results are shown in Fig. 13. A decrease of about 30-40% in 30°C temperature variation is observed.

5. Conclusions

A detailed study of the response of a single crystal of $PbWO_4$ has been performed as function of the temperature variation. This technique allows to determine the relative contribution of Cherenkov and scintillation lights in the signal, due to the fact that Cherenkov light is independent from temperature. Using the already assessed technique of Cherenkov and

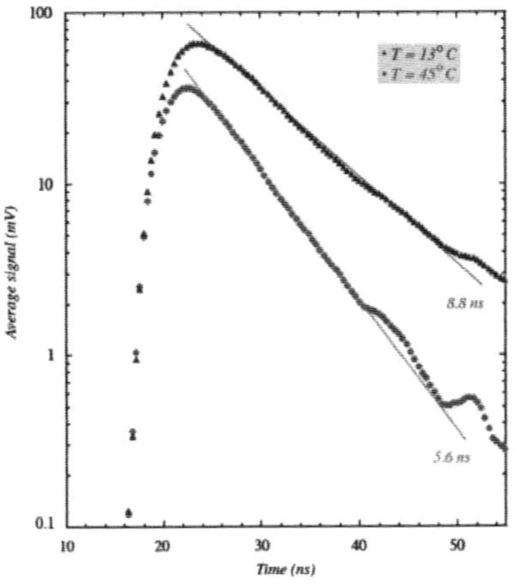

Figure 12. Average time structure of the $PbWO_4$ signal from PMT L at an angle of 30° measured at two different temperatures.

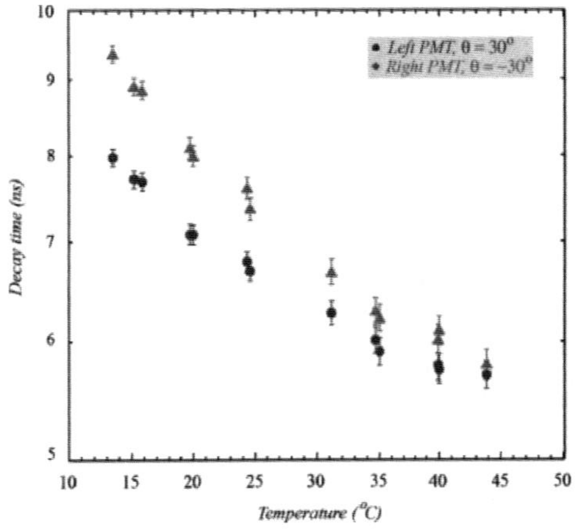

Figure 13. Decay constant of the scintillation time as function of the temperature for angle at which the signal is almost all due to scintillation.

scintillation separation, it has demonstrated that the Cherenkov contribution increases by a factor of two, passing from $13°C$ (left) to $45°C$, and in the same range also the decay time of the scintillation light decay by 30-40%.

References

[1] Ackurin at al., 2007 *Nucl. Instr. and Meth.* A **582** 474
[2] Ackurin at al., 2005 *Nucl. Instr. and Meth.* A **537** 537
[3] Ackurin at al., 2008 *Nucl. Instr. and Meth.* A **593** 530

Journal of Physics: Conference Series **160** (2009) 012069 doi:10.1088/1742-6596/160/1/012069

Time profile analysis of photodetector signals in multi read-out calorimetry with GHz samplers

F Bedeschi[1], **M Bitossi**[1a], **R Carosi**[1], **M Incagli**[1], **R Pegna**[1b] **and F Scuri**[1]

[1]INFN, sez. di Pisa, Largo B. Pontecorvo 3, 56127 Pisa, Italy
[a]Dipartimento di Ingegneria Informatica, University of Pisa, Italy
[b]Dipartimento di Fisica, University of Siena, Italy

E-mail: `roberto.carosi@pi.infn.it`

Abstract. We present possible applications of DAQ systems based on Domino Ring Samplers (DRS) for time profile analysis of photodetector signals used for present and future multiple read-out calorimeters. The example of an 80-channel system in preparation for dual read-out calorimetry (DREAM) is described.

1. Introduction

Calorimeters are crucial components of almost every experiment in high-energy particle physics. The detection of particles that develop electromagnetic showers can be performed with high precision, but this is not true for particles that are subject to the strong interaction.

This is primarily due to following facts:

- the calorimeter typically generates a larger signal per unit deposited energy for the electromagnetic (em) shower component (primarily initiated by the process $\pi^0 \rightarrow \gamma\gamma$) than for the non-electromagnetic one (i.e. $e/h > 1$);
- the fluctuations in the energy sharing between these two components are large and non-Poissonian.

Therefore, in typical calorimeters the hadronic response function is non-Gaussian, the hadronic signals are non-linear, and the hadronic energy exhibits substantial deviations from $E^{1/2}$ scaling [1].

In order to deal with those problems, few ways have been tried, e.g. compensating calorimeters (designed to have $e/h \approx 1$), offline compensation, Particle Flow Analysis.

The DREAM (DUal REad-out Method) collaboration has recently used the Dual Read-out approach [2]. Such approach is based on the measurement of the em shower component, f_{em}, event by event. This is done by comparing the shower signals produced in the form of scintillation light and Cherenkov light in the same detector. In hadronic showers, Cherenkov light production is dominated by their em shower component, with $e/h \approx 5$, as shown by DREAM results. This shows that the signals from the non-em component of the hadron showers are dominated by spallation protons produced in nuclear reactions. These particles are usually not sufficiently relativistic to produce Cherenkov light. The electrons and positrons through which the energy of the em shower component is deposited are relativistic down to a fraction of 1 MeV and thus dominate the production of Cherenkov light in hadron showers.

On the other hand, for the production of scintillation light, the em and non-em components give more similar contributions, $e/h \approx 1.4$.

If E is the energy of the incident particle, the Cherenkov signal (Q) and the scintillation signal (S) are given by

$$Q = E\left[f_{em} + (e/h)_Q^{-1}(1 - f_{em})\right] \tag{1}$$

$$S = E\left[f_{em} + (e/h)_S^{-1}(1 - f_{em})\right] \tag{2}$$

if Q and S are calibrated with electrons.

If the e/h ratios are known, f_{em} can be calculated event by event and the incident energy can be determined. DREAM has shown that using this method, the Gaussian response, the linearity, and the energy resolution can be improved [2].

In principle there are four ways to detect Cherenkov light and separate it from scintillation light. They exploit different properties of S and Q:

- Directionality: S is isotropic, Q is emitted at an angle $\theta_c = \arccos(1/\beta n)$ with respect to the particle momentum;
- Time profile: Q is prompt, S has a characteristic time up to several hundreds ns, depending on the material;
- Wavelength spectrum: Q has a spectrum λ^2, S depends on the material;
- Polarization: Q is polarized, S is not polarized.

DREAM has used the first three methods using a fiber calorimeter, Lead Tungstate crystals ($PbWO_4$) and Bismuth Germanate ($Bi_4Ge_3O_{12}$, or BGO) crystals. Fibers were plastic scintillating fibers (to detect scintillation light), quartz and acrylic plastic fibers (sensitive to Cherenkov light).

Lead Tungstate and BGO have different properties concerning the Q/S separation. The decay times τ for scintillation light are 50 and 300 ns respectively. The λ_{max} of the scintillation light spectrum are 560 and 460 nm. The relative Q/S outputs are roughly 100:1.

DREAM has been able to measure the time structure of the signals using a digital oscilloscope, which provided 5 GSamples/s with an analog bandwidth of 2.5 GHz. The time structure could be studied event by event but for two channels only.

In this paper we propose a method to study the time structure of the signals at DAQ level, event by event and channel by channel.

The separation between Cherenkov and scintillation light is not the end of the story. The kinetic energy of the neutrons is correlated to the nuclear binding energy loss ("invisible energy"). Again, this can be inferred from the time structure of the signals.

In fact, more than 95% of the neutrons are produced in nuclear deexcitation, with average energy $\langle E_n \rangle \approx 3$ MeV. These neutrons lose their energy predominantly through elastic scattering. The energy loss in elastic scattering scales as A^{-1} (A is the atomic mass). Therefore free protons dominate this process.

The density of free protons in DREAM plastic fibers is $8 \cdot 10^{21}$ protons/cm^3. The cross section for elastic $n-p$ scattering goes from 2.2 b (for 3 MeV neutrons) to 12 b (for 0.1 MeV neutrons). In this range of energies, the mean free path between elastic $n-p$ scattering events goes from 56 cm to 10 cm. The average time between subsequent $n-p$ scattering events is about 23 ns. This is independent of E_n. Therefore we expect an exponential tail in the time structure of the signals. Since neutrons lose on average 50% of their kinetic energy in elastic $n-p$ scattering, E_n is reduced to a fraction e^{-1} in about 33 ns if we neglect other processes. If we take into account other processes through which neutrons may lose energy (elastic scattering off C, Si, Cu nuclei, inelastic scattering) we expect an exponential tail with constant $\tau \approx 25$ ns.

The detection of the energy loss by neutrons may lead to an energy resolution down to $15\%/\sqrt{E}$ (E in GeV) [4].

2. Domino Ring Samplers

As discussed, a signal from a hadron calorimeter exhibits a time structure with three components:

- a prompt component due to Cherenkov light;
- a slower component due to scintillation light;
- another slower component due to neutron energy loss.

We propose a DAQ system to make time profile analysis of the individual channels of the DREAM detector. Analog signal sampling of the DREAM read-out for time profile analysis requires some specific characteristics:

- high frequency sampling;
- large bandwidth to avoid smoothing of the prompt Cherenkov component;
- it must cover a large time window to include the slow neutron component;
- it must be cheap and scalable to serve about one hundred channels.

A system based on the Domino Ring Sampler (DRS) chip [5], developed at the Paul Scherrer Institute (PSI), Villigen, Switzerland, for the MEG experiment, is adequate to cope with these requirements.

The main features of the DRS chip are:

- Integration of 10 analog channels in one chip;
- analog bandwidth of about 200 MHz;
- very low power consumption, about 35 mW per chip;
- the response of the sampler cells can be calibrated and the useful input voltage can be extended to 2 Volts (twice the range for commercial Flash ADCs).

In the present version II of the DRS chip, sampling frequency is set to 2.5 GHz with about 200 MHz intrinsic bandwidth. The temporal width of the sampled signal depends on the programmed number of cells (up to 1024) to be sequentially backward converted from trigger time t_0.

The sampler is made of two sections, an analog part for the signal sampling and a digital part for control and multiplexing. The analog signal is stored in a multi capacitor bank (1024 cells in DRS II) that is organized as a ring buffer, in which the single capacitors are sequentially enabled by a shift register driven by a high frequency clock internally generated, called Domino wave. An on-board PLL system locked to an external low frequency reference clock helps in reducing the sampling signal jitter down to less than 200 ps. The chip is housed in a PLCC package and mounted on a mezzanine card.

Once the external trigger has been received, the sampled signal in the ring buffer is delivered to the output stage by a multiplexer and digitized at high resolution (12 bits) at lower frequency (40 MHz). The signal conversion is done by an external ADC.

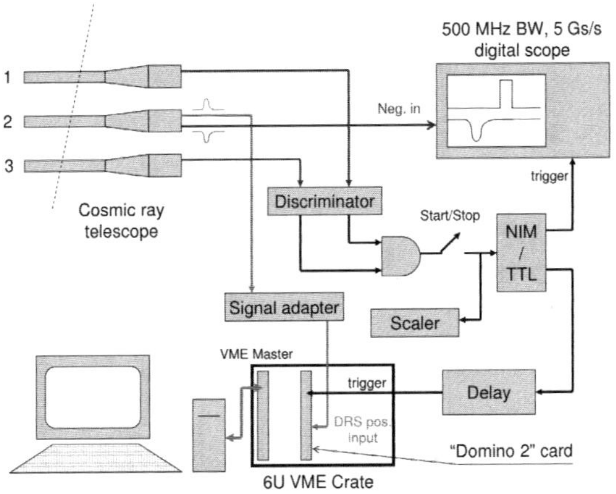

Figure 1. Schematic drawing of the test read-out system.

3. Test setup

A prototype of a board based on DRS II chips has been tested at the Istituto Nazionale di Fisica Nucleare, Pisa, Italy.

The test setup consists of a cosmic ray telescope, as shown in figure 1.

The analyzed PM signal is taken from the central counter 2 of a three scintillator plate telescope. The coincidence of discriminated signals of counters 1 and 3 is used to trigger cosmic ray events. The negative PM signal of counter 2 is fed into the input of a 500 MHz bandwidth, 5 Giga-samples/s digital scope.

The positive output of counter 2 PM is fed into an active signal adapter (50 Ω to differential); no limiting bandwidth is introduced by this device.

The adapted signal is fed to channel 1 of a mezzanine card, hosting a DRS chip and mounted on a motherboard with VME interface.

Stop to the Domino sampling sequence is given by the delayed trigger signal.

In this test the DRS sampling frequency was set at 2.5 GHz (the sampling frequency can be set to a value from 1 through 4 GHz) and all 1024 DRS analog cells converted and acquired (so that the time window was 400 ns).

The DRS sampled signals are acquired with a system based on Struck devices including a SYS3100 interface hosted in the PCI bus of a desktop computer and a VME Crate Controller (Master).

Data acquisition start and stop is handled by a manual switch on the trigger line; on stop the last event in the digital scope is stored and converted to a text file; the corresponding event number in the DRS acquisition chain is counted using a scaler.

The trigger rate is about 8 Hz. It corresponds to a throughput of about 300 kB/s (two Bytes/sample, with 1024 cells/channel, 9 channels/mezzanine, 2 mezzanines). The acquisition time includes DAQ processes, which are not optimized for our setup.

4. Calibration

To transform the DRS output from raw ADC counts to mV, a detailed calibration procedure has been followed.

A signal of constant pulse height is sent to all the 1024 cells and all 8 channels for the 2 mezzanine cards.

The calibration curves, which give the ADC counts as a function of the input pulse height

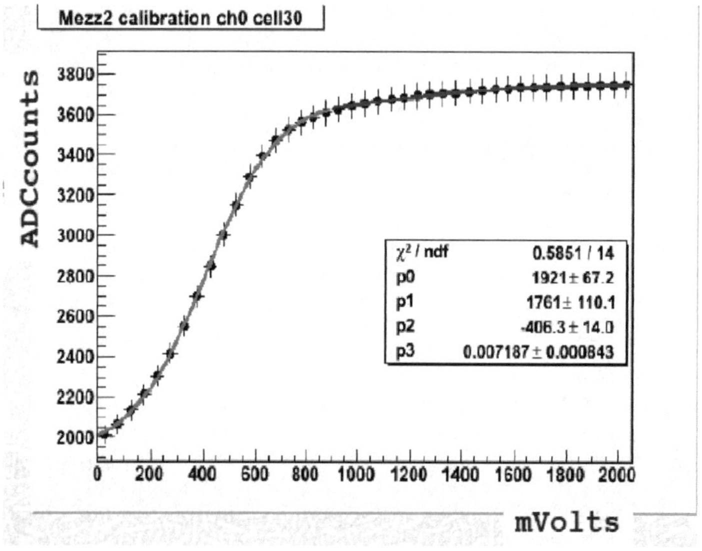

Figure 2. Calibration curve for the cell 30 of channel 0.

in mV, are then obtained. An example is given in figure 2 for the cell no. 20, channel 0, in the second mezzanine card. These curves must be fitted or interpolated and inverted to get the transformation table from ADC counts to millivolts (in fact this has been done with a cubic spline interpolation). From the example figure above we can see that the device is linear with good approximation in the range 100 to 700 mV. For this reason, inside the board it is possible to add an offset in the range -2 V, +2 V to the input signal to have it positive and in the most efficient range. This possibility limits the effective bits to 11 for the ADC range (1 is for the sign). We can see also that the saturation is reached at 1800-1900 ADC counts.

5. Bandwidth

The DRS board has been used to check the bandwidth for the test system. This is important in general to study signal distortions.

We have performed a spectral analysis of the cosmic ray signal at the output of a photomultiplier ("input" signal) and compared with the signal as sampled by the DRS card ("output" signal).

In a preliminary test, using a square wave signal, we already have indications that our prototype readout system has a bandwidth of about 35 Mhz. The rise time t_S of the generated square wave is in fact about 5 ns (from 10% to 90%; this is due to the limited bandwidth of the generator), while the rise time t_O of the signal after sampling by the DRS card is about 11 ns. For a step response of the form

$$V(t) = V_0\left(1 - e^{-t/\tau}\right) \tag{3}$$

where the time constant is τ, the rise time is

$$t_r = t_{90\%} - t_{10\%} = \tau \cdot \log 9 \approx 2.2\tau$$

and the high cutoff frequency $f_H = \log 9/2\pi\tau$.

For a system composed by n cascaded blocks, each having a rise time t_i, whose input signal has a rise time t_S, the output signal has a rise time equal to

$$t_O = \sqrt{t_S^2 + t_1^2 + ... + t_n^2} \tag{4}$$

We can therefore extract the overall rise time of our system

$$t_r = \sqrt{t_1^2 + t_2^2 + ... + t_n^2} = \sqrt{t_O^2 - t_S^2} \approx 9.8 \text{ ns}$$

and the cutoff frequency $f_H \approx 36$ MHz. The overall bandwidth is assumed to be given by the convolution of the cables bandwidth ($\approx$ 150 MHz), the DRS input stage and the DRS chip intrinsic bandwidth ($\approx$ 200 MHz). For a more accurate evaluation of the system bandwidth, a Fourier analysis of the input and output signals have been performed.

Figure 3 shows an input signal in the time domain, as seen from a digital scope.

Figure 4 shows an output signal, sampled by the DRS card. The signal is pedestal subtracted.

The signal analysis in the frequency domain has been done using the Discrete Fourier Transform (DFT).

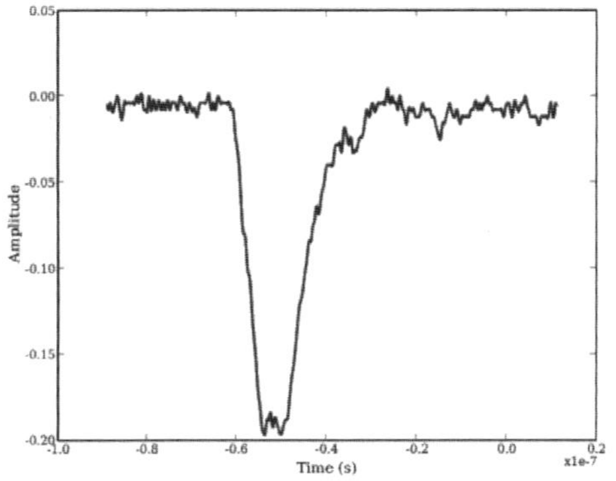

Figure 3. PM signal in the time domain, sampled with the digital scope. Horizontal scale is in seconds.

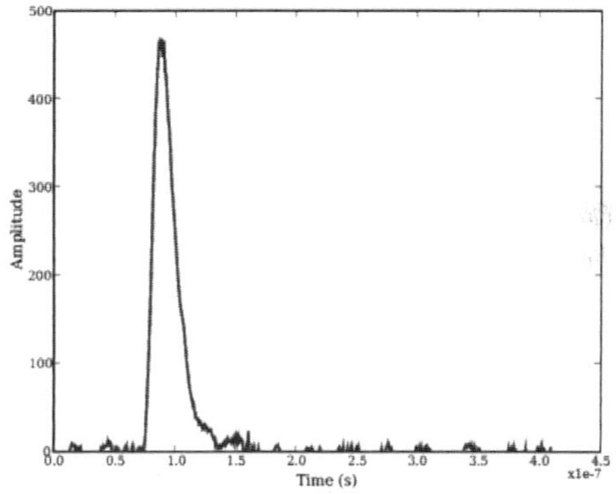

Figure 4. PM signal in the time domain, sampled with the DRS. Horizontal scale is in seconds.

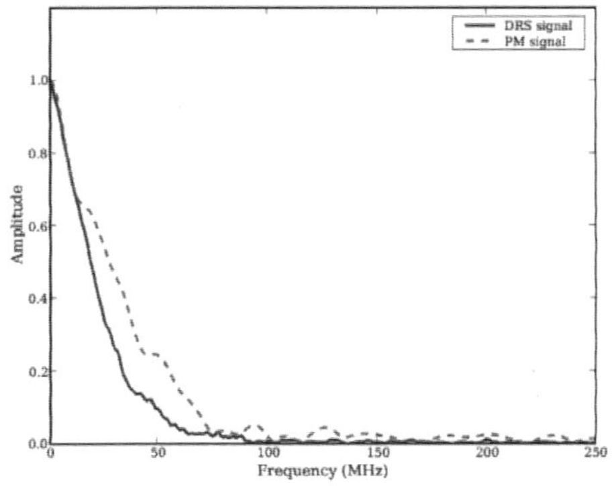

Figure 5. Magnitude of the DFT of the same event PM pulse sampled with the DRS (solid line) and the digital scope (dashed line) signals.

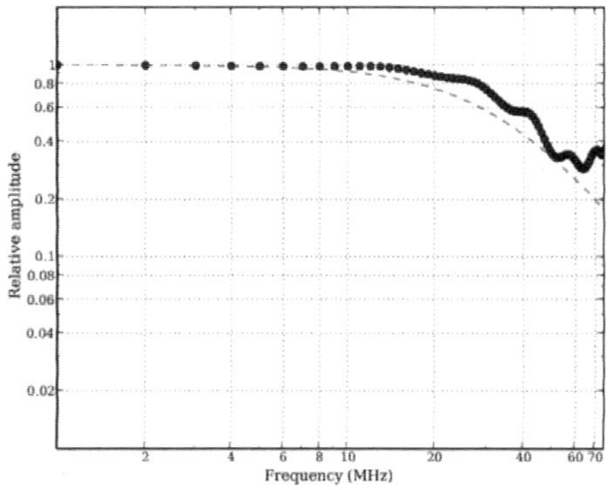

Figure 6. Ratio of the magnitudes of the DFT's of the DRS and digital scope sampled signals. Horizontal scale is in MHz.

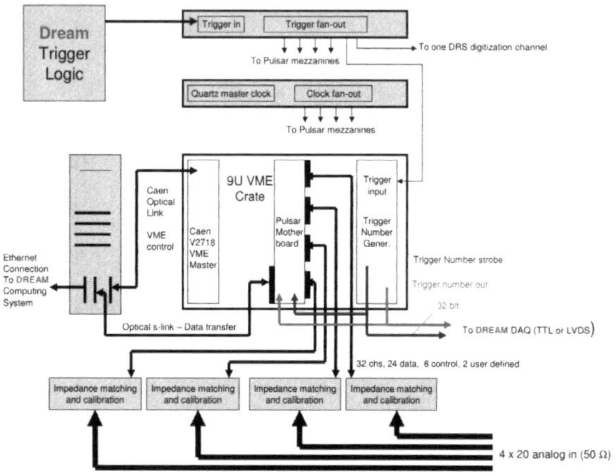

Figure 7. Building block diagram of the new DAQ system for DREAM.

The magnitude of the DFT of the input signal and the magnitude of the DFT of the output signal are shown on the same plot in figure 5. We have normalized the function so that its value is 1 at $f = 0$. The horizontal scale is in Mhz.

Finally, in figure 6 we show the ratio of the magnitudes, i.e. the magnitude of the transfer function of our system (from the photomultiplier to the DRS). We can see that we have a frequency cutoff at about 35 MHz at -3 dB. The dashed line superimposed on the plot represents the magnitude of the DFT for a low-pass filter with cutoff frequency $f_H = 35$ MHz and a second order pole.

The bandwidth we have estimated is the bandwidth of the whole system, which includes cables, board amplification and other limiting devices.

The intrinsic bandwidth of the DRS is about 200 MHz, given by the input capacity of 44 pF in series with a resistor of 22 Ω.

6. Integration in the DREAM system

The DRS chip will be used in the DAQ system for the DREAM test beam at CERN during the Summer 2008. The DAQ main components are shown in figure 7.

Four mezzanine cards mounting two 10 channel DRS chips each are plugged on a 9U VME motherboard. The motherboard, named Pulsar, has been developed by a group of physicists and engineers from Fermilab and the University of Chicago [7]. A picture of a Pulsar board with mezzanine cards is shown in figure 8. The Pulsar has three big Altera APEX20K200 Field Programmable Gate Arrays (FPGA) together with fast large SRAMs. DRS configuration is made via the Pulsar VME interface, while sampled and converted data are sent to the host computer via s-link optical interface. A new design external card is used to match 50 Ω PM signals to the differential input block for DRS and to generate DC levels for the DRS calibration cycle, acting on internal DACs. Auxiliary cards distribute reference and synchronization clock signals and the trigger signal produced by the DREAM logic; another Pulsar board generates the event number, used also by the default DREAM DAQ system in case of two parallel system acquisition.

Data acquisition with the new system is managed with two independent software packages, one used to configure the system before start of data taking, and the other used to handle the data transfer to the host computer in different run modes (calibration, pedestals, physics events). A special task take care to handle local disk occupancy moving data from the host computer

Figure 8. Pulsar board with the four mezzanines with DRS chips.

to the DREAM central computing facility, being relatively high the amount of acquired data: event size is about 160 kB (4 mezzanines, 2 DRS chips/mezzanine, 10 channels/chip, 1024 cells/channels, 2 Bytes/cell) when all DRS cells of all 80 channels are converted. The expected acquisition rate is about 1kHz.

The 80 channels are enough to sample either all PM signals of the fiber detector (quartz and scintillating), or a 40 BGO/$PbWO_4$ crystal matrix (two-side read-out) to be installed upstream of the fiber detector, or a sizeable fraction of both detectors at the same time.

Preliminary results are expected to give first answers on the potentiality of many channel time structure analysis in signal component separation to improve energy resolution. Those results will be the reference for future developments of this DAQ system when higher bandwidth versions (IV) of the DRS chip will be available.

7. Conclusions

We propose a read-out system for the DREAM detector based on the time profile analysis of signals. The core of the system is the DRS chip. It has been successfully tested on a cosmic ray telescope. The system will be used in the DREAM test beam in Summer 2008.

Acknowledgments

We would like to thank the MAGIC group of INFN, Pisa and University of Siena for the support.

References

[1] Wigmans R 2000 *Calorimetry - Energy Measurement in Particle Physics, International Series of Monograph on Physics* vol 107 (Oxford: Oxford University Press) 537
[2] Akchurin N et al 2005 *Nucl. Instrum. Meth.* A **537**
[3] Wigmans R 2007 *Nucl. Instrum. Meth.* A **572** 215
[4] Wigmans R 2007 *New J. Phys.* **10** 025003
[5] Ritt S 2004 *Nucl. Instrum. Meth.* A **518** 407
[6] Pegna R et al 2007 *Nucl. Instrum. Meth.* A **572** 382
[7] Anikeev K at al 2006 *IEEE Trans. Nucl. Sci.* **53** 653

Separation of $PbWO_4$ and BGO signals into Čerenkov and scintillation components

C. Voena, for the DREAM collaboration
Università Sapienza di Roma and INFN Sezione di Roma, Italy
E-mail: `cecilia.voena@roma1.infn.it`

Abstract. We present results from beam tests performed in 2007 on $PbWO_4$ and BGO crystals in the context of the DREAM project. Signals from high energy electrons and pions are analyzed and the possibility of separating the contributions from Čerenkov (Č) and scintillation (S) light for individual events is investigated. Different methods exploiting the difference in timing, in the spectra and in the directionality of the two types of light have been developed to determine the contribution of the two components. In the BGO crystal, Čerenkov signals have been enhanced with the use of optical filters and the ratio Č/S is measured with good precision (~20-30% for energy deposits less than 1 GeV).

1. Introduction

The Dual Readout method allows to improve the performances in hadronic calorimetry by measuring event-by-event the electromagnetic fraction of the hadronic cascades. The principle has been proved by the DREAM calorimeter [1]. The main limitation to the energy resolution of the DREAM detector is the small number of photo-electrons (8 per deposited GeV). This can be improved with the use of homogeneous detectors if Čerenkov light can be separated from scintillation light.

In this paper the results of the measurements performed on a $PbWO_4$ crystal and on a BGO ($Bi_4Ge_3O_{12}$) crystal are described. In order to separate the Čerenkov from the scintillation signals the following differences between the two components are considered:

- Differences in directionality. Scintillation light is emitted isotropically, Čerenkov light is emitted at a characteristic angle with respect to the emitting particle.
- Differences in time structure. Scintillation light is characterized by one or various time constants, Čerenkov light is prompt.
- Differences in spectral properties. Scintillation light has a characteristic spectrum for the crystal in question, Čerenkov light has a λ^{-2} behavior.

2. Crystals and experimental setup

The measurements described in this paper were performed in the H4 beam line of the Super Proton Synchrotron at CERN. The $PbWO_4$ crystal has a length of 18 cm and a cross section of 2.2×2.2 cm^2 (the transverse dimensions correspond to 2.5 radiation lengths, X_0). The BGO crystal has a length of 24 cm and a truncated-pyramidal shape, resulting in a cross section that varied from 2.4×2.4 cm^2 on one side to 3.2×3.2 cm^2 to the opposite side. The transverse dimensions thus varied between 2.2 and 2.9X_0. The index of refractions are 2.2 and 2.15 for the $PbWO_4$ and the BGO, respectively. The $PbWO_4$ scintillation decay time is 10 ns while the BGO decay time is 300 ns. The crystals were read out by two phototubes located on opposite ends and, to reduce the light trapping effects of the large refractive index, in most of the measurements the PMTs were coupled to the crystal by means of silicone "cookies" (n=1.403). In the case of the BGO the light generated in the crystal was filtered before reading it out with a "Yellow" (Y) filter on the small side and a "Ultraviolet" filter (UV) on the large side. The Y filter is highly transparent to the scintillation light, the UV filter is transparent for light in the 300-400 nm range while less than 0.1% of the scintillation light penetrates it (taking into account also the quantum efficiency of the PMTs). The experimental setup is shown in fig. 1.

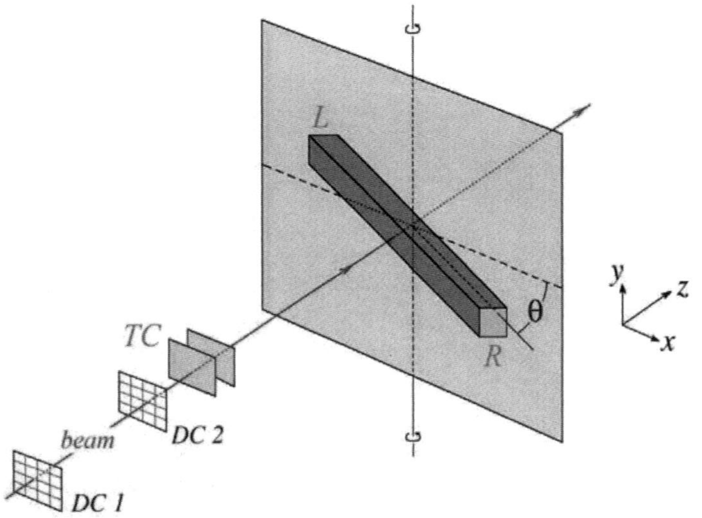

Figure 1. Test beam setup. The scintillation counters (TC) were used to trigger the data acquisition system. The drift chambers (DC1,DC2) were used to reconstruct the impact point of the particles and to select events that entered in a small region (10x10 mm^2) around the crystal center.

The crystal under study was mounted on a platform that could rotate around a vertical axis through its geometrical center. The beam was also steered through this center. The angle θ used through the paper is defined as the angle between the crystal axis and a plane perpendicular to the beam line; the sign is defined such that the crystal orientation in fig. 1 corresponds to $\theta = -30^o$. The crystal signals were acquired with an ADC and with an oscilloscope with a sampling capability of 5 GSample/s at a analog bandwidth of 2.5 GHz.

3. $PbWO_4$ results

Since Čerenkov light is emitted at an angle θ_C=63° by the charged relativistic particles the acceptance of this light for each PMT strongly depends on the angle θ. For the signal from PMT right, this contribution is larger for θ=27°, for which the Čerenkov light goes directly into the PMT. The corresponding angle for PMT left is θ=-27°. Figure 2 (left) shows the average time structure of the signals from 50 GeV electrons in PMT left, for θ=30° and θ=-30°. The trailing edges of the signals are very similar while the leading edge is steeper at θ=-30° and the signal is larger at this angle.

Journal of Physics: Conference Series **160** (2009) 012070 doi:10.1088/1742-6596/160/1/012070

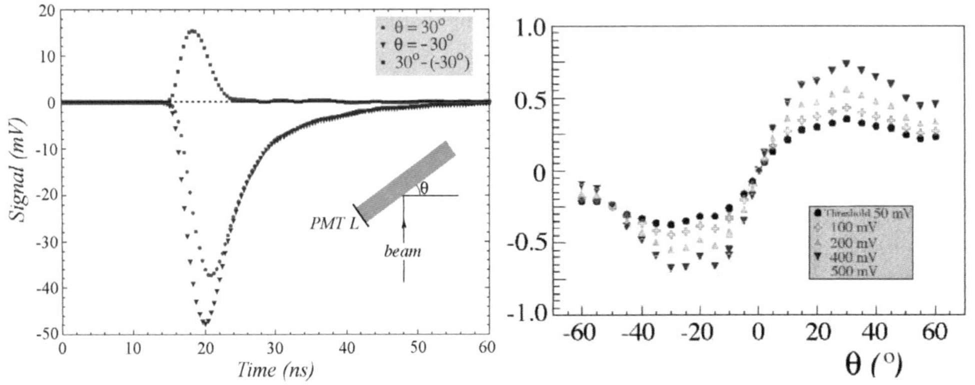

Figure 2. Left: average time structure for PMT left signals from 50 GeV electrons traversing the $PbWO_4$ crystal at different orientations of the crystal with respect to the beam (θ=30° and θ=-30°), and the difference between the two. Right: time difference between the moments the signals from PMT left and right cross a preset threshold level. Results are given as a function of θ for 50 GeV electrons in $PbWO_4$.

The time structure of the signals can be used to determine the relative contribution of the Čerenkov light in different ways. We describe here the most effective method that consists in the measure of the time difference between the moments at which the signals from the two PMTs cross a given threshold level, Δt(L-R). Figure 2 (right) shows Δt(L-R) as a function of θ for 50 GeV electrons and different threshold levels. This difference changes sign at θ=0° as expected and has broad peaks at the angles where acceptance for Čerenkov light is maximal for one of the two PMTs. The performances of this method can be studied by comparing the signal event-by-event distribution with the corresponding distribution at an angle θ for which almost only scintillation light contributes, for example θ=0°. The separation power Π is defined as the ratio of the difference between the mean values and the average width of the two distributions. Figure 3 (left) shows the event-by-event distributions of Δt(L-R) for 50 GeV electrons at θ=0° and θ=30° from which Π = 3.3 is obtained. The separation power as a function of the energy deposit is shown in fig. 3 (right) where the energy scale is derived from Monte Carlo simulation. The expected behavior if photo-electron statistics were the only contribution is also shown for a Čerenkov photo-electron yield of $\sim$50 photo-electrons per deposited GeV. The curve describes the general tendency of the data, but other factors are present: since these have the effect of reducing the separation power, 50 photo-electrons should be interpreted as a lower limit.

In order to investigate different stages of the electromagnetic shower, lead absorber of various thickness was installed upstream of the crystal. As the shower develops the shower particles may not be considered a collection of mips traversing the crystal in the same direction, so it is expected that the directionality of the Čerenkov light is somehow lost while its prompt nature is conserved. The results are in agreement with this expectation and the separation power tends to decrease as the lead thickness increases.

4. BGO results

As stated before, the two PMTs reading the BGO crystal were equipped with two optical filters. Figure 4 shows, for 50 GeV electrons, the average time structure of signals in the Y and UV side with the crystal oriented perpendicular to the beam (θ=0°).

Even if the Čerenkov light represents a very small fraction of the total light it is prominently present in the signals from the UV side.

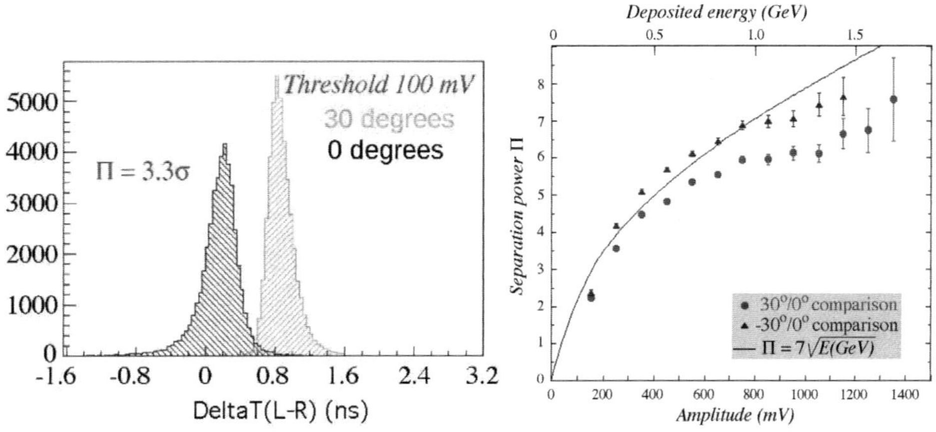

Figure 3. Left: event-by-event distributions of the time difference between the moments the signals from PMT left and right cross a preset threshold level, for 50 GeV electrons in $PbWO_4$ for two different angles. Right: separation power for 50 GeV electrons traversing the $PbWO_4$ crystal as a function of the energy deposit i.e. the amplitude of the signal. The curve describes the expected energy dependence if photo-electron statistics were the only contribution.

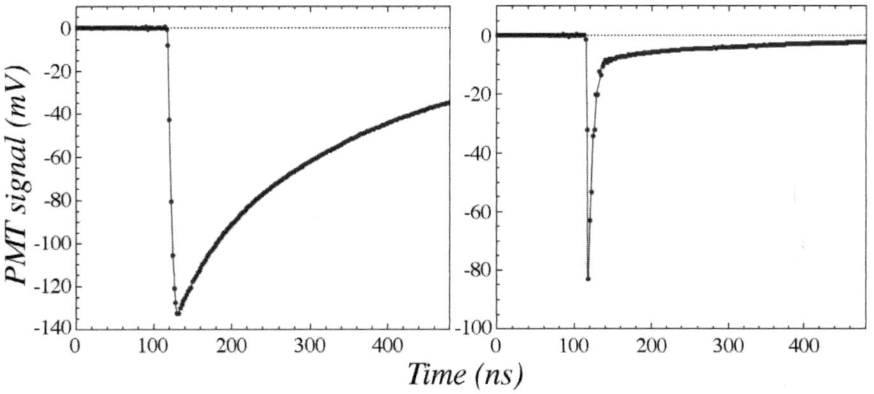

Figure 4. Average time structure of signals from 50 GeV electrons measured in the BGO crystal. Left: Y side. Right: UV side. The crystal was oriented perpendicular to the beam.

The Čerenkov contribution can be measured by integrating the UV signal in a given time window Δt after the start of the pulse. The scintillation contamination in this window is subtracted by taking the shape of the scintillation signal from the Y side and normalizing it to the tail of the UV side distribution. The level of the contamination at a given angle depends on the time window, for example at $\theta=0^o$ in a Δt=20 ns window 20% of the UV signal is due to scintillation light (all the Čerenkov signal is included in such a time window). The ratio Č/S averaged over many events, in a time window Δt=10 ns is shown in fig. 5 (left) as a function of θ. The distribution has a peak at $\theta=30^o$ for which the Čerenkov contribution is maximum in the UV side. The peak is broader for electrons than for pions, probably due to the spread in direction of the particles constituting the shower in the electron signals. Around 20-30 degrees there is a fine structure with peaks for both electrons and pions: since the point-to point fluctuations are very small this should be a systematic effect, probably related to the details of the light collection mechanism.

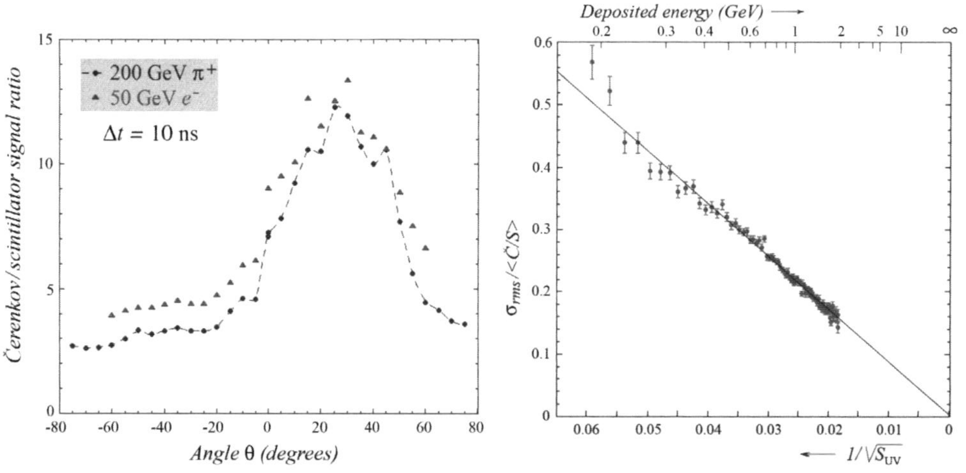

Figure 5. Left: average Č/S ratio in the UV signals from the BGO crystal, for a gate Δt=10 ns from the start of the pulse, as a function of the orientation of the crystal with respect to the beam. Data are for 50 GeV electrons and 200 GeV π^+. Right: fractional width from the event-by-event distribution of the Č/S ratio for 50 GeV electrons traversing the BGO crystal, at θ=30° and Δt= 20 ns.

In order to estimate the Čerenkov light yield 50 GeV electrons have been used. The fractional width of the Č/S event-by-event distribution is shown in fig. 5 (right) as a function of the inverse of the square root of the total UV signal (i.e. the energy deposit in the crystal). The energy scale is obtained on the basis of Monte Carlo studies. A linear fit shows that the resolution on Č/S can be described by the relation:

$$\frac{\sigma_{rms}}{< C/S >} = \frac{0.220}{\sqrt{E(GeV)}} \tag{1}$$

Since the ratio of the numbers of the Č and S photo-electrons in the total UV signals at θ=30° is about 0.44, and assuming that the energy dependence is due entirely to fluctuations in the total number of photo-electrons, the number of Č photo-electrons per deposited GeV is 30.

The precision on the Č/S measurement is shown in fig. 6 (left) as a function of θ, for particles that deposited 1 GeV in the crystal. A precision of 20%-30% is reached.

5. Instrumental effects

5.1. Light attenuation

The short-wavelength Čerenkov light can be attenuated on its way from the production to the light detector and this may give rise to systematic effects. The ratio Č/S is studied as a function of the impact point of the beam particles, i.e. as a function of the distance the light had to travel to the UV filter and its associated PMT, at θ=0°. The measurement were also performed with the filters in the reverse position in order to eliminate effects due to the shape of the crystal. The genuine UV signal attenuation is ~10% over the 24 cm length of the crystal while the ratio Č/S is fairly constant.

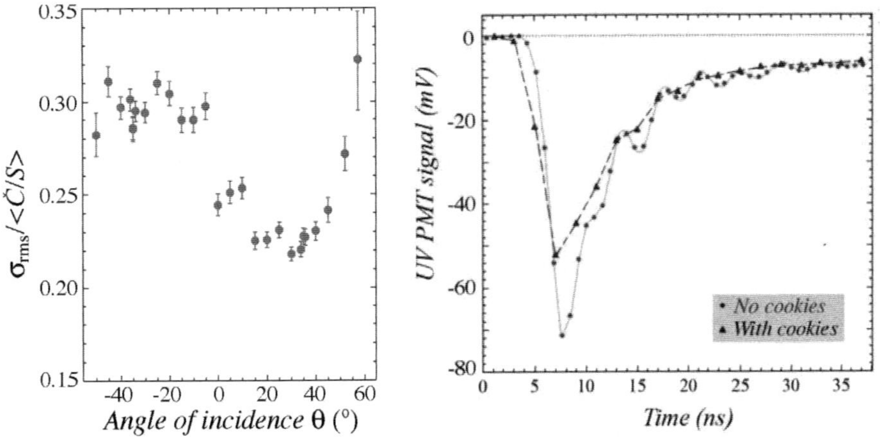

Figure 6. Left: the fractional width of the distribution of the Č/S ratio for 50 GeV electrons traversing the BGO crystal and depositing 1 GeV, as a function of the orientation of the crystal with respect to the beam. Right: average time structure of the prompt component of the UV signals from 50 GeV electrons in BGO, recorded with and without silicone "cookies" mounted between the BGO crystal, the UV filter and the PMT.

5.2. Reflections

Light traveling along the crystal axis experiences a (Fresnel) reflection coefficient of 13% which is reduced to 4% when the silicon cookies are installed between the crystal and the PMT, and the reflection coefficient increases with the angle of incidence to reach 100% at the critical angle. The importance of internal reflections becomes very evident when the silicon cookies are removed, as illustrated in fig. 6 (right) that shows the average time structure of the UV signal component from 50 GeV electrons traversing the BGO crystal at θ=0°, measured with and without cookies. The measurement without cookies reveals an oscillating patter which is absent when the cookies are in place. The equally spaced peaks that can be observed are separated by 4 ns which corresponds to the time light travels a distance of 4*30/2.15=55.8 cm. The light thus travels at an average angle of arcos(48/55.8) = 30°, very close to the critical angle. An exponential fit to these peaks gives 6.1 ns which corresponds to an attenuation length of the crystal of$\sim$80cm.

5.3. Velocity of light in the crystal

The cookie-less measurement has been repeated at different impact points. By measuring the time of the first peak with respect to the trigger of the oscilloscope as a function of the distance traveled by the light (i.e. the impact point of the beam), after correcting for the true longer light path due to reflections, the velocity of light in the crystal can be obtained. The result is 13.3 cm/ns, in good agreement with the expected value, 13.9 cm/ns that can be obtained on the basis of the index of refraction of the crystal.

6. Conclusions

High-Z crystals such as $PbWO_4$ and BGO produce a significant amount of Čerenkov light in addition to scintillation light. We have investigated several methods to extract the Čerenkov contribution to the signals produced by these crystals. BGO turned out to be the more favorable because the spectra of the two lights are quite different, which made possible to produce clean signals of about equal strength with the help of optical filter. Also, the decay time is conveniently long so that the two components can be easily distinguished on the basis of the time structure. We can measure the ratio Č/S in individual events with a relative accuracy of 20-30% for 1 GeV energy deposit in BGO. Altough the Čerenkov fraction of the total signal is considerably larger for $PbWO_4$, the absence of the two advantages listed for BGO made it harder to extract precise Č information. Both crystals produce at least 30 photo-electrons per GeV deposited energy which is a substantial improvement with respect the DREAM calorimeter which had 8 photoelectrons per deposited GeV.

[1] N. Akchurin *at al.*, Nucl. Instr. and Meth. A **537**, 537 (2005).

Journal of Physics: Conference Series **160** (2009) 012071 doi:10.1088/1742-6596/160/1/012071

Dual-readout Calorimetry with Scintillating Crystals

D. Pinci on behalf of the DREAM collaboration
INFN Sezione di Roma, Italy
E-mail: davide.pinci@roma1.infn.it

Abstract. The dual-readout approach, which allows an event-by-event measurement of the electromagnetic shower fraction, was originally demonstrated with the DREAM sampling calorimeter. This approach can be extended to homogeneous detectors like crystals if Cherenkov and scintillation light can be separated. In this paper we present several methods we developed for distinguishing the two components in PWO and BGO based calorimters and the results obtained.

1. Introduction

The possibility of evaluating the amount of the energy released by the electromagnetic part of a hadronic shower (the so called *electromagnetic fraction* f) would allow to account for one of the main sources of the hadronic calorimetry fluctuation. Suppose to have a calorimeter equipped with two sensitive media. For example a medium sensitive to the Cherenkov light and a medium sensitive to the Scintillation light with different response ratios to the hadronic and electromagnetic components of the shower (h/e). After a suitable calibration, the responses provided by the two parts of the calorimeter to a hadronic shower are:

$$C = [f + c(1 - f)]\,E \text{ where } c = (h/e)_C \tag{1}$$

$$S = [f + s(1 - f)]\,E \text{ where } s = (h/e)_S \tag{2}$$

On a event-by-event basis, by simply measuring the C/S ratio, it is possible to evaluate the electromagnetic fraction f as:

$$f = \frac{c - s(C/S)}{(C/S)(1 - s) - (1 - c)} \tag{3}$$

and the energy E released in the calorimeter by the shower, automatically corrected for value of f, is

$$E = \frac{S - \lambda C}{1 - \lambda} \tag{4}$$

where the λ parameter is:

$$\lambda = \frac{1 - q}{1 - c} \tag{5}$$

and it is a “constant” of the calorimeter. Starting from the relation:

$$S = (1 - \lambda)E_0 + \lambda C \tag{6}$$

on a hadron beam of fixed energy E_0, the λ parameter and the beam energy can both be easily obtained from the linear fit of S as a function of C.

Table 1. Main properties of the BGO and PWO.

Material	Scint. light-yield % NaI(Tl)	Scint. decay time (ns)	Scint. spectrum peak (nm)	Transparency cut-off (nm)	Refr. index	Density g/cm^3
BGO	20.0	300	480	320	2.15	7.13
PWO	0.3	10	420	350	2.30	8.28

2. Dual readout method with scintillating crystals

One very promising application of the dual readout calorimetry technique is represented by the DREAM hadronic calorimeter [1]. The main limitation of such a detector is represented by the low Cherenkov photo-electron production (8 ph.e. per deposited GeV). This number arises from the very small sampling fraction and it leads to limited performance on electromagnetic showers. One idea to increase the number of Cherenkov photo-electrons and to improve the performance on electromagnetic showers is to exploit the dual readout method with a homogeneous material. In recent paper ([2]) it was demonstrated the possibility of separating, in a homogeneous scintillating material, the Cherenkov and Scintillation components to the total light yield by exploiting their differences in timing properties, directionality and emission spectrum. Two scintillating materials, the PWO and the BGO were tested, with high energy particle beam, in 2006 and 2007. Their main properties are reported in table 1. For both the materials a complete calorimeter system was built with an electromagnetic section based on scintillating crystals and a hadronic section made by the DREAM detector as shown in fig. 1. The PWO-based calorimeter

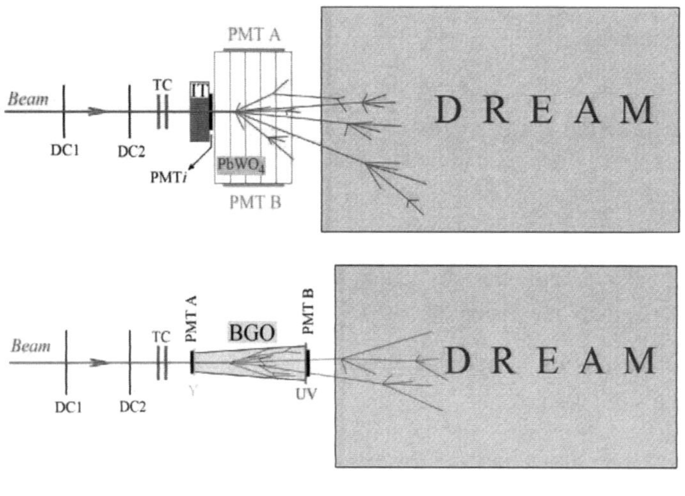

Figure 1. Set-up used for testing the PWO (top) and the BGO crystals (bottom).

was made by a matrix of 19 crystals, readout on the two lateral faces by two fast and low-gain photo-multipliers. In the BGO measurements only one crystal was used, placed parallel to the beam and readout on the two small faces. Between the BGO crystal and the photo-multipliers two optical filters were inserted: a "yellow" filter transparent to the long wavelengths and a "UV" filter transparent to the short wavelengths. The yellow filter transmitted the scintillation light emitted by the crystal, while the "UV" one allowed only the transmission of the Cherenkov light.

In all measurements, the shape of the signals provided by the photo-multipliers were acquired by means of a 5 GS/s oscilloscope. In order to get information on the amount of Cherenkov light produced in each event two different methods were used for the two materials:

- For the *PWO* the asymmetry (B-A)/(B+A) of the light collected by the two PMTs with the matrix rotated at 63° with respect to the beam was evaluated. In this configuration the Cherenkov light (C) reached only the B PMT while the Scintillation one (S) was collected by both the PMTs. It results:

$$\frac{B-A}{B+A} = \frac{C}{C+2S}$$

- For the *BGO* the waveform provided by the PMT placed downstream of the UV filter was off-line integrated in two different gates: *gate-1* around the Cherenkov peak and *gate-2* in the exponential tail (see fig. 2). The analysis of the signal on the "yellow side" allowed the evaluation and subtraction of the contamination of the scintillation light in the *gate-1* window.

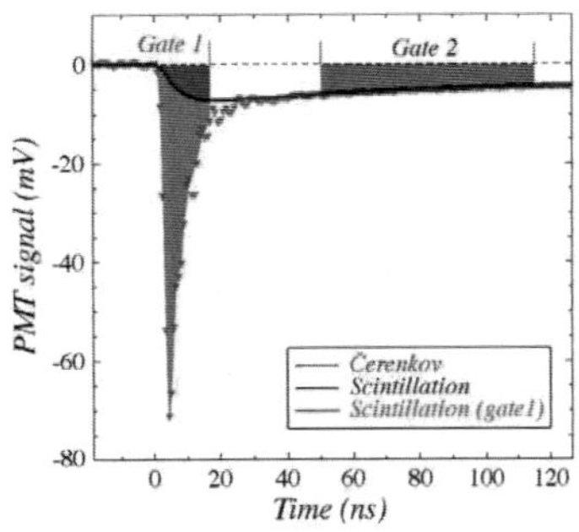

Figure 2. Analysis performed off-line on the "UV-side" signals.

3. Results

In a hadronic shower, the electromagnetic components produced late in the ECAL, will be absorbed by the HCAL. This effect gives rise to a correlation between the electromagnetic fraction measured by the ECAL and the one measured in HCAL by using the formula 3. In fig. 3 the electromagnetic fraction measured in the ECAL and HCAL by means of the above methods are reported.

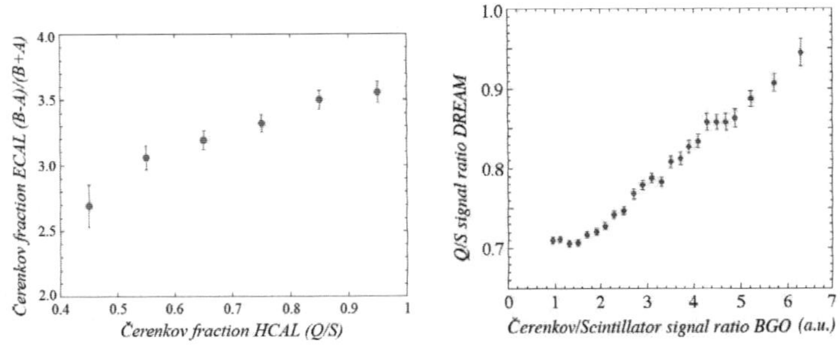

Figure 3. Electromagnetic fraction measured in a section of the calorimeter system as a function of the one measured in the other section for the PWO (left) and the BGO crystals (right).

In particular, for the BGO case, the response of the scintillator fibers in Dream was analyzed as a function of the C/S ratio. For a fixed released-energy value (fig. 4), a large value of f in the shower produces:

- a high scintillation signal;
- a small fractional width in the Dream response because of the lower fluctuations induced by the invisible energy of the non-electromagnetic fraction;

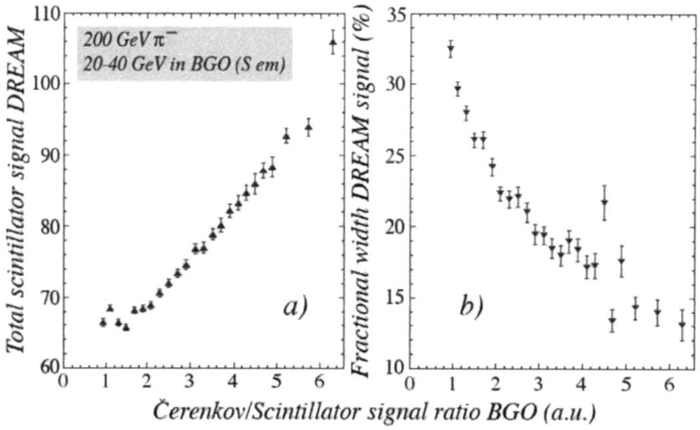

Figure 4. Fractional widths (left) and mean values (right) of the scintillator response distribution in DREAM as a function of the C/S ratio in the BGO-ECAL.

Moreover, as it is shown in fig. 5, events with different f measured in the ECAL have a different distribution in the calorimeter. By choosing events with different values of f in the ECAL it is possible to select different "sub-distributions" in the HCAL scintillator response that are narrower than the global one. A crystal-based ECAL is able to give precious information on the electromagnetic content of the shower and to allow to correct the HCAL response.

4. Conclusion

The separation of Cherenkov and Scintillation components in the signals produced by homogeneous scintillating material was demonstrated to be feasible. This feature gives the possibility of evaluating the electromagnetic fraction of a shower allowing to reduce part of the fluctuations and non-linearities in measuring the Energy released by a hadron. The application of the Dual-Readout method also to the electromagnetic section can be exploited to improve the global performance to electron and pion showers.

References

[1] N. Akchurin *et al.*, "Hadron and jet detection with dual-readout calorimeter",Nucl. Instr. and Meth. A 537 (2005) 537.
[2] N. Akchurin *et al.*. "Contributions of Čerenkov light to the signals from lead tungstate crystals", Nucl. Instr. and Meth. A 582 (2007) 474.
[3] N. Akchurin *et al.* "Dual-readout calorimetry with lead tungstate crystals", Nucl. Instr. and Meth. A. 584 (2008) 273.

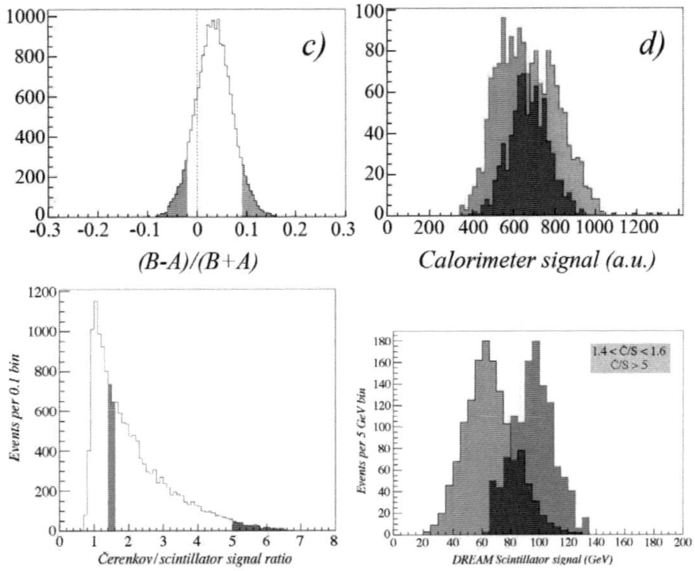

Figure 5. For both the PWO (top) and BGO (bottom) crystals, events with different values of f in ECAL have different "sub-distributions" in the HCAL scintillator response that are narrower than the global one.

Journal of Physics: Conference Series **160** (2009) 012072 doi:10.1088/1742-6596/160/1/012072

Estimate of neutrons event-by-event in DREAM

John Hauptman, for the DREAM Collaboration

Department of Physics and Astronomy, Iowa State University, Ames, IA 50014, USA

E-mail: hauptman@iastate.edu

Abstract. We have measured the contribution of neutrons to hadronic showers in the DREAM module event-by-event as a means to estimate the event-by-event fluctuations in binding energy losses by hadrons as they break up nuclei of the Cu absorber. We make a preliminary assessment of the consequences for hadronic energy resolution in dual-readout calorimeters.

1. Introduction

The DREAM module was designed to measure the electromagnetic content of hadronic showers event-by-event and thereby suppress the effects of these fluctuations in order to not only improve the hadronic energy resolution, but also to achieve a Gaussian response and hadronic linearity in a calorimeter calibrated with electrons.[1] These were achieved in the first DREAM test beam.[1] The next largest fluctuation is the binding energy losses as hadronic particles of the shower lose energy in breaking up the nuclei of the Cu absorber medium. These energy losses are roughly proportional to the MeV neutrons liberated in nuclear break-up, and the energy carried by these neutrons is estimated to be about 20-25% of the total shower energy.

The DREAM module is described in detail in Ref. 1 and consists of 1 tonne of Cu absorber with two types of fibers embedded throughout its volume: scintillating fibers[2] that sample the dE/dx ionization energy loss of all charged particles, and two types of clear fibers (hereafter called Čerenkov fibers) that sample relativistic particles above Čerenkov threshold, $\beta \approx 1/n \approx 0.65$. These particles are predominantly $e^{\pm}$ from the γ-induced showers from the decays of $\pi^0 \to \gamma\gamma$ produced in the hadronic shower. The central seven channels were filled with high-purity quartz fibers[3] and the 12 channels of the outer ring were filled with acrylic plastic fibers.[4]

We modified the DREAM module by summing the PMT signals for the whole DREAM volume for both scintillation and Čerenkov fibers, Fig. 1, and sending these through fast cables to a digital oscilloscope in the counting room. The central scintillating channel (S_0) was sent to the first input; the inner ring of six (6) scintillating channels (S_1) was summed in a standard linear summing unit[5] and send to the second oscilloscope input; the outer ring of 12 scintillating channels (S_2) was sent to the third oscilloscope input; and, 16 of the 19 Čerenkov channels (C) were summed and sent to the fourth oscilloscope input.[6]

[1] For an overview of this and other work by the DREAM Collaboration, see the talk by R. Wigmans, these *Proceedings*.

[2] SCSF-81J produced by Kuraray Co., Ltd, Tokyo, Japan

[3] Polymer-clad fused silica fibers, Polymicro, Phoenix, Arizona, USA.

[4] Raytela PJR-FB750, produced by Toray, Japan.

[5] These LeCroy units from the CERN electronic pool were LeCroy 127EW (or, 127FL) CERN pool ID 4355.

[6] The maximum number of inputs was 16, and since the total number of Čerenkov channels is 19, we chose to leave out three evenly spaced outer channels rather than chain summing units together.

Journal of Physics: Conference Series **160** (2009) 012072 doi:10.1088/1742-6596/160/1/012072

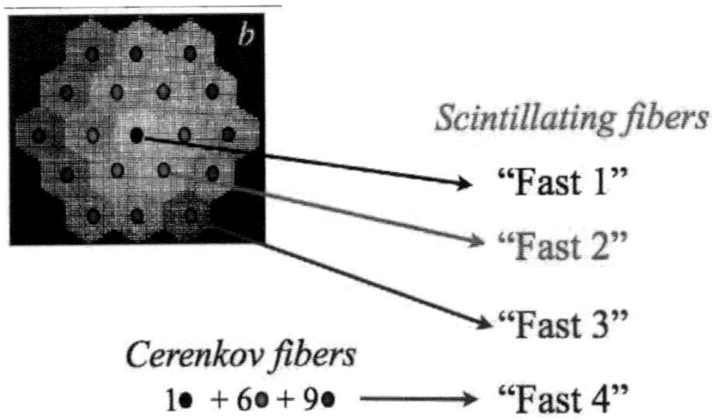

Figure 1. Geometry of the scintillation and Čerenkov channel sums sent to the digitizing oscilloscope. We label the central scintillation channel S_0 (sent to the first input of the fast oscilloscope), and the inner and outer rings S_1 and S_2 (sent to inputs 2 and 3). The Čerenkov sum is labeled C (sent to input 4).

Neutrons liberated from broken-up nuclei will have kinetic energies, T, in the MeV region, velocities of $v \sim \sqrt{2T/M_n} \sim 0.04c \sim 1$ cm/ns, and will therefore persist in the calorimeter volume for tens of nanoseconds as the neutrons elastically scatter from (mostly) Cu nuclei.

The protons in the H-C molecules of the scintillating fibers and the outer ring of Čerenkov fibers are the only free protons in or near the calorimeter. A neutron signal is generated when a neutron elastically scatters from a free proton in a scintillating fiber, $np \rightarrow np$, and the recoil proton ionizes in the scintillator generating scintillation light. This scintillation light luminosity is suppressed by Birks recombination for densely ionizing slow protons by a factor approximately $\sim (1 + k_B \cdot dE/dx)^{-1}$ for dE/dx in MeV/g-cm^{-2} and $k_B \approx 0.008$ g-cm^{-2}/MeV. In addition, the scattering from a free proton is the only effective mechanism for the slowing down of the neutrons from MeV energies to negligible energies below keV over a measurement time of ~ 100 ns.

2. Analysis

The neutron signal is defined as the integral under the $S(t) = S_1(t) + S_2(t)$ curve in Fig. 2, from 20 to 40 ns for each event. This is a very simple definition, but an appropriate start on this problem. The distribution of this neutron signal for 200 GeV "jets"[7] is shown in Fig. 3

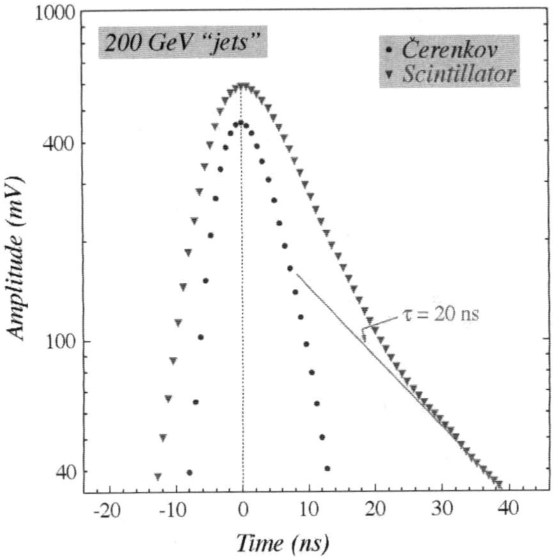

Figure 2. The average Čerenkov pulse, $C(t)$, and the average scintillation pulse, $S_1(t) + S_2(t)$, both centered at $t = 0$. The Čerenkov pulse is very narrow in time, in spite of it being a sum of 16 PMT pulses, and roughly measures the time response of an instantaneous signal. The scintillation pulse is slower due to the finite lifetime of the scintillating fluor (2-3 ns), but also shows a long-lived component with an exponential fall-off with a time constant of 20 ns. This is evidence for slow neutrons in the scintillating fibers.

[7] These are "interaction" jets generated by placing a 0.1-$\lambda_{\rm int}$ lucite block in front of DREAM and accepting events with a charged particle multiplicity above 12 exiting the lucite.

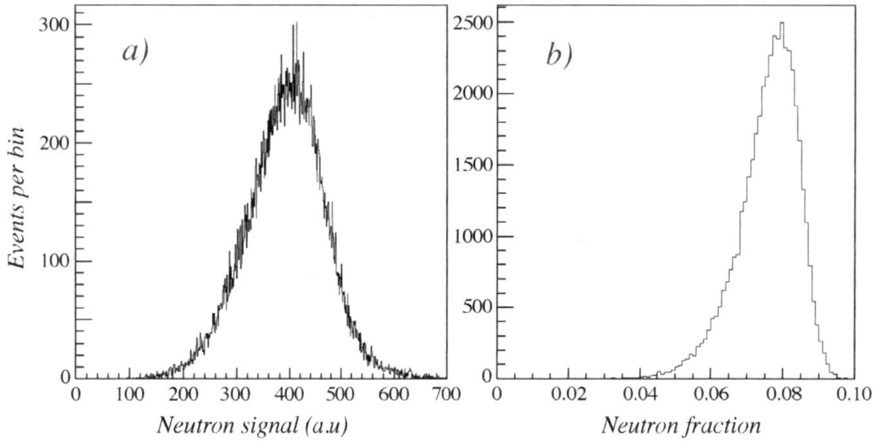

Figure 3. (a) The distribution of the "neutron signal" integral for individual 200 GeV hadronic showers; and, (b) the normalized neutron fraction, f_n.

This neutron fraction is anti-correlated with the Čerenkov pulse displayed in Fig. 4, as it should be, but more interestingly the distribution of the Čerenkov signal is seen to be a superposition of the different slices of the Čerenkov distribution. Selecting on the neutron fraction, f_n, yields the individual nearly Gaussian distributions that make up the total Čerenkov pulse height distribution, as shown in Fig. 5. This decomposition of the total Čerenkov distribution into f_n-selected parts is analogous to the same Čerenkov distribution decomposed into f_{EM}-selected parts in Fig. 34 of the DREAM hadron paper [Ref. 1] and also shown in Figs. 1(a,b) in the talk by Wigmans in these Proceedings.

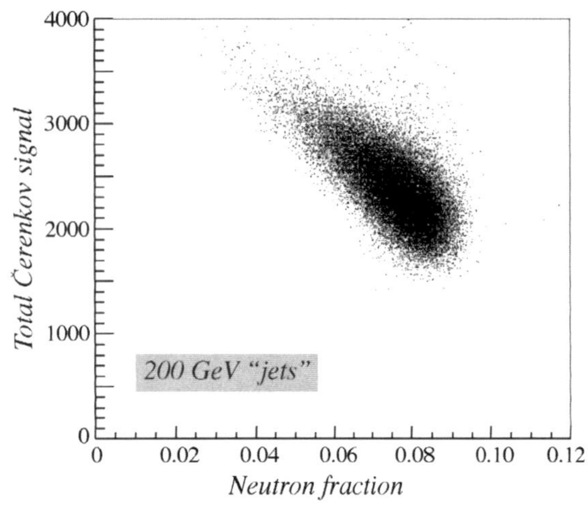

Figure 4. The anti-correlation of the neutron fraction with the Čerenkov signal. This the first and simplest test that f_n is a measure of hadronic content, in this case neutrons. These two variables are as close to the raw measurements as possible.

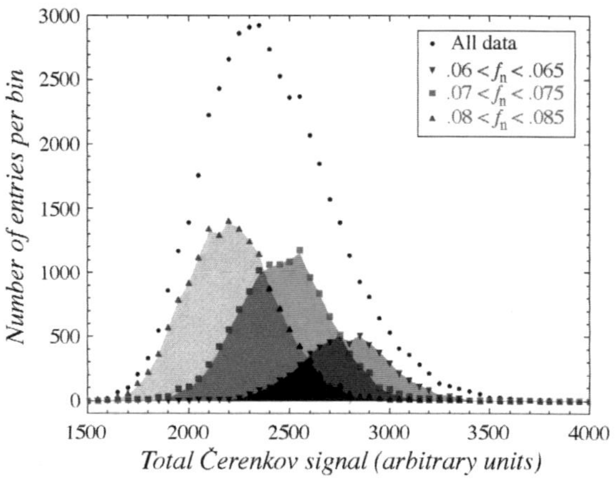

Figure 5. The total distribution is simply the Čerenkov signal for all events. The separate components of the Čerenkov signal distribution are selected as a function of the neutron fraction, f_n, from which it is clear that the wide and asymmetric Čerenkov distribution is just a sum of near-Gaussian distributions for each f_n.

The distribution of the total Čerenkov distribution is shown in Fig. 6(a) for the same 200 GeV hadronic showers, and the rms width of this distribution is shown in the circled points of Fig. 7. This

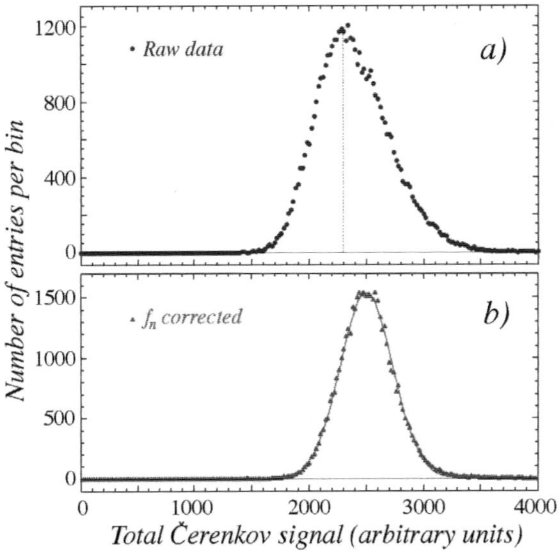

Figure 6. (a) The total Čerenkov distribution from Fig. 5. (b) The separate Čerenkov distributions of Fig. 5, for each small f_n bin, are corrected to a same mean of $\bar{f}_n = 0.07$ and summed.

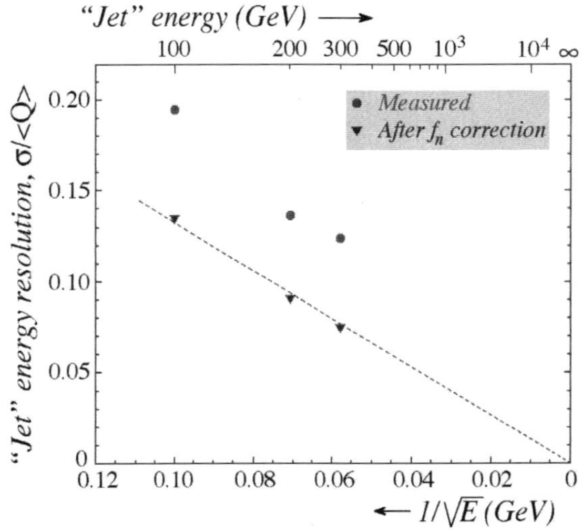

Figure 7. The dependence of the rms widths divided by the means of the distributions in Fig. 6 are plotted for data samples at 100, 200 and 300 GeV beam energy.

width is narrowed by using the f_n-Čerenkov correlation to center each subdistribution at $f_n = 0.07$, shown in Fig. 6(b). The reduction in the rms spread is shown as a function of hadronic shower energy in Fig. 7. The narrowing is about 25% but, maybe more importantly, the f_n-corrected points leave behind no sign of a constant term, as seen by the line drawn through the origin (not a fit). This does not necessarily demonstrate a quantitative improvement in the hadronic energy resolution, but does show that more information leads to a narrowing of the originally wide Čerenkov pulse height distribution. This exercise indicates that there is some useful information in the neutron measurement, f_n, but this might be due in part to its correlation with the EM fraction, f_{EM}.

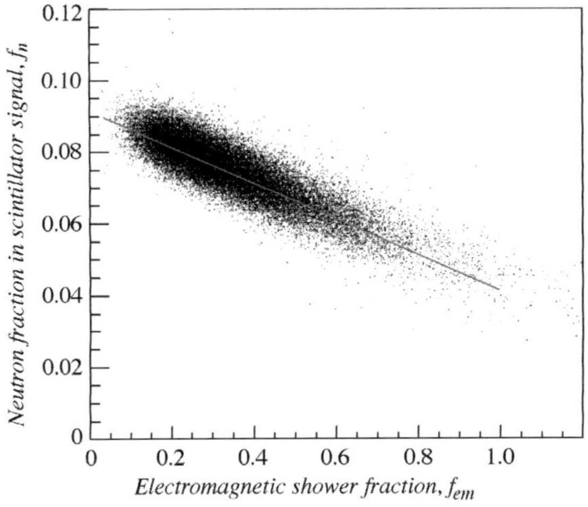

Figure 8. A plot of the neutron fraction, f_n, against the electromagnetic fraction, f_{EM}, is approximately linear, and demonstrates the essential value of this measurement. There are two physical effects which depress the value of f_n: the small 1 kt volume of the DREAM module and the recombination of electrons and ions in the column of dense ionization left by a slow MeV proton, and the consequence suppression of scintillation light yield (Birks mechanism). The EM fraction is linearly related to the ratio $Q/S(C/S)$.

A stronger test is to hold f_{EM} constant and look at the fluctuations in f_n. This is shown using Fig. 8, the nearly linear relationship between f_{EM} and f_n. A selection of showers with $0.70 < Q/S < 0.75$

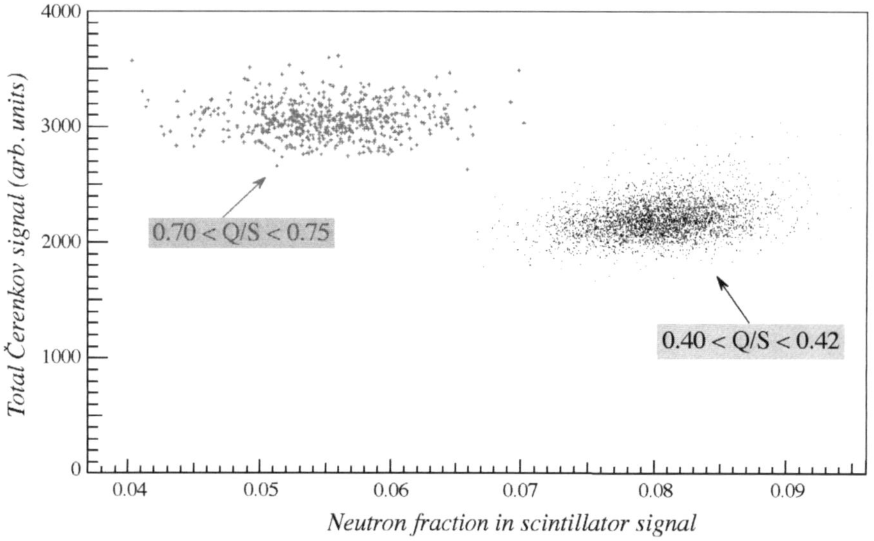

Figure 9. The disperson left in f_n for fixed f_{EM}. The difference in the total Čerenkov signal is due to the difference in the Q/S selections.

from Fig. 8 is shown plotted against f_n and the total Čerenkov signal in Fig. 9. For fixed f_{EM} the neutron fraction varies by approximately 15%, and these are the binding energy loss fluctuations on top of the EM fraction fluctuations. This is also shown for another selection in f_{EM}, $0.40 < Q/S < 0.42$.

The important question is how does an independent measurement of the neutron content of a hadronic shower serve to improve the overall energy resolution.[8] The hadronic shower energy resolution of the DREAM module is limited by lateral leakage fluctuations in this small proof-of-principle module.[9] These leakage fluctuations are estimated at about 4% rms, and therefore any analysis of an unbiased ensemble will be limited to 4% resolution in energy.

With this in mind, we plot the rms width of the Čerenkov signal divided by the mean for showers with fixed EM fraction, $0.40 < Q/S < 0.42$, as a function of the neutron fraction, f_n, in Fig. 10(a). This resolution statistic increases as f_n increases, as it should if the neutron content of showers contributes to the fluctuations in the total energy signal of those showers. If both the EM fraction, f_{EM} and the neutron fraction, f_n, are held fixed, $f_{EM} \approx 0.55$ and $0.045 < f_n < 0.065$, then the width of the Čerenkov signal narrows to about 4.7% as shown in Fig. 10(b). This Čerenkov signal distribution is also nicely Gaussian. For a tighter selection on f_n, $0.050 < f_n < 0.055$, this distribution narrows further to 4.4%, coming interestingly close to our estimate of the leakage fluctuation limit for the DREAM module.

3. Plans for neutrons in DREAM

This measurement is limited by several physical effects: lateral particle leakage, an impedance mismatch in the LeCroy summing units that results in a reflection beginning at about 48ns, low Čerenkov photoelectron statistics, and the somewhat time-incoherent sum of multiple PMT channels into a single (oscilloscope) readout channel. We were possibly lucky that it wasn't worse: the PMT signals went into BNC cable, then into LEMO, then into the 16-input summing units, and the outputs into fast ($0.78c$) cable to the oscilloscope inputs in the counting room.

[8] It certainly serves to improve particle identification of hadronic vs. non-hadronic showers in the calorimeter.

[9] The DREAM module was presciently funded by the ADR (Advance Detector Research) program of the Dept. of Energy with a small grant to study the concept of dual-readout calorimetry. It has achieved this goal with a dozen published papers in *Nucl. Instr. Meths.*, but it was never intended to be a high-performance calorimeter itself.

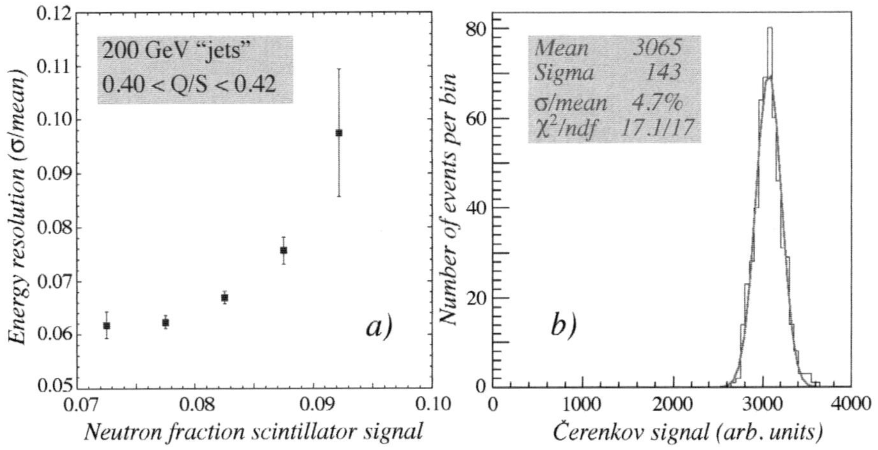

Figure 10. (a) The rms width of the Čerenkov distribution as a function of the neutron fraction, f_n. (b) The Čerenkov signal distribution with an rms width of 4.7% for fixed $f_{EM} \approx 0.55$ and fixed f_n, $0.045 < f_n < 0.065$, fractions.

In the upcoming CERN beam test we expect to read out each DREAM channel independently into a fast circular buffer flash ADC[2] and repeat this neutron analysis, although we will still be limited in overall energy resolution by lateral leakage fluctuations. We will make an attempt at a crude measurement of the leakage event-by-event.

Three decades ago, hadronic calorimeters functioned as counters of penetrating energy rather than as precision devices to measure the energy of a particle or jet of particles. The understanding of the role of neutrons, specifically their role in compensation to achieve equal calorimeter response to hadronic and electromagnetic energy, was established by the SPACAL module[5] 25 years ago. In the present era of a high energy proton collider (LHC) and the prospects for an electron collider (ILC or CLIC), we now seek the ultimate energy resolution possible for hadronic energy. A estimate by Wigmans[4] is that $(13-15)\%/\sqrt{E}$ is the lower limit, essentially determined by unknowable fluctuations in nuclear physics. At a high energy collider, this puts hadronic calorimetry in the "one percent" league with electromagnetic calorimetry.

3.1. Acknowledgments
We acknowledge the US Department of Energy for supporting the original DREAM module and for continued support of these and other measurement in DREAM . We also acknowledge the support of INFN in our plans for improving the neutron measurements. Finally, we acknowledge the excellent quality and variety of beams at CERN.

References
[1] Akchurin N *et al.* 2005, Nucl. Instr. and Meth. **537**, 537. "Hadron and Jet Detection in a Dual-Readout Calorimeter".
[2] Carosi R 2008, Contribution to these Proceedings. "Time profile analysis of photodetector signals in multi read-out calorimetry with GHz samplers".
[3] Barbareschi D *et al.* 2006, Detector Outline Document, 4th Concept, March 2006. The 4th detector has a time-history readout for neutron measurement: `http://physics.uoregon.edu/~lc/wwstudy/concepts/`.
[4] Wigmans R 2000, *Calorimetry, Energy Measurement in Particle Physics*, International Series of Monographs on Physics, Vol. 107, Oxford University Press (2000).
[5] Acosta D *et al.* 1990, Nucl. Instr. and Meth. **A294**, 193.

Journal of Physics: Conference Series **160** (2009) 012073 doi:10.1088/1742-6596/160/1/012073

Progress in hadronic physics modelling in Geant4

John Apostolakis[1], Gunter Folger[1], Vladimir Grichine[1], Aatos Heikkinen[2], Alexander Howard[1], Vladimir Ivanchenko[1], Pekka Kaitaniemi[2], Tatsumi Koi[3], Mikhail Kosov[1,4], Jose Manuel Quesada[5], Alberto Ribon[1], Vladimir Uzhinskiy[1], Dennis Wright[3]

[1] CERN, 1211 Geneve 23, Switzerland
[2] Helsinki Institute of Physics (HIP), P.O.Box 64, FIN-00014 University of Helsinki, Finland
[3] SLAC, 2575 Sand Hill Road, Menlo Park, California, USA
[4] ITEP, B. Cheremushkinskaja 25, Moscow 117259, Russia
[5] Departamento de Física Atómica, Molecular y Nuclear, Universidad de Sevilla, Apartado 1065, E-41080 Sevilla, Spain

E-mail: Gunter.Folger@cern.ch

Abstract. Geant4 offers a set of models to simulate hadronic showers in calorimeters. Recent improvements to several models relevant to the modelling of hadronic showers are discussed. These include improved cross sections, a revision of the FTF model, the addition of quasi-elastic scattering to the QGS model, and enhancements in the nuclear precompound and de-excitation models. The validation of physics models against thin target experiments has been extended especially in the energy region 10 GeV and below. Examples of new validation results are shown.

1. Introduction

The GEANT4 simulation toolkit [1, 2] is widely used for simulation in high energy physics. In particular, LHC experiments use GEANT4 for simulation production and comparison to test beam data. Results from these test beam experiments, showing calorimeter data and GEANT4 predictions, are presented elsewhere in these proceedings.

Both the CMS and ATLAS experiments started by using two main configurations of GEANT4 physics: LHEP and QGSP. The LHEP physics list configuration uses parametrized models to describe the interactions of hadrons. The QGSP physics list uses a string model approach to simulate inelastic interactions of hadrons above about 15-20 GeV, as well as improved cross sections. At lower energies the QGSP physics list is similar to LHEP. The experiments found good agreement of the measured hadronic shower shape, especially longitudinally, to simulation results obtained with LHEP. QGSP describes better the e/π ratio and the resolution, but simulation produces showers which are too short and start too early.

The configuration provided by the QGSP_BERT physics list, together with the addition of quasi-elastic scattering as detailed below, brings the simulated shower shape close to the LHC test beam results. This physics list uses the GEANT4 Bertini cascade model to handle interactions for particles below 10 GeV, reducing the use of parametrized models. This physics list was adopted for production by both the CMS and ATLAS experiments.

To understand the shower shape issue and to improve the simulation, a significant validation effort was started within the GEANT4 hadronic working group. In section 2, we concentrate on the improvements made in the modeling of hadronic interactions with the goal of improving simulation for LHC detectors. Both cross sections and interaction models were studied. The physics performance of models simulating hadronic interactions was examined by comparing to experimental data from thin target experiments.

In section 3 we summarize recent improvements to other GEANT4 hadronic models.

2. Improvements to QGSP and QGSP_BERT physics lists

In the following we will discuss improvements to hadronic models in the physics lists QGSP and QGSP_BERT, as used in production by the LHC experiments. Although an improved modeling of shower shape was an overall goal, the development of models is guided by comparing simulation results of individual models to data of thin target experiments. The main improvement in shower shape results from the addition of quasi-elastic scattering to the modelling of high energy interactions.

2.1. Quasi-elastic scattering

Up to version 8.2 of GEANT4 quasi-elastic scattering, the elastic scattering of the projectile on a nucleon, was ignored for the simulation of high energy interactions by the Quark Gluon string (QGS) model, and the full inelastic cross section was used as the production cross section. Since version 8.3 a simple model for quasi-elastic scattering has been added, and the cross section for quasi-elastic scattering is subtracted from the inelastic cross section, thus reducing the production cross section. Subsequent model refinements added effects of Fermi motion of nucleons on the momenta of the two final state particles.

Quasi-elastic scattering results in longer simulated shower shapes since it reduces the energy and momentum by only a small amount and the scattering angle is typically very forward.

2.2. Review of cross sections

In GEANT4 hadronic cross sections are independent of the model used to generate final states, hence an incorrect cross section for inelastic interactions will have an impact on the shower shape. Therefore an extensive validation effort was undertaken to validate the cross sections used in GEANT4 against experimental data and also versus different evaluations of cross sections. We found that above 100 GeV the adopted relativistic rise was too steep, thus this was corrected for the cross sections in use by the physics lists based on the QGS model.

2.3. Improvements to elastic scattering

Elastic scattering of hadrons on nuclei directly affects the development of hadronic showers in matter. Hadronic calorimeters with scintillators are particularly sensitive to this, due to the energy transfer from low-energy neutrons to recoiling protons by neutron elastic scattering on hydrogen. Initially the elastic model of the GHEISHA package [3] was used in GEANT4 . The elastic scattering process of GEANT4 was recently reviewed and two new models were introduced.

A theory-based model has been developed, which uses the Glauber approach [4] with corrections for inelastic screening [5] and total cross section [6]. The validity range of the Glauber approach starts above 1 GeV.

An alternative phenomenological model was developed in the CHIPS sub-package [7] of GEANT4 . The parameterization of the elastic cross sections for protons and neutrons is based on an analysis of about 600 papers with about 3000 data sets on nucleon-nucleus scattering for targets from hydrogen to lead. As a result, a more precise analytic parameterization of the total and elastic cross sections is obtained.

Neither of these models cover the entire range of particle types and energies, therefore they have been combined in a new GEANT4 process class. In this class, incident protons and neutrons

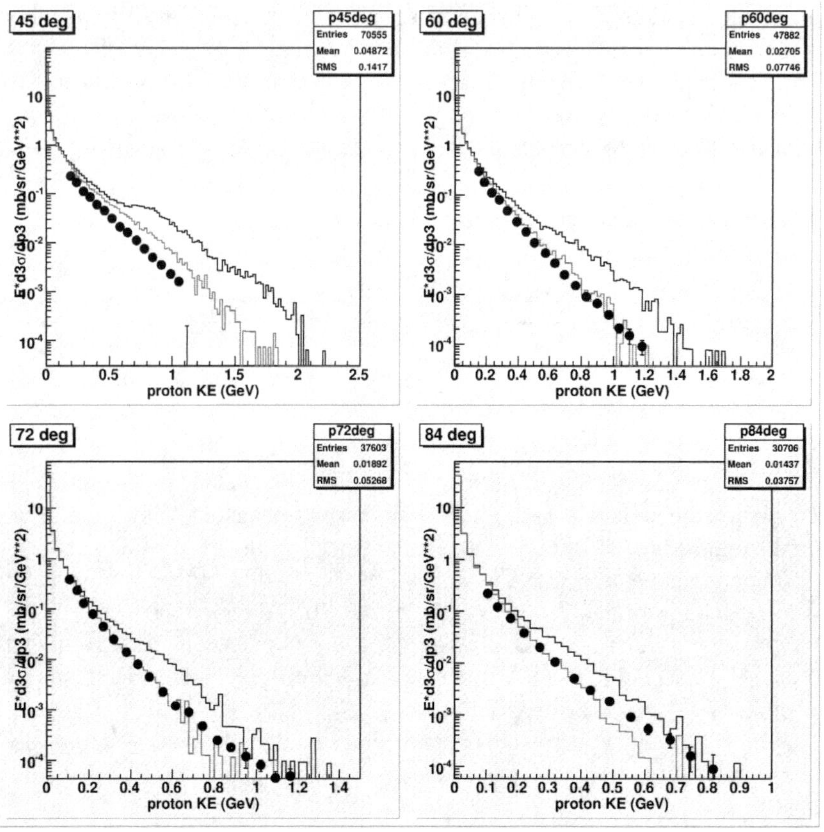

Figure 1. Double differential cross section for inelastic proton scattering from Al at 4 GeV/c. The data from [10] are given by the black points. The Bertini model prediction with the original angular distribution is shown by the black curve and the improved, t-dependent angular distribution is shown by the red curve.

use the CHIPS model, charged pions with incident energy above 1 GeV use the Glauber model, and for all other incident hadron types and energies the GHEISHA-style model is used.

2.4. Improvements to Bertini Cascade

The Bertini-style cascade [8] as implemented in Geant4 has been reasonably successful in describing secondary particle distributions resulting from incident nucleons, pions, kaons and hyperons colliding with nuclei. The applicable energy range of this model is from 100 MeV to 10 GeV. Physics lists using the Bertini cascade are now in use for simulation production by both the ATLAS and CMS experiments.

Improved shower shapes are seen with physics lists using this model [9] and extensive validations at the model level over the full 10 GeV range have in most cases shown reasonable agreement with thin-target data. Significant deviations remain however.

One such deviation is due to the absence of a Coulomb barrier in the evaporation phase of the model. Low energy ($<$ 2-3 MeV) protons were simulated whereas none should exist. The Coulomb barrier has now been included, resulting in the expected low energy behavior for emitted charged hadrons.

Another deviation from data which could have a significant effect on shower shapes has to do with angular distributions. Fig. 1 shows data at four angles for 4 GeV/c protons incident on Al. The upper curve shows the discrepancy with data and the lower curve shows the effect of improving the parameterized angular distribution in the model.

The improvement was made by replacing the relatively flat parameterized angular distribution for incident energies above 2.8 GeV with a simple t-dependent distribution. The net result is more forward-going secondaries in the shower.

Additional improvements to the model are underway and include a better treatment of quasi-elastic scattering and the application of the Coulomb barrier in the precompound phase of the model.

3. Improvements in other models

In addition to improvements in models used by the QGSP_BERT physics configuration, other configurations provide options to assess the systematics of modelling, and to possibly offer new choices in the future.

A significant revision of the GEANT4 FTF model is in progress, and results based on the first set of improvements, presented elsewhere in these proceedings, show that this model should give quite competitive results in the area of shower simulation. As the FTF model can be used down to about 3-5 GeV, there is good overlap with cascade models and the use of parameterized models can be significantly reduced.

3.1. FTF model improvements

The well-known FRITIOF model [11] is now implemented in Geant4 under the abbreviation FTF. It allows the simulation of hadron-hadron and hadron-nucleus interactions starting from laboratory momenta of 3–5 GeV/c. Significant effort has been spent in order to tune the model parameters for a description of hadron-nucleon interactions. As a result, the model describes particle spectra better than several other known models: HIJING, UrQMD, QGSM. The main improvement involves a separate simulation of diffractive and non-diffractive collisions.

Other improvements include the simulation of the elastic scattering of a projectile on the intra-nuclear nucleons and a treatment of the secondary particle formation time. The first improvement allows the low-energy nucleons ejected from a target to be taken into consideration. The second improvement is required for any coupling with low-energy cascade models. Now a combination of the FTF and Binary Cascade models gives a quite satisfactory result for hadron-nucleus interactions, especially for an interpretation of the HARP experimental data.

3.2. Binary Cascade Improvements

The GEANT4 Binary Cascade [12] is a model describing the interactions of protons and neutrons with nuclei. Comparisons to experimental data show that the model is best at low energies, and gives a reasonable description of particle spectra up to about 5-10 GeV. Incident pions up to few GeV were added recently to the list of allowed incident particles. Another development allows the use of the binary cascade as a re-scattering model in combination with the GEANT4 string models. These models use the so-called 'yo-yo' formation time approach [13] to estimate the times and coordinates of secondaries created. A secondary created by the high energy model within the nucleus can interact with the nucleons remaining after the high energy part of the reaction. The main result of including these interactions of secondaries with nucleons is an improved description of the spectra of secondary protons and neutrons.

3.3. Implementation of INCL cascade and ABLA evaporation-fission codes

The INCL4.2 (Liège) intra-nuclear cascade [15], optimized for spallation studies, and the ABLA V3 evaporation [16] with GSI SimFis code for fission are now available in Geant4. The INCL model accepts proton, neutron, pion, deuteron, triton, He3, and α projectiles for energies between 0.2 and 3 GeV. The supported target elements range from carbon through to uranium.

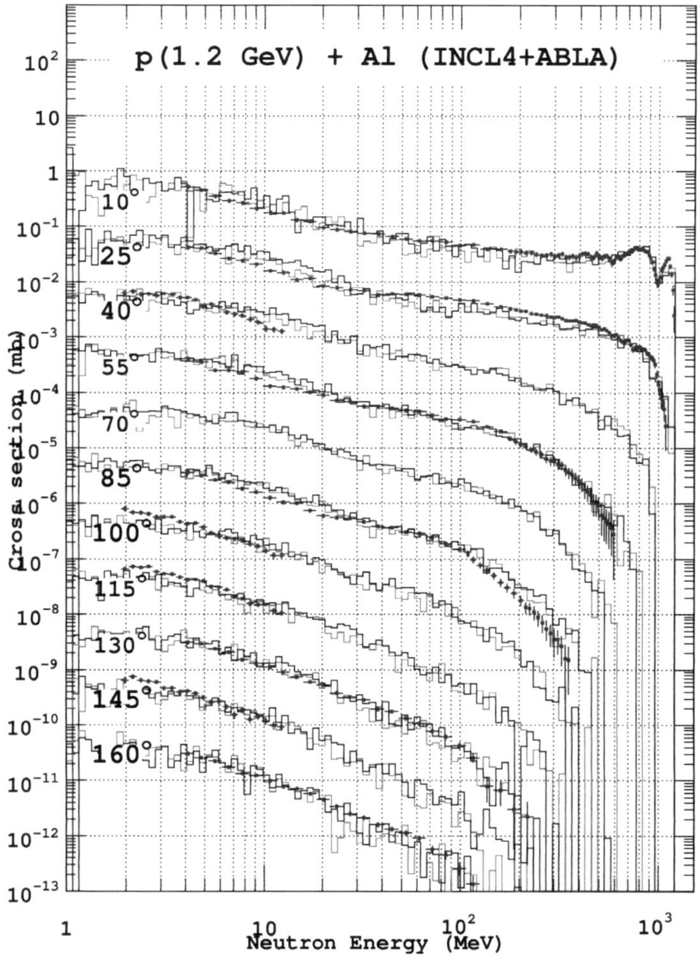

Figure 2. Validation of Geant4 9.1 INCL with ABLA (in red) for 1.2 GeV protons on Al target; black: original Fortran code. Experimental data points are from [18].

A direct translation of the original code has been provided since the Geant4 9.1 release and a verification and validation program was begun to ensure the correctness of the implementation. Fig. 2 demonstrates the neutron production at a typical spallation energy of 1.2 GeV. Good agreement with experiment is observed. The main disadvantage of such a microscopic precision model is the large CPU penalty compared to more phenomenological models.

Two implementations of the PROFI [17] fission model, SimFis3 and SimFis18 (released in 1996 and 2001, respectively) are connected to the Geant4 INCL/ABLA system using an interface that allows direct comparison between the codes. Identical random numbers were used in the unit testing of both the FORTRAN and C++ versions of the SimFis3 to ensure the correctness of the translation.

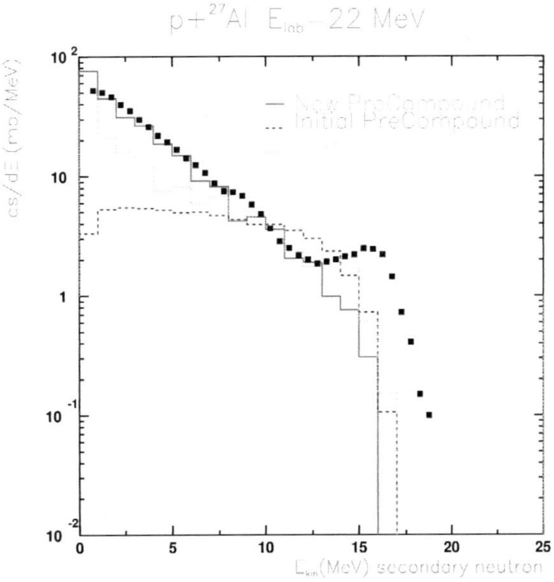

Figure 3. Neutron emission spectrum from ^{22}Al(p,xn) at 22 MeV incident energy

3.4. Improvements of Preequilibrium and Evaporation Models

The precompound stage of nuclear reactions, as described within the framework of the exciton model [20], is used as long as the nuclear system is not in equilibrium. Emission of nuclear fragments or photons from the excited nucleus is simulated using the Weisskopf equilibrium theory [21] , which by itsef is not able to describe the *high energy* tails of the emission spectra.

In the statistical equilibrium state, which is characterised by an equilibrium number of excitons n_{eq}, all three types of transitions are equiprobable. Thus n_{eq} is fixed by $\lambda_{+2}(n_{eq}, U) = \lambda_{-2}(n_{eq}, U)$, where $\lambda_{\pm 2}$ is the transition rate to a state with ± 2 excitons and U is the excitation energy of the compound system. From this condition one can get a *very* rough estimate

$$n_{eq} = \sqrt{2gU}, \tag{1}$$

where g is the single-particle level density.

In practice, the root two factor in eq.1 is considered to be an adjustable parameter in nuclear data evaluations. One must be very cautious with this estimate since in certain circumstances (heavy targets at bombarding energies where pre-equilibrium plays a role) the condition $\lambda_{+2}(n_{eq}, U) = \lambda_{-2}(n_{eq}, U)$ is reached well before $n = n_{eq}$ and therefore the former condition must prevail. In our Monte Carlo simulation, this physically consistent condition has been directly implemented by means of the appropriate algorithm.

The transition probabilities, which are a key ingredient of the model, have been calculated either in a semi-microscopic way [22] or by using a standard parameterisation of the matrix elements [23]. As expected, both of them lead to similar results.

The emission probabilities, either for single nucleons or for complex fragments, have been calculated according to [22], including combinatorial factors R_j, which take into account the availability among the excitons of enough particles to form the one to be emitted; these factors were not included previously.

A proper evaluation of the Coulomb barrier combined with inverse cross sections fitted to experimental data (either directly or indirectly through optical model global fits) is essential for a good description of both the evaporation and preequilibrium stages. Among the different possibilities, we choose for protons the parameterisation of Wellisch[24], and for the rest of the particles we use Chatterjee[14]. The first, unlike the second, gives a good description of the proton inverse reaction cross section at low energies (i.e. in the Coulomb Barrier region) for light systems, such as ^{27}Al, but breaks down for heavy ones; this has been solved by suitable inclusion of the correct Coulomb barrier.

3.5. Improvements to low energy neutron transport

Neutron High Precision Models and Cross Sections (NeutronHP) treat the detailed simulation of neutrons with energies below 20 MeV. It is a data-driven model and depends on the G4NDL evaluated data library.

G4NDL has already been enhanced to additionally include the most-requested elements such as Gd, Nd, Sm, Hf, Ag, Ge, La, Br, Rd and Hg. These have been translated from ENDF-VI[25], ENDF-VII[26] and JENDL-3.3[27] and included in the latest version of G4NDL.

Several other data were also recently introduced to G4NDL. These include data from ENDF-VI used for the special treatment of scattering from chemically bound atoms at thermal energies, and the JENDL high energy cross section data (JENDL-HE) which extends applicability up to 3 GeV. JENDL-HE data for selected materials such as concrete, iron and brass were used in the recent SATIF9 inter-comparison of neutron attenuation calculations[28] where Geant4 shows good agreement with the beam data.

"On-the-fly" calculation of Doppler broadening has long been included in NeutronHP as a means of calculating the effects of thermal motion. It is, however, very CPU-intensive. An option to bypass this calculation and still take thermal motion into account in the final state has thus been added. As a result, a $\sim 400\%$ performance increase has been achieved.

A significant shortfall of the NeutronHP models is the non-conservation of energy and momentum in a reaction. NeutronHP is a data-driven model and not designed for setting up conservation at the single reaction level. However there are many reaction channels which do not require detailed treatments, and for these conservation may be established. Significant improvement in many of these reactions has been made, especially in final state capture, where an option now exists to force the use of the Geant4 Photon Evaporation model whenever a neutron is captured. Previously the less precise G4NDL photon data were used if present.

The above changes have increased the usability and reliability of NeutronHP and efforts for further improvements are ongoing. Since the release of NeutronHP, however, several new evaluated nuclear libraries have become public as well as several new ENDF formats which are not supported by NeutronHP. Rather than make extensive modifications to the NeutronHP code, it was decided to develop a new model based on the Evaluated Nuclear Data Library (ENDL)[29] which is publicly available from Lawrence Livermore National Laboratory. Most evaluated nuclear data libraries including new ones are already translated to ENDL and will be available through the new model.

4. Conclusion

As a result of these studies, we found that a review of several aspects of the GEANT4 modeling of hadronic interactions was required. The main improvements identified for the QGSP physics list were the need to complement the Quark Gluon string (QGS) model by a quasi-elastic model, and the need to use a cascade code for hadrons below 10 GeV. Other improvements include a better description of cross sections at energies above 100 GeV, and an updated elastic scattering model for hadrons. The configuration obtained is the QGSP_BERT physics list. This gives a good description of shower shape, the e/π ratio, and the resolution of hadronic showers.

The effort to improve physics models is ongoing; the Fritiof-based (FTF) model, the pre-compound model and the de-excitation models are being completely revised, and there are refinements and additions to many other hadronic models. Configurations based on these models are availabele as alternate physics lists and allow to estimate systematics, and offer possibilities for future improvements.

5. Acknowledgements

For development of INCL/ABLA code, we are grateful to INCL developers Joseph Cugnon (University of Liege) and Y. Yariv (SphN-DAPNIA, CEA-Saclay and Soreq, Israel) for supporting this project. From the ABLA team Aleksandra Kelic and Karl-Heinz Schmidt kindly provided the code for the Geant4 translation and Christelle Schmitt (IPNL) provided a C++ translation of fission code SimFis18.

References

[1] Aostinelli S *et al* 2003 *Nucl. Instrum. Meth.* A **506** 250-303
[2] J. Allison J *et al* 2006 *IEEE Trans. Nucl. Sci.* **53** 270-278
[3] Fesefeld H 1985 "Simulation of Hadronic Showers, Physics and Applications." *Technical Report PITHA 85-02, Aachen, Germany*
[4] Glauber R J *High Energy Physics and Nuclear Structure.*, edited by S.Devons, Plemium Press, NY, 1970.
[5] Abers E A *et al* 1966 *Nuovo Cimento* **42** 365
Gribov V N 1973 *Zh. Eksp. Theor. Fiz.* **18** (1973) 271.
[6] Nam R A *et al* 1977 *Sov. Jour. of Nuclear Phys.* **26** 550
[7] Kossov V M 2000 Manual for the CHIPS event generator *KEK internal report 2000-17, Feb. 2001 H/R*;
Degtyarenko P V, Kossov V M and Wellisch J P 2000 *Eur. Phys. J.* A **8** (2000) 217
Degtyarenko P V, Kossov V M and Wellisch J P 2001 *Eur. Phys. J.* A **9** 211
Degtyarenko P V, Kossov V M and Wellisch J P 2001 *Eur. Phys. J.* A **9** 221
[8] Geant4 Physics Reference Manual, http://cern.ch/geant4/support/userdocuments.shtml
[9] See other presentations in these proceeedings
[10] En'yo H *et al* 1985 *Phys. Lett.* **159B** 1
[11] Andersson B *et al* 1987 *Nucl. Phys.* B **281** 289
Nilsson-Almquist B and Stenlund E 1987 *Comp. Phys. Commun.* **43** 387
[12] Folger G, Ivanchenko V N and Wellisch J P 2004 *Eur. Phys. J.* A **21** 404-17
[13] Andersson B, Gustafson G, Ingelman G, Sjöstrand T 1983 *Phys. Rep.* **97** 31
[14] Chatterjee A, Murthy K H N and Gupta S K 1981 *Pramana v***16** n5 391
[15] Boudard A *et al* ' 2002 *Phys. Rev.* C **66** 044615
[16] Junghans A R *et al* 1998 *Nuc. Phys.* A **629** 635
[17] Benlliure J *et al* 1998 *Nuc. Phys.* A **628** 458
[18] S. Leray S *et al* 2002 *Phys. Rev.* C **65** 044621
[19] Heikkinen A, Kaitaniemi P, and Boudard A Implementation of INCL cascade and ABLA evaporation codes in Geant4 *Proceedings of CHEP07 Journal of Physics* (to be published)
[20] Griffin J J 1966 *Phys. Rev. Lett.* **17** 478
[21] Weisskopf V E and D. H. Ewing D H 1940 *Phys. Rev.* **57** 472
[22] Gudima K, Mashnik S G and Toneev V D 1983 *Nucl. Phys.* A **401** 329
[23] Gupta S 1981 *Z. Phys.* A **303** 329
[24] Wellisch H P and Axen D 1996 *Phys. Rev.* C **54** 1329
[25] ENDF/B-VI: Cross Section EvaluationWorking Group 1991 *Report BNL-NCS-17541 (ENDF-201)* ed Rose P F (National NuclearData Center, Brookhave National Laboratory, Upton, NY, USA)
[26] Chadwick M B *et al* 2006 *Nuclear Data Sheets* **07** 293-3060
[27] Shibata K *et al* 2002 *J. Nucl. Sci.Technol.* **39** 25
[28] Hirayama H 2008 *Proc. SATIF9, Shielding Aspects of Targets, Accelerators, and Irradiation Facilities* Oak Ridge, USA
[29] http://nuclear.llnl.gov/CNP/Home/Nuclear_Data.htm

Journal of Physics: Conference Series **160** (2009) 012074 doi:10.1088/1742-6596/160/1/012074

The response of the ATLAS Tile Calorimeter to pions and protons in test beam and in GEANT4 Monte Carlo simulation

Margar Simonyan

On behalf of the ATLAS Tile Calorimeter Group, ATLAS Collaboration

Laboratory of Annecy-le-Vieux of Particle Physics, LAPP - 9 Chemin de Bellevue - BP 110 F-74941 Annecy-le-Vieux Cedex - France,

previously Yerevan Physics Institute, Alikhanyan Brothers 2, 375036, Yerevan, Armenia

E-mail: `margar.simonyan@cern.ch`

Abstract. The response of the ATLAS Tile Calorimeter to pions and protons in the energy range of 20 to 180 GeV produced at CERN's Super Proton Synchrotron (SPS) H8 test beam line has been measured. The experimental setup allowed to measure the longitudinal development for pion and proton induced showers up to 20 nuclear interaction lengths. It is found that pion showers penetrate deeper in the calorimeter than the proton induced ones. However, protons induce showers that are wider laterally to the direction of the impinging particle. Including the measured total energy response and the resolution, all observations are consistent with a higher electromagnetic energy fraction in pion induced showers. The data are compared with GEANT4 simulations using several hadronic interaction models.

1. Introduction

The ATLAS Tile Calorimeter (TileCal) is an iron-scintillator hadronic calorimeter located in the central part of the ATLAS detector. The new feature of the TileCal design is the orientation of the scintillating tiles which are placed in planes perpendicular to the LHC colliding beams. However, for the runs analysed here a special non-projective configuration of the test beam setup has been used where the beam direction is perpendicular to the scintillating tiles. This configuration allows an almost full shower containment and makes it possible to measure shower longitudinal profiles up to 20 nuclear interaction lengths[1] (λ).

The measured response of the detector to pions and protons is compared to Monte-Carlo (MC) simulations using several hadronic interaction models available in the GEANT4 [1] framework. Our detailed comparison contributes to test existing simulation packages and to guide their future development so that the stringent requirements to correctly model the calorimeter response and the hadronic shower development, imposed by the LHC physics programs, can be reached.

[1] We use throughout this analysis an effective nuclear interaction length for the Tile calorimeter $\lambda = 20.55$ cm, calculated using the known fraction of materials used in the detector construction and their nuclear interaction lengths.

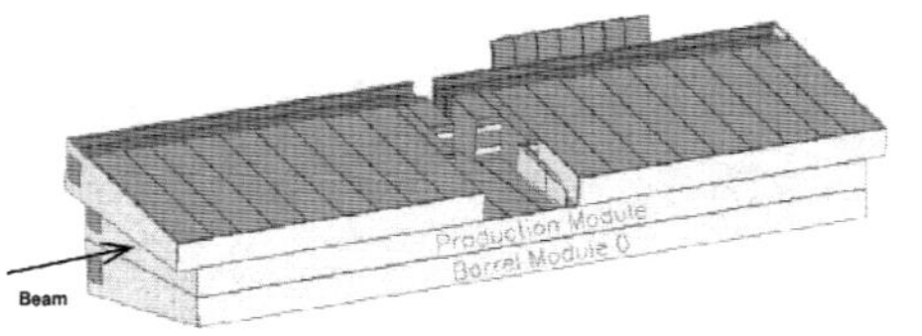

Figure 1. Layout of the calorimeter modules in the test beam set-up. The arrow indicates the direction of the impinging particles.

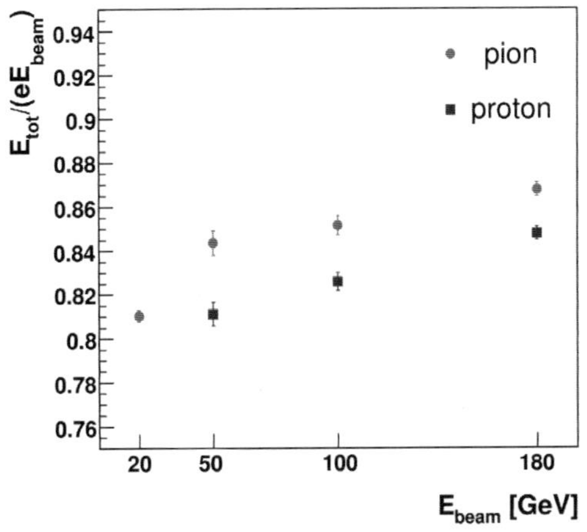

Figure 2. Measured total energy divided by the beam energy for pions and protons as a function of the beam energy. The total energy is presented on the electromagnetic energy scale.

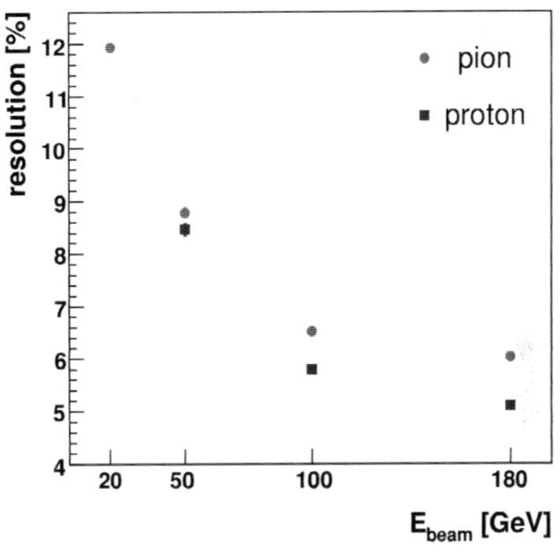

Figure 3. Energy resolution (RMS/E) for pions and protons as a function of the beam energy.

2. Test beam setup

The H8 test beam line at the SPS accelerator of CERN was instrumented with a set of beam detectors: scintillator trigger counters, wire chambers measuring the lateral coordinates of the beam particles and a helium filled threshold Cherenkov counter. The layout of the TileCal modules in the H8 test beam line area is shown in Fig. 1. It consists of two Barrel Modules, 5.64 m long, and 2.93 m long two Extended Barrel modules. As indicated in Fig. 1 the beam impinges from the side of the middle located Barrel Module, perpendicularly to the scintillating tiles. The direction of the beam is defined to be the x-axis.

Longitudinally showers are fully contained in the calorimeter, since no energy is observed in the last layers of the modules. The lateral containment of the showers in the test beam setup is more than 99%.

3. Monte Carlo Simulation Tools

The simulation of the calorimeter modules was performed within the ATLAS software framework ATHENA using the GEANT4 simulation toolkit [1]. Besides purely electromagnetic processes, also hadron and photo-nuclear interactions were simulated. Within the GEANT4 simulation framework several models can be used to simulate the interaction of particles with matter. A "physics list" is a consistent collection of models that covers the interaction of all particles in the whole energy range from thermal energies to the several TeV range. Depending on the application, on the required physics performance and the available computing time, different

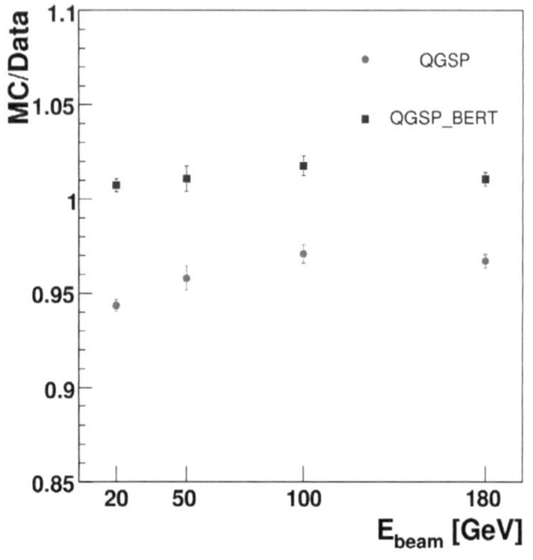

Figure 4. Ratio of the mean energy in the Monte Carlo simulation to the one in the data as a function of the beam energy for pions.

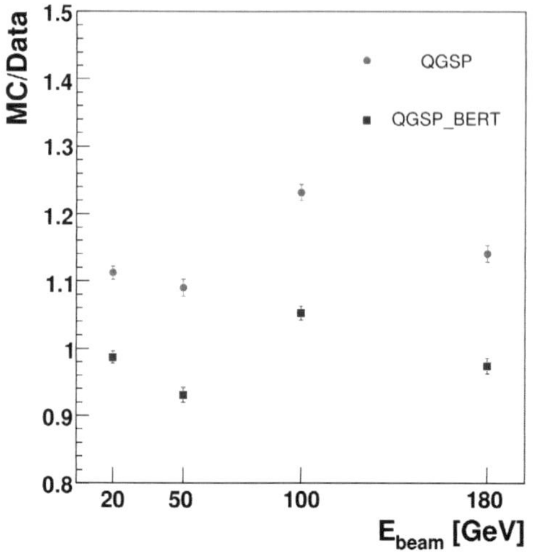

Figure 5. Ratio of the RMS of total energy distribution in the Monte Carlo simulation to the one in the data as a function of the beam energy for pions.

physics lists can be chosen.

In this study different physics lists have been used which are recommended for calorimeter studies at LHC energies. At high energies the Quark-Gluon String (QGS) and Fritiof (FTF) fragmentation models are used. In the intermediate energy region Binary (BIC) and Bertini (BERT) intra-nuclear cascade models are applicable. The Precompound (P) model is used at lower energies.

4. Data Set and Event Selection

Events in the tails of the lateral beam profile are eliminated with the help of the beam chambers. All events within three standard deviations around the mean are accepted. These cuts reduce the momentum spread of the beam and decrease the acceptance for events, where a hadron decays or interacts early in the material of the H8 beam line before entering the calorimeter. In each of the scintillator counters a signal compatible with the one from a minimum ionizing particle is required in order to remove multiple hits from accidental coincidences of beam particles and to reject early showering hadrons.

The Cherenkov counter is used to discriminate pions from protons event-by-event. At 20 GeV only negative beam was available. Electron and muon rejection has been done using calorimeter information. The resulting biases on the measurements are studied and they are found to be negligible.

5. Mean Energy Response and Energy Resolution

The measured mean total energy, normalised to the beam energy is referred here as the response of the calorimeter. It is shown for pions and protons in Fig. 2. For pions, the response increases by about 7% from 20 to 180 GeV. The increase for protons is steeper. The resolution defined as the root mean square of the total energy distribution divided by the mean energy (RMS/E) shown in Fig. 3 is about the same for pions and protons at low energies and is better for

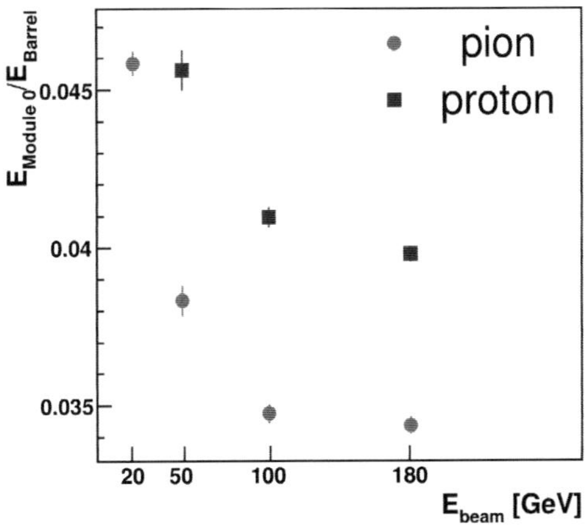

Figure 6. Ratio of the energy deposition in the bottom and central modules (characterising the lateral spread of showers) as a function of the beam energy.

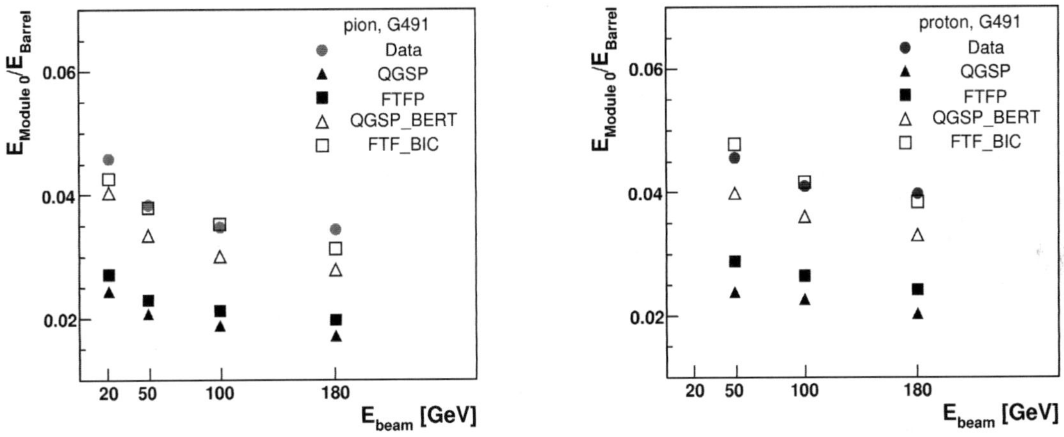

Figure 7. Ratio of energy deposition in the bottom and central modules for data and for various MC simulations as a function of the beam energy.

protons at higher energies ($E \geq 100$ GeV). All these measurements reflect qualitatively the non-compensating nature of the TileCal that leads to a different response to the hadronic and electromagnetic energy depositions during the shower development.

On average, approximately 1/4 of particles produced in each high energy collisions in the hadronic cascade are neutral pions. With increasing beam energy the number of generations in the cascade increases. This leads to a larger fraction of energy transfered into the electromagnetic sector [2]. In case of proton induced hadronic interactions bayron number conservation puts limit on the electromagnetic energy fraction in each collision. In addition, the probability for the leading particle to be a proton in proton induced reactions and to be a pion in pion induced ones is large. The leading pion can well be neutral and therefore can transfer by electromagnetic

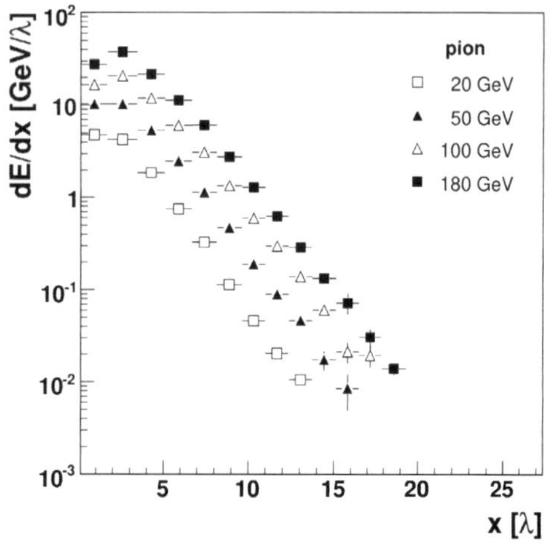

Figure 8. Longitudinal profile of pion induced showers at various energies.

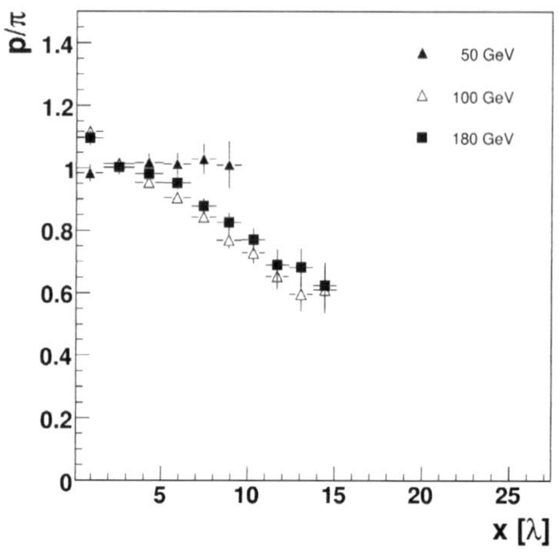

Figure 9. Ratio of the longitudinal profiles of proton induced showers to pion induced ones at various energies.

decay a large fraction of energy into the electromagnetic sector. This explains a higher response of non-compensating calorimeters to pions and a better resolution for protons.

The measured response of pions is compared to the results of the simulations in Fig. 4. The QGSP physics list predicts about 4 % lower response compared to the data. The addition of Bertini's intra-nuclear cascade model leads to an increase of the response in the simulation and results in better agreement with the data. The QGSP_BERT physics list describes the absolute response within 2 %. Similar results are obtained for protons [3].

The RMS of the total energy distribution for pions in the MC is compared to the one in data in Fig. 5. The width of the total energy distribution is larger by about 10 – 20 % in the MC than found in the data. Adding the Bertini cascade brings the simulation to a better agreement with the data (within ±10 %). Similar results have been obtained for proton measurements [4].

6. Lateral Spread

The ratio of the energy deposit in the bottom and central modules is a simple estimator of the lateral spread of the shower. The dependency of this ratio on the beam energy is shown for pions and protons in Fig. 6. Since the measured ratio is larger for pions than for protons, pion induced showers are laterally narrower. This observation might be qualitatively explained by the difference in the electromagnetic energy contents of pion and proton induced showers. Most of the electromagnetic energy is produced near the core of the shower. Since electromagnetic showers are compact, the electromagnetic energy is deposited relatively close to the core of the shower and therefore pion induced showers are relatively narrower.

The same ratio for data and simulation is presented in Fig. 7. Both QGSP and FTFP physics lists predict showers with a lateral spread significantly smaller than the one measured in the data for both pions and protons. The description of QGSP is worse than the one of FTFP. Adding the Bertini and Binary cascade models leads to a good description of the data.

Journal of Physics: Conference Series **160** (2009) 012074 doi:10.1088/1742-6596/160/1/012074

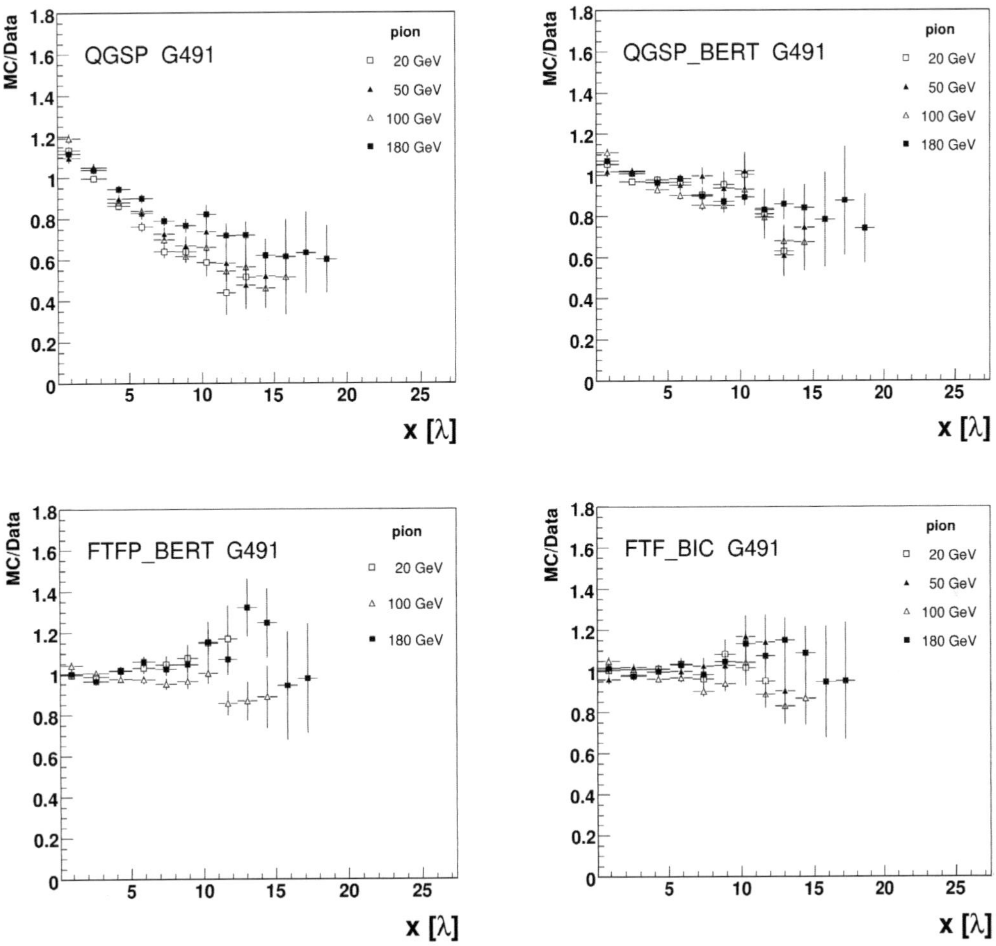

Figure 10. Ratio of various MC simulations to data for the pion longitudinal shower profile description.

7. Longitudinal Shower Profiles

The measured longitudinal profiles of pion induced showers are presented in Fig. 8. The measurements extend up to 20 λ in depth. On average, hadron showers quickly deposit their energy and reach the point within the first few λ in depth where the mean energy deposition is maximal. The average energy deposition then exponentially decreases towards the end of the shower and is lower by approximately four orders of magnitudes at 20 λ.

The ratio of the profiles of showers induced by protons and pions is presented in Fig. 9. Since the statistical uncertainties of the measured longitudinal profiles are relatively higher at the end of the shower, the ratios are only presented for a limited range of depth that depends on the energy. At 50 GeV the ratio is independent of the depth and close to 1, while at higher energies it decreases with depth. This behaviour may be explained by the higher proton interaction cross-section with iron nuclei. Due to this pions, on average, penetrate deeper in the calorimeter. This results in fewer hadronic interactions initiated by pions in the first one or two layers, which, however, are characterised by a higher fraction of neutral pion production by pions. These two differences in the underlying mechanism of shower development for pions and protons have

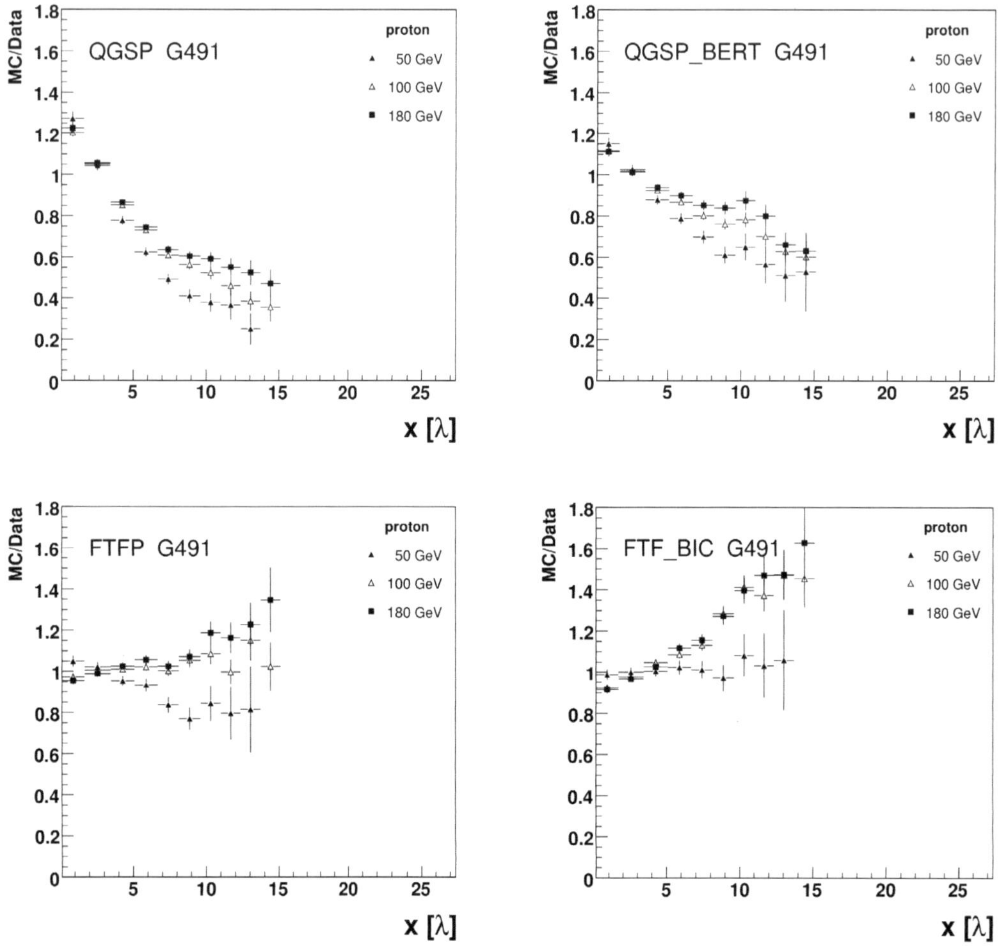

Figure 11. Ratio of various MC simulations to data for the proton longitudinal shower profile description.

opposite effects on longitudinal profiles and happen to balance out at 50 GeV.

The longitudinal profiles of pion induced showers obtained with MC simulation are compared to the measured ones in Fig. 10. The predictions of the hadronic physics lists vary significantly. The QGSP physics list predicts too short showers over the full energy range. Approximately 50% less energy is predicted at a depth of 10 λ as compared to the data. Adding the Bertini inter-nuclear cascade model makes showers longer and brings simulation to a better agreement with the data. Both with Binary and Bertini cascade models the Fritiof based physics list describe longitudinal development of pions induced showers with $\pm 10\%$ precision up to 10λ depth.

Comparison of longitudinal profiles of proton induced showers in the simulation and data is presented in Fig. 11. The QGSP physics list predict very short showers compared to data. The addition of Bertini cascade improves the description but the simulated showers are still too short. Both with Binary and Bertini cascade models the Fritiof based physics lists predict longer showers.

Summary and Conclusions

The response of the Tile calorimeter has been studied for pions and protons within the energy range of 20 to 180 GeV in a special test beam setup where the hadronic showers are almost fully contained in the calorimeter.

The mean response of the detector to pions and protons slowly increases with beam energy. The resolution is better for protons than for pions at higher energies. These observations are consistent with a higher electromagnetic energy fraction in the pion induced showers and larger fluctuations of this fraction.

Longitudinal profiles of pion and proton induced showers have been measured up to a depth of 20 nuclear interaction lengths. The showers induced by protons are longitudinally shorter, but laterally wider. This can be explained by the larger inelastic cross-sections (with iron nucleus) and by the lower fraction of electromagnetic energy depositions for protons.

The experimental data have been compared with the results of GEANT4 simulation, using FTFP and QGSP physics lists, as well as extensions where the Bertini and Binary intra-nuclear cascade models are used. A simple estimator of the lateral shower spread is well reproduced by simulation when Bertini or Binary cascade models are used. Neither of those physics lists is able to reproduce the data both for pions and protons in the whole energy range satisfactorily, although the addition of the intra-nuclear cascade improves the description.

Acknowledgments

We would like to thank all people that helped in the setting-up of the test-beam, in the data taking, in the data reconstruction and analysis, in particular, Alexander Solodkov, Tomas Davidek and Matteo Cavalli-Sforza for their continuous help and useful discussions. We are grateful to our colleagues from the GEANT4 collaboration for their help with the MC simulation and many useful suggestions.

References

[1] S. Agostinelli et al., Nucl. Instr. and Meth. A 506 (2003) 250.

[2] Donald E. Groom, Nucl. Instr. and Meth. A 572 (2007) 633.

[3] H. Hakobyan and M. Simonyan and T. Carli and A. Henriques, Measurement of Pion and Proton Longitudinal Shower Profiles up to 20 Nuclear Interaction Lengths. ATLAS Public note, ATL-TILECAL-PUB-2007-008, `http://cdsweb.cern.ch/record/1004188`.

[4] M. Simonyan, Performance of the ATLAS Tile Calorimeter to pions and protons. Ph.D thesis, `http://cdsweb.cern.ch/record/1098137`.

Journal of Physics: Conference Series **160** (2009) 012075 doi:10.1088/1742-6596/160/1/012075

GEANT4 physics evaluation with testbeam data of the ATLAS hadronic end-cap calorimeter

A E Kiryunin[1], H Oberlack[1], D Salihagić[1], P Schacht[1] and P Strizenec[2]

[1] Max-Planck-Institut für Physik, Werner-Heisenberg-Institut
Föhringer Ring 6, 80805 Munich, Germany
[2] Institute of Experimental Physics of the Slovak Academy of Sciences
Watsonova 47, 04001 Kosice, Slovakia

E-mail: kiryunin@mppmu.mpg.de

Abstract. The validation of GEANT4 physics models is done by comparing experimental data from beam tests of modules of the ATLAS hadronic end-cap calorimeter with GEANT4 based simulations. Various physics lists for the simulation of hadronic showers are evaluated. We present results of studies of the calorimeter performance parameters (like energy resolution and shower shapes) as well as results of investigations of the influence of the Birks' law and of cuts on the time of development of hadronic showers.

1. Introduction

GEANT4 [1] is a package for simulating the passage of particles through matter. It provides variety of tools for detector response simulations, including: geometry description, particle transportation, physics models, visualization and so on. Nowadays it is used by many experiments and in many application areas.

The ATLAS collaboration works with the GEANT4 toolkit since its first production releases in 1999. A lot of efforts have been put over these years by the collaboration to evaluate electromagnetic and hadronic physics models of GEANT4, to follow up their development and to optimise corresponding simulations [2]. Part of the validation work is based on the comparison of GEANT4 predictions with experimental results obtained during various beam tests of ATLAS calorimeter modules. The validation includes comparisons of relevant calorimeter performance parameters like the energy dependence of the response, resolution and shower shapes.

The group of the ATLAS hadronic end-cap calorimeter (HEC) is actively involved in tests and validations of GEANT4 [3, 4]. In this talk the latest results of physics evaluation with HEC testbeam data are presented. Three points will be addressed: a) comparison of different hadronic physics lists, b) study of the saturation of the response in the liquid argon, c) consideration of the time structure of hadronic showers.

2. Hadronic end-cap calorimeter and its beam tests

The ATLAS hadronic end-cap calorimeter [5] is a liquid argon (LAr) sampling calorimeter with parallel copper absorber plates. The HEC is structured in two wheels (HEC1 and HEC2), each wheel consisting of 32 modules. Annular spacers define 8.5 mm gaps between the absorber plates. The thickness of the copper absorber plates is 2.5 cm for HEC1 and 5.0 cm for HEC2. The calorimeter has four longitudinal layers. The total thickness of the calorimeter is ~103 radiation lengths X_0 or ~10 absorption lengths λ.

During the years 2000-2001, the serial production calorimeter modules were exposed to test beams as part of the standard quality control procedure during HEC construction. The beam tests have been carried out in the H6 beam line of the CERN SPS, using secondary and tertiary beams. In total more than 2200 runs were taken at different impact points. Each run consisted of typically 20 000 triggered events. The energy ranged 6-150 GeV for electrons, 10-200 GeV for charged pions and 120, 150, 180 GeV for muons. More details, as well as data analysis results can be found in [6].

3. GEANT4 simulations of the HEC testbeam

For GEANT4 based simulations of the HEC testbeam, the special code (running out of the standard ATLAS software) was prepared. Energy scans with electrons and with negatively charged pions were simulated. Usual statistics was typically 5000 events per beam energy.

GEANT4 version 9.0 (released in June 2007) was used for the present simulations. And the range cut of 30 μm was applied.

3.1. Hadronic physics lists

Different physics lists are provided by the GEANT4 package for simulations of hadronic showers. For the current comparison with the HEC testbeam data, three of them are selected:

- QGSP physics list is based on theory-driven models: it uses the quark-gluon-string model for interactions and a pre-equilibrium decay model for the fragmentation
- QGSP-BERT physics list includes the Bertini cascade model (a theoretical hadronic model for intra-nuclear transport) below ~10 GeV
- QGSP-BERT-HP physics list uses high precision data-driven modeling for low energy neutrons

Estimates of the time needed to simulate charged pions show, that QGSP-BERT based simulations are ~1.7 times slower than simulations with the QGSP physics list. Usage of precise neutron models significantly slows down simulations: QGSP-BERT-HP is almost five times slower than QGSP.

3.2. Birks' law

Recombination effects in the liquid argon lead to the saturation of the response for particles with large dE/dx. Phenomenologically this effect can be described by the Birks' law [7]:

$$\Delta E' = \Delta E \frac{A}{1 + \frac{c}{\rho}\frac{\Delta E}{\Delta x}}, \tag{1}$$

where $\rho = 1.396$ g/cm^3 is the LAr density and parameters of the Birks' formula have the following values: $A = 1$, $c = 0.0045$ g/(MeV cm^2).

To study the influence of this effect on the calorimeter performance parameters, all simulations were done twice: without taking into account the Birks' law and when corresponding energy corrections at each step of simulations were switched on.

3.3. Time structure of hadronic showers

To study the time structure of showers, information about the time of energy depositions in calorimeter cells were kept during simulations. In Figure 1 the visible energies in LAr are shown as functions of the deposition time for 100 GeV pions. It is seen that time profiles of hadronic showers have rather long tails for all studied physics list. And QGSP-BERT clearly predicts slower showers than two other physics lists. Late energy depositions (after 100 ns) amount to 4.1, 1.3 and 0.9 % of the total energy for QGSP-BERT, QGSP and QGSP-BERT-HP, respectively. Long tails of time profiles given by the QGSP-BERT physics lists (if one compares with QGSP-BERT-HP) seem artificial and can be related to neutron modeling as it is implemented in this physics list.

Under conditions foreseen at the Large Hadron Collider (high luminosity, 25 ns interval between bunch crossings), calorimeter signals cannot be integrated over the full time of development of hadronic showers. Readout of calorimeter channels is fast. To model the experimental situation, the following procedure was applied to simulated signals.

For each calorimeter channel in each event a time profile is convoluted with a shaping function corresponding to this channel. (Example of such a function for one of HEC channels is shown in Figure 2.) Reconstructed signal in this case is the amplitude of a convoluted profile at the peak position of a shaping function. Peak positions for different HEC channels are distributed between 50 and 70 ns. Effectively this procedure means the integration of time profiles over a few tens of nanoseconds.

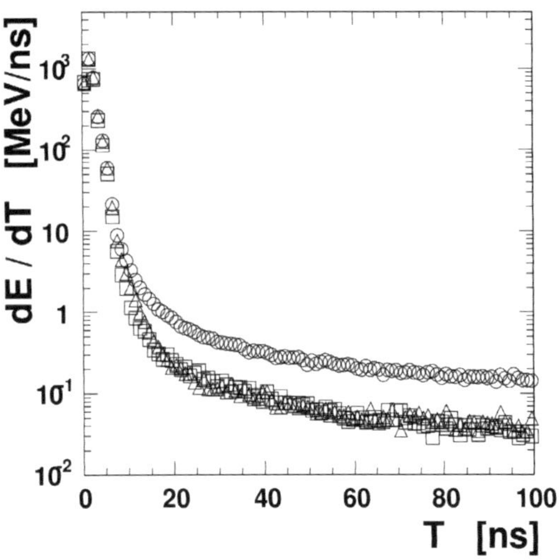

Figure 1. Visible energy in the HEC for 100 GeV charged pions as a function of the time of energy depositions. Squares — QGSP, circles — QGSP-BERT, triangles — QGSP-BERT-HP predictions (with the Birks' law switched on).

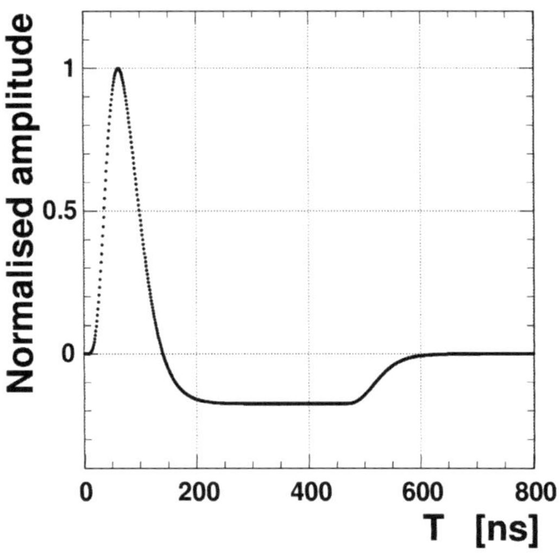

Figure 2. Shaping function for one of HEC channels.

To evaluate the significance of these cuts on the time of shower development, results obtained with the described procedure are compared with results after the integration of signals over the full time interval (i.e. obtained without any time cut).

3.4. Reconstruction and analysis

The analysis of the simulated data followed rather closely the experimental procedure [6]. An energy independent electromagnetic scale factor (obtained by the analysis of electron samples) was applied for the energy reconstruction. Clusters of fixed size were used, i.e. the same set of cells was selected for energy measurements in all events and for all beam energies. The use of clusters of fixed size for the energy reconstruction allowed the evaluation of the electronics noise (including possible correlations between channels). To compare the experimental and simulated data, known spread of the noise was subtracted from the resolution of the experimental data.

4. Results

To evaluate GEANT4 physics models, the following calorimeter performance parameters are studied and presented in this talk: the energy resolution for charged pions, ratio e/π and longitudinal shape of hadronic showers.

4.1. Pion energy resolution

To get the energy resolution and response, Gaussian curves were fitted to the reconstructed energy distributions in the interval $\pm 3\sigma$ around the peak value E_0. The parameters E_0 and σ from this fit were used to determine the resolution σ/E_0.

The pion energy resolution as a function of the beam energy is presented in Figure 3 for the experimental data and for Monte Carlo (MC) simulations with different hadronic physics lists and the Birks' law switch. Comparison of the left-hand and right-hand plots show, that time cuts strongly influence the energy resolution. For the QGSP-BERT physics list, predicting slower hadronic showers, relative increase of 10-30 % (depending on the beam energy) is observed. For the QGSP and QGSP-BERT-HP lists the increase is at the level of ~5 %. This growth brings the energy resolution for QGSP-BERT and QGSP-BERT-HP closer to experimental values.

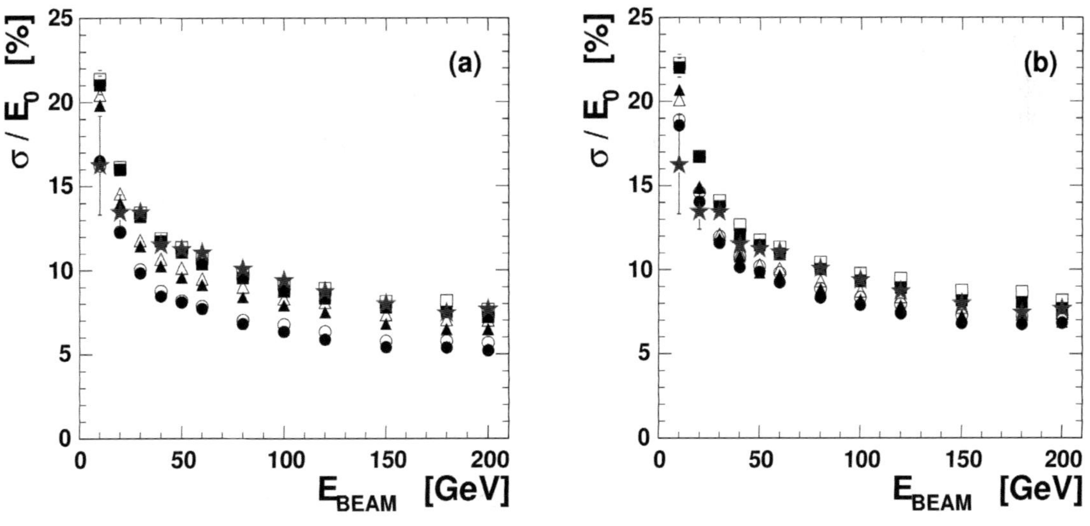

Figure 3. Energy resolution for pions as a function of the beam energy. Stars mark experimental values. Squares — QGSP, circles — QGSP-BERT, triangles — QGSP-BERT-HP predictions. For MC samples: open (full) symbols correspond to simulations with (without) the Birks' law; (a) — no time cut, (b) — after time cuts.

The energy dependence of the resolution can be parametrized by the following two-term formula:

$$\frac{\sigma}{E_0} = \frac{A}{\sqrt{E_{BEAM}}} \oplus B, \tag{2}$$

where A and B are the sampling and the constant terms, respectively. Results, obtained for different simulations after applying time cuts, are summarised in Table 1. They are to be compared with experimental values: $A_{EXP} = 69 \pm 1\ \%\sqrt{GeV}$, $B_{EXP} = 5.8 \pm 0.1\ \%$.

Table 1. Sampling and constant terms of the pion energy resolution, as obtained from simulations after time cuts.

Birks' law	Physics list	A [% $\sqrt{GeV}$]	B [%]
OFF	QGSP	68.1 ± 0.8	6.2 ± 0.1
	QGSP-BERT	57.1 ± 0.7	5.30 ± 0.09
	QGSP-BERT-HP	60.2 ± 0.7	5.4 ± 0.1
ON	QGSP	67.9 ± 0.8	6.9 ± 0.1
	QGSP-BERT	58.6 ± 0.7	5.83 ± 0.09
	QGSP-BERT-HP	59.6 ± 0.7	6.13 ± 0.09

When time cuts are applied and the Birks' law is taken into account, the sampling term of the pion energy resolution is well described by QGSP. At the same time, the constant term is predicted better by the QGSP-BERT and QGSP-BERT-HP physics lists. This means that these two lists give better prediction of the resolution at high beam energies. It is also seen that usage of the Birks' law does not change the sampling term, but leads to the increase of the constant term.

4.2. Ratio e/π

Ratio e/π is defined as a ratio of reconstructed energies E_0 in electron and in pion clusters at the same beam energy. Energy dependences of this ratio are presented in Figure 4 for the experimental data and for different MC simulations. As for the energy resolution, time cuts strongly influence the e/π-ratio for QGSP-BERT: 4-8 % increase is observed for this physics list after applying those cuts. For two other physics list the increase is smaller — 1-2 %. When the Birks' law is taken into account, ratio e/π also becomes larger: by 2-3 % for all physics lists.

To make comparison with experimental data more clear, ratio between simulated and experimental data for e/π can be studied (see Figure 5). After appying time cuts and with usage of the Birk's law, all three hadronic physics lists describe experimental values of the e/π-ratio rather good.

4.3. Longitudinal shape of hadronic showers

Longitudinally the HEC calorimeter is segmented into four layers. (Their thicknesses equal to 1.5, 2.9, 3.0 and 2.8 λ.) This allows to study the longitudinal development of hadronic showers. An appropriate variable is the fraction of energy in a layer with respect to the total energy. It can be defined in the following way:

$$F = \frac{< E_i >}{\sum_{i=1}^{4} < E_i >}, \tag{3}$$

where E_i is the energy in a HEC layer within the reconstructed pion cluster.

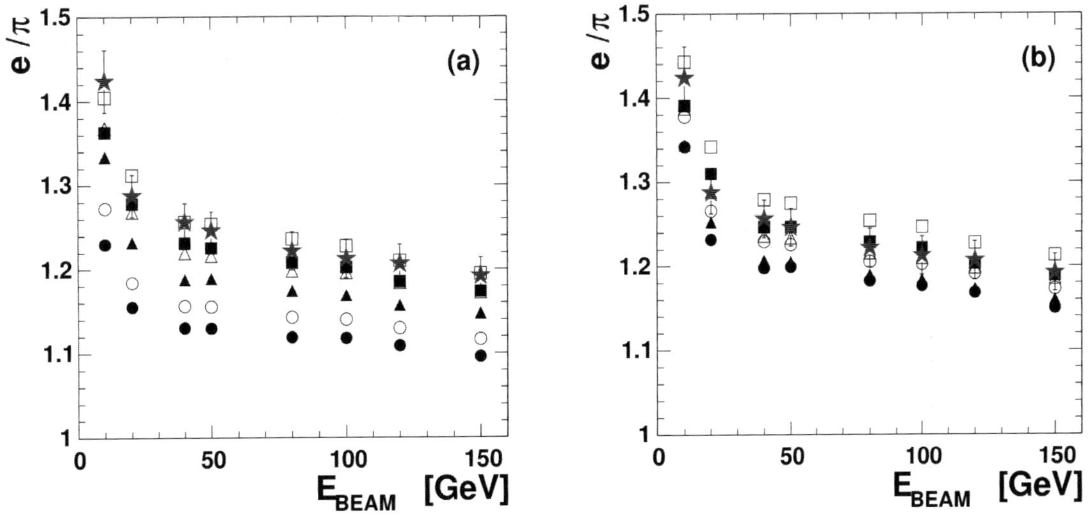

Figure 4. Ratio e/π as a function of the beam energy. Stars mark experimental values. Squares — QGSP, circles — QGSP-BERT, triangles — QGSP-BERT-HP predictions. For MC samples: open (full) symbols correspond to simulations with (without) the Birks' law; (a) — no time cut, (b) — after time cuts.

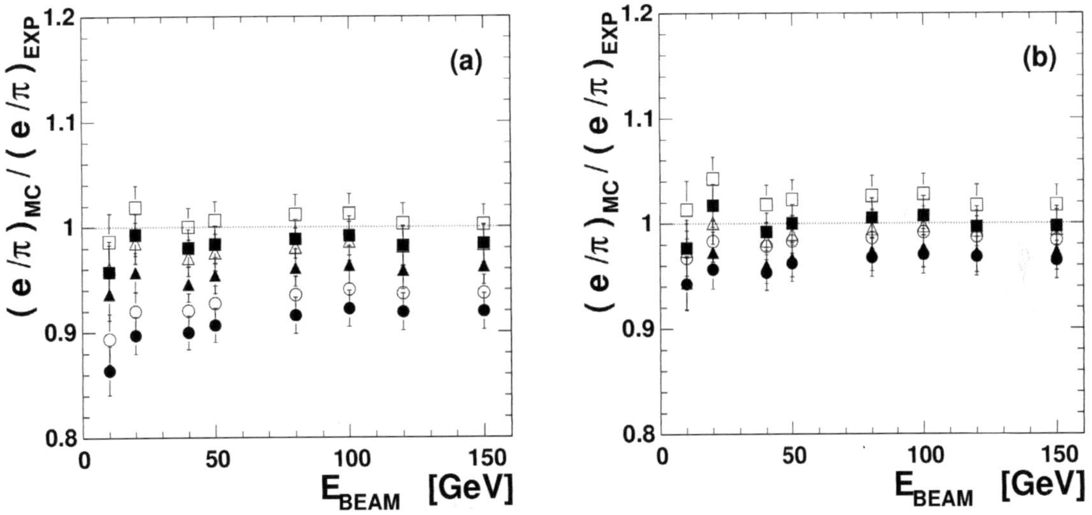

Figure 5. Ratio between simulated and experimental data for e/π as a function of the beam energy. Squares — QGSP, circles — QGSP-BERT, triangles — QGSP-BERT-HP predictions. Open (full) symbols correspond to simulations with (without) the Birks' law; (a) — no time cut, (b) — after time cuts.

These energy fractions are presented in Figure 6 as functions of the beam energy. The ratio between simulated and experimental data is shown as well (Figure 7). Approximately half of the pion signal is deposited in the second layer of the calorimeter. And this is described within a few percent by all three physics lists. At the same time QGSP predicts more energy in the first layer and significantly less energy in the third layer. This means that hadronic showers in this model start earlier and are more compact. The QGSP-BERT and QGSP-BERT-HP physics lists give better description of longitudinal profiles of hadronic showers (except the lowest beam energy of 10 GeV).

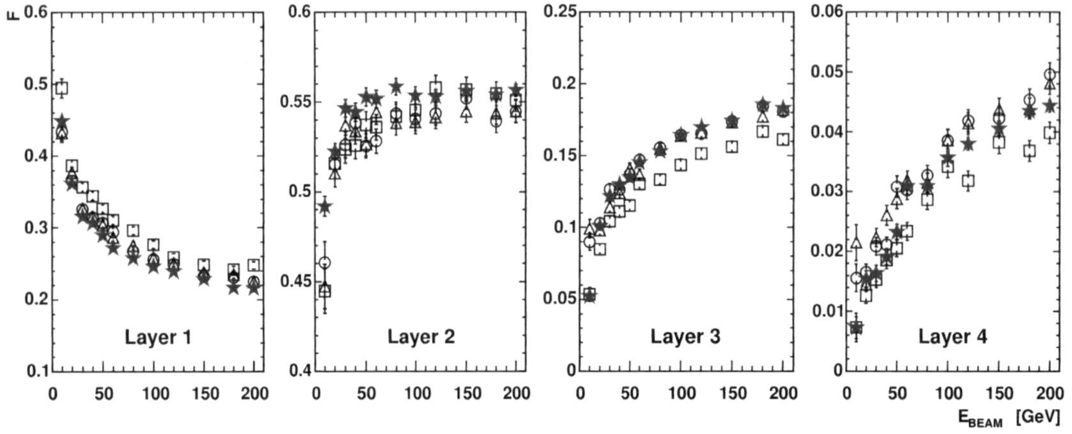

Figure 6. Fractions of the pion shower energy in the various longitudinal layers as a function of the beam energy. Stars mark experimental values. Squares — QGSP, circles — QGSP-BERT, triangles — QGSP-BERT-HP predictions. MC samples: with the Birks' law and after time cuts.

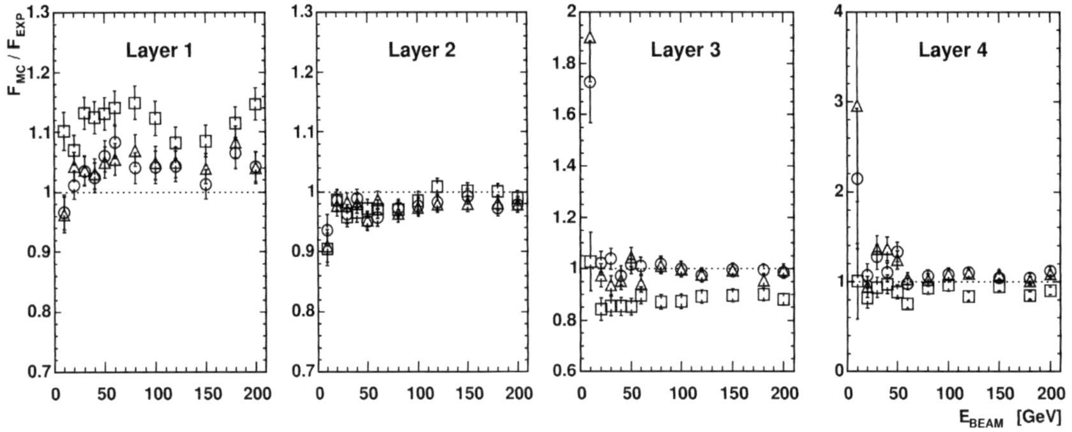

Figure 7. Ratio between simulated and experimental data for the pion energy fractions in the various longitudinal layers as a function of the beam energy. Squares — QGSP, circles — QGSP-BERT, triangles — QGSP-BERT-HP predictions (with the Birks' law and after time cuts).

Additional studies show, that longitudinal profiles do not depend, whether the Birks' law is switched on or off during simulations. Cuts on the time of development of hadronic showers practically do not influence the fraction of energy in HEC layers.

5. Conclusions

GEANT4 based simulations of the HEC testbeam were carried out with different physics lists, namely: QGSP, QGSP-BERT and QGSP-BERT-HP. Influence of the Birks' law and time cuts on the calorimeter performance parameters was investigated. Comparison with experimental results, obtained during beam tests of HEC modules, was done.

Applying cuts on the time of development of hadronic showers (following the experimental procedure of signal measurements in calorimeter cells) has influence on the energy resolution and response for charged pions. Usage of the Birks' law increases the e/π-ratio and the constant term of the pion energy resolution.

After applying time cuts and with the Birks' law switched on: the better description of studied experimental parameters, in total, is given by the QGSP-BERT and QGSP-BERT-HP physics lists. Agreement in the e/π-ratio, close predictions of the resolution at high beam energies and good description of longitudinal profiles of hadronic showers are achieved for these two physics lists.

There are still some open questions. They are addressed to GEANT4 experts. Among them are such requests as a better description of the sampling term of the energy resolution for charged pions with BERT-based physics lists and improvement of the neutron physics in QGSP-BERT.

Acknowledgments

We gratefully acknowledge the encouragement and help of the ATLAS group of the hadronic end-cap calorimeter.

We would like to thank T. Carli for his interest to our work and valuable ideas and suggestions.

We are also grateful to the GEANT4 team, in particular A. Ribon and G. Folger for the help and fruitful discussions.

References

[1] Agostinelli S *et al* 2003 GEANT4 — a simulation toolkit *Nucl. Instr. and Meth.* A **506** 250-303

[2] Rimoldi A *et al* 2008 First Report of the Simulation Optimization Group *ATLAS Note* ATL-SOFT-PUB-2008-002

[3] Kiryunin A E, Oberlack H, Salihagić D, Schacht P and Strizenec P 2006 GEANT4 physics evaluation with testbeam data of the ATLAS hadronic end-cap calorimeter *Nucl. Instr. and Meth.* A **560** 278-90

[4] Kiryunin A E and Salihagić D 2007 Hadronic shower validation experience for the ATLAS end-cap calorimeter *AIP Conf. Proc.* **896** 205-14

[5] Gingrich D M *et al* 2007 Construction, assembly and testing of the ATLAS hadronic end-cap calorimeter *JINST* **2** P05005

[6] Dowler B *et al* 2002 Performance of the ATLAS hadronic end-cap calorimeter in beam tests *Nucl. Instr. and Meth.* A **482** 94-124

[7] Birks J B 1964 *The Theory and Practice of Scintillation Counting* (Pergamon Press)

Comparison of data with Monte Carlo simulations at the ATLAS barrel Combined Testbeam 2004

P. Speckmayer
CERN, CH-1211 Genve 23, Switzerland
E-mail: peter.speckmayer@cern.ch

Abstract. The scheme adopted as baseline by ATLAS for the calibration of hadrons depends strongly on the quality of the description of the data by simulations. In 2004, the calorimeters of the ATLAS barrel region have been exposed to a testbeam in order to evaluate the energy response of pions for the energies ranging from 1 to 350 GeV. For the energy region from 3 to 9 GeV a data analysis with the full systematic uncertainty is available. The data has been compared extensively to GEANT4 simulations. Several combinations of physical models—the so called "physics lists"—are provided by the GEANT4 collaboration and have been evaluated. The best overall description of data is achieved with the physics list QGSP_BERT which describes the energy response of pions within a few percent. QGSP_BERT has been adopted by ATLAS for the simulation of the first data.

Introduction

In 2004 the calorimeters of the ATLAS barrel region have been exposed to a testbeam to evaluate the energy response of pions for energies from 1 to 350 GeV. This data has been compared extensively to simulations. An evaluation of the agreement of the simulations with the data is crucial, since the calibration scheme which ATLAS has adopted as baseline for the hadronic calibration depends strongly on the quality of the simulations since corrections for the non-compensating nature of the calorimeter and for dead material are calculated using MC simulations. An accurate description of the geometry of the testbeam has been simulated with the simulation framework GEANT4. Several physics lists, compilations of models to simulate the interaction of particles of a certain type and a certain energy with matter are provided by the GEANT4 collaboration and their agreement with the data has been evaluated.

1. The Combined Testbeam 2004

In the ATLAS barrel Combined Testbeam 2004 (CTB) a slice of all the sub-detectors of the ATLAS barrel region has been put into the CERN SPS H8 beam line (see fig. 1). Pion beams with energies from 1 GeV to 350 GeV have been shot on the detectors to probe the energy response of the calorimeters. The ATLAS Inner Detector consists of a pixel, a silicon strip detector (SCT) and a transition radiation straw tube tracker (TRT). The calorimeters are liquid argon/lead for the electromagnetic part (LAr) and scintillator tiles/iron for the hadronic part (TileCal). The electromagnetic calorimeter consists of three longitudinal segments (front, middle and back), its electrodes have an accordion shape. A presampler is positioned upstream to correct for the energy which is deposited in the dead material before the calorimeters. The hadronic calorimeter

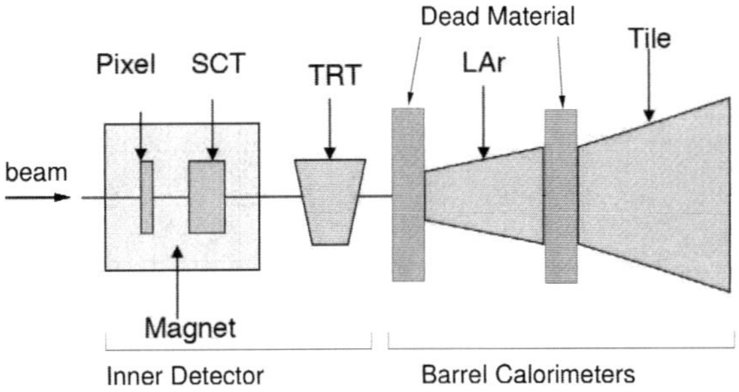

Figure 1: Sketch of the detectors in the ATLAS barrel Combined Testbeam 2004 (CTB). The beam passes the Inner Detector consisting of the Pixel detector, the silicon strip detector (SCT) and the transition radiation straw tube tracker (TRT) before entering the calorimeters. The liquid argon (LAr) calorimeter and the Tile calorimeter are mounted on a rotating table which allows to change the η-position of the calorimeters. The two main areas of dead material (for pions) are before the LAr calorimeter and between the LAr and the Tile calorimeter.

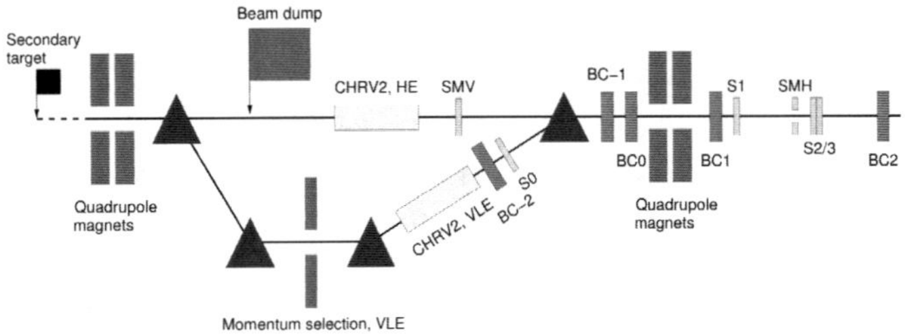

Figure 2: Sketch of the layout of the high energy (HE) and the very low energy (VLE) beam lines for the CTB.

is an iron structure with pockets where scintillator tiles are put. It consists of three longitudinal segments (A, BC and D). The calorimeters were mounted on a rotating table which allowed to change their η-position. For the pion beams, two dead material regions were of importance: the material upstream of the LAr calorimeter (about 0.5λ) for low energy pions, and the dead material region between the LAr and the Tile calorimeter (about 0.5λ) for high energy pions.

In fig. 2 the layout of the high energy (HE) and the very low energy (VLE) beam lines is shown. In the analysis of the VLE data, several analysis cuts have been applied to achieve a good rejection of muons passing through the HE beam line and electron/pion (e/π) separation. The rejection of high energy muons used mainly the timing information of the calorimeters, since these muons are passing directly through the HE beam line and produce their signal earlier or later than the particles which follow the VLE beam line. The effect of muons from pion decays is small and has been included in the systematic uncertainties. The e/π-separation used the signal from the TRT and the Čerenkov detector in the VLE beam line. To achieve a clean pion beam, strong purification cuts had to be applied which lead to a large event rejection of about 100.000 to 1000 events.

In the high energy region, the beam energies 20, 50, 100 and 180 GeV have been analyzed. Electrons are also separated from pions using the TRT. Besides electrons, the pion beam can also

Table 1: Four selected physics lists and their main constituent models. The models are the *quark gluon string* model (QGS), the *Fritiof* model (FTF), the *low energy parametrization* model (LEP) and the *Bertini cascade* model (BERT). For each physics list, the energy region for all models which are used is given. In the overlap region a random decision which model is used is taken.

physics list	low energy region	high energy region
QGSP	0 GeV < LEP < 25 GeV	QGS > 12 GeV
QGSP_BERT	0 GeV < BERT < 9.9 GeV 9.5 GeV < LEP < 25 GeV	QGS > 12 GeV
FTFP	0 GeV < LEP < 5 GeV	FTF > 4 GeV
FTFP_BERT	0 GeV < BERT < 5 GeV	FTF > 4 GeV

contain protons. With the instruments in the beam line pions and protons cannot be separated on an event by event basis. Using the amount of energy deposited in the straws of the TRT can be used to estimate the fraction of the protons in the beam for each beam energy. For the VLE data negative pion beams have been used. There was therefore no proton contamination. The proton fraction has shown to vary from 0% to 75% for the beam energies of 20 GeV to 180 GeV respectively. For the comparisons of data with simulations, simulations of both pions and protons have been produced and mixed according to the estimated fractions of pions and protons. These pion-proton mixes have been compared to data.

2. The simulation framework

The simulations of the CTB have been done with GEANT4, a framework for the simulation of the interaction of particles with matter. Several physics lists are provided by the GEANT4 developers. A physics list defines for all particle types and energy ranges the physical processes and models which are used to simulate the interactions of particles with matter. In table 1 four physics lists and the most important models for the development of hadron showers and their energy ranges are shown. The simulations have been done with GEANT4.7 QGSP_BERT and with GEANT4.91 QGSP_BERT, FTFP and FTFP_BERT. GEANT4.91 is presently the newest version of the simulation framework (Dec. 2007), GEANT4.7 dates from Feb. 2007. GEANT4.91 has improvements of the quasi-elastic and the elastic scattering of nucleons and the multiple scattering.

3. Analysis of the very low energy region for various pseudo-rapidities and comparison with simulations

A full scan from of the beam energies from 3 to 9 GeV and a pseudo-rapidity of $\eta = 0.20$ to 0.65 has been analyzed and a full systematic error analysis has been done[1]. For this data set, the Pixel and the SCT were not available. In this data set the amount of muons from the high energy beam lines was particularly high and therefore it was in addition to the requirements mentioned in sec. 1 necessary to require that in the last layer of the Tile calorimeter a signal compatible with noise is detected. As an example, in fig. 3 the energy response at pseudo-rapidity $\eta = 0.45$ for the beam energies from 3 to 9 GeV and the uniformity for a beam energy of 6 GeV as a function of the pseudo-rapidity are shown. The bars at each point show the uncertainty of the reconstructed energy. The uncertainty ranges from about 8% at 3 GeV to about 1% at 9 GeV. The error is dominated by the statistical uncertainty. The band denotes the uncertainty on the beam energy scale. The energy response is slowly decreasing with the pseudo-rapidity η and

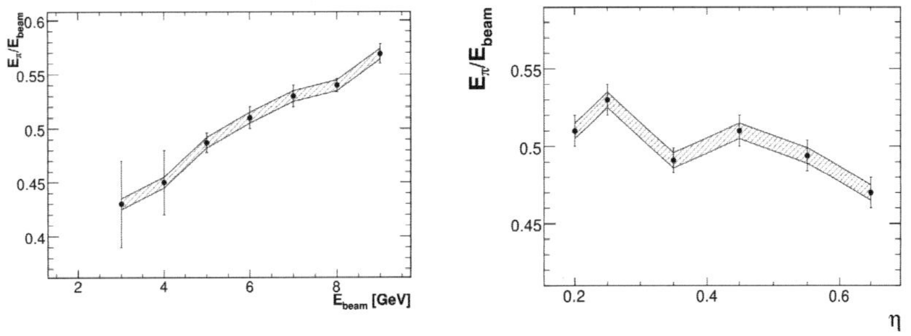

(a) Energy response of pions for $\eta = 0.45$ as a function of the pion energy

(b) Energy response of pions with a beam energy of 6 GeV as a function of η

Figure 3: Examples of the results of the analysis of the VLE data. The energy response for $\eta = 0.45$ as a function of the beam energy (a) and the uniformity over η for a beam energy of 6 GeV (b) are shown. The band around the points denotes the uncertainty on the beam energy scale.

increases with the energy.

For the analyzed η- and energy-scans, the ratio of the energy response of the simulation performed with GEANT4.91 and QGSP_BERT and the data have been calculated. Examples for 3, 6 and 9 GeV are shown in fig. 4. The energy response of the data is described by QGSP_BERT within a few percent.

Examples of the ratio of the resolution of the energy response in simulations and data is shown in fig. 5. The resolution is in the simulations about 10% to 20% too narrow.

4. Analysis of the very low and the high energy data at fixed pseudo-rapidities and comparison with simulations

Fig. 6 shows as example the distributions of the total deposited energy of 100 GeV pions for the data and simulations. The distributions do not have a Gaussian shape. This is not expected, since a part of the energy is deposited in the dead material regions before and between the electromagnetic and hadronic calorimeters. The width of the distribution of GEANT4.7 QGSP_BERT is narrower than in the data. The widths of the distributions of the total deposited energy simulated with GEANT4.91 are wider, such that they agree better with data. The use of the Bertini cascade model causes a shift of the distribution to higher energies which leads to a better agreement with data.

4.1. Linearity and resolution

In fig. 7a the linearities for the energies from 2 to 180 GeV of the simulations done with QGSP, QGSP_BERT, FTFP and FTFP_BERT are compared to data[1]. No systematic uncertainty is included in this comparison. It can be seen, that the Bertini cascade model increases the energy response in the simulations. QGSP_BERT shows the best overall performance for the linearity (within 4%). In particular at low energies, the Fritiof model is not in good agreement with data.

[1] In this analysis a different run period compared to the VLE analysis described in sec 3 has been used. Here, a cut in the last Tile layer was not necessary.

Journal of Physics: Conference Series **160** (2009) 012076 doi:10.1088/1742-6596/160/1/012076

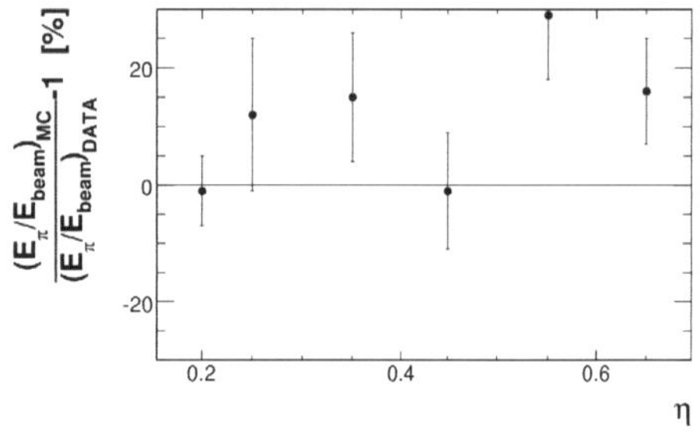

(a) $E_{\text{beam}} = 3$ GeV

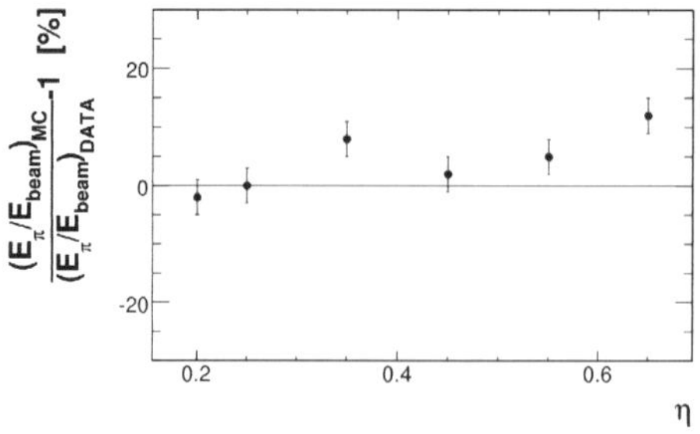

(b) $E_{\text{beam}} = 6$ GeV

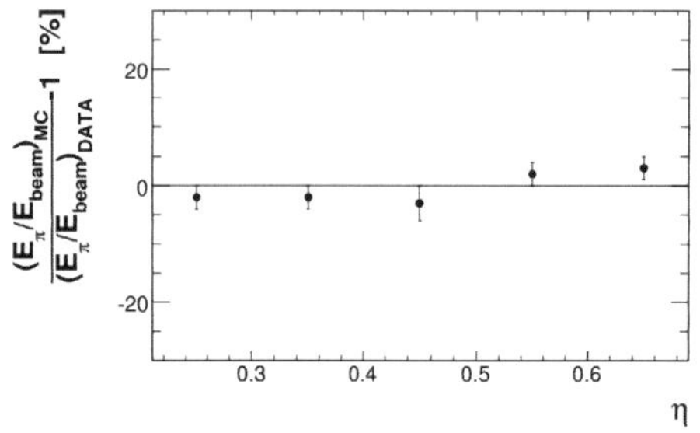

(c) $E_{\text{beam}} = 9$ GeV

Figure 4: Ratio of the energy response of the simulation and the data for η-scans for the beam energies 3, 6 and 9 GeV.

In fig. 7b the resolution for the energies from 2 to 180 GeV are shown for data and simulations with QGSP, QGSP_BERT, FTFP and FTFP_BERT. Physics lists using the Bertini cascade model have a resolution which is too narrow compared to data. The ones without are in better agreement with the data. The disagreement between data and simulations is about 10 to 20%.

4.2. Energy sharing between LAr and Tile calorimeter

Fig. 8 shows the energy sharing between the LAr and the Tile calorimeter for 100 GeV pions. The energy distribution in fig. 8a has a peak at about 0 caused by the events which start to shower after the LAr calorimeter and a broader peak for events which start to shower in the LAr calorimeter. It can be seen that the physics lists using the QGS model are shifted to higher energies. Fritiof describes the energy distribution in the LAr calorimeter better. The Tile calorimeter (see fig. 8b) shows as well two peaks. The one at lower energies comes from events which started to shower before the Tile calorimeter, the peak at higher energies is consists of events which start to shower in the Tile calorimeter. Simulations with QGSP produce too many events with small energies in the Tile calorimeter and too few with high energy. The shower produced with QGSP is too short. Using the Bertini cascade (QGSP_BERT) makes the showers longer, but they are still too short compared to data. FTFP describes well the first peak in the Tile calorimeter, but is shifted to lower energies for the second peak (similar to QGSP). The Bertini cascade shifts the first peak of FTFP_BERT to higher energies compared with data. The high energy peak is well described by both physics lists with the Bertini cascade, FTFP_BERT and QGSP_BERT.

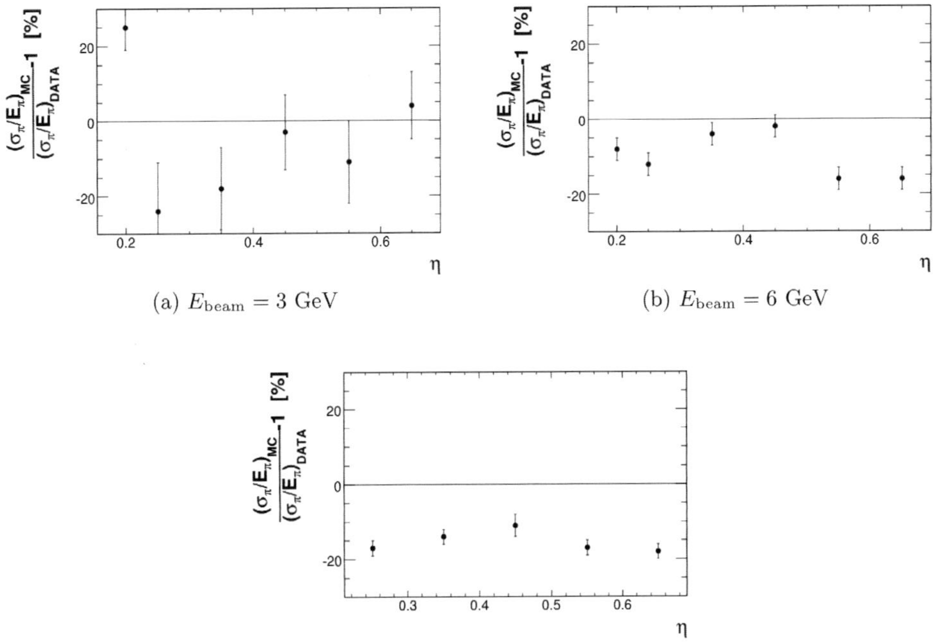

(a) $E_{\text{beam}} = 3$ GeV

(b) $E_{\text{beam}} = 6$ GeV

(c) $E_{\text{beam}} = 9$ GeV

Figure 5: Ratio of the resolution of the energy response of the simulation and the data for η-scans for the beam energies 3, 6 and 9 GeV.

4.3. Barycenter of the showers

In fig. 9 the distribution of the barycenters along the shower axis of the pion showers is shown. Three regions (peaks) can be identified: the first one below 1000 mm, the second one between 1000 and 1500 mm and the third one above 1500 mm. The first peak is formed by events where most of the energy is deposited in the LAr calorimeter. The second peak is formed by events where most of the energy is deposited in the Tile calorimeter. The third peak is created by events where the shower is nearly completely contained in the Tile calorimeter.

Simulations with QGSP produce too many events where the barycenters are too early. Using the Bertini cascade (QGSP_BERT) less events with a barycenter which is too low and more events with a larger barycenter are observed, but it is still not in agreement with the data. FTFP follows the data well, all three peaks are well described. FTFP_BERT shifts the barycenter to higher values, especially for the first peak.

Conclusions

The calorimeters from the barrel region of ATLAS have been exposed to a testbeam (CTB). A data analysis with the full systematic uncertainty is available for 3 to 9 GeV. The pion response is described within a few percent by simulations with the physics list QGSP_BERT, but the energy resolution is about 10 to 20% too narrow compared to data. The Bertini cascade model increases the energy response, but makes the resolution of the energy response narrower. The barycenter of showers simulated with QGSP is too early compared to data. This is improved by using the Bertini cascade (QGSP_BERT), but the barycenter is still too early compared to the

Journal of Physics: Conference Series **160** (2009) 012076 doi:10.1088/1742-6596/160/1/012076

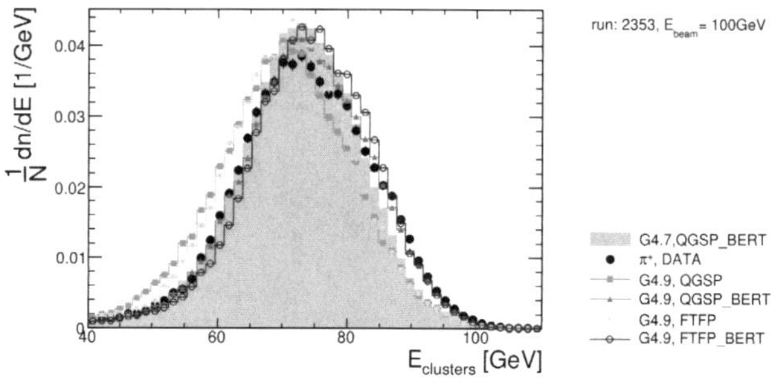

Figure 6: Total deposited energy for 100 GeV pions in the data and in selected physics lists. The width of the distribution of GEANT4.7 QGSP_BERT is narrower than data. With the physics lists from GEANT4.91 the distributions have a width which is comparable to data. The physics lists where cascade models are used the distribution of the total deposited energy is shifted to higher energies, which is in better agreement with the data.

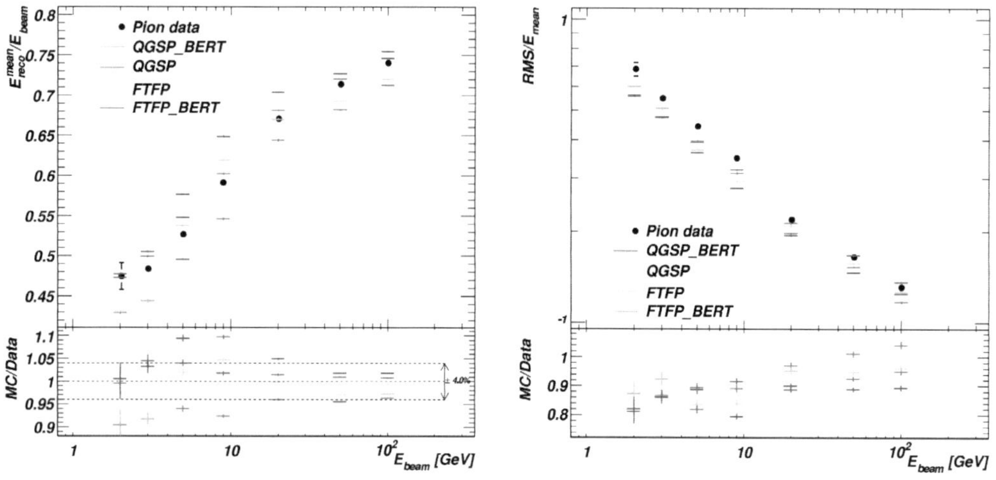

(a) Mean reconstructed energy

(b) Relative resolution

Figure 7: Fig. (a) shows the mean reconstructed energy (a) with respect to the beam energy as reconstructed in data and simulations with the physics lists QGSP, QGSP_BERT, FTFP and FTFP_BERT. The Bertini cascade model increases the energy response. QGSP_BERT shows the best overall performance for the linearity (within 4%). Fig. (b) shows the relative resolution of simulations with the physics lists QGSP, QGSP_BERT, FTFP and FTFP_BERT and of data. The Bertini cascade model makes the resolution narrower (too narrow compared to data). A disagreement of about 10 to 20% in resolution is seen between simulation and data.

data. The Fritiof physics list (FTFP) describes well the distribution of the barycenter of the showers. Adding the Bertini cascade to the Fritiof physics list (FTFP_BERT) produces showers where the barycenter is too late.

As a result from the extensive comparisons of the simulations with the data the physics list

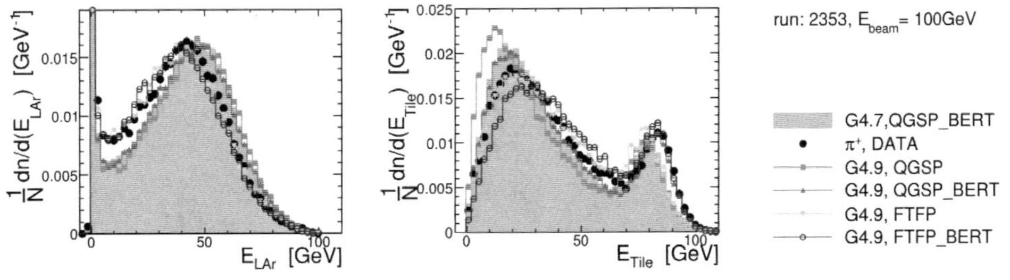

(a) E_{LAr} (b) E_{Tile}

Figure 8: Energy sharing between the LAr and the Tile calorimeter for 100 GeV pions. In (a) the simulations with the physics lists using the QGS model are shifted to higher energies. Fritiof describes the energy distribution in the LAr calorimeter better. Simulations with QGSP produce too many events with small energies in the Tile calorimeter (b) and too few with high energy. Using the Bertini cascade (QGSP_BERT) produces less events with a low energy and more with a high energy, but still too few compared to data. FTFP describes well the first peak in the Tile calorimeter, but is shifted to lower energies for the second peak (similar to QGSP). The Bertini cascade shifts the first peak of FTFP_BERT to higher energies compared with data. The high energy peak is well described by both physics lists with the Bertini cascade, FTFP_BERT and QGSP_BERT.

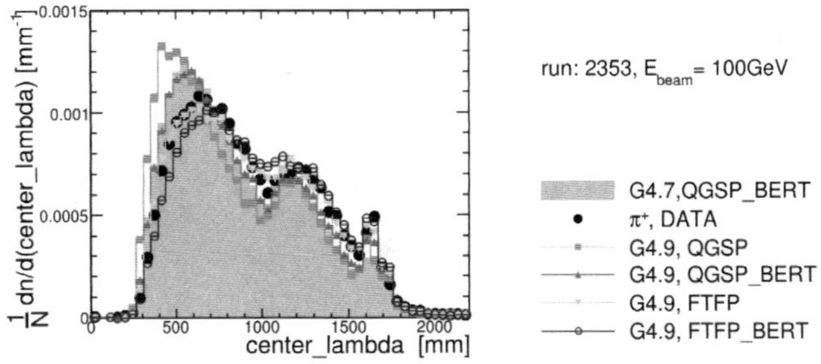

Figure 9: The barycenter (center_lambda) along the axis of showers of the showers produced by 100 GeV pions. The distribution has three peaks. The first is produced by events where most of the energy is deposited in the LAr calorimeter, the second is produced by events depositing most of the energy in the Tile calorimeter and the third peak is produced by events whose shower is nearly completely contained in the Tile calorimeter.

QGSP_BERT has been identified to describe the data best. It therefore has been chosen for the simulation of the first ATLAS data.

References

[1] Cavasinni V, Giangiobbe V and Santoni C 2008 *Study of the response of the central electromagnetic and hadronic calorimeters to pions of energies from 3 to 9 GeV*. Geneva, CERN, ATL-COM-TILECAL-2008-001, 36 p

Journal of Physics: Conference Series **160** (2009) 012077 doi:10.1088/1742-6596/160/1/012077

CALICE scintillator HCAL - electromagnetic and hadronic shower analysis

Erika Garutti (on behalf of the CALICE collaboration)
DESY, Notkestr. 85, 22607 Hamburg, Germany
E-mail: erika.garutti@desy.de

Abstract. The CALICE test beam calorimeters operated at the CERN SPS facility have collected a large sample of hadronic and electromagnetic showers with unprecedented granularity.
The calorimeters and beam line instrumentation are modeled in the MOKKA framework and full simulation of the test beam experiment is performed using GEANT4. Digitization is applied to each calorimeter to account for detector effects and noise. The scintillator-based hadronic calorimeter (AHCAL) is the center of the studies reported in this paper.
The most important validation of the detector modeling and calibration chain is the test of the calorimeter response linearity and resolution for a large range on incident beam energies. Electromagnetic showers are the most demanding test since the energy deposited per single tile in an electromagnetic shower is larger than in a hadronic shower for the same beam energy. Results of the calorimeter response to muons and positrons are discussed and compared to Monte Carlo (MC).
The analysis of single pion showers, recorded with the AHCAL offer the unique possibility to test hadronic shower models using a number of different observables. Total energy and longitudinal profiles have been studied at first and compared with two available shower models in GEANT4. Results of the effect of the shower leakage on the reconstructed energy and energy resolution are presented. Furthermore, studies of shower separation are performed using event mixing techniques. Shower separation is critical for the performance of modern particle flow algorithms, which here for the first time can be tested on experimental data.

1. Introduction

Three highly-granular prototype calorimeters constructed by the CALICE collaboration have been tested at the CERN SPS during 2006 and 2007. The Silicon-Tungsten electromagnetic calorimeter (ECAL, [1]) with 1×1 cm^2 readout pads; the Scintillator-Steel hadron calorimeter (AHCAL, [2]) with 3×3 cm^2 tiles analogically readout; and the Scintillator-Steel Tail-Catcher and Muon Tracker (TCMT, [3]) strip detector have been operated with a common data acquisition providing the readout of $\sim$ 18000 channels. AHCAL and TCMT share not only the same sampling material but also the same photo-detector readout via SiPM [4]. Calibration and corrections of these two detectors are very similar.
The AHCAL is a 38 layers sampling calorimeter, which alternates 2 cm of Steel to active layers assembled with 0.5 cm thick scintillator tiles. About 8000 SiPM are used for the individual readout or each tile. Details on the AHCAL calibration procedure are presented in [5].
In the first year of operation at CERN only 23 layers of the AHCAL have been instrumented. Full commissioning was achieved for the CERN run in 2007. Data from both periods is presented.

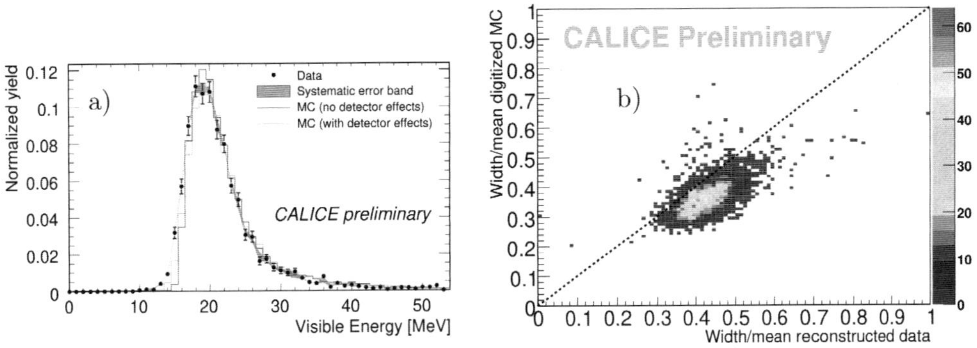

Figure 1. a) Total visible energy deposited by a 120 GeV muon in the AHCAL (points), compared to true and digitized MC. b) Cell-by-cell correlation of the MIP resolution between data and digitized MC.

The validation of the calibration procedure and the MC digitization is obtained by the study of the calorimeter response to muons and electromagnetic showers.
The analysis of single pion showers offers the unique possibility to validate MC hadronic shower models using a large number of well defined observables. Total energy and longitudinal profiles with very high resolution are studied in detail and compared with the available GEANT4 models. The ultimate goal of highly granular calorimeter is shower separation in a single event. To test the CALICE calorimeter system for shower separation quality two single pion events are overlaid and reconstructed with a track-wise clustering algorithm. The quality of shower separation is studied as a function of the distance between the two pions. Furthermore, a naive particle flow algorithm can be applied to the reconstructed clusters assuming one of the two pions to be charged and, therefore, having a perfectly determined momentum from the tracking system. The energy resolution of the second pion (playing the role of the neutral hadron here) is investigated.

2. Validation of MC digitization

The three calorimeters (ECAL, AHCAL and TCMT), as well as the full instrumentation of the SPS-H6 beam-line at CERN, have been implemented in MOKKA [6]. The true GEANT4 response of the detectors is digitized to include the SiPM response function, Poisson smearing and noise. Optical cross-talk between adjacent scintillator tiles is simulated.
In order to validate the digitization steps data is compared to MC for well understood physics processes sensitive the calibration procedure steps. All calorimeter cells are equalized using the signal from minimum ionizing particles (MIP). The response of the whole calorimeter and of the single cells to a MIP is investigated. To test the effect of non-linearity correction and the entire calibration procedure electromagnetic showers are investigated and compared to MC.

2.1. Simulation of muons

The energy deposited by a muon traversing 23 layers of the the AHCAL is shown in Fig. 1a. The distribution is compared to simulation both with and without digitization. In both cases the MC well reproduced the mean value and the spread of the distribution. This indicates that the Landau spread of the MIP energy in the 23 tiles is larger than the spread introduced by detector effects, i.e. Poisson smearing, noise, optical cross-talk. At the MIP energy amplitude the effect of non-linearity of SiPM is negligible.

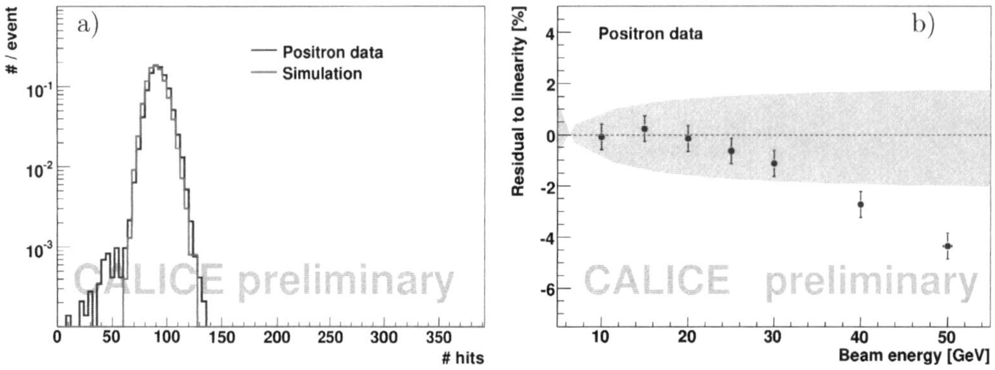

Figure 2. a) Number of reconstructed hits in data and digitized simulation for 10 GeV positrons. Reconstructed data is shown in blue, digitized and reconstructed Monte Carlo in red. The excess in data at ~50 hits is due to muon contamination in the beam. b) Residual to a linear fit to the AHCAL response in the energy range 10-20 GeV. The green band indicates the influence of the calibration uncertainties on the saturation scale on the linearity hypothesis.

Fig. 1b shows cell-by-cell the MIP resolution correlation between data and digitized MC. The MIP width is influenced by Poisson smearing and noise. The correlation is good, but the MC width is about 10% smaller than for the data. This effect may be related to the tile non-uniformity, which is not yet included in the digitization.

2.2. Simulation of electromagnetic showers

For the analysis of electromagnetic showers a threshold cut of 0.5 MIP is applied to the energy deposited in each cell, after calibration. The total number of cells with hits above threshold is very sensitive to the MIP calibration and to the modeling of the noise and optical cross-talk between cells. Fig. 2a shows the total number of hits recorded in the AHCAL for a 10 GeV positron shower. The agreement is very good. The small excess at low number of hits in the data is due to remaining muon contamination in the positron sample, which is not simulated in the MC. For 50 GeV positron momentum the number of hits in MC is systematically 5% lower than in data. Besides the width of the MIP peak, the number of hits above threshold is influenced by light cross-talk between scintillator tiles. This value is assumed to be 10% summed up over all neighbors cells. Further studies will have to show if this disagreement can be improved tuning the amount of light cross-talk.

The integral response to single particles is reconstructed as the energy sum of all hits above threshold over the whole detector. The energy of positrons in the range 10-50 GeV has been reconstructed. Fig. 2b presents the residual to a linear fit in the energy range 10-20 GeV. The error bars include statistics and an uncorrelated systematics of 0.5% from the fit to the total energy distribution. The error band corresponds to the correlated systematic uncertainty of about 5% on the saturation scale. The reconstructed response is linear within errors up to 30 GeV beam momentum. A significant remaining non-linearity of about 4% is observed for 50 GeV positrons.

The energy resolution for positrons is evaluated for data and digitized MC and presented in Fig. 3. A sum in quadrature of noise ($\propto \frac{1}{E}$), stochastic ($\propto \frac{1}{\sqrt{E}}$) and constant terms is used to fit the points. The noise term is fixed from random trigger events to 2 MIP. The stochastic terms in data and MC are $(22.6\pm0.1_{fit}\pm0.4_{calib})\%/\sqrt{E}$ and $(20.9\pm0.3_{fit})\%/\sqrt{E}$, respectively. While

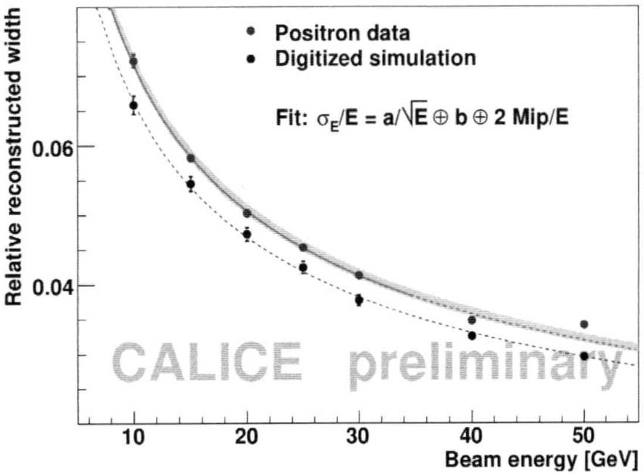

Figure 3. Relative width of the reconstructed energy as a function of beam energy. The resolution is fitted excluding the data points at 40 and 50 GeV because of remaining non-linearity at high energies. The fit results are shown as red (black) line for data (simulation). The green band indicates the influence of calibration uncertainties on the saturation scale. Uncertainties on the MIP scale cancel in the width over mean ratio and are therefore not included.

both constant terms are consistent with zero with an error of 1.5% for data and 2.2% for MC. This preliminary data/MC comparison for electromagnetic processes is considered satisfactory and sufficient to move on to hadronic analysis. The remaining deviations are smaller than 10% and set the scale for the comparison data/MC of hadronic processes.

3. Hadronic showers response in the AHCAL

Hadronic data is calibrated, as electromagnetic data, to the MIP scale.
Pion events collected in October 2006 with a partially instrumented AHCAL are presented. Beam momenta from 6 GeV to 80 GeV have been analyzed. The range 6-20 GeV is a negative pion sample; the range 30-80 GeV is a positive pion sample. No correction for proton contamination in the positive pion sample is applied.
Pions with only MIP-like interaction in the ECAL are selected and the energy deposited in the AHCAL and TCMT are summed to reconstruct contained showers. The response of the calorimeter is shown in Fig. 4a). Data is compared to two MC models. The LHEP and QGSP_BERT models have been chosen, for being the two most discrepant models in terms of total energy deposition. The MC is digitized and reconstructed with the same sampling factors and MIP/GeV conversion factor as for the data.
A discrepancy in the energy scale is observed of about 4% (20%) between data and LHEP (QGSP_BERT).
The residuals to linearity are extracted fitting data and MC in the range 6-20 GeV. Above this range data is consistent with the linear hypothesis at the 2% level, while MC has a positive deviation of 6% (4%) for LHEP (QGSP_BERT). This could be attributed to the non-compensating nature of the calorimeter. The remaining difference between data and MC can be the effect of not properly corrected SiPM non-linearity.
The longitudinal shower profiles of a 20 GeV pion shower is displayed in Fig. 5a and compared to the two MC models. The shower modeled in QGSP_BERT appears to start later than in data and LHEP.

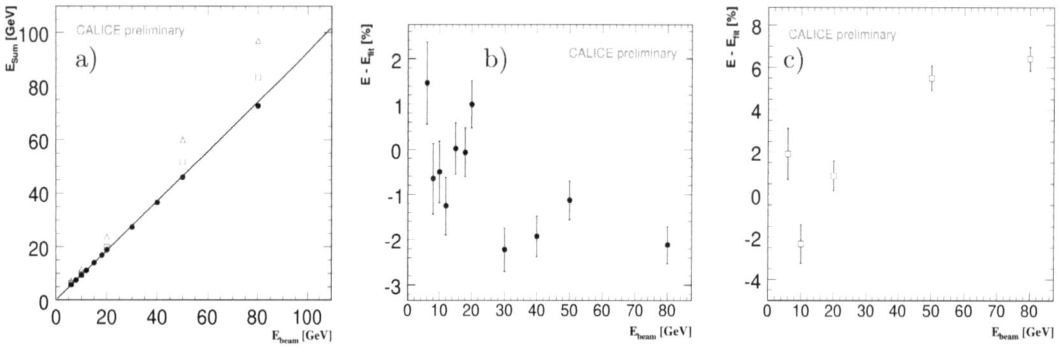

Figure 4. *a) Linearity of the AHCAL and TCMT combined response to pions. Black circles are data, open squares digitized LHEP and open triangles are digitized QGSP_BERT. Residual from a linear fit to the total energy in the range 6-20 GeV for data (b) and LHEP (c).*

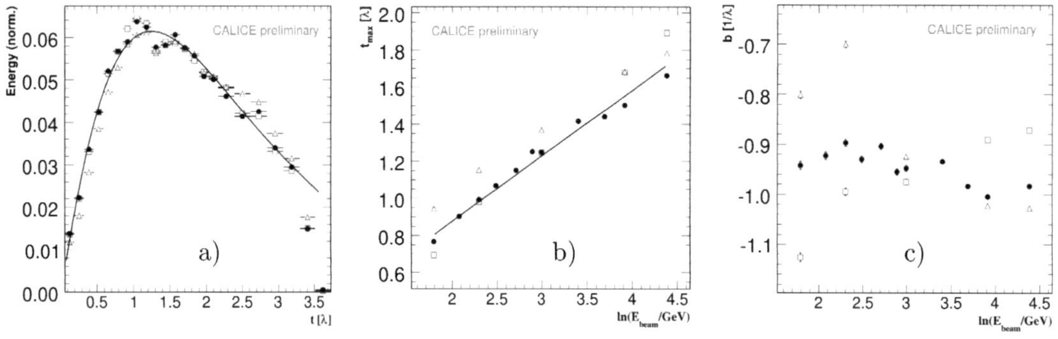

Figure 5. *a) Longitudinal shower profile of 20 GeV pions. The curve is a fit to the data points, excluding the last two layers which are known to have low efficiency.*
Energy dependence of the longitudinal shower maximum (b) and shower attenuation parameter (c). Black circles are data, open squares digitized LHEP and open triangles are digitized QGSP_BERT.

The shower maximum is determined with a fit to this distribution using the parameterization

$$\frac{dE}{dt} = kt^{a-1} \cdot e^{bt}, \tag{1}$$

where t is the depth in the calorimeter expressed in units of λ_0. The last two layers of the AHCAL are not included in the fit since they are known to have low efficiency. The energy dependence of the shower maximum $t_{max} = \frac{a-1}{b}$, and the shower attenuation b, are presented in Fig. 5b and Fig. 5c. The expected logarithmic increase of t_{max} with energy is observed. The value of $b < 1$ is expected for pion showers.
The energy resolution obtained with the combined AHCAL and TCMT is presented in Fig. 6 for data and MC. The data lies in between the two models chosen for comparison, illustrating the

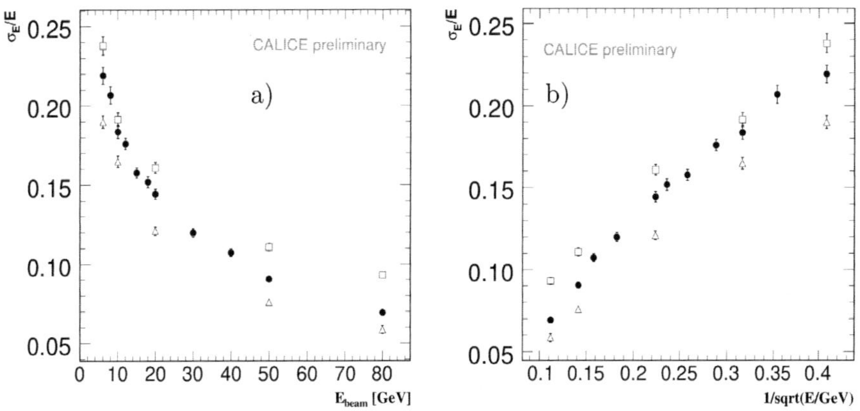

Figure 6. *Energy resolution for pions in the combined AHCAL and TCMT detector as a function of the beam energy, E (a) and of $1/\sqrt{E}$ (b). Black circles are data, open squares digitized LHEP and open triangles are digitized QGSP_BERT. The errors are statistical only.*

large variations in the MC predictions. This plot is, at present, only a qualitative comparison.

4. Shower leakage

Hadronic shower development suffers from much larger fluctuations than electromagnetic ones. In particular the shower starting point varies over a much larger range in the calorimeter depth from event to event. The determination of the shower starting point is crucial to study hadronic shower shapes in detail and to address the problem of shower leakage. The definition of shower starting point is based on the topological resolution of the AHCAL. The track of the incoming pion is followed through the first layers of the detector until its first interaction takes place and an energy larger than minimum ionizing is released. The starting point identification method can be generalized to neutral hadron showers, but this study is restricted to charged hadrons.
Data is presented which was collected in a combined run of AHCAL and TCMT in 2007 at CERN. The fully instrumented AHCAL was rotated at an angle of 60 degrees with respect to the incoming beam. This gives the maximum depth of active detector achievable with the setup.
Fig. 7a shows the position of shower start in units of interaction length, λ_0. The decreasing exponential slope of 0.88 λ_0 agrees with the value extracted using a completely independent method and shown in Fig. 5c.
Fig. 7b shows the longitudinal profile for 10 GeV pion showers. Knowing for each event the shower starting point, it is possible to shift the single event profile to the same starting point. This is shown by the open symbols. The longitudinal shower profile becomes much shorter after correcting event-by-event for the variation of the shower starting point. A more meaningful and independent data to MC comparison will be possible using these profiles.
The knowledge of the shower starting point has been used to study the effect of shower leakage on the energy reconstructed and energy resolution of the calorimeter, shown in Fig. 8a and Fig. 8b, respectively. Fig. 8a has been used to determine a correction for the energy deposited as a function of the shower starting point. If this correction is applied to pion showers of 10 GeV

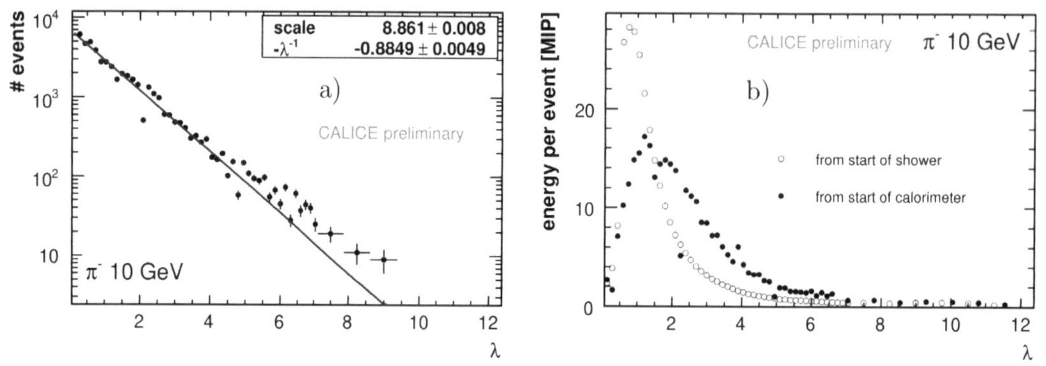

Figure 7. *a) Position of shower start in the calorimeter as a function of the number of interaction lengths. b) Longitudinal shower profile in the calorimeter determined before and after the shift, event by event, to the shower starting point, for 10 GeV pion showers.*

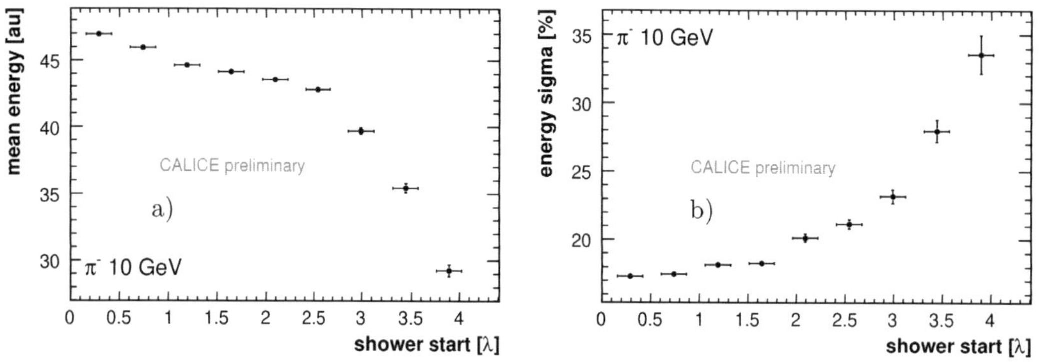

Figure 8. *a) Mean energy deposited in the calorimeter as a function of the shower starting point. b) Energy resolution as a function of the shower starting point, for 10 GeV pion showers.*

the mean value of the reconstructed energy better agrees with the expected beam momentum, but the energy resolution does not improve. This correction has no effect on the resolution of single particles, but can be potentially useful in the reconstruction of jets.

5. Showers separation

During test beam data taking single pion events are recorded. These events are selected and overlaid offline for the study of showers separation in the calorimeter.

Calibrated pion events which interact in the ECAL as MIP-like particles are selected, which have no energy leaking to the TCMT. These fully contained showers in the AHCAL are selected according to their relative distance and overlaid on a cell basis. The energy of the overlaid event in each cell is the sum of the single pion energies. This method over-estimates the noise of the calorimeter by double counting it from the two events.

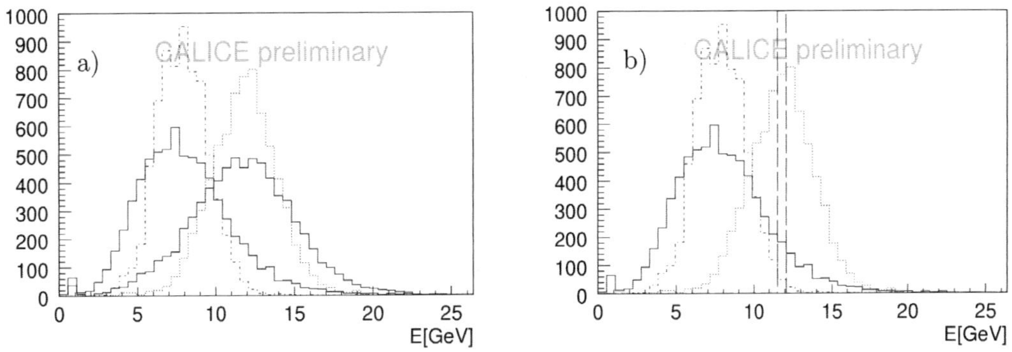

Figure 9. *Energy distribution of reconstructed clusters $E_{cluster}$ (solid lines) for and overlaid event with a 8 GeV and a 12 GeV pions. The dash-dotted and dotted lines are the calorimeter energy distributions for single events, E^1_{calo} and E^2_{calo}. In b) the cluster associated with the charged hadron has been replaced by the beam momentum, E_{track} (dashed line).*

The overlaid events are then input to a clustering algorithm which identifies the two clusters and measures their total energy. A track-wise clustering algorithm is used, which is an algorithm developed for particle flow in the full ILC detector [7]. The parameters of the code have been tuned for the AHCAL geometry to maximize the number of two identified clusters and minimize the difference between reconstructed and true cluster energy in the calorimeter.
An example is presented where two pions of 8 GeV and 12 GeV respectively are overlaid. The reconstructed cluster energies $E_{cluster}$ are shown in Fig. 9a as compared to the energy reconstructed in the calorimeter for the single pions, E_{calo}. One can see a broadening of the energy spectra of both pions due to the clustering procedure. The efficiency of two cluster separation with this method is more than 90% for two particles at a distance of 10 cm.

5.1. Naïve Particle Flow

Given the reconstructed clusters, an assumption can be made to test a naïve Particle Flow approach. One of the two pions is considered to be a charged hadron; its energy measured in the calorimeter is replaced by the tracker information, see Fig. 9b. For each overlaid event a "track" is defined by choosing the position of one of the mip stubs in the ECAL. The track's energy, E_{track}, is the beam momentum of the corresponding event. The track is assigned to the cluster with cluster position (projection of the center of gravity to the $x - y$ plane) closest to the track.
To quantify the result of the particle flow approach compared to the ideal particle flow a quantity has been defined, the efficiency of shower separation. For events where exactly two clusters have been found the cluster energy $E_{cluster}$ in an interval of $\pm\, 3\sigma$ of E^1_{calo} is integrated and compared to E^1_{calo}. The ratio of these two quantities defines the particle flow efficiency,

$$eff_{Pflow} = \frac{\int_{-3\sigma}^{+3\sigma} E_{cluster}}{\int_{-\infty}^{+\infty} E^1_{calo}}. \quad (2)$$

The ideal particle flow case corresponds to infinitely separated clusters for which $E_{cluster} = E^1_{calo}$. Fig. 10 shows the efficiency of particle flow for 6 GeV charged particle (track). Neutral particles

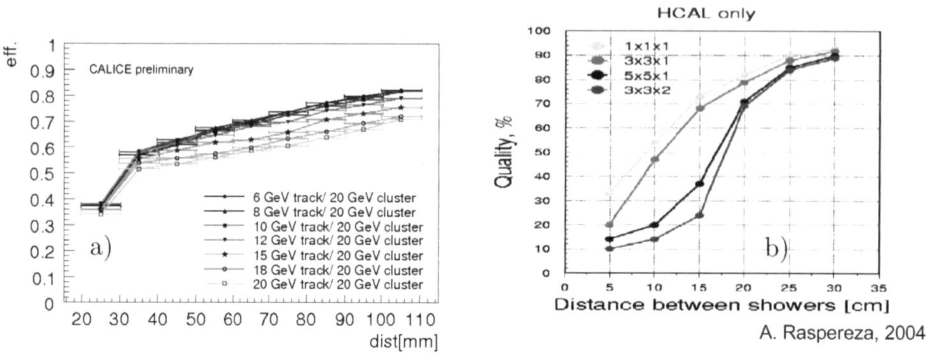

Figure 10. *a) Efficiency of particle flow (defined in text) as a function of particles separation, for 20 GeV cluster. b) MC studies of shower separation for various AHCAL tile geometries.*

with energy between 6 and 20 GeV are separated from the charged one. The efficiency increases for larger distances between particles and for lower energy of the neutral particle. The test of larger distances is not possible using the currently analyzed data set.

This result can be compared to MC simulation studies [8] performed for the optimization of the AHCAL cell size. Comparing to the 3x3x1 cm^3 configuration (which is the one the AHCAL has been realized with) there is good qualitative agreement between data and MC trend. It has to be noted that in these studies one neutral and one charged particle were simulated, whereas in data we use two charged pions. This may be an advantage for the clustering algorithm since the MIP-like stab of the charged particle considered neutral is not removed before clustering. Detailed comparison with MC is still needed to validate particle flow in simulation with real data.

6. Conclusions

The calibration procedure of the AHCAL CALICE prototype has been tested on electromanetic data. A linear response up to 30 GeV is obtained after correction of the SiPM non-linear response. A deviation of 4% from linearity is still visible for 50 GeV positrons, which will require more detailed studies.

The MC digitization procedure includes the effects of the SiPM response function, Poisson smearing and noise, and optical cross-talk between adjacent scintillator tiles. The digitization has been validated with muon and positron data at the level of better than 10%.

This analysis shows a first comparison of CALICE hadron data to MC models including digitization. First results indicate that the digitized MC models agree in trend with the data in all observables studied. For the extraction of the total energy from the visible events it is observed that the two MC models have different MIP to GeV conversion factors or different sampling factors for hadrons. This effect will have to be investigated in more detail.

The deviation from linear behaviour at high energy is different in data and MC. While data is consistent with linearity up to 80 GeV at the 2.5% level, both MC models indicate a positive deviation. This effect could be due to an increasing e/π at high energy, which is either not observed in data or compensated by a remaining non-linearity effect.

The two independent analysis of the longitudinal shower profile and the shower starting point determination yield consistent values for the interaction length of the calorimeter for pions to

be 0.88 λ_0.
The first application of a Particle Flow type of reconstruction to real data has been presented. This type of analysis is a powerful tool to compare various clustering algorithms in a realistic hadronic shower environment. Once these results have been validated by comparison to MC prediction they will be an important step in the verification of Particle Flow performance for physics events.
The shower separation efficiency obtained from these preliminary studies is consistent with the MC studies which have driven the design of the AHCAL protorype.

Acknowledgments
This work is supported by DESY, by the University of Hamburg and by the Helmholtz-Nachwuchsgruppen fond VH-NG-206.

References
[1] J. Repond *et al.*,"Design and Commmissioning of the Physics Prototype of a SiW Electromagnetic Calorimeter for the International Linear Collider", submitted to JINST.
[2] G. Eigen, "The Calice scintillator HCAL testbeam prototype", in Proc. of "XII International Conference on Calorimetry in High Energy Physics", Chicago, IL, 2006.
[3] D. Chakraborty, "The Tail-Cathcer/Muon Tracker for the CALICE Test Beam", in Proc. of the "2005 International Linear Collider Workshop (LCWS05)", Stanford, CA, 2005.
[4] P. Buzhan, B. Dolgoshein, L. Filatov, A. Ilyin, V. Kantzerov, V. Kaplin, A. Karakash, F. Kayumov, S. Klemin, E. Popova, S. Smirnov, Nucl. Instrum. Meth. A **504** (2003) 48 (2003).
[5] A. Lucaci-Timoce, these proceedings.
[6] "MOKKA - a detailed Geant4 simulation for the International Linear Collider detectors", http://polywww.in2p3.fr/activites/physique/geant4/tesla/www/mokka/mokka.html
[7] V. Morgunov, A. Raspereza, "Novel 3D clustering algorithm and two particle separation with tile HCAL", ILC note: LC-TOOL-2004-022, http://www-flc.desy.de/lcnotes/.
[8] V. Morgunov, A. Raspereza, ECFA workshop, Montpellier, 13-16 Nov. 2004.

Journal of Physics: Conference Series **160** (2009) 012078 doi:10.1088/1742-6596/160/1/012078

Performance of the ATLAS Liquid Argon Endcap Calorimeter in Beam Tests

Pavol Strizenec
IEP SAS, Watsonova 47, SK-04001 Kosice, Slovakia
E-mail: strizene@saske.sk

Andrey Minaenko
IHEP, RU-142 284 Protvino, Russian Federation
E-mail: minaenko@ihep.ru

(for the ATLAS LAr Endcap group)

Abstract. The pseudo-rapidity region $2.5 < |\eta| < 4.0$ in ATLAS is a particularly complex transition zone between the endcap and forward calorimeters. A set-up consisting of 1/4 respectively 1/8 of the full azimuthal acceptance of the ATLAS liquid argon endcap and forward calorimeters has been exposed to beams of electrons, pions and muons in the energy range $E \leq 200$ GeV at the CERN SPS. Data have been taken in the endcap and forward calorimeter regions as well as in the transition region. This beam test set-up corresponds very closely to the geometry and support structures in ATLAS. A detailed study of the performance in the endcap and forward calorimeter regions is described. The data are compared with MC simulations based on GEANT 4 models.

1. Introduction

This talk summarizes the main performance parameters of LAr Endcap calorimeters in a beam-test, which was carried out in 2004. It is a continuation of calibration studies of individual parts - electromagnetic (EMEC) [1, 2, 3, 4], hadronic endcap (HEC) [5, 6] and forward calorimeters (FCal) [7]. This beam-test cover particularly difficult forward region $2.5 < |\eta| < 4.0$ (the transition from the electromagnetic endcap calorimeter EMEC and hadronic endcap calorimeter HEC to the forward calorimeter FCal). The set-up was as close as possible to the final ATLAS detector, also with respect to the support structures and dead material distribution. It should be stressed that the set-up reproduced the ATLAS projective geometry at one $|\eta|$ point, but not over the entire $|\eta|$ region. The beam was incident at an angle of 4.2° onto the face of the calorimeters, corresponding to an average $|\eta|$ of 3.2.

2. Setup and Data

The beam tests have been carried out in the H6 beamline at the CERN SPS, which provides hadrons, electrons or muons in the energy range $6\,\text{GeV} \leq \text{E} \leq 200\,\text{GeV}$. The particle trigger was based on the coincidence of three scintillation counters S1, S2 and S3, and in order to be able

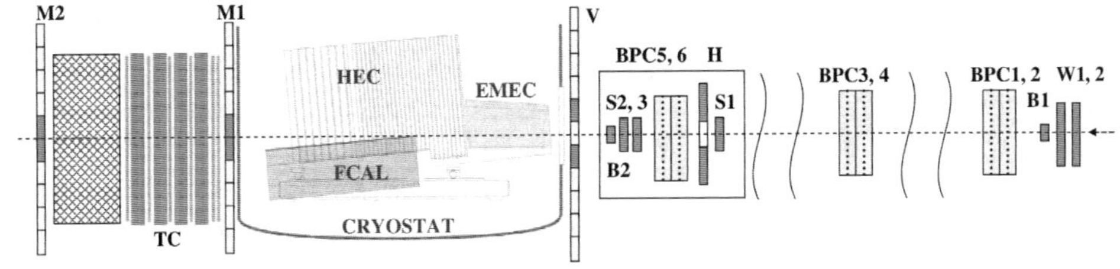

Figure 1. Schematic view of the general beam test set-up. Shown are the modules in the cryostat and the beam instrumentation used: multiwire proportional chambers (BPC), scintillation counters (S, B) and scintillator walls (V, M1, M2). The beam moves from right to left.

to adjust for the various vertical beam positions, these counters - as well as the beam chambers - were mounted on a vertically movable table. The impact position and angle of beam particles were derived using hit information from 6 multiwire proportional chambers (BPC-1 - BPC-6), each with a vertical (y) and horizontal (x) read-out plane. The positions of the BPC wire planes along the beamline were surveyed prior to the beam test. Track fits were performed separately in the $x - z$ and $y - z$ planes. Average track residuals calculated run-by-run for the full run period are typically below $\sim 200\ \mu$m, in both coordinates. This tracking system allowed determination of the particle impact angle with a precision of $\sim 10\ \mu$rad and the impact point with a precision of $\sim 100\ \mu$m (at high energies).

The beam enters through the cryostat window from right at a nominal position of $y = 0$ cm. Shown are the inner EMEC (front), the HEC front wheel and rear wheel modules as well as the FCal1 and FCal2 modules (below the HEC modules) inside a cryostat. In addition a cold tail catcher is placed right behind the last FCal module in order to measure any leakage beyond the FCal2 module. In a similar way the energy leakage for the rear wheel HEC modules is measured in a warm tail-catcher (TC) placed outside the cryostat.

The electromagnetic endcap calorimeter (EMEC) is a liquid argon (LAr) sampling calorimeter with lead as absorber material. The liquid argon gap between the absorber plates increases with the radius and the accordion wave amplitude. For this beam test a module of the inner wheel has been rebuilt using leftover electrodes available. Except for the missing outermost electrode in azimuth (ϕ), it corresponds exactly to a module from the series production.

The hadronic endcap calorimeter (HEC) [5, 6] is a liquid argon sampling calorimeter with flat copper absorber plates, structured longitudinally in two separate wheels. Longitudinally each wheel has two read-out sections. Due to the limitation in space, a special set of 8 front and 8 rear HEC modules has been built for this beam test. These had a reduced coverage in $|\eta|$, corresponding to $2.1 < |\eta| < 3.2$. Also the rear modules were of half size in depth.

The forward calorimeters (FCal) [8] cover in ATLAS the region $3.1 < |\eta| < 4.9$. Each endcap is made of three modules, one behind the other, providing three depth segments. All modules have cylindrical electrodes made of a solid rod within a thin-walled tube. The gap between the rod and the tube fills with liquid argon. The electrode rods, tubes, and matrix of the electromagnetic module are made of copper. The next two modules are hadronic calorimeters. Their electrode rods are solid tungsten, the tubes are copper, and the matrix is sintered tungsten alloy. The FCal modules used in this beam test are engineering prototypes of the FCal1 and FCal2 ATLAS modules and are those used in the 1998 beam tests [7].

In the two run periods more than 4000 runs have been taken with electrons, pions or muons in the energy range $6\,\mathrm{GeV} \leq \mathrm{E} \leq 200\,\mathrm{GeV}$ with about 80 million triggers in total. Energy scans have been taken at a standard set of impact points. In addition, horizontal and vertical scans have been done at fixed particle energies. The frequently used impact points D and H for the EMEC/HEC and FCal data correspond to $\eta = 2.8$ and $\eta = 3.65$ in ATLAS.

We focus in this talk on the electron and pion performance results in the EMEC/HEC and FCal regions. First results from the vertical scan covering the full crack region are given as well. Here it is only relevant to compare the data with MC simulation in order to verify the geometry and inactive material distribution. Also a discussion of noise cut influence to a energy reconstruction performance is added.

The noise subtraction for the data has been done using the exact list of read-out cells in a data cluster for the next random trigger event in the same data file. Thus for the noise estimate a close correlation to the real event with respect to time and actual cluster is kept. Table 1 shows the typical noise, in MeV, at the appropriate electromagnetic scale, for cells in the different longitudinal sections of the three calorimeter systems. Table 2 shows the corresponding mean reconstructed noise, for a number of different cone sizes R. For the electromagnetic sections of the EMEC and FCal, the noise is reconstructed at the electromagnetic scale, using cone sizes appropriate for electromagnetic clusters. The right half of the Table 2 shows the reconstructed noise expected for the reconstruction of pion data. These use the larger cone sizes more appropriate to hadronic reconstruction and sum over all longitudinal sections. These results, in GeV, have been calibrated to the hadronic energy scale.

Table 1. Example of typical mean cell noise (MeV) on the electromagnetic scale

Calorimeter section	**EM2**	**EM3**	**HEC0**	**HEC1**	**HEC2**	**FCal1**	**FCal2**
Mean Noise [MeV]	80	60	200	280	460	240	370
Expected Noise [MeV]	80	60	190	250	420	180	315

Table 2. Example of mean noise (GeV) in typical cone for electrons ($R = 0.15$ and $R = 0.25$) and hadrons ($R = 0.30$, $R = 0.40$ and $R = 0.50$). The energy is given on the electromagnetic scale for electrons and on the hadronic scale for hadrons.

Cone	$R = 0.15$	$R = 0.25$	$R = 0.30$	$R = 0.40$	$R = 0.50$
EMEC/HEC [GeV]	0.55	1.3	4.2	5.6	7.1
FCal [GeV]	0.59	1.1	3.1	4.6	6.7

To compare data with MC expectation the simulation code GEANT 4 [9], version 7.1 has been used. From the physics lists for hadronic shower simulations available in GEANT 4 the physics lists QGSP-GN 2.6 and QGSP-BERTINI 2.6 have been used. The QGSP-BERTINI option is expected to increase the shower size somewhat in comparison to the QGSP-GN version and thus should give a better description of the shower shapes. Therefore both options have been used for the study of the pion shower shape.

3. Electron and Pion Results

Figure 2 shows the response for electrons of 193 GeV when performing a vertical scan at $x = 0$ covering almost the full acceptance. A cone size of $R = 0.15$ has been used. Shown is the total response as well as the response in the main longitudinal section of either the EMEC or FCal. In all figures in this chapter the open symbols represents the MC, the solid one a data. The small deviation of the data from MC prediction at $y \sim -115$ mm is due to a weak response of a single electronic channel in the data.

Figure 3 shows the response as function of energy for electromagnetic clusters for the EMEC impact point D (corresponding to $|\eta| = 2.8$, and for FCal impact point H. For the cone size of

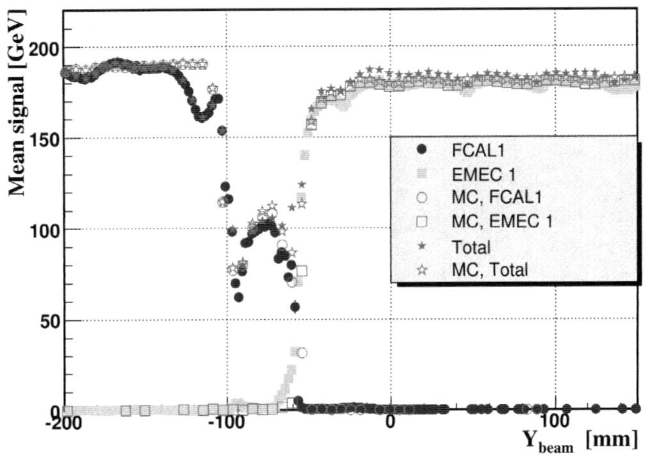

Figure 2. Energy response for electrons of 193 GeV when performing a vertical scan covering almost the full acceptance. Shown is the total response as well as the response in the main longitudinal section of either the EMEC or FCal.

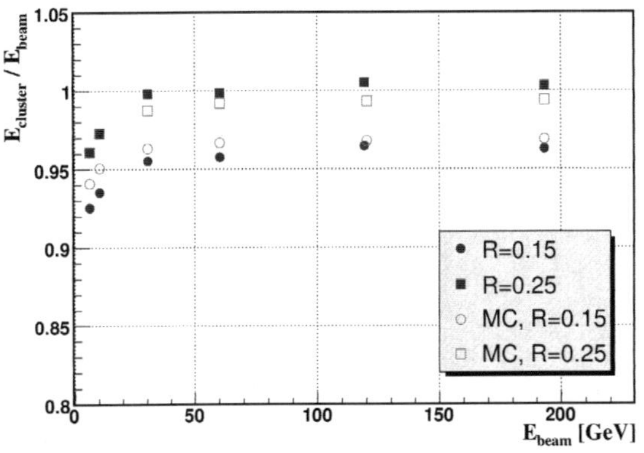

(a)

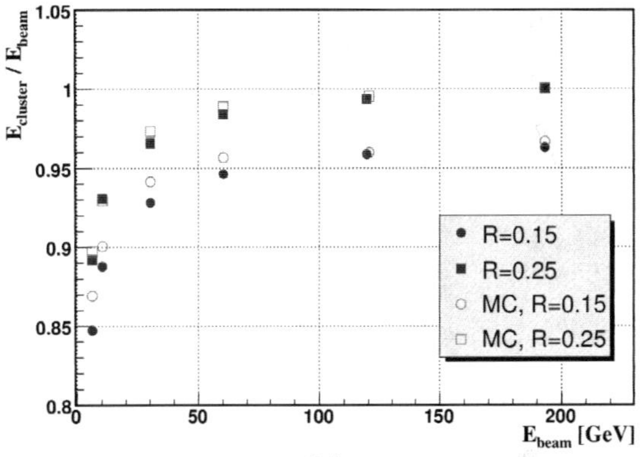

(b)

Figure 3. Energy dependence of the response to electrons with the impact point in EMEC (a) and FCal (b). Shown are the results for a cone size $R = 0.15$ and $R = 0.25$.

$R = 0.15$ the leakage out of cluster is $\sim 4\%$ at high energies in point D and $\sim 5\%$ in point H. The linearity is well described by the MC, the deviations at low energies are to a large extent due to the dead material in the beam in front of the active calorimeter, as expected from and seen in MC simulations.

Similarly the energy resolution has been studied for electrons reconstructed in a cone. The results are shown in figure 4 after noise subtraction and compared with MC expectations. The resolution has been parametrized using

$$\frac{\sigma(E)}{E} = \frac{a}{\sqrt{E}} \oplus b. \tag{1}$$

For the larger cone option the data yield typically a sampling term of $a = (13.5 \pm 0.5)\,\%\sqrt{\text{GeV}}$ and a constant term $b = (0.7{\pm}0.1)\,\%$ in point D and $a = (29.3{\pm}0.7)\,\%\sqrt{\text{GeV}}$ and a constant term $b = (3.0 \pm 0.1)\,\%$. in point H. In general, the energy resolution expected from MC is somewhat better than seen in the data. This holds for both cone sizes, particularly at low energies. For the larger cone option the MC prediction is approaching the data at higher energies. But for the smaller cone option the data yield a somewhat larger constant term.

As was the case for electrons, a vertical scan of the response to pions can reveal the fine structure of the hadronic calorimeters of the set-up. Figure 5 show the response on the

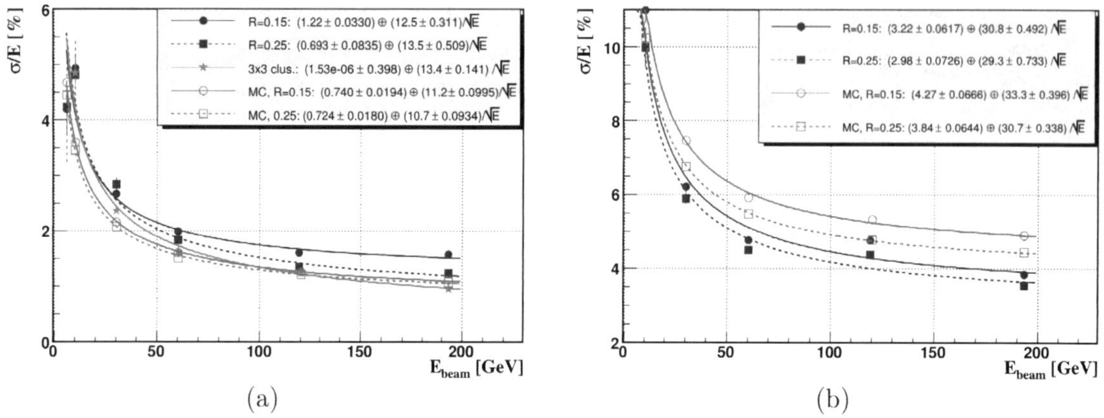

Figure 4. Energy dependence of the energy resolution for electrons with the impact point in EMEC (a) and FCal (b). Shown are the results for a cone size of $R = 0.15$ and $R = 0.25$.

electromagnetic scale for pions of 200 GeV. In general, the MC predicts a larger signal in the electromagnetic sections and a smaller one in the hadronic sections compared to the data. This is a well known problem in the GEANT 4 QGSP and QGSP-GN simulations, yielding somewhat more compact hadronic showers than seen in the data. Except for this overall scale factor, the MC shows a rather good agreement with the data, in particular for the details of the response when passing the crack region.

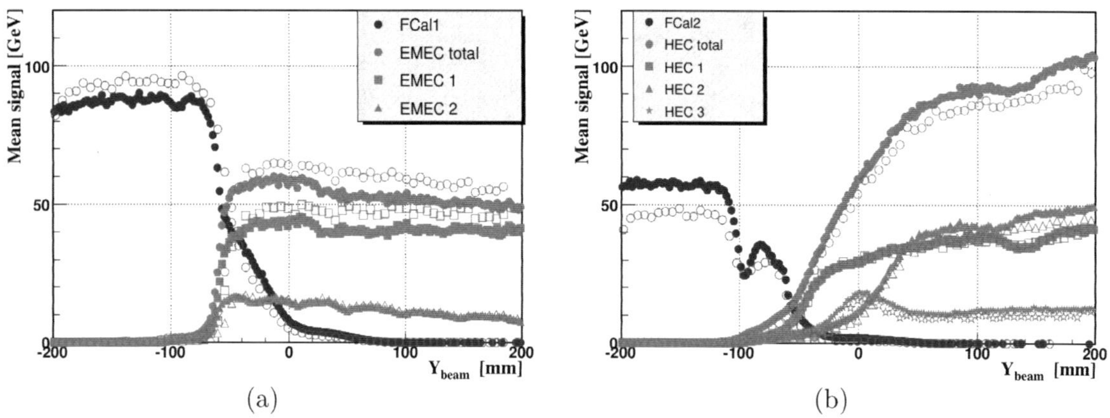

Figure 5. Energy response on the electromagnetic scale for pions of 200 *GeV* when performing a vertical scan. Shown is the response in the FCal1, total EMEC, as well as in the two main longitudinal sections of the EMEC on part (a), and FCal2, total HEC, as well as in the individual longitudinal sections of the HEC (HEC1, HEC2, HEC3) on part (b). For the energy reconstruction a cone size of $R = 0.30$ has been used.

The response as function of energy has been studied for pions using energy deposits on electromagnetic scale within a cone of $R = 0.3$ and $R = 0.5$. Figure 6 shows the response as function of energy. The expectations from MC simulations are shown as well. The data for the smaller cone size are somewhat below the MC expectations. Part of the difference might be the rather compact lateral shower size in GEANT 4 QGSP with respect to the data, an effect which seems to be more pronounced at low energies (see e.g. [10]). For the larger cone size the MC describes the data reasonably well.

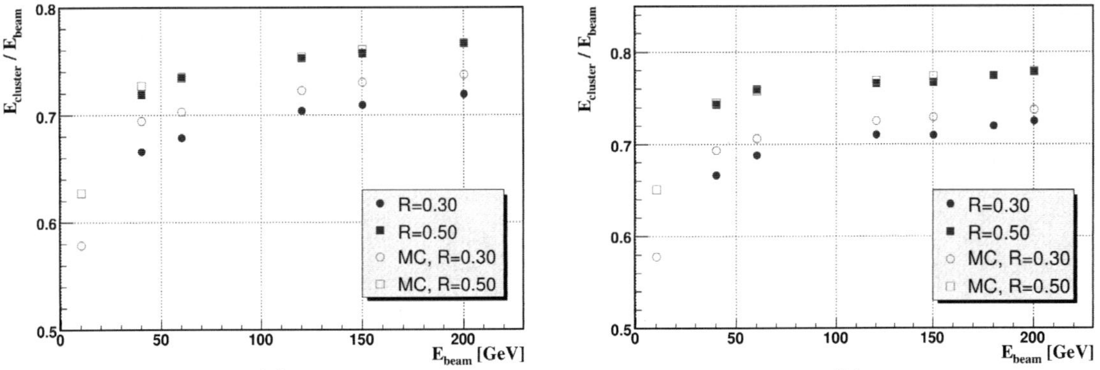

(a) (b)

Figure 6. Energy dependence of the response to pions with the impact point in EMEC (a) and FCal (b), using the electromagnetic scale. The energy has been reconstructed within a cone of $R = 0.3$ and $R = 0.5$ respectively.

The energy resolution for pions is affected by the different and in addition energy dependent response to pions in the different calorimeter sections. Therefore we tried to compare the data with MC simulations using the 'bench mark approach', where one calibration constant per longitudinal section and energy is used, rather than one overall calibration constant at each energy as done in the HEC stand alone beam tests (for details see [5]). Figure 7 shows the energy resolution as function of energy. The expectations from MC simulations are shown as well. The noise has been subtracted. The energy resolution expected from the MC simulation

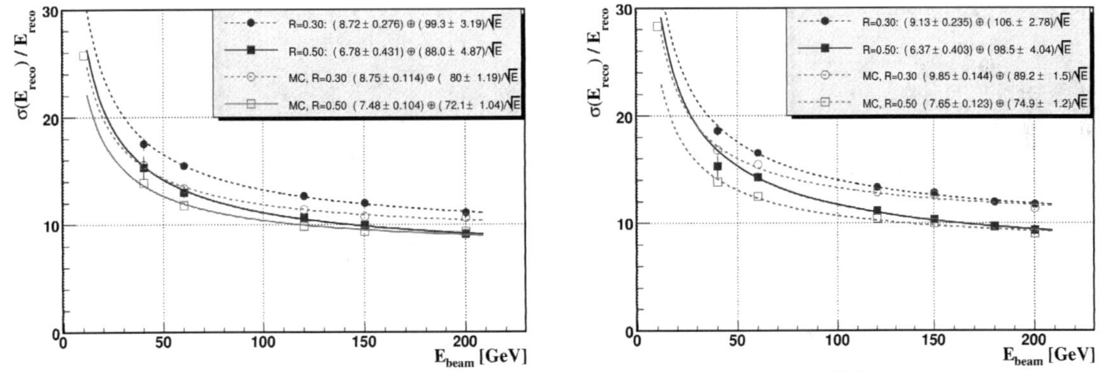

(a) (b)

Figure 7. Energy dependence of the energy resolution for pions with the impact point in EMEC (a) and FCal (b), using the bench mark approach. Here one calibration constant per longitudinal section has been determined from a fit minimizing the energy resolution. The energy has been reconstructed within a cone of $R = 0.3$ resp. $R = 0.5$.

(GEANT 4 QGSP-BERTINI) is in rather good agreement for the higher energies and for the larger cone, but at low energies is somewhat better than what is seen in the data. The resolution has been parametrized using equation (1). For point D the data for the cone size $R = 0.5$ yield typically a sampling term of $a = (88 \pm 5)\,\%\sqrt{\text{GeV}}$ and a constant term $b = (6.8 \pm 0.4)\,\%$ whereas the MC expectations are $a = (72 \pm 1)\,\%\sqrt{\text{GeV}}$ and $b = (7.5 \pm 0.1)\,\%$ respectively. The differences in the terms obtained are largely driven by the somewhat different energy dependence, yielding a better resolution for the MC at low energies, but being in agreement with the data at high energies. The conclusions are similar for point H, where data yield typically a sampling term of $a = (98.5 \pm 4.0)\,\%\sqrt{\mathrm{GeV}}$ and a constant term $b = (6.4 \pm 0.4)\,\%$ whereas the MC expectations give $a = (74.9 \pm 1.2)\,\%\sqrt{\mathrm{GeV}}$ and a constant term $b = (7.7 \pm 0.1)\,\%$.

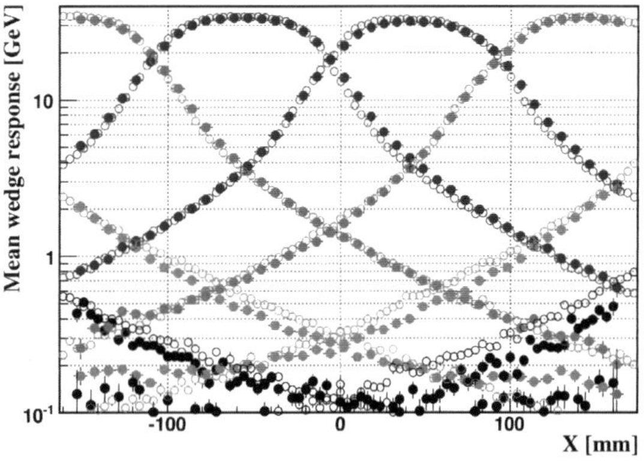

Figure 8. Lateral shower shape for pions at 60 GeV. Shown is the energy in individual ϕ wedges of $\Delta\phi = 0.2$, integrated in η and summed for the EMEC and HEC calorimeters. At each pion impact point the six relevant contributing signals using the electromagnetic scale are shown.

The lateral shower shape has been studied using a horizontal scan with pions at 60 GeV (see Fig. 8). Shown is the energy in individual ϕ wedges of $\Delta\phi = 0.2$, integrated in η and summed for the EMEC and HEC calorimeters. At each pion impact point the eight contributing signals (8 ϕ wedges) using the electromagnetic scale have been studied, two of them are rather small. The six most relevant signals are shown for the data and MC prediction using GEANT 4 with the QGSP-BERTINI physics list. In general the MC prediction describes the data well.

4. Noise Cuts Analysis

Because of usage of topological clustering as a default in ATLAS reconstruction, we have also tried to check the influence of various noise cut on response and resolution. In the figure 9(a), the response of the electrons with cone cluster of $R = 0.25$ is shown, only cells passing the noise cut are taken into energy sum. The cut -3σ means no cut at all, and cut is asymmetric, only positive signals are taken into account. In figure 9(b) the resolution dependence of the same cluster is shown. Similar plots for pions are shown in figure 10. We see, that the optimal cut in

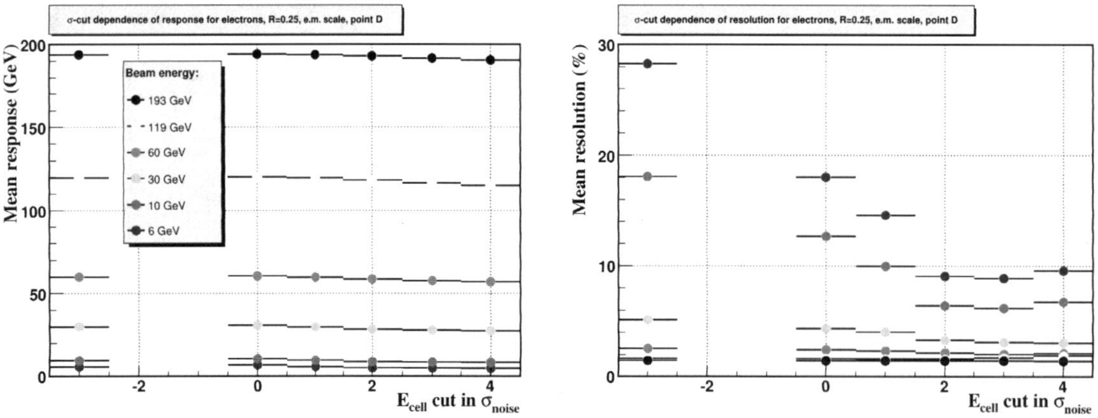

(a) (b)

Figure 9. Response (a) and resolution (b) dependence on σ cut used on cells in cone $R = 0.25$ for electrons in point D.

the case of electrons is a 3σ, whereas for pions 0σ brings the best performance.

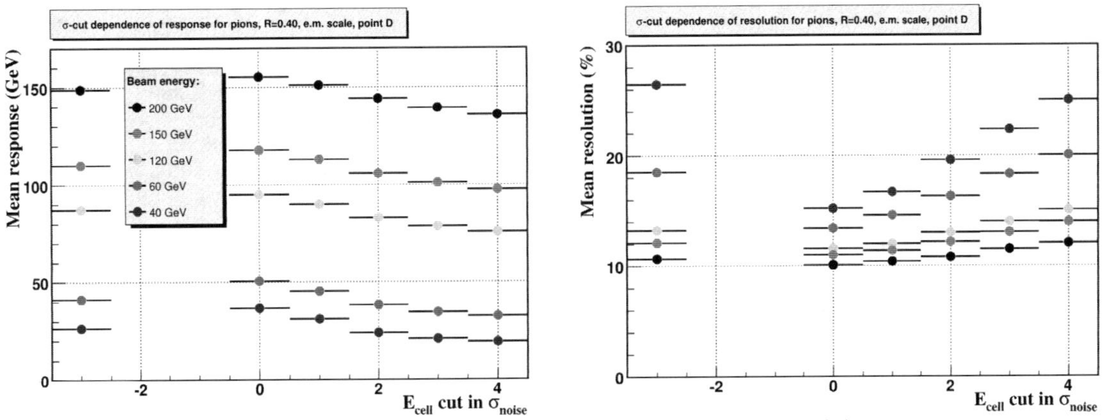

(a) (b)

Figure 10. Response (a) and resolution (b) dependence on σ cut used on cells in cone $R = 0.25$ for pions in point D.

5. Conclusions

The region of the ATLAS endcap (EMEC/HEC) and forward (FCal) calorimeters in the pseudorapidity range $2.5 < |\eta| < 4.0$ has been studied in beam test runs with electrons and pions. The performance of the EMEC/HEC as well FCal for electrons and pions has been assessed. The results have been compared in detail with MC simulations (GEANT 4). In general, the data show agreement with MC predictions at higher energies. At low energies the GEANT 4 physics list QGSP-GN predicts a somewhat larger pion response (at the electromagnetic scale), coupled with a better energy resolution and more compact shower size than seen in the data. Here QGSP-BERTINI yields a better agreement, in particular for the shower shape.

Acknowledgments

This work has been supported by the Bundesministerium für Bildung, Wissenschaft, Forschung und Technologie, Germany, under Contract numbers 05 HA 8EX1 6, 05 HA 8UMA 8 and 05 HA 8PXA 2, by the Slovak funding agency VEGA under Contract number 2/0061/08 and by ISTC under Contract number 1621. We thank all funding agencies for financial support.

References

[1] Colas J et al 2007 Response Uniformity of the ATLAS Liquid Argon Electromagnetic Calorimeter *Nucl. Instr. and Meth.* A **582** 429

[2] Aharrouche M et al 2006 Energy linearity and resolution of the ATLAS electromagnetic barrel calorimeter in an electron test-beam *Nucl. Instr. and Meth.* A **568** 601

[3] Colas J et al 2005 Position resolution and particle identification with the ATLAS EM calorimeter *Nucl. Instr. and Meth.* A **550** 96

[4] Aubert B et al 2003 Performance of the ATLAS electromagnetic end-cap module 0 *Nucl. Instr. and Meth.* A **500** 178

[5] Dowler B et al 2002 Performance of the ATLAS Hadronic End-Cap Calorimeter in Beam Tests *Nucl. Instr. and Meth.* A **482** 94

[6] Gingrich D M et al 2007 Conctruction, Assembly and Testing of the ATLAS Hadronic End-Cap Calorimeter *Journal of Instrumentation* JINST02(2007)P05005

[7] Archambault J P et al 2008 Energy Calibration of the ATLAS Liquid Argon Forward Calorimeter *Journal of Instrumentation* JINST03(2008)P02002

[8] Artamonov A et al 2008 The ATLAS Forward Calorimeter *Journal of Instrumentation* JINST03(2008)P02010

[9] Agostinelli S et al 2003 GEANT 4 - a simulation toolkit *Nucl. Instr. and Meth.* A **506** 250

[10] Kiryunin A E et al 2006 GEANT 4 physics evaluation with testbeam data of the ATLAS hadronic end-cap calorimeter *Nucl. Instr. and Meth.* A **560** 278

Journal of Physics: Conference Series **160** (2009) 012079 doi:10.1088/1742-6596/160/1/012079

The overview of the ATLAS local hadronic calibration.

Guennadi Pospelov, for the ATLAS hadronic calibration group.
Max-Planck-Institut für Physik, Werner-Heisenberg-Institut
Föhringer Ring 6, 80805 Munich, Germany

E-mail: pospelov@mppmu.mpg.de

Abstract.
We present the concept of the local hadronic calibration for the ATLAS calorimeter system as a tool to get calibrated jets at particle level for any jet algorithm. The procedure is based on detailed Geant4 simulations providing information on energy deposits in all parts of the ATLAS detector. Calibration starts from topological clusters reconstructed and calibrated at the electromagnetic scale. First classification step tags clusters as mainly electromagnetic, or hadronic depending on cluster shape variables. Clusters which are identified as being electromagnetic are kept at the electromagnetic scale. Hadronic clusters receive cell weights to correct for the invisible energy deposits of hadrons. Next steps are out-of-cluster corrections for lost energy deposited in calorimeter cells outside of reconstructed clusters and dead material corrections for energy deposited outside of active calorimeter volumes. Finally out-of-jet corrections are applied to get the final jet energy scale.

1. Introduction

The jet energy scale is one of the main systematic uncertainties in many physics studies foreseen with the ATLAS detector [1]. Top mass reconstruction or measurements of inclusive jet cross-section are examples relevant for the first data taking phase. The initial parton energy differs from the energy of the hadronic shower because of detector effects, like calorimeter non-compensation and energy losses in dead material, jet effects, like bending of secondary particles out of the jet cone by the magnetic field or jet reconstruction algorithm specifics, and effects from the collision physics environment.

In this paper we present the concept of the local hadronic calibration for jets to the particle level independent on any jet algorithm. The ultimate goal of the procedure is to provide jet algorithms with constituents — calibrated clusters with energies corresponding to the stable particle energies. The key feature of the approach is to factorize corrections in several sequential steps to disentangle detector effects of different types and to correct them independently.

Starting point for the local hadronic calibration is topological clustering [2] in the calorimeter cells which have been calibrated at the electromagnetic (*em*) scale. Cluster shape variables are then used to classify clusters as having electromagnetic or hadronic nature. The hadron-like clusters are subject of a cell weighting procedure to compensate for the lower responce of the calorimeter to hadronic deposits, while clusters classified as electromagnetic-like, are kept at the original scale. Next out-of-cluster (OOC) corrections for lost energy deposited in calorimeter cells outside of reconstructed clusters in the tail of hadronic shower are applied. Finally, dead material

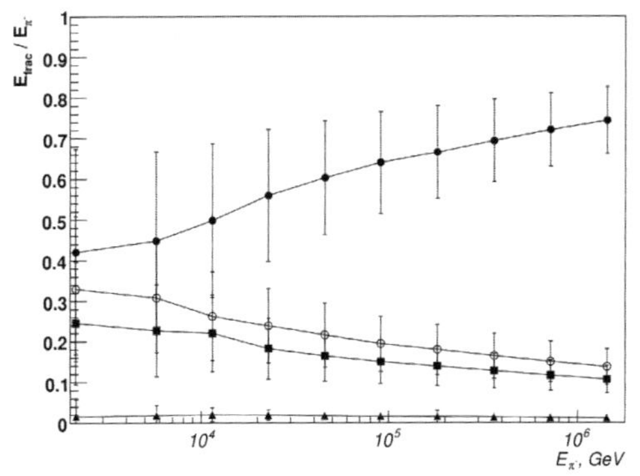

Figure 1. Calibration energy fractions (• - visible *em*, ∘ - visible *non-em*, ▪ - invisible, ▴ - escaped) released in active and inactive material in the endcap region ($1.5 < |\eta| < 3.0$) by single charged pions vs. the pion energy.

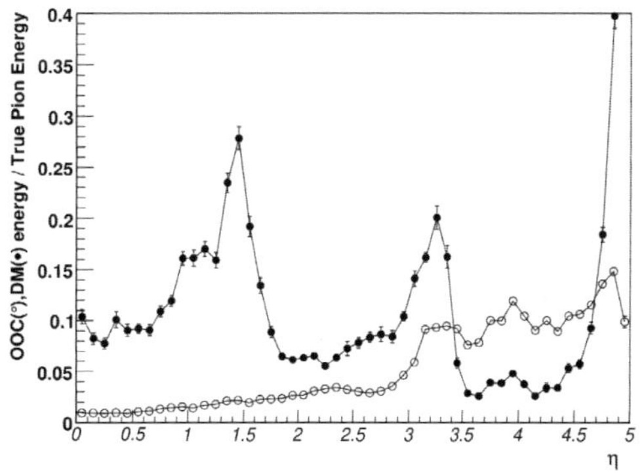

Figure 2. Fractions of true calibration hit energies outside topological clusters but inside the calorimeters (∘) and outside calorimeters (•) as function of pion pseudo-rapidity for 200 GeV pion energy.

(DM) corrections are applied on the cluster level to account for energy deposits outside of active calorimeter volumes, e.g. in the cryostat, the magnetic coil and calorimeter intermodular cracks.

To derive the correction coefficients for all calibration steps we use actual energy deposits in active and inactive volumes known from detailed `Geant4` [3] simulations via the so-called *calibration hits* mechanism.

2. Simulation and Calibration Hits

The classification and calibration methods described in this paper are based on detailed simulations of single pions in the standard ATLAS environment. The data set consist of 10^6 neutral and $2 \cdot 10^6$ charged pions with initial η, ϕ coordinates randomly distributed across the entire detector. The initial pion energies were generated in 1000 logarithmically equidistant steps from 150 MeV to 2 TeV.

Calibration hits provide access to the energy deposits in active (liquid argon, scintillator), inactive (absorber) and dead material regions for the following four categories of energy deposits:

- visible *em*-energy: The energy released by electrons or positrons via ionization.
- visible *non-em*-energy: The energy released by charged particles other than electrons or positrons via ionization.
- in-visible energy: The energy released by non-ionizing processes such as break-up of nuclear bindings.
- escaped energy: The energy leaving the mother volume in form of non-interacting or escaping particles like neutrinos or muons.

Figure 1 shows the average fraction of each of these 4 categories summed over the active and inactive regions in the endcap calorimeters as a function of the pion energy. Error bars on the figure represent *rms* values. Figure 2 shows the total out-of-cluster and dead material energy fractions known from calibration hits as function of pion pseudorapidity for 200 GeV charged pions.

The aim of the calibration is to reconstruct the lost invisible and escaped energy on an event-by-event basis with measured energy depositions inside calorimeter cells.

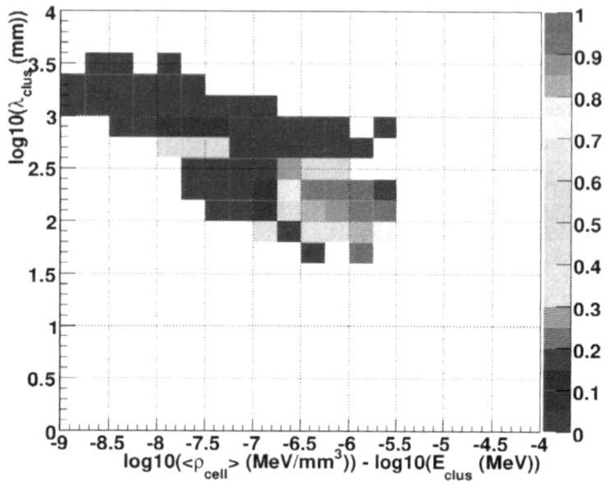

Figure 3. Probability weights to observe a neutral pion cluster for 8 GeV $< E_{\mathrm{cluster}} <$ 16 GeV.

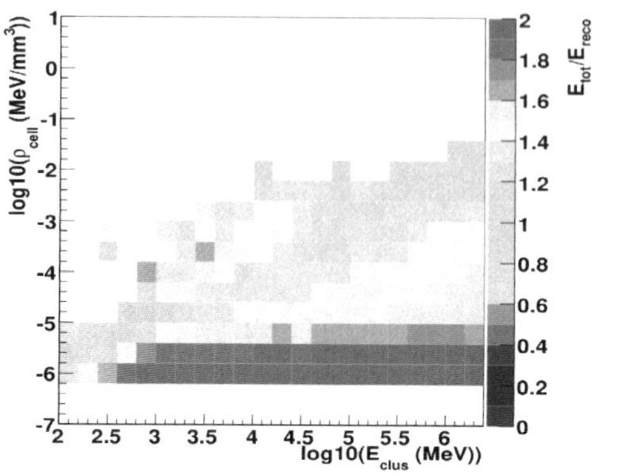

Figure 4. Hadronic cell weights for Tile barrel sampling 1 at $0.2 < |\eta| < 0.4$.

3. Local Hadronic Calibration

3.1. Cluster Classification

Aim of the cluster classification is to distinguish between clusters seeded by electromagnetic or hadronic interactions. For topological clusters where the correspondence of cluster and particle level is close this would lead to a separation of electrons, photons and neutral pions on one side from charged hadrons (mainly pions) and neutrons on the other side such that both particle classes can be optimally calibrated.

The current default procedure in ATLAS is to classify topological clusters basing on the predicted phase-space population of clusters stemming from charged and neutral pion simulations. The separation of hadronic from electromagnetic showers relies on the following cluster shape variables: the average cell energy density $\langle\rho\rangle$ and the depth of the shower center $\lambda_{\mathrm{center}}$. For both particle species the 4-dimensional phase space for clusters in $|\eta|$, E_{cluster}, $\log_{10}(\lambda_{\mathrm{center}})$ and $\log_{10}(\langle\rho\rangle) - \log_{10}(E_{\mathrm{cluster}})$ is binned and filled with simulated pions in the energy range $E_{\mathrm{cluster}} = 200\,\mathrm{MeV} - 2\,\mathrm{TeV}$ and $|\eta| < 5$. Assuming an a-priori probability ratio of $1 : 2$ for observing a neutral pion over a charged pion the probability weight in each phase-space bin for observing a neutral pion is given by: $w_i = n_i^{\pi^0}/(n_i^{\pi^0} + 2n_i^{\pi^\pm})$, where $n_i^{\pi^{0,\pm}}$ is the fraction of clusters at given energy and η of neutral or charged pions in a given phase-space bin i from the simulation. A cluster is classified as electromagnetic if the lookup value for $w_i > 0.5$ at a given cluster phase-space bin.

Figure 3 shows the classification weights obtained with this procedure for the region $0.2 < |\eta| < 0.4$ and $8\,\mathrm{GeV} < E_{\mathrm{cluster}} < 16\,\mathrm{GeV}$ on a colored scale from 0 (purely hadronic) to 1 (purely electromagentic as a function of the two moments $\langle\rho\rangle$ and $\lambda_{\mathrm{center}}$. It is clearly visible that electromagnetic showers dominate the region of high energy density and small cluster depth. It is also apparent from the plots that classification is most difficult for clusters with smaller energies as the distributions for charged and neutral pions have larger overlaps in that regime.

The classification performance in terms of the energy fractions classified as electromagnetic and hadronic for neutral and charged single pions is shown in figure 5. For neutral pions the method reaches classification efficiencies of $80 - 85\%$ above 50 GeV, but below that energy the electromagnetic fraction drops linearly with the logarithm of the pion energy. For charged pions the hadronic fraction is rather stable between $80 - 90\%$ over the entire energy range – at low energies due to the low classification efficiency of neutral pions and at large energies due to the better separation of neutral and charged showers.

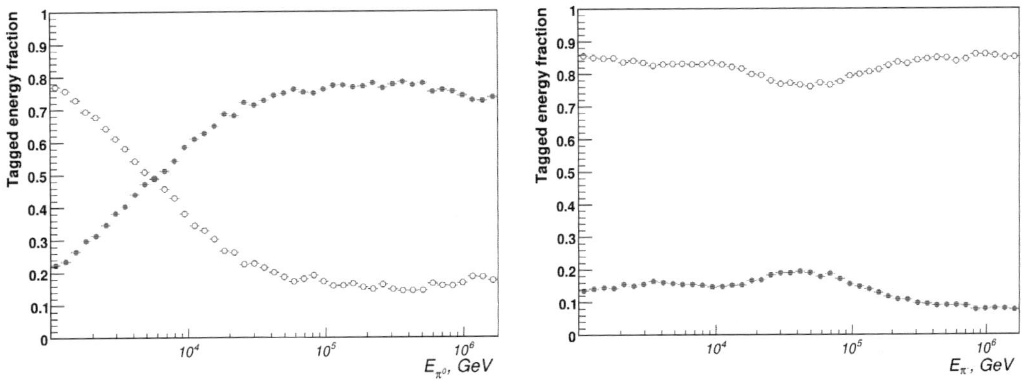

Figure 5. Energy fractions of single pions classified as electromagnetic (•) or hadronic (○) for π^0's (left), $\pi^\pm$'s (right) as function of the pion energy averaged over all η and ϕ.

3.2. Cluster Weighting

For clusters classified as hadronic a H1-type cell-based weighting procedure [4, 5] is applied to correct for invisible and escaped energy fractions. The weight tables are filled with the ratios $E_{\text{cell}}^{\text{calib}}/E_{\text{cell}}^{\text{reco}}$, where $E_{\text{cell}}^{\text{reco}}$ is the regular simulated energy based on the visible energy in the active media including noise, while $E_{\text{cell}}^{\text{calib}}$ is the sum of the four calibration energies deposited in the given cells active and inactive parts. Average weights are kept as a function of the cluster energy the cell belongs to, the energy density $\rho_{\text{cell}} = E_{\text{cell}}^{\text{reco}}/V_{\text{cell}}$ of the cell, and the $|\eta|$ of the cell center. There are 25 equidistant $|\eta|$-bins from 0.0 to 5.0, 20 logarithmic energy-density bins from $-7 < \log_{10}(\rho_{\text{cell}}(\text{MeV/mm}^3)) < 1$ and 22 logarithmic cluster energy bins from $\log_{10}(100(\text{MeV})) < \log_{10}(E_{\text{clus}}(\text{MeV})) < \log_{10}(2 \cdot 10^6(\text{MeV}))$. Examples of the actual weights for the first sampling of the Tile calorimeter obtained this way from $2 \cdot 10^6$ charged pions are shown in figure 4.

3.3. Out-of-Cluster corrections

Out-of-cluster corrections aim for recovering from the lost low energetic deposits at the tails of hadronic showers. This energy is deposited in calorimeter cells which are not part of any reconstructed cluster due to noise thresholds in the clustering.

The total fraction of out-of-cluster energy from single pion simulations w.r.t. the weighted total energy inside clusters is averaged in bins of pion energy, pion $|\eta|$, and the pion depth λ in the calorimeter. The λ values of all clusters in a given event are combined with energy proportional weights to obtain the latter. 12 energy bins with the boundaries $\{0, 1, 2, 4, 8, 16, 32, 64, 128, 256, 512, 1024, 2048\}$ GeV, 25 equidistant $|\eta|$ bins from 0 to 5 and 20 equidistant bins in $\log_{10}(\lambda/\text{mm})$ from 0 to 4 define the lookup table.

However, out-of-cluster energy estimated with the help of tables of that kind turned out to be overrated in multi particle event. This is due to the fact, that energy which is outside of one cluster could actually be inside the other cluster. To resolve this problem the level of isolation of clusters is used [6]. It is determined by the fraction of cells on its outer perimeter not included in other clusters. An isolation of 0 means that all cells on its outer perimeter are included in neighboring clusters, a value of 1 indicates that the cluster is totally isolated – i.e. that all its neighbor cells are not part of any cluster. This isolation fraction is multiplied with the expected out-of-cluster energy fraction for each cluster as obtained from the out-of-cluster tables described above. This accounts for the fact that clusters tend to be less isolated in full events compared to single pion events from simulations or beam tests.

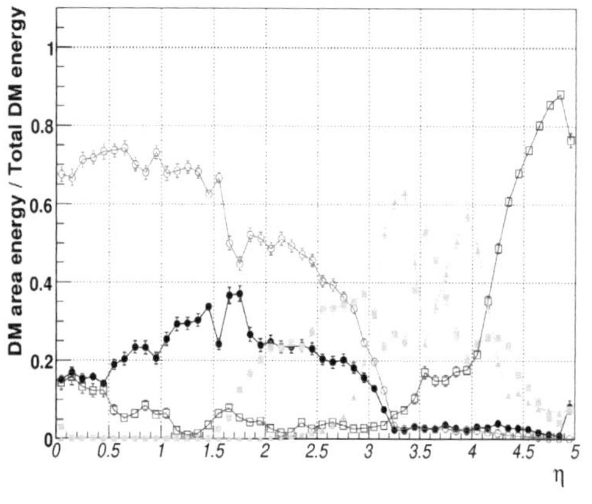

Figure 6. Ratio of energy deposited in different DM regions (● - before *em* calorimeter, ○ - between *em* and *had*, ■ - intermodular cracks, □ - leakages, ▲ - before forward calorimeter) to the total DM energy for charged pion with energy 200 GeV.

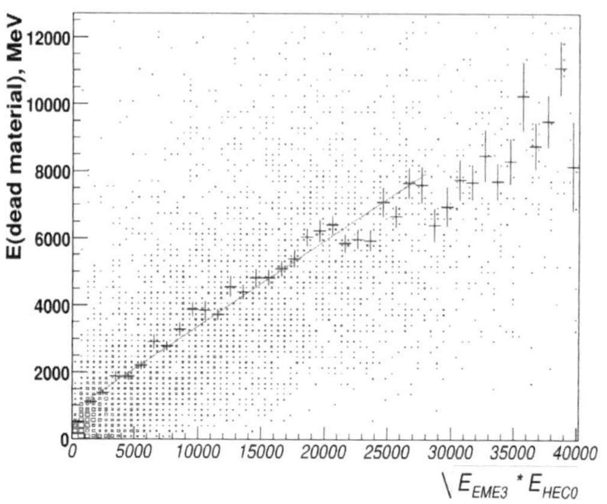

Figure 7. Dependence of energy deposited between *em* and *had* endcap calorimeters on $\sqrt{E_{EME3} \cdot E_{HEC0}}$, where E_{EME3} energy in last *em* sampling, E_{HAD0} energy in first *had* samplings.

3.4. Dead material corrections

Dead material corrections are accounting for energy deposited outside of active zones of the electromagnetic and hadronic calorimeters. This includes energy deposited in upstream material (Inner Detector material, magnetic coil, cryostat walls), different crack regions between calorimeter modules and material behind calorimeter system (leakages). Figure 6 shows the ratio of energy deposited in different DM regions to the total DM energy for charged pions with 200 GeV energy as a function of the pion pseudorapidity. These regions are treated variously to get and to apply DM correction coefficients. For upstream material the good correlations of DM deposits with the preshower signals are used. Deposits between electromagnetic and hadronic calorimeters are corrected for by weighting the geometric average of signals in the last *em* and first *had* sampling. (see figure 7). Correction coefficients then are parametrised as a function of cluster energy and pseudorapidity. For intricate DM regions where there are no cluster quantitities correlating well with DM energy deposits, 3 dimensional lookup tables on $E_{cluster}$, $\eta_{cluster}$, $\lambda_{cluster}$ are used to store average ratios of DM energy to the cluster energy.

4. Performance

Figure 8 shows the ratio of reconstructed over simulated pion energy at the *em*-scale, and after applying 3 sequential steps of local hadronic calibration: hadronic weighting, out-of-cluster and dead material corrections. The plot has been obtained for 100 GeV single charged pions in the region $0.2 < |\eta| < 0.4$. It is clearly seen that each step shifts the mean of the reconstructed energy toward the truth energy and, at the end, gives the correct value. The energy resolution is improving on each step as well.

The performance of local hadronic calibration on di-jet events is presented in figure 9. Jets are formed by the K_T jet algorithm with R = 0.6 from topological clusters, calibrated with the local hadronic calibration scheme, and compared to the truth matched jets. Thus, for the barrel and endcap regions the local hadronic calibration recovers linearity for 100 GeV jets up to 92%, while for 10 GeV jets missed energy reaches 25%. Main sources of remaining energy correction and their impact on the energy underestimation are presented below (calculation has been done

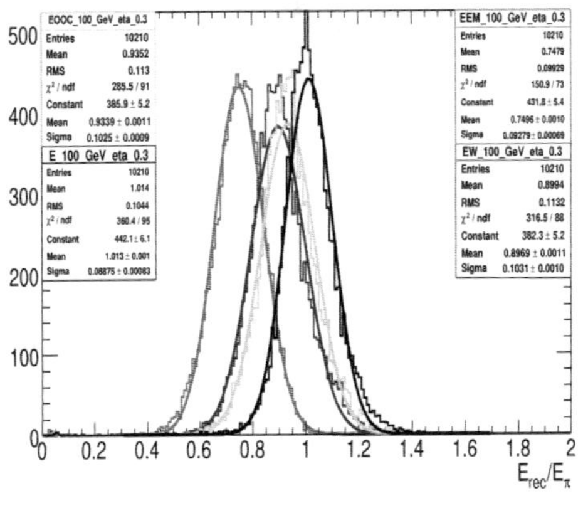

Figure 8. The ratio of reconstructed over simulated pion energy at various stages of the calibration.

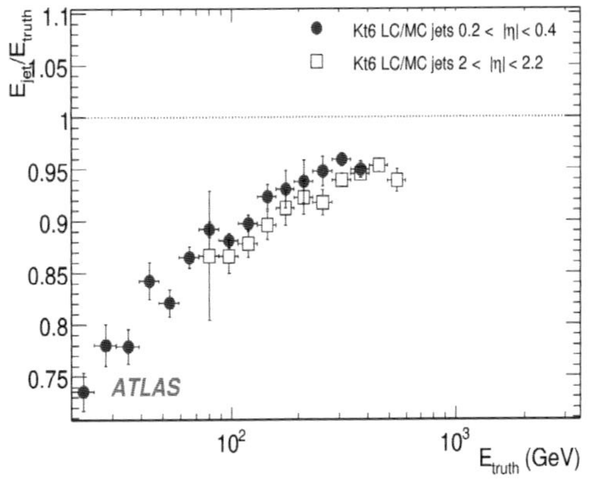

Figure 9. Linearity for K_T jet algorithm with R = 0.6 for 2 different η regions as a function of jet energy.

for $p_\perp \simeq 150GeV$):

- missclassification (hadronic deposits are treated as electromagnetic and vice versa) - 3%.
- calorimeter inefficiency (particle leaves signal below noise threshold) - 3%.
- lost particles (bending of particles outside the acceptance cone by magnetic field) - 2%.

All these effects are subject of jet level corrections which are currently under development.

5. Conclusion

In this article the local hadron calibration as a tool to calibrate topological clusters from *em* scale to the particle level is discussed. It will be used to provide jet correction for the ATLAS detector. The method includes four consecutive steps. In a first step clusters are classified as electromagnetic or hadronic based on cluster energy density and calorimeter cluster depth. In the second step hadronic-like clusters are corrected for the invisible energy released in hadronic interactions. Then energy deposited in calorimeter cells not belonging to any cluster and in dead material outside calorimeter is corrected for. The correction coefficients are derived from single neutral and charged pion simulations. After application of the last calibration steps, the method is able to improve energy resolution and to restore true energy for single pions. For jets reconstruction the method is proven to recover linearity up to 92% in di-jet events with $p_\perp \simeq 150\,\mathrm{GeV}$. The main sources of remaining missing energy are: missclasification, calorimeter inefficiency for low energetic pions and bending of particles out of the jet cone. They are the subject of the very last jet level correction step, which is currently under development.

References

[1] ATLAS Collaboration 2008 The ATLAS experiment at the CERN Large Hadron Collider *submitted to JINST 06/05/2008*

[2] Lampl W *et al* 2008 Calorimeter clustering algorithms: Description and Performance *ATLAS Note* ATL-LARG-PUB-2008-002

[3] Agostinelli S *et al* 2003 GEANT4 — a simulation toolkit *Nucl. Instr. and Meth.* A **506** 250-303

[4] H1 Collaboration 1994 *Nucl. Instr. and Meth.* A 344 (1994) 492

[5] H.P. Wellisch, J.P. Kubenka, H. Oberlack, P. Schacht 1994 *Preprint* MPI-PhE/94-03

[6] Bergeaas E *et al* 2008 Topological cluster classification and weighting *ATLAS Note* ATL-COM-LARG-2008-006

Validation of the ATLAS hadronic calibration with the LAr End-Cap beam tests data

Teresa Barillari

Max-Planck-Inst. fuer Physik Werner-Heisenberg-Institut, Foehringer Ring 6, D-80805 Muenchen, Germany

E-mail: barilla@mppmu.mpg.de

Abstract. The high granularity of the ATLAS calorimeter and the large number of expected particles per event require a clustering algorithm that is able to suppress noise and pile-up efficiently. Therefore the cluster reconstruction is the essential first step in the hadronic calibration. The identification of electromagnetic components within a hadronic cluster using cluster shape variables is the next step in the hadronic calibration procedure. Finally the energy density of individual cells is used to assign the proper weight to correct for the invisible energy deposits of hadrons due to the non-compensating nature of the ATLAS calorimeter and to correct for energy losses in material non instrumented with read-out.

The weighting scheme employs the energy density in individual cells. Therefore the validation of the monte carlo simulation, which is used to define the weighting parameters and energy correction algorithms, is an essential step in the hadronic calibration procedure.

Pion data, obtained in a beam test corresponding to the pseudorapidity region $2.5 < |\eta| < 4.0$ in ATLAS and in the energy range 40 GeV $\leq E \leq$ 200 GeV, have been compared with monte carlo simulations, using the full ATLAS hadronic calibration procedure.

1. Introduction

The primary goal of the ATLAS calorimeter is to measure the energy and the direction of the electrons, photons and jets. A good calorimeter performance is achieved by also having a good missing transverse energy and particle identification determination.

The beam tests [1] in the particularly difficult forward region $2.5 < |\eta| < 4.0$ (the transition from the electromagnetic end-cap calorimeter EMEC [2] and hadronic endcap calorimeter HEC [3] [4] to the forward calorimeter FCal [5]) was carried out in 2004 in the H6 beam-line at the CERN SPS which provides hadrons, electrons or muons in different energy range.

The beam test set-up has been chosen to be as close as possible to the final ATLAS detector, both with respect to the calorimeter modules, the support structures, and the dead material distribution. The set-up was also reproducing the ATLAS projective geometry but only at one $|\eta|$ point, and not over the entire $|\eta|$ region. The load in the liquid argon (LAr) cryostat [1] consisted of the inner section of the EMEC module, eight specially constructed front wheel HEC modules, eight purpose-built rear wheel HEC modules and the FCal modules. Limited by cryostat dimensions the depth of the rear wheel HEC modules was half of the ATLAS modules.

The goal of the study is to obtain the hadronic calibration in the forward region $2.5 < |\eta| < 4.0$, including corrections for dead material effects. The validation of the monte carlo (MC)

simulation in ATLAS using beam test data is an important step in the calibration studies. The default ATLAS hadronic calibration procedure is based on MC simulations.

In the following the initial studies done to validate the Geant4 monte carlo simulation using the ATLAS hadronic calibration procedure [6] with the 2004 EMEC-HEC-FCAL beam test data [1] are presented.

2. Data and monte carlo samples

During the beam tests more than 4000 runs have been taken with electrons, pions or muons in the energy range $6\,\text{GeV} \leq E \leq 200\,\text{GeV}$. Energy scans have been taken at a standard set of impact points. The impact points used in the analysis here presented, point D and point H for the EMEC/HEC and FCal regions, correspond to $\eta = 2.8$ and $\eta = 3.65$ in ATLAS. Five pion runs in the energy range $40\,\text{GeV} \leq E \leq 200\,\text{GeV}$ have been used to study the EMEC/HEC region and other five pion runs in the same energy range were used to test the FCal region. Six of these runs are negative pion data at 120, 180 and 200 GeV, and four are positive pion runs at 40 and 60 GeV. The beam test MC simulation used in the analysis consists of charged and neutral single pions obtained using Geant4 version 8.3 and the hadronic physics list QGSP_EMV. The default ATLAS hadronic calibration corrections, weights, dead material corrections etc., have been applied to the beam test monte carlo simulation.

3. Local hadronic calibration in ATLAS

The local hadronic calibration [6] [7] is one of the calibration approaches studied in the ATLAS collaboration. It consists of four main calibration steps. First the cluster are reconstructed in the calorimeters with an algorithm to optimize noise suppression and particle separation. Shower shape variables and other cluster characteristics are used to classify the clusters as hadronic or electromagnetic. The hadronic clusters are then subject to a cell weighting procedure to compensate the different response to hadrons compared to electrons. After the weighting, out-of-cluster corrections are applied to the clusters to correct for the energy deposited in the calorimeter but outside calibrated clusters. Finally the dead-material corrections are applied on both electromagnetic and and hadronic clusters to compensate for the energy losses in materials outside the calorimeter.

4. Performance of the hadronic calibration

The application of the ATLAS local hadronic calibration scheme to beam tests data, pions at 200 GeV in the EMEC/HEC region, is shown in figure 1. The reconstructed energy over the nominal beam energy (E_{Reco}/E_{π}) distribution at electromagnetic scale (EM) is indicated in the figure by the open triangles and the dash-dotted line Gaussian fit. The reconstructed energy distribution after applying the weighting step is indicated by the open circles and its dotted line Gaussian fit. The full triangles and the dashed line Gaussian fit in the figure show the distribution obtained after adding the out-of-cluster correction (OOCC). The reconstructed energy after all corrections, included the dead material (DM) corrections, is shown by the full circles and the full line Gaussian fit. Figure 1 shows that the linearity (E_{Mean}/E_{π}) computed as the mean of the Gaussian fitted to the energy distribution (E_{Mean}) divided by the pion beam energy (E_{π}), improves at every step: at EM scale the linearity is 75.95%, after weighting it is 87.44%, after adding the OOCC the linearity is 90.98%, and after all corrections the E_{Mean}/E_{π} reaches a value of 99.92%. The resolution of the energy computed as σ/E_{Mean}, at EM scale is 11.64%, after applying weighting the resolution gets to a value of 10.1%, after OOCC it is 10.0%, and after applying the DM correction the resolution is 9.9%. We see here that an improvement of the resolution is achieved after applying weighting corrections.

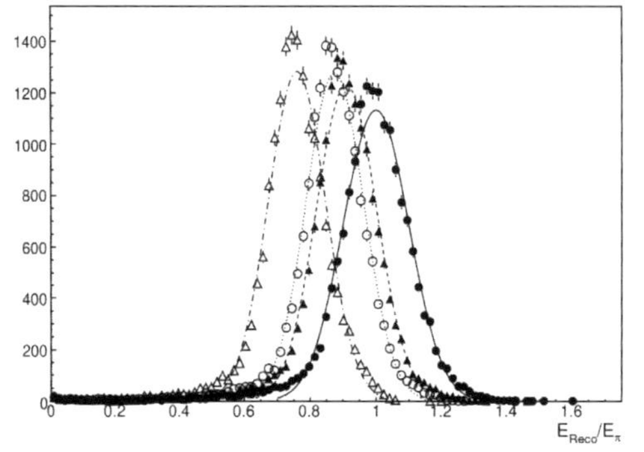

Figure 1. Plot shows the ratio on the reconstructed energy over the 200 GeV nominal pion beam energy (E_{Reco}/E_{π}), at various steps of the hadronic calibration procedure.

5. Linearity and resolution of data and simulation

The linearity in the EMEC/HEC region at EM scale and after all the calibration corrections for data and MC simulation is shown in figure 2 (left plot). The linearity values are given in table 1. The full circles in this figure indicate data at EM scale, open squares show the simulation results obtained with Geant4 QGSP_EMV also at EM scale, the full triangles and the open stars are here the data and the MC results obtained after calibration. This figure shows an overall agreement (within 3%) between data and monte carlo simulation.

The linearities values obtained in the FCal region for data and MC at EM scale and after all calibration corrections are given in table 1. The values in this table show that in the FCal region at higher beam energies the monte carlo has about 1% better linearity than the data. It should be emphasized here that the dead material in front of the FCal in the beam test set-up [1] is different than in ATLAS, this difference has not being corrected yet in the beam test monte carlo simulation.

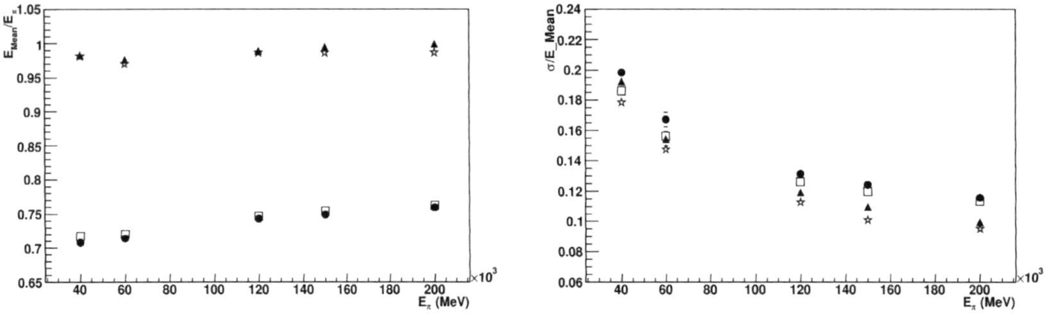

Figure 2. Linearity (left plot) and resolution (left plot) obtained in the EMEC/HEC region for data and MC as function of the beam energy. Full circles (open squares) are data (MC) at EM scale, full triangles (open stars) are data (MC) after calibration. Values are given in table 1.

The right plot in figure 2 shows the resolution obtained with data and monte carlo at EM scale and after the calibration corrections for the EMEC/HEC region. The resolution values are given in table 1. The symbols in this figure are the same as the ones used in the linearity plot. This plot shows that in the EMEC/HEC region the resolution improves after all calibration corrections, and the ATLAS hadronic calibration corrections are here working.

The resolution values obtained in the FCal region for data and MC at EM scale and after all calibration corrections are given in table 1. The values in this table show that in the FCal region, the resolution improves more in the data than in the monte carlo. The differences in the resolution values observed here, like in the linearity case, are dued to the fact that the beam test analysis is using the default ATLAS calibration corrections. ATLAS is also using a monte carlo simulation, Geant4 QGSP_EMV, that doesn't describe well the hadronic cascade model, Geant4 QGSP_BERTINI need to be used instead. Once the new beam test monte carlo simulation is available, with better weights, better beam tests DM corrections, and better hadronic cascade simulation, further improvements can be done for the whole hadronic calibration procedure.

Table 1. Linearities (E_{Mean}/E_π) and resolutions (σ/E_{Mean}) values in percentage obtained in the EMEC/HEC (first 5 rows) and in the FCal (last 5 rows) regions for data and MC at EM scale and after calibration corrections (Calib.). E_π is the nominal beam energy. The uncertainties values here are the statistical errors from the Gaussian fits.

E_π (GeV)	E^{Data}_{Mean}/E_π EM	E^{MC}_{Mean}/E_π EM	E^{Data}_{Mean}/E_π Calib.	E^{MC}_{Mean}/E_π Calib.
40	70.78 ± 0.24	71.73 ± 0.14	98.28 ± 0.32	98.13 ± 0.18
60	71.40 ± 0.40	72.00 ± 0.11	97.63 ± 0.52	97.01 ± 0.15
120	74.31 ± 0.08	74.69 ± 0.10	98.92 ± 0.09	98.65 ± 0.11
150	74.90 ± 0.08	75.41 ± 0.10	99.49 ± 0.09	98.63 ± 0.10
200	75.95 ± 0.07	76.25 ± 0.10	99.92 ± 0.08	98.67 ± 0.11
40	70.06 ± 0.27	70.49 ± 0.20	97.57 ± 0.33	97.27 ± 0.27
60	73.05 ± 0.29	73.61 ± 0.17	96.91 ± 0.36	95.71 ± 0.22
120	75.90 ± 0.09	77.04 ± 0.14	95.70 ± 0.11	94.93 ± 0.18
150	76.47 ± 0.13	77.61 ± 0.14	93.89 ± 0.15	94.53 ± 0.16
200	78.05 ± 0.08	78.89 ± 0.13	92.92 ± 0.09	94.02 ± 0.15
E_π (GeV)	σ^{Data}/E_{Mean} EM	σ^{MC}/E_{Mean} EM	σ^{Data}/E_{Mean} Calib.	σ^{MC}/E_{Mean} Calib.
40	19.83 ± 0.26	18.61 ± 0.15	19.23 ± 0.25	17.85 ± 0.15
60	16.72 ± 0.48	15.62 ± 0.13	15.44 ± 0.47	14.76 ± 0.12
120	13.14 ± 0.09	12.62 ± 0.10	11.93 ± 0.08	12.26 ± 0.10
150	12.42 ± 0.09	11.97 ± 0.10	10.95 ± 0.08	10.08 ± 0.09
200	11.56 ± 0.08	11.33 ± 0.11	9.95 ± 0.07	9.50 ± 0.10
40	22.01 ± 0.30	22.97 ± 0.25	19.24 ± 0.29	22.64 ± 0.25
60	18.55 ± 0.35	18.80 ± 0.20	16.75 ± 0.30	19.38 ± 0.21
120	14.34 ± 0.10	15.14 ± 0.16	14.54 ± 0.11	15.18 ± 0.16
150	13.52 ± 0.16	14.32 ± 0.14	12.50 ± 0.15	13.50 ± 0.15
200	12.29 ± 0.09	13.46 ± 0.14	11.72 ± 0.09	13.16 ± 0.15

[1] Aderholz R et al., Performance of the ATLAS liquid argon endcap calorimeter in the pseudorapidity region $2.5 < |\eta| < 4.0$ in beam tests, Preprint submitted to *Elsevier Science*

[2] ATLAS EM end-cap collaboration, Constraction, assembly and tests of the ATLAS electromagnetic end-cap calorimeter, to be published in *Nucl. Instr. and Meth.* **A**

[3] Dowler B et al. Performance of the ATLAS hadronic end-cap calorimeter in beam tests 2002 *Nucl. Instr. and Meth.* A **482** p94

[4] Gingrich D M et al., Construction, assembly and aesting of the ATLAS hadronic end-cap calorimeter 2007 *JINST* 2 P05005

[5] Artamonov A et al., The ATLAS forward calorimeter, 2008 *Journal of Instrumentation JINST* **3** P02010

[6] Bergeaas E, et al. 2008 Topological cluster classification and weighting *ATL-COM-LARG-2008-006* (Geneva, CERN)

[7] Menke S, Vivarelli I 2008 Detector Level Jet Corrections *ATL-COM-PHYS-2008-095* (Geneva, CERN) p35

Test of the ATLAS pion calibration scheme in the ATLAS combined test beam

Francesco Spanò [1]

Columbia University, Nevis Laboratories, 136 South Broadway, P. O. Box 137, Irvington, NY 10533, USA

E-mail: `francesco.spano@cern.ch`

Abstract. Pion energy reconstruction is studied using the data collected during the 2004 ATLAS combined test beam. The strategy to extract corrections for the non-compensating nature of the ATLAS calorimeters for dead material losses and for leakage effects is discussed and assessed. The default ATLAS strategy based on a weighting technique of the energy deposits in calorimeter cells is presented and compared to a novel technique exploiting correlations among energy deposited in calorimeter layers.

1. Introduction

Non-linearity and resolutions degradation in energy reconstruction of hadrons by calorimeters result from non-compensation effects compounded by unmeasured energy deposited in non-instrumented (dead) material. Calibration techniques are used to recover linearity and improve resolution.

In the year 2004 the ATLAS collaboration carried out a test-beam where a full central [2] slice of the ATLAS detector was exposed to beams of electrons and pions a large energy range. One of the main purposes of this combined test-beam was to test the ATLAS strategy to use a calibration based on simulation to reconstruct the correct energy of pions.

The general long-term ATLAS strategy to use simulation to calibrate the detector response to hadrons is shown in [1], while the quality of the simulation for the combined test beam data is discussed in [2]. The result of these two inputs is shown in the present report.

The experimental setup and the real and simulated data samples are briefly illustrated in sections 2 and 3. The pion calibration techniques are illustrated in section 4. The performance on simulation and data is discussed in terms of linearity and resolution in section 5.

2. Experimental setup

The 2004 ATLAS combined test beam is shown in the sketch of figure 1. It was composed of a full central slice of the ATLAS detector extending for about three units in pseudo-rapidity [3] and for 0.3 radians the azimuthal direction, ϕ, around the beam axis. The central semiconductor pixel

[1] On behalf of the Combined Test Beam and Hadron Calibration groups of the ATLAS collaboration.

[2] The part of the ATLAS detector that is closest to the proton-proton interaction point in terms of pseudo-rapidity (defined in footnote 3).

[3] Pseudo-rapidity (η) is defined at η = -log(tan($\theta/2$)) where θ is the polar angle in the detector, shown in fig 1.

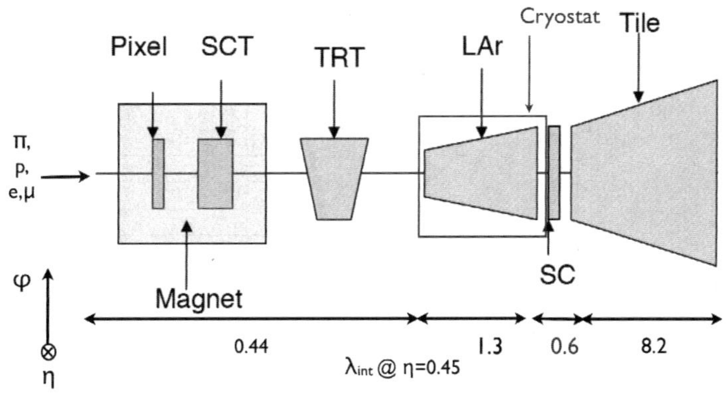

Figure 1. The experimental layout for the 2004 ATLAS combined test beam. See text for details.

Table 1. Properties of positive pion data samples used in the analysis. The sample fraction of proton events is also reported. At 180 GeV, pion events are selected from a sample of positron events.

Positive pion data samples		
Selected events	Energy (GeV)	Proton contam. (%)
8000	20	0
15000	50	41
7000	100	59
5000	180	75

and strip detectors were housed in a bending magnet and followed by the straw-tube transition radiation tracker. The ATLAS central sampling calorimeters followed: one barrel module of the liquid argon-lead electromagnetic calorimeter (LAr) with accordion shape was housed in its cryostat and put in front of three hadronic iron-scintillator modules (Tile) stacked in the azimuthal direction, orthogonally to the incoming test beam axis. The setup was exposed to beams of particles (pions, protons, electrons and muons) in the energy range 1 to 350 GeV. At $\eta = 0.45$ the material in front of the calorimeters is estimated to consist of 0.44 λ_{int} [4] and the calorimetry stretches for about 9.5 λ_{int} : 1.3 λ_{int} for LAr and 8.2 λ_{int} for Tile. The LAr cryostat accounts for additional 0.6 λ_{int} in between the LAr and the Tile.

3. Data and simulation samples

The data consists of samples of events in which positive pions impinge on the experimental setup at $\phi = 0$ and $\eta = 0.45$. They are summarized in table 1.

The pion beams are generated from proton primary beams extracted from the CERN SPS

[4] The interaction length is calculated for protons.

accelerator: the resulting proton contamination is measured by estimating the fraction of proton events that are necessary to reproduce the observed probability of generating a high energy hit in the transition radiation detector. The pion selection is documented in [2]. The proton contamination required simulation of samples of pions and protons in the range 15 to 230 GeV with GEANT 4.7 [3] using the QGSP_BERT physics [5] list and a consistent description of the test-beam set-up. The 4 million events simulated were split in two statistically independent sets of samples, the first one used for deriving simulation corrections and the second for testing the expected performance.

4. Pion calibration techniques

An incoming pion in the ATLAS detector causes a shower that is sampled by the seven calorimeter layers of the combined electromagnetic and hadronic sections. Any hadronic calibration scheme has to recover the intrinsic losses due to the invisible energy lost in nuclear interactions. Additional imperfections in the reconstruction also need to be accounted for: corrections are required for the imperfect energy collection of the clustering algorithm (out-of cluster) and incomplete shower containment (leakage of neutrons, muons and neutrinos). Finally corrections for energy deposits in non-instrumented material have an important role. In particular, in the ATLAS central region (barrel) a non-negligible amount of dead material is present between the electromagnetic and hadronic compartments i.e. in the midst of the longitudinal development of most hadronic showers.

Two calibration techniques are considered. The default ATLAS local hadronic calibration (LH in the following) is described in detail in [1]. A novel technique (described in section 4) is also considered: it is based on the use of correlations between the signals at the layer level, summing the clustered energy in each calorimeter longitudinal segment. It is called layer correlation calibration (LC in the following). The ansatz is that hadronic and electromagnetic energy deposits have different fluctuations properties and, consequently, variables that are sensitive to fluctuations in the total energy can be used both to derive all the corrections and improve the resolution of the total energy measurement.

The two techniques result in different outputs: LH produces calibrated clusters that will be used to form calibrated jets. On the other hand LC provides calibrated layer energies: such scheme is technically extendible to jets, but it will be aimed at calibrating the given jet energy depositions in a layer.

5. Performance of the hadronic calibration schemes

Performance is assessed in terms of linearity and relative resolution.

The total energy is fitted with a Gaussian in the $[\mu\text{-}\ 2\sigma,\ \mu + 2\sigma]$ interval where μ is the mean of the initial energy distribution and σ is its standard deviation. Linearity is defined as the ratio of the expected fitted average to the beam energy as a function of the beam energy. Relative resolution is defined as the ratio of the fitted standard deviation to the fitted average as a function of the beam energy.

5.1. Expected linearity and resolution for the LH scheme

Linearity obtained with LH from simulated positive pion events is shown in the upper plot of figure 2. At the electromagnetic scale the typical linearity shape for non-compensating calorimeters is observed: about 75% of the beam energy is measured and the linearity ratio increases with beam energy due to the increasing electromagnetic fraction of the shower. The compensation weights recover about 10% of the total beam energy. The small out-of-cluster corrections account for about 1% of the beam energy. The remaining 10% is recovered by adding the dead material corrections. Linearity is finally recovered within 2% for beam energy larger than 20 GeV.

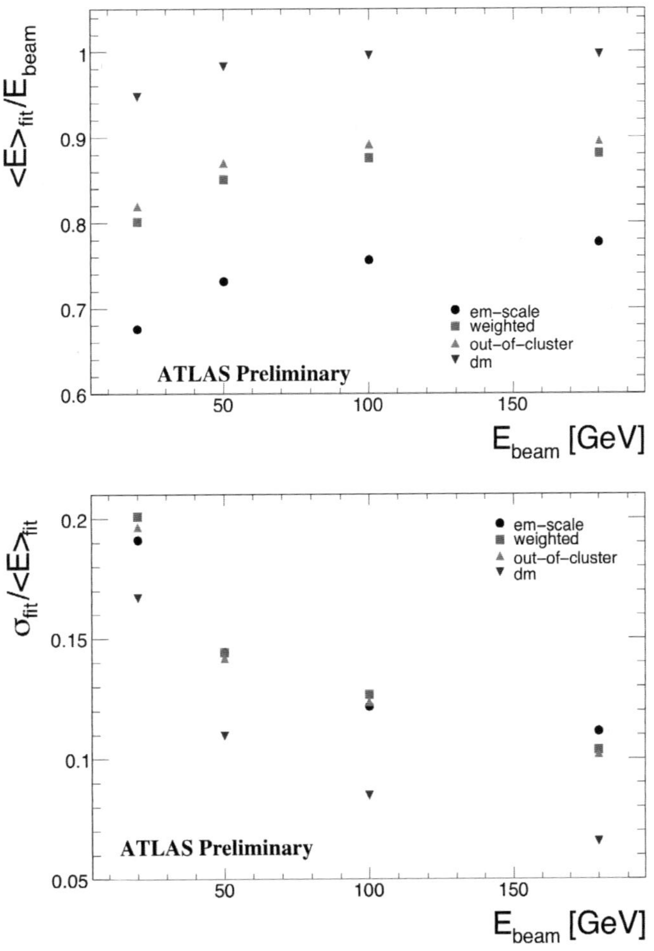

Figure 2. LH scheme: linearity (upper plot) and relative resolution (lower plot) for simulated positive pion events impinging on the test-beam setup at $\eta = 0.45$. The various stages of correction are shown (see text for details).

The relative resolution is shown in the lower plot of figure 2. Dead material effects are expected to play a dominant role. The improvement in relative resolution deriving from suppressing the various fluctuations is expected to reach 11% to 40%.

5.2. LC scheme

The LC technique defines the total pion energy as the sum of clustered energy for each calorimeter layer. The event-by-event layer energy corrections are defined as a function of a specific pair of linear combinations of layer energies. Such combinations are the components of the seven-dimensional vector of layer energies along the vector space basis derived by a principal component analysis (PCA) [4]: the two components are used along the basis vectors whose associated PCA variance gives the largest contributions to the fluctuations of the total energy. Intuitively, the corrections depend on the "directions of largest independent fluctuations", $E_{eig,i}$, where i is an integer from one to seven.

Both the weights aimed at compensating the invisible energy and the corrections aimed at

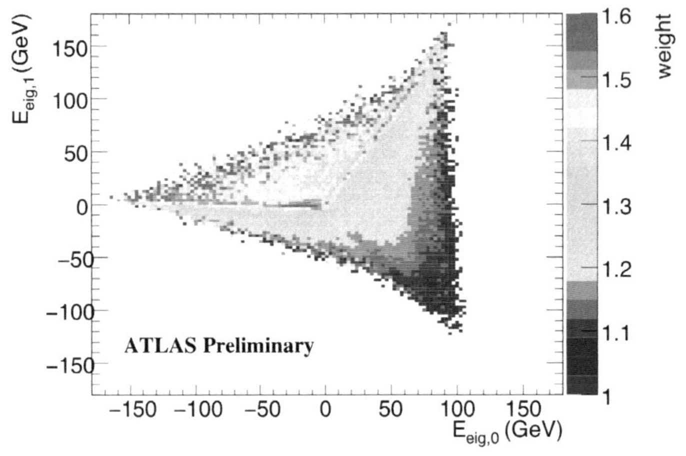

Figure 3. LC scheme: two dimensional look-up table of compensation weights for the first layer of the Tile hadronic calorimeter (see details in the text). Pions and protons samples are mixed to achieve a 41% proton contamination.

recovering dead material losses are derived as two dimensional look-up tables (seven for the layer energies, one for the dead material correction). For both types of tables, $E_{eig,0}$, the linear combination of layer energies with the largest contributions to the total energy fluctuation is one of the two dimension.

$E_{eig,1}$, the linear combination with the second largest contributions to the total energy fluctuation, is the second variable for the calculation of the compensation weights. The weight for a given layer energy in a given bin of the two dimensional table is defined as the average $E_{true}^{layer}/E_{rec}^{layer}$ over all the events in the bin [5]. An example of the weight table for the first Tile layer is shown in figure 3. The separation between the high weight region, dominated by invisible energy, and the low-weight region, dominated by visible energy, is evident.

For the dead material correction, the second dimension is represented by, $E_{eig,2}$,the layer energy combination with the third largest contribution to the total energy fluctuations [6]. The look-up table is shown in figure 4 where the high dead material correction region is well separated form the rest. The correction is derived as a function of the normalized linear combinations mentioned above (each combination is divided by the best estimate of the total energy) and it is expressed as a fraction of the total energy itself.

A small correction for the leakage, dead material energy losses upstream of the calorimeters and in between the first (presampler) and second ("strips") LAr layers is calculated by a parametrization obtained from simulation as a function of the total energy estimate E_{tot} [7].

An iterative procedure is then applied: a given total energy estimate provides a new dead material correction which can in turn be used to determine the total energy. A few iterations are required to obtain a stable result.

All look-up tables are filled by the full set of simulated samples from 15 to 230 GeV so as to reduce the beam energy dependence of the correction as much as possible.

[5] Such correction includes out-of-cluster effects as the numerator is the true total energy deposited in the layer.
[6] For dead material corrections this combination is found to have the best expected performance.
[7] The formula is $E_{func}^{DM}(E) = C_{upstream}(E)E_{LArpresampler} + C_{LArpresampler-strips}(E)\sqrt{|E_{LArpressampler}E_{LArstrips}|} + C_{leakage}(E)$.

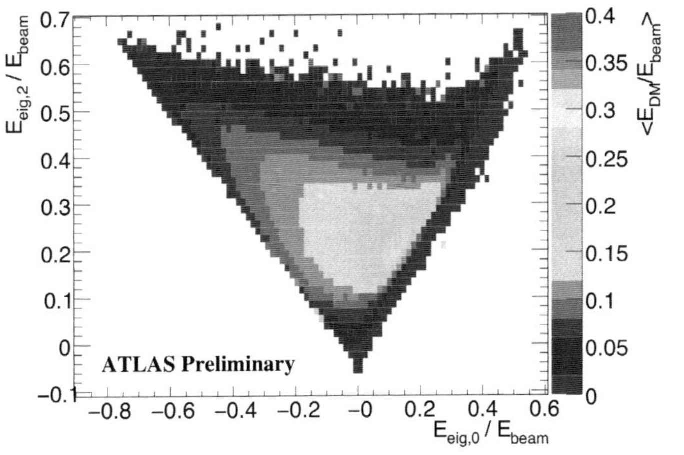

Figure 4. LC scheme: two dimensional look-up table of corrections for energy loss in dead material (as a fraction of total energy) between TileCal and LAr electromagnetic calorimeter (see details in the text). Pions and prootn samples are mixed to achieve a 41% proton contamination.

5.3. Linearity and resolution for the LC scheme

In the case of the LC scheme a mix of pions and protons was used to derive the corrections and to simulate the data. The contamination values are those from table 1.

The upper plot of figure 5 shows the linearity obtained for both data and simulation: the agreement is within 2% at all stages of calibration. The resulting picture is similar to that outlined in section 5.1 for the LH scheme. The reconstruction at the electromagnetic scale is accounting for 75% of the beam energy. The compensation weights recover about 12% of the beam energy while the dead material correction accounts for about 10%. The dead material correction for losses between Tile and LAr represents about 80% of the total dead material corrections. The LC method recovers linearity within 3% over the whole energy range.

The relative resolution is shown the lower plot of figure 5. The simulation foresees a relative improvement of 17 to 24%: the data behave consistently showing an improvement of 17 to 21%. Even though the relative behaviour is the same, the simulation underestimates the resolution in the data by about 25%. GEANT4.9 [2] is expected to improve the data description.

5.4. LH vs LC scheme: linearity

Figure 6 shows the comparison for the resulting linearity when applying both calibration techniques to the same data set [8] The result is quite consistent: the linearity is recovered within 2 to 5% by both techniques.

[8] The electromagnetic scale result is taken from the LC method: a slightly different event selection was applied to the data when applying the LH scheme.

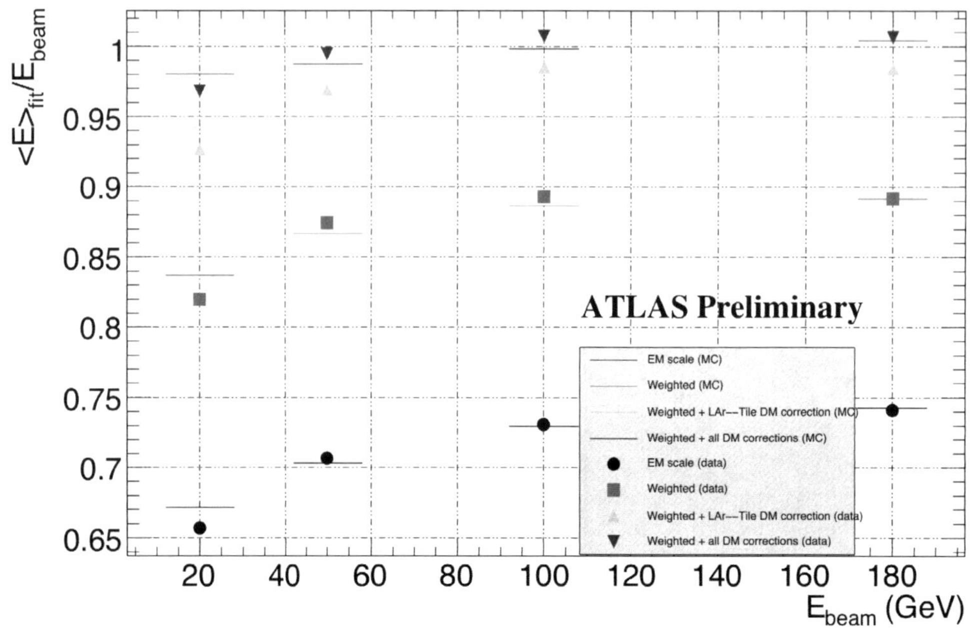

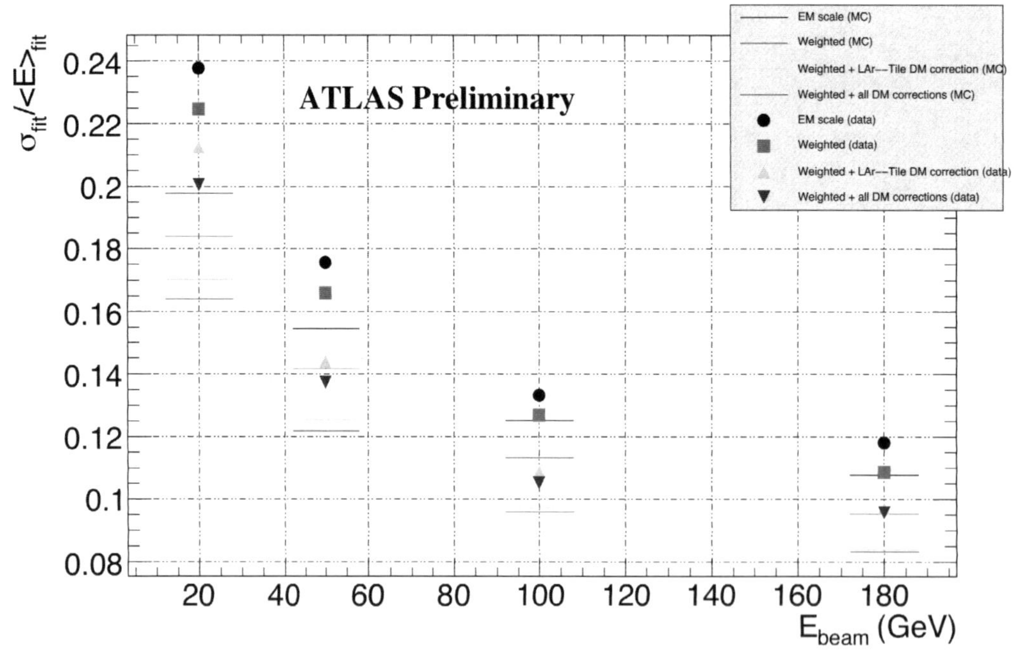

Figure 5. LC scheme: linearity (upper plot) and relative resolution (lower plot) for positive pion events impinging on the test-beam setup at $\eta = 0.45$ at the various stages of correction (see text for details). Data are shown in filled symbols, simulated events are represented by horizontal lines.

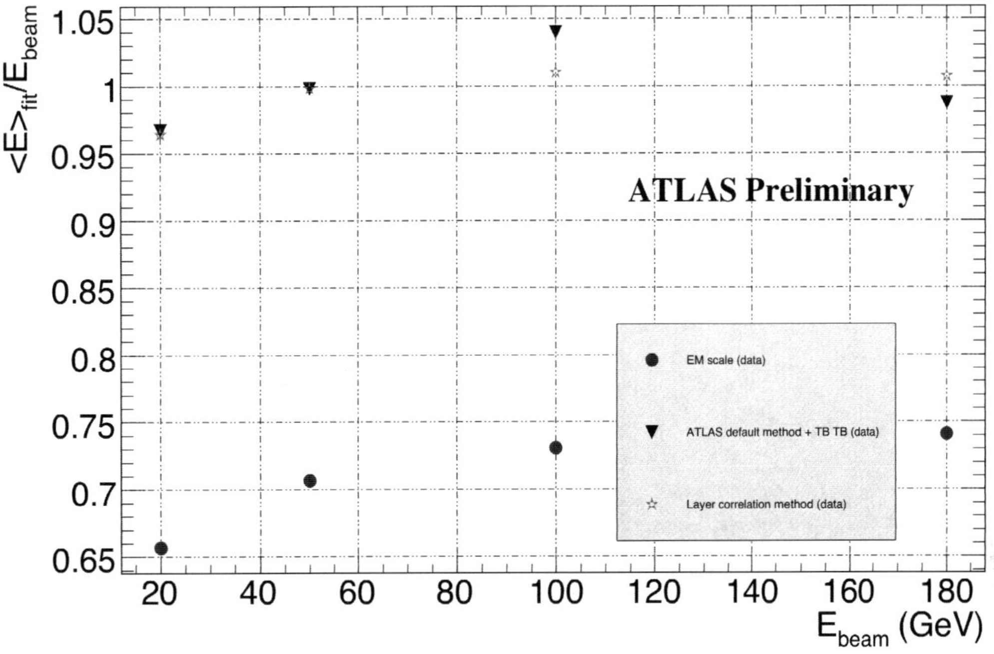

Figure 6. Comparison of linearity for the LH and LC schemes. Only the electromagnetic scale and the fully corrected scale are shown.

6. Conclusions

A simulation-based cell-weighting technique for hadronic signal calibration was applied to pion energy reconstruction in 2004 ATLAS combined test beam for beam energy in the range 20 to 180 GeV. A novel technique based on the correlation amongst layer energies was also used.

The linearity of response to charged pions is recovered within 2 to 5% by both approaches in good agreement between data and simulation; compensation weights and dead material effects have similar impact. According to simulation, the relative energy resolution is expected to improve (by 20-30% to 40%). LC actually achieves an improvement of 17 to 21%. Simulation underestimates data resolution by 10 to 25%; dead material effects are dominant.

Data-simulation discrepancies at the electromagnetic scale keep their size at all stages of calibration, thus simulation performance is the limiting factor.

Acknowledgments

The essential collaboration of Tancredi Carli, Karl-Johan Grahn and Peter Speckmayer is gratefully acknowledged.

References

[1] Pospelov G 2008, these proceedings
[2] Speckmayer P 2008, these proceedings
[3] Agostinelli S *et al.* 2003 *Nucl. Instr. and Meth.* A **506** 250-303
Allison J it et al. 2006 *IEEE Transactions on Nuclear Science* **53 No. 1** 270-278.
[4] Jackson J E 2005 *A User's Guide to Principal Components* (Newark, NJ : Wiley) p 505
[5] Guthrie M P, Alsmiller R G and Bertini H W 1968 *Nucl. Instr. and Meth.* A **66** 29
Guthrie M P, Bertini H W 1971 *Nucl. Phys.* A **169** 670

Fast Simulation of Electromagnetic Showers in the ATLAS calorimeter: Frozen Showers

E. Barberio[1], J. Boudreau[2], B. Butler[3], S.L. Cheung[4], A. Dell'Acqua[5], A. Di Simone[5], E. Ehrenfeld[6,7], M. V. Gallas[5], A. Glazov[7], Z. Marshall[8,9], J. Mueller[2], R. Plačakytė[7], A. Rimoldi[10], P. Savard[4], V. Tsulaia[2], A. Waugh[11], C.C Young[3]

[1] School of Physics, University of Melbourne, Parkvill, Victoria 3010, Australia
[2] University of Pittsburgh, Department of Physics and Astronomy, 3941 O'Hara Street, Pittsburgh, PA 15260, USA
[3] Stanford Linear Accelerator Center, 2575 Sand Hill Road, Menlo Park, CA 94025, USA
[4] Department of Physics, University of Toronto, 60 Saint George Street, Toronto M5S 1A7, Ontario, Canada
[5] European Laboratory for Particle Physics, CERN, 1211 Geneva 23, Switzerland
[6] Institut für Experimentalphysik, Universität Hamburg, Luruper Chaussee 149, 22761 Hamburg, Germany
[7] Deutsches Elektronen Synchrotron, Notkestrasse 85, 22603 Hamburg, Germany
[8] Physics Department, California Institute of Thechnology, Pasadena, CA 91125,USA
[9] Columbia University, Nevis Laboratory, 136 So. Broadway, Irvington, NY 10533, USA
[10] Instituto di Fisica Nucleare, Universita di Pavia & I.N.F.N., Via A. Bassi 6, 27100 Pavia, Italy
[11] Falkiner High Energy Physics Group, School of Physics, University of Sydney, Sydney NSW 2006, Australia

E-mail: `ringaile@mail.desy.de`

Abstract. One of the most time consuming process simulating *pp* interactions in the ATLAS detector at LHC is the simulation of electromagnetic showers in the calorimeter. In order to speed up the event simulation several parametrisation methods are available in ATLAS. In this paper we present a short description of a frozen shower technique, together with some recent benchmarks and comparison with full simulation.

1. Introduction

An expected high rate of proton-proton collisions in ATLAS detector at LHC [1] requires large samples of simulated events (Monte Carlo) to study various physics processes. A detailed simulation of particle reactions ("full simulation") in the ATLAS detector is based on GEANT4 [2] and is very accurate. However, due to complexity of the detector, high particle multiplicity and GEANT4 itself, the average CPU time spend to simulate typical QCD event in *pp* collision is 20 or more minutes for modern computers.

During detector simulation the largest time is spend in the calorimeters (up to 70%) most of which is required for electromagnetic particles in the electromagnetic (EM) part of the calorimeters. This is the motivation for fast simulation approaches which reduce the simulation time without affecting the accuracy.

Journal of Physics: Conference Series **160** (2009) 012082 doi:10.1088/1742-6596/160/1/012082

Several of fast simulation methods available within the ATLAS simulation framework (standard Athena based simulation program) are discussed here with the focus on the novel frozen shower library (FS) technique. The results obtained with FS are presented here as well.

2. Electromagnetic ATLAS calorimeters

There are three main liquid-argon (LAr) EM sampling calorimeters in the ATLAS detector which cover pseudorapidity range $|\eta| < 1.475$ (barrel), $1.375 < |\eta| < 3.2$ (two end-cap components) and $|\eta| < 4.9$ (forward calorimeters FCAL), see Figure 1. Both barrel and end-cap calorimeters have accordion shaped electrodes and absorber plates over the whole coverage thus providing the full ϕ symmetry. The accordion waves vary with radius and η in both calorimeters which is important for optimisation of the performance.

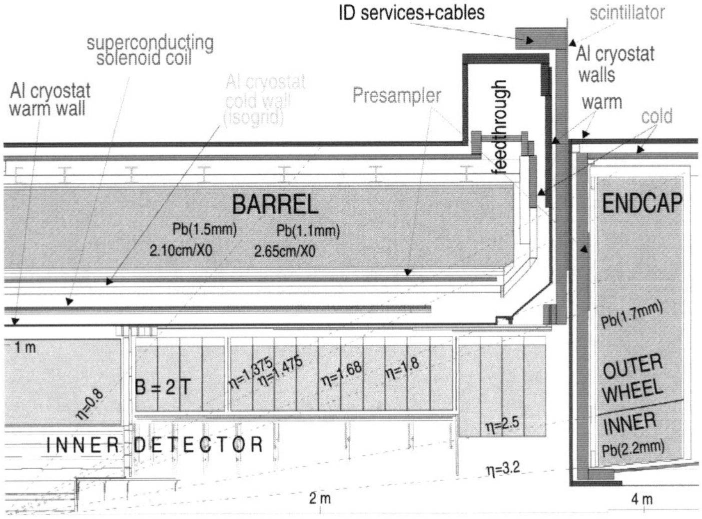

Figure 1. Layout of ATLAS EM calorimeters (FCAL is not shown).

The forward calorimeter (Figure 2) has a different layout which is motivated by the fact that the FCAL is located at high η and is exposed to a high flux of particles, thus is optimised for the best performance in high rate environment. The active material is LAr in gaps between a hexagonal array of copper tubes and rods. More information about the EM calorimeters can be found elsewhere, e.g. in [3].

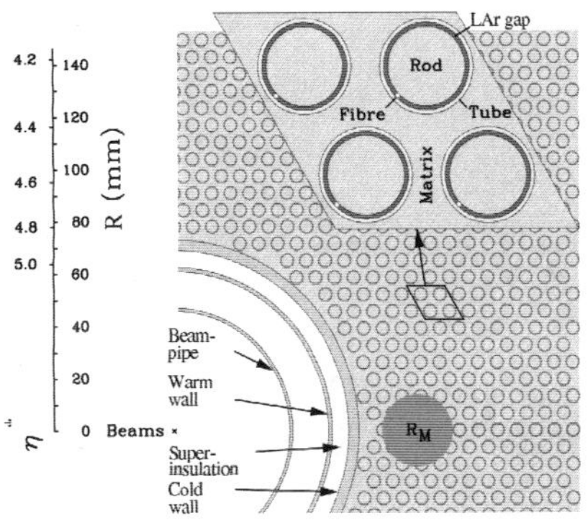

Figure 2. Schematic few of forward calorimeter module illustrating the arrangement of copper tubes and rods with LAr gaps in between.

3. Fast shower simulation techniques for ATLAS calorimeter

There are three basic approaches for the fast shower simulation in different energy ranges: High energy approach which uses parametrisation method, low energy (below 1 GeV) approach for which frozen showers are used and very low energy particle "spot" (or "killing") method.
The basic idea of parametrisation is that two, longitudinal and radial, energy distributions are sufficient to describe the spatial energy distribution of EM showers in the calorimeter. This method performs best for $e^{\pm}$ in high energy region, i.e. at energies of 10 GeV and above where calorimeter is sufficiently uniform and shower is well described by parametrised functions. The full description of the parametrisation method can be found in [4] while in [5] and [6] more details about the application of this method are given. The lower energy - frozen shower - method is an alternative approach where showers are terminated by substituting the initial low energy EM particle by the shower from the library. This method is described in the following section. The "spot" approach is applicable only for very low energy particles where single particles ($e^{\pm}$) below 10 MeV are substituted by a single spot ("hit") with the same as the particle energy.

3.1. Frozen Showers

The frozen shower approach is well described by its name. Prior to the simulation of physics events, showers of electromagnetic particles below a certain energy cut-off (typically 1 GeV) are frozen and stored in a library for use at simulation time. Then, during simulation, when electromagnetic particles fall below that energy cut-off, the particle is stopped and a frozen shower from the library is substituted. The main steps of shower library creation are discussed in the following paragraphs.

Compression

For the library creation the incident particle of fixed energy and position (in η and ϕ) is started from the calorimeter surface. The energy of the incident particle in the calorimeter is deposited in "spots" or hits, which are subsequently recorded. For the hits spatial coordinates, energy fraction and time information are saved. The sampling fraction, the sampling fluctuations and charge-collection effects in the calorimeter are applied to the energy deposits before they are frozen. Furthermore, in order to save disk space as well as memory consumption, hit information is compressed. The compression involves two processes, *hit merging* and *truncation*:

- if the distance between any two hits in the calorimeter is smaller then a given parameter R_{min}, then hits are merged into one deposit at the energy weighted center between them. The process is repeated until no hits with distance $< R_{min}$ are left. Typical value of R_{min} is 5 mm.
- hits whose energies are below the fraction f of the total energy sum of all hits, are truncated (f is typically 95%). The energy of the remaining hits is rescaled back to preserve the total deposited energy.

In Figure 3 the shower compression steps are illustrated.

Binning

In order to follow the most important variations in shower properties the library is binned in *energy* and *position*. The same *energy* bins for all three ATLAS EM calorimeters are used, they are placed at: 1, 2, 5, 10, 20, 50, 100, 200, 500, 1000 MeV. Figure 4 shows distribution of the collected energy for each generated energy bin.
The *position* bins however are different for different calorimeters. η binning is chosen for the barrel and end-cap calorimeters since they properties, and thus the shower properties, vary strongly with η. The bins are optimised to closely follow the effective sampling variations in calorimeters. Since the variation of the response to EM particles with ϕ (due to the accordion structure of

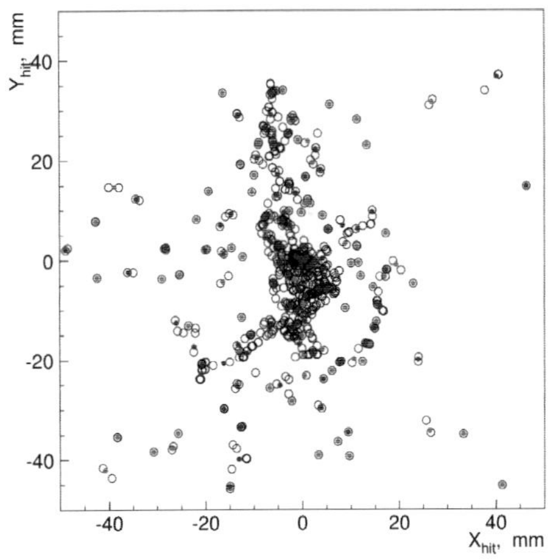

Figure 3. The transverse view of the energy deposits (hits) from a 0.5 GeV energy electron shower. The open points are original hits, red points are hits after the merging, blue points are left after the merging and truncation.

the calorimeter) is smaller than the the lateral shower size, during the creation of the shower libraries we average over ϕ position. As an example, the variation of the sampling fraction in end-cap calorimeter (caused by η dependent high voltage corrections introduced to compensate changing detector geometry) and the corresponding η binning are shown in Figure 5.

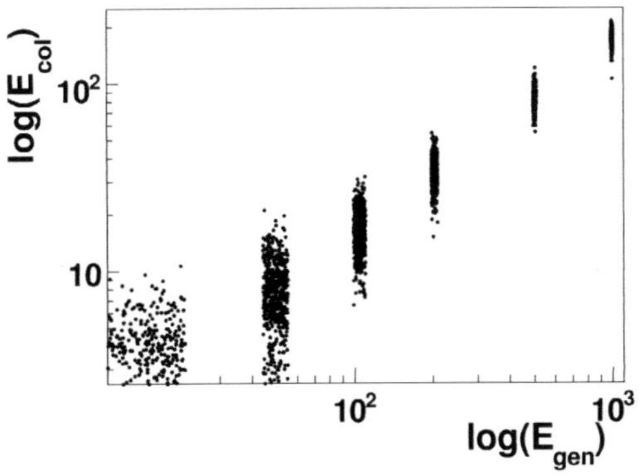

Figure 4. The correlation of the energy collected in the sensitive calorimeter detector (E_{col}) with the generated energy (E_{gen}) as recorded in the frozen shower library.

As previously mentioned, the forward calorimeter differs from other detectors by its design: contrary to barrel and end-cap, FCAL has very small active material layers (LAr gaps within copper tubes and rods). Because of this, the shower shapes in the forward calorimeter have almost no η dependence but strongly depend on the position inside the detector, i.e. the distance to the active material. Therefore the position binning for FCAL is chosen as the distance of the particle impact position from the center of the closest rod.
With a typical 1000 events in each position-energy bin, the size of the frozen shower library is about 50 MBytes for one calorimeter.

Usage of FS library
During the simulation each particle is checked for its type, energy and containment in the calorimeter. When all conditions are fulfilled (for the FS method it is: Particle is $e^{\pm}$ or γ, it is in a valid energy range and fully contained in the calorimeter), a shower from the library of corresponding energy and position bin is taken. For the particle at pseudorapidity η, the shower bin is chosen randomly from two closest pseudorapidity bins (η_1 and η_2) with a probability to take η_1 bin given as $(\eta_2 - \eta/\eta_2 - \eta_1)$. For FCAL the distance from the rod center is checked.

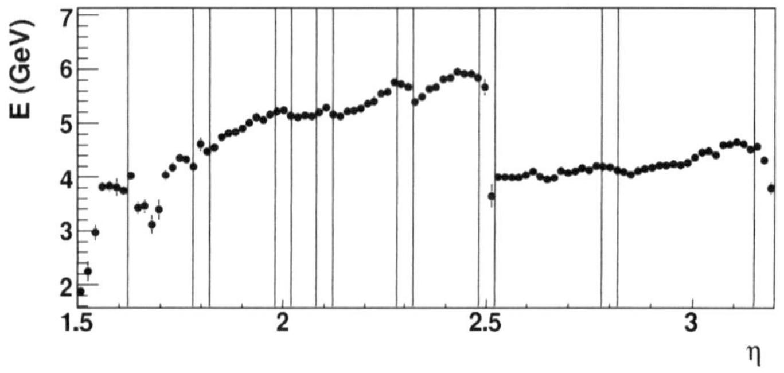

Figure 5. The energy deposited by 64 GeV electrons in the end-cap calorimeter. The library bins in η are shown as vertical lines.

The energy bin is chosen from the two adjacent energy bins (the random energy distribution follows a logarithmic distribution). Once the shower is chosen from the library, the shower hits are transformed into the ATLAS coordinate system and the total deposited energy is rescaled to the energy of the substituted particle.

3.2. Results

In this section the comparison of full and fast (FS) simulations for 64 GeV electrons simulated in various pseudorapidity regions is presented at both, simulation (before digitisation and reconstruction) and reconstruction level.

The performance of FS at simulation level is shown in Figure 6 and 7: In Figure 6 the energy response from the full simulation as a function of η (covering range from the central to forward detector region) is compared to the energy response from the FS. A good agreement with the full simulation is observed in the whole η range. Small deviations at the cracks and calorimeter edges (η = 0, 1.4, 2.5, 3.2) are due to incomplete shower containment where the description of showers becomes more complicated.

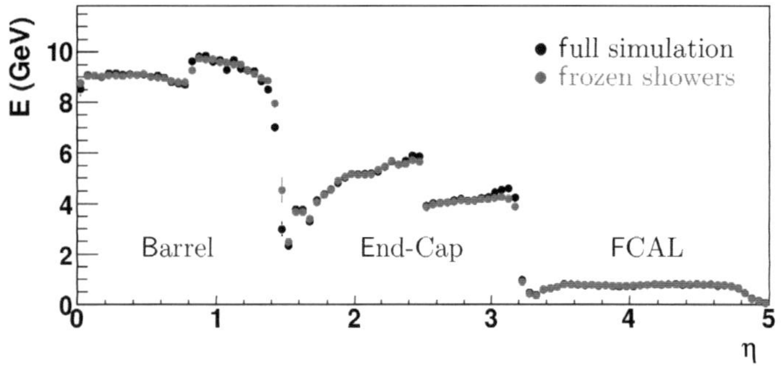

Figure 6. The comparison of energy in sampling fraction from full simulation and FS as a function of η.

In Figure 7 the CPU time in arbitrary units for full simulation and FS is shown as a function of η. As can be seen from the picture, the improvement in speed for the 64 GeV electron simulation using frozen showers can reach factor of ten and more compared to the full simulation[1].

Comparison of full and fast simulation for the shower shapes at reconstruction level for 64 GeV electrons with η = 1.0 are shown in Figure 8. Here the shower width in 3 strips of first EM barrel compartment and the lateral shower width in the second barrel compartment calculated in the window of 3x5 cells are presented. As can be seen from these examples, a good agreement between full simulation and FS at the reconstruction level is achieved.

[1] The factor of ten improvement in simulation of EM particles in the calorimeter corresponds to about a factor of two for the full simulation of a typical QCD event in ATLAS detector.

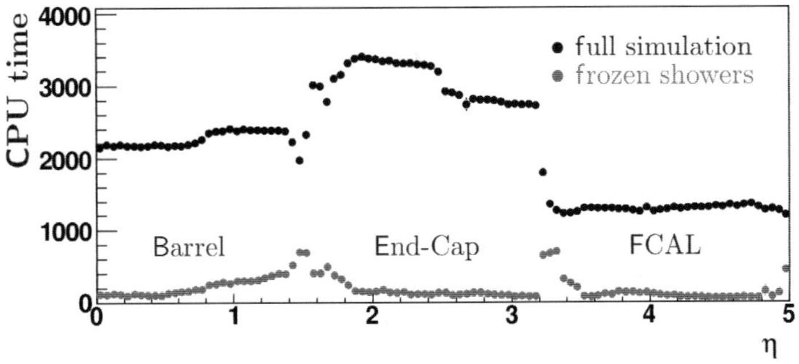

Figure 7. The comparison of simulation time from full simulation and FS as a function of η.

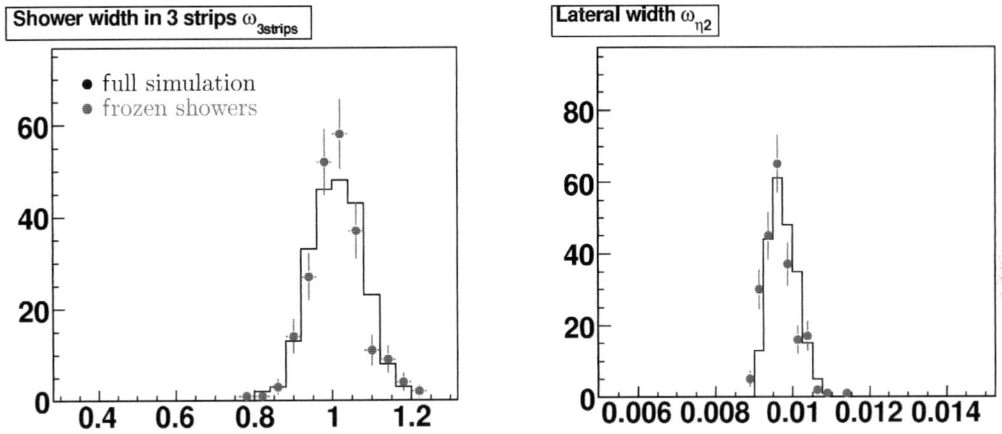

Figure 8. Shower width calculated in 3 strips in the first compartment of barrel EM calorimeter (left) and lateral shower width in the second compartment (right) for the full and fast (FS) simulation. Distributions are shown for 64 GeV and $\eta = 1.0$ electrons.

4. Conclusions

A fast simulation approach including such techniques as shower parametrisation and frozen showers has been implemented and is available for use in the ATLAS simulation framework. A short description of these methods was presented here. A significant gain in speed is achieved using the fast simulation with a good agreement of energy distributions and shower shapes over the large pseudorapidity range at both, simulation and reconstruction, level.

References

[1] ATLAS Collaboration, *The ATLAS Experiment at the CERN Large Hadron Collider*, 2007, submitted to JINST.

[2] S. Agostinelli et al., *GEANT4 - A Simulation Toolkit*, Nucl. Inst. Meth. A506, 2003, 250303.

[3] ATLAS Collaboration, *ATLAS Liquid Argon Calorimeter: Technical design report* 1996, CERN-LHCC-96-4.

[4] Grindhammer G, Rudowicz M and Peters S *The parametrized simulation of electromagnetic showers in homogeneous and sampling calorimeters*, 2000, hep-ex/0001020.

[5] Ehrenfeld W., *Fast Shower Simulation in the ATLAS calorimeter*, contributed paper to International Conference on Computing in High Energy and Nuclear Physics (CHEP'07), Victoria BC Canada, 2007.

[6] Marshall Z., *The GEANT4-based ATLAS Fast Electromagnetic Shower Simulation*, contributed paper to 10th ICATPP Conference on Astroparticle, Particle, Space Physics, Detectors and Medical Physics Applications, Villa Olmo, Como, 2007.

Journal of Physics: Conference Series **160** (2009) 012083 doi:10.1088/1742-6596/160/1/012083

Fine shashlik simulation from tests results

M. Prokudin, I. Korolko

SSC RF ITEP, Bolshaya Cheremushkinskaya 25, 117218 Moscow, Russia

E-mail: `Mikhail.Prokudin@cern.ch`

Abstract. We have studied the response uniformity of the LHCb electromagnetic calorimeter modules in a beam of high energy muons. Tests of prototype module built with thin (0.5 mm) absorber and scintillator plates were performed as well as transverse scans of an ECAL module with 50 GeV electrons. To understand the light collection uniformity a dedicated Monte Carlo program for the the light propagation in different scintillator tiles was developed. Parameters of this program are tuned and checked using experimental data

1. Introduction

During last 25 years sampling calorimeters of "shashlik" type have proven their applicability for a wide range of high energy physics experiments. This type of calorimeter was successfully operated at PHENIX [1] at BNL, DELPHI [2] at CERN and HERA-B [3] at DESY. Wave length shifting (WLS) fibers used for the scintillator light readout allow us to build hermetic calorimeters with minimal amount of dead material. Fast response, adequate energy resolution, radiation resistance, flexible transverse and longitudinal granularity of these detectors and relatively moderate cost make them very attractive to use in large modern experiments. Last years we are developing the "shashlik" technology trying to improve the calorimeter energy resolution and to decrease the Moliere radius.

The energy resolution of "shashlik" calorimeter is described with a following formula:

$$\frac{\sigma_E}{E} = \frac{\boldsymbol{a}}{\sqrt{E}} \oplus \boldsymbol{b},$$

where $\boldsymbol{a}$ and $\boldsymbol{b}$ stand for the sampling and constant terms respectively. The sampling term depends on a thickness of absorber plates and photostatistics. An obvious wish to minimize the sampling term dictates the choice of thin (down to 0.3 — 0.5 mm) absorber plates. At the same time the very high multiplicities of modern high energy physics and nuclear physics experiments have to be taken into account. Overlapping showers from neighbour tracks considerably dilute the intrinsic energy resolution and identification power of a calorimeter with poor segmentation (large Moliere radius). To minimize Moliere radius one should also use thin scintillator tiles. On the other hand, the constant term of energy resolution is determined by transverse nonuniformity of light collection efficiency in scintillator tiles. This nonuniformity rises with decreasing the scintillator thickness worsening the constant term significantly. We have performed detailed studies of light collection efficiency in "shashlik" calorimeter modules built with thick (4mm) and thin (0.5mm) scintillator tiles. Test beam measurements of outer, middle and inner modules of the LHCb electromagnetic calorimeter [4] were complemented with investigation

Journal of Physics: Conference Series **160** (2009) 012083 doi:10.1088/1742-6596/160/1/012083

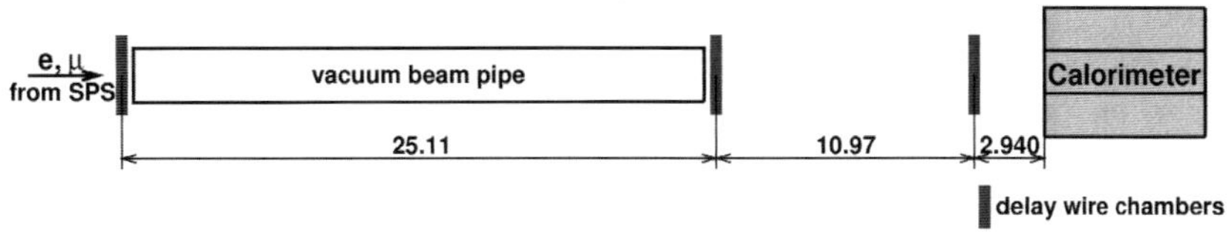

Figure 1. Sketch of the LHCb calorimeter test beam installation.

of new "shashlik" module, which consists of 280 alternating layers of 0.5mm thick absorber and scintillator.

To understand the light collection quantitatively a dedicated ray-tracing program and corresponding optical model of scintillator tile readout with WLS fibers were developed. The parameters of the optical model were extracted from the experimental data collected during muon scan of the LHCb calorimeter inner and outer modules. Independent measurements performed with new module were used to study the dependence of light collection uniformity on scintillator tile thickness. Finally all Monte Carlo parameters were crosschecked by comparison of simulation predictions with electron scan data obtained for the LHCb inner module.

2. Experimental measurements

Experimental studies of the calorimeter modules were performed at the X7 beam which is a part of the West Area test beam complex at CERN. The beam momentum settings were used in the range between 5 and 100 GeV/c. Beam polarity was negative, and particles available were π^-, e^- and μ^-. The electron beam was produced using a lead sheet to convert photons from neutral pion decays, while the muon beam used a beam dump to absorb all secondary particles except muons.

The beam line was instrumented with scintillator counters to form the beam trigger. Three sets of delay wire chambers [5] with a typical spatial resolution of 0.2 mm were used to reconstruct the beam particle trajectory and to determine the particle entry point at the calorimeter module surface. The arrangement of tested calorimeter modules and delay wire chambers at the beam line is shown in Figure 1.

The calorimeter modules were assembled in a 3×3 array: the tested module was put in a center of assembly, while other 8 modules were used to control the leakage of electromagnetic shower. Uniformity of the calorimeter response for muons was measured with central module only because the net energy leakage to all neighbor cells was less than electronics noise in these channels. High statistical samples of 100 GeV/c muons were used to scan the surface of the calorimeter cells with 1×1 mm^2 step. The obtained at every step signal distribution was then fitted with Landau function. Determined most probable values were stored in a 2D histogram representing the response uniformity of a given calorimeter cell to muons. For the outer LHCb module 1×2 mm^2 steps were used because of a limited statistics.

For electrons the energy was collected in a central module and 4 modules on top, bottom, right and left from the center. The net signal in four corner modules was comparable with the noise of electronics. The front face of the calorimeter cell was scanned with 1×1 mm^2 step. Individual signal distributions were then fitted with Gaussian. The positions of distribution maximum were stored in a 2D histogram representing the response uniformity of a given calorimeter cell to electrons.

3. Light collection modelling

A dedicated ray-tracing program was developed for modelling light collection inside the scintillator cell. Optical processes handled by the ray-tracer involve refraction, mirror and diffuse reflection, attenuation in medium and absorption on plastic surface and in tyvek which is used to interlay plastic and absorber tiles. Optical properties of all surfaces and media depend on photon wavelength to model the dispersion. Additional dependence on local tile coordinate was added for modelling light masks. The geometry of plastic tiles is defined by a set of primitives (cuboids, cone and cylinder) and Boolean operations with them. A dedicated voxelization [6] procedure is used to speed up calculations. The photon is assumed to propagate along the straight line between two interactions neglecting diffuse scattering in the scintillator medium.

Tiles have cuboid shape with cylindrical holes for WLS fibers and one clear fiber in the centre of every cell. The edges of the tile are assumed to be coated with special white paint simulating the diffuse reflection due to the special chemical treatment used during mass production. The same white paint also covers thin (1.0 — 1.5 mm) bands at the very edge of top and bottom surfaces of the tile. A scintillator tile with all fibers is covered with tyvek sheets from both sides. Photons are generated uniformly inside tile volume with isotropic momentum distribution. Photon wave length is generated according to the POPOP emission spectra. Each photon is propagating inside described setup until its stop in scintillator or absorption on the tile surface, in the tyvek or in the fiber core. Photon stopped in the fiber core is reemitted isotropically with shifted wavelength corresponding to the WLS fiber reemission spectrum. The direction of reemitted photon is then checked to ensure that it is not able to leave the fiber and will reach the photomultiplier sensitive window. Small attenuation in the light mixers located between fibers bundle and PMT is neglected. Computed 2D map of light collection efficiency is finally convoluted with the measured map of the tile thicknesses which determines the energy deposition for minimum ionizing particle crossing the tile in a given point perpendicular to its surface. The last step is important because plastic tile thickness is not constant with a current mass production technology. The result of such convolution is referred later as an effective 2D map of light collection efficiency.

The results of ray-tracing program depend on certain number of parameters such as: photon attenuation length in scintillator, whiteness of tyvek and quality of white paint covering tile edges, probability of photon absorption on the surface of plastic tile (so called surface quality) and some others. All these parameters could be extracted from the comparison of simulation results with the experimental measurements of light collection uniformity. Such comparison should take into account the transverse shape of incoming particle energy deposition in the calorimeter module which dilute the intrinsic variation of light collection. Muon data is therefore most suitable for our purposes. Effective 2D maps of light collections were used in GEANT3.21 [7] program for weighting of each elementary energy deposition depending on the local coordinate in calorimeter cell. The similar simulation chain was then repeated for electrons with selected set of effective 2D maps as a cross-check. The inner module of the LHCb calorimeter was also modeled with Geant4 [8] transport code. The results of two GEANT versions are identical. Algorithms of incoming particle energy reconstruction after GEANT simulation correspond exactly to those used in the analysis of test beam data. Tuned set of ray-tracing program parameters allow us to reproduce the experimental data almost perfectly.

4. Comparison of simulation with experimental data

Experimental results obtained for the inner calorimeter cell are presented in Figure 2 for two 1 mm wide bands — passing over the fiber positions (lower plot) and located between two fiber rows (upper plot). Hatched histogram shows the GEANT predictions convoluted with effective 2D light collection efficiency maps from our dedicated simulations. Global nonuniformity is negligible. Special chemical treatment of tile edges increased the light collection efficiency, compensating effects of dead material between modules. Results for the middle calorimeter module are very similar. In the outer modules the inter-fiber distance is significantly higher (15 mm instead of 10 mm in other modules). Corresponding experimental results for the outer calorimeter module are presented in Figure 3 together with GEANT simulation predictions. Visible variation of the MIP particle response is diluted with worse spatial resolution of the tracking system used at that time period. Increased inter-fiber distance smoothed the behavior of light collection efficiency between the fibers. The quality of diffuse light reflection and the width of painted bands were found to be the most important parameters of optical model. The last important parameter was the quality of scintillator tile surface. The dependence of light collection efficiency variations on several other factors (like refraction index of the scintillator, attenuation length of the scintillator) was found to be almost negligible.

Electromagnetic showers are much wider in comparison with almost point-like energy depositions of minimum ionizing particles. Therefore the variations of reconstructed energy for electrons and photons are considerably reduced. The results of a lateral scan of the LHCb inner module with 50 GeV/c^2 electrons are shown in Figure 4. In the cell of the inner module we observed 0.5% variation of response. This cell is located in the centre of the module and therefore the effects of dead material are diminutive. The variation of the ECAL response to electrons and photons from the LHCb events would be significantly decreased due to the nonzero impact angles.

The response uniformity of the prototype module is shown in Figure 5. In comparison with the LHCb inner module the prototype has significantly thinner scintillator tiles, where the light collection uniformity is much worse. This effect is explained by a finite probability for photon to be captured on the scintillator tile boundary. In thin tiles a number of reflections photon undergoes to travel till absorption if WLS fiber is much higher than in thick tiles. So the probability of absorption is also higher.

The optical model with extracted parameters was used to calculate the overall light yield

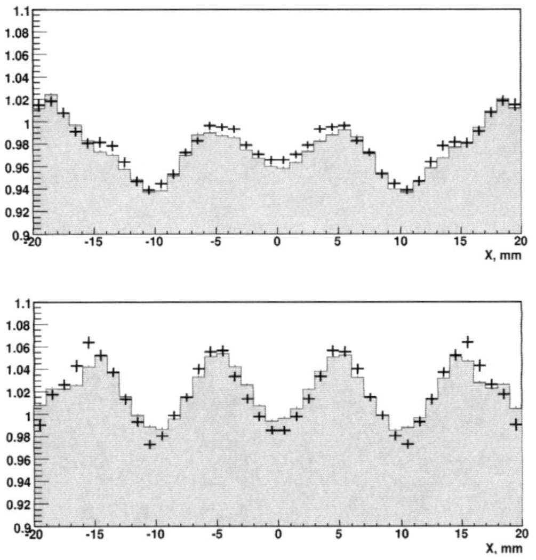

Figure 2. The response uniformity of the inner LHCb module measured with muons (error bars) and simulated (hatched histogram). The scan was made in 1mm wide bands between two fiber rows (upper) and through the fiber positions (bottom).

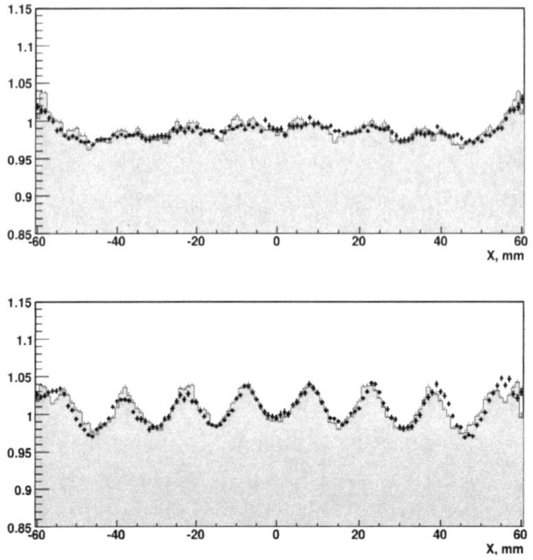

Figure 3. The response uniformity of the outer LHCb module measured with muons (error bars) and simulated (hatched histogram). The scan was made in 2mm wide bands between two fiber rows (upper) and through the fiber positions (bottom).

for all tested modules. Ray-tracing program was not tuned to reproduce the absolute value of light yield; therefore we have used the inner module simulation results for normalization

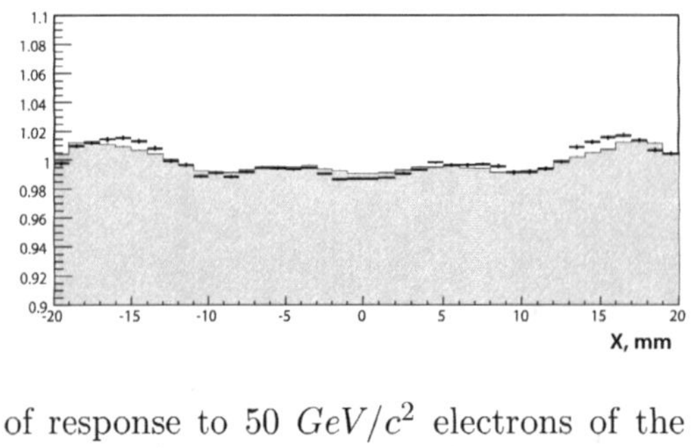

Figure 4. Uniformity of response to 50 GeV/c^2 electrons of the inner module (error bars) compared with simulation results (hatched histogram). The scan was made in 1mm wide bands between two fiber rows (upper) and near bottom fiber raw (bottom). The average response of the module for experimental data and MC simulation was normalized to unity.

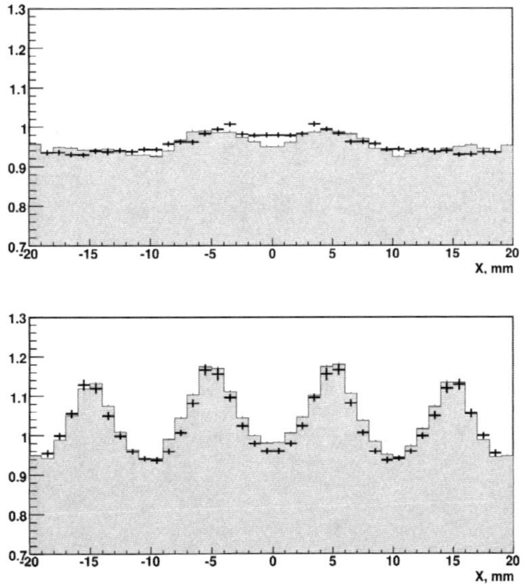

Figure 5. The response uniformity of the prototype module measured with muons (error bars) and simulated (hatched histogram). The scan was made in 1mm wide bands between two fiber rows (upper) and through the fiber positions (bottom).

on experimental data. The comparison of Monte Carlo predictions with data is presented in Table 1. The data for cosmic stand are taken from [9].

Table 1. Light yield for different modules.

	LHCb inner	LHCb middle	LHCb outer	prototype
Test beam	3000 ± 150	3600 ± 180	2500 ± 130	700 ± 60
Cosmic stand	3100 ± 130	3500 ± 170	2600 ± 120	no data
Modeling	3000	3600	2570	600

5. Conclusion

Using the tuned ray-tracing program we have calculated the 2D maps of the light collection efficiency for three different technologies of "shashlik" scintillator tiles production:

- tiles with transparent edges;
- tiles with mirrored edges (ideal mirror);
- tiles with matted edges and bounds near edges (LHCb like).

The results are presented in Figure 6, averaging over narrow (1 mm wide) bands precisely in between the fiber rows. Tiles with transparent edges have the lowest and most non-uniform light collection efficiency. Mirroring of the tile edges allows to increase the light yield by a factor of 3 and improves the uniformity, but does not compensate dead material effects. Chemical treatment of tile edges and bounds near edges resulting in a diffuse reflection of the scintillating light allows a factor of 2 increase of overall light yield and compensates the decrease in detector response on the outer borders of the calorimeter modules due to dead material.

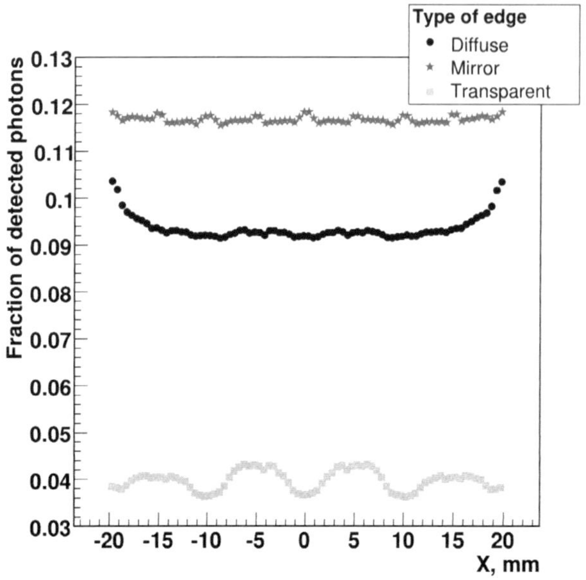

Figure 6. Light collection efficiency simulated for transparent (squares), mirrored (stars) and matted (solid dots) tiles

The uniformity of the LHCb calorimeter response was measured at the test beam and modeled. The model is able to describe the response uniformity with unique set of parameters. It also describes the overall light yield data with the same set of parameters. This model will be used in simulation of LHCb calorimeter response and for the development of new "shashlik" calorimeters with better energy resolution and small Moliere radius.

References

[1] Adcox K *et al.* (PHENIX) 2003 *Nucl. Instrum. Meth.* **A499** 469–479
[2] Benvenuti A C *et al.* Given at 4th International Conference on Calorimetry in High-energy Physics, La Biodola, Italy, 19-25 Sep 1993
[3] Hartouni, E and others DESY-PRC-95-01
[4] Amato S *et al.* 2000 *LHCb calorimeters Technical Design Report* Technical Design Report LHCb (Geneva: CERN)
[5] Spanggaard J 1998 Delay Wire Chambers — A User Guide (*Preprint* `CERN-SL-98-023-BI`)
[6] Arie Kaufman and Daniel Cohen and Roni Yagel 1993 *Computer* **26** 51–64 ISSN 0018-9162
[7] 1993 Geant3. CERN Program Library Long Writeup (*Preprint* `W5013`)
[8] Agostinelli et al 2003 *NIM* **A 506**
[9] Aref'ev A *et al.* 2008 Design, construction, quality control and performance study with cosmic rays of modules for the LHCb electromagnetic calorimeter Tech. Rep. LHCb-2007-148. CERN-LHCb-2007-148 CERN Geneva

AUTHOR INDEX

AUTHOR INDEX

AUTHOR INDEX

AUTHOR INDEX

AUTHOR INDEX